Plant Microbes Symbiosis: Applied Facets

Naveen Kumar Arora
Editor

Plant Microbes Symbiosis: Applied Facets

Springer

Editor
Naveen Kumar Arora
Department of Environmental Microbiology
Babasaheb Bhimrao Ambedkar University
Lucknow, Uttar Pradesh, India

ISBN 978-81-322-2067-1 ISBN 978-81-322-2068-8 (eBook)
DOI 10.1007/978-81-322-2068-8
Springer New Delhi Heidelberg New York Dordrecht London

Library of Congress Control Number: 2014951212

Printed on acid-free paper

Springer is part of Springer Science+Business Media (www.springer.com)

Preface

Plant associated microorganisms (PAMs) play a number of very important roles to better the life of their host plant. It is believed that the origin of plant-microbe interactions is as old as the arrival of plants on land, and both the mutual partners selected each other after a long evolutionary process. These useful microbes provide the plants with nutrients, battle pathogens, combat abiotic stresses, and can play several other roles in the plant ecosphere. Plant-microbe interactions in tandem can perform acts such as reclamation of barren soils, degradation of recalcitrant pollutants, and removal of dangerous compounds from the ecosystems.

Injudicious use of chemicals in agriculture and industries has led to major environmental and health related issues for humans and other organisms inhabiting the planet. These chemicals are accumulating in the food chain and ecosystems. In fact, so many of them reach into the food products which we consume for our sustenance. The indiscriminate use of chemicals is in fact responsible for the reduced fertility of soils by killing several important and useful PAMs. In case of microorganisms we even do not know what we have lost over the years. But one thing is certain that in barren soils the population of useful plant growth promotory (PGP) microbes is very low or even more alarmingly they may altogether be absent from such soils. Hence there is an urgent need to re-introduce these beneficial microbes in the soil along with their compatible plants. The book discusses the various applications of PGP microbes and the importance and utility of plant-microbe interactions in sorting out the issues related to soil degradation and pollution.

The volume is in the form of 19 chapters contributed by experts, from around the globe, involved in research on applied utility of plant-microbe interactions. The tome discusses in great detail the present scenario of utilizing the PGP microbes for enhancement of crop yields by providing nutrients through biocycling or mobilization or fixation, fighting phytopathogens and other stresses such as salinity, drought, pH, etc. An integrated management of nutrients, diseases, and stresses can be achieved by application of diverse PGP microbes in combination. However, to provide these benefits to the plant the PGP microbes have to be augmented in the soil, in particular in the habitats where they are deficient in numbers. This can only be achieved by uplifting the quality of current bioinoculants by evaluating the possible bottlenecks and taking cue from latest research for the development of novel futuristic bioinoculants. Utilization of advanced biotechnological tools such as genomics, proteomics, and genetic engineering can go a long way in achieving the targets. Bt crops are already proving to be a big success story for the case. Exploitation of plant-microbe symbiosis for reclamation of barren or polluted lands, degradation of pollutants, and combating the impact of climate change are the future directions to be explored. The book focuses on these very issues which will be extremely important for fulfilling the food security aspirations of swelling population on reducing arable land in a sustainable manner.

Researchers involved in the field of bioinoculant technology, PGPRs, plant-microbe interactions, rhizosphere biology, bioremediation, biotic and abiotic stress management for crop production, and related areas will find the compilation extremely useful. The researchers will not only get a comprehensive insight on what is happening in this field around the globe but will also be able to know about the future directions of research. The book will be very useful for faculty and graduate and post-graduate students of life sciences, specifically those pursuing career in agriculture microbiology or biotechnology. The book discusses in detail how the plant-microbe symbiosis have been utilized so far and how we can utilize this wonderful relationship in future so as to use fewer chemicals and ensuring for future generations safe and sustainable agri-ecosystems.

I would like to thank all the contributors for providing great knowledge on diverse aspects of the subject. The applications of plant-microbe associations have been comprehensively covered in the tome due to the vast expertize of the authors, and I can say that it is due to their effort that the book has become unique in itself. I thank Dr. Mamta Kapila, Senior Editor, Springer (India), for her continuous support and pursuance. It is because of her that the project could stick to the time frame and see the light of day. My special thanks to Prof. R. C. Sobti, Vice Chancellor, BBA University, Lucknow, for his incessant support and encouragement. I would like to thank the team of my research scholars including Sakshi Tewari, Jitendra Mishra, Rachna Singh, Maya, Shweta and Jay Prakash for helping in compilation of the manuscript. Most importantly, I have to acknowledge the support and care of my wife, Preeti Arora, for looking after me and my kids, Pranay and Nav, who by their presence make all the tensions and fatigue go burst.

I hope that the readers will find the book very beneficial. For any further suggestions or corrections please get back to yours truly.

Lucknow, Uttar Pradesh, India Naveen Kumar Arora

Contents

About the Editor

Dr. Naveen Kumar Arora, Ph.D. Microbiology, Associate Professor and Head, Department of Environmental Microbiology, Babasaheb Bhimrao Ambedkar University (A Central University), Lucknow, Uttar Pradesh, India, is a renowned researcher in the field of Environmental Microbiology and Biotechnology. His specific area of research is rhizosphere biology and PGPRs. He has 41 research papers published in premium international journals and several articles published in magazines and dailies. He is an editor of the book *Plant Microbe Symbiosis: Fundamentals and Advances*, published by Springer. He is a member of several national and international societies and reviewer of several international journals. He has delivered lectures in conferences and seminars around the globe. He has a long-standing interest in teaching at the PG level and is involved in taking courses in bacteriology, microbial physiology, environmental microbiology, agriculture microbiology, and industrial microbiology. He has been an advisor to 57 post-graduate and 3 doctoral students. Recently, he was awarded for excellence in research by the Honorable Governor of Uttar Pradesh. Although an academician and researcher by profession, he has a huge obsession for the wildlife and its conservation and has authored a book, *Splendid Wilds*. He has a dedicated website www.naveenarora.co.in for the cause of wildlife and environment conservation.

Rhizosphere Bacteria for Crop Production and Improvement of Stress Tolerance: Mechanisms of Action, Applications, and Future Prospects

1

Sajid Mahmood Nadeem, Muhammad Naveed, Maqshoof Ahmad, and Zahir Ahmad Zahir

Abstract

Rhizosphere bacteria associated with plant roots can enhance crop productivity through a number of direct and indirect mechanisms. These beneficial bacteria attracted the scientists around the globe due to their significant contribution to mitigate adverse effects of environmental stresses on plants. These plant growth-promoting rhizobacteria (PGPR) have the potential to improve crop production under stress conditions solely and/or in combination with other microbes. The use of PGPR as co-inoculants with symbiotic bacteria is a potential biotechnological approach to promote nodulation for improving crop biomass and soil health. Multi-strain bacterial consortia are also proved useful for enhancing plant growth and development particularly in conditions where single inoculation was not so effective. The objectives of present review are to highlight the basic mechanisms used by such bacteria in general and the applied aspects of these bacteria for improving plant stress tolerance and ultimately crop productivity particularly. Certain examples highlighting their significant role for enhancing plant growth under biotic and abiotic stress conditions have been reviewed. The role of PGPR for improving nodulation when used with nitrogen-fixing bacteria has been discussed. The potential of genetically engineered rhizobacteria that possess the required trait necessary under certain environmental conditions has also been evaluated. The areas that need further research and future perspectives of this technology have been discussed in detail.

S.M. Nadeem
Sub-Campus Burewala, University of Agriculture, Faisalabad, Pakistan

M. Naveed • Z.A. Zahir (✉)
Institute of Soil and Environmental Sciences, University of Agriculture, Faisalabad, Pakistan
e-mail: zazahir@yahoo.com

M. Ahmad
University College of Agriculture and Environmental Sciences, The Islamia University of Bahawalpur, Bahawalpur, Pakistan

N.K. Arora (ed.), *Plant Microbes Symbiosis: Applied Facets*,
DOI 10.1007/978-81-322-2068-8_1, © Springer India 2015

Introduction

Although plants are well adapted to a number of environmental conditions, certain biotic and abiotic stresses cause significant reduction in plant growth and yield. When plants are subjected to environmental stresses, a number of plant processes are affected that ultimately lead to improper growth and development. Under stress, plants reduce their vegetative growth to conserve and redistribute resources that decrease their chances of survival under severe conditions (Skirycz and Inze 2010). These stresses affect the plant growth and development in one way or the other. For example, in water-limited conditions, reduction in initial plant growth occurs that results in plant leaves with fewer and smaller cells (Aguirrezabal et al. 2006; Skirycz et al. 2010). Saline conditions affect the plant growth by a number of negative impacts including ion toxicity, osmotic effect, and specific ion effect (Munns 1993; Ashraf 1994; Sosa et al. 2005; Conde et al. 2011; Habib and Ashraf 2014). The presence of heavy metals and other contaminants also reduces the crop productivity by disturbing various plant processes (Aydinalp and Marinova 2009; Yadav 2010).

In the context of climate change and mounting population pressure, increasing plant tolerance against these stresses is a dire need. A number of techniques have been adopted to understand the mechanisms and providing tools to enhance plant tolerance against environment stresses. Some of these techniques are based on the use of chemicals that may be toxic for environment. For example, there is environmental concern about the application of aminoethoxyvinylglycine (a chemical inhibitor of ethylene) for protecting the plant from stress-induced impact of ethylene. Similarly, there are also concerns about the use of fertilizers, herbicides, fungicides, and pesticides, which provide nutrition as well as protect the plant from pest and diseases.

One of the alternatives and emerging technologies to solve this problem is the use of naturally occurring PGPR. There are a number of reports that show the effectiveness of PGPR for enhancing plant growth and development (Zahir et al. 2004; Paul and Nair 2008; Glick 2012). This environment-friendly population is equally effective for promoting crop productivity under normal and stress conditions (Zahir et al. 2004; Nadeem et al. 2010b; Saharan and Nehra 2011). There are a number of reports revealing the mechanism of growth promotion by PGPR that may be direct or indirect (Nadeem et al. 2010b; Glick 2012; Ahemad and Kibret 2014).

The use of this environment-friendly approach could be one of the alternatives to minimize the use of chemicals, which cause a number of negative impacts on human health directly and indirectly. The use of this technology could also be useful for improving yield of agricultural crops that otherwise looks difficult due to the limited availability of agricultural inputs as well as adverse environmental conditions. The present review highlights the significant impact of PGPR on plant growth and development under stress environment. Some important aspects of PGPR that make it a useful approach for sustainable agriculture have been discussed in detail.

Rhizosphere: A Nutrient-Rich Niche for Beneficial Bacteria

The zone surrounding the plant roots in which complex relations exist between plant, microorganisms, and the soil is called rhizosphere. According to Hiltner (1904), it is a region of soil around the plant root where maximum microbial activity is present compared to surrounding soil. Plant root exudates are a source of food for microbes and these exudates support the growth and development of microbial population. In view of Hiltner, not only the bacteria present in the rhizosphere are attracted by the root exudates but there are also "uninvited guests" that use these exudates for their survival (Hartmann et al. 2008). The chemical changes occurring at soil–root interface have marked influence on plant microbial interactions as well as availability and acquisition of nutrients. The rhizosphere soil is a favorable habitat for microbial growth which exerts a potential impact on plant health as well as soil fertility (Antoun and Prevost 2005; Podile and Kishore 2006).

In the rhizosphere, a number of important and intensive interactions occur among microorganisms, plant, as well as soil microfauna (Antoun and Prevost 2005). As diverse interactions between organisms take place in this region, it is also described as "microbial hot spot" (Whipps 2001). These interactions are based on complex exchanges between roots and microorganisms, and the beneficial, detrimental, and neutral nature of these relationships is all regulated by complex molecular signaling (Dardanelli et al. 2010). The beneficial interactions can significantly influence plant growth and development, and therefore have a positive influence on crop production.

Root exudates contain adequate amount of carbon and energy that is readily available for microbial growth and development and the physiology of microbial cell (Somers et al. 2004). These exudates consist of a number of organic compounds, auxins, sugars, vitamins, and flavonoids (Neumann and Romheld 2001; Somers et al. 2004; Dardanelli et al. 2009; Raaijmakers et al. 2009). Due to the presence of a variety of compounds in the root exudates, microorganisms like PGPR are present in high concentration around the roots and utilize these compounds as food to promote their growth and development (Whipps 1990; Glick et al. 1999).

The length of the rhizosphere soil may vary and range from less than 1 mm to several millimeters depending upon the presence of root exudates (Darrah 1993; Hubel and Beck 1993; Bringhurst et al. 2001). The microbial population in the rhizosphere is different from surrounding environment owing to the presence of large amounts of nutrients released by plant roots (Burdman et al. 2000). According to an estimate, the number of microorganisms in the rhizosphere is 19–32 times higher than surrounding bulk soil, i.e., free of roots (Bodelier et al. 1997). Due to the presence of exudates in this area, most microorganisms occur within 50 μm of root surface. About 50–100 mg of exudates are released from per gram of root that is enough to support 2×10^{10} bacterial cells (Foster 1983). Although a large number of microbial population are present in the rhizosphere, only 7–15 % of the total root surface is generally occupied by microbial cells (Foster 1983; Pinton et al. 2001).

Due to the nutrient-rich environment compared to surrounding bulk soil, the competition among microbial population for nutrients is very high in this environment (Berg et al. 2005). The rhizosphere microorganisms having the ability to grow rapidly by degrading substances like chitin, cellulose, and root and seed exudates control the rhizosphere competition (Baker 1991), and being the most abundant microorganisms in the rhizosphere, bacteria are highly competitive and aggressively colonize the plant roots (Antoun and Prevost 2005; Bouizgarne 2013).

Plant Growth-Promoting Rhizobacteria: An Overview

The term plant-growth promoting rhizobacteria (PGPR) was first used by Kloepper and Schroth (1978) to describe the bacteria that colonize the plant roots. These microorganisms generally exist more or less near the roots due to the presence of root exudates that are a source of food for their growth (Whipps 1990), and most of these microbes depend on root exudates for their survival (Glick et al. 1998; Khalid et al. 2006). Some of these PGPR belong to important genera including *Serratia*, *Arthrobacter*, *Azospirillum*, *Pseudomonas*, *Burkholderia*, *Enterobacter*, *Bacillus*, *Azoarcus*, *Paenibacillus*, *Klebsiella*, *Erwinia*, *Beijerinckia*, *Flavobacterium*, and *Gluconacetobacter* (Okon and Labandera-Gonzalez 1994; Glick 1995; Podile and Kishore 2006; Dardanelli et al. 2009).

Some researchers have described PGPR on the basis of their growth-promoting characteristics. For example, Bashan and Holguin (1998) categorized the PGPR into two classes, i.e., biocontrol PGPB (plant growth-promoting bacteria) and PGPB. Khan (2005) on the basis of the relationship of rhizobacteria with plants divided them into two groups, i.e., symbiotic rhizobacteria and free-living rhizobacteria. However, Gray and Smith (2005) classified PGPR as extracellular and intracellular. According to their view, extracellular PGPR exist in the rhizosphere, on the rhizoplane, or in the spaces between cells of the root cortex, while intracellular PGPR exist inside the root cells. Sayyed et al. (2010) termed beneficial microorganisms as yield-increasing bacteria (YIB) or PGPR. Burr and Caesar (1984) reported these bacteria as plant health-promoting rhizobacteria (PHPR) and nodule-promoting rhizobacteria that are present in the rhizosphere. In view of Vessey (2003), PGPR are bacteria which flourish in the rhizosphere and may grow in, on, or around plant tissues and enhance growth and development. Gray and Smith (2005) mentioned the gradient of root proximity among rhizobacteria, including bacteria living near roots, bacteria in the rhizoplane, bacteria within the root tissues, and bacteria that reside inside the cell in specialized structures called nodule.

There are a number of ways through which PGPR are helpful for promoting plant growth and development. According to Glick (1995), PGPR facilitate plant growth by three different ways that include the synthesis of compounds for plant uptake, facilitating the nutrient uptake, and disease prevention (Garcia de Salamone et al. 2001; Raj et al. 2003; Guo et al. 2004). PGPR use various mechanisms which may take place simultaneously or sequentially at different plant growth stages. These mechanisms are divided into two major categories, i.e., direct and indirect growth-promoting mechanisms (Scher and Baker 1982; Shanahan et al. 1992; Glick 1995; Boddey and Dobereiner 1995; Glick et al. 1995; Flaishman et al. 1996). Nitrogen fixation, phosphate solubilization, and phytohormone and siderophore production are some examples of direct growth promotion by the PGPR (Kloepper et al. 1989; Glick et al. 1995). PGPR also play an important role indirectly in plant growth promotion by inhibiting the growth of plant pathogens (Glick and Bashan 1997; Persello-Cartieaux et al. 2003). This can be achieved by the production of antibiotic and antifungal metabolites, lytic enzymes, and hydrogen cyanide, inducing systemic resistance and also by reducing the availability of certain nutrients such as iron required by the pathogen for its growth (Kloepper and Mariano 2000; Antoun and Kloepper 2001; Van Loon 2007; Labuschagne et al. 2010).

In addition to these general mechanisms of growth promotion, PGPR also protect the plant from the deleterious effects of various stresses. Stress conditions affect not only plant growth but are also harmful for the microbial community. However, under stress conditions, certain PGPR maintain their growth by using particular mechanisms. For example, *Pseudomonas* sp. has the ability to withstand drought stress environment by the production of exopolysaccharides (EPS) (Sandhya et al. 2009). PGPR also enhance plant growth under stress conditions like salinity, drought, heavy metals, flooding, temperature, and petrochemicals like diesel (Saleem et al. 2007; Nadeem et al. 2010b). Under stress conditions PGPR work through particular mechanisms like lowering of stress-induced ethylene, production of EPS, regulating nutrient uptake, enhancing the activity of antioxidant enzymes, and suppressing the growth of pathogens (Glick et al. 2007; Sandhya et al. 2009; Belimov et al. 2009a; Nadeem et al. 2010a; Saharan and Nehra 2011).

Rhizosphere Bacteria for Plant Biomass Production

For maintaining soil function, the presence of microorganisms is very essential because they play critical role in decomposition, nutrient cycling, toxin removal, and improving soil structure (van Elsas and Trevors 1997). Their use for improving nutrient status of soil is very essential and important for agricultural practices (Freitas et al. 2007). There are several reports showing the remarkable role of these microbes for improving plant growth and yield of a number of crops (Kloepper et al. 1980; Biswas et al. 2000; Asghar et al. 2002; Zahir et al. 2004; Gray and Smith 2005; Figueiredo et al. 2008; Nadeem et al. 2010b).

It has been observed that certain PGPR strains have advantage to protect the plant from more than one stress. This growth promotion under different environments indicates that these strains utilize more than one mechanism to promote growth in that particular environment. This might be due to their various growth-promoting traits like 1-aminocyclopropane-1-carboxylate (ACC) deaminase activity and EPS production, inducing systemic resistance (ISR) and better root colonization (Bakker et al. 2007; Sandhya et al. 2009; Zahir et al. 2009; Nadeem et al. 2010b). Some of the important aspects of PGPR that are useful for plant biomass have been reviewed and discussed in the following sections.

Biofertilizers

With increase in population, the demand for food is also increasing. No doubt, the use of modern technologies has increased crop production many times compared to old conventional approaches. However, this increase in crop yield is generally due to the use of excessive amount of fertilizers (i.e., the requirement of high-yielding varieties), pesticides, and other commercial growth-promoting substances. In addition to increase in crop yield, most of these substances also cause negative impact on environment. The excessive losses of chemical fertilizer, pesticides, and other waste products cause environmental and social problems throughout the world (Singh et al. 2011). Although certain physical and chemical methods are being adopted to solve such problems, it has been observed that it cannot be done effectively without applying microbial methods and techniques (Singh et al. 2011).

One of the environment-friendly techniques is the use of biofertilizers. A biofertilizer is a product that contains living microorganisms, which enhance plant growth and yield through direct or indirect mechanisms (Fuentes-Ramirez and Caballero-Mellado 2005). According to Vessey (2003), biofertilizer is a microbial product, which when applied colonizes the rhizosphere or plant interior and enhances the growth of plant by increasing the availability or supply of nutrients.

Biofertilizers are important components of integrated nutrient systems, and their use (containing beneficial microorganisms) instead of synthetic chemicals is known to improve plant growth and development and may help to sustain environmental health and soil productivity (O'Connel 1992; Wu et al. 2005). The mechanisms of biofertilizer for improving plant growth are not still understood clearly; however, some growth-promoting attributes have been mentioned in the literature, which include nitrogen fixation, phytohormone production, solubilization, and availability of nutrients, reducing stress impact by enzymes like ACC deaminase, chitinase, etc. (Vessey 2003; Glick 2012; Ahemad and Kibret 2014). Some of the selected examples of the effect of biofertilizers on different crops have been reviewed in the following sections.

Cereals

Cereals like wheat, maize, and rice are the main staple food for most of the world population. Due to an increase in population, their demand is also increasing. The extensive use of chemical fertilizers for increasing yield of such valuable crops results in a negative impact on environment. Moreover, these nonleguminous crops do not fix nitrogen, and therefore, application of nitrogen to these crops is another expensive input. However, the use of PGPR can improve the nitrogen uptake of nonleguminous crops as it was observed in maize (31 % N) and palm seedlings where inoculated plants showed 31 and 40 % more nitrogen, respectively, compared to uninoculated control (El-Komy et al. 1998; Amir et al. 2001). Similarly, inoculation of nitrogen-fixing and phosphate-solubilizing bacteria (PSB), in combination, proved more effective for enhancing nutrient uptake in various crops including sorghum, barley, and wheat (Alagawadi and Gaur 1992; Belimov et al. 1995; Galal 2003).

Yazdani et al. (2009) in a field experiment observed that application of manure as well as biofertilizer enhanced the yield of maize. They further concluded that application of PSB along with other PGPR strains reduced the application of phosphorus up to 50 % without any reduction of yield. They concluded that up to 50 % nitrogen fertilizer could be saved by inoculation with

Azospirillum and phosphobacteria. However, in earlier study conducted on millet, Chandrasekar et al. (2005) demonstrated that the effectiveness of biofertilizer was more when used with chemical fertilizers. The efficiency of biofertilizer also increased when used with other microbial populations. For example, the use of mycorrhizal fungi along with bacteria enhances the effectiveness of biofertilizers. Wu et al. (2005) developed two biofertilizers containing arbuscular mycorrhizal fungi (*Glomus mosseae* and *Glomus intraradices*) with or without bacterial strains (nitrogen fixer, *Azotobacter chroococcum*; PSB, *Bacillus megaterium*; and K solubilizer, *Bacillus mucilaginous*). Effect of biofertilizer was evaluated on soil particles and maize (*Zea mays*) growth. They also compared the efficiency of these biofertilizers with chemical and organic fertilizers. The results showed that the biofertilizer containing mycorrhizal fungi with three strains of rhizobacteria significantly enhanced the growth of maize. The application of biofertilizer not only improved the plant growth but also caused positive effect on soil properties. The organic matter and nitrogen contents of soil increased with the application of biofertilizer. It was also observed that half the amount of biofertilizer caused similar effect when compared with organic and chemical fertilizer.

Legumes

Legumes are well known for their ability to fix atmospheric nitrogen, which may be a good source of nitrogen for the succeeding crop. To fulfill the phosphorus requirements of plants, soluble phosphatic fertilizers are preferably used for crop production; however, the manufacturing cost of chemical fertilizers makes them economically unaffordable. Stamford et al. (2007) conducted a study to minimize this problem by using biofertilizers. Experiments were carried out on yam bean (*Pachyrhizus erosus*) in low-P available soil. Biofertilizer was prepared from rock phosphate and sulfur and then inoculated with *Acidithiobacillus*. An increase in soil acidity and available P was observed where biofertilizers were applied. This study indicates better performance of biofertilizer for enhancing growth.

Vermicompost, a biofertilizer that contains worms, beneficial bacteria, and a number of major and micronutrients, proves very useful for enhancing plant growth. Jeyabal and Kuppuswamy (2001) studied the response of vermicompost prepared from agricultural and industrial waste by using earthworms (*Eudrilus eugeniae*) along with fertilizer nitrogen and another biofertilizer prepared from *Azospirillum* and phosphobacteria on rice–legume system. The study showed that application of vermicompost, *Azospirillum*, and phosphobacteria biofertilizer increased rice yield compared to fertilizer nitrogen alone. The residual effect of these fertilizers was studied on the succeeding legume. The residual effect of N supplied through vermicompost and *Azospirillum* produced higher nodule number and nodule weight than fertilizer N alone. The yield was also higher in case of biofertilizer compared to nitrogen fertilizer. Similarly, Sinha et al. (2010) studied the efficacy of vermicompost against DAP (diammonium phosphate) and FYM (farm yard manure) on *Cicer* and *Pisum*. A significant improvement of growth parameters like root length, shoot length, number of pods, and number of nodules in both crops was observed compared to untreated control. The applications of biofertilizer also increased the population of nitrogen-fixing bacteria.

The application of biofertilizers has proved not only useful for enhancing growth and yield of grain legumes but also the biomass of fodder legume crops. *Galega orientalis* is a perennial leguminous fodder crop. Biofertilizer caused positive effect on the growth of this fodder crop (Arkhipchenko et al. 2005). Three biofertilizers were prepared from pig farm sludge and poultry dung from aerobic and poultry dung from anaerobic conditions and then checked for quality, nutrient content, and bacterial communities. *Bacillus* spp. represented maximum microbial population, i.e., 20–35 % of the viable heterotrophic microorganisms in the final product. The population of pathogenic fungi was suppressed in the final product which might be due to *Bacillus* sp. because of their ability to produce antibiotics to suppress the growth of phytopathogenic fungi (Foldes et al. 2000; Knox et al. 2000). Application of these fertilizers to *G. orientalis* significantly increased the yield. It was also observed that biofertilizer also reduced the damage of insects to plants. Moreover, biofertilizers prepared from aerobic conditions were more effective than anaerobic ones.

Miscellaneous

The use of biofertilizers has also been reported to have positive results on other crops. The field study of Shehata and El-Khawas (2003) on sunflower (*Helianthus annuus*) showed the effectiveness of biofertilizers in natural conditions. Two biofertilizers (Biogien and Microbien) were applied alone and in combination. The application of biofertilizers whether applied in combination or separately significantly enhanced the growth and yield of sunflower. Biofertilizer also enhanced oil and nutrient content of seed as well as nitrogen and nitrogenous compounds and, however, decreased the saturated fatty acids and increased the unsaturated fatty acids.

Nutrition requirement of banana crop is high and use of chemical fertilizers increases the production cost of this crop. Literature showed the beneficial effects of biofertilizer on banana crop. For example, Rivera-Cruz et al. (2008) conducted a study to compare the effectiveness of poultry manure and banana waste as inoculant carrier of bacterial strain and also to establish efficient dose of biofertilizer. They demonstrated that application of biofertilizer not only increased the plant growth but also enhanced the phosphate solubilization, enzyme activities, and aggregate stability. They found that both fertilizers were useful carriers for PGPR. Keeping in view the huge requirement of fertilizers for banana, which is not only costly but has negative impact on environment also, Mia et al. (2010) conducted studies to evaluate the role of PGPR for improving the nutritional requirements of banana. It has been observed that PGPR inoculation with 33 % nitrogen not only enhanced the nutrient supply but also improved root and shoot growth.

It is evident from the above discussion that the use of PGPR as biofertilizers is very useful for promoting plant growth. These biofertilizers are not only useful for fulfilling the nutrient requirement of the plant but also enhance the plant growth by various mechanisms. It has also been observed that although the application of biofertilizers alone can promote plant growth, combined application of biofertilizer with chemical fertilizers can give better results. Therefore, the application of biofertilizer can reduce the use of chemical fertilizer and significant yield could still be obtained.

Phytostimulators

As it was discussed earlier, PGPR use different mechanisms for improving plant growth and development. One of the well-established and common mechanisms used by PGPR is the production of plant growth regulators (PGRs), also called as phytohormones. These PGRs stimulate many plant processes and ultimately enhance growth of plant. Therefore, PGPR owing to their ability to produce phytohormones also act as phytostimulators. The five major plant growth-promoting hormones that play important role in plant growth promotion include auxins, abscisic acid, gibberellins, cytokinins, and ethylene (Arshad and Frankenberger 1998; Garcıa de Salamone et al. 2001; Gutierrez-Manero et al. 2001; Egamberdiyeva 2005).

Although a number of workers have shown that plants have the ability to synthesize these phytohormones endogenously (Moore 1989; McKeon et al. 1995; Walton and Li 1995), PGPR, particularly rhizosphere bacteria, are another potential source of these growth-promoting substances (Frankenberger and Arshad 1995; Patten and Glick 1996). Several workers have reported the production of PGRs from a wide variety of rhizobacteria (Arshad and Frankenberger 1998; Lugtenberg and Kamilova 2009; Ahemad and Kibret 2014). No doubt, chemically synthesized PGRs can also be applied for promoting plant growth; however, microbial-derived PGRs are more effective and economical. Due to low thresholds between inhibitory and stimulatory levels of chemically synthesized PGRs, microbial-produced hormones are more effective by virtue of their continuous and slow release (Khalid et al. 2006).

These phytohormones enhance plant growth not only in normal conditions, but some hormones also play important role by promoting plant growth in unfavorable environment (Barazani and Friedman 1999; Patten and Glick 2002; Raghavendra et al. 2010). For example, indoleacetic acid (IAA) is helpful for alleviating the impact of drought stress on plant growth and development (Leinhos and Bergmann 1995). San-Francisco et al. (2005) while studying the effect of IAA and its precursor on the development, mineral nutrition, and free polyamine content of pepper plants found an increase in spermine and a decrease in putrescine in pepper leaves by the application of IAA. Similarly, abscisic acid (ABA) also considered as an important PGR, playing an important role in the photoperiodic induction of flowering (Wilmowicz et al. 2008), plays central role in stress (Zhang et al. 2006).

In earlier studies, it has been observed that high nitrate concentration is inhibitory to root growth (Zhang et al. 1999; Scheible et al. 1997; Forde 2002); however, the mechanism of this inhibition was not known. Tian et al. (2008) reported that reduced maize growth due to elevated levels of nitrate was due to the low IAA level in the plant root. Salinity is a major constraint to sustainable agriculture, and a decrease in cytokinin supply from root to shoot was also observed under salinity stress (Naqvi and Ansari 1974). Egamberdieva (2009) attributed better growth of inoculated wheat under salinity stress to the production of phytohormones by PGPR. He used PGRs including gibberellins, auxins, zeatin, and ethephon as well as auxin-producing *Pseudomonas* spp. The results showed that PGRs and *Pseudomonas* spp. significantly enhanced the seedling growth under saline conditions compared to control.

Arkhipchenko et al. (2005) observed a positive response on growth of legume fodder by the application of PGPR. They found a significant quantity of L-tryptophan that was the major precursor of IAA (Barazani and Friedman 1999). They demonstrated that this growth promotion was due to IAA. It is also evident from the work of Cassan et al. (2009) who observed that bacterial strains *Azospirillum brasilense* and *Bradyrhizobium japonicum* enhanced the germination and growth of corn and soybean. The effect was positive whether the strains were used singly or in combination. The strains had the ability to produce IAA, gibberellic acid, and zeatin (member of cytokinins). They observed that both strains were able to secrete these hormones into the culture medium, at a concentration that was sufficient to produce morphological and physiological changes in young seed tissues.

Barbieri et al. (1991) and Burdman et al. (1996) reported the effectiveness of *Azospirillum* sp. for enhancing root growth in grasses and legume, respectively, through phytohormone production. Also, IAA-producing *Bradyrhizobium* sp. enhanced total biomass and grain yield in chickpea (Alagawadi and Gaur 1988). The co-inoculated effect of these two strains, *Azospirillum* and *Bradyrhizobium* sp., was evaluated by Molla et al. (2001). In a laboratory study, soybean was co-inoculated with *Azospirillum* spp. (*A. brasilense* and *A. lipoferu m*) and *Bradyrhizobium* sp., and its potential for enhancing root growth and nodulation was evaluated. They observed that total root length, root hair, and shoot dry weight significantly increased by co-inoculation. Co-inoculated plants also showed more number of nodules and weight. It was also observed that *A. brasilense* performed better than *A. lipoferum*.

Phytoextraction is an important and attractive approach to clean up the soil environment from toxic heavy metals. The ability of a plant to take up heavy metals depends upon its better root system. The PGRs are helpful in phytoextraction by influencing the root growth and metal availability. IAA is known to induce root growth by enhancing cell division and cell extension (Taiz and Zeiger 2000). It has been observed that chelating agents used to extract metals increase the metal concentration that may become harmful for plant growth (Tandy et al. 2006); however, this negative impact could be eliminated by the application of IAA (Liu et al. 2007). Tassi et al. (2008) found that exogenous PGR based on cytokinins positively enhanced phytoextraction of Pb and Zn and improved biomass production of *Helianthus annuus*.

From the above discussion, it is clear that phytohormones produced by PGPR play an important role for plant growth promotion. These phytohormones accelerate a number of plant processes. A large number of PGPR strains produce different types of growth hormones. These phytohormones are not only helpful for normal growth processes but also protect the plants from negative impact of stress. Owing to the ability of producing hormones, these PGPR strains enhance the plant growth and development under adverse conditions.

Synergistic Application with Symbiotic Bacteria (Co-inoculation)

The effectiveness of PGPR for enhancing plant growth is well documented. These beneficial bacteria can also be used effectively with other microbes for promoting their growth and yield. This co-inoculation is equally effective under normal as well as stress conditions. The study of Mishra et al. (2011) showed the effectiveness of co-inoculation of *Pseudomonas* and *Rhizobium* sp. for growth enhancement and nutrient acquisition of lentil. Although inoculation of bacterial strain singly caused significant effect on growth, nodulation, and nutrient uptake, co-inoculation of *Pseudomonas* with *Rhizobium* recorded maximum increase in nodulation, chlorophyll content, and total iron, nitrogen, and P uptake. This study indicates the synergistic relationship between the strains that is helpful for improving plant growth and also suggested the use of such strains for biofertilizer production. Previous studies have also revealed that combination of microorganisms showed synergistic interaction and was found helpful for enhancing plant growth (Khammas and Kaiser 1992; Belimov et al. 1995; Bashan and Holguin 1997a, b; Bashan 1998). For example, *Azospirillum* is considered as a *Rhizobium* helper, which enhances nodulation and nodule activity and provides resistance in plants against unfavorable conditions (Itzigsohn et al. 1993; Fabbri and DelGallo 1995). Egamberdieva et al. (2010) found that co-inoculation of *Pseudomonas* sp. capable of producing IAA and *Rhizobium* (not producing IAA) enhanced shoot and root dry weight of fodder *Galega* (*Galega orientalis* Lam.) compared to *Rhizobium* alone.

The co-inoculation becomes more essential when a single strain is not so effective in some particular environment. For example, under salinity stress, reduction in nodulation has been observed (Zahran 1999). Nitrogenase enzyme that is a major component of nitrogen fixation is affected by salt concentration (Jofre et al. 1998). Also, under stress environment, elevated level of ethylene is produced that is inhibitory for the overall process of nitrogen fixation (Taiz and Zeiger 2000). This elevated level of ethylene is due to ACC that is the immediate precursor of ethylene. It has been observed that certain PGPR strains contain ACC deaminase enzyme, which could be helpful for reducing the ethylene concentration that causes negative impact on nodulation in legumes under stress environment (Ahmad et al. 2011). It was observed that co-inoculation of *Pseudomonas* and *Rhizobium phaseoli* proved very effective for enhancing growth and nodulation of mung bean under salinity stress. It has been observed that this growth enhancement was equally effective under natural salt-affected field condition (Ahmad et al. 2013a, b). Co-inoculation of *Rhizobium* with two different PGPR strains enhanced the growth of mung bean in saline environment under axenic conditions (Fig. 1.1). This growth enhancement was also observed when same study was conducted in greenhouse (Fig. 1.2). Similarly, Malik and Sindhu (2011) also observed that co-inoculation of *Mesorhizobium* sp. with IAA-producing *Pseudomonas* increased the nodule number and nodule biomass of chickpea (*Cicer arietinum*). The positive results of co-inoculation are not only observed in case of legume but also in nonlegumes. The study of Bano and Fatima (2009) showed that inoculation of *Pseudomonas* and *Rhizobium* proved useful for promoting maize growth under saline conditions by increasing water content and decreasing electrolyte leaking of inoculated plants.

However, for getting significant results, it is necessary to use the partners that have compatibility with each other. In case of noncompatible

Fig. 1.1 Effect of co-inoculation on root/shoot growth of mung bean seedlings under axenic conditions at 6 dS m^{-1} (**Mk20** and **Mk25**: PGPR containing ACC deaminase; **M9** and **M6**: *Rhizobium*)

Fig. 1.2 Effect of co-inoculation on growth of mung bean at 6 dS m^{-1} in pot trial (**Mk25**: PGPR containing ACC deaminase; **M9**: *Rhizobium*)

partner, effective results may not be obtained. Moreover, the kind of plant may also have influence on the performance of inoculation. For example, the study of Rosas et al. (2006) showed that co-inoculation of *Pseudomonas putida* with *Sinorhizobium meliloti* was more effective in case of soybean than alfalfa where *Bradyrhizobium japonicum* was more successful with *P. putida*.

Rhizosphere Bacterial Consortia

Although PGPR inoculation has been proved useful for enhancing plant growth and development in normal and stress conditions, in certain cases particularly in field, inconsistent results are obtained with single inoculum (Lucy et al. 2004; Smyth et al. 2011). This inability of bacteria to show their full potential might be due to certain factors including competition with indigenous population and low-quality inoculum. The use of multi-strain inoculum could be one of the best solutions to this problem. A multi-strain bacterial consortium is a group of different species of bacteria, which act together as a community. The inoculation of *Withania somnifera* (Ashwagandha) with multi-strain consortia of *Azospirillum*, *Azotobacter*, *Pseudomonas*, and *Bacillus* showed significant increase in plant height and root length when compared to uninoculated control (Rajasekar and Elango 2011). For obtaining significant results, it is necessary to use such species which have compatibility with each other in addition to having good growth-promoting traits. Choure and Dubey (2012) suggested some characteristics of a good bacterial consortium. These include bacterial species which are resistant to environmental stress, have natural enzymatic activity and long shelf life, and are nonpathogenic besides having good sustainability and economic feasibility.

The performance of bacterial consortium was more effective when used in a stress environment. The study of Upadhyay et al. (2011) conducted under salinity stress condition showed that dual inoculation enhanced wheat growth by increasing sugar and proline content as well as decreasing sodium content of inoculated wheat. Similarly, Annapurna et al. (2011) demonstrated that single and dual inoculation showed variable response under stress condition. Although single inoculation also proved effective for diluting the negative impact of salinity, the bacterial consortium was more effective for inducing salinity tolerance in wheat. Similarly, the use of more than one bacterial species was also effective for promoting nodulation when bean plants were co-inoculated with *Rhizobium* spp. and *Bacillus polymyxa* (Petersen et al. 1996).

The susceptibility of plant to pathogens is one of the important factors that adversely affect plant growth and development and ultimately cause yield reduction. Although plant pathogens can be controlled effectively by synthetic chemicals, there are certain environmental concerns related to these toxic substances. It has been observed that the use of bacterial consortium is also very effective for promoting plant growth by protecting them from pathogenic diseases. The combined application of *Sinorhizobium fredii* KCC5 and *Pseudomonas fluorescens* LPK2 significantly reduced the wilt disease as reported by Gupta et al. (2002). They demonstrated that the use of microbial consortia could be an effective way for the management of fungal disease. The study of Choure and Dubey (2012) showed that the multi-strain consortium of three bacterial species *P. fluorescens* LPK2, *S. fredii* KCC5, and *A. chroococcum* AZK2 inhibited the conidial and chlamydospore germination of *Fusarium udum* and reduced the *Fusarium* wilt in *Cajanus cajan*. They demonstrated that multi-strain bacterial inoculum can be used effectively for protecting the plant from fungal diseases.

Precursor–Inoculum Interaction

As discussed in the earlier section, PGPR produce a number of phytohormones that stimulate plant growth and development. These growth hormones are also very effective for enhancing plant growth under stress environment. It has been observed that same strains may produce variable amount of phytohormones that could be the reason of inconsistent results obtained under certain conditions. This might be due to certain factors; among them the availability of suitable precursor is one of the primary factors affecting the secretion of these hormones. This drawback can be overcome by the use of specific precursor in addition to inoculation. It has been observed that exogenous application of a precursor increased the magnitude of phytohormone production many fold in culture as well as in soil (Asghar et al. 2000, 2002). This precursor–inoculum interaction is useful for promoting plant growth. A number of studies conducted by various workers showed that precursor–inoculum interaction significantly enhanced plant growth (Sarwar and Kremer 1995; Zahir et al. 1997; Arshad and Frankenberger 1998). This approach is also effective for controlling the production of a hormone by providing suitable concentration of a particular substrate.

The approach is effective in a number of ways particularly in conditions where the application of a particular hormone is difficult. For example, ethylene is a gaseous hormone that stimulates various physiological processes in plants. The application of ethylene due to its gaseous nature is difficult. However, its precursor L-methionine (L-MET) can release ethylene in the presence of soil microflora therefore stimulating plant growth (Arshad and Frankenberger 1998; Khalid et al. 2006; Mahmood et al. 2008).

L-tryptophan is considered as an efficient precursor for microbial biosynthesis of auxins (Frankenberger and Arshad 1991). Zahir et al. (2005) conducted a study to evaluate the effectiveness of precursor–inoculum interaction for enhancing maize growth. Auxin production by *Azotobacter* was measured in the presence and absence of its precursor L-tryptophan. Results of field trials showed that application of *Azotobacter* alone or in combination with L-tryptophan significantly enhanced maize growth. However, combined application of *Azotobacter* and L-tryptophan caused pronounced effect on grain yield, fresh biomass, and total nitrogen uptake compared to uninoculated control and where sole inoculation was used. This study showed that the application of a precursor along with microbial strain could be more effective. Prior to this, similar results were also been observed in other crops (Zahir et al. 1997, 2000). In an earlier study, Khalid et al. (1999) also reported the growth and yield enhancement of wheat by the application of *Azotobacter* in the presence of L-tryptophan.

Precursor–inoculum interaction is not only effective under normal conditions but is also useful for promoting plant growth under stress conditions. To evaluate the performance of *Rhizobium* for improving growth and yield of mung bean (*Vigna radiata* L.), a pot study was conducted under saline conditions in the presence and absence of L-tryptophan (Zahir et al. 2010). Although salinity reduced the growth of mung bean, application of L-TRP and *Rhizobium* appeared to mitigate the adverse effects of salt stress. It was further observed that combined application of *Rhizobium* and L-tryptophan produced more pronounced effect.

The above discussion shows that although PGPR can enhance plant growth by phytohormone production, in certain cases the expected results are not obtained. This might be due to their inefficient production of these hormones. This low production of phytohormones occurs due to unavailability of a particular substrate. It has been observed that such limitation can be removed by using suitable substrate during inoculation. The use of this technology could be beneficial in the environment where sole bacterial inoculation cannot perform efficiently. The significant results can also be obtained by using multi-strain consortia.

Rhizosphere Bacteria for Improving Stress Tolerance

In arid and semiarid climate, different environmental stresses affect plant growth by affecting a number of physiological processes positively or negatively. These stresses may be abiotic (salinity, drought, temperature, and heavy metal stress) or biotic, i.e., pathogen and disease stress. Stress induces different physiological disorders such as nutrient deficiency, hormonal imbalance, increase in concentration of certain osmolytes, and higher levels of ethylene in plants. High ethylene concentration is one of the contributing factors for premature senescence of different plant parts. Synthesis of ethylene under stress can be suppressed biologically through the action of rhizosphere bacteria that can degrade the ethylene precursor ACC through their enzyme ACC deaminase (Glick et al. 1998). Rhizosphere bacteria have the potential to colonize plant roots under stressed environment. These bacteria lower the endogenous level of ethylene and thus promote plant growth. In addition, these bacteria possess several other traits like production of different growth regulators such as auxins, gibberellins, cytokinins, and/or polyamines, which directly promote plant growth or indirectly induce systemic resistance. In this section, significant improvement in plant biomass under stress conditions through inoculation with different strains of PGPR has been reviewed and discussed.

Abiotic Stress Regulation

Salinity, drought, temperature, and heavy metals are some of the important abiotic stresses that adversely affect the plant growth one way or the other. Some of the common effects of these stresses on plant growth include hormonal and nutritional imbalances, ion toxicity, and osmotic effect. PGPR have shown to facilitate plant growth indirectly by making the plant more resistant to stress (Glick 2004) or directly by facilitating the uptake of nutrients from the environment, by influencing phytohormone production (e.g., auxin, cytokinin, and gibberellins), by enzymatic lowering of plant ethylene levels, and/or by production of siderophores (Patten and Glick 1996; Kohler et al. 2006; Nadeem et al. 2009; Ahmad et al. 2011). The production of exopolysaccharides and disaccharides by some rhizosphere bacteria may act as protectants against abiotic stresses such as drought, salinity, and temperature stress. For example, some bacteria have the ability to produce trehalose which acts as protectant in plants against abiotic stresses. Trehalose is resistant to acid and high temperature and is a highly stable molecule that can form a gel phase as cells dehydrate, replacing water and, as a result, decreasing damage from drought and salt. Moreover, trehalose also prevents degradation and aggregation of proteins under different stresses. Thus, inoculation of plants with trehalose-producing rhizobacteria confers resistance against stresses (Suarez et al. 2008; Rodriguez-Salazar et al. 2009). The effectiveness of PGPR inoculation to reduce the negative impact of these stresses on plant growth has been reviewed in the following sections.

Salinity Stress

Salinity is one of the most critical constraints hampering agricultural productions in many areas around the world. However, utilization of salt-affected soils for agricultural production is also indispensable to feed the burgeoning population over the globe. Reducing salinity-induced ethylene by any mechanism could decrease the negative impact of salinity on plant growth. Recent studies have revealed that plants inoculated with rhizobacteria were better able to thrive through the salinity stress while demonstrating a normal growth pattern. Hydrolysis of ACC by rhizobacterial strains carrying ACC deaminase activity diluted the adverse effect of salinity by decreasing the ethylene production thus improving seedling growth of plants (Shaharoona et al. 2006; Nadeem et al. 2007, 2009; Ahmad et al. 2011). Reduction in ethylene production due to inoculation with PGPR containing ACC deaminase has been reported in various studies as well (Glick et al. 1998; Indiragandhi et al. 2008).

Inoculation with PGPR containing ACC deaminase has been reported to eliminate or at least reduce the stress-induced ethylene-mediated negative impact on plants (Glick 2005; Safronova et al. 2006). It has been well documented that different bacterial species belonging to different genera such as *Agrobacterium*, *Achromobacter*, *Azospirillum*, *Burkholderia*, *Enterobacter*, *Pseudomonas*, and *Ralstonia*, possessing variable ACC deaminase activity, increased the growth of different crops (Arshad et al. 2008; Nadeem et al. 2010b). Many researchers have reported that exogenous application of ethylene and/or salinity had adverse effect on plant growth; however, inoculation with PGPR containing ACC deaminase improved the plant growth (Shaharoona et al. 2006; Cheng et al. 2007; Nadeem et al. 2009; Ahmad et al. 2011). Improvement in dry matter and salt tolerance of tomato (Mayak et al. 2004a), canola (Cheng et al. 2007), mung bean (Ahmad et al. 2011), wheat (Zahir et al. 2009; Nadeem et al. 2010a), and chickpea (Patel et al. 2012) has been reported due to inoculation with PGPR containing ACC deaminase under stress.

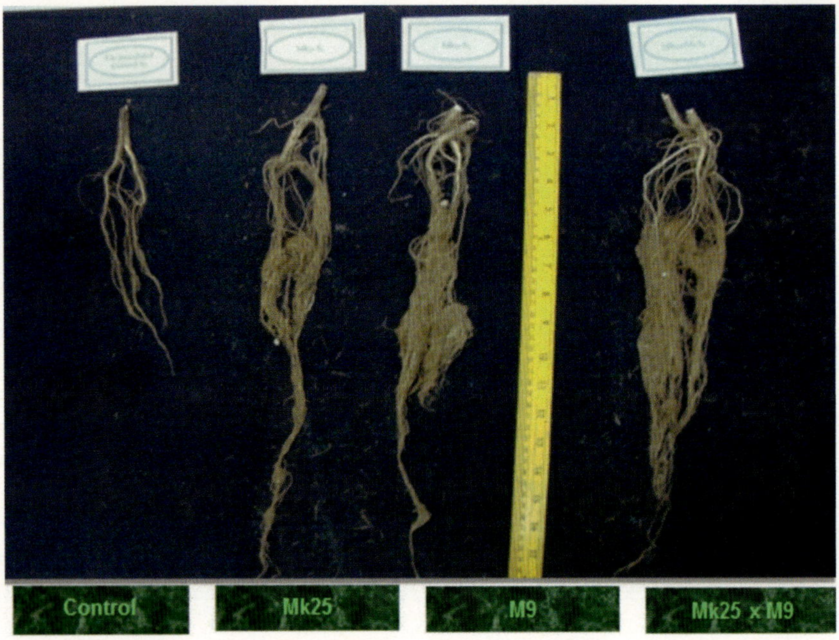

Fig. 1.3 Effect of co-inoculation on root length of mung bean at 6 dS m^{-1} in pot trial (**Mk25**: PGPR containing ACC deaminase; **M9**: *Rhizobium*)

The worth of PGPR strains containing ACC deaminase, i.e., *P. fluorescens*, has been reported by Shaharoona et al. (2008), where they studied that the efficacy of these strains carrying ACC deaminase may likely be due to the reduction of stress-induced ethylene level. It is also evident from a pot study where co-inoculation of *Rhizobium* with a PGPR strain containing ACC deaminase significantly enhanced the root length (Fig. 1.3) might be due to lowering of stress-induced ethylene (Ahmad et al. 2012).

The microbial treatments also reduce the Na^+ uptake by plants and increase the K^+ uptake thus increasing the K^+/Na^+ ratio which otherwise is reduced by salinity (Giri et al. 2007). The rhizobacterial species *Variovorax paradoxus* effectively colonized the plant roots under stressed conditions and hence can be used efficiently for attenuating the growth inhibition of plants under stressed conditions (Belimov et al. 2009a). These PGPR containing ACC deaminase can decrease the stress-induced endogenous ethylene level (Mayak et al. 2004a). The inoculation with *A. brasilense* increases the relative water content in the leaf due to the production of ABA by the bacterial strains thus reducing the ACC level in plants (Cohen et al. 2008).

Mayak et al. (2004a) reported that *Achromobacter piechaudii* having ACC deaminase activity significantly increased the fresh and dry weights of tomato seedlings grown in the presence of NaCl (up to 172 mM). The bacterium reduced the production of ethylene by tomato seedlings, which otherwise was stimulated when seedlings were grown with increasing salt concentrations. However, the sodium content of the plant was not decreased, whereas the uptake of phosphorous and potassium was slightly increased, which might have contributed in part to the activation of processes involved in the alleviation of the adverse effects of salts on plant growth. The bacterium also increased the water use efficiency (WUE) in saline soil and helped in alleviating the salt suppression of photosynthesis. *Pseudomonas fluorescens* strain TDK1 containing ACC deaminase activity enhanced the saline resistance in groundnut plants and increased yield as compared with that inoculated with *Pseudomonas* strains lacking ACC deaminase activity (Saravanakumar and Samiyappan 2007).

Similarly, ACC deaminase bacteria have also been reported to increase the salt tolerance in plants by lowering the synthesis of salt-induced stress ethylene and promote the growth of canola in saline environment (Cheng et al. 2007). Nadeem et al. (2009) observed similar results in case of maize grown under salt stress in response to inoculation with PGPR containing ACC deaminase activity. Hamaoui et al. (2001) reported that inoculation of chickpea with *A. brasilense* significantly reduced the negative effects on plant growth caused by irrigation with saline water (EC, 4.8 dS m^{-1}; 714 mg L^{-1} NaCl) under controlled conditions. They conducted an experiment in a greenhouse and found that the inoculation with *A. brasilense* significantly increased the nodulation and root and shoot development under both normal and saline water conditions. It has been well documented that the PGPR containing ACC deaminase have the ability to tolerate high concentrations of salinity under a wide range of temperature variations (Hontzeas et al. 2005).

Chlorophyll contents in the plant shoot are increased when inoculated with *Pseudomonas* stains having ACC deaminase activity (Glick et al. 1997; Nadeem et al. 2009). This increase in chlorophyll content may be due to the increased photosynthetic leaf area of the plant by inoculation compared to uninoculated control where leaf area was reduced due to salinity stress (Marcelis and van Hooijdonk 1999). Thus, improvement in chlorophyll contents might have resulted in increased photosynthetic rate which ultimately produced more photosynthates, leading to increased yield of crops.

Increased nutrient concentration in maize plants has been reported under salt stress due to bacterial inoculation in crop plants (Hamdia et al. 2004; Nadeem et al. 2009). Inoculation with PGPR containing ACC deaminase significantly increased P and K contents in salt-stressed tomato plants (Mayak et al. 2004a). Similarly, enhanced uptake of N, P, K, and Ca content of soybean due to inoculation with PGPR under salinity stress has been reported by Han and Lee (2005). It has been reported that the root elongation rate and mineral N, P, and K and microelement uptake are consequently improved after *Azospirillum* inoculation (Dobbelaere and Okon 2007). This could result in a generally better mineral nutrition of the plant (such as phosphorous, iron, and molybdenum) essential for rhizobia–nodule formation and nitrogen fixation activities (Burdman et al. 1998). Thus, enhanced nodulation with improved mineral nutrition might have improved the yield of mung bean under salt-affected conditions.

Inoculation with PGPR containing ACC deaminase reduces the absorption and uptake of Na^+. The reduction in Na^+ uptake might be due to the decreased passive flow of Na^+ into the vascular tissues due to more proliferation of roots in inoculated/co-inoculated plants. Similarly, Nadeem et al. (2009) observed that PGPR containing ACC deaminase significantly increased K^+ in maize under salinity stress. Increase in K^+ uptake and decreased Na^+ uptake due to inoculation with PGPR having ACC deaminase have also been reported by Yue et al. (2007). It has been reported that an increase in concentration of K^+ in plants under salt stress could ameliorate the deleterious effects of salinity on growth and yield (Giri et al. 2007). This improvement in ionic/nutritional balance might have resulted into more growth and yield of crop plants.

Different PGPR strains differ in their ability to promote plant growth due to difference in ACC deaminase activity (Shaharoona et al. 2006; Nadeem et al. 2007, 2009). This difference may also be due to the presence of other growth-promoting characters, i.e., chitinase activity, phosphate solubilization, root colonization, etc., in addition to ACC deaminase activity. Mechanisms used by rhizobacteria to induce salinity stress have been summarized in Table 1.1.

Drought Stress

Drought is one of the most important abiotic stresses that have drastic effects on plant growth and productivity by creating dehydration and osmotic stress (Boyer 1982; Glombitza et al. 2004). There are a number of chemicals (Mascher et al. 2005; Jaleel et al. 2007; Liu et al. 2009), PGRs (Lu et al. 2009; Yuan et al. 2010), and alternative products which can be used to reduce the deleterious effects of drought on plant growth

Table 1.1 Abiotic stress tolerance in plants induced through rhizosphere bacteria

Crop	Rhizobacterial strain	Mechanism used/processes regularized	Reference
Salt stress			
Chickpea	*P. putida* MSC1 and *Pseudomonas pseudoalcaligenes* MSC4	Phosphate solubilization, siderophore production, and IAA production	Patel et al. (2012)
Pepper	*Bacillus aryabhattai* RS341 and *Brevibacterium epidermidis* RS15	Suppression of ethylene production due to ACC deaminase activity	Siddikee et al. (2012)
Eggplant	*Xanthobacter autotrophicus* BM13, *Enterobacter aerogenes* BM10, and *Bacillus brevis* FK2	Increase in potassium, K^+–Na^+ ratio, and K^+–Na^+ selectivity in the eggplant shoots	Abd-El-Azeem et al. (2012)
Maize	*P. fluorescens* S20, and *Enterobacter aerogenes* S14	ACC deaminase activity, siderophore production, enhancing microbial growth in the rhizosphere	Ahmad et al. (2014)
	A. chroococcum	Phosphate solubilization, auxin production, and catalase activity	Rojas-Tapias et al. (2012)
	E. aerogenes and *P. fluorescens*	ACC deaminase activity	Zahir et al. (2012)
Cotton	*P. putida* Rs-198	Increased absorption of Mg^{2+}, K^+, and Ca^{2+} and decreased uptake of Na^+ from the soil and also improved production of endogenous indoleacetic acid (IAA) content and reduced abscisic acid (ABA) content of cotton	Yao et al. (2010)
Arabidopsis thaliana	*B. subtilis* GB03	Soil bacteria-induced salt tolerance in plants by tissue-specific regulation of the sodium transporter HKT1	Zhang et al. (2008)
Cucumber and tomato	*Pseudomonas chlororaphis* TSAU13	Protecting plants against soilborne pathogens and regulation of phytohormone production	Egamberdieva (2012)
Sunflower	*P. fluorescens* biotype F and *P. fluorescens* CECT 378^T	Indoleacetic acid and siderophore production and improvement in K^+–Na^+ ratio	Shilev et al. (2012)
Mung bean	*Pseudomonas syringae* Mk1, *P. fluorescens* Mk20, and *P. fluorescens* Biotype G, Mk25	Inhibition of ethylene due to ACC deaminase activity, catalase activity, chitinase activity, phosphate solubilization, root colonization, siderophore production	Ahmad et al. (2012, 2013a, b)
Wheat	*P. putida* (W2) and *P. fluorescens* (W17)	Inhibition of ethylene due to ACC deaminase activity, phosphate solubilization, root colonization, siderophore production	Nadeem et al. (2010a)
Drought stress			
Wheat	*Burkholderia phytofirmans* PsJN	Modulation of metabolism, improving ionic balance	Naveed et al. (2014a)
	Bacillus amyloliquefaciens 5113 and *A. brasilense* NO40	Improvement in homeostatic mechanisms and may be due to a combination of morphological, physiological, and metabolic effects	Kasim et al. (2012)
	Azospirillum	Morphological modifications of the coleoptile xylem architecture, upregulation of its own indole-3-pyruvate decarboxylase gene, and enhanced bacterial IAA synthesis	Pereyra et al. (2012)

(continued)

Table 1.1 (continued)

Crop	Rhizobacterial strain	Mechanism used/processes regularized	Reference
Rice	*A. brasilense*	AM colonization through promotion of fungal propagule germination, stimulation of mycelial growth, or changes in the root architecture through the production of growth factors	Ruiz-Sanchez et al. (2011)
Maize	*B. phytofirmans* PsJN, *Enterobacter* sp. FD17	Modulation of metabolism through endophytic colonization	Naveed et al. (2014b)
	P. putida	Suppression of ethylene production due to ACC deaminase activity	Vilchez and Manzanera (2011)
	Pseudomonas entomophila BV-P13, *Pseudomonas stutzeri* GRFHAP-P14, *P. putida* GAP-P45, *P. syringae* GRFHYTP52, and *Pseudomonas monteilii* WAPP53	Regulation of activities of antioxidant enzymes, ascorbate peroxidase (APX), catalase (CAT), glutathione peroxidase (GPX)	Sandhya et al. (2010)
Chickpea	*Paenibacillus lentimorbus* B-30488	Enhanced growth due to biofilm formation	Khan et al. (2011)
Cucumber	*Bacillus cereus* AR156, *B. subtilis* SM21, and *Serratia* sp. XY21	Reductions in the leaf monodehydroascorbate (MDA) content and extent of the peroxidation of plasma lemma, protection of roots from detrimental effects of drought stress, and increase in activity of antioxidant enzymes and chlorophyll contents	Wang et al. (2012)
Ornamental plants	*V. paradoxus* 5C-2	Reduction in endogenous production of stressed ethylene due to ACC deaminase activity	Sharp et al. (2011)
Green gram (*V. radiata*)	*P. fluorescens* Pf1	Induced systemic tolerance by regulating the activity of catalase and peroxidase	Saravanakumar et al. (2011)
Sugarcane	*Azospirillum* sp.	Production of IAA and enhanced microbial activity in the rhizosphere of inoculated plants	Moutia et al. (2010)
Heavy metal stress			
Brassica juncea	*Enterobacter* sp. CBSB1	The introduced glutathione synthase gene *gcsgs* into endophytic strain CBSB1 improved phytoremediation efficiency of host plant	Qiu et al. (2014)
Spartina densiflora	*Aeromonas aquariorum* SDT13, *Pseudomonas composti* SDT3, and *Bacillus* sp. SDT14	Consortium was able to twofold increase seed germination and to protect seeds against fungal contamination in polluted estuaries	Andrades-Moreno et al. (2014)
Agrostis capillaris	*Pantoea* sp. and *Bacillus* sp.	Inoculation with seed endophytes significantly improved plant growth and Cd uptake	Truyens et al. (2014)
Wheat	*P. putida* KT2440	Expression of phytochelatin synthase gene in plants resulted in high heavy metal resistance and high accumulation capacity of Cd	Yong et al. (2013)

(continued)

Table 1.1 (continued)

Crop	Rhizobacterial strain	Mechanism used/processes regularized	Reference
Mirabilis jalapa	*P. fluorescens* strains AA27, MO49, CD1, and CC45	Significantly promoted leaf uptake for Cr, Cu, and Zn	Petriccione et al. (2013)
Pearl millet	*Pseudomonas*, *Citrobacter*, *Acinetobacter*, *Serratia*, and *Enterobacter*	Phosphorus solubilization, IAA production, tolerance to Zn toxicity, temperature stress, and salinity stress	Misra et al. (2012)
Chinese brake (*Pteris vittata* L.)	*Rhodococcus* sp. TS1, *Delftia* sp. TS33, *Comamonas* sp. TS37, *Delftia* sp. TS41, and *Streptomyces lividans* sp. PSQ22	Enhanced population of rhizosphere microbes and increase in arsenic bioavailability through reduction of arsenate	Yang et al. (2012)
Canola and barley	*P. putida* 11, *P. putida* 4, and *P. fluorescens* 169	IAA production, siderophore production, and ACC deaminase activity	Yancheshmeh et al. (2011)
Temperature stress			
Vitis vinifera L.	*B. phytofirmans* strain PsJN	The endophytic colonization of PsJN improved plant photosynthesis and regulated carbohydrate metabolism compared to control	Fernandez et al. (2012)
	B. phytofirmans PsJN	Stress-related gene expression, modulating carbohydrate metabolism and metabolite accumulation	Theocharis et al. (2012)
Wheat	*Pseudomonas lurida* M2RH3	Production of siderophores, IAA, and HCN and solubilization of phosphorus	Selvakumar et al. (2011)
	Pseudomonas sp.	Bacterization improved the root/shoot length and biomass and the level of cellular metabolites compared to control	Mishra et al. (2011)

and physiology. However, work has been carried out, and a number of rhizobacterial species have been explored which have the ability to elucidate the deleterious effects of drought on plant growth due to ACC deaminase activity (Zahir et al. 2008; Belimov et al. 2009a, b) and production of PGRs along with some other mechanisms (Benabdellah et al. 2011). Literature which supports the various mechanisms used by rhizosphere bacteria to induce drought stress tolerance in plants has been reported in Table 1.1.

Rhizosphere bacteria induce abiotic stress tolerance in plants by imposing physical and chemical changes which is called as "induced systemic tolerance" (IST) (Yang et al. 2008) through various mechanisms. Induced drought tolerance in *Arabidopsis thaliana* has been reported due to stimulation of transcription of a drought-response gene, Early Responsive to Dehydration 15 (ERD15), and of an ABA-responsive gene, RAB18, through inoculation with rhizobacterial strain *Paenibacillus polymyxa* (Timmusk and Wagner 1999). Induction of systemic tolerance to drought in many plant species by the action of ACC deaminase-containing rhizobacteria has also been a well-established fact. For example, a significant increase in growth of tomato and pepper seedlings has been reported through inoculation with ACC deaminase-containing PGPR strain *Achromobacter piechaudii* ARV8 which otherwise was decreased due to drought stress. Moreover, ethylene production in tomato seedlings under drought was also decreased due to inoculation (Mayak et al. 2004b). An increase in N, P, and K concentrations in lettuce plants inoculated with *Bacillus* sp. under drought stress has been reported by Vivas et al. (2003).

Fig. 1.4 Impact of inoculation on cucumber growth under drought stress (**Level 2**: water stress by applying water of 70 % field capacity)

Rhizosphere bacteria also induce drought tolerance in plants when applied in combination with mycorrhizal and rhizobial strains. For example, co-inoculation of two rhizobacterial strains *P. polymyxa* (DSM 36) and *P. polymyxa* Loutit (L) with *Rhizobium tropici* (CIAT 899) has pronounced effects in alleviating the drought stress in the common bean (*Phaseolus vulgaris* L.) as compared to sole inoculation (Marcia et al. 2008). Similarly, co-inoculation of rhizobacterial strain *Pseudomonas mendocina* with two arbuscular mycorrhizal strains *G. intraradices* and *G. mosseae* was more effective to increase phosphatase activity, osmolytes regulation, and the activity of other stress enzymes in lettuce leaves grown under drought stress (Kohler et al. 2008). Similarly, some bacteria have the ability to induce drought stress in crop plants due to production of EPS. For example, Grover et al. (2011) reviewed the use of various strains of *Pantoea*, *Rhizobium*, *Pseudomonas*, *Achromobacter*, *Azospirillum*, and *Bacillus* in wheat and sunflower which showed their ability to induce drought stress due to EPS production ability. Rhizobacteria could play a significant role in stress management. Thus, these bacteria can successfully be used to improve drought stress tolerance in crop plants. In a greenhouse study conducted under water stress conditions, inoculation with PGPR enhanced the growth of cucumber compared to uninoculated control (Fig. 1.4).

Temperature Stress

Plants are also sensitive to fluctuation in temperature and heat stress (Mendelsohn et al. 1994; Robertson et al. 1998). Like other environmental stresses, temperature stress also affects plant growth, which might be due to the production of elevated level of ethylene (Cheikh and Jones 1994). Reports are available in literature (Table 1.1) showing that ACC deaminase also protects the plant from temperature stress (Bensalim et al. 1998; Ait Barka et al. 2006). Ethylene and its precursor ACC also have a potential role in flower senescence, leaf abscission, and wilting of flowering species (Woltering and van Doorn 1988; Reid and Wu 1992). Ethylene production decreases the shelf life of flowers that is a major impediment in the success of flowering business (Saleem et al. 2007). Rhizobacteria containing ACC deaminase can also enhance the shelf life of flowers by ameliorating temperature stress (Nayani et al. 1998).

Heavy Metal Stress

Restoration and remediation of metal-polluted soils through phytoremediation is a safe and environment-friendly approach which has gained the attention of scientists (Wu et al. 2006). Reports are available on the use of hyperaccumulating plants for the remediation of heavy metal-polluted soils (Belimov et al. 2005; Zaidi et al. 2006). However, with an increase in metal concentration in plants, the growth decreases due to impaired metabolic activity. Heavy metals enhance the ethylene synthesis that ultimately affect the plant growth in a number of ways (Safronova et al. 2006; Arshad et al. 2008). In addition to an increase in ethylene concentration, heavy metals also cause iron deficiency and evolution of active oxygen species (Burd et al. 1998, 2000; Buchanan 2000). Heavy metal stress tolerance in plants can be induced by a number of abiotic and biotic sources. Biotic elicitors, however, are more beneficial than abiotic as these approaches are safe and environmental friendly. Therefore, the use of rhizosphere bacteria to induce the heavy metal tolerance in plants has gained the attention of scientists as an emerging field of research. Plant–microbe interactions in the rhizosphere have the potential to improve biomass production and heavy metal stress tolerance of crop plants (Glick 2003).

Rhizobacteria may improve plant growth through siderophore production, synthesis of phytohormones and enzymes, as well as phosphate solubilization (Verma et al. 2001). The modulation of high ethylene concentration in plants produced under stressed conditions by some rhizobacteria having ACC deaminase activity may also mediate in heavy metal stress tolerance in crop plants. It is well documented that rhizobacteria containing ACC deaminase activity can improve plant growth and heavy metal stress tolerance in plants in metal-contaminated soils (Belimov et al. 2005; Madhaiyan et al. 2007; Kumar et al. 2008; Belimov and Wenzel 2009).

The rhizosphere bacteria also promote nodulation in contaminated soils. The bacterial strain isolated from Cd-contaminated soil increased the nodulation in clover plants (Vivas et al. 2005) due to accumulation of heavy metal by PGPR in their cells and, thus, reduced the bioavailable Cd concentrations, thereby reducing its uptake by plants and rhizobia. This eliminated Cd toxicity and increased nodulation. The other aspects of this increased nodulation may be because of the enhanced enzymatic activity in the soil (phosphatase, dehydrogenase, etc.) and auxin production by PGPR strains.

Biotic Stress Regulation

Rhizosphere bacteria indirectly influence the plant growth by inhibiting the growth of certain pathogenic organisms by producing antagonistic substances, inducing resistance against pathogens, or releasing compounds against some pests (Glick 1995). The indirect role of rhizobacteria for growth promotion and as defensive agents against diseases and insect pests has been of great importance in modern agriculture in addition to their function as direct growth promoters which has already been well documented (Table 1.2).

Disease Control

Bacteria that are used to control plant diseases and insect pests are called as biocontrol agents, while those having antagonistic properties against certain pathogenic microbes are called as antagonists (Beattie 2006; Glick 2012). As antagonists to plant pathogens, these bacteria can synthesize certain hydrolytic enzymes such as proteases, glucanases, chitinases, cellulases, and lipases. These enzymes hydrolyze the fungal cells (Singh et al. 1999; Frankowski et al. 2001; Kim et al. 2008; Neeraja et al. 2010; Maksimov et al. 2011). Also, they colonize the root surface and rhizosphere where they compete with pathogenic organisms for nutrients and space (Stephens et al. 1993; Kamilova et al. 2005; Compant et al. 2005; Lugtenberg and Kamilova 2009). Another important mechanism is that they cleave the ACC, the immediate precursor of ethylene, into ammonia and α-ketobutyrate through their ACC deaminase activity, thus regulating the production of ethylene (Glick et al. 1998; Mayak et al. 2004a). Furthermore, these bacteria also produce siderophores and antibiotics (Glick 2012). So, rhizosphere bacteria can successfully be used for

Table 1.2 Biotic stress tolerance in plants induced through rhizosphere bacteria

Crop	Stress	Rhizobacterial strain	Mechanism used/processes regularized	Reference
Tobacco	Blue mold	*S. marcescens* 90–116, *Bacillus pumilus* SE 34	Induced systemic resistance due to increased level of salicylic acid and higher activity of β-1,3-glucanase	Zhang et al. (2002)
	Tobacco necrosis virus	*P. fluorescens*	Salicylic acid-dependent activation of PR-1 gene	Park and Kloepper (2000)
Groundnut	*Rhizoctonia bataticola*	*Pseudomonas* sp.	Production of hydroxamate-type siderophores, hydrocyanic acid, and IAA	Gupta et al. (2002)
Chickpea	*Fusarium avenaceum*	*Enterobacter* sp.	ACC deaminase activity and production of IAA and siderophores	Hynes et al. (2008)
Tea	Brown root rot disease	*Ochrobactrum anthropi* TRS-2	Production of siderophores, chitinase, β-1,3-glucanase, peroxidase and phenylalanine ammonia lyase activity	Chakraborty et al. (2009)
Pepper	*Myzus persicae*	*B. subtilis* G803	Suppression of growth due to production of antibiotics	Kokalis-Burelle et al. (2002)
	M. persicae	*B. cereus* MJ-1	Production of gibberellins	Joo et al. (2005)
Tomato	*Phytophthora infestans*	*Bacillus* spp. and *Paenibacillus* sp.	Inhibition of fungal growth due to production of antifungal metabolites and catalase activity	Lamsal et al. (2013)
	Tomato wilt disease caused by *Fusarium oxysporum* f. sp. *lycopersici*	*B. subtilis* S2BC-1 *B. subtilis* GIBC-Jamog	Enhanced activities of chitinases and β-1,3-glucanase and higher intensity of peroxidase isoforms	Shanmugam and Kanoujia (2011)
	Botrytis cinerea	*Bacillus subitus* S499	Activation of lipoxygenase induced by surfactin-overproducing bacteria	Ongena et al. (2007)
Wheat	*Microconidia nivale*	*P. fluorescens*	Production of siderophores and antibiotics	Amein et al. (2008)
	F. oxysporum, *Fusarium culmorum*, *Fusarium solani*, *Pythium ultimum*, *Alternaria alternata*, *B. cinerea*, *Phytophthora cryptogea*	*Acinetobacter*, *Pseudomonas*, *Staphylococcus*, *Bacillus*, *Enterobacter*, *Pantoea*, *Alcaligenes*	Antagonism	Egamberdieva et al. (2008)
	Take-all disease caused by *Gaeumannomyces graminis*	*B. pumilus* 7 km	Stimulation of plant's defensive mechanisms due to increased activities of soluble peroxidase (SPOX), ionically cell wash peroxidase (CWPOX), β-1,3-glucanase, β-1,4-glucanase and enhanced phenolic contents	Sari et al. (2007)

(continued)

Table 1.2 (continued)

Crop	Stress	Rhizobacterial strain	Mechanism used/processes regularized	Reference
Canola	*Sclerotinia sclerotiorum*	*P. chlororaphis*	Enhanced activities of chitinases and β-1,3-glucanase	Fernando et al. (2007)
	Blackleg disease caused by *Leptosphaeria maculans*	*B. cereus* DFE4 *B. amyloliquefaciens* DFE16	Production of lipopeptide antibiotics iturin A, bacillomycin D, and surfactin	Ramarathnam et al. (2011)
Potato	*Fusarium sambucinum*, dry rot	*Serratia plymuthica 5–6*	Production of antifungal metabolites, competition for nutrients and space, and induction of resistance mechanisms in the host	Gould et al. (2008)
Cotton	Cotton root rot caused by *F. oxysporum*	*M. phlei* MbP18 *B. polymyxa* BcP26 *B. amyloliquefaciens* BcA12 *P. alcaligenes* PsA15	Production of phytohormones, antibiosis, and competition for nutrients and niche	Adesemoye and Egamberdieva (2013)
Cucumber	Cotton aphids	*Bacillus* sp.	Production of phytohormone and alteration of metabolic pathways	Stout et al. (2002)
Common bean	*B. cinerea*, *Colletotrichum lindemuthianum*	*P. putida* WCS358	Siderophore-mediated competition for iron and induced systemic resistance	Meziane et al. (2005)
Potato, canola	*P. infestans* US-8	*Bacillus*, *Pseudomonas*, *Rahnella*, and *Serratia*	Production of antibiotics and induction of plant defense systems	Daayf et al. (2003)

the biocontrol of soilborne plant pathogens (Sturz et al. 2000; Compant et al. 2005). These bacteria can also be used in combination with mycorrhizal fungi. For example, the growth of root pathogens like *Fusarium* and *Cylindrocarpon* spp. was inhibited when *Bacillus subtilis* was inoculated in combination with *Laccaria proxima* and *Suillus granulatus* (Schelkle and Peterson 1996).

Indirectly, rhizosphere bacteria have the ability to prevent the proliferation of plant pathogens through the synthesis of different antibiotics (Raaijmakers et al. 2002; Compant et al. 2005; Mazurier et al. 2009). Some strains also have the ability to synthesize hydrogen cyanide in addition to the production of antibiotics which can react synergistically with antibiotics in controlling plant pathogens. Some of the mechanisms that are used by PGPR as biocontrol of plant pathogens have been reviewed in Table 1.2.

Pest Resistance

Rhizosphere bacteria also have the ability to induce systemic resistance in plants. The growth and development of insects is affected by *P. fluorescens* at all growth stages. The growth of larvae of corn earthworm, *Helicoverpa zea*, was affected by *Pseudomonas maltophila* that causes more than 60 % reduction in adult emergence; also, the emerged pupae and adults from infected larvae were smaller and weak (Bong and Sikorowski 1991). In another study, conducted on cotton plants treated with *Pseudomonas gladioli*, the growth and development of *Helicoverpa armigera* was greatly affected which may be due to an increase in their polyphenol and terpenoid content induced by rhizobacteria (Qingwen et al. 1998). Zehnder et al. (1997) while studying the effect of rhizobacterial strains as inducers of systemic resistance in cucumber reported that *P. putida* strain 89B-27, *Serratia marcescens* strain 90–166, *Flavimonas oryzihabitans* strain INR-5, and *Bacillus pumilus* strain INR-7 significantly reduced the population of *Acalymma vittatum* and *Diabrotica undecimpunctata howardi* on cucumber. But these strains varied in their efficacy to control the insect pests which might be due to difference in effectiveness of these strains as rhizosphere colonizers. Genetically engineered strains with induced insecticidal crystal protein have more potential as biocontrol agents of Lepidoptera insect pests. Transgenic *Pseudomonas cepacia* strain 526, when genetically modified with the crystal protein gene, showed insecticidal activity against tobacco hornworm (Stock et al. 1990). Thus, rhizobacteria can be used as effective biotechnology to control insect pests, and this technology has a prosperous future in sustainable agriculture. Certain bacteria can also be used for inhibiting weed growth owing to the production of certain compounds like hydrogen cyanide (Kremer 2007).

The work reviewed in the above sections indicates that PGPR are using different mechanisms to enhance plant growth and development. Some of these mechanisms are general, while others are specific. By these and other unknown mechanisms, PGPR promote plant biomass and yield under stress conditions by creating a favorable environment that is conducible for plant growth in adverse conditions. Due to root colonization ability and their interaction with plant and other microbial populations, PGPR have great potential to improve root growth, enhance biomass yield, and increase crop yields substantially.

Genetically Engineered Rhizobacteria for Biomass Production and Stress Tolerance Improvement

A number of studies reviewed in the above sections showed the tremendous performance of PGPR for improving plant growth and development under normal as well as stress conditions. This growth enhancement is due to their number of well-defined growth-promoting traits. The occurrence of these growth-promoting traits in bacterial strains is common; however, the presence of all these traits in a single species may be a rare one. Although a number of bacterial strains are considered as PGPR owing to the presence of one or more traits, these bacterial strains may also lack some particular traits that could be beneficial for promoting plant growth under a particular environment (Nadeem et al. 2012). Through biotechnological and molecular approaches, genetically engineered bacteria can be produced expressing some particular traits.

The significant results obtained through microbial inoculation under laboratory and natural field conditions motivate the scientists to develop genetically engineered bacteria expressing a gene of the particular trait. The performances of these bacteria have been evaluated for promoting growth and yield. It has been observed that not only the wild strain but also the genetically engineered strain significantly enhanced the plant growth (Shah et al. 1998; Nascimento et al. 2012). For example, the rhizobacteria containing ACC deaminase are very helpful for promoting plant growth under stress conditions by reducing the negative impact of stress-induced ethylene (Glick et al. 2007; Nadeem et al. 2010a). However, certain bacterial strains do not have such activity and are not able to promote plant growth due to the inability to reduce the elevated level of ethylene that is produced under stress conditions. Efforts have been made to introduce ACC deaminase gene in such bacteria and then observe their ability for promoting plant growth. The work of Shah et al. (1998) showed that cloned ACC deaminase gene can be expressed in *Escherichia coli* enabling this bacterium to grow on ACC as a sole source of nitrogen and having ability to promote root elongation of canola seedlings. The presence of heavy metals and other toxic elements affects the plant growth. Even in certain cases it has been observed that in the presence of toxic compounds, PGPR are unable to promote plant growth (Davison 2005; Pimentel et al. 2011). The growth-promoting ability of PGPR in contaminated environment can be enhanced if inoculated strains have both growth-promoting traits and contaminant-degrading ability. This can be achieved by constructing genetically engineered bacteria having contaminant-degrading ability and well-established growth-promoting traits. Keeping in view of these, Yang et al. (2011) studied the growth-promoting ability of genetically modified bacteria for enhancing maize growth in the presence of phenol. They used two strains for the construction of a recombinant. One was phenol-degrading *Pseudomonas aeruginosa* SZH16 but did not have growth-promoting ability, and the other was *P. fluorescens* having growth-promoting ability but did not have phenol-degrading ability. The modified strain not only enhanced maize growth but also degraded phenol. Similar results were also observed by Taghavi et al. (2005) where genetically modified endophyte *B. cepacia* BU0072 enhanced the root and leaf mass of yellow lupin in the presence of toluene. Also, endophytic *P. putida* VM1441 protected the pea plant from toxic effect of naphthalene as well as enhanced the efficiency of naphthalene removal (Germaine et al. 2009).

The genetically modified bacteria have also proved useful for protecting the plants from diseases. The biocontrol ability of a PGPR strain can be enhanced by insertion of a functional gene. For example, genetically modified *P. fluorescens* SBW25 was more efficient in reducing damping-off disease in pea than the wild-type strain (Timms-Wilson et al. 2000). This strain was not previously known to produce antimicrobial substances. Similar results were also observed by Bainton et al. (2004) where genetically modified *P. fluorescens* SBW25 enhanced the disease resistance ability of pea plant, although the wild type also had biocontrol ability but reduced number of emerged pea seedlings that did not occur in the case of modified strain.

Concluding Remarks and Future Prospects

The above-discussed review indicates that PGPR can be used effectively for promoting plant growth and development. It has been observed that due to effective root colonization and their interaction with plant and other microbial populations, PGPR have great potential for improving root growth, enhancing biomass yield, and increasing crop yields. The PGPR are equally effective for enhancing growth under control as well as natural conditions. In addition to normal growth-promoting traits, the PGPR also protect the plant from certain biotic and abiotic stresses by using some particular mechanisms, for example, the presence of ACC deaminase enzyme and the production of EPS are one of the very important mechanisms that play a key role for reducing the negative impact of salinity and drought. Through these and other unknown mechanisms, PGPR not only promote plant biomass yield in normal conditions but also create a favorable environment that is conducible for microbes themselves as well as for plant growth in adverse conditions.

One of the drawbacks of this technology which is reported by some of the workers is the lack of reproducible results. It is evident from the above discussion that such limitation can be minimized through precursor–inoculum interaction and by multi-strain bacterial consortium. Significant growth enhancement can be obtained through precursor–inoculum interaction in an environment where a particular substrate is missing or present in low quantity to show full potential of bacterial strains. Similarly, where a single strain was not so effective for providing desired results particularly under stress environment, the use of multi-strain inocula could be one of the better strategies to promote plant growth.

Although the above-discussed review clearly indicates the effectiveness of PGPR for enhancing plant growth and development under stress conditions, certain areas need further research in this regard. The selection of a suitable strain keeping in view the particular problem is very important. For example, the negative impact of stress-induced ethylene can be minimized by the inoculation of PGPR containing ACC deaminase activity. Similarly, a strain containing ACC deaminase activity and EPS-producing ability can be used effectively for protecting the plant from stress-induced impact of salinity and drought. However, a biocontrol strain can be considered as the best if it has fungal-degrading enzymes or other related mechanisms like inducing systemic resistance.

The use of multi-strain inocula is a good strategy where a single strain is not able to give desired results; however, the compatibility of strains with each other is a very important aspect to obtain better results. The selection of suitable and compatible strains is one of the prerequisites for the success of this technology. The significant results can only be obtained by selecting PGPR strains having multiple traits as well as using these strains in proper combination. It is also important to examine the availability of suitable substrates for microbial growth. Significant results can only be obtained if suitable substrates are available for bacteria; otherwise, precursor–inoculum interaction could be a better strategy under such conditions.

The literature shows that genetically engineered bacteria can also be used effectively for promoting plant growth under normal and stress conditions; however, this aspect needs some comprehensive research. The effectiveness of this population under natural field conditions is not much evaluated. Further, the behavior of such genetically modified bacteria in soil environment and their impact on microbial community structure need comprehensive research.

In spite of a number of studies that show the effectiveness of PGPR for enhancing plant growth and development, the field application of these bacteria is limited. This might be due to nonavailability of a suitable carrier material and/or method of application. This aspect needs much research so that their application can be made easily as well as effectively. At present, one of the better ways of PGPR application is their use through biofertilizers. Although certain studies have shown the effectiveness of these biofertilizers for enhancing crop production, this aspect still needs much research. There are certain queries about the use of biofertilizers that are related to their shelf life and effectiveness in the natural field conditions. Moreover, the use of more than one strain which have compatibility with each other is another important aspect that should be kept in mind during biofertilizer preparation. This aspect also needs research so that suitable strains can be used effectively to maximize the benefits that can be obtained on sustainable basis.

References

Abd El-Azeem SAM, Elwan MWM, Sung JK, Ok YS (2012) Alleviation of salt stress in eggplant (*Solanum melongena* L.) by plant-growth-promoting rhizobacteria. Commun Soil Sci Plant Anal 43:1303–1315

Adesemoye AO, Egamberdieva D (2013) Beneficial effects of plant growth–promoting rhizobacteria on improved crop production: prospects for developing economies. In: Maheshwari DK, Saraf M, Aeron A (eds) Bacteria in agrobiology: crop productivity. Springer, Berlin, pp 45–63

Aguirrezabal L, Bouchier–Combaud S, Radziejwoski A, Dauzat M, Cookson SJ, Granier C (2006) Plasticity to soil water deficit in *Arabidopsis thaliana*: dissection of leaf development into underlying growth dynamic and cellular variables reveals invisible phenotypes. Plant Cell Environ 29:2216–2227

Ahemad M, Kibret M (2014) Mechanisms and applications of plant growth promoting rhizobacteria: current perspective. J King Saud Univ Sci 26:1–20

Ahmad M, Zahir ZA, Asghar HN, Asghar M (2011) Inducing salt tolerance in mung bean through coinoculation with *Rhizobium* and PGPR containing ACC deaminase. Can J Microbiol 57:578–589

Ahmad M, Zahir ZA, Asghar HN, Arshad M (2012) The combined application of rhizobial strains and plant growth promoting rhizobacteria improves growth and productivity of mung bean (*Vigna radiata* L.) under salt–stressed conditions. Ann Microbiol 62:1321–1330

Ahmad M, Zahir ZA, Khalid M, Nazli F, Arshad M (2013a) Efficacy of *Rhizobium* and *Pseudomonas* strains to improve physiology, ionic balance and quality of mung bean under salt–affected conditions on farmer's fields. Plant Physiol Biochem 63:170–176

Ahmad M, Zahir ZA, Nadeem SM, Nazli F, Jamil M, Khalid M (2013b) Field evaluation of *Rhizobium* and *Pseudomonas* strains to improve growth, nodulation and yield of mung bean under salt-affected conditions. Soil Environ 32:158–165

Ahmad M, Zahir ZA, Jamil M, Nazli F, Latif M, Akhtar MF (2014) Integrated use of plant growth promoting rhizobacteria, biogas slurry and chemical nitrogen for sustainable production of maize under salt–affected conditions. Pak J Bot 46:375–382

Ait Barka E, Nowak J, Clement C (2006) Enhancement of chilling resistance of inoculated grapevine plantlets with a plant growth–promoting rhizobacterium, *Burkholderia phytofirmans* strain PsJN. Appl Environ Microbiol 72:7246–7252

Alagawadi AR, Gaur AC (1988) Associative effect of Rhizobium and phosphate solubilizing bacteria on the yield and nutrient uptake of chickpea. Plant Soil 105:241–246

Alagawadi AR, Gaur AC (1992) Inoculation of *Azospirillum brasilense* and phosphate–solubilizing bacteria on yield of sorghum (*Sorghum bicolor* L. Moench) in dry land. Trop Agric 69:347–350

Amein T, Omer Z, Welch C (2008) Application and evaluation of Pseudomonas strains for biocontrol of wheat seedling blight. Crop Prot 27:532–536

Amir HG, Shamsuddin ZH, Halimi MS, Ramlan MF, Marziah M (2001) Effects of Azospirillum inoculation on N_2 fixation and growth of oil palm plantlets at nursery stage. J Oil Palm Res 13:42–49

Andrades-Moreno L, del Castillo I, Parra R, Doukkali B, Redondo-Gómez S, Pérez-Palacios P, Caviedes MA, Pajuelo E, Rodríguez-Llorente ID (2014) Prospecting metal–resistant plant–growth promoting rhizobacteria for rhizoremediation of metal contaminated estuaries using Spartina densiflora. Environ Sci Pollut Res 21:3713–3721

Annapurna K, Ramadoss D, Vithal L, Bose P, Sajad A (2011) PGPR bioinoculants for ameliorating biotic and abiotic stresses in crop production. In: Proceedings of the 2nd Asian PGPR conference, Beijing, pp 67–72

Antoun H, Kloepper JW (2001) Plant growth–promoting rhizobacteria (PGPR). In: Brenner S, Miller JH (eds) Encyclopedia of genetics. Academic, New York, pp 1477–1480

Antoun H, Prevost D (2005) Ecology of plant growth promoting rhizobacteria. In: Siddiqui ZA (ed) PGPR: biocontrol and biofertilization. Springer, Dordrecht, pp 1–38

Arkhipchenko IA, Salkinoja-Salonen MS, Karyakina JN, Tsitko I (2005) Study of three fertilizers produced from farm waste. Appl Soil Ecol 30:126–132

Arshad M, Frankenberger WT Jr (1998) Plant growth regulating substances in the rhizosphere, microbial production and function. Adv Agron 62:46–51

Arshad M, Shaharoona B, Mahmood T (2008) Inoculation with plant growth promoting rhizobacteria containing ACC-deaminase partially eliminates the effects of water stress on growth, yield and ripening of *Pisum sativum* L. Pedosphere 18:611–620

Asghar HN, Zahir ZA, Khaliq A, Arshad M (2000) Assessment of auxin production from rhizobacteria isolated from different varieties of rapeseed. Pak J Agric Sci 37:101–104

Asghar HN, Zahir Z, Arshad M, Khaliq A (2002) Relationship between in vitro production of auxins by rhizobacteria and their growth–promoting activities in *Brassica juncea* L. Biol Fertil Soils 35:231–237

Ashraf M (1994) Breeding for salinity tolerance in plants. Crit Rev Plant Sci 13:17–42

Aydinalp C, Marinova M (2009) The effects of heavy metals on seed germination and plant growth on alfalfa plant (*Medicago sativa*). Bulg J Agric Sci 15:347–350

Bainton NJ, Lynch JM, Naseby D, Way JA (2004) Survival and ecological fitness of *Pseudomonas fluorescens* genetically engineered with dual biocontrol mechanisms. Microb Ecol 48:349–357

Baker R (1991) Diversity in biological control. Crop Prot 10:85–94

Bakker PAHM, Raaijmakers JM, Bloemberg GV, Hofte M, Lemanceau P, Cooke M (2007) New perspectives and approaches in plant growth–promoting rhizobacteria research. Eur J Plant Pathol 119:241–242

Bano A, Fatima M (2009) Salt tolerance in *Zea mays* (L) following inoculation with Rhizobium and Pseudomonas. Biol Fertil Soils 45:405–413

Barazani O, Friedman J (1999) Is IAA the major root growth factor secreted from plant-growth-mediating bacteria? J Chem Ecol 25:2397–2406

Barbieri P, Baggio C, Bazzicalupo M, Galli E, Nuti MP ZG (1991) *Azospirillum*-Gramineae inter-action: effect of indole-3-acetic acid. In: Polsinelly M, Materassi R, Vincenzini M (eds) Developments in plant and soil sciences; nitrogen fixation. Kluwer Academic Publishers, Dordrecht, pp 161–168

Bashan Y (1998) Inoculants of plant growth–promoting bacteria for use in agriculture. Biotechnol Adv 16:729–770

Bashan Y, Holguin G (1997a) *Azospirillum*-plant relationships: environmental and physiological advances. Can J Microbiol 43:103–121

Bashan Y, Holguin G (1997b) Short– and medium– term avenues for *Azospirillum* inoculation. In: Ogoshi A, Kobayashi K, Homma Y, Kodama F, Kondo N, Akino S (eds) Plant growth–promoting rhizobacteria-present status and future prospects. Faculty of Agriculture, Hokkaido University, Sapporo, pp 130–149

Bashan Y, Holguin G (1998) Proposal for the division of plant growth-promoting rhizobacteria into two classification: biocontrol-PGPB (plant growth-promoting bacteria) and PGPB. Soil Biol Biochem 30:1225–1228

Beattie GA (2006) Plant-associated bacteria: survey, molecular phylogeny, genomics and recent advances. In: Gnanamanickam SS (ed) Plant-associated bacteria. Springer, Dordrecht, pp 1–56

Belimov AA, Wenzel WW (2009) The role of rhizosphere microorganisms in heavy metal tolerance of higher plants. Asp Appl Biol 98:81–90

Belimov AA, Kojemiakov AP, Chuvarliyeva CV (1995) Interaction between barley and mixed cultures of nitrogen fixing and phosphate-solubilizing bacteria. Plant Soil 173:29–37

Belimov AA, Hontzeas N, Safronova VI, Demchinskaya SV, Piluzza G, Bullitta S, Glick BR (2005) Cadmium-tolerant plant growth-promoting bacteria associated with the roots of Indian mustard (*Brassica juncea* L. Czern.). Soil Biol Biochem 7:241–250

Belimov AA, Dodd IC, Hontzeas N, Theobald JC, Safronova VI, Davies WJ (2009a) Rhizosphere bacteria containing 1-aminocyclopropane-1-carboxylate deaminase increase yield of plants grown in drying soil via both local and systemic hormone signaling. New Phytol 181:413–423

Belimov AA, Dodd IC, Safronova VI, Davies WJ (2009b) ACC-deaminase-containing rhizobacteria improve vegetative development and yield of potato plants grown under water–limited conditions. Asp Appl Biol 98:163–169

Benabdellah K, Abbas Y, Abourouh M, Aroca R, Azcon R (2011) Influence of two bacterial isolates from degraded and non–degraded soils and arbuscular mycorrhizae fungi isolated from semi–arid zone on the growth of *Trifolium repens* under drought conditions: mechanisms related to bacterial effectiveness. Eur J Soil Biol 47:303–309

Bensalim S, Nowak J, Asiedu SK (1998) A plant growth promoting rhizobacterium and temperature effects on performance of 18 clones of potato. Am J Potato Res 75:145–152

Berg G, Eberl L, Hartmann A (2005) The rhizosphere as a reservoir for opportunistic human pathogenic bacteria. Environ Microbiol 7:1673–1685

Biswas JC, Ladha JK, Dazzo FB, Yanni YG, Rolfed BG (2000) Rhizobial inoculation influences seedling vigor and yield of rice. Agron J 92:880–886

Boddey RM, Dobereiner J (1995) Nitrogen fixation associated with grasses and cereals: recent progress and perspectives for the future. Fertil Res 42:241–250

Bodelier PLE, Wijlhuizen AG, Blom CWPM, Laanbroek HJ (1997) Effects of photoperiod on growth of and denitrification by Pseudomonas chlororaphis in the root zone of *Glyceria maxima*, studied in a gnotobiotic microcosm. Plant Soil 190:91–103

Bong CFJ, Sikorowski PP (1991) Histopathology of cytoplasmic polyhedrosis virus (Reoviridae) infection in corn earworm *Helicoverpa zea* (Boddie) larvae (Insecta: Lepidoptera: Noctuidae). Can J Zool 69:2121–2127

Bouizgarne B (2013) Bacteria for plant growth promotion and disease management. In: Maheshwari DK (ed) Bacteria in agrobiology: disease management. Springer, Berlin

Boyer JS (1982) Plant productivity and environment. Science 218:443–448

Bringhurst RM, Cardon ZG, Gage DJ (2001) Galactosides in the rhizosphere: utilization by *Sinorhizobium meliloti* and development of a biosensor. Proc Natl Acad Sci U S A 98(8):4540–4545

Buchanan KL (2000) Stress and the evolution of condition dependent signals. Tree 15:156–160

Burd GI, Dixon DG, Glick BR (1998) A plant growth promoting bacterium that decreases nickel toxicity in seedlings. Appl Environ Microbiol 64:3663–3668

Burd GI, Dixon DG, Glick BR (2000) Plant growth promoting bacteria that decrease heavy metal toxicity in plants. Can J Microbiol 46:237–245

Burdman S, Volpin H, Kigel J, Kapulnik Y, Okon Y (1996) Promotion of nodulation and growth of common bean (*Phaseolus vulgaris* L.). Soil Biol Biochem 29:923–929

Burdman S, Vedder D, German M, Itzigsohn R, Kigel J, Jurkevitch E, Okon Y (1998) Legume crop yield promotion by inoculation with *Azospirillum*. In: Elmerich C, Kondorosi A, Newton WE (eds) Biological nitrogen fixation for the 21st century. Kluwer Academic Publishers, Dordrecht, pp 609–612

Burdman S, Jurkevitch E, Okon Y (2000) Recent advances in the use of plant growth promoting rhizobacteria (PGPR) in agriculture. In: Subba Rao NS, Dommergues YR (eds) Microbial interactions in agriculture and forestry. Science Publishers, Inc., Enfield, pp 229–250

Burr TJ, Caesar A (1984) Beneficial plant bacteria. Crit Rev Plant Sci 2:1–20

Cassan F, Perrig D, Sgroy V, Masciarelli O, Penna C, Luna V (2009) *Azospirillum brasilense* Az39 and *Bradyrhizobium japonicum* E109, inoculated singly or in combination, promote seed germination and early seedling growth in corn (*Zea mays* L.) and soybean (*Glycine max* L.). Eur J Soil Biol 45:28–35

Chakraborty U, Chakraborty BN, Basnet M, Chakraborty AP (2009) Evaluation of *Ochrobactrum anthropi* TRS–2 and its talc based formulation for enhancement of growth of tea plants and management of brown root rot disease. J Appl Microbiol 107:625–634

Chandrasekar BR, Ambrose G, Jayabalan N (2005) Influence of biofertilizers and nitrogen source level on the growth and yield of *Echinochloa frumentacea* (Roxb.), Link. J Agric Technol 1:223–234

Cheikh N, Jones RJ (1994) Disruption of maize kernel growth and development by heat stress, role of cytokine/abscisic acid balance. Plant Physiol 106:45–51

Cheng Z, Park E, Glick BR (2007) 1-Aminocyclopropane-1-carboxylate deaminase from *Pseudomonas putida* UW4 facilitates the growth of canola in the presence of salt. Can J Microbiol 53:912–918

Choure K, Dubey RC (2012) Development of plant growth promoting microbial consortium based on interaction studies to reduce wilt incidence in *Cajanus cajan* L. Var. Manak. World J Agric Sci 8:118–128

Cohen AC, Bottini R, Piccoli PN (2008) *Azospirillum brasilense* Sp245 produces ABA in chemically–defined culture medium and increases ABA content in *Arabidopsis* plants. Plant Growth Regul 54:97–103

Compant S, Brion D, Jerzy N, Christophe C, Essaïd AB (2005) Use of plant growth–promoting bacteria for biocontrol of plant diseases: principles, mechanisms of action and future prospects. Appl Environ Microbiol 71:4951–4959

Conde AM, Chaves M, Geros H (2011) Membrane transport, sensing and signaling in plant adaptation to environmental stress. Plant Cell Physiol 52:1583–1602

Daayf F, Adam L, Fernando WGD (2003) Comparative screening of bacteria for biological control of potato late blight (strain US–8), using in vitro, detached–leaves, and whole–plant testing systems. Can J Plant Pathol 25:276–284

Dardanelli MS, Manyani H, Gonzalez-Barroso S, Rodrıguez-Carvajal MA, Gil–Serrano AM, Espuny MR, Lopez–Baena FJ, Bellogın RA, Megıas M, Ollero FJ (2009) Effect of the presence of the plant growth promoting rhizobacterium (PGPR) *Chryseobacterium balustinum* Aur9 and salt stress in the pattern of flavonoids exuded by soybean roots. Plant Soil 328:483–493

Dardanelli MS, Manyani H, Gonzalez–Barroso S, Rodriguez–Carvajal MA, Gil–Serrano AM, Espuny MR, López–Baena FJ, Bellogin RA, Megías M, Ollero FJ (2010) Effect of the presence of the plant growth promoting rhizobacterium (PGPR) *Chryseobacterium balustinum* Aur9 and salt stress in the pattern of flavonoids exuded by soybean roots. Plant Soil 328:483–493

Darrah PR (1993) The rhizosphere and plant nutrition: a quantitative approach. Plant Soil 155–156(1):1–20

Davison J (2005) Risk mitigation of genetically modified bacteria and plants designed for bioremediation. J Ind Microbiol Biotechnol 32:639–650

Dobbelaere S, Okon Y (2007) The plant growth promoting effect and plant responses. In: Elmerich C, Newton WE (eds) Associative and endophytic nitrogen-fixing bacteria and cyanobacterial associations. Kluwer Academic Publishers, Dordrecht, pp 1–26

Egamberdieva D (2009) Alleviation of salt stress by plant growth regulators and IAA producing bacteria in wheat. Acta Physiol Plant 31:861–864

Egamberdieva D (2012) *Pseudomonas chlororaphis*: a salt-tolerant bacterial inoculants for plant growth stimulation under saline soil conditions. Acta Physiol Plant 34:751–756

Egamberdieva D, Kamilova F, Validov S, Gafurova L, Kucharova Z, Lugtenberg B (2008) High incidence of plant growth–stimulating bacteria associated with the rhizosphere of wheat grown on salinated soil in Uzbekistan. Environ Microbiol 10:1–9

Egamberdieva D, Berg G, Lindstrom K, Rasanen LA (2010) Co-inoculation of *Pseudomonas* spp. with *Rhizobium* improves growth and symbiotic performance of fodder galega (*Galega orientalis Lam.*). Eur J Soil Biol 46:269–272

Egamberdiyeva D (2005) Plant growth promoting rhizobacteria isolated from a calcisol in semi-arid region of Uzbekistan: biochemical characterization and effectiveness. J Plant Nutr Soil Sci 168:94–99

El-Komy HMA, Moharram TMM, Safwat MSA (1998) Effects of *Azospirillum* inoculation on growth and N2 fixation of maize subjected to different levels of FYM using 15 N–dilution method. In: Malik KA (ed) Nitrogen fixation with non-legumes. Kluwer Academic Publishers, London, pp 49–59

Fabbri P, DelGallo M (1995) Specific interaction between chickpea (*Cicer arietinum*) and three chickpea–Rhizobium strains inoculated singularly and in combination with *Azospirillum brasilense* Cd. In: Fendrik I, Del Gallo M, Vanderleyden J, de Zamaroczy M (eds) Azospirillum VI and related microorganisms, genetics – physiology –ecology, vol G37, NATO ASI Series, Series G: Ecological Sciences. Springer, Berlin, pp 207–212

Fernandez O, Theocharis A, Bordiec S, Feil R, Jasquens L, Clement C, Fontaine F, Ait Barka E (2012) *Burkholderia phytofirmans* PsJN acclimates grapevine to cold by modulating carbohydrate metabolism. Mol Plant Microbe Interact 25:496–504

Fernando WGD, Zhang Y, Nakkeeran S, Savchuk S (2007) Biological control of *Sclerotinia sclerotiorum* (Lib.) de Bary by *Pseudomonas* and *Bacillus* species on canola petals. Crop Prot 26:100–107

Figueiredo MVB, Burity HA, Martinez CR, Chanway CP (2008) Alleviation of water stress effects in common bean (*Phaseolus vulgaris* L.) by co-inoculation *Paenibacillus* x *Rhizobium tropici*. Appl Soil Ecol 40:182–188

Flaishman MA, Eyal Z, Zilberstein A, Voisard C, Hass D (1996) Suppression of *Septoria tritici* blotch and leaf rust of wheat by recombinant cyanide producing strains of *Pseudomonas putida*. Mol Plant Microbe Interact 9:642–645

Foldes T, Banhegyi I, Herpai Z, Varga L, Szigeti J (2000) Isolation of Bacillus strains from the rhizosphere of cereals and in vitro screening for antagonism against phytopathogenic, food–born pathogenic and spoilage microorganisms. J Appl Microbiol 89:840–846

Forde BG (2002) Local and long-range signaling pathways regulating plant responses to nitrate. Annu Rev Plant Physiol Plant Mol Biol 53:203–224

Foster RC (1983) The fine structure of epidermal cell mucilages of roots. New Phytol 91:727–740

Frankenberger WT Jr, Arshad M (1991) Yield response of *Capsicum annuum* to the auxin precursor, L-tryptophan applied to soil. PGRSA Q 19:231–240

Frankenberger WT Jr, Arshad M (1995) Phytohormones in soils: microbial production and function. Marcel Dekker, New York, p 503

Frankowski J, Lorito M, Scala F, Schmid R, Berg G, Bahl H (2001) Purification and properties of two chitinolytic enzymes of *Serratia plymuthica* HRO–C48. Arch Microbiol 176:421–426

Freitas ADS, Vieira CL, Santos CERS, Stamford NP, Lyra MCCP (2007) Characterization of isolated rhizobia of *Pachyrhizus erosus* cultivated in saline soil of the state of Pernambuco, Brazil. Bragantia 66:497–504

Fuentes-Ramirez LE, Caballero-Mellado J (2005) Bacterial biofertilizers. In: Siddiqui ZA (ed) PGPR: biocontrol and biofertilization. Springer, Dordrecht, pp 143–172

Galal YGM (2003) Assessment of nitrogen availability to wheat (*Triticum aestivum* L.) from inorganic and organic N sources as affected by *Azospirillum brasilense* and *Rhizobium leguminosarum* inoculation. Egypt J Microbiol 38:57–73

Garcia de Salamone IE, Hynes RK, Nelson LM (2001) Cytokinin production by plant growth promoting rhizobacteria and selected mutants. Can J Microbiol 47:404–411

Germaine KJ, Keogh E, Ryan D, Dowling DN (2009) Bacterial endophyte mediated naphthalene phytoprotection and phytoremediation. FEMS Microbiol Lett 296:226–234

Giri B, Kapoor R, Mukerji KG (2007) Improved tolerance of *Acacia nilotica* to salt stress by arbuscular mycorrhiza, *Glomus fasciculatum*, may be partly related to elevated K^+/Na^+ ratios in root and shoot tissues. Microb Ecol 54:753–760

Glick BR (1995) The enhancement of plant growth by free–living bacteria. Can J Microbiol 41:109–117

Glick BR (2003) Phytoremediation: synergistic use of plants and bacteria to clean up the environment. Biotechnol Adv 21:383–393

Glick BR (2004) Bacterial ACC–deaminase and the alleviation of plant stress. Adv Appl Microbiol 56:291–312

Glick BR (2005) Modulation of plant ethylene levels by the bacterial enzyme ACC deaminase. FEMS Microbiol Lett 251:1–7

Glick BR (2012) Plant growth–promoting bacteria: mechanisms and applications. In: Ano T, Comi G, Shoda M (eds) Scientifica, Hindawi Publishing Corporation, pp 1–15

Glick BR, Bashan Y (1997) Genetic manipulation of plant growth–promoting bacteria to enhance biocontrol of fungal phytopathogens. Biotechnol Adv 15:353–378

Glick BR, Karaturovic DM, Newell PC (1995) A novel procedure for rapid isolation of plant growth promoting Pseudomonas. Can J Microbiol 41:533–536

Glick BR, Liu C, Ghosh S, Dumbroff EB (1997) Early development of canola seedlings in the presence of the plant growth promoting rhizobacterium *Pseudomonas putida* GR 12–2. Soil Biol Biochem 29:1233–1239

Glick BR, Penrose DM, Li J (1998) A model for the lowering of plant ethylene concentrations by plant growth promoting rhizobacteria. J Theor Biol 190:63–68

Glick BR, Patten CL, Holguin G, Penrose DM (1999) Biochemical and genetic mechanisms used by plant growth promoting bacteria. Imperial College Press, London

Glick BR, Cheng Z, Czarny J, Cheng Z, Duan J (2007) Promotion of plant growth by ACC deaminase– producing soil bacteria. Eur J Plant Pathol 119:329–339

Glombitza S, Dubuis PH, Thulke O, Welzl G, Bovet L, Götz M, Affenzeller M, Geist B, Hehn A, Asnaghi C, Ernst D, Seidlitz HK, Gundlach H, Mayer KF, Martinoia E, Werck-Reichhart D, Mauch F, Schäffner AR (2004) Crosstalk and differential response to abiotic and biotic stressors reflected at the transcriptional level of effector genes from secondary metabolism. Plant Mol Biol 54:817–835

Gould M, Nelson LM, Waterer D, Hynes RK (2008) Biocontrol of *Fusarium sambucinum*, dry rot of potato, by *Serratia plymuthica* 5—6. Biocontrol Sci Tech 18:1005–1016

Gray EJ, Smith DL (2005) Intracellular and extracellular PGPR: commonalities and distinctions in the plant-bacterium signaling processes. Soil Biol Biochem 37:395–412

Grover M, Ali SZ, Sandhya V, Rasul A, Venkatesvarlu B (2011) Role of microorganisms in adaptation of agriculture crops to abiotic stress. World J Microbiol Biotechnol 27:1231–1240

Guo JH, Qi HY, Guo YH, Ge HL, Gong LY, Zhang LX (2004) Biocontrol of tomato wilt by plant growth promoting rhizobacteria. Biol Control 29:66–72

Gupta CP, Dubey RC, Maheshwari DK (2002) Plant growth enhancement and suppression of *Macrophomina phaseolina* causing charcoal rot of peanut by fluorescent pseudomonas. Biol Fertil Soils 35:399–405

Gutierrez-Manero FJ, Ramos–Solano B, Probanza A, Mehouachi J, Tadeo FR, Talon M (2001) The plant-growth-promoting rhizobacteria *Bacillus pumilus* and *Bacillus licheniformis* produce high amounts of physiologically active gibberellins. Physiol Plant 111:206–211

Habib N, Ashraf M (2014) Effect of exogenously applied nitric oxide on water relations and ionic composition of rice (*Oryza sativa* L.) plants under salt stress. Pak J Bot 46:111–116

Hamaoui B, Abbadi JM, Burdman S, Rashid A, Sarig A, Okon Y (2001) Effects of inoculation with *Azospirillum brasilense* on chickpeas (*Cicer arietinum*) and faba beans (*Vicia faba*) under different growth conditions. Agronomie 21:553–560

Hamdia MA, Shaddad MAK, Doaa MM (2004) Mechanism of salt tolerance and interactive effect of *Azospirillum brasilense* inoculation on maize cultivars grown under salt stress conditions. Plant Growth Regul 44:165–174

Han HS, Lee KD (2005) Physiological responses of soybean– inoculation of *Bradyrhizobium japonicum* with PGPR in saline soil conditions. Res J Agric Biol Sci 1:216–221

Hartmann A, Rothballer M, Schmid M (2008) Lorenz Hiltner, a pioneer in rhizosphere microbial ecology and soil bacteriology research. Plant Soil 312:7–14

Hiltner L (1904) Uber neuere Erfahrungen und Problem auf dem Gebiet der Bodenbakteriologie und unter besonderer Berucksichtigung der Grundungung und Brache. Arb Dtsch Landwirtsch Ges 98:59–78

Hontzeas N, Richardson AO, Belimov A, Safronova VI, Abu-Omar MM, Glick BR (2005) Evidence for horizontal transfer of 1-aminocyclopropane-1-carboxylate deaminase genes. Appl Environ Microbiol 71:7556–7558

Hubel F, Beck E (1993) In-situ determination of the P-relations around the primary root of maize with respect to inorganic and phytate-P. Plant Soil 157:1–9

Hynes RK, Leung GC, Hirkala DL, Nelson LM (2008) Isolation, selection, and characterization of beneficial rhizobacteria from pea, lentil and chickpea grown in western Canada. Can J Microbiol 54:248–258

Indiragandhi P, Anandham R, Kim KA, Yim WJ, Madhaiyan M, Sa TM (2008) Induction of defense responses in tomato against *Pseudomonas syringae* pv. tomato by regulating the stress ethylene level with *Methylobacterium oryzae* CBMB20 containing 1-aminocyclopropane-1-carboxylate deaminase. World J Microbiol Biotechnol 4:1037–1045

Itzigsohn R, Kapulnik Y, Okon Y, Dovrat A (1993) Physiological and morphological aspects of interactions between *Rhizobium meliloti* and alfalfa (*Medicago sativa*) in association with *Azospirillum brasilense*. Can J Microbiol 39:610–615

Jaleel CA, Manivannan P, Sankar B, Kishorekumar A, Gopi R, Somasundaram R, Panneerselvamet R (2007) Induction of drought stress tolerance by ketoconazole in

Catharanthus roseus is mediated by enhanced antioxidant potentials and secondary metabolite accumulation. Colloids Surf B Biointerfaces 60(2):201–206

Jeyabal A, Kupuswamy G (2001) Recycling of organic wastes for the production of vermicompost and its response in rice legume cropping system and soil fertility. Eur J Agron 15:153–170

Jofre E, Rivarola V, Balegno H, Mori G (1998) Differential gene expression in *Azospirillum brasilense* under saline stress. Can J Microbiol 44:929–936

Joo GJ, Kin YM, Kim JT, Rhee IK, Kim JH, Lee IJ (2005) Gibberellins–producing rhizobacteria increase endogenous gibberellins content and promote growth of red peppers. J Microbiol 43:510–515

Kamilova F, Validov S, Azarova T, Mulders I, Lugtenberg B (2005) Enrichment for enhanced competitive plant root tip colonizers selects for a new class of biocontrol bacteria. Environ Microbiol 7:1809–1817

Kasim WA, Osman ME, Omar MN, Abd-El-Daim IA, Bejai S, Meijer J (2012) Control of drought stress in wheat using plant–growth–promoting bacteria. J Plant Growth Regul 32:122–130

Khalid M, Zahir ZA, Waseem A, Arshad M (1999) Azotobacter and L–tryptophan application for improving wheat yield. Pak J Biol Sci 2:739–742

Khalid A, Arshad M, Zahir ZA (2006) Phytohormones: microbial production and applications. In: Uphoff N, Ball AS, Fernandes E, Herren H, Husson O, Laing M, Palm C, Pretty J, Sanchez P, Sanginga N, Thies J (eds) Biological approaches to sustainable soil systems. Taylor & Francis, Boca Raton, pp 207–220

Khammas KM, Kaiser P (1992) Pectin decomposition and associated nitrogen fixation by mixes cultures of *Azospirillum* and *Bacillus* species. Can J Microbiol 38:794–797

Khan AG (2005) Role of soil microbes in the rhizosphere of plants growing on trace metal contaminated soils in phytoremediation. J Trace Elem Med Biol 18:355–364

Khan N, Mishra A, Chauhan PS, Nautiyal CS (2011) Induction of *Paenibacillus lentimorbus* biofilm by sodium alginate and $CaCl_2$ alleviates drought stress in chickpea. Ann Appl Biol 159:372–386

Kim YC, Jung H, Kim KY, Park SK (2008) An effective biocontrol bioformulation against Phytophthora blight of pepper using growth mixtures of combined chitinolytic bacteria under different field conditions. Eur J Plant Pathol 120:373–382

Kloepper JW, Mariano RLR (2000) Rhizobacteria to induce plant disease resistance and enhance growth – theory and practice. In: International symposium on biological control for crop protection, Rural Development Administration, Suwon, pp 99–116

Kloepper JW, Schroth MN (1978) Plant growth promoting rhizobacteria on radishes. In: Proceedings of the 4th international conference on plant pathogenic bacteria. Angers, pp 879–882

Kloepper JW, Leong J, Teintze M, Schroth MN (1980) Pseudomonas siderophores: a mechanism explaining disease suppressive soils. Curr Microbiol 4:317–320

Kloepper JW, Lifshitz R, Zablotowicz RM (1989) Free living bacterial inocula for enhancing crop productivity. Trends Biotechnol 7:39–44

Knox OGG, Killham K, Leifert C (2000) Effects of increased nitrate availability on the control of plant pathogenic fungi by the soil bacterium *Bacillus subtilis*. Appl Soil Ecol 15:227–231

Kohler J, Caravaca F, Carrasco L, Roldan A (2006) Contribution of *Pseudomonas mendocina* and *Glomus intraradices* to aggregates stabilization and promotion of biological properties in rhizosphere soil of lettuce plants under field conditions. Soil Use Manag 22:298–304

Kohler J, Hernandez JA, Caravaca F, Roldan A (2008) Plant-growth-promoting rhizobacteria and arbuscular mycorrhizal fungi modify alleviation biochemical mechanisms in water-stressed plants. Funct Plant Biol 35:141–151

Kokalis-Burelle N, Vavrina CS, Rosskopf EN, Shelby RA (2002) Field evaluation of plant growth–promoting rhizobacteria amended transplant mixes and soil solarization for tomato and pepper production in Florida. Plant Soil 238:257–266

Kremer RJ (2007) Deleterious rhizobacteria. In: Gnanammanickam SS (ed) Plant associated bacteria. Springer, Amsterdam, pp 335–357

Kumar KV, Singh N, Behl HM, Srivastava S (2008) Influence of plant growth promoting bacteria and its mutant on heavy metal toxicity in *Brassica juncea* grown in fly ash amended soil. Chemosphere 72:678–683

Labuschagne N, Pretorius T, Idris AH (2010) Plant growth promoting rhizobacteria as biocontrol agents against soil–borne plant diseases. In: Maheshwari DK (ed) Plant growth and health promoting bacteria. Springer, Berlin

Lamsal K, Kim SW, Kim YS, Lee YS (2013) Biocontrol of late blight and plant growth promotion in tomato using rhizobacterial isolates. J Microbiol Biotechnol 23:1–8

Leinhos V, Bergmann H (1995) Changes in yield, lignin content and protein pattern of barley (*Hordeum vulgare* cv. Alexis) induced by drought stress. J Appl Bot 69:206–210

Liu D, Li T, Yang X, Islam E, Jin X, Mahmood Q (2007) Enhancement of lead uptake by hyperaccumulator plant species *Sedum alfredii* Hance using EDTA and IAA. Bull Environ Contam Toxicol 78:280–283

Liu RZ, Jiang XL, Guan HS, Li XX, Du YS, Wang P, Mou H (2009) Promotive effects of alginate–derived oligosaccharides on the inducing drought resistance of tomato. J Ocean Univ China (Ocean Coast Sea Res) 8(3):303–311

Lu S, Su W, Li H, Guo ZF (2009) Abscisic acid improves drought tolerance of triploid bermudagrass and involves H_2O_2– and NO–induced antioxidant enzyme activities. Plant Physiol Biochem 47:132–138

Lucy M, Reed E, Glick BR (2004) Application of free living plant growth promoting rhizobacteria. Antonie Van Leeuwenhoek 86:1–25

Lugtenberg B, Kamilova F (2009) Plant-growth-promoting rhizobacteria. Annu Rev Microbiol 63:541–555

Madhaiyan M, Poonguzhali S, Sa TM (2007) Characterization of 1-Aminocyclopropane-1-carboxylate deaminase (ACC) deaminase containing *Methylobacterium oryzae* and interactions with auxins and ACC regulation of ethylene in canola (*Brassica campestris*). Planta 226:867–876

Mahmood MH, Khalid A, Khalid M, Arshad M (2008) Response of etiolated pea seedlings and cotton to ethylene produced from L–methionine by soil microorganisms. Pak J Bot 40:859–866

Maksimov IV, Abizgil'dina RR, Pusenkova LI (2011) Plant growth promoting rhizobacteria as alternative to chemical crop protectors from pathogens (review). Appl Biochem Microbiol 47:333–345

Malik DK, Sindhu SS (2011) Production of indole acetic acid by *Pseudomonas* sp.: effect of coinoculation with *Mesorhizobium* sp. Cicer on nodulation and plant growth of chickpea (*Cicer arietinum*). Physiol Mol Biol Plants 17:25–32

Marcelis LFM, Van Hooijdonk HV (1999) Effect of salinity on growth, water use and nutrient use in radish (*Raphanus sativus* L.). Plant Soil 215:57–64

Marcia VBF, Burity HA, Martinez CR, Chanway CP (2008) Alleviation of drought stress in the common bean (*Phaseolus vulgaris* L.) by co–inoculation with *Paenibacillus polymyxa* and *Rhizobium tropici*. Appl Soil Ecol 40:182–188

Mascher R, Nagy E, Lippmann B, Hornlein S, Fischer S, Scheiding W, Neagoe A, Bergmann H (2005) Improvement of tolerance to paraquat and drought in barley (*Hordeum vulgare* L.) by exogenous 2–aminoethanol: effects on superoxide dismutase activity and chloroplast ultrastructure. Plant Sci 168:691–698

Mayak S, Tirosh T, Glick BR (2004a) Plant growth–promoting bacteria confer resistance in tomato plants to salt stress. Plant Physiol Biochem 42:565–572

Mayak S, Tirosh T, Glick BR (2004b) Plant growth–promoting bacteria that confer resistance to water stress in tomato and pepper. Plant Sci 166:525–530

Mazurier S, Corberand T, Lemanceau P, Raaijmakers JM (2009) Phenazine antibiotics produced by fluorescent pseudomonads contribute to natural soil suppressiveness to *Fusarium wilt*. ISME J 3:977–991

McKeon TA, Fernandez-Maculet JC, Yang SF (1995) Biosynthesis and metabolism of ethylene. In: Davies PJ (ed) Plant hormones physiology, biochemistry and molecular biology. Kluwer Academic Publishers, Dordrecht, pp 118–139

Mendelsohn R, Nordhaus W, Shaw D (1994) The impact of global warming on agriculture: a Ricardian analysis. Am Econ Rev 84:753–771

Meziane H, Van der Sluis I, Van Loon LC, Hofte M, Bakker PAHM (2005) Determinants of *Pseudomonas putida* WCS358 involved in inducing systemic resistance in plants. Mol Plant Pathol 6:177–185

Mia MAB, Shamsuddin ZH, Wahab IZ, Marziah M (2010) Effect of plant growth promoting rhizobacterial (PGPR) inoculation on growth and nitrogen incorporation of tissue-cultured Musa plantlets under nitrogen–free hydroponics condition. Aust J Crop Sci 4:85–90

Mishra PK, Bisht SC, Ruwari P, Joshi GK, Singh G, Bisht JK, Bhatt JC (2011) Bioassociative effect of cold tolerant Pseudomonas spp. and *Rhizobium leguminosarum* -PR1 on iron acquisition, nutrient uptake and growth of lentil (*Lens culinaris* L.). Eur J Soil Biol 47:35–43

Misra N, Gupta G, Jha PN (2012) Assessment of mineral phosphate–solubilizing properties and molecular characterization of zinc–tolerant bacteria. J Basic Microbiol 52:1–10

Molla AH, Shamsuddin ZH, Halimi MS, Morziah M, Puteh AB (2001) Potential for enhancement of root growth and nodulation of soybean co–inoculated with *Azospirillum* and *Bradyrhizobium* in laboratory systems. Soil Biol Biochem 33:457–463

Moore TM (1989) Biochemistry and physiology of plant hormones, 2nd edn. Springer, New York, p 330

Moutia JFY, Saumtally S, Spaepen S, Vanderleyden J (2010) Plant growth promotion by *Azospirillum* sp. in sugarcane is influenced by genotype and drought stress. Plant Soil 337:233–242

Munns R (1993) Physiological processes limiting plant growth in saline soils: some dogmas and hypotheses. Plant Cell Environ 16:15–24

Nadeem SM, Zahir ZA, Naveed M, Arshad M (2007) Preliminary investigations on inducing salt tolerance in maize through inoculation with rhizobacteria containing ACC deaminase activity. Can J Microbiol 53:1141–1149

Nadeem SM, Zahir ZA, Naveed M, Arshad M (2009) Rhizobacteria containing ACC deaminase confer salt tolerance in maize grown on salt affected soils. Can J Microbiol 55:1302–1309

Nadeem SM, Zahir ZA, Naveed M, Asghar HN, Arshad M (2010a) Rhizobacteria capable of producing ACC–deaminase may mitigate salt stress in wheat. Soil Sci Soc Am J 74:533–542

Nadeem SM, Zahir ZA, Naveed M, Ashraf M (2010b) Microbial ACC deaminase: prospects and applications for inducing salt tolerance in plants. Crit Rev Plant Sci 29:360–393

Nadeem SM, Shaharoona B, Arshad M, Crowley DE (2012) Population density and functional diversity of plant growth promoting rhizobacteria associated with avocado trees in saline soils. Appl Soil Ecol 62:147–154

Naqvi SM, Ansari R (1974) Estimation of diffusible auxin under saline growth conditions. Experientia 30:350–354

Nascimento F, Brigido C, Alho L, Glick BR, Oliveira S (2012) Enhanced chickpea growth–promotion ability of a *Mesorhizobium* strain expressing an exogenous ACC deaminase gene. Plant Soil 353:221–230

Naveed M, Hussain MB, Zahir ZA, Mitter B, Sessitsch A (2014a) Drought stress amelioration in wheat through inoculation with *Burkholderia phytofirmans* strain PsJN. Plant Growth Regul 73(2):121–131. doi:10.1007/s10725-013-9874-8

Naveed M, Mitter B, Reichenauer TG, Krzysztof W, Sessitsch A (2014b) Increased drought stress resilience of maize through endophytic colonization by *Burkholderia phytofirmans* PsJN and *Enterobacter* sp. FD17. Environ Exp Bot 97:30–39

Nayani S, Mayak S, Glick BR (1998) Effect of plant growth-promoting rhizobacteria on senescence of flower petals. Indian J Exp Biol 36:836–839

Neeraja C, Anil K, Purushotham P, Suma K, Sarma P, Moerschbacher BM, Podile AR (2010) Biotechnological approaches to develop bacterial chitinases as a bioshield against fungal diseases of plants. Crit Rev Biotechnol 30:231–241

Neumann G, Romheld V (2001) The release of root exudates as affected by the plant's physiological status. In: Pinto R, Varanini Z, Nannipieri P (eds) The rhizosphere: biochemistry and organic substances at the soil-plant interface. Dekker, New York, pp 41–93

O'Connel PF (1992) Sustainable agriculture–a valid alternative. Outlook Agric 21(1):5–12

Okon Y, Labandera–Gonzalez C (1994) Agronomic applications of *Azospirillum*: an evaluation of 20 years of worldwide field inoculation. Soil Biol Biochem 26:1591–1601

Ongena M, Jourdan E, Adam A, Paquot M, Brans A, Joris B, Arpigny JL, Thonart P (2007) Surfactin and fengycin lipopeptides of *Bacillus subtilis* as elicitors of induced systemic resistance in plants. Environ Microbiol 9:1084–1090

Park K, Kloepper J (2000) Activation of PR–1a promoter by rhizobacteria that induce systemic resistance in tobacco against *Pseudomonas syringae* pv. tabaci. Biol Control 18:2–9

Patel D, Jha CK, Tank N, Saraf M (2012) Growth enhancement of chickpea in saline soils using plant growth-promoting rhizobacteria. J Plant Growth Regul 31:53–62

Patten CL, Glick BR (1996) Bacterial biosynthesis of indole-3-acetic acid. Can J Microbiol 42:207–220

Patten CL, Glick BR (2002) Role of Pseudomonas putida indoleacetic acid in development of the host plant root system. Appl Environ Microbiol 68:3795–3801

Paul D, Nair S (2008) Stress adaptations in a plant growth promoting rhizobacterium (PGPR) with increasing salinity in the coastal agricultural soils. J Basic Microbiol 48:378–384

Pereyra MA, Garcia P, Colabelli MN, Barassi CA, Creus CM (2012) A better water status in wheat seedlings induced by *Azospirillum* under osmotic stress is related to morphological changes in xylem vessels of the coleoptile. Appl Soil Ecol 53:94–97

Persello-Cartieaux F, Nussaume L, Robaglia C (2003) Tales from the underground: molecular plant-rhizobacterial interactions. Plant Cell Environ 26:189–199

Petersen DJ, Srinivasan M, Chanway CP (1996) *Bacillus polymyxa* stimulates increased *Rhizobium etli* populations and nodulation when co–resident in the rhizosphere of *Phaseolus vulgaris*. FEMS Microbiol Lett 142:271–276

Petriccione M, Di Patre D, Ferrante P, Papa S, Bartoli G, Fioretto A, Scortichini M (2013) Effects of *Pseudomonas fluorescens* seed bioinoculation on heavy metal accumulation for *Mirabilis jalapa* phytoextraction in smelter–contaminated soil. Water Air Soil Pollut 224:1645

Pimentel MR, Molina G, Dionísio AP, Marostica MR Jr, Pastore GM (2011) The use of endophytes to obtain bioactive compounds and their application in biotransformation process. Biotechnol Res Int 2011:1–11

Pinton R, Varanini Z, Nannipieri P (2001) The rhizosphere: biochemistry and organic substances at the soil-plant interface. Marcel Dekker, New York

Podile AR, Kishore GK (2006) Plant growth promoting rhizobacteria. In: Gnanamanickam SS (ed) Plant associated bacteria, 1st edn. Springer, Dordrecht, pp 195–230

Qingwen Z, Ping L, Gang W, Qingnian C (1998) On the biochemical mechanism of induced resistance of cotton to cotton bollworm by cutting off young seedling at plumular axis. Acta Phytophylacica Sin 25:209–212

Qiu Z, Tan H, Zhou S, Cao L (2014) Enhanced phytoremediation of toxic metals by inoculating endophytic *Enterobacter* sp. CBSB1 expressing bifunctional glutathione synthase. J Hazard Mater 267:17–20

Raaijmakers JM, Vlami M, de Souza JT (2002) Antibiotic production by bacterial biocontrol agents. Antonie Van Leeuwenhoek 81:537–547

Raaijmakers JM, Paulitz CT, Steinberg C, Alabouvette C, Moenne–Loccoz Y (2009) The rhizosphere: a playground and battlefield for soilborne pathogens and beneficial microorganisms. Plant Soil 321:341–361

Raghavendra AS, Gonugunta VK, Christmann A, Grill E (2010) ABA perception and signalling. Trends Plant Sci 15:395–401

Raj SN, Deepak SA, Basavaraju P, Shetty HS, Reddy MS, Kloepper JW (2003) Comparative performance of formulations of plant growth promoting rhizobacteria in growth promotion and suppression of downy mildew in pearl millet. Crop Prot 22:579–588

Rajasekar S, Elango R (2011) Effect of microbial consortium on plant growth and improvement of alkaloid content in *Withania somnifera* (Ashwagandha). Curr Bot 2:27–30

Ramarathnam R, Fernando WGD, de Kievit T (2011) The role of antibiosis and induced systemic resistance, mediated by strains of *Pseudomonas chlororaphis*, *Bacillus cereus* and *B. amyloliquefaciens*, in controlling blackleg disease of canola. Biol Control 56:225–235

Reid MS, Wu MJ (1992) Ethylene and flower senescence. Plant Growth Regul 11:37–43

Rivera-Cruz MC, Narcia AT, Ballona GC, Kohler J, Caravaca F, Roldan A (2008) Poultry manure and banana waste are effective biofertilizer carriers for promoting plant growth and soil sustainability in banana crops. Soil Biol Biochem 40:3092–3095

Robertson MJ, Bonnett GD, Hughes RM, Muchow RC, Campbell JA (1998) Temperature and leaf area expansion of sugarcane: integration of controlled-environment, field and model studies. Aust J Plant Physiol 25:819–828

Rodriguez-Salazar J, Suarez R, Caballero-Mellado J, Iturriaga G (2009) Trehalose accumulation in *Azospirillum brasilense* improves drought tolerance and biomass in maize plants. FEMS Microbiol Lett 296:52–59

Rojas-Tapias D, Moreno–Galvan A, Pardo-Diaz S, Obando M, Rivera D, Bonilla R (2012) Effect of inoculation with plant growth-promoting bacteria (PGPB) on amelioration of saline stress in maize (*Zea mays*). Appl Soil Ecol 61:264–272

Rosas SB, Andres JA, Rovera M, Correa NS (2006) Phosphate *solubilizing Pseudomonas putida* can influence the rhizobia-legume symbiosis. Soil Biol Biochem 38:3502–3505

Ruiz-Sanchez M, Armada E, Munoz Y, de Salamone IEG, Aroca R, Ruiz–Lozano JM, Azcon R (2011) *Azospirillum* and arbuscular mycorrhizal colonization enhance rice growth and physiological traits under well–watered and drought conditions. J Plant Physiol 168:1031–1037

Safronova VI, Stepanok VV, Engqvist GL, Alekseyev YV, Belimov AA (2006) Root-associated bacteria containing 1-aminocyclopropane-1-carboxylate deaminase improve growth and nutrient uptake by pea genotypes cultivated in cadmium supplemented soil. Biol Fertil Soils 42:267–272

Saharan BS, Nehra V (2011) Plant growth promoting rhizobacteria: a critical review. Life Sci Med Res LSMR- 21

Saleem M, Arshad M, Hussain S, Bhatti AS (2007) Perspective of plant growth promoting rhizobacteria (PGPR) containing ACC deaminase in stress agriculture. J Ind Microbiol Biotechnol 34:635–648

San Francisco S, Houdusse F, Zamarreno AM, Garnica M, Casanova E, Garcia-Mina JM (2005) Effects of IAA and IAA precursors on the development, mineral nutrition, IAA content and free polyamine content of pepper plants cultivated in hydroponic conditions. Sci Hortic 106:38–52

Sandhya V, Ali SKZ, Grover M, Reddy G, Venkateswarlu B (2009) Alleviation of drought stress effects in sunflower seedlings by the exopolysaccharides producing *Pseudomonas putida* strain GAP-P45. Biol Fertil Soils 46:17–26

Sandhya V, Ali SZ, Grover M, Reddy G, Venkateswarlu B (2010) Effect of plant growth promoting Pseudomonas spp. on compatible solutes, antioxidant status and plant growth of maize under drought stress. Plant Growth Regul 62:21–30

Saravanakumar D, Samiyappan R (2007) ACC deaminase from *Pseudomonas fluorescens* mediated saline resistance in groundnut (*Arachis hypogea*) plants. J Appl Microbiol 102:1283–1292

Saravanakumar D, Kavino M, Raguchander T, Subbian P, Samiyappan R (2011) Plant growth promoting bacteria enhance water stress resistance in green gram plants. Acta Physiol Plant 33:203–209

Sari E, Hetebarian HR, Aminian H (2007) The effects of *Bacillus pumilus*, isolated from wheat rhizosphere, on resistance in wheat seedling roots against the take-all fungus, *Gaeumannomyces graminis* var. *tritici*. J Phytopathol 155:720–727

Sarwar M, Kremer RJ (1995) Enhanced suppression of plant growth through production of L-tryptophan–derived compounds by deleterious rhizobacteria. Plant Soil 172:261–269

Sayyed RZ, Gangurde NS, Patel PR, Joshi SA, Chincholkar SB (2010) Siderophore production by *Alcaligenes faecalis* and its application for growth promotion *Arachis hypogaea*. Indian J Biotech 9:302–307

Scheible WR, Lauerer M, Schulze ED, Caboche M, Stitt M (1997) Accumulation of nitrate in the shoot acts as a signal to regulate shoot–root allocation in tobacco. Plant J 11:671–691

Schelkle M, Peterson RL (1996) Suppression of common root pathogens by helper bacteria and ectomycorrhizal fungi in vitro. Mycorrhiza 6:481–485

Scher FM, Baker R (1982) Effect of Pseudomonas putida and a synthetic iron chelator on induction of soil suppressiveness to *Fusarium wilt* pathogens. Phytopathology 72:1567–1573

Selvakumar G, Joshi P, Suyal P, Mishra PK, Joshi GK, Bisht JK, Bhatt JC, Gupta HS (2011) *Pseudomonas lurida* M2RH3 (MTCC 9245), a psychrotolerant bacterium from the Uttarakhand Himalayas, solubilizes phosphate and promotes wheat seedling growth. World J Microbiol Biotechnol 27:1129–1135

Shah S, Li JP, Moffatt BA, Glick BR (1998) Isolation and characterization of ACC deaminase genes from two different plant growth-promoting rhizobacteria. Can J Microbiol 44:833–843

Shaharoona B, Arshad M, Zahir ZA (2006) Effect of plant growth promoting rhizobacteria containing ACC-deaminase on maize (*Zea mays* L.) growth under axenic conditions and on nodulation in mung bean (*Vigna radiata* L.). Lett Appl Microbiol 42:155–159

Shaharoona B, Naveed M, Arshad M, Zahir ZA (2008) Fertilizer-dependent efficiency of Pseudomonads for improving growth, yield, and nutrient use efficiency of wheat (*Triticum aestivum* L.). Appl Mirobiol Biotechnol 79:147–155

Shanahan P, O'Sullivan DJ, Simpson P, Glennon JD, O'Gara F (1992) Isolation of 2,4-diacetylphloroglucinol from a fluorescent pseudomonad and investigation of physiological parameters influencing its production. Appl Environ Microbiol 58:353–358

Shanmugam V, Kanoujia N (2011) Biological management of vascular wilt of tomato caused by *Fusarium oxysporum* f. sp. *lycospersici* by plant growth-promoting rhizobacterial mixture. Biol Control 57:85–93

Sharp RG, Chen L, Davies WJ (2011) Inoculation of growing media with the rhizobacterium *Variovorax paradoxus* 5C-2 reduces unwanted stress responses in hardy ornamental species. Sci Hortic 129:804–811

Shehata MM, El–Khawas (2003) Effect of two biofertilizer on growth parameters, yield characters, nitrogenous components, nucleic acids content, minerals, oil content, protein profiles and DNA banding pattern of sunflower (*Helianthus annus* L. cv. Vedock) yield. Pak J Biol Sci 6:1257–1268

Shilev S, Sancho ED, Benlloch-Gonzalez M (2012) Rhizospheric bacteria alleviate salt-produced stress in sunflower. J Environ Manage 95:S37–S41

Siddikee MA, Chauhan PS, Sa T (2012) Regulation of ethylene biosynthesis under salt stress in red pepper (*Capsicum annuum* L.) by 1-aminocyclopropane-1-carboxylic acid (ACC) deaminase-producing halotolerant bacteria. J Plant Growth Regul 31:265–272

Singh PP, Shin YC, Park CS, Chung YR (1999) Biological control of *Fusarium wilt* of cucumber by chitinolytic bacteria. Phytopathology 89:92–99

Singh JS, Pandey VC, Singh DP (2011) Efficient soil microorganisms: a new dimension for sustainable agriculture and environmental development. Agric Ecosyst Environ 140(3–4):339–353

Sinha J, Biswas CH, Ghosh A, Saha A (2010) Efficacy of vermicompost against fertilizer on Cicer and Pisum and on population diversity of N_2 fixing bacteria. J Environ Biol 31(3):287–292

Skirycz A, Inze D (2010) More from less: plant growth under limited water. Curr Opin Biotechnol 21:197–203

Skirycz A, De Bodt S, Obata T, De Clercq I, Claeys H, De Rycke R, Andriankaja M, Van Aken O, Van Breusegem F, Fernie AR, Inze D (2010) Developmental stage specificity and the role of mitochondrial metabolism in the response of Arabidopsis leaves to prolonged mild osmotic stress. Plant Physiol 152:226–244

Smyth EM, McCarthy J, Nevin R, Khan MR, Dow JM, O'Gara F, Doohan FM (2011) *In vitro* analyses are not reliable predictors of the plant growth promotion capability of bacteria; a *Pseudomonas fluorescens* strain that promotes the growth and yield of wheat. J Appl Microbiol 111:683–692

Somers E, Vanderleyden J, Srinivasan M (2004) Rhizosphere bacterial signalling: a love parade beneath our feet. Crit Rev Microbiol 30:205–240

Sosa L, Llanes A, Reinoso H, Reginato M, Luna V (2005) Osmotic and specific ion effect on the germination of *Prosopis strombulifera*. Ann Bot 96(2):261–267

Stamford NP, Santos PR, Santos CERS, Freitas ADS, Dias SHL, Lira Junior MA (2007) Agronomic effectiveness of biofertilizers with phosphate and *Acidithiobacillus* in a Brazilian tableland acidic soil. Bioresour Technol 98:1311–1318

Stephens PM, Crowley JJ, O'Connell C (1993) Selection of pseudomonad strains inhibiting *Pythium ultimum* on sugar-beet seeds in soil. Soil Biol Biochem 25:1283–1288

Stock CA, Mcloughlin TJ, Klein JA, Adang M (1990) Expression of a *Bacillus thuringiensis* crystal protein gene in *Pseudomonas cepacia* 526. Can J Microbiol 36:879–884

Stout MJ, Zehnder GW, Baur ME (2002) Potential for the use of elicitors of plant defence in arthropod management programs. Arch Insect Biochem Physiol 51:222–235

Sturz AV, Christie BR, Nowak J (2000) Bacterial endophytes: potential role in developing sustainable systems of crop production. Crit Rev Plant Sci 19:1–30

Suarez R, Wong A, Ramirez M, Barraza A, Orozco Mdel C, Cevallos MA, Lara M, Hernandez G, Iturriaga G (2008) Improvement of drought tolerance and grain yield in common bean by overexpressing trehalose–6–phosphate synthase in rhizobia. Mole Plant Microbe Interact 21:958–966

Taghavi S, Barac T, Greenberg B, Borremans B, Vangronsveld J, van der Lelie D (2005) Horizontal gene transfer to endogenous endophytic bacteria from poplar improves phytoremediation of toluene. Appl Environ Microbiol 71:8500–8505

Taiz L, Zeiger E (2000) Plant physiology, 2nd edn. Benjamin Cumings Publishing Company, Redwood City

Tandy S, Ammann A, Schulin R, Nowack B (2006) Biodegradation and speciation of residual SS-ethylenediaminedisuccunic acid (EDDS) in soil solution after soil washing. Environ Pollut 142:191–199

Tassi E, Pouget J, Petruzzelli G, Barbafieri M (2008) The effects of exogenous plant growth regulators in the phytoextraction of heavy metals. Chemosphere 71:66–73

Theocharis A, Bordiec S, Fernandez O, Paquis S, Dhondt-Cordelier S, Baillieul F, Clement C, Barka EA (2012) *Burkholderia phytofirmans* PsJN primes *Vitis vinifera* L. and confers a better tolerance to low nonfreezing temperatures. Mol Plant Microbe Interact 25:241–249

Tian Q, Chen F, Liu J, Zhang F, Mi G (2008) Inhibition of maize root growth by high nitrate supply is correlated to reduced IAA levels in roots. J Plant Physiol 165:942–951

Timms-Wilson TM, Ellis RJ, Renwick A, Rhodes DJ, Mavrodi DV, Weller DM, Thomashow LS, Bailey MJ (2000) Chromosomal insertion of phenazine-1-carboxylic acid biosynthetic pathway enhances efficacy of damping-off disease control by *Pseudomonas fluorescens*. Mol Plant Microbe Interact 13:1293–1300

Timmusk S, Wagner EGH (1999) The plant-growth-promoting rhizobacterium *Paenibacillus polymyxa* induces changes in Arabidopsis thaliana gene expression: a possible connection between biotic and abiotic stress responses. Mol Plant Microbe Interact 12:951–959

Truyens S, Jambon I, Croes S, Janssen J, Weyens N, Mench M, Carleer R, Cuypers A, Vangronsveld J (2014) The effect of long-term Cd and Ni exposure on seed endophytes of *Agrostis capillaris* and their potential application in phytoremediation of metal-contaminated soils. Int J Phytoremediation 16:643–659

Upadhyay SK, Singh JS, Singh DP (2011) Exopolysaccharide–producing plant growth- promoting rhizobacteria under salinity condition. Pedosphere 21:214–222

Van Elsas JD, Trevors JT (1997) Modern soil microbiology. Marcel Dekker, New York, pp 1–20

Van Loon LC (2007) Plant responses to plant growth-promoting rhizobacteria. Eur J Plant Pathol 119:243–254

Verma SC, Ladha JK, Tripathi AK (2001) Evaluation of plant growth promoting and colonization ability of endophytic diazotrophs from deep water rice. J Biotechnol 91:127–141

Vessey JK (2003) Plant growth promoting rhizobacteria as biofertilizers. Plant Soil 255:571–586

Vilchez S, Manzanera M (2011) Biotechnological uses of desiccation-tolerant microorganisms for the rhizoremediation of soils subjected to seasonal drought. Appl Microbiol Biotechnol 91:1297–1304

Vivas A, Marulanda A, Ruiz–Lozano JM, Barea JM, Azcon R (2003) Influence of a *Bacillus* sp. on physiological activities of two arbuscular mycorrhizal fungi and on plant responses to PEG–induced drought stress. Mycorrhiza 13:249–256

Vivas A, Barea JM, Azcon R (2005) Interactive effect of *Brevibacillus brevis* and *Glomus mosseae*, both isolated from Cd contaminated soil, on plant growth, physiological mycorrhizal fungal characteristics and soil enzymatic activities in Cd polluted soil. Environ Pollut 134:257–266

Walton DC, Li Y (1995) Abscisic acid biosynthesis and metabolism. In: Plant hormones: physiology, biochemistry and molecular biology. Dordrecht, pp 140–157

Wang CJ, Yang W, Wang C, Gu C, Niu DD, Liu HX, Wang YP, Guo JH (2012) Induction of drought tolerance in cucumber plants by a consortium of three plant growth-promoting rhizobacterium strains. PLoS One 7(12):e52565

Whipps JM (1990) Carbon economy. In: Lynch JM (ed) The rhizosphere. Wiley, Chichester, pp 59–99

Whipps J (2001) Microbial interactions and biocontrol in the rhizosphere. J Exp Bot 52:487–511

Wilmowicz E, Kesy J, Kopcewicz J (2008) Ethylene and ABA interactions in the regulation of flower induction in *Pharbitis nil*. J Plant Physiol 165:1917–1928

Woltering EJ, van Doorn WG (1988) Role of ethylene in senescence of petals–morphological and taxonomical relationships. J Exp Bot 39:1605–1616

Wu SC, Caob ZH, Lib ZG, Cheung KC, Wong MH (2005) Effects of biofertilizer containing N-fixer, P and K solubilizers and AM fungi on maize growth: a greenhouse trial. Geoderma 125:155–166

Wu SC, Cheung KC, Luo YM, Wong MH (2006) Effects of inoculation of plant growth–promoting rhizobacteria on metal uptake by *Brassica juncea*. Environ Pollut 140:124–135

Yadav SK (2010) Heavy metals toxicity in plants: an overview on the role of glutathione and phytochelatins in heavy metal stress tolerance of plants. S Afr J Bot 76:167–179

Yancheshmeh JB, Khavazi K, Pazira E, Solhi M (2011) Evaluation of inoculation of plant growth–promoting rhizobacteria on cadmium and lead uptake by canola and barley. Afr J Microbiol Res 5:1747–1754

Yang JW, Kloepper JW, Ryu CM (2008) Rhizosphere bacteria help plants tolerate abiotic stress. Trends Plant Sci 14:1–4

Yang L, Wang Y, Song J, Zhao W, He X, Chen J, Xiao M (2011) Promotion of plant growth and in situ degradation of phenol by an engineered *Pseudomonas fluorescens* strain in different contaminated environments. Soil Biol Biochem 43:915–922

Yang Q, Tu S, Wang G, Liao X, Yan X (2012) Effectiveness of applying arsenate reducing bacteria to enhance arsenic removal from polluted soils by *Pteris vittata* L. Int J Phytoremediation 14:89–99

Yao LX, Wu ZS, Zheng YY, Kaleem I, Li C (2010) Growth promotion and protection against salt stress by *Pseudomonas putida* Rs-198 on cotton. Eur J Soil Biol 46:49–54

Yazdani M, Bahmanyar MA, Pirdashti H, Esmaili MA (2009) Effect of phosphate solubilization microorganisms (PSM) and plant growth promoting rhizobacteria (PGPR) on yield and yield components of corn (*Zea mays* L.). Proc World Acad Sci Eng Technol 37:90–92

Yong X, Chen Y, Liu W, Xu L, Zhou J, Wang S, Chen P, Ouyang P, Zheng T (2013) Enhanced cadmium resistance and accumulation in *Pseudomonas putida* KT2440 expressing the phytochelatin synthase gene of *Schizosaccharomyces pombe*. Lett Appl Microbiol 58:255–261

Yuan GF, Jia CG, Li Z, Sun B, Zhang LP, Liu N, Wang QM (2010) Effect of brassinosteroids on drought resistance and abscisic acid concentration in tomato under water stress. Sci Hortic 126:103–108

Yue H, Mo W, Li C, Zheng Y, Li H (2007) The salt stress relief and growth promotion effect of RS-5 on cotton. Plant Soil 297:139–145

Zahir ZA, Arshad M, Azam M, Hussian A (1997) Effect of an auxin precursor L-tryptophan and *Azotobacter* inoculation on yield and chemical composition of potato under fertilized conditions. J Plant Nutr 20:745–752

Zahir ZA, Abbas SA, Khalid M, Arshad M (2000) Substrate-dependent microbially derived plant hormones for improving growth of maize seedlings. Pak J Biol Sci 3:289–291

Zahir ZA, Arshad M, Frankenberger WT Jr (2004) Plant growth promoting rhizobacteria application and perspectives in agriculture. Adv Agron 81:96–168

Zahir ZA, Asghar HN, Akhtar MJ, Arshad M (2005) Precursor (L-tryptophan)-inoculum (*Azotobacter*) interactions for improving yields and nitrogen uptake of maize. J Plant Nutr 28:805–817

Zahir ZA, Munir A, Asghar HN, Shahroona B, Arshad M (2008) Effectiveness of rhizobacteria containing ACC-deaminase for growth promotion of peas (*Pisum sativum*) under drought conditions. J Microbiol Biotechnol 18:958–963

Zahir ZA, Ghani U, Naveed M, Nadeem SM, Asghar HN (2009) Comparative effectiveness of *Pseudomonas* and *Serratia* sp. containing ACC deaminase for improving growth and yield of wheat (*Triticum aestivum* L.) under salt-stressed conditions. Arch Microbiol 191:415–424

Zahir ZA, Shah MK, Naveed M, Akhter MJ (2010) Substrate-dependent auxin production by *Rhizobium phaseoli* improves the growth and yield of *Vigna radiata* L. under salt stress conditions. J Microbiol Biotechnol 20:1288–1294

Zahir ZA, Akhtar SS, Ahmad M, Saifullah NSM (2012) Comparative effectiveness of *Enterobacter aerogenes* and *Pseudomonas fluorescens* for mitigating the depressing effect of brackish water on maize. Int J Agric Biol 14:337–344

Zahran HH (1999) *Rhizobium*-legume symbiosis and nitrogen fixation under severe conditions and in an arid climate. Microbiol Mol Biol Rev 63:968–989

Zaidi S, Usmani S, Singh BR, Musarrat J (2006) Significance of *Bacillus subtilis* strain SJ-101 as a bioinoculant for concurrent plant growth promotion and nickel accumulation in *Brassica juncea*. Chemosphere 64:991–997

Zehnder G, Kloepper JW, Yao C, Wei G (1997) Induction of systemic resistance in cucumber against cucumber beetles (Coleoptera: Chrysomelidae) by plant growth-promoting rhizobacteria. J Econ Entomol 90:391 396

Zhang HM, Jennings A, Barlow PW, Forde BG (1999) Dual pathways for regulation of root branching by nitrate. Proc Natl Acad Sci U S A 96:6529–6534

Zhang S, Reddy MS, Kloepper WJ (2002) Development of assays for assessing induced systemic resistance by plant growth–promoting rhizobacteria against blue mold of tobacco. Biol Control 23:79–86

Zhang J, Jia W, Yang J, Ismail AM (2006) Role of ABA in integrating plant responses to drought and salt stresses. Field Crop Res 97:111–119

Zhang H, Kim MS, Sun Y, Dowd SE, Shi H, Pare PW (2008) Soil bacteria confer plant salt tolerance by tissue–specific regulation of the sodium transporter HKT1. Mol Plant Microbe Interact 21:737–744

Biopesticides: Where We Stand?

2

Jitendra Mishra, Sakshi Tewari, Sachin Singh, and Naveen Kumar Arora

Abstract

Chemical pesticides are well known for their effective role in disease management because not only they act on a broad host range but production technology is also less expensive. However, the devastating part is their huge negative impact on the environment including the living beings of the planet. In spite of this, in the absence of suitable alternative, the use of synthetic pesticides has dominated around the globe. By the advent of greener approach of developing and using biopesticides, the situation is gradually changing but in fact can move far more swiftly in this direction which will be sustainable and eco-friendly. Although biopesticides are slowly replacing the chemical pesticides, a complete global look at the scenario indicates that the former and particularly the industries based on them are still in an insecure position in comparison to the chemicals which rule the agriculture. We can say that the biopesticides, although show a great promise, have not come up to the desired level so as to displace the dominance of chemicals. In this chapter, the global scenario of biopesticides is discussed emphasizing upon the current demand, use, constraints, and remedies.

Introduction

Two-thirds of today's world population depends upon agriculture for livelihood, but nowadays, growth and production of agricultural crops are getting hampered day by day (Elumalai and Rengasamy 2012). When farmers see their agricultural crops declining in yield and production, they often expect a dramatic, magical treatment to make them lush, green, and healthy again, so that the productivity increases. As a result, they start using chemical pesticides, disregarding their future effects. The extensive use of these synthetic organic chemicals in the past decades has led to a number of long-term environmental problems (Arora et al. 2012). Keeping all these

J. Mishra • S. Tewari • S. Singh • N.K. Arora (✉)
Department of Environmental Microbiology, BBA University, Lucknow 226025, Uttar Pradesh, India
e-mail: nkarora_net@rediffmail.com

N.K. Arora (ed.), *Plant Microbes Symbiosis: Applied Facets*,
DOI 10.1007/978-81-322-2068-8_2, © Springer India 2015

facts in mind, a very big challenge in the new millennium is to produce more and more food from shrinking per capita arable land, keeping the environment safe and innocuous.

As agricultural production intensified over the past few decades, producers became more and more dependent on agrochemicals. The conventional chemical pesticides have although enhanced the food production, but have also adversely affected the environment and nontarget organisms. Chemical fertilizers and pesticides are continuously accumulating in the environment, harming the ecosystem, causing pollution, and inflicting diseases at alarming levels (Gerhardson 2002; Arora et al. 2010). The heavy use of pesticides has already caused grave damage to health, ecosystems, and groundwater. Many of the pesticides currently being used have the tendency to survive in plants for a long time. They also enter the food chain and are found in meat and dairy products and remain as residue in the soil and ecosystem for long durations.

Therefore, it is very urgent to identify alternatives to chemical pesticides for plant protection without sacrificing the productivity and profitability of agriculture. Due to the side effects of chemical pesticides, sustainable crop production through eco-friendly management is essentially required in the present scenario.

Biopesticides offer powerful tools to create a new generation of sustainable agriculture products. They are the most likely alternatives to some of the most problematic chemical pesticides currently in use. Biopesticides offer solutions to concerns such as pest resistance, traditional chemical pesticides, and public concern about side effects of pesticides on the surrounding environment and ultimately on human health. The overriding challenge for the biopesticide industry is to live up to the promises and expectations of the end users or the market and public as whole. There are unanswered questions and unexamined assumptions about these biological and eco-friendly alternatives. Challenges to biopesticides stem from questions about their efficacy and safety, public and grower confusion about the spectrum of biopesticide products in the market, and current market conditions that paradoxically both hinder and favor the field's growth.

The aim of the review is to highlight the present global scenario of biopesticides, including its market, availability, and usefulness. Besides this, the compilation also addresses the possible constraints and dilemmas associated with biopesticides throughout the globe and the direction to what should be explored in the future to uplift the stature. Biopesticides offer an environmentally sustainable approach to increase crop production and health, contributing substantially in making the twenty-first century the age of biotechnology hence every effort should be made to enhance their use and popularity.

Shift of Wheel from Pesticides to Biopesticides

It is well known that there have been some discoveries in the past which not only have changed the life of man but also had major influence on the globe, and a very well-known chemical pesticide para-dichlorodiphenyltrichloroethane (DDT) was one of them (West and Campbell 1946). After World War II, its discovery was considered as important as that of penicillin (Felix 1958). In one way when penicillin was saving the life of millions from life-threatening bacterial infections, DDT was protecting crops from a variety of pests. Although for a few years the glory of DDT was extolled by everyone, but soon the whole picture started to change. For the first time, Rachel Carson, a marine biologist and conservationist, stated in her book *Silent Spring* that we are paying much more than what we are getting to get rid of mosquitoes (Carson 1962). Finally in the USA, DDT was banned in 1972 (Griswold 2012). There are myriads of incidences dealing with DDT poisoning that are already known and some are needed to be further explored (Hill and Robinson 1945; Dresdend 1948; Keane 1972; Tschirley 1973; Longnecker et al. 1997; Conis et al. 2010; Qiu 2013). DDT was not the only culprit; other categories of synthetic pesticides such as organophosphates (OP), carbamates, and pyrethroids were also launched, and by 1980, their impact on pest control and environment was also well recognized (Aktar et al. 2009).

OP are advanced over organochlorines being nonpersistent in environment and do not cause bioaccumulation in the food chain; however, they are much more acutely toxic to humans and other mammals than organochlorines, and their exposure by inhalation, swallowing, or absorption through the skin can lead to immediate health problems (Baird and Cann 2008). Mostly people residing in agricultural areas, farmworkers, and small children are more frequently exposed to hazards of organophosphates (Landrigan et al. 1999; Eskenazi et al. 1999; Fenske et al. 2000; McCauley et al. 2001; Quandt et al. 2004; Eskenazi et al. 2008). Acute toxicity of organophosphates and carbamates is a severe problem in developing countries, and ignorance about their hazards and the lack of information have led to many deaths among agricultural workers (Konradsen et al. 2003). A high proportion of pesticide poisonings and deaths occur in developing countries, and the main victims are farming workers especially those having insufficient knowledge of pesticide hazards (Pimentel and Greiner 1996). Intensive use of synthetic pesticides created more harsh ecological conditions that resulted in reduced soil fauna and habitat loss of micro- and macroflora (Edwards and Thompson 1973; Tripathi and Sharma 2005; Frampton et al. 2006; Bezchlebová et al. 2007). Recent studies showed that the increased use of pesticides is responsible for the vanishing population of bees and several other useful insects involved in pollination of flowers of agriculturally important crops (Gill et al. 2012). In a similar study, pesticide exposure raised question on global decline of frog population (Brühl et al. 2013). Even several species of birds have become extinct or are on the verge of it because of the pesticides. It can also be concluded that the negative impact of pesticides is much more than what is visible by the aid of the present technology.

By 2001, over 100 nations in Stockholm Convention agreed to sign an international treaty to phase out completely persistent organic pollutants ("POPs"), including DDT (Downie 2003). Albeit the synthetic pesticides have showed devastating effect on the ecosystems, in the absence of an alternative, it was impossible to diminish their utility and effects (Wu and Chen 2004; Aktar et al. 2009). Meanwhile, a phrase by Roger Ascham "Necessitie, the inuentour of all goodnesse" proved to be true for some workers in the field of biological control. Pioneer workers such as A. Bassi in Italy, V. Auduoin in France, J. Le Conte in the USA, and E. Metchnikoff in Russia proposed that antagonistic microbes might be useful in controlling crop pests and may be alternatives to synthetic pesticides (Le Conte 1874; Steinhaus 1949, 1957, 1975; McCoy et al. 1988). Later, these pioneer discoveries and further researches proved to be milestones in development of microbe-based pesticides (Schönbeck and Dehne 1986; Sundheim and Tronsmo 1988).

In the eighteenth century and even in the beginning of the nineteenth century, the focus of biological control was to use animals such as birds and entomophagous insects; microbes were not even properly known at that time. The discovery of *Bacillus thuringiensis* (Bt) showed a wider aspect of microbe-based biological control (Aronson et al. 1986; Martin and Traverse 1989; Siegel and Shadduck 1990; Marrone 1994; Joung and Co^te' 2000). Microbial pest control was a very new concept, and its selective action on pest attracted the concentration of researchers and industrialists equally, and soon the first commercial Bt product, Thuricide, was registered in the USA in 1961 (USEPA 1998). Since then, different subspecies, varieties, and strains of Bt have been identified that are effective against a variety of insects (Gonzales et al. 1982; Carlton 1988). In a span of a very few years, Bt has covered up to 90 % of the whole biopesticide market (Chapple et al. 2000; Chattopadhyay et al. 2004; Romeis et al. 2006), and several Bt strains are now registered as biopesticides throughout the world (Glare and O'Callaghan 2000). In 1965, the first fungal product "Boverin" was developed in the former Union of Soviet Socialist Republics (USSR). In Boverin, *Beauveria bassiana* was used to control the Colorado potato beetle and the codling moth (De Faria and Wright 2007). In

1973, *Heliothis* nuclear polyhedrosis virus (NPV) was first declared as a viral biopesticide (Szewczyk et al. 2011). In the last few years, although several new biopesticides have been registered and served as better alternatives to synthetic pesticides, they still lack far behind the chemicals and the desired levels (Clemson 2007). Biopesticides have emerged as alternatives to chemical pesticides, and we also know their importance and benefits, but in reality to date, the goals have not been attained and chemicals still rule throughout the globe.

Concept of Biopesticides

In very general terms, according to the US Environmental Protection Agency (USEPA), biopesticides are pesticides derived from natural materials such as animals, plants, bacteria, and minerals. Biopesticides also include living organisms that destroy agricultural pests. The EPA separates biopesticides into three major classes based on the type of active ingredient used, namely, biochemical, plant-incorporated protectants, and microbial pesticides (USEPA 2008). Biochemical pesticides are chemicals either extracted from natural sources or synthesized to have the same structure and function as the naturally occurring chemicals. Biochemical pesticides are distinguished from conventional pesticides both by their structure (source) and mode of action (mechanism by which they kill or control pests) (O'Brien et al. 2009). At a global level, there is an inconsistency in understanding the term biopesticide as aforementioned operative definition of the term biopesticide given by USEPA is not followed in the entire world and that is why International Biocontrol Manufacturer's Association (IBMA) and the International Organization for Biological Control (IOBC 2008) promote to use the term biocontrol agents (BCAs) instead of biopesticide (Guillon 2003). IBMA classifies biocontrol agents into four groups: (1) macrobials, (2) microbials, (3) natural products, and (4) semiochemicals (insect behavior-modifying agents). Among all the biocontrol agents, the most important products are microbials (41 %), followed by macrobials (33 %), and, finally, other natural products (26 %) (Guillon 2003). This review focuses on microbe-based biopesticides.

Microbial Pesticides

Microbial pesticides are also known as BCAs. They offer the advantages of higher selectivity and lower or no toxicity in comparison to conventional chemical pesticides (MacGregor 2006). The active ingredient of a microbial pesticide is typically the microorganism. Microbial ingredients can be either the spores or the organisms themselves. The most commonly used microbial biopesticides are living organisms, which are pathogenic for the pest of interest. These include biofungicides (*Trichoderma*, *Pseudomonas*, *Bacillus*), bioherbicides (*Phytophthora*), and bioinsecticides (*Bt*) (Gupta and Dikshit 2010). Microbial pesticides come from naturally occurring or genetically altered bacteria, fungi, algae, viruses, or protozoans. They suppress pests either by producing toxic metabolites specific to the pest, causing disease, preventing establishment of other microorganisms through competition, or various other modes of action (Clemson 2007). Of the total biopesticide market for all crop types, bacterial biopesticides claim about 74 %; fungal biopesticides, about 10 %; viral biopesticides, 5 %; predator biopesticides, 8 %; and "other" biopesticides, 3 % (Thakore 2006). By 2008, there were approximately 73 microbial active ingredients that were registered by the USEPA. The registered microbial biopesticides included 35 bacterial products, 15 fungi, 6 nonviable (genetically engineered) microbial pesticides, 8 plant-incorporated protectants, 1 protozoan, 1 yeast, and 6 viruses (Steinwand 2008). Microbial biopesticides may be delivered to crops in many forms including live organisms, dead organisms, and spores, and the next subsection presents various forms of microbe-based pesticides that are being used presently.

Bacterial Biopesticides

Bacterial biopesticides are the most common form of microbial pesticides that function in multiple ways. Generally, they are used as insecticides, although they can be used to control the growth of plant pathogenic bacteria and fungi. As an insecticide, they are generally specific to individual species of moths and butterflies or species of beetles, flies, and mosquitoes. To be effective, they must come into contact with the target pest and may be required to be ingested. In insects, bacteria disrupt the digestive system by producing endotoxins that are often specific to the particular insect pest. When used to control pathogenic bacteria or fungus, the bacterial biopesticide colonizes on the plant and crowds out the pathogenic species (O'Brien et al. 2009).

The members of the genus *Bacillus* are often considered as microbial factories for the production of vast array of biologically active molecules, some of which are potentially inhibitory for fungal growth (Schallmey et al. 2004). The most widely used microbial pesticides are subspecies and strains of *B. thuringiensis* (Bt), accounting for approximately 90 % of the biopesticide market in the USA (Chattopadhyay et al. 2004). Since its discovery in 1901, Bt has been widely used to control insect pests important in agriculture, forestry, and medicine (Mazid and Kalita 2011). Its principal characteristic is the synthesis, during sporulation, of crystalline inclusions containing proteins known as δ endotoxins or Cry proteins, which have insecticidal properties. To date, over one hundred *B. thuringiensis*-based bioinsecticides, biopesticides, and biofungicides have been developed. Microbial pesticides containing *B. thuringiensis* var. *kurstaki* kill the caterpillar stage of a wide array of butterflies and moths. In addition, the genes that code for the insecticidal crystal proteins have been successfully transferred into different crop plants including cotton, tomato, brinjal, etc. that lead to significant economic benefits. Due to their high specificity and safety in the environment, *B. thuringiensis* and Cry proteins are efficient, safe, and sustainable alternatives to chemical pesticides for the control of insect pests (Roy et al. 2007; Kumar 2012).

Plant pathogenic fungi and oomycetes are major threats for crop and plant production. Therefore, the control of fungal diseases by *Bacillus*-based biopesticides represents an interesting opportunity for agricultural biotechnology. Indeed, several commercial products based on various *Bacillus* species such as *B. amyloliquefaciens*, *B. licheniformis*, *B. pumilus*, and *B. subtilis* have been marketed as biofungicides (Fravel 2005). These *Bacillus*-based products have been developed especially for the control of fungal diseases. A high number of reports have described the beneficial effects of several *Bacillus* species against diseases elicited by oomycetes and fungal pathogens. Some examples are the suppression of root diseases (such as avocado root rot, tomato damping off, and wheat take-all), foliar diseases (such as cucurbit and strawberry powdery mildews), and postharvest diseases (such as green, gray, and blue molds) (Cazorla et al. 2007; Pertot et al. 2008; Arrebola et al. 2010). Certain strains of *B. subtilis* are being used against a range of plant pathogens that cause damping off and soft rots (Kloepper et al. 2004; Haas and Defago 2005; Berg 2009). Apart from this *B. pumilus* QST 2808, *B. subtilis* QST GB03 are used for designing biopesticides, namely, Ballad®Plus and Kodiak®, for commercial purposes in the USA (Stewart et al. 2011).

Due to their catabolic versatility and excellent root-colonizing capability, pseudomonads are also being investigated extensively for the use in biocontrol of pathogens in agriculture (Ganeshan and Kumar 2006). They are known to enhance plant growth and yield, reduce severity of many diseases, and are considered to be among the most prolific PGPRs (Hoffland et al.1996; Wei et al.1996). Several species of *Pseudomonas* are being used for designing biopesticides that include *P. fluorescence*, *P. aeruginosa*, *P. syringae*, etc. Certain strains of *Pseudomonas aureofaciens* are being used against a range of plant pathogens including damping off and soft rots (Kloepper et al. 2004; Haas and Defago

2005; Berg 2009). The cell suspensions of pseudomonads are immobilized on certain carriers and are prepared as formulations for easy application, storage, commercialization, and field use. In India, *P. fluorescens* biopesticide is effectively being used against late blight of potato; it is available commercially under diverse brand names such as Krishi bio rahat, Krishi bio nidan, Mona, etc. Virulent cells of bacterial antagonist *P. fluorescens* are taken to prepare a biopesticide formulation that is effective against phytopathogen *Ralstonia solanacearum* (Bora and Deka 2007; Chakravarty and Kalita 2011).

P. syringae strains ESC-10 and ESC-11 were initially registered (licensed for sale and distribution) in 1995; at the end of April 2000, there were three end products containing ESC-10 and 2 end products containing ESC-11 in the USA (Bull et al. 1997). An attractive role of fluorescent pseudomonads in biological control of fungal plant pathogens has been illustrated against *Aspergillus*, *Alternaria*, *Fusarium*, *Macrophomina*, *Pythium*, *Sclerotinia*, and *Rhizoctonia* (Dunne et al. 1998; Gupta et al. 2001). Several different commercially available biopesticides in the USA that are developed from *Pseudomonas* and effective against fungal phytopathogens are Spot-Less, At-Eze, Bio-Save 10LP, and Bio-Save 11LP (Vargas 1999; Nakkeeran et al. 2005; Khalil et al. 2013). Bioformulation, biopesticides, and bioinoculants developed from fluorescent pseudomonads can serve multifaceted functions of plant growth promotion, bioremediation, and disease management (Arora et al. 2008, 2013; Khare and Arora 2011; Tewari and Arora 2013).

Certain other bacterial strains like that of *Agrobacterium radiobacter* are also used to control pests such as *Agrobacterium tumefaciens*. Other PGPRs like *Pantoea agglomerans* strain E325, *Streptomyces lydicus* WYEC 108, and *Coniothyrium minitans* strain CON/M/91-08 are also used nowadays for designing new biopesticides such as Bloomtime Biological™[3], Actinovate®SP, and Contans®WG (in the USA) that are proving to be boon in the field conditions (Chunxue et al. 2010).

Fungal Biopesticides

The fungal pathogens play a major role in the development of diseases on many important field and horticultural crops, resulting in severe plant yield losses (Khandelwal et al. 2012). Intensified use of fungicides has resulted in accumulation of toxic compounds potentially hazardous to human and environment and also in the buildup of resistance in the pathogens. Fungal biopesticides can be used to control insects and plant diseases including other fungi, bacteria, nematodes, and weeds. The mode of action is varied and depends on both the pesticidal fungus and the target pest. One advantage of fungal biopesticides in comparison with many of the bacterial and all of the viral biopesticides is that they do not need to be eaten to be effective. However, they are living organisms that often require a narrow range of conditions including moist soil and cool temperatures to proliferate. Biocontrol agents like *Trichoderma* are acclaimed as effective, eco-friendly, and cheap, nullifying the ill effects of chemicals. Therefore, of late, these biocontrol agents are identified to act against an array of important soil-borne plant pathogens causing serious diseases of crops (Bailey and Gilligan 2004).

Fungal biopesticides used against plant pathogens include *T. harzianum*, which is an antagonist of *Rhizoctonia*, *Pythium*, *Fusarium*, and other soil-borne pathogens (Harman 2005). *Trichoderma* is a fungal antagonist that grows into the main tissue of a disease-causing fungus and secretes enzymes that degrade the cell walls of the other fungus and then consumes the contents of the cells of the target fungus and multiplies its own spores. *Trichoderma* is one of the common fungal biocontrol agents being used worldwide for suitable management of various foliar- and soil-borne plant pathogens like *Ceratobasidium*, *Fusarium*, *Rhizoctonia*, *Macrophomina*, *Sclerotium*, *Pythium*, and *Phytophthora* spp. (Dominguesa et al. 2000; Anand and Reddy 2009). *Trichoderma viride* has proved to be very promising against soil-borne plant parasitic fungi (Khandelwal et al. 2012). A specific strain *Muscodor albus* QST 20799 is a naturally occurring fungus originally isolated from the bark of a cinnamon tree in Honduras. When hydrated, this *M. albus* strain is reported to produce a number of volatiles, mainly alcohols, acids, and esters, which inhibit and kill certain bacteria and other organisms that cause soil-borne and postharvest diseases. Products containing QST 20799 can be used in fields, greenhouses, and warehouses (USEPA 2008).

B. bassiana (Balsamo) Vuillemin and *Metarhizium anisopliae* (Metchnikoff) Sorokin are naturally occurring entomopathogenic fungi that infect sucking pests including *Nezara viridula* (L) (green vegetable bug) and *Creontiades* sp. (green and brown mirids) (Sosa-Goméz and Moscardi 1998). Fungi have the unique ability to attack insects by penetrating through the cuticle making them ideal for the control of sucking pests. *B. bassiana* is currently registered in the USA as Mycotrol ES ® (Mycotech, Butte) and Naturalis L® (Troy Biosciences). These products are registered against sucking pests such as whitefly, aphids, thrips, mealybugs, leafhoppers, and weevils. Studies also show that *B. bassiana* is virulent against *Lygus hesperus* Knight (Hemiptera: Miridae), a major pest of alfalfa and cotton in the USA (Noma and Strickler 2000).

Viral Biopesticides

Like bacteria and fungi, viral biopesticides play a significant role in antagonizing pathogens especially bacteria in the form of bacteriophages. Apart from it, viruses are host specific, infecting only one or a few closely related species, thus offering minimal off-target impacts (Cory and Myers 2003; England et al. 2004; Raymond et al. 2005; Hewson et al. 2011). A bacteriophage is a virus that infects bacterial cell walls. If the virus attacks bacteria that cause plant disease, it can be used as a pesticide. A large number of phage pesticides are currently used under commercial trade names and sold in the markets. Patent protection and intellectual property are important factors in the commercialization of phage pesticides. The concept of phage therapy has existed for over 90 years, and multiple companies have acquired patents and established commercial platforms for using them (Gill et al. 2007). A leading company of the USA, Omnilytics, has developed a range of phage products for the control of *Xanthomonas campestris* pv. vesicatoria, for the treatment of bacterial spot of tomatoes and peppers, and *P. syringae* pv. tomato, which is the causative agent of bacterial speck on tomatoes (Frampton et al. 2012; Schofield et al. 2012).

Baculovirus is the main virus that is commercially used for designing phage pesticides. Since the start of their commercial use, baculoviruses have been tested extensively to assess their safety in order to meet registration requirements (reviewed in Burges et al. 1980a, b; Gröner 1986; Ignoffo 1975). Baculoviruses develop in the nuclei of the host insect cells. When ingested by the host insect, infectious virus particles are liberated internally and become active. Once in the larval gut, the virus's protein overcoat quickly disintegrates, and the viral DNA proceeds to infect digestive cells. Within a few days, the host larvae are unable to digest food and so weaken and die (Thakore 2006). Baculoviruses are particularly attractive for use as biopesticides due to their high host specificity. Each virus only attacks particular species of insects, and they have been shown to have no negative impacts on plants, mammals, birds, fish, or nontarget insects (D'Amico 2007). Baculoviruses can also cause sudden and severe outbreaks within the host population for complete control of the disease (Sylvar 2008). Another major advantage of baculoviruses is that in some cases, they can replace and serve as an alternative to the antibiotics and chemical pesticides (O'Brien et al. 2009).

As of 2010, over 24 baculovirus species have been reported to be registered for use in insect pest management throughout the world (Kabaluk et al. 2010). The market share of baculoviruses is 6 % of all microbial pesticides (Quinlan and Gill 2006; Marrone 2007), and millions of hectares have been treated with registered baculovirus products over the years (Szewczyk et al. 2009; Kabaluk et al. 2010; Moscardi et al. 2011). Despite many years of use and testing against nontarget organisms, no adverse effects have ever been attributed to baculoviruses (McWilliam 2007).

Biopesticides: Global Scenario

A large number of crop pests cause about 40 % reduction in the world's crop yield (Oerke et al. 1994), and control measures adopted by using synthetic chemicals alone have remained formidable as the data suggests that approximately 5.6 billion pounds of pesticide are used worldwide and are responsible for the unbalancing of our environment (Alavanja 2009). Whereas application of biopesticides showed lesser or no toxicity to crops and is considered environment friendly, these are not globally as dominating as synthetics. Application and development trend of biopesticides has been reviewed by Leng et al. (2011). Worldwide, approximately 1,400 biopesticide products were being sold (Marrone 2007). Table 2.1 depicts a comprehensive list of commercially available biopesticides in the markets around the globe. These products are commercially successful and widely available as liquid concentrates, wettable powders, and ready-to-use dusts and granules. Among them, bacterial products are more frequently used (Fig. 2.1) especially those from Bt (Lisansky 1997). Production of Bt always remained on priority in biopesticide industry, and currently it is the main bacterium being used in agricultural pest control (Brar et al. 2006; Ali et al. 2008). Its firm position in biopesticide industry is indicated by the fact that more than 53 % of the world biopesticide market is occupied by about 200 Bt-based products (CABI 2010), and almost 50 % of this is consumed by America particularly in the USA and Canada (Guerra et al. 2001). Data on microbial biopesticide agents from Agriculture and Agri-Food Canada (Kabaluk and Gazdik 2005) and the USEPA indicate that more than 200 products are being sold in the USA, compared to only 60 comparable products in the EU. In the UK, only 5 microbial products were reported to be sold, compared to 10 in Germany and 15 each in France and the Netherlands (Chandler et al. 2008). Till 2003, the largest market shares in biopesticide belonged to North America with 44 %, followed by Europe with 20 %, Asia (13 %), Oceania (11 %), Latin America (9 %), and lowest for Africa (3 %) (Fig. 2.2). In this section, we discuss the continent-wise scenario of biopesticides.

North America

North America leads in market share of biopesticides and covers 44 % of it. In Canada alone, end-user sales of pesticides were valued at $1.4 billion in 2010, and of this, microbial pesticides represented about 0.5 % ($7.4 million), with 88 % of microbials represented by Bt ($6 million for use in forestry; $500,000 for use in agriculture), 6.7 % ($500,000) for other bacteria, 0.67 % ($50,000) for viruses, 0.67 % ($50,000) for fungi, and 4.1 % ($300,000) for nematodes (CPL Business Consultants 2010). Numerous universities across Canada are actively engaged in the research and development of biopesticides ranging from botanical to microbial pesticides for control of insects, weeds, and fungi. In the USA, the first registered microbial pesticide was prepared using *Bacillus popilliae* (*Paenibacillus popilliae*) in 1948 for the control of the Japanese beetle (Schneider 2006). Whereas a commercially unsuccessful entomopathogenic fungus (*Hirsutella thompsonii*)-based product Mycar was first registered for control of citrus rust mites, for which approval was granted to Abbott Laboratories in 1981 (McCoy 1996). In the USA, EPA is the sole agency responsible for encouraging development and use of biopesticides. Biopesticides and Pollution Prevention Division (under the Pesticide Programs) was established in 1994 in the USA with the goal of reducing risks associated with pesticide use in agricultural and nonagricultural settings and advocates adoption of biopesticides. In the USA, pesticide has to pass stringent regulations before its marketing, and the Federal Insecticide, Fungicide, and Rodenticide Act (FIFRA) does this job along with EPA, assessing carcinogenic risks or short-term mutagenicity assays to ensure that pesticide can be used with a reasonable certainty, and it will not harm human health or the environment. Such stringency in regulation created a positive pressure on growers to adopt biopesticides.

Table 2.1 Commercially available biopesticides in the global market

Category of biopesticide	Products common name or trade name	Targets
The USA		
Bactericides		
A. radiobacter k84	Galltrol – A	Crown gall disease
P. agglomerans C9-1	BlightBan C9-1	Fire blight
P. agglomerans E325	Bloomtime	Fire blight
Bacteriophage of *P. syringae* pv. Tomato	AgriPhage	Bacterial speck
Bacteriophage of *X. campestris* pv. Vesicatoria	AgriPhage	Bacterial spot
Fungicides		
B. licheniformis SB3086	EcoGuard	Fungal diseases
Bacillus mycoides isolate J	BacJ	Cercospora
B. pumilus GB 34	GB34	Seedling diseases – *Pythium* and *Rhizoctonia*
B. pumilus QST 2808	Sonata Ballad Plus	Powdery mildew, downy mildew, and rusts
B. subtilis GB03	Companion Kodiak	*Fusarium*, *Pythium*, *Rhizoctonia*
B. subtilis MBI 600	Histick N/T Pro-Mix with Biofungicide	Damping off
B. subtilis subsp. amyloliquefaciens FZB24	Taegro	*Fusarium* and *Rhizoctonia* wilt diseases
P. aureofaciens Tx-1	Spot-Less	Turf fungal diseases
Pseudomonas chlororaphis 63–28	At-Eze	Soil and seed-borne fungi
P. syringae ESC 10	Bio-Save 10LP	Postharvest diseases
P. syringae ESC 11	Bio-Save 11LP	Postharvest diseases
Streptomyces griseoviridis K61	Mycostop Biofungicide Mycostop Mix	Fungi causing damping off, stem, and crown rots
S. lydicus WYEC108	Actinovate Actino-Iron	Fungi causing damping off, stem and crown rots
Ampelomyces quisqualis M10	PowderyGon	Powdery mildew
Aspergillus flavus AF36	*Aspergillus flavus* AF36	*Aspergillus flavus* producing aflatoxin
A. flavus NRRL 21882	Afla-guard	*Aspergillus flavus producing aflatoxin*
C. minitans CON/M/91–08	Contans	*Sclerotinia minor*, *Sclerotinia sclerotiorum*
Gliocladium catenulatum J1446	Prestop	Seed-borne and soil-borne diseases
M. albus QST 20799	Arabesque	Postharvest diseases
Pseudozyma flocculosa PF-A22 UL	Sporodex	Powdery mildew
Trichoderma asperellum ICC 012 and *T. harzianum* (*gamsii*) ATCC080	Tenet Bioten Remedier	Soil-borne diseases
T. harzianum ATCC 20476	Binab	Wound healing
T. harzianum Rifai T-22	PlantShield RootShield T-22 Planter box	Seed and foliar diseases
T. harzianum T-39	Trichodex	Soil and foliar diseases

(continued)

Table 2.1 (continued)

Category of biopesticide	Products common name or trade name	Targets
Trichoderma polysporum ATCC 20475	Binab T	Soil and foliar diseases
Ulocladium oudemansii U3	BOTRY-Zen	Botrytis and *Sclerotinia*
Verticillium albo-atrum WC S850	DutchTrig	Dutch elm disease
Bacteriophage of *P. syringae* pv. tomato	AgriPhage	Tomato leaf spot
Candida oleophila strain O	NEXY	Postharvest fruit molds
Fungicides/bactericides		
B. subtilis QST713	Serenade	Foliar fungal and bacterial diseases
Herbicides		
Bacillus cereus BP01	MepPlus	Plant growth regulator
Alternaria destruens 059	Smolder	Herbicide – dodder
Chondrostereum purpureum PFC 2139	Chontrol Paste	Herbicide – stump sprout inhibitor
Colletotrichum gloeosporioides f.sp. aeschynomene ATCC 202358	LockDown	Herbicide – northern Jointvetch
Puccinia thlaspeos woad (dyer's woad rust)	Woad Warrior	Herbicide – Dyer's woad
Insecticides		
B. popilliae	Milky Spore Powder	Japanese beetle grubs
Bacillus sphaericus Serotype H5a5b strain 2362 ATCC1170	VectoLex	Mosquito larvae
B. thuringiensis subsp. aizawai NB200	Florbac	Moth larvae
B. thuringiensis subsp. israelensis	BMP	Mosquito and blackflies
B. thuringiensis subsp. israelensis EG2215	Gnatrol	Mosquito, flies
	Aquabac	
B. thuringiensis subsp. aizawai delta-endotoxin in killed *P. fluorescens*	M-Trak	Colorado potato beetle
B. thuringiensis subsp. aizawai GC-91	Agree WG	Plutella
B. thuringiensis subsp. kurstaki	Thuricide Forestry	Lepidopteran larvae
	Wilbur-Ellis BT 320	
	Dust	
	Dipel	
	Deliver	
	Biobit HP	
	Foray	
	Javelin WG	
	Green Light	
	Hi-Yield Worm Spray	
	Ferti-Lome	
	Bonide	
	Britz BT	
	Worm Whipper	
	Security Dipel Dust	
B. thuringiensis subsp. kurstaki BMP 123	BMP123	Lepidopteran larvae
B. thuringiensis subsp. kurstaki EG2348	Condor	Lepidopteran larvae
B. thuringiensis subsp. tenebrionis	Novodor	Colorado potato beetle

(continued)

Table 2.1 (continued)

Category of biopesticide	Products common name or trade name	Targets
B. thuringiensis subsp. kurstaki EG7826	Lepinox WDG	Lepidopteran larvae
B. bassiana 447	Baits Motel Stay-awhile	Ants
B. bassiana ATCC 74040	Naturalis L	Various insects
B. bassiana GHA	Mycotrol ES	Various insects
	Mycotrol O	
	Botanigard 22WP	
	BotaniGard ES	
B. bassiana HF23	balEnce	Housefly
M. anisopliae F52	Tick-Ex	Ticks and grubs
Paecilomyces fumosoroseus Apopka 97	PFR-97	Whitefly and thrips
Nosema locustae	Nolo-Bait	Grasshopper and crickets
	Semaspore Bait	
Anagrapha falcifera NPV	CLV-LC	Lepidopteran larvae
Cydia pomonella GV	CYD-X	Virus codling moth
Gypsy moth NPV	Gypchek	Gypsy moth
H. zea NPV (previously *Heliothis zea* NPV)	GemStar	Cotton bollworm, tobacco, budworm
Indian meal moth GV (*Plodia interpunctella* GV)	FruitGuard	Indian meal moth
Mamestra configurata NPV (107308)	Virosoft	Bertha armyworm
Spodoptera exigua NPV	Virus Spod-X	Beet armyworm
Saccharomyces cerevisiae	Bull Run	Fly attractant
Nematicides		
Bacillus firmus I-1582	BioNem	Nematodes
Pasteuria usgae	Econem	Nematodes
Myrothecium verrucaria	DiTera	Nematodes
Paecilomyces lilacinus 251	MeloCon WG	Nematodes
Virucides		
Zucchini yellow mosaic virus – weak strain	AgroGuard-Z	Zucchini yellow mosaic
Europe		
Aureobasidium pullulans	Blossom Protect	Fire blight, postharvest diseases
Phlebiopsis gigantea (several strains)	Rotstop	Conifer root rots
P. chlororaphis	Cedomon, Cerall	*Pyrenophora teres*, *P. graminea*, *Tilletia caries*, *Septoria nodorum*, *Fusarium* spp.
Pseudomonas sp. DSMZ 13134	Proradix	Root rots
S. griseoviridis K61	Mycostop	*Fusarium* wilt, *Botrytis* gray mold, root rot, stem rot, stem end rot, damping off, seed rot, soil-borne damping off, crown rot, *Rhizoctonia*, *Phytophthora*, wilt, seed damping off, early root rot
A. quisqualis AQ10	AQ10	*Leaf disease*
C. oleophila strain O		Postharvest disease
C. minitans CON/M-91–05	Contans WG	*Sclerotinia sclerotiorum*, *S. minor*
G. catenulatum J1446	Prestop, Prestop mix	Damping off, gummy stem blight, gray mold, root rot, stem rot, wilt, storage diseases, foliar diseases, seed rot

(continued)

Table 2.1 (continued)

Category of biopesticide	Products common name or trade name	Targets
P. flocculosa PF-A22 UL	Sporodex	Powdery mildew
Pythium oligandrum	Polyversum	Root rots
Trichoderma asperellum (ICC012) (T25) (TV1) (formerly *T. harzianum*)	Tenet	Fungal infections (*Pythium*, *Phytophthora*, *Botrytis*, *Rhizoctonia*)
Trichoderma atroviride IMI 206040 (formerly *T. harzianum*)	Binab T Pellets	*Botrytis cinerea*, pruning wound infection *Chondrostereum purpureum*
T. atroviride I-1237	Esquive	Fungal infections (*Pythium*, *Phytophthora*, *Botrytis*, *Rhizoctonia*)
Trichoderma gamsii (formerly *T. viride*) (ICC080)	Remedier	Fungal infections (*Pythium*, *Phytophthora*, *Botrytis*, *Rhizoctonia*)
T. harzianum Rifai T-22 ITEM 108 or KRL-AG2	Trianum P	Root diseases
T. harzianum Rifai T-39 (IMI 206039)	Trichodex Rootshield	*Botrytis cinerea*, *Colletotrichum* spp., *Fulvia fulva*, *Monilia laxa*, *Plasmopara viticola*, *Pseudoperonospora cubensis*, *Rhizopus stolonifer*, *Sclerotinia sclerotiorum*
T. polysporum and *T. harzianum*	Binab T Vector	Fungal pathogens, fairy ring, *Botrytis*, *Verticillium*, *Pythium*, *Fusarium*, *Phytophthora*, *Rhizoctonia*, *Didymella*, *Chondrostereum*, *Heterobasidion*
V. albo-atrum (WCS850) (formerly *Verticillium dahliae*)	Dutch Trig	Dutch elm disease
Fungicides/bactericides		
B. subtilis QST 713	Serenade	Fire blight, *Botrytis* spp.
Insecticides		
B. thuringiensis subsp. *aizawai* GC-91	Turex	Lepidoptera pests
B. thuringiensis subsp. *israelensis* AM65	VectoBac	Sciarids
B. thuringiensis subsp. *kurstaki* HD-1	Dipel WP	Lepidoptera pests
B. thuringiensis subsp. kurstaki ABTS 351, PB 54, SA 11, SA12, and EG 2348	Batik Delfin	Lepidoptera pests
B. thuringiensis subsp. *kurstaki* BMP 123	BMP 123 Prolong	Lepidoptera pests
B. thuringiensis subsp. *tenebrionis* NB 176	Novodor	Coleoptera pests
B. bassiana ATCC 74040	Naturalis L	Thrips, whitefly, mites
B. bassiana GHA Fungus	Botanigard	Whiteflies, aphids, thrip
Lecanicillium muscarium (Ve6) (former *Verticillium lecanii*)	Mycotal, Vertalec	Whiteflies, thrips, aphids (except the Chrysanthemum aphid: *Macrosiphoniella sanborni*)
P. fumosoroseus Apopka 97	Preferal WG	Greenhouse whiteflies (*Trialeurodes vaporariorum*)
P. fumosoroseus Fe9901	Nofly	Whiteflies
Adoxophyes orana BV-0001 GV	Capex	Summer fruit tortrix (*Adoxophyes orana*)
Cydia pomonella GV	BioTepp	Codling moth (*Cydia pomonella*)
Spodoptera exigua NPV	Spod-X GH	*Spodoptera exigua*

(continued)

Table 2.1 (continued)

Category of biopesticide	Products common name or trade name	Targets
Nematicides		
P. lilacinus PL 251	BioAct WG	Common plant parasitic nematodes
Virucides		
Zucchini Yellow Mosaic Virus, weak strain Virus	Curbit	Yellow mosaic virus
China		
Bactericides		
A. radiobacter	Trade name not available	Crown gall
Bacillus polymyxa	Trade name not available	Crown gall
Bacillus sphaericus	Trade name not available	Crown gall
Fungicides		
B. cereus	Trade name not available	Bacterial wilt, sheath blight/rice false smut, bacterial wilt
B. licheniformis		Downy mildew, *Fusarium* wilt
B. subtilis	Trade name not available	Bacterial wilt, root rot, tobacco black shank, rice blast, rice false smut
Trichoderma spp.		Fungus downy mildew, *Rhizoctonia cerealis*, gray mold
Fungicides/bactericides		
P. fluorescens	Trade name not available	Bacterial wilt, root rot
Insecticides		
B. thuringiensis subsp. aizawa	Trade name not available	Lepidopteran pests
B. thuringiensis subsp. israelensis	Trade name not available	Lepidopteran pests
B. thuringiensis subsp. kurstaki		Lepidopteran pests
Pseudomonas alcaligenes	Trade name not available	Locusts, grasshoppers
B. bassiana		*Monochamus alternatus*, *Dendrolimus punctatus*
Conidobolus thromboides	Trade name not available	Aphids
M. anisopliae		Cockroaches, grasshoppers, locusts
P. lilacinus	Trade name not available	Nematodes
Pochonia chlamydosporia		Nematodes
Dendrolimus cytoplasmic polyhedrosis virus	Trade name not available	Virus Caterpillars
NPV, *Ectropis obliqua hypulina* NPV, *Laphygma exigua* NPV, *Prodenia litura* NPV, *Buzura suppressaria* NPV, *Gynaephora ruoergensis* NPV, *Mythimna separata* NPV	Trade name not available	Virus Beet armyworm, lepidoptera, looper, *H. armigera*, *Laphygma exigua*
Periplaneta fuliginosa densovirus virus	Trade name not available	Cockroaches
Pieris rapae GV, *Mythimna separata* GV, *Plutella xylostella* GV	Trade name not available	*Pieris rapae*, *Plutella xylostella*
Japan		
Insecticides (bacterial, fungal, viral, parasitic nematodes)		
B. thuringiensis kurstaki	Toarowaa Esmark Guardjet, Dipol, Tuneup Fivestar BioMax DF	Lepidopteran larvae

(continued)

Table 2.1 (continued)

Category of biopesticide	Products common name or trade name	Targets
B. thuringiensis aizawai	Quark XenTari Florbac Sabrina	Lepidopteran larvae
B. thuringiensis aizawai +kurstaki	Bacilex	Lepidopteran larvae
B. thuringiensis japonensis	BuiHunter	Cockchafers and white grubs
B. bassiana	BotaniGard	Thrips, whiteflies, diamondback moth
P. fumosoroseus	Preferd	Whitefly, aphids
Lecanicillum longisporum	Vertalec	Aphids
Adoxophyes orana GV+Homona magnanima GV	Hamaki-Tenteki	*Adoxophyes honmai* and *Homona magnanima*
Steinernema carpocapsae	Bio Safe	Weevils, black cutworm, common cutworm, peach fruit moth
India		
Fungicide		
P. fluorescens	ABTEC Pseudo	Plant soil-borne diseases
	Biomonas	
	Esvin Pseudo	
	Sudo	
	Phalada 104PF	
	Sun Agro Monus	
	Bio-cure-B	
A. quisqualis	Bio-Dewcon	Powdery mildew
T. harzianum	Biozim	Soil-borne pathogens
	Phalada 105	
	Sun Agro Derma H	
T. viride	Monitor, Trichoguard	Soil-borne pathogens
	NIPROT	
	Bioderma	
	Biovidi	
	Eswin Tricho	
	Biohit	
	Tricontrol	
	Ecoderm	
	Phalada 106TV	
	Sun Agro Derma	
	Defense SF	
Fungicides/bactericides		
B. subtilis		Soil-borne pathogens
Insecticides		
B. thuringiensis subsp. israelensis	Tacibio, Technar	Lepidopteran pests
B. thuringiensis subsp. Kurstaki	Bio-Dart	Lepidopteran pests
	Biolep	
	Halt	
	Taciobio-Btk	

(continued)

Table 2.1 (continued)

Category of biopesticide	Products common name or trade name	Targets
B. bassiana	Myco-Jaal	Coffee berry borer, diamondback moth, thrips, grasshoppers, whiteflies, aphids, codling moth
	Biosoft	
	ATEC Beauveria	
	Larvo-Guard	
	Biorin	
	Biolarvex	
	Biogrubex	
	Biowonder	
	Veera	
	Phalada 101B	
	Bioguard	
	Bio-power	
M. anisopliae	ABTEC	Coleoptera and lepidoptera, termites, mosquitoes, leafhoppers, beetles, grubs
	Verticillium	
	Meta-Guard	
	Biomet	
	Biomagic	
	Meta	
	Biomet	
	Sun Agro Meta	
	Bio-Magic	
P. fumosoroseus	Nemato-Guard	Whitefly
	Priority	
P. lilacinus	Yorker	Whitefly
	ABTEC	
	Paceilomyces	
	Paecil	
	Pacihit	
	ROM biomite	
	Bio-Nematon	
Verticillium lecanii	Verisoft	Whitefly, coffee green bug, homopteran pests
	ABTEC	
	Verticillium	
	Vert-Guard	
	Bioline	
	Biosappex	
	Versitile	
	Ecocil	
	Phalada 107 V	
	Biovert Rich	
	ROM Verlac	
	ROM Gurbkill	
	Sun Agro Verti	
	Bio-Catch	

(continued)

Table 2.1 (continued)

Category of biopesticide	Products common name or trade name	Targets
H. armigera NPV	Helicide	*H. armigera*
	Virin-H	
	Helocide	
	Biovirus-H	
	Helicop	
	Heligard	
Spodoptera litura NPV	Spodocide	*S. litura*
	Spodoterin	
	Spodi-cide	
	Biovirus-S	
Nematicides		
Verticillium chlamydosporium		Nematodes
Australia		
Fungicide		
P. fluorescens	ABTEC Pseudo	Plant soil-borne diseases
	Biomonas	
	Esvin Pseudo	
	Sudo	
	Phalada 104PF	
	Sun Agro Monus	
	Bio-cure-B	
A. quisqualis	Bio-Dewcon	Powdery mildew
T. harzianum	Biozim	Soil-borne pathogens
	Phalada 105	
	Sun Agro Derma H	
T. viride	Monitor, Trichoguard	Soil-borne pathogens
	NIPROT	
	Bioderma	
	Biovidi	
	Eswin Tricho	
	Biohit	
	Tricontrol	
	Ecoderm	
	Phalada 106TV	
	Sun Agro Derma	
	Defense SF	
Fungicides/bactericides		
B. subtilis		Soil-borne pathogens
Insecticides		
B. thuringiensis subsp. israelensis	Tacibio, Technar	Lepidopteran pests
B. thuringiensis subsp. Kurstaki	Bio-Dart	Lepidopteran pests
	Biolep	
	Halt	
	Taciobio-Btk	

(continued)

Table 2.1 (continued)

Category of biopesticide	Products common name or trade name	Targets
B. bassiana	Myco-Jaal	Coffee berry borer, diamondback moth, thrips, grasshoppers, whiteflies, aphids, codling moth
	Biosoft	
	ATEC Beauveria	
	Larvo-Guard	
	Biorin	
	Biolarvex	
	Biogrubex	
	Biowonder	
	Veera	
	Phalada 101B	
	Bioguard	
	Bio-power	
M. anisopliae	ABTEC	Coleoptera and lepidoptera, termites, mosquitoes, leafhoppers, beetles, grubs
	Verticillium	
	Meta-Guard	
	Biomet	
	Biomagic	
	Meta	
	Biomet	
	Sun Agro Meta	
	Bio-Magic	
P. fumosoroseus	Nemato-Guard	Whitefly
	Priority	
P. lilacinus	Yorker	Whitefly
	ABTEC	
	Paceilomyces	
	Paecil	
	Pacihit	
	ROM biomite	
	Bio-Nematon	
V. lecanii	Verisoft	Whitefly, coffee green bug, homopteran pests
	ABTEC	
	Verticillium	
	Vert-Guard	
	Bioline	
	Biosappex	
	Versitile	
	Ecocil	
	Phalada 107 V	
	Biovert Rich	
	ROM Verlac	
	ROM Gurbkill	
	Sun Agro Verti	
	Bio-Catch	

(continued)

Table 2.1 (continued)

Category of biopesticide	Products common name or trade name	Targets
H. armigera NPV	Helicide	*H. armigera*
	Virin-H	
	Helocide	
	Biovirus-H	
	Helicop	
	Heligard	
S. litura NPV	Spodocide	*S. litura*
	Spodoterin	
	Spodi-cide	
	Biovirus-S	
Nematicides		
Verticillium chlamydosporium		Nematodes
Bactericides		
A. radiobacter	NoGall	Crown gall disease
Fungicides		
T. harzianum	Trichodex	*Botrytis* spp.
Insecticides		
B. sphaericus	VectoLex	Mosquito larvae
B. thuringiensis subsp. a*izawai*	Agree, Bacchus, XenTari	Lepidoptera larvae
B. thuringiensis subsp. israelensis	Aquabac, BTI, Teknar, Vectobac	Mosquito larvae
B. thuringiensis subsp. kurstaki	Biocrystal, Caterpillar, Killer, DiPel, Costar, Delfin, Full-Bac WDG	Lepidoptera larvae
M. anisopliae	BioCane, Granules	Gray-backed cane grub (scarabs)
M. anisopliae subsp. *acridum*	Green Guard	Locusts and grasshoppers
M. flavoviride	Chafer Guard	Redheaded pasture cockchafer
H. armigera NPV	Heliocide	*Helicoverpa* spp.
	Vivus Gold	
	Vivus Max	
H. zea NPV	Gemstar	*Helicoverpa* spp.
	Vivus	
Africa		
Bactericides	**Products**	**Targets**
A. radiobacter	Crown Gall Inoculant	Crown gall
Fungicides		
B. subtilis 101	Shelter	Root and leaf diseases
B. subtilis 102	Artemis	Root and leaf diseases
B. subtilis 246	Avogreen	Root and leaf diseases
B. subtilis QST 713	Serenade	*Botrytis* spp.
A. quisqualis AQ10	Bio-Dewcon	Powdery mildew

(continued)

Table 2.1 (continued)

Category of biopesticide	Products common name or trade name	Targets
T. harzianum	Eco-77	Root diseases
	Eco-T	
	Promot	
	Romulus	
	Rootgard	
	Trichoplus	
	Trykocide	
T. harzianum 39	Trichodex	Root diseases
T. harzianum DB103	T-Gro	Root diseases
Fungicides/bactericides		
B. subtilis	Defender	Soil-borne fungi and bacteria
Insecticides		
B. thuringiensis subspp. aizawai and kurstaki	Agree	Lepidoptera larvae
B. thuringiensis subsp. israelensis	VectoBac	Mosquito
B. thuringiensis subsp. kurstaki	DiPel	Lepidoptera larvae
	Rokur	
	Thuricide	
B. thuringiensis subsp. kurstaki H7	Florbac WG	Lepidoptera larvae
B. bassiana	Bb Plus, Bb weevil, Sparticus	Thrips, weevils, whiteflies
M. anisopliae subsp. acridum IMI 330 189	Green Muscle	Locust
GV	Trade name not available	Lepidoptera larvae
Pseudomonas resinovorans bacteriophage	Agriphage	Insect pest control
Nematicides		
P. lilacinus	Bio-Nematon	Nematodes
P. lilacinus 251	PL Plus	Nematodes

Source: Kunimi (2007) and Kabaluk et al. (2010)

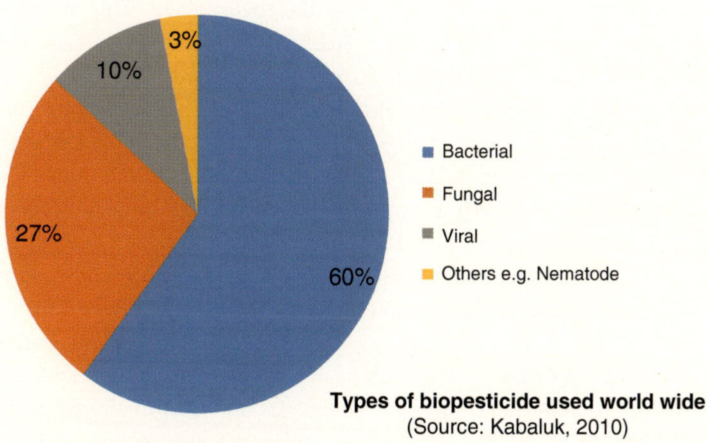

Fig. 2.1 Global biopesticide market based on types of microbes used

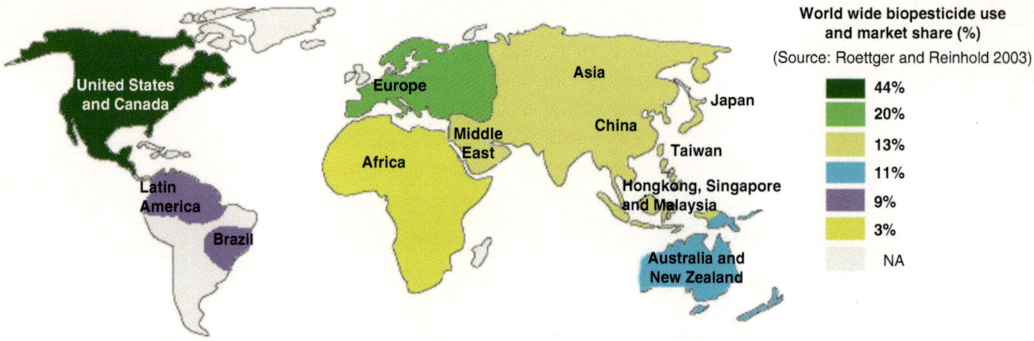

Fig. 2.2 Global biopesticide use and market

Keeping in mind the importance of minor crops (defined as any crop grown on 300,000 acres or less), the US Department of Agriculture (USDA) initiated the Interregional Research Project No. 4, commonly known as IR-4. The USDA IR-4 program has provided grant funding to key influencers such as land-grant university extension specialists to demonstrate with the end user the performance of biopesticides in realistic programs with conventional pesticides (Radcliffe et al. 2009). Overall progress of biopesticides in the USA can be considered as satisfactory. According to EPA data in the USA, 102 microbials, 52 biochemicals, and 48 semiochemicals are being used as biopesticides (USEPA 2011).

Europe

Europe ranks second in biopesticide production. The largest individual European biopesticide market is Spain, followed by Italy and France (Business Wire 2010). The first Bt-based product (Thuricide) was approved in Europe in 1964, whereas first registration for an entomopathogenic fungus *L. longisporum* was given in 1981 to Tate and Lyle in the UK (Quinlan 1990). Europe also belongs to continents where the MRL (Maximum Residue Level) regulations in food products have been strictly decided (Regulation (EC) 396/2005), and the European Food Safety Authority (EFSA) has an important role in this system. EFSA provides independent scientific advice to support the risk managers (EU institutions and Member States) in defining appropriate regulatory frameworks and making decisions to protect consumers. These may involve adopting or revising European legislation on food or feed safety or deciding whether to approve regulated substances such as pesticides and food additives, and if so in which foods or crops and at what levels. This resulted in farmers to produce crop having very less concentration of pesticide or with no detectable limit. In this residue-controlling movement the Ministry of Food, Agriculture, and Fisheries in France launched the program "Ecophyto 2018," in 2008, for a 50 % reduction in the use of pesticides by 2018 (agriculture.gouv.fr/IMG/pdf/PLAN_ECOPHYTO_2018_eng.pdf). In 2011, Europe implemented the Plant Protection Products Regulation (www.eppo.int/PPPRODUCTS/information/new_eu_regulations.htm), and according to that, any crop protection product validates not only that it is effective but also that it should be risk free to humans and to the environment. This made 74 % of all commercial pesticides illegal, as only 26 % of substances passed the new stringent tests, biopesticides being among them (www.pangaeaventures.com/blog/biopesticides-the-next-crop-of-cleantech-home-runs). Currently, most of the European consumers are demanding production of food that hasn't been treated with conventional pesticides; that is why in the present time, Europe has set the standard for biopesticides use. In spite of the stark law on residue control, development of new biopesticide faces hurdle of registration which may take at least 5 years and cost up to € 0.5 million. Presently there are 68 biopesticide active substances registered in

Europe that consist of 34 microbials, 11 biochemicals, and 23 semiochemicals (EUPD 2010). The proportion of the European microbial pesticide market taken by *B. thuringiensis*-based products has declined from an estimated 90 % in 2000 to 50.6 % in 2008 mainly because of the arrival of other BCAs (Business Wire 2010). The largest increases since 2005 were seen in non-Bt bacteria, notably *B. subtilis*, and in fungal-based products, including *C. minitans* and *Trichoderma*. There were also significant increase in viral sales and a steady rise in the nematode market (Business Wire 2010).

Asia

Asia is the largest continent in the world and known for its diverse biodiversity. In Asia, the economy depends heavily on agriculture. Using biopesticides in food crops such as rice, maize, and vegetables is increasing gradually. Rice is the main staple hence much of the biocontrol experimentation and practical use was done on this crop. China is one of the most populated countries in the world, and it is producing biopesticides since 1960, most of them being in the form of unformulated dried cultures (Xu et al. 1987). Biopesticide application was done in 800,000 ha in 1972 which reached 27,000,000 ha in 2000 (Ye and Chen 2002; Zhang 2002; Yang 2007). Till 2008, there were 327 biopesticides registered in China. Among these, 270 bacterial biopesticides are obtained from 11 microbial species (mostly Bt), 22 registered fungal biopesticides from 6 fungal species, and 35 registered viral biopesticides, 14 of which are developed from *Heliothis armigera* NPV (ICAMA 2008).

Japan is one of the pioneer countries as far as the use of biopesticides is considered. In Japan, registration process for Bt was a typical case (Aizawa and Ishiwata 2001) as the earlier application had posed a negative impact on sericulture. In 1972, a study committee on Bt products was established for studying the effects of Bt products on silkworm rearing in sericulture (Aizawa and Fujiyoshi 1973), and after 7 years, the committee concluded that Bt products would not pose a threat to silkworm rearing if farmers were prevented from spraying Bt products on mulberry fields (Study Committee on *Bacillus thuringiensis* Products 1984). Finally, the first Bt registered product was launched in 1981 (Kunimi 2007) and later formulated products of Bt aizawai, Bt kurstaki, and Bt japonensis were also introduced (Ohba et al. 1992). NPV was first time used to control cabbage armyworm, *Mamestra brassicae*, in 1962 (Kunimi 1998). Entomopathogenic fungus *B. bassiana* was also tried to control pine caterpillar (*Dendrolimus spectabilis*) in 1933 by Hidaka (1933). In the last few years, Japanese research in biocontrol field has resulted in the identification and characterization of several new insect pathogens, delivery systems, and formulation development, but the sale of microbial-based products remained less satisfactory, even less than 2 % of all insecticides sold (Kunimi 2007).

In India, biocontrol concept was in practice ever since neem was used as an alternative to chemical pesticides. Farmers have been using neem not only for vegetable protection but also in various other medically imported applications. Evidence of using insects and birds for pest eradication is also found (Subramaniam 1952). In India, microbial-based pesticides evolved as an emergent need when chemical insecticides failed to control *Helicoverpa armigera*, *S. litura*, and other pests of cotton (Armes et al. 1992; Kranthi et al. 2002). In India, the first time commercial production of biocontrol agents was started by Bio-Control Research Laboratories (BCRL), a division of Pest Control (India) Limited, under contract with Plant Protection Research Institute (PPRI) (Manjunath et al. 1992). The rise of biopesticides in India is being encouraged by the government as part of the integrated pest management (IPM) program. The Ministry of Agriculture and the Department of Biotechnology are largely responsible for supporting the production and application of biopesticides, and most of biopesticides are being supplied free of cost by the research agencies to farmers through extension services (Alam 1994). In the last few years, microbes exhibiting good biocontrol potential have been discovered by many workers and are commercially exploited for large-scale biopesticide development (Rabindra 2001; Ignacimuthu et al. 2001; Koul et al. 2003; Ranga Rao et al. 2007). Most of the products are from antagonistic fungi (especially *Trichoderma* spp.) and bacteria (especially Bt and *P. fluorescens*), whereas viral biopesticides consist of NPV and granuloviruses (GV) (Rabindra 2005). But large-scale production poses certain difficulties, and lesser developed technology is the main hurdle for industrial production. Hence, most of the products are being produced at small-scale facilities for local use only, mostly by sugar mills and cooperatives, state agricultural departments, IPM centers, and agricultural universities. By 2006, only 12 biopesticides (that of Bt, *Trichoderma*, *Pseudomonas*, and *Beauveria* species) had been registered, but 194 substances were listed as chemical pesticides (Gupta 2006). However, biopesticides consumption has increased in India as there was 219 metric tons (MT) of biopesticide used in 1996–1997 which increased to 683 MT in 2000–2001, whereas load of chemical pesticides had declined from 56,114 MT to 43,584 MT in the same years (Shukla and Shukla 2012). India is still facing problems associated with production, and quality of the products are not up to the desired level because of which farmers are not very enthusiastic for biopesticides in comparison to chemical pesticides (Gupta and Dikshit 2010).

Bt was the first biopesticide product to be introduced and commercialized into the Thai market in 1965 (Rushtapakomchai 2003). But it was largely neglected by farmers as they did not know much about Bt due to its slow and highly selective action (Rushtapakomchai 2003). Finally in 1969, Bt kurstaki was introduced to control lepidopterous larvae of cruciferous crops (Prasetphol et al. 1969; Vattanatangum 1989). Active research on biopesticide started in 1980 to produce products containing Bt, NPV, and fungi (Jones et al. 1993). The promotion of biopesticide is going on in Thailand, and by 2003 there were six major biopesticides in the market from Bt (subspecies kurstaki, aizawai, and tenebrionis), fungal products of *T. harzianum*, entomopathogenic fungi, entomopathogenic nematodes, and NPV (Rushtapakomchai 2003). Although the production is increasing, it is not enough to fulfill the demand of the whole country, whereas poor quality of some of the noncommercial biopesticides is also a cause for concern (Warburton et al. 2002).

In South Korea, microbial pest control was initiated in 1970 and involved the use of entomopathogenic viruses, bacteria, fungi, and nematodes to control pests in forestry, agriculture, and golf courses, but it took many years when the first commercial biopesticide, "Solbichae" (a Bt subsp. aizawai), was registered in 2003 to control diamondback moth and beetle armyworm in Chinese cabbage (Jeong et al. 2010). By 2009, 34 microbial pesticide products were registered to control insect pests and plant diseases in Korea (Jeong et al. 2010).

Australia

Microbial control first began in the late 1960s with the GV of codling moth (*Cydia pomonella*) and nucleopolyhedrosis virus (NPV) of *Helicoverpa zea*. Initially, limited success was observed in field trials in comparison with their chemical counterparts, but GV proved to be effective against potato tuber moth (*Phthorimaea operculella*) (Reeda and Springetta 1971). A vast variety of products have been registered, but among them *B. thuringiensis* subsp. *kurstaki* (Btk) is the most famous product, and in 1987–1988 when there was increased incidence of cotton pest, Btk-based products accounted for a 2 % market share of all the insecticides (Powles and Rogers 1989; Fitt 1994, 2004). Fungal biocontrol research was mainly directed toward *Metarhizium* to control a wide range of insects (Milner and Jenkins 1996) but especially its effectiveness against locust and grasshopper was of much interest and its oil-based formulation with high persistence in soil made it very popular in Australia (Ibrahim et al. 1999; Milner et al. 1997; Lomer 2001). BioGreen was first registered and commercially produced *Metarhizium*-based product. After the success of BioGreen, other *Metarhizium*-based products also came up and showed remarkable response in biocontrol of crop diseases (Milner 2000).

Latin America

There is a growing trend of using biopesticides in organic culture in Latin America, especially at the local level production for indigenous use has increased meaningfully by the involvement of the local government and NGOs. In Latin America, a proportion of Bt-based products cover about 40 % of the market (CPL Business Consultants 2010). In Argentina, Bt products were first used in 1950 against *Colias lesbia* in alfalfa (Botto 1996). The first virus-based product was registered in 2000 by Agro Roca, and the fungal product based on *B. bassiana* for controlling *Triatoma infestans* and *Musca domestica* was first registered by Alves et al. (2008). The use of biopesticides has risen in Brazil, and by 2010 approximately 3 million hectares of agricultural cropland were being treated annually with microbial pesticides (Kabaluk et al. 2010). The use of Bt in Brazil started in the early 1990s, and at that time only three commercial products were available in the Brazilian market, all based on Btk (Dipel, Thuricide, and Bactospeine) (Habib and Andrade 1991). Approximately 40 commercial mycoinsecticides available in the Brazilian market are registered by 19 for-profit companies. More than 20 laboratories operated by sugar/ethanol mills produce *M. anisopliae* for their own use to control cercopids in cane fields (Kabaluk et al. 2010). Universities, research institutes, nonprofit organizations, rubber tree farms, and cattle farms also produce various fungal microbial control agents. By 2011, there were only 26 BCAs registered that is far less in comparison to 1,352 chemical pesticides (formulations and mixtures) registered. However, at the same time a number of unregistered BCA are also sold, and their number is higher in comparison to registered products, as the registration process involves high cost and length of time (Bettiol 2011). However, Argentina and Brazil have showed potential in utilizing biocontrol agents, and Cuba is also one of the leads in the pest management by biological means in Latin America (Sinclair and Martha 2001).

Africa

In Africa, biological control has been used for a long time to control invasive alien plant species (IAPs) (Olckers 1999). Besides this, devastating effects of locusts and grasshoppers on African agriculture had also realized the requirement of biological control (Groote et al. 2001). Application of fungal-based products of *M. anisopliae* has proven to be effective in pest management (Lomer and Prior 1992; Bateman 1997; Hunter et al. 2001). Studies of using viruses for control of insect pests are also reported, but their long-term effective commercial use is not well documented (Cherry 2004). In Africa, despite broad interest in the use of BCA, their availability to growers has until recently been very restricted because of limited demand, technical and financial constraints, and in-country regulatory frameworks (Cherry and Gwynn 2007). According to estimation, the annual biocontrol sales for the whole of Africa in 2003 were approximately $23 million, including $5 million for bacterial products (Guillon 2003). In Kenya, for example, in 2002, the total pesticide sales were valued at approximately $57.4 million, of which $1.15 million (2 %) are accounted for by BCA sales, predominantly Bt-based products (Wabule et al. 2004). Overall, the data are not very well documented for Africa.

Global Market

Biopesticide companies have invested billions of dollars to develop a variety of microbial products so as to eradicate crop diseases. It is impossible to starkly tell the market trends for biopesticides, and there is a considerable discrepancy in both predication of global sales and selecting category of biopesticides. Yet some reliable data sources are market survey websites and biopesticide companies. The biopesticide market is growing at more than 20 % per year, and there may be a tremendous increase possible in the next 5 years (Market and

Table 2.2 Worldwide biopesticide sales

Biopesticide	Estimated sales figures (in $US million) North America	Europe	Asia and Australia	Latin America	Africa and Middle East	Total
Total Bt (products based on *B. thuringiensis* serotypes)	72.0	27.57	74.75	30.19	6.28	201.79
Other bacteria	23.94	6.30	14.05	4.56	0.40	49.25
Viruses	5.57	7.47	23.90	3.80	0.48	41.22
Fungi	15.85	5.64	18.85	35.96	0.78	77.08
Nematodes and other	9.4	7.50	0.95	0.16	0.13	18.14
Total	126.76	54.48	132.5	74.67	8.07	396.48

Source: CPL Business Consultants (2010)

Market 2013). In comparison, there is a gradual decline in synthetic, and the overall market for the biopesticides had increased from $672 million in 2005 to over $1 billion in 2010, at an annual average growth rate (AAGR) of 9.9 % (Industrial Equipment News 2013). A more detailed report by BCC Research (2012) expected biopesticides market to total $2.1 billion in 2012 which may exceed beyond $ 3.7 billion in 2017, with a compound annual growth rate (CAGR) of 12 %. An estimation of global biopesticide sales by CPL Business Consultants (2010) is given in Table 2.2.

Biopesticide Resources

In the field of biocontrol, modern technology and digitalization of print media offered quick access to several resources dealing with the matter of interest. There are myriads of literature available free of cost in the World Wide Web (www). Although this e-service has been very popular in developed countries, various databases and online services have been developed with the motive to benefit farmers and researchers so as to provide ampule amount of information indiscriminately. A summary of the databases and services are mentioned in Box 2.1.

Constraints with Biopesticides and Possible Remedies

Since biopesticide is a product generally with live organism(s), utmost care is needed, at all the steps, beginning from the production till the end use to maintain the microbial load and vigor. Production technology of biopesticide requires proper care and aid of sophisticated equipments to ensure availability of quality products in the market. As discussed again and again, biopesticides although offer a great promise are still not able to perform up to the mark, and in fact it will not be wrong to mention that these eco-friendly products have not taken market by storm and are performing below par. Kabi (1997) gave stress on the production of quality biopesticides, since they are important in rendering sustainability to farming systems. Insufficient knowledge, lack of adequate machinery, inappropriate handling and improper distribution, importation laws for live inoculants, and several other issues can lead to lack of quality products and loss of market. Major constraints associated with biopesticide development and growth are discussed in this section.

Lack of Awareness

Agriculture market is witnessing an increase in demand for environment friendly, chemical residue-free organic products. Growth in some regions is however hindered due to well-established chemical pesticide markets, lack of awareness about benefits of biopesticides, and uneven efficiency of biopesticides. The lack of awareness, knowledge, and confidence in farmers is one of the chief reasons for the lagging of these eco-friendly pest control alternatives. There are lots of ifs that farmers observe while using biopesticides; the results are sometimes not homogenous or consistent, and hence the users find themselves

Box 2.1: Biopesticide Resources

International Biocontrol Manufacturers Association (IBMA)

IBAMA is the worldwide association of biocontrol-based industries producing microorganisms, macroorganisms, semiochemicals, and natural pesticides for plant protection and public health. IBMA has 180 members and participates in the activities of international organizations such as OECD, FAO, WHO, the International Forum for Chemical Safety, the European Commission, etc. The main objective of IBMA is to associate the international organizations and manufacturers that are involved in the development and use of biocontrol agents. IBMA also maintains the product quality of the biocontrol agents and also forms and maintains the ethical professional rules. IBMA transfers the information for biocontrol agents to the interested parties and also organizes training program (details periodically updated at website: http://www.ibma.ch/news.html) to improve the skill of the company staff members for the better research. IBMA has four divisions; they are microbial biocontrol agents, natural and biochemical products, semiochemicals, and invertebrate biological control agents. IBMA provides a platform for the biocontrol products companies to share the knowledge and accordingly improve their business performance.

Online Information Service for Non-Chemical Pest Management in the Tropics (OISAT) was launched by the Pesticide Action Network (PAN) Germany in 2003, with the aim of limiting the use of hazardous pesticides and providing safer alternatives to poor farmers (Carina Webber 2008). OISAT info is a web-based information tool offering trainers, extension workers, and farmers a quick access to up-to-date information in the form of illustrations, photographs, and glossary terms in order to minimize pest damage in a safe, effective, and ecologically sound way. Every report or articles before publication undergoes a peer review by experts working on related fields then can be publically accessed at www.oisat.org

The International Organization for Biological Control (IOBC) was established in 1955 as a global organization affiliated to the International Council of Scientific Unions (ICSU). This nonprofit organization is known for its leadership in environmentally safe methods of pest and disease control and establishing a quality standard in the field of classical biological control (Cock et al. 2009).

Biopesticide Industry Alliance (BPIA) founded in 2000 with the mission of increasing awareness of biopesticide and to deliver full range of benefits of biopesticides in pest management program. BPIA is also involved in facilitating global acceptance, successful development, and commercialization of biopesticides. BPIA is a standards committee that evaluates and recommends quality and efficacy standards for biopesticides in agriculture, forestry, turf and ornamental, public health, consumer, and other target markets, even industry.

The Bio-Pesticides Database (BPDB) was developed by the Agriculture and Environment Research Unit (AERU) based at the University of Hertfordshire, the UK, is a comprehensive relational database of basic identification, physicochemical, toxicological, ecotoxicological, and other related data for both the more traditional agricultural pesticides (PPDB) and veterinary substances (VSDB). This database provides around 450 records of biocontrol agents ranging from naturally derived substances to insect predators (website http://sitem.herts.ac.uk/aeru/bpdb/index.html). BPDB retrieved information at global level from various resources such as scientific literature and databases, manuals, registration databases dossiers, company technical datasheets, and research projects. Before uploading data on the website, its quality is assessed that involves cross checking and peer review by experts.

confused in adopting these greener alternatives. Microbial pest control products require more attention of farmers than chemical products. Condition is worst in developing countries where most of the farmers are even not familiar with the term "biopesticide" and lack efficient skills to practice and use them (Alam 2000).

Research and development of biological pest control methods must be given priority, and people in general and agriculturists in particular must be educated about the handling and use of such control measures. It is absolutely necessary to create awareness, among the end users (farmers), regarding the use, efficacy, benefits, and importance of biopesticides, and also harmful effects of chemical pesticides must be known and explained. The responsiveness could be achieved by introducing certain extension activities such as organizing teaching programs, workshops, and entrepreneurs dealing with the idea of promoting sustainable agriculture using biological products (Amin 2013). Guidance, explanation, and monitoring should be done regularly by proceeding for interactive questionnaire sessions that will inculcate knowledge and learning in the farmers regarding the use, application, and handling of biopesticides. Farmers should also be trained for the methods of application in fields, and some encouraging activities can be introduced such as providing rewards to the farmers who applied biopesticides or promoted their use (Halim and Ali 1998). The efforts of various government agencies to popularize the use of biopesticides will definitely have impact in elevating current status and application of biopesticides all over the world. The National Farmer Policy (2007) in India has strongly recommended the promotion of biopesticides for increasing agricultural production and sustaining the health of farmers and environment. It also includes the clause that biopesticides would be treated at par with chemical pesticides in terms of support and promotion.

The use of transgenic crops like Bt crops is also not up to the mark especially in developing countries. In transgenic crops, functional foreign gene is incorporated by biotechnological tools. Objections to development and deployment of transgenic crops by farmers rest on several issues related to ethnic matters, associated risks, lack of confidence, costs, market control, etc. Farmers generally lack faith in the use of transgenic crop. Hence, it is essential that training and teaching should be given to the farmers in regard to the use of transgenic crops. Special schools should be established particularly in the rural areas so as to train and teach the farmers to efficiently use the biopesticides. Proper application of biopesticides is very important to achieve the optimum results. Farmers need to be made aware of the implications of chemicals on the current and future prospects of soil and yields. These goals can be achieved by the support of government and corporate houses.

Lack of Faith and Inconsistent Field Performance

Lack of faith in the use of biopesticides was found to be one of the major factors responsible for their lagging behind (Arora et al. 2010). Many farmers who stopped using biopesticides reported that it was mainly because the supply was extremely unreliable and the performance very inconsistent (Alam 2000). A key factor involved in the lack of success has been the rapid decline in the size of populations of active cells, to levels ineffective to achieve the objective, following introduction into soil. Abiotic soil factors (e.g., textural type, pH, temperature, and moisture) exert their (direct) effect on inoculant population dynamics by imposing stresses of various natures on the living cells introduced in the fields (Evans et al. 1993). They can also act indirectly by affecting the activity of the indigenous soil microflora. Hence, maintenance of sufficient activity of inoculated populations over a prolonged period after release often represents the main hurdle in the successful use of microbes as biopesticides (Arora et al. 2010). Furthermore, efficient introduction into soil during the growing season is a major technical constraint. It is extremely important that a minimum effective threshold population of the introduced biopesticide is maintained in the soil/rhizosphere so as to combat the pests and pathogens (Arora et al. 2010).

Building farmer's faith and confidence by developing appropriate stress-tolerating formulation can increase product stability, constancy, and viability that can reduce inconsistency in natural field conditions. Production technology, employed for designing biopesticides, should be improvised along with sophisticated quality control measures and monitoring facilities. The designed biopesticide should be reliable, specific, indigenous, and replicable in its activity. Extensive research should be conducted in the fields to develop appropriate formulation working efficiently under diverse in vivo conditions (Retchelderfer 1984; Greaves 1993). Technical and chemical compatibility along with innovative application methods is a prerequisite for the success of a new biopesticide product in the agricultural industry. Developed formulation should be compatible with crop production practices and equipment. The host range and abiotic conditions under which the formulation is most effective must be clearly mentioned on the packets and if possible explained to the farmers.

Poor Quality and Shelf Life

Poor quality and performance are also one of the serious problems that hindered biopesticides takeover on the market. Several workers reported that the biopesticides being sold in the market are contaminated and have a low count of microorganisms (Singleton et al.1996; Alam 2000; Arora et al. 2010). Due to low bacterial count, it is not surprising that their performance is poor, deprived, and uneven. Due to which the shelf life is low and inconsistent in performance resulting in decline in the demand.

The inconsistent and seasonal nature of the existing demand requires efficient storage. The storage of biopesticides requires special facilities and skills, which most producers, shopkeepers, and farmers do not possess. Shelf life is a cessation (end) of several factors like production technology, carrier and packaging material used, and mode and distance of transport. All these levels are desired to sustain the shelf life. Bacterial survival in the desired formulation is affected by several variables: the culture medium used for bacterial cultivation, the physiological state of the bacteria when harvested from the medium, the use of protective materials, the type of drying technology used, the presence or absence of contaminants, and the rate of dehydration (Paul et al. 1993). Extreme care should be taken throughout the designing process to minimize the chances of contamination. It is also important that precautions should be taken to avoid adulteration during packaging, storage, and application of biopesticides.

Future research efforts in formulation technology should emphasize processes that will achieve viable and stable biological products. The most suggested solutions to the problem of survival time are air-dried and lyophilized preparations of biopesticides (Nakkeeran et al. 2005). Decrease in the water content in the biopesticides can be used for long-term survival during storage. In this way, the bacteria in the formulation remain inactive, resistant to environmental stresses, insensitive to contamination, and thus become more compatible with chemical pesticide applications (Bashan 1998). However, dehydration phase is also the most sensitive part of the entire formulation process, especially for nonspore-forming bacteria (Shah-Smith and Burns 1997).

Hence, it could be suggested that the main factors that have potential to affect economic feasibility of the biopesticide product are designing and optimizing perfect formulation technology. Good formulation can be reflected by the long product storability. Several commonly used biopesticide formulations with extended shelf life include granules, pellets, and dry powder based. Granules can protect the active agent from desiccation and also provide basic food for the agent. Powder is easy to apply by suspending it in water and also can cover a wide area of application (Urquhart and Punja 1997; Amin 2013).

High Budget of Production and Lesser Agribusiness

Hi-tech instrumentation required for producing biopesticides under completely sterile conditions is not getting acceptance. Screening of suitable strains and research and development issues add to the budget. Large-scale screening of strains with biological activity is still required (comparable to more than 1:20,000 screened molecules for a new chemical product) (Bashan 1998). High sensitivity to temperature and other external conditions of these "living" inputs calls for enormous caution at the stage of manufacture/culture, transportation/distribution, and application. This involves investment in packaging, storage, and use of suitable carrier materials (Arora et al. 2001, 2010). In general, firms with larger production facilities are expected to invest more on networks to understand and access the market. All these factors prove that the raw material and instrumentation facility initially required for the biopesticide production are costly, and companies will only develop these products if there is a long-term profit in doing so. However, consistency and long-term returns can reduce the cost and enhance the profits.

A number of features of the agricultural economy make it difficult for companies to invest in developing new biopesticide products and, at the same time, make it hard for farmers to decide about adopting the new technology (Chandler et al. 2011). Most of the established companies relinquished their wish to do business in microbial pesticides but finally left the field due to huge losses in the agribusiness (CPL Business Consultants 2006). Commercial aspects of biopesticide industries were studied by various workers (Warrior 2000; Benuzzi 2004; Gelernter 2005) that confirm limited success and huge expenditure (Stewart 2001; Hallett 2005; CPL Business Consultants 2007; Droby et al. 2009). The market's potential in business decision of biopesticide industry is now being forecasted by leading global management consulting and market research firms that suggest that the agribusiness dealing with biopesticides requires huge intake of money, high-risk factors, and less profit (Leng et al. 2011).

Henceforth, the profit in biopesticide business could be made only by using novel techniques and tools. Agribusiness companies need to use certain innovative techniques and cheap raw material to capture or protect market shares by offering new products that buyers want (product innovations) and also by cutting costs (process innovations) and minimizing risk factors. Multifaceted bioformulations based on microbial consortia with diverse activities can be useful in bringing down the costs. Biopesticides designed from consortia will have multiple and holistic applicability in promoting plant growth; protecting plant health; strengthening plant-microbe associations under stress, pollutant, or contaminant-affected regions; and protecting plants from the attack of phytopathogens through biological control (Arora et al. 2013). Advanced countries take advantage of technology, using a variety of procedures such as licensing, buying, and accessing what others develop (Pray and Nagarajan 2010, 2012). Product protection by patenting and support by government in taxation and infrastructure development can also help.

Regulatory Framework

Registration of biopesticide is the main hurdle in the development, and most of the times registration is much more expensive than the production. Registration is not only expensive but also time-consuming (Ehlers 2006). The main problem is that biopesticides contain active cells (live organisms), and these live forms are treated like pathogens by the government agencies. Another issue is regarding the import and export of biopesticides; again, it should be pointed that export and import of chemical pesticide is much easier (as no one doubts on its integrity) in comparison to the use of biopesticides.

The assessment of risks is important because it provides the basis for governments' decisions whether to approve or register new biological pesticides and whether to renew the registration of old ones. Registration requires collation of data and preparation of dossier for submission to a national regulatory authority. By these efforts, governments can speed up the process of approving safer new pesticides and stopping use of riskier ones. Some major countries and their registration details are provided in Table 2.3. USEPA encourages development and use of

Table 2.3 Biopesticide regulation structures in different countries

Name of country/ continent	Registration/governing bodies
The USA	The USEPA is the main authority to regulate the use, sale, and distribution of conventional chemical and biological pesticides. USEPAs do this primarily from three statutes: (1) the Federal Insecticide, Fungicide, and Rodenticide Act (FIFRA) as amended, (2) the Federal Food, Drug, and Cosmetic Act (FFDCA), and (3) the Food Quality Protection Act (FQPA) of 1996. All ensure that pesticides use does not cause unreasonable adverse effects to humans or the environment
Europe	Here, the term "biopesticide" generally includes products with active substances based on microorganisms, botanicals (or biochemicals or plant extracts), and semiochemicals (including pheromones) and regulated under the Council Directive 91/414/EEC. This directive provides a list of active substances authorized for incorporation in plant protection products and lays down the requirements for application dossiers for new active substances and new plant protection products
China	In 1997, The Regulation on Pesticide Administration was introduced (revised in 2001 and 2004) to supervise and control manufacturing, marketing, and use of pesticides/biopesticide and agrochemicals in China or import to China
Australia	Registration of pesticides is governed by the Agricultural and Veterinary Chemicals Code Act 1994 and administered by the Australian Pesticides and Veterinary Medicines Authority (APVMA). The importation of a biological agent also requires authorization from the Australian Quarantine Inspection Service (AQIS) prior to introduction. If the organism has been genetically altered, approval from the Office of the Gene Technology Regulator (OGTR) is required prior to importation or release
India	Biopesticides fall under the Insecticide Act (1968). Central Insecticides Board (CIB) and the Registration Committee (RC) are two "high-powered" bodies under this Act. CIB is the Apex Advisory. It comprises eminent scientists of all disciplines/fields concerned. Whereas, the RC grants registrations to the persons desiring to import or manufacture insecticides, after scrutinizing their formulae and verifying claims with respect to their bio-efficacy and safety to human beings and animals
South Korea	Registration at first level is governed by the Agromaterials Management Division (AMD) and the Rural Development Administration (RDA), and then a new record is evaluated by Pesticide Safety Evaluation Division (PSED) and National Academy of Agricultural Science (NAAS). Once this dossier is deemed to contain sufficient information, the PSED holds two technical expert committees for product management and safety management. The committees provide the examination results to the PSED, who adjust the results and reports to the AMD and RDA. The AMD have a dedicated council for agrochemical safety for the final decision and for reporting to the applicant's dossier

Source: Kabaluk et al. (2010)

biopesticides. Since biopesticides tend to pose fewer risks than chemical pesticides, EPA generally requires much less data to register a biopesticide than to register a conventional pesticide (Kumar 2012).

The governments can frame regulations at the global level by organizing meetings, workshops, and conferences regarding uplifting the status of biopesticides/bioformulation. Governments should set up regulatory framework that could be accepted globally. Presently, different countries have different rules, and regulations due to which problems related to registration, use, import, and export do occur. The regulation can set up uniform acts or laws that could be accepted globally, so that there is a common policy regarding the use of biopesticides.

Health and Ecological Risks

There may be some chances of adverse health effects if biopesticides are not used according to the instructions mentioned on the product, but in comparison to chemical pesticides, risks are far lower. Biopesticides containing Bt as active ingredient are not reported to show any major adverse effects on human health, but in some cases, occupational exposure confirmed health risks (Green et al. 1990; Bernstein et al. 1999; Doekes et al. 2004). Studies on fungal biopesticides suggest that spore of entomopathogenic fungi such as *Trichoderma*, *M. anisopliae*, and *B. bassiana* may cause allergy to farmworkers (Iida et al. 1994; Darbro and Thomas 2009). Recently, studies on mice confirmed as a robust fungal allergen from biopesticides of *M. anisopliae* (Ward et al. 2009, 2011). *M. anisopliae* is also reported to affect survival of nontarget pests (Thungrabeab et al. 2006).

Thus, it is necessary that before developing a biopesticide, strain monitoring should be extensively done. Doubtful strains should be screened out, as adding such microbes may result in the loss of confidence. However, care must be undertaken to ensure that any newly introduced natural product, being a microbial agent or secondary metabolite, should possess no threat to the operator, the environment, or the consumer before it is introduced into crop protection systems (Copping and Menn 2000). Apart from it, governments should also set up defined standards and permissible limits in regard to using biopesticides so that it diminishes the health risks. Essential data regarding the composition, toxicity, degradation, and other characteristics of the biopesticides is important and should be submitted to the respective agencies of various countries. In the USA, the registrants have to submit to EPA so as to ensure that the pesticide is safe for use. EPA conducts rigorous reviews to ensure that pesticides do not have adverse effects on human health or the environment (Kumar 2012).

Problems Associated with Viral Biopesticides

The main problem associated with viral biopesticides is the requirement of live host, tissue, or cell line culture for the proliferation and cultivation. Development of tissue culture laboratory in an institute (that works on viruses) again needs ethical clearances from the government organizations (Lapointe et al. 2012). The main cons of viral biopesticides are the culturing techniques, handling errors, and upsurge costs that ultimately hamper growth, production, and manufacture.

In all viral pathogens, including the most common biopesticides, the baculoviruses, replication is dependent upon the availability of permissive host cells. The accessibility and susceptibility of host cells to viral invasion and replication are classified into three categories: permissive, semi-permissive, and nonpermissive. In semi-permissive, infections result in limited viral progeny resulting from defects in some replication events, such as gene expression or viral DNA replication. In nonpermissive infections, cells do not support viral replication, and the process does not yield infectious progeny (Lapointe et al. 2012). Determining what factors influence the level of permissiveness of an insect cell to a particular baculovirus has proven to be challenging because baculovirus host range is affected not only by the interactions between the baculovirus and the host cell at the molecular level but also by aspects of insect behavior and physiology (Miller and Lu 1997; Cory and Hoover 2006; Thiem and Cheng 2009). Research and innovative techniques are required so as to ease the production, reduce the cost, and simplify the application of viral pesticides.

Competition with Chemical Pesticides

All over the world, chemical pesticides are used in very high amounts (Donaldson et al. 1995), and one-third of the agricultural production is dependent on pesticides (Liu et al. 2002). According to USEPA, over 1 billion tons of pesticides are used in the USA every year, and this is 22 % of the estimated 5.2 billion pounds of pesticides used worldwide (USEPA 2011). Consumption of pesticides in some of the developed countries is almost 3,000 g ha^{-1} (Khater 2012). Practically, biopesticides are not as effective as chemicals. Ahmad et al. (2007) compared the effectiveness of *B. thuringiensis* with, megamos (chemical pesticide) in controlling thrips

(*Thrips tabaci*) on garlic (*Allium sativum*) and found maximum yield using megamos treatment. In case of chemical pesticides, lesser quantity is suffice to kill a vast quantity of pests which is the main reason why farmers choose chemical pesticides over biopesticides. There have been several cases where chemical pesticides reduced losses of many crops, and in their absence it is reported that global losses would have risen from present levels of around 42 % to close to 70 % (Oerke et al. 1994). Knuston et al. (1990) provided further detail that in the absence of chemical control of weeds in wheat production, US yields would fall by 30 % and 5 % in the absence of fungicides and herbicides (Knutson et al. 1990). Similar studies also confirmed chemical dependence for the production of major crops around the globe (Farah 1994; Warren 1998; Webster et al. 1999; Aktar et al. 2009). Some workers are working on synergistic action of microbial biopesticides and chemical pesticides by IPM programs (Irigaray et al. 2003; Koppenhöfer and Fuzy 2003). IPM can also result in gradual decrease in the use of chemicals leading to development of confidence among farmers for the biopesticides. Research on combining microbial biopesticides with synthetic pesticides has showed improvement in control of some pest species including pesticide-resistant varieties (Khalique and Ahmed 2001; Cuthbertson et al. 2005). By removing the previously mentioned hurdles and constraints, confidence for biopesticides can be developed. This will only enhance the market for them. Quality products with the ability to act in field conditions will be able to compete with the chemicals and gradually overtake the market.

Conclusion

Since inception of biopesticides, their position and situation still remains in dilemma. Farmers find themselves confused and less confident in selecting biopesticides over the synthetics. Despite the fact that presently biopesticides are being used everywhere in the world, it is also known that developed countries seem to be ahead in their wider application (Chandler 2011). Developing countries have huge possibilities for using biopesticides as the production can be less expansive and labor is cheap in comparison to developed nations (Roettger and Reinhold 2003). Also countries like India are vastly dependent upon agriculture for not only feeding their populations but also for the economy which depends majorly on this sector. However, most of the challenges faced for the upliftment of biopesticides are fundamental and cosmopolitan. These include the efficacy of the microbial activity, survival of microorganisms, delivery systems, determining host range, and avoiding injury to nontarget organisms, consistency, performance in field conditions, economics, government regulations, and confidence among the end users. Gelernter (2007) has described the future of biocontrol in Asia, and according to him unreasonable expectations for performance, inappropriate regulatory guidelines, lack of documentation on the uptake of microbial control strategies, difficulties in implementing local production schemes, and inhibition of scientific exchange are the main hurdles in establishment of biocontrol. Biopesticide production also faces problem of quality control, and at the global level there is no uniformity in processes and methods. In this regard, Van Lantern (2003) starkly emphasized that the characteristics that affect overall quality have to be identified and must be quantifiable and relevant for the field performance of the parasitoid or predator. Though a lot of research is going on biopesticide development, but still it is further needed to be emphasized and explored (Gaind and Kaushik 2008). This exploration could be done by developing strong policy and encouragement from governments to the industry as well as end user by means of liberal tax benefits, incentives, etc. Creating awareness among the farmer community about the beneficial effects of BCAs and harmful effects of chemical protectants will certainly create a congenial and long-standing effect which can lead to commercial success of the biopesticides (Swati and Adholeya 2008).

In spite of all these limitations, biopesticides are gradually becoming popular, especially among local farmers, and that is why a statement of David Cary, executive director of the IBMA, provides a hope of betterment that the biological control market "that is only 3 %, or $1.3 billion, of the $44 billion global crop protection business, is growing 10 % a year" (Patrick and Kaskey 2012). The growth of biopesticides is indicative of its importance in the sustainable agriculture by producing food crop with lesser chemical use. There is prediction that the world population will exceed nine billion by 2050, and efforts have to be made to meet the demand of 70–100 % more food from the same land area without the extensive use of chemicals (http://unsdsn.org/files/2013/05/130112-HLP-TG7-Solutions-for-sustainable-food production.pdf). Currently, the ultimate requirement for biopesticides to develop is to overcome the shortcomings which are associated with them. This reformation will certainly create similar acceptance as the synthetic pesticides and importantly without any adverse effect on the environment. A concerted effort of research institutes, universities, nongovernment organizations (NGO), and government organizations is required to elevate the stature of biopesticides. Determination at global level is required to strengthen these green alternatives and push off the red poisonous chemicals from our platter and of course the ecosystem as a whole.

References

Ahmad S, Khan IA, Hussain Z, Shah SIA, Ahmad M (2007) Comparison of a biopesticide with some synthetic pesticides against aphids in rapeseed crop. Sarhad J Agric 23:1117–1120

Aizawa K, Fujiyoshi N (1973) Development of bacterial insecticides in Japan. J Ferment Technol 51:363–365

Aizawa K, Ishiwata S (2001) His discovery of sottokin (*Bacillus thuringiensis*) in 1901 and subsequent investigations in Japan. In: Ohba M, Nakamura O, Mizuki E, Akao T (eds) Proceedings of a centennial symposium commemorating Ishiwata's discovery of *Bacillus thuringiensis*. Kurume, Japan, pp 1–14

Aktar MW, Sengupta D, Chowdhury A (2009) Impact of pesticides use in agriculture: their benefits and hazards. Inter Discip Toxicol 2:1–12

Alam G (1994) Biotechnology and sustainable agriculture: lessons from India, Technical Paper No. 103. OECD Development Centre, Paris

Alam G (2000) A study of biopesticides and biofertilizers in Haryana, India Gatekeeper Series No. 93. IIED, UK

Alavanja MC (2009) Pesticides use and exposure extensive worldwide. Rev Environ Health 24:303–309

Ali S, Zafar Y, Ali MG, Nazir F (2008) *Bacillus thuringiensis* and its application in agriculture. Afr J Biotechnol 9:2022–2031

Alves SB, Lopes RB, Vieira S, Tamai MA (2008) Fungos entomopatogénicos usados no controle de pragasna America Latina. In: Alves SB, Biaggioni LR (eds) Controle microbiano de pragasna America Latina. FEALQ, Piracicaba, pp 69–110

Amin N (2013) Teaching of biopesticide development as a technoprenuership opportunity in plant protection. J Biol Agric Healthc 3:2224–3208

Anand S, Reddy J (2009) Biocontrol potential of *Trichoderma* sp against plant pathogens. Inter J Agri Sci 2:30–39

Armes NJ, Jadhav DR, Bond GS, King ABS (1992) Insecticide resistance in *Helicoverpa armigera* in south India. Pest Sci 34:355–364

Aronson A, Beckman W, Dunn P (1986) *Bacillus thuringiensis* and related insect pathogens. Microbiol Rev 50:1–24

Arora NK, Kumar V, Maheshwari DK (2001) Constraints, development and future of the inoculants with special reference to rhizobial inoculants. In: Maheshwari DK, Dubey RC (eds) Innovative approaches in microbiology. Singh and Singh, Dehradun, pp 241–245

Arora NK, Khare E, Naraian R, Maheshwari DK (2008) Sawdust as a superior carrier for production of multipurpose bioinoculant using plant growth promoting rhizobial and pseudomonad strains and their impact on productivity of *Trifolium repense*. Curr Sci 95: 90–94

Arora NK, Khare E, Maheshwari DK (2010) Plant growth promoting rhizobacteria: constraints in bioformulation, commercialization, and future strategies. In: Maheshwari DK (ed) Plant growth and health promoting bacteria. Springer, Berlin, pp 97–116

Arora NK, Tewari S, Singh S, Lal N, Maheshwari DK (2012) PGPR for protection of plant health under saline conditions. In: Maheshwari DK (ed) Bacteria in agrobiology: stress management. Springer, Berlin, pp 239–258

Arora NK, Tewari S, Singh R (2013) Multifaceted plant-associated microbes and their mechanisms diminish the concept of direct and indirect PGPRs. In: Arora NK (ed) Plant microbe symbiosis-fundamentals and advances. Springer, India, pp 411–449

Arrebola E, Jacobs R, Korsten L (2010) Iturin A is the principal inhibitor in the biocontrol activity of *Bacillus amyloliquefaciens* PPCB004 against postharvest fungal pathogens. J Appl Microbiol 108:386–395

Bailey DJ, Gilligan CA (2004) Modeling and analysis of disease induced host growth in the epidemiology of take all. Phytopathology 94:535–540

Baird C, Cann M (2008) Pesticides. In: Environmental chemistry. W H Freeman and Company, New York

Bashan Y (1998) Inoculants of plant growth promoting bacteria use in agriculture. Biotech Adv 6:729–770

Bateman RP (1997) Methods of application of microbial pesticide formulations for the control of locusts and grasshoppers. Memoires Entomol Soc Canada 171: 69–81

Benuzzi M (2004) What will be the future for BCAs? The industry's point of view on problems in developing BCAs. IOBC/WPRS Bull 27(8):429–431

Berg G (2009) Plant-microbe interactions promoting plant growth and health: perspectives for controlled use of microorganisms in agriculture. Appl Microbiol Biotechnol 84(1):11–18

Bernstein IL, Bernstein JA, Miller M, Tierzieva S, Bernstein DI, Lummus Z, Selgrade MK, Doerfler DL, Seligy VL (1999) Immune responses in farm workers after exposure to *Bacillus thuringiensis* pesticides. Environ Health Persp 107:575–582

Bettiol W (2011) Biopesticide use and research in Brazil. Outlooks Pest Manag 22:280–283

Bezchlebová J, Cernohlávková J, Lána J, Sochová I, Kobeticová K, Hofman J (2007) Effects of toxaphene on soil organisms. Ecotoxicol Environ Saf 68:326–334

Bora LC, Deka SN (2007) Wilt disease suppression and disease enhancement in (Lycopersicon esculentum) by application of *Pseudomonas fluorescens* based biopesticide (Biofor-Pf) in Assam. Indian J Agr Sci 77: 490–494

Botto EN (1996) Control biológico de plagas en La Argentina: informe de la situación actual. In: Zapater C (ed) El control biológico en América Latina. Buenos Aires, pp 1–8

Brar SK, Tyagi VRD, Valéro JR (2006) Recent advances in downstream processes and formulations of *Bacillus thuringiensis* based biopesticide. Process Biochem 41:323–342

Brühl CA, Schmidt T, Pieper S, Alscher A (2013) Terrestrial pesticide exposure of amphibians: an underestimated cause of global decline? Sci Rep 3:1135

Bull CT, Stack JP, Smilanick JL (1997) *Pseudomonas syringae* strains ESC-10 and ESC-11 survive in wounds on citrus and control green and blue molds of citrus. Biol Control 8:81–88

Burges HD, Croizier G, Huber J (1980a) A review of safety tests on baculoviruses. Entomophaga 25: 329–340

Burges HD, Huber J, Croizier G (1980b) Guidelines for safety tests on insect viruses. Entomophaga 25:341–348

Business Wire (2010) Research and markets: the 2010 biopesticides market in Europe & company index – opportunities exist which could raise the total market to $200 million by 2020. New York

CABI (2010) The 2010 worldwide biopesticides: market summary. CPL Business Consultants, London, p 40

Carina Webber (ed) (2008) How to grow crops without endosulfan, PAN Germany

Carlton B (1988) Development of genetically improved strains of *Bacillus thuringiensis*. In: Hedin P, Menn J, Hollingworth R (eds) Biotechnology for crop protection. American Chemical Society, Washington, DC, pp 260–279

Carson R (1962) The silent spring. Houghton Mifflin, USA

Cazorla FM, Romero D, Garcia AP, Lugtenberg BJJ, Vicente A, Bloemberg G (2007) Isolation and characterization of antagonistic *Bacillus subtilis* strains from the avocado rhizoplane displaying biocontrol activity. J App Microbiol 103:1950–1959

Chakravarty G, Kalita MC (2011) Management of bacterial wilt of brinjal by *P. fluorescens* based bioformulation. ARPN J Agri Biol Sci 6(3):1–11

Chandler D, Davidson G, Grant WP, Greaves J, Tatchell GM (2008) Microbial biopesticides for integrated crop management: an assessment of environmental and regulatory sustainability. Trends Food Sci Technol 19:275–283

Chandler D, Bailey AS, Tatchell GM, Davidson G, Greaves GWP (2011) The development, regulation and use of biopesticides for integrated pest management. Philos Trans R Soc Lond B Biol Sci 1573: 1987–1998

Chapple AC, Downer RA, Bateman RP (2000) Theory and practice of microbial insecticide application. In: Lacey LA, Kaya HA (eds) Field manual of techniques in invertebrate pathology. Kluwer, Dordrecht, pp 5–37

Chattopadhyay A, Bhatnagar NB, Bhatnagar R (2004) Bacterial insecticidal toxins. Crit Rev Microbiol 30:33–54

Cherry AJ (2004) Public-private partnerships for development and implementation of entomopathogenic viruses as bioinsecticides for key lepidopteran pests in Ghana and Benin, West Africa, Final Technical Report, Project R7960. Natural Resources Institute, Chatham, p 42

Cherry AC, Gwynnn RL (2007) Perspective on the development of biocontrol in Africa. Biocontrol Sci Technol 17:665–676

Chunxue C, Sunjeong P, McSpadden Gardener BB (2010) Biopesticide controls of plant diseases: resources and products for organic farmers in Ohio. Fact Sheet Agri Nat Res 1–10

Clemson HGIC (2007) Organic pesticides and biopesticides, Clemson extension, home and garden information center. Clemson University, Clemson

Cock MJW, van Lenteren JC, Brodeur J, Barratt BIP, Bigler F, Bolckmans K, Co^nsoli FL, Haas F, Mason PG, Parra JRP (2009) In: The use and exchange of biological control agents for food and agriculture, Report prepared for the FAO genetic resources commission by the IOBC global commission on biological control and access and benefit sharing, IOBC, Bern

Conis E, MS, MJ (2010) Debating the health effects of DDT: Thomas Jukes, Charles Wurster and the fate of an environmental pollutant, Public Health Rep, 125:337–342

Copping LG, Menn JJ (2000) Biopesticides: a review of their action, applications and efficacy. Pest Manag Sci 56:651–676

Cory JS, Hoover K (2006) Plant-mediated effects in insect–pathogen interactions. Trends Ecol Evol 21(5):278–286

Cory JS, Myers JH (2003) The ecology and evolution of insect baculoviruses. Annual Rev Ecol Evol Syst 34:239–272

CPL Business Consultants (2006) Biopesticides 2007. CPL Business Consultants, Wallingford

CPL Business Consultants (2007) Biopesticides 2007, a how to do it guide to biopesticides, vol 5. CPL Business Consultants, Wallingford

CPL Business Consultants (2010) The 2010 worldwide biopesticides market summary, vol 1. CPL Business Consultants, Wallingford

Cuthbertson AGS, Walters KFA, Northing P (2005) The susceptibility of immature of *Bemisia tabaci* to the entomopathogenic fungus *Lecanicillium muscarium* on tomato and verbena foliage. Mycopathologia 159:23–29

D'Amico V (2007) Baculovirus in biological control: a guide to natural enemies in North America. http://www.nysaes.cornell.edu/ent/biocontrol/pathogen/baculoviruses

Darbro JM, Thomas MB (2009) Spore persistence and likelihood of aeroallergenicity of entomopathogenic fungi used for mosquito control. Am J Trop Med Hyg 80:992–997

De Faria MR, Wright SP (2007) Mycoinsecticides and mycoacaricides: a comprehensive list with worldwide coverage and international classification of formulation types. Biol Control 43:237–256

Doekes G, Larsen P, Sigsgaard T, Baelum J (2004) IgE sensitization to bacterial and fungal biopesticides in a cohort of Danish greenhouse workers: the BIOGART study. Am J Ind Med 46:404–407

Dominguesa FC, Queiroza JA, Cabralb JMS, Fonsecab LP (2000) The influence of culture conditions on mycelial structure and cellulose production by *Trichoderma reesei* rut C-30. Enz Microb Technol 26:394–401

Donaldson D, Kiely T, Grube A (1995) Pesticide's industry sales and usage 1998–1999 market estimates, report no. EPA-733-R-02-OOI. US Environmental Protection Agency, Washington, DC

Downie D (2003) Global POPs policy: the 2001 Stockholm convention on persistent organic pollutants. In: Downie D, Fenge T (eds) Northern lights against POPs: combating toxic threats in the Arctic. McGill-Queens University Press, Montreal

Dresdend D (1948) Site of action of D.D.T. and cause of death after acute D.D.T. poisoning. Nature 162:1000–1001

Droby S, Wisniewski M, Macarisin D, Wilson C (2009) Twenty years of postharvest biocontrol research: is it time for a new paradigm? Postharvest Biol Technol 52:137–145

Dunne C, Moënne-Loccoz Y, McCarthy J, Higgins P, Powell J, Dowling DN, O'Gara F (1998) Combining proteolytic and phloroglucinol-producing bacteria for improved biocontrol of Pythium-mediated damping-off of sugar beet. Plant Pathol 47:299–307

Edwards CA, Thompson AR (1973) Pesticides and the soil fauna. Residue Rev 45:1–79

Ehlers RU (2006) Einsatz der Biotechnologie im biologischen Pflanzenschuz. Schnreihe dtsch Phytomed Ges 8:17–31

Elumalai LK, Rengasamy R (2012) Synergistic effect of seaweed manure and *Bacillus sp.* on growth and biochemical constituents of *Vigna radiata L.* J Biofertil Biopestici 3:121–128

England LS, Vincent ML, Trevors JT, Holmes SB (2004) Extraction, detection and persistence of extracellular DNA in forest litter microcosms. Mol Cell Probes 18:313–319

Eskenazi B, Bradman A, Castorina R (1999) Exposures of children to organophosphate pesticides and their potential adverse health effects. Environ Health Perspect 107:409–419

Eskenazi B, Lisa GR, Amy RM, Asa B, Kim H, Nina H, Caroline J, Laura F, Dana BB (2008) Pesticide toxicity and the developing brain. Basic Clin Pharmacol Toxicol 102:228–236

EUPD (2010) European union pesticides database. http://ec.europa.eu/food/plant/protection/evaluation/database_act_subs_en.htm

Evans J, Wallace C, Dobrowolski N (1993) Interaction of soil type and temperature on the survival of *Rhizobium leguminosarum* bv Viciae. Soil Biol Biochem 25:1153–1160

Farah J (1994) Pesticide policies in developing countries: do they encourage excessive use? Discussion paper no. 238. IBRD/World Bank, Washington, DC

Felix MI (1958) Men, molds and history. MD Publication, New York

Fenske RA, Lu C, Simcox NJ, Loewenherz C, Touchstone J, Moate TF, Allen EH, Kissel JC (2000) Strategies for assessing children's organophosphorus pesticide exposures in agricultural communities. J Expo Anal Environ Epidemiol 10:662–671

Fitt GP (1994) Cotton pest management: part 3, an Australian perspective. Ann Rev Entomol 39: 543–562

Fitt GP (2004) Implementation and impact of transgenic Bt cottons in Australia. In: Cotton production for the new millennium, Proceedings third world cotton research conference, Cape Town, South Africa, Agricultural Research Council – Institute for Industrial Crops, Pretoria, South Africa, pp 371–381

Frampton GK, Jansch S, Scott-Fordsmand JJ, Römbke J, Van den Brink P (2006) Effects of pesticides on soil invertebrates in laboratory studies: a review and analysis using species sensitivity distributions. J Environ Toxicol Chem 25:2480–2489

Frampton RA, Pitman AR, Fineran PC (2012) Advances in bacteriophage-mediated control of plant pathogens. Int J Microbiol 2012:326452

Fravel DR (2005) Commercialization and implementation of biocontrol. Ann Rev Phytopathol 43:337–359

Gaind S, Kaushik BD (2008) Biofertilizers for sustainability, agroresources and technology. In: Maheshwari DK, Dubey RC (eds) Potential microorganisms for sustainable agriculture: a techno-commercial perspective. IK International Publishing House Pvt Ltd, New Delhi, pp 67–87

Ganeshan G, Kumar MA (2006) *Pseudomonas fluorescens*, a potential bacterial antagonist to control plant diseases. J Plant Interact 1:123–134

Gelernter WD (2005) Biological control products in a changing landscape. In: Proceedings of the BCPC international congress, vol 1, Glasgow, Scotland, The British Crop Protection Council, Hampshire, UK, pp 293–300

Gelernter WD (2007) Microbial control in Asia: a bellwether for the future? J Invertebr Path 95:161–167

Gerhardson B (2002) Biological substitutes for pesticides. Trends Biotechnol 20:338–343

Gill JJ, Hollyer T, Sabour PM (2007) Bacteriophages and phage-derived products as antibacterial therapeutics. Expert Opin Ther Pat 17:1341–1350

Gill RJ, Rodriguez OR, Raine NE (2012) Combined pesticide exposure severely affects individual and colony level traits in bees. Nature 491:105–108

Glare TR, O'Callaghan M (2000) *Bacillus thuringiensis*: biology, ecology and safety. Wiley, Chichester

Gonzalez JM Jr, Brown BJ, Carlton BC (1982) Transfer of *Bacillus thuringiensis* plasmids coding for δ-endotoxin among strains of *B. thuringiensis* and *B. cereus*. Proc Natl Acad Sci 79:6951–6955

Greaves MP (1993) Formulation of microbial herbicides to improve performance in the field. In: Proceedings of 8th EWRS symposium quantitative approaches in weed and herbicide research and their practical application. Braunschweig, Germany, pp 219–225

Green M, Heumann M, Sokolow R, Foster LR, Bryant R, Skeels M (1990) Public health implications of the microbial pesticide *Bacillus thuringiensis*: an epidemiological study, Oregon, 1985–86. Amer J Public Health 80(7):848–852

Griswold E (2012) How 'silent spring' ignited the environmental movement. http://www.nytimes.com/2012/09/23/magazine/how-silent-spring-ignited-the-environmental movement.html?pagewanted=all&_r=0

Gröner A (1986) Specificity and safety of baculoviruses. In: Granados RR, Federici BA (eds) The biology of baculoviruses, vol 2, Practical application for insect control. CRC Press, Boca Raton, pp 177–202

Groote HD, Douro-Kpindou O-K, Ouambama Z, Gbongboui C, Müller D, Attignon S, Lomer C (2001) Assessing the feasibility of biological control of locusts and grasshoppers in West Africa: incorporating the farmers' perspective. Agric Hum Values 18(4):413–428

Guerra PT, Wong LJG, Roldán HM (2001) Bioinseticidas: Su empleo, produción y commercialization en México. Ciencia UANL 4:143–152

Guillon ML (2003) Regulation of biological control agents in Europe. In: Roettger U, Reinhold M (eds) International symposium on biopesticides for developing countries. CATIE, Turrialba, pp 143–147

Gupta PK (2006) Status of biopesticides-Indian scene. Toxicol Int 13:65–73

Gupta S, Dikshit AK (2010) Biopesticides: an ecofriendly approach for pest control. J Biopest 3:186–188

Gupta CP, Dubey RC, Kang SC, Maheshwari DK (2001) Antibiosis mediated necrotrophic effect of *Pseudomonas* GRC2 against two fungal pathogens. Curr Sci 81:91–94

Haas D, De'fago G (2005) Biological control of soil-borne pathogens by fluorescent pseudomonads. Nat Rev Microbiol 3:307–319

Habib MEM, de Andrade CFS (1991) Controle microbiano de insetos com o uso de bactérias. Informe Agropecuário 15:21–26

Halim H, Ali MM (1998) Training and professional development. In: Swanson BE, Bentz RP, Sofranko AJ (eds) Improving agricultural extension: a reference manual. Food and Agriculture Organization of the United Nations, Rome

Hallett SG (2005) Where are the bioherbicides? Weed Sci 53:404–415

Harman GE (2005) Overview of mechanisms and uses of *Trichoderma* spp. 648. Phytopathology 96:190–194

Hewson I, Brown JM, Gitlin SA, Doud DF (2011) Nucleopolyhedrovirus detection and distribution in terrestrial, freshwater, and marine habitats of Appledore Island, Gulf of Maine. Microbial Ecol 62:48–57

Hidaka Y (1933) Utilization of natural enemies for control of the pine caterpillar. J Jpn Forest Soc 15: 1221–1231

Hill KR, Robinson G (1945) Fatal D.D.T. poisoning. Br Med J 2:845–847

Hoffland E, Hakulinen J, van Pelt JA (1996) Comparison of systemic resistance induced by avirulent and non-pathogenic *Pseudomonas* species. Phytopathology 86:757–762

Hunter DM, Milner RJ, Spurgin PA (2001) Aerial treatment of the Australian plague locust, *Chortoicetes terminifera* (Orthoptera: Acrididae) with *Metarhizium anisopliae* (Deuteromycotina: Hyphomycetes). Bull Entomol Res 91:93–99

Ibrahim L, Butt TM, Beckett A, Clark SJ (1999) The germination of oil formulated conidia of the insect pathogen, *Metarhizium anisopliae*. Mycol Res 103: 901–907

ICAMA (2008) Pesticide manual, the institute for the control of agrochemicals. Ministry of agriculture, China (in Chinese)

Ignacimuthu S, Sen A, Janarthanan S (eds) (2001) Microbials in insect pest management. Science Publishers, Enfield, p 188

Ignoffo CM (1975) Evaluation of in vivo specificity of insect viruses. In: Summers M, Engler R, Falcon LA, Vail PV (eds) Baculoviruses for insect pest control. Amerc Soc Microbiol, Washington, DC, pp 52–57

Iida A, Sanekata M, Fujita T, Tanaka H, Enoki A, Fuse G, Kanai M, Rudewicz PJ, Tachikawa E (1994) Fungal metabolites XVI structures of new peptaibols, trichokindins I–VII, from the fungus *Trichoderma harzianum*. Chem Pharm Bull 42:1070–1075

Industrial Equipment News (2013) Biopesticides market to reach $1 billion in 2010, West Afton Ave, Yardley, PA

IOBC (2008) International organization for biological control. IOBC Newsletter 84:5–7

Irigaray FJSC, Marco-Mancebon V, Perez-Moreno I (2003) The entomopathogenic fungus *Beauveria bassiana* and its compatibility with triflumuron: effects on the two-spotted spider mite *Tetranychus urticae*. Biol Control 26:168–173

Jeong JK, Sang GL, Siwoo L, Hyeong JJ (2010) South Korea. In: Kabaluk, JT, Antonet MS, Mark SG, Stephanie GW (eds) The use and regulation of microbial pesticides in representative jurisdictions worldwide. IOBC Global

Jones KA, Westby A, Reilly PJA, Jeger MJ (1993) Exploitation of microorganisms in the developing countries of the tropics. In: Jones DG (ed) Exploitation of microorganisms. Chapman and Hall, London, pp 343–370

Joung KC, Co^te' JC (2000) A review of the environmental impacts of the microbial insecticide *Bacillus thuringiensis*. In: Agriculture and Agri- Food Canada, Technical Bulletin No. 29

Kabaluk T, Gazdik K (2005) Directory of microbial pesticides for agricultural crops in OECD countries, Agriculture and Agri-Food Canada. http://www.agr.gc.ca/env/pest/index_-e.php??s1=pub&page=micro

Kabaluk JT, Svircev AM, Goette lMS, Woo SG (eds) (2010) The use and regulation of microbial pesticides in representative jurisdictions worldwide. IOBC Global, p 99

Kabi MC (1997) Impact of biofertilizer on rural development. In: Proceedings of National Conference on impact of biotechnology and modern horticulture in rural development. Jadavpur University, Calcutta

Keane WT (1972) Eliminate DDT? Quest for an advantageous benefit: risk ratio. Sci Total Environ 2:141–163

Khalil IAIM, Appanna V, Rick DP, Ronald JH, Lucie G, Tharcisse B, Kelvin L, René P, Kathy AD, Ian KM, Sharon LIL, Kithsiri EJ (2013) Efficacy of Bio-Save 10LP and Bio-Save 11LP (*Pseudomonas syringae*) for management of potato diseases in storage. Biol Control 64:315–322

Khalique D, Ahmed K (2001) Synergistic interaction between *Bacillus thuringiensis* (Berliner) and Lambda-cyhalothrin (Pyrethroid) against, chickpea pod borer, *Helicoverpa armigera* (Huebner). Pakistan J Biol Sci 4:1120–1123

Khandelwal M, Datta S, Mehta J, Naruka R, Makhijani K, Sharma G, Kumar R, Chandra S (2012) Isolation, characterization and biomass production of *Trichoderma viride* using various agro products- A biocontrol agent. Adv Appl Sci Res 3:3950–3955

Khare E, Arora NK (2011) Dual activity of pyocyanin from *Pseudomonas aeruginosa* –antibiotic against phytopathogen and signal molecule for biofilm development by rhizobia. Can J Microbiol 57:708–713

Khater HF (2012) Prospects of botanical biopesticides in insect pest management. Pharmacologia 3:641–656

Kloepper JW, Ryu CM, Zhang S (2004) Induced systemic resistance and promotion of plant by *Bacillus* spp. Phytopathology 94:1259–1266

Knutson RD, Taylor CR, Penson JB, Smith EG (1990) Economic impacts of reduced chemical use. Knutson and Associates, College Station

Konradsen F, Hoekb CDC, Hutchinson G, Daisley H, Singh S, Eddleston M (2003) Reducing acute poisoning in developing countries-options for restricting the availability of pesticides. Toxicology 192:249–261

Koppenhöfer AM, Fuzy AM (2003) Biological and chemical control of Asiatic garden beetle *Maladera castanea (Coleoptera Scarabaeidae)*. J Econ Entomol 96:1076–1082

Koul O, Dhaliwal GS, Marwaha SS, Arora JK (2003) Future perspectives in. In: Koul O, Dhaliwal GS, Marwaha SS, Arora JK (eds) Biopesticides and pest management, vol 1. Campus Books International, New Delhi, pp 386–388

Kranthi KR, Jadhav DR, Kranthi S, Wanjari RR, Ali S, Russell D (2002) Insecticide resistance in five major insect pests of cotton in India. Crop Prot 21: 449–460

Kumar S (2012) Biopesticides: a need for food and environmental safety. J Biofert Biopest 3:1–3

Kunimi Y (1998) Japan. In: Hunter-Fujita HR, Entwistle PF, Evans HF, Crook NF (eds) Insect viruses and pest management. Wiley, Chichester, pp 269–279

Kunimi Y (2007) Current status and prospects on microbial control in Japan. J Invertebr Pathol 95:181–186

Landrigan PJ, Claudio L, Markowitz SB, Berkowitz GS, Brenner BL, Romero H, Wetmur JG, Matte TD, Gore AC, Godbold JH, Wolff MS (1999) Pesticides and inner-city children: exposures, risks, and prevention. Environ Health Perspect 107:431–437

Lapointe R, Thumbi D, Lucarotti CJ (2012) Recent advances in our knowledge of Baculovirus: molecular biology and its relevance for the registration of Baculovirus-based products for insect pest population control. In: Soloneski S (ed) Integrated pest management and pest control – current and future tactics. Intech, Europe, pp 481–522

Le Conte JL (1874) Hints for the promotion of economic entomology. Am Assoc Adv Sci 22:11–22

Leng P, Zhang Z, Guangtang P, Zhao M (2011) Applications and development trends in biopesticides. Afr J Biotechnol 10:19864–19873

Lisansky S (1997) Microbial biopesticides, In: Evans HF (ed) Microbial insecticides, Novel or necessity? Proceedings No. 68. British crop protection council, Farnham, UK, pp 3–10

Liu CJ, Men WJ, Liu YJ (2002) The pollution of pesticides in soils and its bioremediation. Syst Sci Compr Stud Agric 18:295–297

Lomer C (2001) Assessing the feasibility of biological control of locusts and grasshoppers in West Africa: incorporating the farmers' perspective. Agric Hum Value 18:413–428

Lomer CJ, Prior C (eds) (1992) Biological control of locusts and grasshoppers. CAB International, Wallingford

Longnecker MP, Rogan WJ, Lucier G (1997) The human health effects of DDT (dichlorodiphenyltrichloroethane) and PCBS (polychlorinated biphenyls) and an overview of organochlorines in public health. Annu Rev Public Health 18:211–244

MacGregor JT (2006) Genetic toxicity assessment of microbial pesticides: needs and recommended approaches. Intern Assoc Environ Mutagen Soc 1–17

Manjunath TM, Kumar N, JB, Nagaraj DN (1992) A report on the survey for natural enemies of the Mauritius thorn, *Caesalpinia decapetala* in India, Unpublished Report of Bio-Control Research Laboratories, Bangalore, India

Market and Market (2013) Report code: CH 1266 Global biopesticides market – trends and forecasts (2012–2017), India

Marrone PG (1994) Present and future use of *Bacillus thuringiensis* in integrated pest management systems: an industrial perspective. Biocon Sci Technol 4: 517–526

Marrone PG (2007) Barriers to adoption of biological control agents and biological pesticides, CAB reviews: perspectives in agriculture, veterinary science, nutrition and natural resources 2(51). CAB International, Wallingford

Martin PAW, Traverse RS (1989) Worldwide abundance and distribution of *Bacillus thuringiensis* isolates. Appl Environ Microbiol 55:2437–2442

Mazid S, Kalita JC (2011) A review on the use of biopesticides in insect pest management. Int J Sci Adv Technol 1:169–178

McCauley L, Beltran M, Phillips J, Lasarev M, Sticker D (2001) The Oregon migrant farmworkers community: an evolving model for participatory research. Environ Health Perspect 109:449–455

McCoy CW (1996) Pathogens of eriophyoid mites. In: Lindquist EE, Sabelis MW, Bruin J (eds) World crop pests: eriophyoid mites, their biology, natural enemies and control, vol 6. Elsevier, Amsterdam, pp 481–490

McCoy CW, Samson RA, Boucias DG (1988) Entomogenous fungi. In: IgnoVo CM, Mandava NB (eds) Handbook of natural pesticides, vol 5, Microbial pesticides part A, Entomogenous protozoa and fungi. CRC Press, Boca Raton, pp 151–236

McWilliam A (2007) Environmental impact of baculoviruses, FAO.R7299_FTR_anx3. http://www.fao.org/docs/eims/upload/agrotech/2003/R7299_FTR_anx3.pdf

Miller LK, Lu A (1997) The molecular basis of baculovirus host range. In: Miller LK (ed) The baculoviruses. Plenum Press, New York, pp 217–235

Milner RJ (2000) Current status of *Metarhizium* as a mycoinsecticide in Australia. Biocontrol News Inf 21:47N–50N

Milner RJ, Jenkins K (1996) *Metarhizium*: a versatile mycoinsecticide of the future. Prof Pest Manag 1:32–36

Milner RJ, Baker GL, Hooper GHS, Prior C (1997) Development of a mycoinsecticide for the Australian plague locust. In: Krall S, Peveling R, Diallo DB (eds) New strategies in locust control. Birkhäuser, Basel, pp 177–183

Moscardi F, de Souza Lobo M, de Castro Batista ME, Moscardi LM, Szewczyk B (2011) Baculovirus pesticides – present state and future perspectives. In: Ahmad I, Ahmad F, Pichtel P (eds) Microbes and microbial technology. Springer, New York, pp 415–445

Nakkeeran S, Dilantha Fernando WG, Zaki A (2005) Plant growth promoting rhizobacteria formulations and its scope in commercialization for the management of pests and diseases. In: Siddiqui ZA (ed) PGPR: biocontrol and biofertilization. Springer, Dordrecht, pp 257–296

National farmers Policy (2007) Department of Agriculture and Cooperation, Ministry of Agriculture Government of India, India

Noma T, Strickler K (2000) Effects of *Beauveria bassiana* on Lygus *Hesperus* (Hemiptera: Miridae) feeding and oviposition. Environ Entmol 29:394–402

O'Brien KP, Franjevic S, Jones J (2009) Green chemistry and sustainable agriculture: the role of biopesticides, advancing green chemistry. http://advancinggreenchemistry.org/wp-content/uploads/Green-Chem-and-Sus.-Ag.-the-Role-of-Biopesticides.pdf

Oerke EC, Dehne HW, Schnbeck F, Weber A (1994) Crop production and crop protection: estimated losses in major food and cash crops. Elsevier, Amsterdam

Ohba M, Iwahana H, Asano S, Suzuki N, Sato R, Hori H (1992) A unique isolate of *Bacillus thuringiensis* serovar japonensis with a high larvicidal activity specific for scarabaeid beetles. Lett App Microbiol 14:54–57

Olckers T (1999) Introduction: biological control of weeds in South Africa (1990–1998), African Entomology, Memoir No 1

Patrick W, Kaskey J (2012) Biopesticide: killer bugs for hire. Bloomberg Business Week. http://www.businessweek.com/articles/2012-07-19/biopesticides-killer-bugs-for-hire

Paul E, Fages J, Blanc P, Goma G, Pareilleux A (1993) Survival of alginate-entrapped cells of *Azospirillum lipoferum* during dehydration and storage in relation to water properties. Appl Microbiol Biotechnol 40:34–39

Pertot I, Gobbin D, De Luca F, Prodorutti D (2008) Methods of assessing the incidence of Armillaria root rot across viticultural areas and the pathogen's genetic diversity and spatial–temporal pattern in northern Italy. Crop Prot 27:1061–1070

Pimentel D, Greiner A (1996) Environmental and socio-economic costs of pesticide use. In: Pimentel D (ed) Techniques for reducing pesticides: environmental and economic benefits. Wiley, Chichester

Powles RJ, Rogers PL (1989) *Bacillus* toxin for insect control – a review. Aust J Biotechnol 3:223–228

Prasetphol S, Areekul P, Buranarerk A, Kritpitayaavuth M (1969) Life history of orange dog butterfly and its microbial control, technical bulletin no.10. Department of Agriculture, Bangkok

Pray CE, Nagarajan L (2010) Price controls and biotechnology innovation: are state government policies reducing research and innovation by the Ag biotech industry in India? Ag Bio Forum 13:297–307

Pray CE, Nagarajan L (2012) Innovation and research by private agribusiness in India, IFPRI distribution paper 118. IFPRI, Washington, DC

Qiu J (2013) Organic pollutants poison the roof of the world: accumulation of DDT in Himalayas exceeds that seen in Arctic, Nature News

Quandt SA, Arcury TA, Rao P, Snively BM, Camann DE, Doran AM, Yau AY, Hoppin JA, Jackson DS (2004) Agricultural and residential pesticides in wipe samples from farmworker family residences in North Carolina and Virginia. Environ Health Perspect 112:382–387

Quinlan RJ (1990) Registration requirements and safety considerations for microbial pest control agents in the European economic community. In: Laird M, Lacey LA, Davidson EW (eds) Safety of microbial insecticides. CRC Press, Boca Raton, pp 11–18

Quinlan RJ, Gill A (2006) The world market for microbial biopesticides, overview volume. CPL Business Consultants, Wallingford, p 26

Rabindra RJ (2001) Emerging trends in microbial control of crop pests. In: Rabindra RJ, Kennedy JS, Sathiah N, Rajasekaran B (eds) Microbial control of crop pests. Tamil Nadu Agriculture University, Coimbatore, pp 110–127

Rabindra RJ (2005) Current status of production and use of microbial pesticides in India and the way forward. In: Rabindra RJ, Hussaini SS, Ramanujam B (ed) Microbial biopesticide formulations and application, Technical document No.55, Project directorate of biological control, pp 1–12

Radcliffe EB, Hutchison WD, Cancelado RE (eds) (2009) Integrated pest management: concepts, tactics, strategies and case studies. Cambridge University Press, New York

Ranga Rao GV, Rupela OP, Rameshwar Rao V, Reddy YVR (2007) Role of biopesticides in crop protection: present status and future prospects. Ind J Plant Prot 35:1–9

Raymond B, Hartley SE, Cory JS, Hails RS (2005) The role of food plant and pathogen-induced behavior in the persistence of a nucleopolyhedrovirus. J Invert Patho 88:49–57

Reeda EM, Springetta BP (1971) Large-scale field testing of a granulosis virus for the control of the potato moth (*PhthorimaeaOperculella* (Zell.) (Lep., Gelechiidae)). Bull Ent Res 61:223–233

BCC Research (2012) Global markets for biopesticides, Report code: CHM029D, Business Communication Company Research, LLC, USA

Retchelderfer K (1984) Factors affecting the economic feasibility of the biological control of weeds. In: Delfoss ES (ed) Proceedings of VI international symposium on biological control of weeds, Agr Can Bull, p 135–144

Roettger U, Reinhold M (eds) (2003) International symposium on biopesticides for developing countries, CATIE, Turrialba, Costa Rica

Romeis J, Meissle M, Bigler F (2006) Transgenic crops expressing *Bacillus thuringiensis* toxins and biological control. Nat Biotechnol 24:63–71

Roy A, Moktan B, Sarkar PK (2007) Characteristics of *Bacillus cereus* isolates from legume-based Indian fermented foods. Food Contr 18:1555–1564

Rushtapakomchai W (2003) Use and production of biopesticides in Thailand. In: Roettger U, Reinhold M (eds) International symposium on biopesticides for developing countries. CATIE, Turrialba, pp 126–130

Schallmey M, Singh A, Ward OP (2004) Developments in the use of *Bacillus* species for industrial production. Can J Microbiol 50:1–17

Schneider W (2006) US EPA Regulation of biopesticides, microbial and biochemical pesticide regulation. In: REBECA workshop on current risk assessment and regulation practice. Salzau, Germany

Schofield DA, Bull CT, Rubio I, Wechter WP, Westwater C, Molineux IJ (2012) Development of an engineered bioluminescent reporter phage for detection of bacterial blight of crucifers. Appl Environ Microbiol 78:3592–3598

Schönbeck F, Dehne HW (1986) Use of microbial metabolites inducing resistance against plant pathogens. In: Fokkema NJ, Van den Heuvel J (eds) Microbiology of the Phyllosphere. Cambridge University Press, UK, pp 363–377

Shah-Smith DA, Burns RG (1997) Shelf-life of a biocontrol *Pseudomonas putida* applied to sugar beet seeds using commercial coating. Biocontrol Sci Technol 7:65–74

Shukla R, Shukla R (2012) Market potential for biopesticides: a product for agricultural. IJMRR 2:91–99

Siegel JP, Shadduck JA (1990) Clearance of *Bacillus sphaericus* and *Bacillus thuringiensis* ssp. israelensis from mammals. J Econ Entomol 83:347–355

Sinclair M, Martha T (2001) Going against the grain: agricultural crisis and transformation. Oxfam Americas, Boston

Singleton PW, Boonkerd N, Carr TJ, Thompson JA (1996) Technical and market constraints limiting legume inoculant use in Asia. In: Extending nitrogen fixation research to farmers' fields: proceedings of an international workshop on managing legume nitrogen fixation in the cropping system of Asia. ICRISAT Asia Centre, India

Sosa-Gomez DR, Moscardi F (1998) Laboratory and field studies on the infection of stink bugs, Nezara viridula, Piezodorus guildinii, and Euschistus heros (Hemiptera: Pentatomidae) with Metarhizium anisopliae and Beauveria bassianain. Brazil J Invertebr Pathol 2:115–120

Steinhaus EA (1949) Principles of insect pathology. McGraw Hill Co, New York

Steinhaus EA (1957) Microbial diseases of insects. Annu Rev Microbiol 11:165–182

Steinhaus EA (1975) Disease in a minor chord. Ohio State University Press, Columbus

Steinwand B (2008) Biopesticide ombudsman (personal communication). US Environmental Protection Agency, Washington, DC

Stewart A (2001) Commercial biocontrol – reality or fantasy? Australasian Plant Pathol 30:127–131

Stewart A, Hill R, Stark C (2011) Desktop evaluation on commercially available microbial-based products for control or suppression of *Pseudomonas syringae* pv. Actinidiae. Bio Prot Res Centre, pp 1–26

Subramaniam VK (1952) Control of the fluted scale in peninsular India. Ind J Entomol 16:391–394

Sundheim L, Tronsmo A (1988) Hyperparasites in biological control. In: Mukerji KG, Garg KL (eds) Biocontrol of plant diseases, vol 1. CRC Press Boca Raton, USA, pp 53–70

Swati S, Adholeya A (2008) Biological control: alternative paradigms for commercialization. In: Maheshwari DK, Dubey RC (eds) Potential microorganisms for sustainable agriculture: a techno-commercial perspective. IK Publishing House Pvt Ltd, New Delhi

Sylvar Technologies (2008) Research. http://www.sylvar.ca/content/13636

Szewczyk B, Rabalski L, Krol E, Sihler W, Lobo de Souza M (2009) Baculovirus biopesticides – a safe alternative to chemical protection of plants. J Biopesticides 2:209–216

Szewczyk B, Lobo de Souza M, Batista de Castro ML, Moscardi ML, Moscardi F (2011) Baculovirus biopesticides. In: Stoytcheva M (ed) Pesticides – formulations, effects, fate. InTech, doi:10.5772/13219

Tewari S, Arora NK (2013) Transactions amongst microorganisms and plant in the composite rhizosphere habitat. In: Arora NK (ed) Plant microbe symbiosis-fundamentals and advances. Springer, India pp 411–449

Thakore Y (2006) The biopesticide market for global agricultural use. Ind Biotechnol 2:192–208

Thiem S, Cheng X-H (2009) Baculovirus host-range. Virol Sin 24:436–457

Thungrabeab M, Blaeser P, Sengonca C (2006) Effect of temperature and host plant on the efficacy of different entomopathogenic fungi from Thailand against *Frankliniella occidentalis* (Pergande) and Thrips tabaci Lindeman (Thysanoptera:Thripidae) in the laboratory. J Plant Dis Protect 113:181–187

Tripathi G, Sharma M (2005) Effects of habitats and pesticides on aerobic capacity and survival of soil fauna. Biomed Environ Sci 18:169–175

Tschirley FH (1973) Pesticides, relation to environmental quality. JAMA 224:1157–1166

Urquhart EJ, Punja ZK (1997) Epiphytic growth and survival of *Tilletiopsu pallescens*, a potential biological control agent of *Sphaerotheca jilznea*, on cucumber leaves. Can J Bot 75:892–901

USEPA (1998) Reregistration Eligibility Decision (RED), *Bacillus thuringiensis*, USA

USEPA (2008) What are biopesticides? http://www.epa.gov/pesticides/biopesticides/whatarebiopesticides.htm

USEPA (2011) Pesticide news story: EPA releases report containing latest estimates of pesticide use in the United States, USA

Van Lantern JC (2003) Need for quality control for mass produced biological control. In: Van Lantern JC (ed) Quality control and production of biological control agents theory and testing procedures. CABI International, UK

Vargas JM (1999) Biological control: a work in progress. Golf Course Manag 67:1–4

Vattanatangum A (1989) Diamondback moth control with *Bacillus thuringiensis* products in Thailand. Isr J Entomol 23:131–139

Wabule MN, Ngaruiya PN, Kimmins FK, Silverside PJ (ed) (2004) Registration for biocontrol agents in Kenya. In: Proceedings of the pest control products board/Kenya agricultural research institute/department for international development crop protection programme workshop. Natural Resources International Limited, Aylesford, UK

Warburton H, Ketunuti U, Grzywacz D (2002) A survey of the supply, production and use of microbial pesticides in Thailand, NRI Report 2723. Natural Resources Institute, University of Greenwich, Chatham, p 100

Ward MDW, Chung YJ, Haykal-Coates N, Copeland LB (2009) Differential allergy responses to *Metarhizium anisopliae* fungal component extracts in BALB/c mice. J Immunotoxicol 6(1):62–73

Ward MWD, Chung YJ, Copeland LB, Doerfler DL (2011) Allergic responses induced by a fungal biopesticide *Metarhizium anisopliae* and house dust mite are compared in a mouse model. J Toxicol 1–13

Warren GF (1998) Spectacular increases in crop yields in the United States in the twentieth century. Weed Technol 12:752–760

Warrior P (2000) Living systems as natural crop-protection agents. Pest Manag Sci 56:681–687

Webster JPG, Bowles RG, Williams NT (1999) Estimating the economic benefits of alternative pesticide usage scenarios: wheat production in the United Kingdom. Crop Prot 18:83–89

Wei G, Kloepper JW, Tuzun S (1996) Induced systemic resistance to cucumber diseases and increased plant growth by plant growth-promoting rhizobacteria under field conditions. Phytopathology 86:221–224

West TW, Campbell GA (1946) DDT, the synthetic insecticide. Chapman & Hall, London

Wu C, Chen X (2004) Impact of pesticides on biodiversity in agricultural areas. Ying Yong Sheng Tai Xue Bao 15(2):341–344

Xu QF, Song YL, Du CX, Zun SL, Wang WX, Xu BS (1987) An investigation of culturing the fungus pathogen, *Beauveria bassiana* in maize whorl against corn borer, *Ostrinia furnacalis*. J Jilin Agric Sci 4:25–27

Yang HW (2007) Advances in bio-control of plant diseases and pests in China. Sci Technol Rev 25:56–60

Ye ZC, Chen JF (2002) Biological control, Chinese encyclopedia of academic research: 20th century. Fujian Education Press, pp 328–331

Zhang GZ (2002) Research and development of biopesticides in China. J Hubei Agri Coll 22:472–475

Biotrophic Plant-Microbe Interactions for Land Reclamation and Sustainable Agriculture Development

3

Vivek Kumar, Priyanku Teotia, Sandeep Bisht, and Shivesh Sharma

Abstract

Anthropogenic undesired actions intended at agricultural and technological advancement have led to the non-judicious creation and usages of various chemicals. Contamination of soil and formation of barren lands are a worldwide crisis, and reclamation of this using chemical or physical means is not a solution. The negative aspects of pollutants in the soil and environment lead to diverse impact on human beings, flora and fauna also. This undesirable facet relies on the pollution type, its severity and nature. The hunt for alternative methods for digging and incineration to clean contaminated sites resulted in the application of bioremediation techniques, but this is not cost-effective. The cost-effective and viable mode could be efficient utilization of plant-microbe interaction (PMI) pair in agricultural land reclamation. In the process of active rhizosphere functioning, root exudates of plant lead to proliferation, survival, and working of microorganisms, which subsequently results in a more efficient degradation of contaminants. The plant root system actually helps to spread microbes in the soil and assists in penetrating otherwise hard soil layers and surfaces. The inoculation of pollutant-degrading bacteria on plant seed can be an important additive to improve the efficacy of bioremediation or plant bioaugmentation. Biotrophic PMI is promising, a relatively novel technique employed in reclamation of the contaminated or degraded agricultural soils. It may be defined as the exploitation of efficient microbes along with their host plants to utilize or remove, obliterate, or

V. Kumar (✉)
Amity Institute of Microbial Technology, AMITY University, Noida 201303, India
e-mail: vivekbps@gmail.com

P. Teotia
Department of Biotechnology, Division of Biosciences, College of Applied Education & Health Sciences, Meerut 250003, India

S. Bisht
Department of Basic Science, VCSG College of Horticulture, Uttarakhand University of Horticulture & Forestry, Bharsar, Pauri 246123, Uttarakhand, India

S. Sharma
Department of Biotechnology, MLN National Institute of Technology, Allahabad, UP, India

N.K. Arora (ed.), *Plant Microbes Symbiosis: Applied Facets*,
DOI 10.1007/978-81-322-2068-8_3,

impound hazardous chemicals at a particular site. This technology has so far been used experimentally to take away toxic heavy metals and other pollutants from contaminated soil; expansion of its capacity for applications to remove and degrade organic pollutants in the environment is the next phase. This chapter presents an overview of present aspects of microbes-plant relations in reclamation for feasible and viable augmentation of agriculture land biodiversity.

Introduction

Soil is an inadequate reserve, and its well-being is vital for any sustainable development in a particular area. Many developmental or unwanted anthropogenic activities have resulted in soil deprivation, and in the case of agricultural lands, there is successive loss of soil fertility. The characteristic of soil depends not only on its physical or chemical properties but also on the microbial diversity and activity of its flora and fauna (Doran and Linn 1994). From the last many years, enormous area of cultivable agricultural land has been degraded, neglected, and deserted. It has been estimated by the Food and Agricultural Organization that around 25–28 % of the world's land has developed some sort of dereliction, and this land is no more utilized for any sort of production. Countries in the Latin American region have developed heavily degraded land (around 27 %), all due to human activities (Bot et al. 2000). Among Latin American countries, only Brazil has around 236 million hectares of degraded land or approximately four times the area dedicated to arable crops. Degradation of these grazing lands, which were earlier rich and fertile agricultural areas, occurred chiefly owing to overgrazing, unattended vegetation burning, and non-application of chemical fertilizers in the soil. These soils were once fertile agricultural soil, which were uncovered in gullies, and did not have any organic matter to support microbial and plant growth, therefore resulted in poor soil health and physicochemical structure (Sharma et al. 2006). Comparable circumstances also happened in soils uncovered by road and rail corridors, land leveling on construction sites, and mining activities. Under these conditions, the revegetation of these areas by native vegetation is almost extremely limited. Moreover, the seeds from native forest species may be deposited by wind, water, or animal vectors, and many may germinate, they do not thrive, and spontaneous revegetation of the areas normally does not occur. In other situations these areas are invaded by persistent, fire-prone grasses that are known to preclude or severely slow subsequent forest regeneration (Parrota et al. 1997).

Reclamation of Land Polluted with Heavy Metals

Heavy metals are elements having some metallic characteristics such as tensile stress, conductivity, stability as anions, cations, ligand specificity, etc., and should have an atomic number more than 20. In any ecosystem, the most widespread and general heavy metals responsible for contamination are cadmium, copper, lead, mercury, nickel, arsenic, and chromium. Metals occur as natural components in soil, and many heavy metals are required by plants as micronutrients. As a result of human activities such as mining and smelting of metals, electroplating, gas exhaust, energy and fuel production, fertilizer, sewage and pesticide application, municipal waste generation, etc. (Raymond and Felix 2011), soil metal pollution has become one of the most severe environmental problems today. Heavy metal contamination of water and soils poses a major environmental and human health concern.

Efficient microbes and native plant interaction can be a gorgeous approach to attain the proficient taking away of pollutants in contaminated soils. Plants can be used to get rid of an ample range of pollutants, because they can be monitored easily and many agriculture technologies are existing to lessen the costs of the treatment. The choice of micro-phytoremediation is usually cost-effective since it is innately and automatically driven (Rajkumar et al. 2012). Once this process is established, the contaminated sites typically involve slight financial input with regard to continuation. As heavy metals stay on place, the cost of transportation connected to shifting the soil contaminated with heavy metals to a treatment facility could be abolished. Scientists and agriculturist estimate that, on an average, the micro-phytoremediation technique is cost-effective compared to other available remediation methods (Pilon 2005). Abolition of transportation of heavy metal-contaminated soils also diminishes the chances of contaminants proliferating to other uncontaminated environments.

Micro-phytoremediation of heavy metals in contaminated soils may take one of several forms: micro-phytoextraction, micro-rhizofiltration, phytostabilization, and micro-phytovolatilization. Micro-phytoextraction is a method in which plants along with associated microbes help to uptake metals from soil into the roots and shoots of the plant. Micro-rhizofiltration is the employment of plant roots to uptake, concentrate, or precipitate metals from industrial effluents, and micro-phytostabilization is the use of microbes and plants to lessen the movement of heavy metals through absorption and precipitation by plant roots, therefore decreasing their bioavailability. On the other hand micro-phytovolatilization is the uptake and liberation of volatile compounds like mercury- or arsenic-containing components into the environment. The plant roots interrelate with a great number of diverse microbes, and these interactions play a major role in phytoremediation of heavy metals (Jing et al. 2007). The working efficacy of associated plant-microbe mutualism in a heavy-metal contaminated land can be influenced by associated microorganisms and type of plant. Microbes in soil play very important roles in plant nutrient recycling, upholding of soil structure, detoxification of harmful chemicals, biocontrolling of plant pests, and also regulation of plant growth (Kumar et al. 2000; Sandaa et al. 2001). Thus, the native potential bacteria can supplement the remediation aptitude of plants or diminish the phytotoxicity of the heavy metal-contaminated soil. Additionally, microbes and plants can form precise associations in which the plant provides the microbes with an explicit carbon basis that persuades the microbes to trim down the phytotoxicity of the contaminated soil. Plants and microbes can form nonspecific relations in which common plant activities encourage the associated microbial community, which results in the usual microbial metabolic activity, which biodegrades contaminants in the soil. Plants' root can supply exudates, which results in increasing ion solubility and then availability. These biochemical devices boost the remediation functions of microbes associated with roots (Marques-Ana et al. 2009). In a nutshell, the variation and adaptation abilities of plant and microbe as well as the bioremediation capability of the microbes are very crucial in reducing the damaging consequences of heavy metal pollution in any agricultural soil system.

Reclamation of Lands Polluted with Petroleum Hydrocarbons

Contamination of soil by petroleum hydrocarbons is an extensive, prevalent, and well-known worldwide environmental concern. While the majority of these hydrocarbon contaminants occur due to natural resource, release and accumulation of hydrocarbon contaminants through human act have led to noteworthy spoilage of agricultural soil as well as groundwater. A good number of common petroleum hydrocarbons in the environment are related with the handling and use of petrol, diesel, crude oil, and other oil products in automobiles and machineries employed in transportation and other industrial activities. The examples of hydrocarbons which are of environmental concern are gases (e.g., methane and propane, though these do not play a significant role in agricultural soil contamination or degradation), liquids (hexane and benzene, play a role as major contaminants), waxes (paraffin wax and naphthalene), and other polymers (synthetic rubber, polyethylene, polypropylene, polyvinyl butyral, polyacrylonitrile, silicon, neoprene, and plastic or polystyrene). The three key classes of aromatic hydrocarbons of environment pollution concern are (i) polycyclic aromatic hydrocarbons, (ii) heterocyclic aromatic hydrocarbons, and (iii) alkyl polyaromatic hydrocarbons (McGuinness and Dowling 2009).

Although various forms of petroleum hydrocarbons are employed in modern industrial society, for exploration, extraction, purification, and transportation, use of petroleum products, however, poses inevitable environmental risks and menace (Pena-Castro et al. 2006). The presence of petroleum hydrocarbons (PHCs) in soils is not only an adverse factor for human health but also a negative impetus for plant growth (Meudec et al. 2007; Euliss et al. 2008). During microbe phytoremediation of organic contaminants, plants can further benefit from associated bacteria possessing appropriate degradation pathways and metabolic capabilities, leading to more efficient contaminant degradation and reduction of both phytotoxicity and evapotranspiration of volatile contaminants. Apart from environmental conditions the inoculation method and physicochemical properties of the soil are the main factors influencing the survival and activity of an inoculated strain as well as the growth of plants. Soil microorganisms act as biological catalysts for the bulk bioconversion processes and in few cases operate as exclusive and particular bioreactors to biodegrade the organic contaminants. An organic contaminant in the soil could be biodegraded by the process of oxidation or reduction, by the elimination of functional group of the molecule, or by the partial degradation of the crux structure of the hydrocarbon molecules. Mineralization is a method through which the natural and original organic molecules are degraded to generate carbon dioxide, microbial biomass, and inorganic compounds containing N, P, K, and S (Turco and Sadowsky 1995). Fruitful inoculation and use of selected potential soil microbes in feasible means is the initial action to improve the micro-phytoremediation of a contaminated soil. It is equally important that the microorganisms remain viable and can colonize roots and rhizosphere in order to continue improving plant growth and potentially degrade soil contaminants (Ramos et al. 2010).

Prakash et al. (2014) isolated, characterized, and evaluated the potential of PHC-degrading bacterial strains from oil-contaminated soil in Meerut region, India. Among 59 oil-degrading bacterial cultures isolated from oil-contaminated soil samples, one *Bacillus*, two species of *Pseudomonas*, and one species of *Micrococcus* were used for in vitro remediation of benzene, diesel, toluene, anthracene, and naphthalene. Maximum biodegradation of selected hydrocarbon, i.e., benzene and toluene, was done by consortium formulation. Moreover, *Bacillus* sp., APHP6; *Pseudomonas* sp., APHP9; *Pseudomonas* sp., APBP1; *Micrococcus* sp., APIO4; and their consortium resulted in 54.8 %, 60.2 %, 40.9 %, 32.5 %, and 66.2 % decrease in benzene concentration and 61.2 %, 68.4 %, 53.7 %, 39.3 %, and 75.4 % decrease in diesel concentration, respectively, after 6 days of incubation.

Bisht et al. (2010) isolated several naphthalene- and anthracene-degrading bacteria from the rhizosphere of *Populus deltoides*, which were growing in non-contaminated soil. Among these, four isolates, i.e., *Kurthia* sp., *Micrococcus varians*, *Deinococcus radiodurans*, and *Bacillus circulans*, utilized chrysene, benzene, toluene, and xylene, in addition to anthracene and naphthalene. In another study, Bisht et al. (2014) isolated eight endophytic bacteria from the roots of *Populus deltoides* growing in non-contaminated sites at the natural vegetation of Garhwal Himalayas, India. Among these eight, only *Bacillus* sp. SBER3 isolate was able to metabolize a wide range of polyaromatic hydrocarbons (PAH) and other hydrocarbons, i.e., anthracene, naphthalene, benzene, toluene, and xylene. Furthermore, quantitatively *Bacillus* sp. SBER3 was able to produce 22 mg ml^{-1} of IAA after 4 days and solubilized 0.96 mg ml^{-1} of available phosphorus, respectively, after 120 h; SBER3 also produced siderophore and 1 aminocyclopropane-1-carboxylate (ACC) deaminase. Along with these traits, SBER3 under in vitro condition inhibited the phytopathogenic fungi *Rhizoctonia solani*, *Macrophomina phaseolina*, *Fusarium oxysporum*, and *Fusarium solani* with percent growth inhibition (PGI) of 60 %, 61.5 %, 64.3 %, and 12 %, respectively. Rhizoremediation potential of *Bacillus* sp. SBER3 was demonstrated in PAH-contaminated soil model system. Significant enhancement in shoot, root length, and root and shoot biomass including the stem girth of *P. deltoides* in comparison to control was also recorded, and concurrently endophytic isolate Bacillus sp. SBER3 degraded 45.6 % of PAH in the soil model system after 120 days. Such potentially PHC-biodegrading bacterial strains could be effectively employed in the rhizosphere of plants growing in contaminated soil for reclamation and reviving nearly biological dead soil. It is becoming increasingly clear that the mutualistic relationships that occur between plants and microbes especially in the rhizospheric zone could play a vital role in improving the efficiency of phytoremediation (Bisht et al. 2014). The addition of efficient microbes enhances the PHC elimination from soil, possibly by increasing plant germination and by stimulating the plants to grow easily and build up wider root biomass for more nutrients and water uptake.

Reclamation of Lands Degraded Due To Soil Erosion

A lot of many landmasses including fertile soil have suffered from soil erosion. In the broad background, the land not in its original undisturbed state is referred as eroded land. In the utilization milieu, eroded land means loss of biological agricultural productivity or erosion in the land's ability to hold enviable vegetation. Man-made fracas such as tunnels, roads, and building erections result in the diversified land erosion problems, soil fertility loss, slope creation, loose sedimentation and displacement of soil by air waterways, and finally disruption of vital indigenous habitat. The cost of development has to be paid in terms of land degradation, but, the gravity of the damaging impact may be diminished by the speedy reestablishment of aboriginal plants (Mackenzie and Mackenzie 1995).

To increase variety and diversity of plant species in any soil, the type of flora management or revegetation depends upon the status of soil disturbance. Microorganisms are highly distinct in the soil and offer an extraordinary challenge to scientists to understand their biodiversity. It becomes very challenging for researchers to assure a successful restoration of microbes, as compared to flora and fauna. Microbes should be given utmost priority of any restoration of biodiversity as they represent the vast majority of the minute but important diversity of an ecosystem at every site, but still the role of microbes is rarely considered in terms of maintaining an ecological diversity and stability (Blanco and Lal 2008). Native and potential microbes should be a center of attention, for restoring disturbed agriculture lands. Without most species of plants and animals, disturbed lands can become steady, and the majority of ecosystems would function normally, but without naturally occurring or introduced microorganisms, the natural ecosystem would stop functioning. For evaluation of microbial restoration and their diversity, there could be three critical points for measurement. The first point is to define the space and sequential allocation of different species and their affiliation to ecosystem progression. The second worry is to review the affluence of microorganisms within the different ecosystems. Finally, the third one is finding the ability for spreading and usual reestablishment for the requirement of man-made introduction of efficient microbes in the rhizosphere (Toy et al. 2002).

The role of symbiotic or associative microorganisms is much less varied and clearly plays decisive roles in the creation and perseverance of crop or forest plants including weed plant composition. Essentially there are two types of activities: one is plant pathogens that hinder plant growth, and the second one are symbionts that remove nutrient resources and exchange those resources with plants for energy, protection, and enriched plant root exudates. Many reforestation and replanting methods successfully established the native flora which reduced soil erosion. The purpose of these methods was to establish the native vegetation in eroded area with the help of microbes and also not allowing the exotic plants to acclimatize in that area. The best tactic has been the planting of competitive and fast-growing native plants inoculated with effective microbes, and it could be bacteria or endophytic fungi. These symbiotic microorganisms augment plant growth by providing unavailable nutrients and other growth factors. The severe soil erosion results in consequent fine dust pollution. Fine dust directly or indirectly leads to chronic respiratory diseases (Tiittanen et al. 1999; Samet et al. 2000). This method has been successfully employed all over the world to combat land degradation by erosion (Rapp 1986). Since long, it has been presumed that degraded and deforested agricultural lands would relapse to their original vegetation in due course of time, and this practice commonly happens in temperate lands where reforestation occurs in the absence of domestic grazing and no man degradative activities (Weber 1986). In such soils natural microflora helps to rebuild the degraded ecosystem.

For reclamation of desert, plants, especially xerophytes, are exceptional top soil binders. They have the potential of stopping soil erosion and lessening dirt spread in deserted agricultural and town areas. Hence revegetation programs have to be correctly implemented to allow the growth of xerophytes in degraded areas. To cope up the demand of medicinal plants in the international market, it will be a better option to use degraded or eroded lands for their cultivation. This exercise will assist to rescue the lands from the overload of different unwanted activities besides being the economic exchequer for the growers (USDA 2014). However, the role of associated rhizosphere microbes should be monitored so as to utilize them along with the plants in reclamation activities.

Reclamation of Lands Polluted with Fly Ash

Combustion of coals in thermal power plants is one of the major sources of environmental pollution due to generation of huge amounts of ashes, which are disposed of in large ponds in the vicinity of the thermal power plants. This problem is of particular significance in India, which utilizes coals of very high ash content (approximately. 40–55 wt.%). Fly ash is a diverse complex combination of noncrystalline and crystalline solid phases and is generally a fine-powdered ferroaluminosilicate material with Al, Ca, Fe, Na, and Si as the predominant elements. Certain other metal elements like B, Mo, S, and Se are characteristically enriched in fly ash particles (Adriano et al. 1980). Presence of fly ash metals disturbs the natural characteristics of soil, which adversely affects microbial population and plant growth. High amount of fly ash causes hostile situation for flora and fauna to thrive in any ecosystem. Adjustment or correction of soil with some amount of fly ash is likely to influence the natural properties. Therefore, for land amendment purpose, if fly ash is to be employed, it is essential to seriously assess its possible consequences on the free-living as well as rhizospheric microbes, which play a pivotal role in sustaining the fertility and productivity of the agriculture land. Literature is very scant regarding impact of fly ash on biological activities of soil micro/macro flora and fauna. Limited and incongruous information is available on this subject. Garau et al. (1991) reported that application of fly ash to agricultural soil restrains the microbial respiration activity, nitrification, and mineralization rate. Pichtel and Hayes (1990) informed that microbial population and other biological activities are hindered by fly ash application.

Conversely, it has also been observed that the application of less amount of fly ash augments the biomass and variety of the soil microflora and enhances the aptitude of the agricultural soil to sustain plant growth and proliferation. Modifications in the soil characteristics using fly ash are also shown during microbial enzyme activity study, which increased up to 20–50 % in the presence of earthworms (*Drawida willsi Michaelsen*). It could be due to the fact that earthworm excreta acted as substrates for microbial growth and proliferation (Pati and Sahu 2004). Enrichment of vermicompost from nutrients points of view using N-fixing and P-solubilizing *Azotobacter chroococcum* has been studied and found that the enriched vermicompost could be used to dilute the effects of unwanted factors of disturbed soil (Kumar and Singh 2001). Presence of simple organic compounds and N source is vital for proliferation of plentiful microbial population in fly ash-amended soils. Therefore, application of organic additives like press mud, sewage sludge, and bioorganic waste into fly ash-contaminated soil will certainly augment microorganism population by providing essential micro- and macronutrients along with a carbon source. Organic matter will mitigate the lethal effects of fly ash metals, balance the C:N ratio, and encourage the microbial population and their diversity (Pichtel and Heys 1990).

Fertility of various agricultural soils could be decreased at a shocking rate owing to the loss of organic substance as a result of poor soil biological health, soil erosion, and compaction and due to broad variety of metal pollutants present in fly ash. The physical, chemical, and biological characteristics of an agricultural ecosystem are influenced by the micro and macro flora. Presence of organic matter and sufficient number of active microbial population will recycle the nutrients and prevent the soil compaction and toxicity due to accumulation of chemicals by fly ash. However, in many cases adverse circumstances in troubled soils may lead to no or poor vegetation growth. Such disturbed soils can be restored using active rhizospheric microbial populations. Here plant growth-promoting microorganisms (PGPM) play a critical role by providing essential growth factors to the plants, overcoming abiotic and biotic effects and mitigating the toxic effects of contaminants in the soil. Disturbed soils with fly ash all over the world necessitate biological remediation approach (Schützendübel and Polle 2002). This approach is connected with a passionate hunt for aboriginal or exotic microbes, which can adapt in fly ash-amended environment and help the plant to grow, and directly or indirectly phytoremediation of the disturbed place will eventually be remediated. The arbuscular mycorrhizae (AM) fungi and plant growth-promoting bacteria can carry out this job efficiently in inconsiderate soil environment and encourage the bioremediation of bothered agricultural soils (Fomina et al. 2005; Baum et al. 2006).

The knowledge of efficacy and working aspects of soil inoculation of competent microbes is inadequate mainly because of lack of analyzing methods and also by unavailability of quick and efficient location-specific selection techniques. The greatest location-specific information will be generated by using active, efficient, and competitive free-living, associative, and endophytic microbes which are compatible with location-specific plants (Wenzel 2009). Well-suited potentially competent microbes can surely and significantly be a part of bioremediation of inorganic contaminants. This is the reason that screening of efficient microbial inoculums is a key foundation of triumphant micro-biotechnological application in the soil. Nevertheless, a number of impediments should be conquered to attain the unbeaten commercialization of microbial technology. The most pertinent and important study of soil microbes for consequent applications in bioremediation of fly ash-contaminated soils is the research trials and observations done in natural environmental conditions. In spite of this, only thorough genetic and molecular exploration of an individual competent microbe in highly accurate and proscribed environmental circumstances can explicate the fundamental and vital means of their influence on plants in fly ash soils. For this reason, all aspects of procedural ladder are significant for uncovering, elucidation, and following controlled exploitation of communications among plant roots and their associated soil microbes (Read 2002).

Role of PMI in Sustainable Agriculture

In any good soil, a distinctive equilibrium of chemical, physical, and biological aspects maintains perfect soil characteristic. A healthy agri-environment structure and performance is dominated mainly by soil microbial dynamic system. A balanced and vigorous soil microbial community that decomposes organic matter and recycles macro- and micronutrients results in a sustainable and productive agriculture ecosystem (Solanki et al. 2011). Therefore, effective

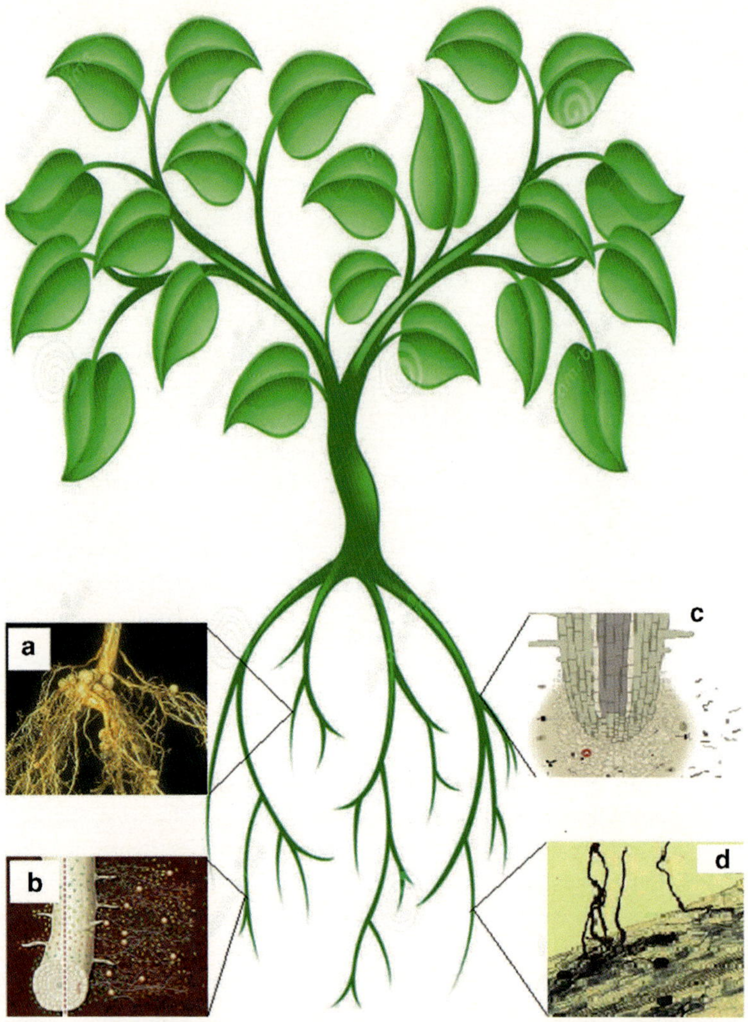

Fig. 3.1 Advantageous microorganisms in the soil, rhizosphere, and roots: (**a**) root nodule bacteria, (**b**) free-living and rhizospheric bacteria, (**c**) root exudates attract useful bacteria, and (**d**) AM fungi exploring soil for nutrient uptake

communications between the plants and saprophytic microbes are an important aspect to figure out the degradation of the eco-environment system and have key impact on the working and development of agricultural soil structure (Kumar et al. 2000). Modification in land use or types of cropping system can alter the microbial population constitution of the rhizospheric soil. Knowledge of influence of dissimilar crop plants growing individually or as a mixed cropping system, which could play a vital role in decisive composition or type of microbial population, able to effectively recycle the available nutrients and production of phytohormones and repression of plant disease is essential. An optimal balance and management of soil fertility is required for enhancement of agricultural sustainability and productivity (Johansson et al. 2004). The role of naturally occurring and competitive advantageous free-living, associative, endophytic microbes and arbuscular mycorrhizal fungi (AMF) needs to be fully exploited in agriculture, regardless of its sizeable stimulus on crop production (Fig. 3.1).

Many workers have reported that beneficial microbes have manifold functions in soil ecosystem (Johansson et al. 2004; Kumar et al. 2013). These microbial communities are perfect means in soil for sustainable agriculture and restoration of degraded lands, forestry, and horticulture. On the other hand still there are research gaps as well in our understanding of microbial functions especially in the rhizosphere. Even without understanding and concentrating on managing the role of microorganisms, the growth of agriculture took place in the last several decades. It was not known that rhizospheric microbial community and their actions are indispensable for soil biological vigor and plant growth. Free-living as well as rhizospheric microbes are accountable for fundamental functions such as recycling of nutrients, N_2 fixation, P solubilization, biocontrol activities, and finally decomposition of organic, dead, and decay materials (Kumar et al. 2000, 2001).

The supervision of soil microorganisms in soil, through screening and inoculation of potential microbial species or by simply encouraging naturally accessible microbes, embraces immense assurance for sustainable and balanced agriculture system. For this, there are many reasons; firstly, microorganisms are versatile in functioning, ranging from biocontrol agents to plant growth promoter. Secondly, once introduced, microbes can perpetuate and survive well, with availability of favorable conditions. Finally, managing the microbial population system may encourage soil flexible systems. Existing agricultural ecosystems are insolvent in terms of their biodiversity and physiological and functional variety (Kumar et al. 2009, 2011). For maximum profitable utilization of microbial strains, cautiously controlled field experiments of crop plants are essential. An augmented understanding of plant-microbe association could impart the way for developing sustainable agriculture in order to guarantee food and feed production with nominal threat to the environmental system. To accomplish food, feed security, and sustainable agriculture production system, abovementioned aspects need promotion and awareness among common man and farmers.

Role in Soil Health

The health of soil ecosystem depends upon the diverse microbial community present, which contributes by transforming unavailable atmospheric N_2 into available form, solubilizes P, and provides other nutrients to plant for its optimum growth. Microbes also contribute a central role in nutrient recycling within the soil. Any uncontaminated soil contains plentiful genera of fungi, protozoa, and bacteria, which play an active part as biocontrol agents, phagocytizing pathogenic microbes. Using manifold means, plant growth-promoting microbes (PGPM) promote the development and maturity of crop plants exactly and implicitly. PGPM produce secondary metabolites, i.e., plant growth-promoting substances, which result in modifications of root morphology consequential in larger root exterior zone for maximum absorption and transport of nutrients to plants. Excretion of siderophores, by soil microbes, causes chelation of iron leading to competition for iron by soilborne plant pathogens. The larger root surface region for uptake of macro- and micronutrients and production of secondary metabolites may facilitate to optimize nutrient recycling during abiotic and biotic stresses due to unsuitable weather or soil conditions or due to competition with pathogenic microbes (Yang and Crowley 2000).

Many bioinoculums for leguminous and nonleguminous crops have been developed, which have resulted in awareness all over the globe. Commonly used PGPM bioinoculants of *Rhizobium*, *Azospirillum*, *Azotobacter*, *Bacillus*, *Pseudomonas*, AM fungi, etc., are commercially available for many cultivated crops. The purpose of PGPM is to recycle the available and unavailable nutrients in soil with minimum application of chemicals. Application of PGPM under field as well as in greenhouse conditions has encouraging responses; however, under field conditions, due to some unspecified reasons, applied bioinoculants do not perform as per expectations. For healthy soil system and development of competent microbes, the inoculation of PGPM is advisable in every cropping season after thorough research of soil and climatic conditions (Carvalhais et al. 2011).

Aboriginal plant species and their associative microbes due to their adaptation and acclimatization are doing well in environmental treatment as well as in healthy soil formation. Introduction of exotic species in the Thar Desert, India, and Sahel regions, Africa, for restoration of ecosystem and soil health did not generate encouraging results as exotic species could not fit into the new environment (Catterson et al. 1987; Sinha 1997) as compared to the aboriginal species. For rehabilitation of acidic or saline-contaminated soils especially in desert regions, importance should be given to those plants, which can survive or tolerate these harsh conditions and finally help in soil building and its productivity. Isolation and regular reinoculation of potentially competent microbes from such regions will certainly help to restore the soil health for many years to come.

Role in Abiotic and Biotic Stress

A variety of abiotic factors such as climate change, drought stress, pollution, and heavy metal contamination can influence the development of microbial relationships in the rhizosphere. In almost most of the cases, presence of mycorrhizae can alleviate the stresses to plants caused by these external influences. Inefficient mining processes, treatment of soil with sewage sludge or industrial effluents, and overuse of heavy metal-containing fertilizers or gas exhausts have, among other factors, contributed to the creation of large areas contaminated by heavy metals, radionuclides, and persistent organic pollutants. Such areas may jeopardize human health and ecology (Mathur et al. 2007) and, therefore, should not only be under strict control but should also be managed in a way which attenuates the possibility of any risk concerning the contamination of the food chain and the groundwater or air by wind erosion.

Soil Salinity Reclamation

In arid and semiarid regions, soil salinity is considered the most significant constituent of land degradation. It is an active process, spreading globally in more than 100 countries and covering on an average more than one billion hectares. Excess soil salinity leads to noteworthy degradation in fertile and irrigated agriculture lands due to poor understanding and mismanagement of natural resources. Recently, some noteworthy technological developments are available to evaluate soil salinity by using remote sensing technologies. The geographical information system, geophysical methods, and modeling, from regional, national to farm levels, also provide a major overlook to monitor land degradation due to salinity. The diverse types of soil salinity which influence agricultural productivity, like irrigation-induced salinity and "transient" dryland salinity, have been observed in desert regions of the world (Munns and Tester 2008). Obviously, soil salinity is one of the key ecological stresses that lead to reduction in agricultural productivity all over the world. Land degradation due to soil salinity and population growth has led agricultural scientists to develop salt-tolerant or resistant crops using genetic techniques (Yamaguchi and Blumwald 2005). On the other hand, the biochemical, molecular, and physiological means of salt resistance or tolerance in agricultural crops are yet to be understood in detail, and this is the reason that advancement in developing salt-tolerant crops has been sluggish. In the saline agricultural environments, soil salinity is often accompanied by additional abiotic and biotic stresses. For example, metals like high boron concentrations often happen in saline environments. Therefore, one has to understand the interactions between soil salinity and high boron toxicity in crops (Jahromi et al. 2008).

Plants are affected by salinity in various manners such as heavy metal or precise ion toxicity, physiological disorders, and also osmotic effects (Esechie et al. 2002). The optional approach to develop salt tolerance in plants could be by introducing salt-tolerant beneficial microbes that may augment crop production or help plants to thrive in adverse environment. Important role has been played by symbiotic relationships such as AM fungi, whose hyphal networks increase the soil surface area for the plant cells. The ectomycorrhizal fungi, which form a network layer around the root hair system and also in to the root intercellular spaces, and rhizospheric plant growth-promoting rhizobacteria are known to play a very significant role under saline conditions (Manchanda and Garg 2008; Ben Salah et al. 2009). One of the main storage carbohydrates in the extraradical mycelium as well as in spores of AM fungi is trehalose. The biosynthesis of trehalose is reliant upon colonization of AM fungi with plant roots (Hoekstra et al. 1992; Schubert et al. 1992). Accordingly, the accumulation of trehalose might help to defend plants against salt stress. Romero et al. (2006) reported that silicon improves salt tolerance in plants by limiting Na^+ uptake through an unknown means, thus, improving photochemical efficiency and therefore, protecting cell membrane integrity, increasing antioxidant enzyme activity, and reducing water loss. Fascinatingly, the AM fungi *Glomus intraradices* is able to build up silicon and can improve its concentration in the host plant (Hammer et al. 2011), but the exact role of silicon in preventing water loss is yet to be deciphered.

The regulation of phytohormone production by microbes has been known in controlling plant vegetative growth during the osmotic stress by regulating the plant growth and plant leaf senescence (Kumar et al. 2001; Ghanem et al. 2011). Aroca et al. (2008) reported that AM fungi infection increases leaf abscisic acid (ABA) accumulation in drying soil, but limited information is available on whether AM fungi modify the ABA relations of plants growing in saline soil. The effect of AM fungi on plant ABA level seems contradictory, but ABA level in plants appears important in determining mycorrhizal fungi colonizing the plant root system. In one experiment, ABA-deficient tomato mutant, which accumulates at least 50 % lesser ABA concentration in roots compared to its wild type (Cornish and Zeevaart 1988), exhibited less root colonization by AM fungi *Glomus intraradices* DAOM197198 (Herrera et al. 2007). Some AM fungal species such as *Scutellospora calospora* attain maximum spore germination under high salt (300 mM NaCl) concentrations (Juniper and Abbott 2006; Jahromi et al. 2008). Application of potential microbes to enhance plant growth and yield in saline soil conditions entails that the inoculants should also be salt tolerant or resistant, which highlights the prospective of using microbes from saline habitats. Certain plant growth-promoting microbes, which can survive, proliferate, and colonize the root system in saline environment (Paul and Nair 2008), certainly offer substantial promise as bioinoculants.

It is a long-term objective to reclaim saline soil using improved and efficient plant-microbe pair for salinity-affected land. Plant survival can be accomplished with an understanding of the means of osmoadaptation in microbes and plants. Accumulation of amino acids and carbohydrates in microbes such as proline, glycine betaine, glutamate, and trehalose in response to soil salinity or high osmolarity results in better adaptation. Here proline plays a very prominent role in osmoregulation through change or adaptation in osmotic stress that shifts the dominant osmolyte from glutamate to proline in *Azospirillum brasilense* (Tripathi et al. 2002). Upadhyay et al. (2012) isolated ten species of *Bacillus* from wheat rhizosphere of saline soil. Proline content increased with NaCl stress, and maximum production was recorded with isolate SU8- 2.73 and 11.95 g mg protein at 0 % and 10 % NaCl (w/v), respectively.

Species of *Bacillus* have evolved highly sophisticated regulatory networks for protection against sudden unfavorable environmental changes, including nutrient malnourishment, changes in conditions such as temperature and humidity, oxidative stress, and abrupt increase in soil or surrounding salinity. Spore-forming bacteria such as species of *Bacillus* are one of the major types of soil bacteria that help in survival and proliferation of plants in saline environment mycorrhizal fungi being the other ones.

Drought Resistance

Due to excess and uncontrolled irrigation by farmers, water table has gone down, which has resulted in arid or semiarid conditions. In such conditions the role of microbes becomes very important to help the plant overcome the drought. The transpiration and conductance rate reduces in plant leaf due to water scarcity, the water use effectiveness increases, and using this technique the plants survive over the drought period (Aroca and Ruiz 2009). An increase in ABA in plant leaf, which reduces water loss by stomata closing and leading to production of activated oxygen species, results in prevention of water loss (Cho et al. 2008). Additional plant-signaling components, for instance, salicylic acid, methyl jasmonate, and ethylene, also play a vital role in stomata closure and reduce the rate of transpiration. Techniques that permit plants to handle drought conditions are synchronized by gene expression alteration. Genes which manage drought conditions are of two types: (a) functional genes, which encode for transporter proteins and detoxify enzymes and biosynthesis of osmolyte enzymes, and (b) regulatory genes, which encode for transcriptional factors (Aroca and Ruiz 2009). In general, the favorable impacts of rhizospheric PGPM on plant water stress conditions are caused by alteration in phytohormonal contents, specially ABA, ethylene, and cytokinins. Inoculation of *Arabidopsis* with *A. brasilense* Sp245 exhibited higher ABA amount compared to uninoculated plants (Cohen et al. 2008). Arzanesh et al. (2011) used different strains of *A. lipoferum* (under drought conditions) to inoculate wheat seedlings. Inoculation with bacteria having properties of phytohormone production and phosphate solubilization alleviated wheat plant drought stress and improved plant development and growth yield (Kumar and Narula 1999; Narula et al. 2000). Bacterial production of hormone-like substances and their ability to stimulate plant endogenous hormone levels were believed to play a key role in this process (Dobbelaere et al. 1999). Under aerobic conditions, it has been observed that *A. brasilense* generates noteworthy amounts of diffusible gas, nitric oxide. These biomolecules have been reported to work as signaling molecules in an indole acetic acid-stimulated pathway, which is implicated in adventitious root growth (Creus et al. 2005; Molina et al. 2008). Bacterium, *P. polymyxa*, isolated from drought-prone *Arabidopsis thaliana* induced drought-responsive genes at a transcriptional point, which shows a role in overcoming drought stress. Plant inoculated with bacterium was further resistant to water stress compared to uninoculated (Timmusk and Wagner 1999).

PMI and Nutrient Uptake

The availability of nutritional material present in soil greatly influences the plant growth and yield in an ecosystem. Normally, a plant could face noteworthy remonstrance in harvesting a sufficient amount of these nutrients to suffice the requirement of fundamental and essential metabolic course owing to their unavailability and fixation in soil (Liu et al. 2011). A scarcity of a few one of them might consequence in poor or less plant growth. Indications of nutrient shortage could result in diminutive growth, plant tissue death, and small size and yellowing of leaves due to decreased amount of chlorophyll production, an essential chrome material for photosynthesis. The deficiency of nutrient supply to the plants will have adverse effects on agriculture, resulting in fewer yields with poor quality. It can also affect the overall eco-environment of a region including the soil microflora, whose survival and growth greatly depend upon the root exudates. There are chiefly two types of plant nutrients: macro- and micronutrients or trace elements. The macronutrients are the construction lump of vital cellular machinery, and parts such as nucleic acids, proteins, and carbohydrates therefore are needed in huge amount by the plant cells. Nitrogen, phosphorous, potassium, sulfur, and magnesium are crucial and vital macro elements for a plant. Micronutrients, such as Zn, Fe, Mn, Cu, Co, and Ni, are requisite in very little amounts, because these play an important role in enzyme functions and activity. Plant adopts different strategies for the uptake of Fe from soil, as depicted in Figs. 3.2 and 3.3.

Two major nutrients, N and P, are considered most important for the plant growth, yield, and development since they are frequently present in minute quantities or are available in a form that cannot be taken up by the plant (Vessey 2003). Consequently, the development and progression of most plant species resulted in the improvement of reciprocally symbiotic, helpful, and advantageous

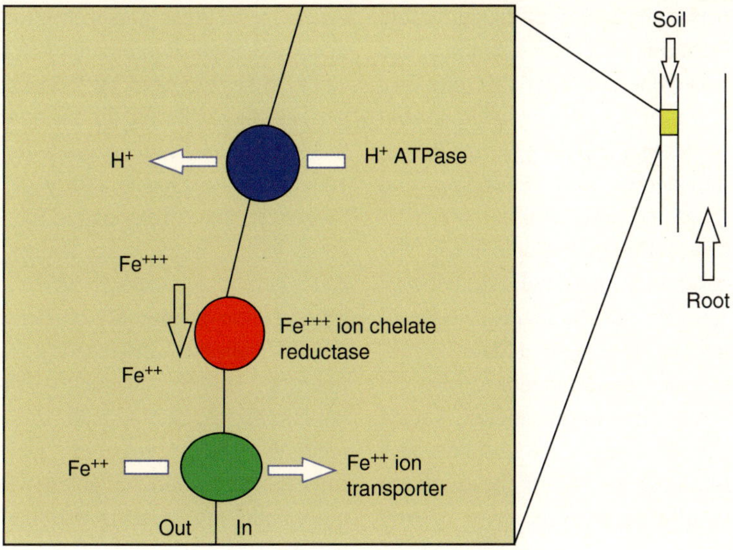

Fig. 3.2 Plants induce the activity of a proton ATPase, a ferric chelate reductase, and a ferrous iron transporter during iron limitation (without bacteria)

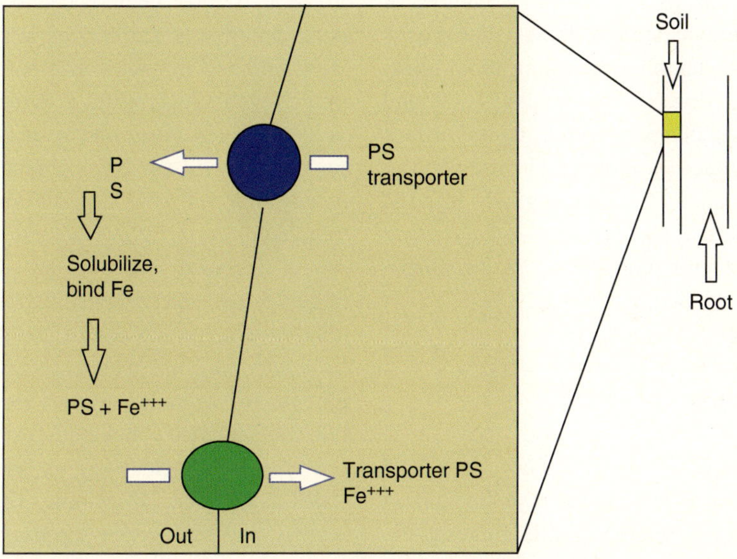

Fig. 3.3 Plants in association with microbes synthesize and secrete phytosiderophores (PS) into the soil during iron deficiency; the PS bind ferric iron complexes that are transported to plant root

relationships with soil microbes. The result of this relationship developed the dependency of plants and microbes on each other. Many associative, free-living, or endophytic microbes fix atmospheric N for the plants and also solubilize the fixed P making it available to the plant (Kumar and Narula 1999; Kumar et al. 2009, 2011, 2013). Diverse bacterial and fungal species have the potential to solubilize diverse insoluble inorganic P forms, such as tricalcium phosphate, dicalcium phosphate, hydroxyapatite, and rock phosphate. Such microorganisms are present in most agricultural lands and constitute about 40 % of the culturable population (Richardson 2001). Among the common bacterial genera having this property are *Pseudomonas*, *Bacillus*, *Rhizobium*, *Burkholderia*, *Achromobacter*, *Agrobacterium*, *Micrococcus*, *Aerobacter*, *Flavobacterium*, and *Erwinia* (Rodríguez and Fraga 1999).

The limitations of diverse nutrients in soil also affect the rhizospheric microbial populations; this effect is due to the modification of root micro-rhizosphere environment (Chen et al. 2006). These differences are thought to relate to modifications in rhizodeposition patterns. A wide range of deficiency symptoms are observed in a plant in nutrient-limiting conditions. These nutrient limitations also influence the secretion of root exudates (Kumar and Narula 1999). The symbiotic microbes can noticeably play a vital function in approaching complicated organic nitrogen, nullifying the role of saprophytes in soil nutrient mineralization. In spite of this, in a number of mycorrhizal organizations, these saprotrophs cooperate with plants by providing them nitrogen. The mycorrhizal fungi increase N, P, and other nutrients in plants by increasing the surface area using hyphal extensions (Jonsson et al. 2001; Lerat et al. 2003). This augment in N, P, and other nutrient uptake is related to the encouragement of bacteria proliferating in the micro-rhizosphere of the plant.

Future Aspects

Novel technologies and approaches should be implemented that will aid to improve the biodegradation efficacy of microbes in the rhizosphere. Consequently, additional studies of the selection of potent rhizospheric microorganisms or communities, which can survive and proliferate in the root zone of a plant which is suitable for rhizoremediation or phytoremediation, could result in better and extended approach. This plant microbe scheme then can be an interesting means to develop and expand bioremediation into a widely accepted land area. The inoculation of genetically engineered microorganisms with unique pathways will create novel or improved functions, which can seize a great prospective and promise for increased phyto-/bioremediation. Bioremediation can also be improved by the employment of the genes responsible for the biosynthetic pathways that can augment biodegradation processes by enhancing the bioavailability of the substrates. Moreover, genes responsible for encoding resistance to important stress factors present in the ecosystem may increase both the survival and the proliferation of specially created microbes. As a result, genetic engineering of aboriginal microbes, suitably adapted to native environmental conditions, could make the bioremediation methods more feasible and eco-friendly. Importance should be given while developing bioremediation systems using a plant-microbe pair to select a native type of microorganism, which can enhance the process using natural gene transfer mechanism. This is also useful to evade the complications of national as well as international legislation matters regarding restricting, using, and monitoring the application of genetically modified microbes in a particular ecosystem. Nevertheless, with a universal political budget towards green and sustainable bioremediation methods, the employment of plant-associated microbes to biodegrade toxic man-made compounds or overcome natural inhibiting factors in degraded land may provide competent, cost-effective, and sustainable green technology. Another exhilarating choice could be to exploit the potential of plant-associated microbial genome for degraded ecosystem or biodegradable contaminated compounds, using the application of rDNA technology to create transgenic plants using bacterial genome. These plants can express bacterial genes resulting in enhanced plant lenience and metabolism of noxious organic compounds in a particular area. Some transgenic plants have been developed for phyto/bioremediation of petroleum hydrocarbons and heavy metals. Genetically engineered poplar plantlets excreting bacterial mercuric reductase enzyme have been shown to germinate and successfully grow in mercury-contaminated soil. Based on this line several other native plants can be modified using microbial genomes to grow well in contaminated or degraded soils.

Conclusion

Biotrophic plant-microbe interaction process is a hopeful approach to reinstate degraded lands. However, there are certain challenging issues which still linger on to be conquered. Certain amount of understanding is required to recognize how plants and microbes will react in the non-favorable or unfertile environment. There is still a lot of work to be carried out, more in open fields, especially in degraded ecosystems based on the laboratory and greenhouse studies. One has to provide commercially feasible systems for using plant-associative free-living, endophytic, and rhizospheric microbes to biodegrade a wide range of contaminated compounds and overcome abiotic and biotic factors in any environmental system. Beyond doubt, since long the soils have been built by plants and their associated microbes, the association of both of these has contributed in improving the physicochemical characteristics of soils to permit better development of plants and their associated microbes. This association has led to increased plant growth, depth and loosening of soil, and structuring; better water holding capacity; better aeration and gaseous exchange; increased availability of inorganic and organic compounds to roots; and enhanced rock weathering and nutrient recycling. There are also other key questions to work for novel crops and better cropping practices for efficient and better harvesting of nutrients from soil, especially P, which gets fixed in soil. The application of P-solubilizing microbes solves this problem to a great extent. The complex ecology of the rhizosphere is a potentially interesting research discipline, having numerous exhilarating defiance and remonstrance, equally with a basic and a strategically applied perception. Recognition of dissimilar techniques related to rhizospheric plant-microbes communication opened novel potential and chances to devise a tactic for increased crop growth and yields. Many researchers explain numerous advantageous effects of PGPM on plant yield and growth by changing the microenvironment of root rhizospheric system including nutrient supply and controlling pathogens. After this the use of micro-biotechnology techniques can be employed to make better PGPM strains by combining many beneficial features in one organism. Finally, one could question that modification of micro-rhizosphere by inoculation of competent PGPM will influence the natural microbes. Whether it will adversely affect the natural population or have synergistic effect is not clear at present. One has to increase the knowledge and information of such microbial interactions and their behavior before we exactly plan to introduce the competent or all in one PGPM in the rhizosphere for sustainable growth of the plant.

References

Adriano DC, Page AL, Elseewi AA, Chang AC, Straughan I (1980) Utilization and disposal of fly ash and other coal residues in terrestrial ecosystems: a review. J Environ Qual 9:333–344

Aroca R, Ruíz LJM (2009) Induction of plant tolerance to semi-arid environments by beneficial soil microorganisms (a review). In: Lichtfouse E (ed) Climate change, intercropping, pest control and beneficial microorganisms, sustainable agriculture reviews 2. Springer, Dordrecht, pp 121–135

Aroca R, Vernieri P, Ruiz LJM (2008) Mycorrhizal and non-mycorrhizal Lactuca sativa plants exhibit contrasting responses to exogenous ABA during drought stress and recovery. J Exp Bot 59:2029–2041

Arzanesh M, Alikhani H, Khavazi K, Rahimian H, Miransari M (2011) Wheat (*Triticum aestivum* L.) growth enhancement by *Azospirillum* sp. under drought stress. World J Microbiol Biotcchnol 27:197–205

Baum C, Hrynkiewicz K, Leinweber P, Meißner R (2006) Heavy-metal mobilization and uptake by mycorrhizal and nonmycorrhizal willows (*Salix dasyclados*). J Plant Nutr Soil Sci 169:516–522

Ben SI, Albacete A, Marti'nez AC, Haouala R, Labidi N, Zribi F, Martinez V, Pe'rez-Alfocea F, Abdelly C (2009) Response of nitrogen fixation in relation to nodule carbohydrate metabolism in Medicago ciliaris lines subjected to salt stress. J Plant Physiol 166:77–88

Bisht S, Pandey P, Sood A, Sharma S, Bisht NS (2010) Biodegradation of naphthalene and anthracene by chemo-tactically active rhizobacteria of *Populus deltoids*. Braz J Microbiol 41:922–930

Bisht S, Pandey P, Kaur G, Aggarwal H, Sood A, Sharma S, Kumar V, Bisht NS (2014) Utilization of endophytic strain *Bacillus* sp. SBER3 for biodegradation of polyaromatic hydrocarbons (PAH) in soil model system. Eur J Soil Biol 60:67–76

Blanco H, Lal R (2008) Principles of soil conservation and management [E-book]. Springer, Dordrecht, pp 1–19

Bot AJ, Nachterhaele RO, Young A (2000) Land resource potential and constraints at regional and country levels. Land and Water Development Division, FAO, UN, Rome

Carvalhais LC, Dennis PG, Fedoseyenko D, Hajirezaei MR, Borriss R, Von WN (2011) Root exudation of sugars, amino acids, and organic acids by maize as affected by nitrogen, phosphorus, potassium, and iron deficiency. J Plant Nutr Soil Sci 174:3–11

Catterson TM, Gulick FA, Resch T (1987) Rethinking forestry strategy in Africa: experiences drawn from USAID activities. Desertification Control Bull 33:31–37

Chen YP, Rekha PD, Arun AB, Shen FT, Lai WA, Young CC (2006) Phosphate solubilizing bacteria from subtropical soil and their tricalcium phosphate solubilizing abilities. Appl Soil Ecol 34:33–41

Cho SM, Kang BR, Han SH, Anderson AJ, Park JY, Lee Y, Cho BH, Yang KY, Ryu CM, Kim YC (2008) 2R,3R-butanediol, a bacterial volatile produced by *Pseudomonas chlororaphis* O6, is involved in induction of systemic tolerance to drought in *Arabidopsis thaliana*. Mol Plant Microbe Interact 21:1067–1075

Cohen A, Bottini R, Piccoli P (2008) *Azospirillum brasilense* Sp. produces ABA in chemically-defined culture medium and increases ABA content in *Arabidopsis* plant. Plant Growth Regul 54:97–103

Cornish K, Zeevaart JAD (1988) Phenotypic expression of wild-type tomato and three wilty mutants in relation to abscisic acid accumulation in roots and leaflets of reciprocal grafts. Plant Physiol 87:190–194

Creus CM, Graziano M, Casanovas EM, Pereyra MA, Simontacchi M, Puntarulo S, Arassi CA, Lamattina L (2005) Nitric oxide is involved in the *Azospirillum brasilense*-induced lateral root formation in tomato. Planta 221(2):297–303

Dobbelaere S, Croonenborghs A, Thys A, Vande Broek A, Vanderleyden J (1999) Phytostimulatory effect of *Azospirillum brasilense* wild type and mutant strains altered in IAA production on wheat. Plant Soil 212:153–162

Doran JW, Linn DM (1994) Microbial ecology of conservation management systems. In: Hatfield JL, Stewart BA (eds) Soil biology: effects on soil quality. (Advances in soil science). Lewis, Boca Raton, pp 1–27

Esechie HA, Al-Saidi A, Al-Khanjari S (2002) Effect of sodium chloride salinity on seedling emergence in chickpea. J Agron Crop Sci 188:155–160

Euliss K, Ho C, Schwab AP, Rock S, Banks MK (2008) Greenhouse and field assessment of phytoremediation for petroleum contaminants in a riparian zone. Bioresour Technol 99:1961–1971

Fomina MA, Alexander IJ, Colpaert JV, Gadd GM (2005) Solubilization of toxic metal minerals and metal tolerance of mycorrhizal fungi. Soil Biol Biochem 37:851–866

Garau ML, Dalmau JL, Felipo MT (1991) Nitrogen mineralization in soil amended with sewage sludge and fly ash. Biol Fert Soils 12:199–204

Ghanem ME, Albacete A, Smigocki AC, Frébort I, Pospíšilová H, Martínez-Andújar C, Acosta M, Sánchez-Bravo J, Lutts S, Dodd IC, Alfocea FP (2011) Root-synthesised cytokinins improve shoot growth and fruit yield in salinised tomato (*Solanum lycopersicum* L.). J Exp Bot 62:125–140

Hammer EC, Nasr H, Pallon J, Olsson PA, Wallander H (2011) Elemental composition of arbuscular mycorrhizal fungi at high salinity. Mycorrhiza 21:117–129

Herrera MMJ, Steinkellner S, Vierheilig H, Ocampo JA, Garcı'a-Garrido JM (2007) Abscisic acid determines arbuscule development and functionality in tomato arbuscular mycorrhiza. New Phytol 175:554–564

Hoekstra FA, Crow JH, Crowe LM, Van Roekel T, Vermeer T (1992) Do phospholipids and sucrose determine membrane phase transitions in dehydrating pollen species? Plant Cell Environ 15:601–606

Jahromi F, Aroca R, Porcel R, Ruiz LJM (2008) Influence of salinity on the *in vitro* development of *Glomus intraradices* and on the *in vivo* physiological and molecular responses of mycorrhizal lettuce plants. Microbial Ecol 55(1):45–53

Jing YD, He ZL, Yang XE (2007) Role of soil rhizobacteria in phytoremediation of heavy metal contaminated soils. J Zhejiang Univ Sci B 8(3):192–207. doi:10.1631/jzus.2007.B0192

Johansson JF, Paul LR, Finlay RD (2004) Microbial interactions in the mycorrhizosphere and their significance for sustainable agriculture. FEMS Microbiol Ecol 48:1–13

Jonsson LM, Nilsson LC, Wardle DA, Zackrisson O (2001) Context dependent effects of ectomycorrhizal species richness on tree seedling productivity. Oikos 93:353–364

Juniper S, Abbott LK (2006) Soil salinity delays germination and limits growth of hyphae from propagules of arbuscular mycorrhizal fungi. Mycorrhiza 16:371–379

Kumar V, Narula N (1999) Solubilization of inorganic phosphates by *Azotobacter chroococcum* mutants and their effect on seed emergence of wheat. Biol Fertil Soil 28:301–305

Kumar V, Singh KP (2001) Enriching vermicompost by nitrogen fixing and phosphate solubilizing bacteria. Bioresour Technol 76:173–175

Kumar V, Aggarwal NK, Singh BP (2000) Performance and persistence of P-solubilizing *Azotobacter chroococcum* in wheat rhizosphere. Folia Microbiol 45(4):342–346

Kumar V, Behl RK, Narula N (2001) Effect of P-solubilizing *Azotobacter chroococcum* on yield traits and their survival in the rhizosphere of wheat genotypes under field conditions. Acta Agron Hungarica 49(2):141–149

Kumar V, Solanki AS, Sharma S (2009) Yield and economics of *Withania somnifera* influenced by dual inoculation of *Azotobacter chroococcum* and *Pseudomonas putida*. Turk J Biol 33:219–223

Kumar V, Singh AS, Sharma S (2011) AM fungi, *A. chroococcum*, yield, nutrient uptake and economics of

Plantago ovata in Indian arid region. Thai J Agric Sci 44(1):53–60

Kumar V, Bisht S, Teotia P, Sharma S, Solanki AS (2013) Interaction between *G. fasciculatum* and *A. chroococcum* for yield, nutrients uptake and cost economy of *Lepidium sativum* in Indian arid region. Thai J Agric Sci 46(1):21–28

Lerat S, Lapointe L, Gutjahr S, Piché Y, Vierheilig H (2003) Carbon partitioning in a split-root system of arbuscular mycorrhizal plants is fungal and plant species dependent. New Phytol 157:589–595

Liu JJ, Wang GH, Jin J, Liu JD, Liu XB (2011) Effects of different concentrations of phosphorus on microbial communities in soybean rhizosphere grown in two types of soils. Ann Microbiol 61:525–534

Mackenzie FT, Mackenzie JA (1995) Our changing earth: an introduction to earth system science and global environmental change. Prentice Hall, Englewood Cliffs, p 580

Manchanda G, Garg N (2008) Salinity and its effects on the functional biology of legumes. Acta Physiol Plantarum 30:595–618

Marques-Ana PG, Rangel-Anto' Nio OSS, Castro-Paula ML (2009) Remediation of heavy metal contaminated soils: phytoremediation as a potentially promising clean-up technology. Crit Rev Environ Sci Technol 39:622–654

Mathur J, Bizzoco RW, Ellis DG, Lipson DA, Poole AW, Levine R, Kelley ST (2007) Effects of abiotic factors on the phylogenetic diversity of bacterial communities in acidic thermal springs. Appl Environ Microbiol 73(8):2612–2623

McGuinness M, Dowling D (2009) Plant-associated bacterial degradation of toxic organic compounds in soil. Int J Environ Res Public Health 6:2226–2247

Meudec A, Poupart N, Dussauze J, Deslandes E (2007) Relationship between heavy fuel oil phytotoxicity and polycyclic aromatic hydrocarbon contamination in *Salicornia fragilis*. Sci Total Environ 381:146–156

Molina FC, Creus CM, Simontacchi M, Puntarulo S, Lamattina L (2008) Aerobic nitric oxide production by *Azospirillum brasilense* Sp245 and its influence on root architecture in tomato. Mol Plant Microbe Interact 21:1001–1009

Munns R, Tester M (2008) Mechanisms of salinity tolerance. Annu Rev Plant Biol 59:651–681

Narula N, Kumar V, Behl RK, Deubal A, Gransee A, Merbach W (2000) Effect of P-solubilizing *Azotobacter chroococcum* on N, P, K uptake in P-responsive wheat genotypes grown under green house conditions. J Plant Nutr Soil Sci 163:393–398

Parrota JA, Knowles OH, Wunderle JM Jr (1997) Development of floristic diversity in 10-year-old restoration forests on a bauxite mined site in Amazonia. For Ecol Manag 99:21–42

Pati SS, Sahu SK (2004) CO_2 evolution and enzyme activities (dehydrogenase, protease and amylase) of fly ash amended soil in the presence and absence of earthworms (*Drawida willsi Michaelsen*) under laboratory conditions. Geoderma 118:289–301

Paul D, Nair S (2008) Stress adaptations in a plant growth promoting rhizobacterium (PGPR) with increasing salinity in the coastal agricultural soils. J Basic Microbiol 48:378–384

Peña-Castro JM, Barrera-Figueroa BE, Fernández-Linares L, Ruiz-Medrano R, Xoconostle-Cázares B (2006) Isolation and identification of up-regulated genes in bermuda grass roots (*Cynodon dactylon* L.) grown under petroleum hydrocarbon stress. Plant Sci 170:724–731

Pichtel JR, Heys JM (1990) Influence of fly ash on soil microbial activity and population. J Environ Qual 19:596–597

Pilon SE (2005) Phytoremediation. Annu Rev Plant Biol 56:15–39

Prakash A, Bisht S, Singh J, Teotia P, Kela R, Kumar V (2014) Biodegradation potential of Petroleum Hydrocarbons (PHCs) by bacteria and mixed bacterial consortium isolated from contaminated sites. Turk J Eng Environ Res 38:31–40 (Accepted)

Rajkumar M, Sandhya S, Prasad MN, Freitas H (2012) Perspectives of plant-associated microbes in heavy metal phytoremediation. Biotechnol Adv 30(6):1562–1574

Ramos JL, Duque E, van Dillewjin P, Daniels C, Krell T, Espinosa-Urgel M, Ramos-González MI, Rodríguez S, Matilla M, Wittich R, Segura A (2010) Removal of hydrocarbons and other related chemicals via the rhizosphere of plants. In: Timmis KN, McGenity TJ, van der Meer JR, de Lorenzo V (eds) Handbook of hydrocarbon and lipid microbiology. Springer, Berlin, pp 2575–2581

Rapp A (1986) Introduction to soil degradation processes in drylands. Clim Chang 9:19–31

Raymond AW, Felix EO (2011) Heavy metals in contaminated soils: A review of sources, chemistry, risks and best available strategies for remediation. ISRN Ecol 20

Read DJ (2002) Towards ecological relevance-progress and pitfalls in the path towards an understanding of mycorrhizal functions in nature. In: van der Heijden MGA, Sanders IR (eds) Mycorrhizal ecology. Springer, Berlin, pp 3–29

Richardson AE (2001) Prospects for using soil microorganisms to improve the acquisition of phosphorus by plants. Aust J Plant Physiol 28:897–906

Rodríguez H, Fraga R (1999) Phosphate solubilizing bacteria and their role in plant growth promotion. Biotechnol Adv 17:319–339

Romero AMR, Jurado O, Cuartero J (2006) Silicon alleviates the deleterious salt effect on tomato plant growth by improving plant water status. J Plant Physiol 163:847–855

Samet JM, Dominici F, Curriero FC, Coursac I, Zeger SL (2000) Fine particulate air pollution and mortality in 20 U.S. Cities, 1987–1994. N Engl J Med 343:1742–1749

Sandaa RA, Torsvik V, Enger'e (2001) Influence of long-term heavy-metal contamination on microbial communities in soil. Soil Biol Biochem 33:287–295

Schubert A, Wyss P, Wiekman A (1992) Occurrence of trehalose in vesicular-arbuscular mycorrhizal fungi and in mycorrhizal roots. J Plant Physiol 140:41–45

Schützendübel A, Polle A (2002) Plant responses to abiotic stresses: heavy metal-induced oxidative stress and protection by mycorrhization. J Exp Bot 53:1351–1365

Sharma NK, Sharma S, Kumar V, Poonam A (2006) Species specific *Rhizobium-Albizia lebbeck* interaction. Ind J For 29(2):175–179

Sinha RK (1997) Towards ecological regeneration, biodiversity conservation and environmental restoration of the Thar Desert ecosystem in India. Desertification Control Bull 30:28–36

Solanki AS, Kumar V, Sharma S (2011) AM fungi, *A. chroococcum*, yield, nutrient uptake and economics of *Chlorophytum borivilianum* in Indian arid region. J Agric Tech 7(4):983–991

Tiittanen P, Timonen KL, Ruuskanen J, Mirme A, Pekkanen J (1999) Fine particulate air pollution, resuspended road dust and respiratory health among symptomatic children. Eur Respir J 13:266–273

Timmusk S, Wagner EG (1999) The plant-growth-promoting rhizobacterium *Paenibacillus polymyxa* induces changes in *Arabidopsis thaliana* gene expression: a possible connection between biotic and abiotic stress responses. Mol Plant Microbe Interact 12:951–959

Toy TJ, Foster GR, Renard KG (2002) Soil erosion, processes, prediction, measurement, and control [E-book]. Wiley, New York, pp 25–43

Tripathi AK, Nagarajan T, Verma SC, Le Rudulier D (2002) Inhibition of biosynthesis and activity of nitrogenase in *Azospirillum brasilense* Sp7 under salinity stress. Curr Microbiol 44(5):363–367

Turco RF, Sadowsky MJ (1995) The microflora of bioremediation, In: Skipper HD, Turco RF (eds) Bioremediation: science and application. Soil Science Society of America special publication 43, Madison, pp 87–103

United States Department of Agriculture (USDA), Natural Resources Conservation Service, A Soil profile. Retrieved 21 Feb 2014, from United State Department of Agriculture, Web Site: http://soils.usda.gov/education/resources/k_12/lessons/profile/

Upadhyay SK, Maurya SK, Singh DP (2012) Salinity tolerance in free living plant growth promoting Rhizobacteria. Ind J Sci Res 3(2):73–78

Vessey JK (2003) Plant growth promoting rhizobacteria as biofertilizers. Plant Soil 255:571–586

Weber FR (1986) Reforestation in Arid lands. VITA Publishing, Riverside

Wenzel WW (2009) Rhizosphere processes and management in plant-assisted bioremediation (phytoremediation) of soils. Plant Soil 321:385–408

Yamaguchi T, Blumwald E (2005) Developing salt-tolerant crop plants: challenges and opportunities. Trends Plant Sci 10:615–620

Yang CH, Crowley DE (2000) Rhizosphere microbial community structure in relation to root location and plant iron nutritional status. Appl Environ Microbiol 66:345–351

Plant-Microbe Partnerships for Enhanced Biodegradation of Polychlorinated Biphenyls

4

Prameela Jha and Prabhat Nath Jha

Abstract

Our modern civilization rests on synthetic organic compounds. Some of these compounds which have become indispensable for human society are termed as persistent organic pollutants (POPs). They enter into the food chain where they bioaccumulate in fat tissues of organisms at various trophic levels and usually biomagnify the physiological effects at higher levels resulting in severe health hazards. Civil and mechanical engineering remediation methods deployed to clean such environmental contaminants have serious shortcomings of being expensive and environmentally invasive. Thus, there is an urgent need to search for alternative strategy(ies) which does not suffer from abovementioned shortcomings. The past decade has witnessed a momentum in the use of plants and its associated microflora partnerships as an alternative option. Plant and associated bacteria interact synergistically to lower down the levels of contaminants present in the immediate vicinity of root systems. Plants modulate root secretions and exudations to orchestrate recruitment of catabolically active bacteria in their rhizosphere. This recruitment in turn biodegrades contaminant(s) present in the rhizospheric soil by bacterial communities which subsequently reduces environmental stress on the plant. Thus, it is highly recommended to explore novel plant microbe pairs naturally operating in organic contaminated ecosystems and study the members of microbial communities to get more insight into a phenomenon of phytoremediation so that the technology may be optimized and extended from lab to land. The present review highlights the synergistic effects of plant-microbe interactions in the phytoremediation of organic pollutants with special reference to polychlorinated biphenyls (PCBs). Also, the state-of-the-art techniques used to study microbial diversity within PCB contaminated environments are discussed.

P. Jha (✉) • P.N. Jha
Environmental & Microbial Biotechnology
Laboratory, Department of Biological Sciences,
Birla Institute of Technology and Science,
Pilani 333031, India
e-mail: drprameelajha@gmail.com

N.K. Arora (ed.), *Plant Microbes Symbiosis: Applied Facets*,
DOI 10.1007/978-81-322-2068-8_4, © Springer India 2015

Introduction

Comfort and convenience are the words that best describe modern-day human society. But to achieve comfort and convenience in every walk of life, we are synthesizing certain organic compounds which are not suitable for both humans and the environment. A wide variety of toxic chemicals including heavy metals and POPs are being indiscriminately dispensed into the environment. POPs are chemical substances that persist in the environment, bioaccumulate through the food web, and pose a risk of causing adverse effects to human health and the environment. These compounds have the capacity to travel long ranges from the source of origin and reach the places where they were never used or synthesized, thus, making them more dangerous. The Stockholm Convention on POPs, under the auspices of the United Nations Environment Programme (UNEP), an international agreement enforced in 2004 (Stockholm convention 2004), enlisted 12 organic compounds as POPS of major concern. These 12 POPs include PCBs, 9 chlorinated organic pesticides (aldrin, chlordane, dichlorodiphenyltrichloroethane (DDT), dieldrin, endrin, mirex, heptachlor, hexachlorobenzene, and toxaphene), and dioxins and furans. Other organic pollutants of major concern are industrial solvents, total petroleum hydrocarbons (TPH), polyaromatic hydrocarbons (PAHs), halogenated hydrocarbons, volatile organic compounds (VOCs), and brominated flame retardants (Table 4.1). The health hazards of these organic pollutants include skin irritation, endocrine disruption, increased risks of developing cancer and developmental disabilities, and increased incidence of Parkinson's disease (Falck et al. 1992; Wolff et al. 1993; Faroon et al. 2001; Flower et al. 2004; Ascherio et al. 2006).

For public health safety and environment protection, various measures have been taken with concomitant investments of resources by government in most countries, but remediation of contaminated sites still remains a serious problem to be addressed. Various physicochemical processes including soil washing, electrokinetic methods, chemical reduction or oxidation of contaminant, incinerations, etc., appear to be expensive, environment invasive, and many times nonspecific, producing secondary contaminants (Pandey et al. 2009; Weyens et al. 2009, 2010a). Thus, there is a constant thrust to develop alternative strategies for remediating contaminated sites. During the last two decades, phytoremediation has emerged as a potential tool for the cleanup of polluted environment. Phytoremediation is the use of plants to remediate polluted soils which is an eco-friendly and cost-effective technology and currently receiving considerable global attention (Glick 2010; Yousaf et al. 2011; Khan et al. 2013). It is an *in situ* solar-powered remediation technology, requiring minimal site disturbance and maintenance resulting in high public acceptance (Weyens et al. 2009, 2010a). Since conventional remediation options currently available are frequently expensive and environmentally invasive, phytoremediation emerges out to be a valuable alternative, especially for the treatment of large contaminated areas with diffused pollution (Weyens et al. 2010a, b; Nesterenko-Malkovskaya et al. 2012).

Rhizoremediation: An Edge Over Phytoremediation

Plant uptake is the first and crucial step in whole plant metabolism of organics. In case of constant plant and environmental features, the lipophilicity of the compound, expressed as its octanol-water partition coefficient (K_{ow}), is the determining factor for root entry and translocation. Organic contaminants with a log K_{ow} <1 are considered to be very water soluble, and plant roots do not generally accumulate them at a rate surpassing passive influx into the transpiration stream (Cunningham and Berti 1993). Contaminants with a log K_{ow} >3.5 show high sorption to the roots but slow or no translocation to the stems and leaves (Trapp et al. 2001). However, plants readily take up organic contaminants with a log K_{ow} between 0.5 and 3.5, as well as weak electrolytes. After plant uptake, the organic compound may be metabolized and/

Table 4.1 Persistent organic pollutants of major concern enlisted by the Stockholm Convention 2004

POP	Use	Structure
Aldrin	Crop insecticide (corn, cotton)	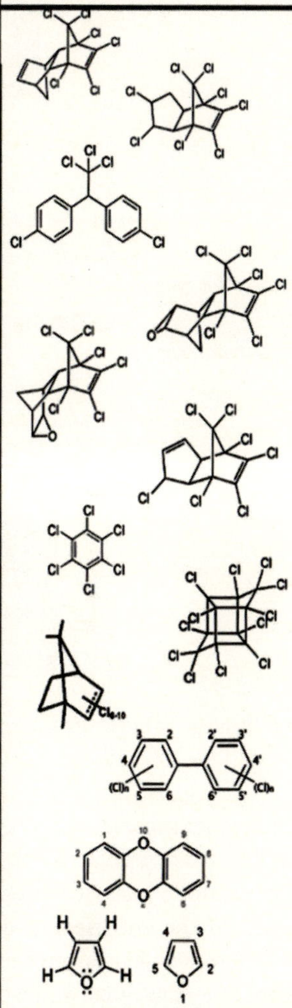
Chlordane	Crop insecticide (vegetables, citrus, cotton, potatoes)	
DDT (dichloro-diphenyl-trichloroethane)	**Crop insecticide (cotton)**	
Dieldrin	Crop insecticide (cotton, corn)	
Endrin	Crop insecticide (cotton, grains)	
Heptachlor	Insecticide (termites and soil insects)	
Hexachlorobenzene	Fungicide for seed treatment	
Mirex	Insecticide (termites, fire ants)	
Toxaphene	Insecticide (livestock, crops)	
PCBs	**Industrial chemical (paint and plastic additive)**	
Dioxins	**Unintentionally produced during combustion**	
Furans	Unintentionally produced during combustion	

or released into the atmosphere via evapotranspiration through the stem and/or leaves. Although plants often metabolize or sequester organics, they have significant inabilities in two ways, as being photoautotrophic; plants do not rely on organic molecules as a source of energy or carbon. By consequence, unlike microorganisms, during evolution plants were not under selective pressure to develop the capacity to degrade chemically recalcitrant molecules, leading to a much more limited spectrum of chemical structures that they can metabolize (Burken 2004). To avoid accumulation and potential toxicity to sensitive organelles, plant metabolizes organic molecules (other than photosynthates), converts them into more water-soluble forms, and sequesters into vacuole or cell walls. The process is described as the "Green Liver Model" (Sandermann 1992). By contrast, microbial metabolism often ends with the organics being converted into CO_2, water, and cellular biomass. Therefore, in order to obtain a more efficient degradation of organic compounds, plants depend on their associated microorganisms. Plants themselves have a positive effect on the microbial degradation of

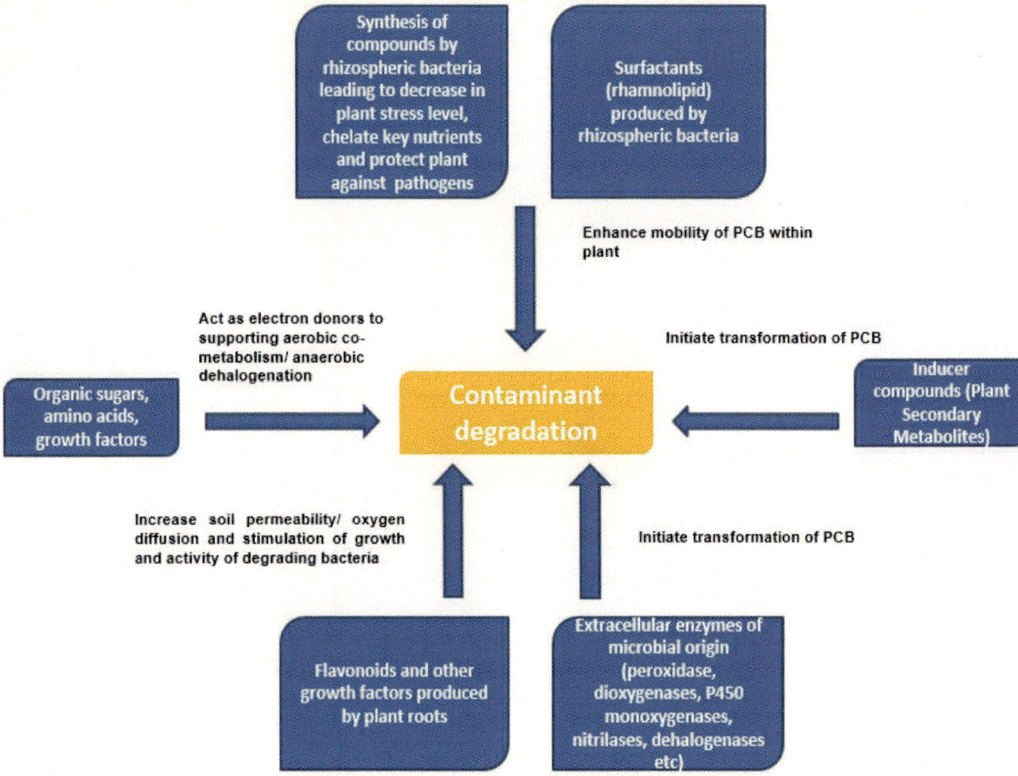

Fig. 4.1 Synergistic effect of plant and rhizospheric microbes during rhizoremediation and phytoremediation

organic contaminants (Crowley et al. 1997). This increased degradation potential is the result of higher microbial densities and metabolic activities in the rhizosphere due to microbial growth on root exudates and cell debris originating from the plant roots. Some of the synergistic activities are summarized in Fig. 4.1. Moreover, dense populations of diverse heterotrophic microorganisms are living in the rhizosphere, phyllosphere, and inside the plant (endophytes). These microbial associations increase the capacity for a stepwise transformation of organic contaminants by consortia and provide a habitat that is conducive to genetic exchange and gene rearrangements. The emerging picture suggests that plants draw pollutants, including PAHs, into their rhizosphere to varying extents via the transpiration stream (Harvey et al. 2002). Subsequent degradation can occur in the plant itself, or in the rhizosphere, or both. However, plants readily absorb compounds with a log K_{ow} between 0.5 and 3.5. These compounds seem to enter the xylem faster so that rhizosphere microflora cannot degrade them, even if the latter is enriched with bacteria capable of degrading the compound (Trapp et al. 2000). Therefore, after this class of compounds is taken up by the plants, endophytes seem to be especially suitable for the degradation of these compounds. Endophytic bacteria can be defined as bacteria colonizing the internal tissues of plants without showing symptoms of infection or negative effects on their host, and of the nearly 300,000 plant species that exist on the earth, each individual plant is host to one or more endophytes (Strobel et al. 2004). The association of endophytes with their host is varied and complex, and we have just started understanding these interactions. Several plant growth promotory effects of endophytes on their host ranging from N_2 fixation, biocontrol against phytopathogens, phytohormone production, etc., are well documented (Parmeela and Johri 2004; Compant et al. 2010; Gupta et al. 2013). The combined use of plants and endophytic bacteria to clean up environmental pollutants has gained momentum in the past years as they have high potential to promote plant growth and contribute to degradation of pollutants (Germaine et al. 2006; Compant et al. 2010; Khan et al. 2013).

Large-scale applications of phytoremediation still face a number of obstacles, including the levels of contaminants (being toxic for the organisms involved in remediation), bioavailable fraction of the contaminants (being too low), and, in some cases, evapotranspiration of volatile organic pollutants from soil or groundwater to the atmosphere. Exclusively derived processes by plants alone (phytoremediation) are time-consuming. Further, high levels of pollutants pose toxicity to the remediating plants. This process can be accelerated by exploring the plant-microbe partnership, which would improve the plant growth by facilitating the sequestration of contaminants modified by microbial communities in rhizosphere, which is more precisely called as plant-assisted rhizoremediation (Gerhardt et al. 2009; Afzal et al. 2011; Teng et al. 2011; Yousaf et al. 2011). Plant root systems efficiently absorb surrounding water, hence bringing dissolved pollutants into the rhizosphere zone for degradation. Contaminant-degrading bacteria can be conveniently included in plant growth-promoting rhizosphere bacteria as they colonize root vicinity and utilize root exudates as sole carbon and energy source facilitating concomitant degradation of organic contaminants (Singer et al. 2003). Also, plants positively affect the size of colonizing populations and exhibit certain preferences towards particular group/genera of microbial community. This preference is usually induced by shifts in chemical composition of root exudations (Jha et al. 2014). This strategy is referred as ex planta means of phytoremediation. Interestingly, plant roots not only secrete enzymes but also improve the degrading ability of microorganisms present in rhizo-/endorhizosphere (Gianfreda and Rao 2004; Kuiper et al. 2004).

Aminocyclopropane-1-Carboxylic Acid (ACC) Deaminase-Producing Bacteria and Phytoremediation

Roots are the most important part of a plant. The extensive root system not only provides anchorage but also helps in absorption of essential nutrients from soil so that the plant may flourish well. Sometimes roots play a larger-than-life role by hyperaccumulating metals and other organic contaminants from rhizosphere, thus, participating in phytoremediation. However, the stress induced due to the presence of contaminants exerts negative effect on plants by elevating ethylene production, a phytohormone known to reduce root growth, thus, posing a major limitation to phytoremediation. The phytohormone ethylene (C_2H_4) has a central role in modulating the growth and cellular metabolism of plants. This is believed to be involved in disease resistance, biotic/abiotic stress tolerance, plant-microbe partnerships, and plant nutrient cycle. The enhanced level of ethylene is usually plant's response to stress induced by the presence of metals, organic and inorganic chemicals, damage (mainly by insects and nematodes) or mechanical wounds, and most importantly, presence of phytopathogens (as bacteria and fungi) (Abeles et al. 1992). The synthesis of "stress ethylene" occurs in two peaks (Glick et al. 2007; van Loon et al. 2006). The first peak is relatively small and consumes the existing 1-aminocyclopropane-1-carboxylic acid (ACC), the immediate precursor of ethylene. This consumption of ACC in stressed plants is known to initiate the transcription of spectrum of genes responsible for encoding plant defensive proteins (Robinson et al. 2001). The second peak of ethylene in plant occurs by the synthesis of additional ACC in plant, and it is known to be detrimental to the plant. The overproduction of ethylene can cause the inhibition of root elongation and lateral root growth, and root hair formation consequently reduced phytoremediation efficiency. However, some bacteria are capable of alleviating the stress-mediated impact on plants by enzymatic hydrolysis of ACC. ACC is involved in the biosynthetic pathway of ethylene, as an intermediate in the conversion of methionine to ethylene following biosynthetic sequence: methionine-S-adenosylmethionine (SAM)–ACC– C_2H_4 (Adam and Young 1979). In general, ACC is exuded from plant roots or seeds and then taken up by the ACC-utilizing bacteria before being oxidized by the plant ACC oxidase and cleaved by ACC deaminase to α-ketobutyrate (αKB) and ammonia. The bacteria utilize the ammonia evolved from ACC as a sole nitrogen source and thereby decrease ACC within the plant with the concomitant reduction of plant ethylene. The decreased ethylene levels in plants hosting ACC-utilizing bacteria derive benefits by stress alleviation and enhanced plant productivity. Since SAM is converted by ACC synthase to ACC, the ACC synthase protein seems to play a main controlling role in the ethylene biosynthesis pathway. In the absence of ACC-utilizing bacteria, ACC is oxidized by ACC oxidase to form ethylene, cyanide, and CO_2.

Plant: Microbe Partnerships for Enhanced Phytoremediation of Organic Pollutants

In order to understand, optimize, and monitor bio/rhizoremediation, a process occurring naturally in soil ecosystems, it is most important to explore and understand specific plant-microbe pairs which carry out these degradations efficiently in nature. Several plant species including *Salix caprea, Willow, Birch, Pinus, Poplus, Solanum nigrum, Nicotiana tabacum, Armoracia rusticana,* among others, are well known to harbor bacterial populations capable of degrading organic compounds especially PCBs (Leigh et al. 2007; Ionescu et al. 2009; Uhlik et al. 2012). Siciliano et al. (2001) showed that plants growing in xenobiotic contaminated soils have natural ability to recruit endophytes having catabolic genes (Siciliano et al. 2001). In a groundbreaking study, Leigh and coworkers (2006) investigated abundances, identities, and degradation capabilities of microbial populations associated with mature tree species (*Pinus nigra, Salix caprea*) naturally growing in PCB-contaminated sites and identification of plants that are capable of preferentially promoting growth of PCB-degrading bacteria in their rhizosphere. Authors found that plant species like *P. nigra* and *S. caprea* were having good rhizoremediation potentials as they were harboring increased number of PCB degraders in their root zones (Leigh et al. 2006). Both plants produced terpenoids, tannins, some phenolic compounds, among others, bearing marked analogy with aromatic pollutants that serve as a substrate and induced PCB degradation pathway (Singer et al. 2003). In another study conducted by Ionescu and coworkers (2009), four different plant species, viz., *S. caprea, S. nigrum, N. tabacum*, and *A. rusticana* were used to study the influence of plant growth on total microbial counts and plant-mediated effects on total counts of bacteria. Authors found that the most efficient plant species promoting the growth of PCB degraders were *N. tabacum* and *A. rusticana*. It is also proven experimentally that mixed cropping helps in inducing enzyme systems involved in the degradation of pollutants. Li et al. (2013) conducted an experiment to see the effect of mixed plantation on PCB removal mediated by two well-known plant species (alfalfa and tall fescue). Authors concluded that even mixed plantation contributes to significant increase in phytoremediation activity of associated microbial populations. Most common PCB degraders identified were *Pseudomonas mendocina, Sphingobacterium mizutae, Agrobacterium radiobacter, Ochrobactrum anthropi,* and *Burkholderia cepacia*. Other plants as *Festuca arundinacea, Morus* sp., *Panicum virgatum (switchgrass), Cucurbita pepo, Arabidopsis thaliana, Lolium multiflorum*, etc., are also well known as home to several PCB and other aromatic hydrocarbon-degrading microorganisms (Li et al. 2013). Some commonly used plant and bacteria combinations used successfully by various authors has been complied in (Table 4.2).

Rhizoengineering: Manipulating Root Exudations

Plants actively participate in recruiting or exhibiting preferences towards a specialized group of population to colonize its rhizosphere. This microbial community is often referred as the

Table 4.2 Plant-microbe combinations exploited in PCB degradation

Plant used	Microbes used	Origin	Reference
Medicago sativa	*Pseudomonas*	Rhizosphere	Ionescu et al. (2009)
	Rhodococcus		
	Rhizobium		
N. tabacum	*Achromobacter* sp.	Rhizosphere	Ionescu et al. (2009)
S. caprea	*Sphingobacterium mizutae*	Rhizosphere	Ionescu et al. (2009)
	Burkholderia cepacia		
A. rusticana	*Bacillus pumilus*	Rhizosphere	Ionescu et al. (2009)
S. nigrum	*Pseudomonas mendocina*	Rhizosphere	Ionescu et al. (2009)
	P. fluorescens		
Populus	*Burkholderia cepacia*	Endophytic	Taghavi et al. (2005)
Triticum spp.	*Herbaspirillum* sp. K1	Endophytic	–
P. nigra	*Microbacterium oxydans* type strain	Rhizosphere	Siciliano and Germida (1998)
S. caprea	*Achromobacter* sp.	Rhizosphere	Leigh et al. (2006)
Fraxinus excelsior	*Rhodococcus erythreus* type strain	Rhizosphere root zone	Schell (1985)
Betula pendula	*R. erythreus* type strain	Rhizosphere root zone	Schell (1985)
Robinia pseudoacacia	*Rhodococcus* sp., *Arthrobacter oxydans*	Rhizosphere root zone	Schell (1985)

second genome of plant (Berendsen et al. 2012). The vast spectrum of structural and functional diversities of plant root exudates and secondary plant metabolites (SPMs) is an eternal source to an interdisciplinary researcher to explore the dynamic microcosm of the rhizosphere. Hence, it has been envisaged as an ideal site to modify microbial populations to suit various applications in the soil. This manipulation of plant root secretions and associated microbes to improve plant health and productivity is termed as rhizoengineering (O'Connell et al. 1996). Rhizoengineering exploits the strategies based on favoring the growth of the targeted microbes that possess the ability to metabolize exotic nutrients exuded by plants (Lugtenberg et al. 2001). A nutritional bias or favoritism is created, and population of bacteria can be further enriched which can preferentially utilize the given nutrients. Savka and Farrand (1997) published an early report to achieve success in rhizoengineering based on favorably partitioning the exotic nutrients, opines, which were produced by the transgenic plants leading to improved and competitive growth of the metabolizing strains in comparison with the microbes unable to metabolize opines (opine concept). This strategy provides selective nutritional advantage to *Agrobacterium* and other selective bacteria considering genetic exchanges (horizontal gene transfer). In yet another approach, various species of *Pseudomonas* selectively using salicylate as carbon source (Colbert et al. 1993) and detergents (Igepal Co-720 and dioctyl sulfosuccinate) from maize rhizosphere had been isolated. Flavonoids, most studied as chemicals involved in molecular cross talk between plant roots and specific microorganisms, are known to induce transcription of nodulation genes in nitrogen-fixing rhizobia (Brencic and Winans 2005). *Rhizobium* sp. and *Bradyrhizobium* sp. transformed the flavonoids quercetin, daidzein, and genistein through novel biochemical pathways in contrast to modes of degradation favored by other rhizobacteria (Toussaint et al. 2012). Flavonoids, naringenin, and apigenin have been shown to support growth of PCB-degrading bacteria (Fletcher and Hegde 1995). Plant production of substrates that cause nutritional bias on the one hand and enzyme induction for the biodegradation of xenobiotic compounds on the other offers intriguing and very attractive opportunities for rhizosphere manipulations (rhizoengineering) for enhancing rhizodegradation of xenobiotic contaminants beyond the inherent capabilities of the plant-microbe system. Further researches focused on profiling of root exudates

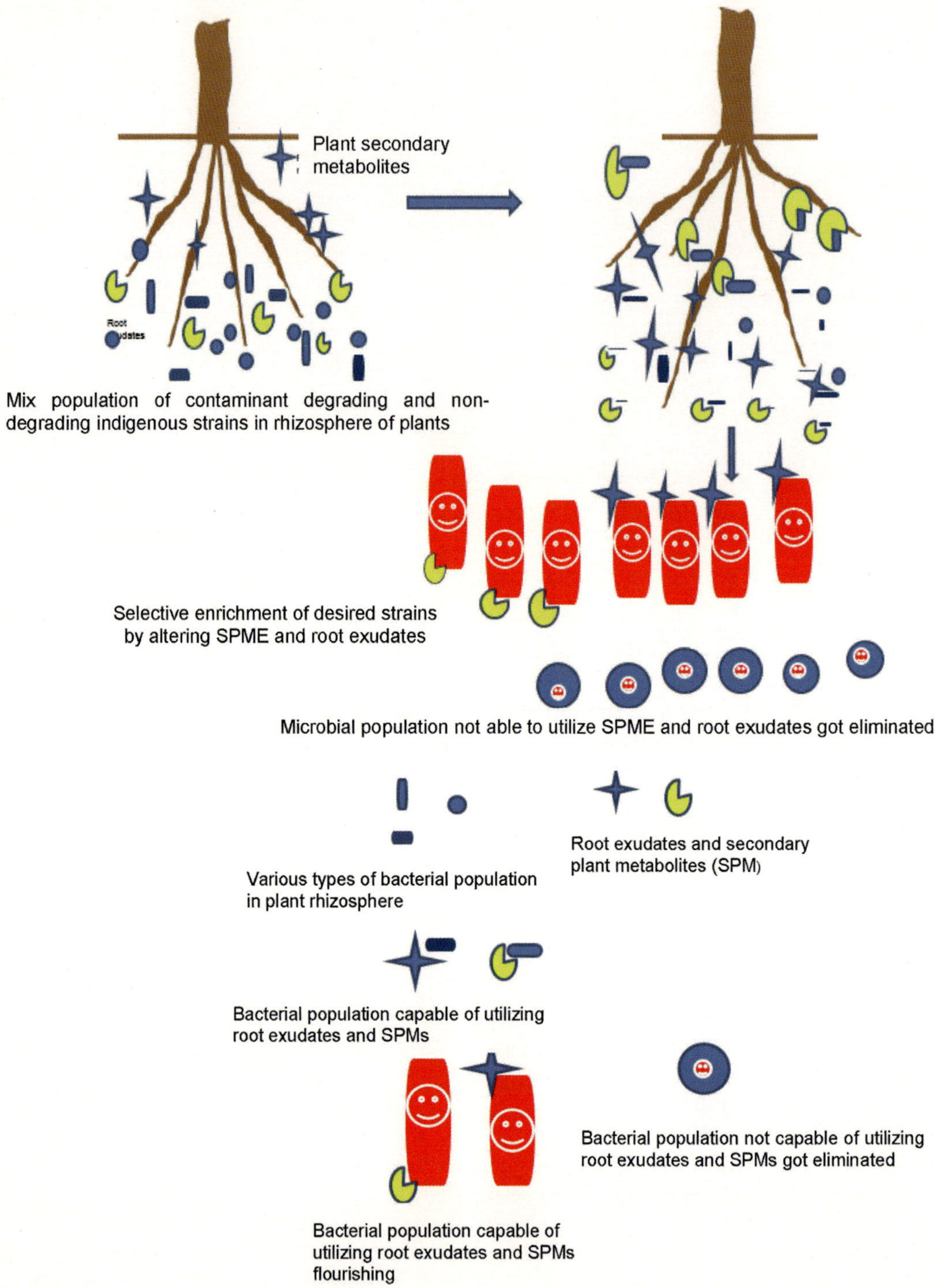

Fig. 4.2 Role of root exudates and SPMs in contaminant (polychlorinated biphenyls, polyaromatic hydrocarbons) degradation

to identify compounds that may be exploited to create nutritional bias for enhancing rhizodegradation of persistent xenobiotic contaminants such as PCB have immense potential (Narasimhan et al. 2003). A schematic representation of rhizoengineering is demonstrated in Fig. 4.2.

Approaches to Study Diversity and Activity of Microbes in Contaminated Soil

The advents of molecular biology techniques bypassing cultivation steps have enabled us to generate information on the total microbial composition at a particular site and time (Table 4.3). These molecular phylogenetic techniques are based on isolating DNA directly from environmental samples as soil, water, sediments, etc. However, there are certain advantages and disadvantages associated with each culture-dependent and culture-independent techniques. Few important techniques currently being used are discussed in brief here.

Biochemical Identification

It represents one of the oldest means of microbial identification and characterization. Several commercial kits are available in facilitating easy and rapid identification of pure cultures on the basis of phenotypes (Busse et al. 1996). NEFERM test 24, API 20, BIOLOG plates, among others, are most routinely used in microbiological laboratories. However, identification based on biochemical tests suffers major drawbacks of being labor intensive, requirement for specific consumables, and having lengthy processes.

Whole-Cell MALDI-TOF MS

This is relatively a recent and revolutionary technique, providing accurate measurement of molecular weight of a compound and giving information on molecules and its structure. Difference in mobility of current carriers with different molecular weight under electromagnetic field is the main principle for this technique (Lay 2000). Fragments of proteins, higher molecular weight fragments of lipid oligosaccharide components of the cell wall, serve as ionized molecules and biomarkers conserved within a particular group of microorganisms and, hence, allow identification, easy grouping, and characterization. Kurzawova et al. (2012) successfully used the technique to identify bacterial species based on the composition of ribosomal proteins in bacterial cells isolated from two plant species *N. tabacum* and *S nigrum*, well known to house PCB degraders. The technique is quick and requires less effort in sample preparation. Relatively low sample size requirement makes it popular among research fraternity (Krishnamurthy and Ross 1996). In a recent variation of the technique, whole-cell instead of extracted proteins or other molecules from environmental samples is used. Koubek et al. (2012) in an interesting study compared three different methods, viz., NEFERM test 24, MALDI-TOF MS, and sequencing of 16S rRNA for taxonomic identification of bacterial isolates recovered from sediments contaminated with PCB. Authors found that each method has its own limitation and advantage for resolution. However, MALDI-TOF MS appeared as the most reliable method with immense potential for quick screening of clinical and environmental samples (Koubek et al. 2012). Restriction patterns of PCR-amplified fragments of particular gene can also be analyzed with MALDI-TOF MS (Taranenko et al. 2002).

Genetic Fingerprinting Techniques

Genetic fingerprinting techniques provide a pattern of genetic diversity in a microbial community. There are several fingerprinting techniques including denaturing gradient gel electrophoresis (DGGE)/thermal gradient gel electrophoresis (TGGE), terminal restriction fragment length polymorphism (T-RFLP), amplified ribosomal DNA restriction analysis (ARDRA), DNA sequencing, and real-time PCR among others. A genetic fingerprinting technique is based on

Table 4.3 Tools and techniques used to identify microorganisms metabolizing various organic pollutants

Technique	Environmental sample	Target compound	Microorganism recovered	Reference
Biochemical tests	Sediment	PCB	*P. fluorescens*	Dercova et al. (2008)
			P. aeuroginosa	
			Ochrobactrum anthropi	
			A. radiobacter	
Mass spectrometry (MALDI-TOF)	Horseradish rhizosphere soil	Biphenyl	*Rhizobium radiobacter*	Uhlik et al. (2011)
			Arthrobacter sp.	
			Achromobacter sp.	
			Rhodococcus sp.	
			Serratia fonticola	
16S rDNA sequencing	Alfalfa and tall fescue rhizosphere soil	PCB	*Burkholderia* sp.	Li et al. (2013)
			Pseudomonas sp.	
			Comamonas testosteroni	
DGGE/TGGE	Soil	PCB	*Flexibacter* sp.	Tang et al. (2013)
			Bdellovibrio sp.	
			Bacillus niabensis *Bradyrhizobium elkanii*	
			Sphingomonas sp.	
T-RFLP	Soil	Chlorobiphenyl	*Hydrogenophaga*	Uhlik et al. (2013)
			Pseudoxanthomonas	
			Azoarcus	
			Burkholderia	
SIP	Soil	Biphenyl	*Burkholderia*	Uhlik et al. (2012)
			Pseudomonas	
			Rhodanobacter	
SIP	Soil	Benzoate	*Rhodanobacter*	Uhlik et al. (2012)
			Pseudomonas	
			Azotobacter	
SIP	Soil	Naphthalene	*Pseudomonas*	Uhlik et al. (2012)
			Rhodanobacter	
			Pandoraea	
SIP	Soil	Chlorobiphenyl		Uhlik et al. (2013)

deducing polymorphism in total or a part of genetic material. DGGE is based on the identification of differences in base sequences even if the size of DNA fragments is the same. The principle behind the technique is decreased electrophoretic mobility of DNA in polyacrylamide gels containing a linear gradient of DNA denaturants (DGGE) or a linear temperature gradient (TGGE). In both DGGE and TGGE, DNA is PCR-amplified with primers targeting part of the studied gene, mostly 16S rRNA gene or some specific catabolic activity gene. The 5′ of the forward primer is attached with 35–40 base pair GC clamp which facilitates a certain portion of the DNA to remain double stranded during electrophoresis. In the following denaturation step, hydrogen bonds are disrupted while holding together double strands of DNA (Marsh and Nakatsu 2007). Hence, the separation of DNA occurs proportional to G + C content and distribution. Watanabe and coworkers beautifully exploited the technique to detect and characterize dominant phenol-degrading bacteria in activated sludge (Watanabe et al. 1998). In a similar technique, single-strand conformational polymorphism (SSCP), the separation of single-stranded

DNA is achieved on the polyacrylamide gel and differences in mobility results due to differences in the folded secondary structure. Here no denaturants are used; hence, DNA molecules of equal size fold and hence move differently according to the DNA base sequences (Schiwieger and Tebbe 1998). Ribosomal intergenic spacer analysis (RISA) is yet another technique being exploited in microbial community analysis of a contaminated site. The heterogeneity of length and sequences of tRNAs encoded in intergenic spacer region between 16S and 23S rRNA genes is studied for differentiation between bacterial strains and closely related species. Again, in RISA, the intergenic spacer region is PCR-amplified, denatured, and separated on polyacrylamide gel. The sequence polymorphisms are detected by silver stain or in an advanced version, i.e., automated ribosomal intergenic spacer analysis (ARISA), and the forward primer is fluorescently labeled and is automatically detected, which increases the sensitivity of the method. In an interesting study, Watts and coworkers (2001) while studying PCB-dechlorinating communities in enrichment culture did comparative analysis among three molecular techniques ARDRA, DGGE, and T-RFLP. Authors found that the methods have different biases as the same enrichment culture gave different clone frequencies (ARDRA), band intensities (DGGE), and peak heights (T-RFLP). However, all the techniques proved effective in qualitative analysis and identification of organism (Watts et al. 2001). Also, authors indicated that of all the techniques DGGE appeared to be rapid and the most effective tool for studying microorganisms in highly enriched cultures.

Stable Isotope Probing (SIP)

This is a relatively new technique that relates microbial diversity to the metabolic activity, initially reported by Boschker et al. (1998) to investigate oxidation of greenhouse gas methane in freshwater sediment (Boschker et al. 1998). There exist several variations including DNA-SIP, RNA-SIP, SIP of amino sugars, and SIP of methyl esters of fatty acids (PLFA-SIP). SIP directly allows tracking of stable isotope (C^{13}) from an environmental sample in order to probe which microbes are functionally active in utilizing a given pollutant. This is done usually through incubating a sample with a heavier isotope of carbon-, nitrogen-, or oxygen-labeled compound. This compound is usually a substrate for that community. The labeled compound if utilized is inserted into the cellular components after which phylogenetic markers as DNA, RNA, phospholipid fatty acids, or proteins can be extracted and analyzed. Uhlik et al. (2012) in an interesting study investigated shifts in bacterial community composition induced by naringenin, caffeic acid, and limonene, due to their degradative capabilities on long-term PCB-contaminated soil. They coupled pyrosequencing of 16s rRNA gene tag-encoded amplicons and DNA-stable isotope probing (SIP) to analyze and identify populations actively involved in 4-chlorobiphenyl catabolism. Authors concluded that out of three PSMs used, application of naringin showed efficient degradation when compared to the control. Also, they identified that *Hydrogenophaga*, *Terrimonas*, *Paucimonas*, and *Pseudorhodoferax* were the most dominant genera.

Radajewski et al. (2000) enhanced the phylogenetic resolution of SIP by demonstrating that stable isotope-labeled DNA could be used to identify individual members of mixed microbial communities. Since isotopic enrichment increases buoyant density of DNA, subsequent density centrifugation in CsCl gradients is exploited to separate "heavy" (labeled) from "natural" (unlabeled) DNA. As a final step 16S rDNA clone libraries are constructed from "heavy" DNA and sequenced to obtain the identity of organisms assimilating the given ^{13}C-labeled substrate used. Another major variation of the technique is RNA-SIP where SSU (small subunit) rRNA molecule can be used (Manefield et al. 2002). Huang et al. (2009) successfully used mRNA-SIP to recover the key phylogenetics and functional RNAs and concluded that RNA-SIP can be used for studying transcriptomes together with taxonomic identities (Huang et al. 2009). However, there are certain serious shortcomings to the tool, being low mRNA yields from environmental samples, difficulties in separating it from other types of RNAs, and its relative instability.

Biosensors

Biosensors can be defined as a coupling of any biological system with microelectronic or any other device which facilitates the rapid, accurate, and detection of very low levels of contaminants in body fluids, water, or soil. Biosensors based on microorganisms are emerging as candidates of choice being easy to maintain, cost-effective, high population sizes, and accurate in activity and functioning. Additionally, they can be tailored to meet the specific requirements when monitoring specific contaminants. Casavant et al. (2003) used a *gfp*-labeled *Pseudomonas fluorescens* A506 to detect low concentrations of toluene (0.2 mm) and trichloroethylene in the rhizosphere and observed an increased *gfp* expression in the root-colonizing biosensor population when the plant rhizosphere was exposed to toluene. Interestingly, they noted that 14 % more cells were induced in uncontaminated rhizosphere soil than bulk soil. It was further confirmed that the promoter is induced by variety of alkyl-substituted benzene derivatives and branched alkenes (Singer et al. 2003).

Biosensors also appear to be useful tools when carrying out environmental risk assessments of polluted sites while monitoring the efficacy of a remediation strategy because of their ability to detect only the biologically relevant (bioavailable) fraction of the contaminant. Liu et al. (2010) in a study constructed three biosensor strains: *P. fluorescens* F113rif*gfp*, *P. fluorescens* F113rifPCB*gfp*, and *P. fluorescens* F113L::1180*gfp* designed to detect the bioavailability and biodegradation of PCBs in real contaminated soil. As the F113rifPCB*gfp* and F113L::1180*gfp* strains degraded PCBs, they produced chlorinated benzoate end products which in turn induced the expression of *gfp*. Therefore, these strains could actively report degradation of PCBs by their own due to the fluorescence in the presence of an end product (benzoate). This approach is equally suited to study both soil bioremediation and plant rhizoremediation. Some authors suggested that by immobilizing the biosensor cells, ease of detection, accuracy, and reproducibility could be improved. There exists a positive correlation between the PCB levels within the samples and the percentage of fluorescing biosensors cells when immobilized PCB-degrading biosensors (F113L::1180*gfp* and F113rifPCB*gfp*) were introduced into PCB-contaminated soil and sludge. The greater number of fluorescent cells is a direct manifestation of greater bioavailability of PCBs, and their degradation by the GM (genetically modified) biosensors which in turn indicated that the soil/sludge posed a greater risk to human health. Demnerova et al. (2005) carried out PCB degradation experiments using three different plant species. Their results supported previous existing plethora of data that the number of natural PCB degraders on roots was at least ten times higher than that in bulk soil. Gfp-expressing biosensor cells were easy to detect, and they were in higher densities in soils planted with pea, indicating increased PCB degradation activity compared with unplanted soil. Authors stated that this study demonstrated that biosensors could be used at two major fronts, firstly, to evaluate a specific contaminated site for bioremediation potential, and secondly to monitor PCB degradation in real PCB-contaminated soil.

qRT-PCR

Use of real-time PCR increases the probability of accurately determining population sizes of PCB-degrading microbes. Biosensor cell numbers detected by the qRT-PCR method are tenfold higher than the numbers detected using the plate counting method. The high sensitivity of real-time PCR is because it could detect both culturable and non-culturable inoculants together with dead inoculants (Cubero and Graham 2005). It was also documented previously by Wang et al. (2004) that by using a qRT-PCR method, a tenfold higher level of the engineered *Pseudomonas putida* could be detected compared to the plate counting method, during 2-chlorobenzoate degradation in soil.

Now next-generation sequencing, such as pyrosequencing along with various bioinformatics software tools, is emerging as a potential tool for identification of functional communities of microbes present in polluted sites.

Conclusion

Rhizoremediation has successfully occupied center stage in the remediation of sites contaminated by metals and especially organic pollutants. Several genera of bacteria have already proved their potentials in lowering the levels of organic pollutants along with their plant partner. Thus, it becomes immensely important to explore some new such partnerships to optimize remediation strategies in long-term contaminated soils. Recently, molecular biology-based techniques have broadened our scope in not only identifying the phylogeny of the individual members in bacterial communities engaged in polluted sites but also in linking their phylogeny with metagenomics. Thus, in order to fully understand and optimize the bioremediation processes, much deeper researches based on exploring diversity in metabolic pathways and understanding the mutual relationships between or among plants and their associated microbes should be of paramount importance to help achieve goals. Use of biosensors, another complementary tool, has enabled us to have deeper insight in monitoring the efficiency of various environmentally oriented bioassays.

Acknowledgments The corresponding author thanks the Department of Science and Technology (DST), India, for their financial support as Women Scientist scheme, DST No. SR/WOS-A/LS-275/2011 (G).

References

Abeles FB, Morgan PW, Saltveit ME Jr (1992) Ethylene in plant biology, 2nd edn. Academic, New York, pp 414

Adams DO, Yang SF (1979) Ethylene biosynthesis: identification of l-amino cyclopropane carboxylic acid as an intermediate in the conversion of methionine to ethylene. Proc Natl Acad Sci U S A 76:170–174

Afzal M, Yousaf S, Reichenauer TG, Kuffner M, Sessitsch A (2011) Soil type affects plant colonization, activity and catabolic gene expression of inoculated bacterial strains during phytoremediation of diesel. J Hazard Mater 186:1568–1575

Ascherio A, Chen H, Weisskopf MG, O'Reilly E, McCullough ML, Calle EE, Schwarzschild A, Thun MJ (2006) Pesticide exposure and risk for Parkinson's disease. Ann Neurol 60:197–203

Berendsen RL, Pieterse CMJ, Bakker PAHM (2012) The rhizosphere microbiome and plant health. Trends Plant Sci 17:478–486

Boschker HTS, Nold SC, Wellsbury P, Bos D, de Graaf W, Pel R, Parkes RJ, Cappenberg TE (1998) Direct linking of microbial populations to specific biogeochemical processes by ^{13}C- labeling of biomarkers. Nature 392:801–805

Brencic A, Winans SC (2005) Detection of and response to signals involved in host-microbe interactions by plant-associated bacteria. Mol Biol Rev 69:155–194

Burken JG (2004) Uptake and metabolism of organic compounds: green-liver model. In: McCutcheon SC, Schnoor JL (eds) Phytoremediation: transformation and control of contaminants. Wiley, Hoboken, p 59

Busse HJ, Denner EB, Lubitz W (1996) Classification and identification of bacteria: current approaches to an old problem. Overview of methods used in bacterial systematics. J Bacteriol 47:3–38

Casavant NC, Thompson D, Beattie GA, Phillips GJ, Halverson LJ (2003) Use of a site-specific recombination based biosensors for detecting bioavailable toluene and related compounds in roots. Environ Microbiol 5:238–249

Colbert SF, Schroth MN, Weinhold AR, Hendson M (1993) Enhancement of population densities of *Pseudomonas putida* PpG7 in agricultural ecosystem by selective feeding with carbon source salicylate. Appl Envrion Microbiol 59(7):2064–2070

Compant S, Clément C, Sessitsch A (2010) Plant growth-promoting bacteria in the rhizo- and endosphere of plants: their role, colonization, mechanisms involved and prospects for utilization. Soil Biol Biochem 42:669–678

Crowley DE, Alvey S, Gilbert ES (1997) Rhizosphere ecology of xenobiotic degrading microorganisms. In: Kruger EL, Anderson TA, Coats JR (eds) Phytoremediation of soil and water contaminants, vol 664, ACS symposium series. American Chemical Society, Washington, DC, pp 19–34

Cubero J, Graham JH (2005) Quantitative real-time polymerase chain reaction for bacterial enumeration and allelic discrimination to differentiate *Xanthomonas* strains on citrus. Bacteriology 95:1333–1340

Cunningham SD, Berti WB (1993) Remediation of contaminated soils with green plants: an overview. In Vitro Cell Dev Biol 29:207–212

Demnerova K, Mackova M, Spevakova V, Beranova K, Kochankova L, Lovecka P, Ryslava E, Maceket T (2005) Two approaches to biological decontamination of ground-water and soil polluted by aromatics-characterization of microbial populations. Int Microbiol 8:205–211

Dercova K, Cicmanova LP, Demnerova K, Mackova M, Hucko P, Kushnir P (2008) Isolation and identification of PCB-degrading microorganisms from contaminated sediments. Int Biodeter Biodeg 62:219–225

Falck F Jr, Ricci A Jr, Wolff MS, Godbold J, Deckers P (1992) Pesticides and polychlorinated biphenyl residues in human breast lipids and their relation to breast cancer. Arch Environ Health 47:143–146

Faroon O, Jones D, de Rosa C (2001) Effects of polychlorinated biphenyls on the nervous system. Toxicol Ind Health 16:305–333

Fletcher JS, Hegde RS (1995) Release of phenols by perennial plant roots and their potential importance in bioremediation. Chemosphere 31:3009–3016

Flower KB, Hoppin JA, Lynch CF, Blair A, Knott C, Shore DL, Sandler DP (2004) Cancer risk and parental pesticide application in children of agricultural health study participants. Environ Health Perspect 112:631–635

Gerhardt KE, Huang XD, Glick BR, Greenberg BM (2009) Phytoremediation and rhizoremediation of organic soil contaminants: Potential and challenges. Plant Sci 176:20–30

Germaine KJ, Liu X, Cabellos GG, Hogan JP, Ryan D, Dowling DN (2006) Bacterial endophyte-enhanced phytoremediation of the organochlorine herbicide 2,4-dichlorophenoxyacetic acid. FEMS Microbiol Ecol 57:302–310

Gianfreda L, Rao MA (2004) Potential of extracellular enzymes in remediation of polluted soils: a review. Enz Microb Biotechnol 35:339–354

Glick BR (2010) Using soil bacteria to facilitate phytoremediation. Biotechnol Adv 28:367–374

Glick BR, Cheng Z, Czarny J, Duan J (2007) Promotion of plant growth by ACC deaminase-containing soil bacteria. Eur J Plant Pathol 119:329–339

Gupta G, Panwar J, Jha PN (2013) Natural occurrence of *Pseudomonas aeruginosa*, a dominant cultivable diazotrophic endophytic bacterium colonizing *Pennisetum glaucum* (L.). R Br Appl Soil Ecol 64:252–261

Harvey PJ, Campanella BF, Castro PM, Harms H, Lichtfouse E, Schäffner AR, Smrcek S, Werck-Reichhart D (2002) Phytoremediation of polyaromatic hydrocarbons, anilines and phenols. Environ Sci Pollut Res Int 9:29–47

Huang WE, Ferguson A, Singer AC, Lawson K, Thompson IP, Kalin RM, Larkin MJ, Bailey MJ, Whiteley AS (2009) Resolving genetic functions within microbial populations: in situ analyses using rRNA and mRNA stable isotope probing coupled with single-cell raman fluorescence in situ hybridization. Appl Environ Microbiol 75:234–241

Ionescu M, Beranova K, Dudkova V, Kochankova L, Demnerova K, Macek T, Mackova M (2009) Isolation and characterization of different plant associated bacteria and their potential to degrade polychlorinated biphenyls. Int Biodeter Biodeg 63:667–672

Jha P, Panwar J, Jha PN (2014) Secondary plant metabolites and root exudates: guiding tools for polychlorinated biphenyl biodegradation. Int J Environ Sci Technol. doi:10.1007/s13762-014-0515-1,1-14

Khan S, Afzal M, Iqbal S, Khan QM (2013) Plant–bacteria partnerships for the remediation of hydrocarbon contaminated soils. Chemosphere 90: 1317–1332

Koubek J, Uhlik O, Jecna K, Junkovaa P, Vrkoslavova J, Lipov J, Kurzawova V, Macek T, Mackova M (2012) Whole –cell MALDI- TOF: rapid screening method in environmental microbiology. Int Biodeter Biodeg 69:82–86

Krishnamurthy T, Ross PL (1996) Rapid identification of bacteria by direct matrix-assisted laser desorption/ionization mass spectrometric analysis of whole cells. Rapid Commun Mass Spectrom 10:883–888

Kuiper I, Lagendijk EL, Bloemberg GV, Lugtenberg BJ (2004) Rhizoremediation: a beneficial plant–microbe interaction. Mol Plant Microbe Interact 17:6–15

Kurzawova V, Stursa P, Uhlik O, Norkova K, Strohalm M, Lipov J, Kochankova L, Mackova M (2012) Plant – microorganism interactions in bioremediation of polychlorinated biphenyl-contaminated soil. New Biotechnol 30:15–22

Lay JO (2000) MALDI-TOF mass spectrometry and bacterial taxonomy. Anal Chem 19:507–516

Leigh MB, Prouzová P, Macková M, Macek T, Nagle DP, Fletcher JS (2006) Polychlorinated biphenyl (PCB)-degrading bacteria associated with trees in a PCB-contaminated site. Appl Environ Microbiol 72:2331–2342

Leigh MB, Pellizari VH, Uhlik O, Sutka R, Rodrigues J, Osrom NE, Zhou J, Tiedje J (2007) Biphenyl utilizing bacteria and their functional genes in a pine root zone contaminated with polychlorinated biphenyls (PCBs). ISME J 1:134–148

Li Y, Liang F, Zhu Y, Wang F (2013) Phytoremediation of a PCB contaminated soil by alfalfa and tall fescue single and mixed plants cultivation. J Soil Sed. doi:10.1007/s11368-012-0618-6,13:925-931

Liu X, Germaine KJ, Ryan D, Dowling DN (2010) Genetically modified Pseudomonas biosensing biodegraders to detect PCB and chlorobenzoate bioavailability and biodegradation in contaminated soils. Bioeng Bugs 1(3):198–206

Lugtenberg BJ, Dekkers L, Bloemberg GV (2001) Molecular determinants of rhizosphere colonization by Pseudomonas. Annu Rev Phytopathol 39:461–490

Manefield M, Whiteley AS, Griffiths RI, Bailey MJ (2002) RNA stable isotope probing, a novel means of linking microbial community function to phylogeny. Appl Environ Microbiol 68:5367–5373

Marsh TL, Nakatsu CH (2007) Analysis of microbial communities with denaturing gradient gel electrophoresis and terminal restriction fragment length polymorphism. In: Reddy CA, Beveridge TJ, Breznak JA, Marzluf G, Schmidt TM, Snyder LR (eds) Methods for general and molecular microbiology. ASM Press, Washington, DC, pp 909–923

Narasimhan K, Basheer C, Bajic VB, Swarup S (2003) Enhancement of plant-microbe interactions using a rhizosphere metabolomics-driven approach and its application in the removal of polychlorinated biphenyls. Plant Physiol 132:146–153

Nesterenko-Malkovskaya A, Kirzhner F, Zimmels Y, Armon R (2012) *Eichhornia crassipes* capability to remove naphthalene from wastewater in the absence of bacteria. Chemosphere 87:1186–1191

O'Connell KP, Goodman RM, Handelsman J (1996) Engineering the rhizosphere: expressing a bias. Trends Biotechnol 14:83–88

Pandey J, Chauhan A, Jain RK (2009) Integrative approaches for assessing the ecological sustainability of in situ bioremediation. FEMS Microbiol Rev 33:324–375

Parmeela, Johri BN (2004) Phylogenetic analysis of rice endophytes showing antagonism against *Rhizoctonia solani*. Curr Sci 87(5):687–692

Radajewski S, Ineson P, Parekh NR, Murrell JC (2000) Stable-isotope probing as a tool in microbial ecology. Nature 403:646–649

Robinson MM, Griffith M, Pauls KP, Glick BR (2001) Dual role of ethylene in susceptibility of tomato to Verticillium wilt. J Phytopathol 149:385–388

Sandermann H Jr (1992) Plant metabolism of xenobiotics. Trends Biochem Sci 17:82–84

Savka MA, Farrand SK (1997) Modification of rhizobacterial populations by engineering bacterium utilization of novel plant-produced resource. Nat Biotechnol 15:363–368

Schell MA (1985) Transcriptional control of the nah and sal hydrocarbon- degradation genes from plasmid NAH7. J Bacteriol 26:2049–2057

Schiwieger F, Tebbe CC (1998) A new approach to utilize PCR-single-strand-conformation polymorphism for 16S rRNA gene-based microbial community analysis. Appl Environ Microbiol 64:4870–4876

Siciliano SD, Germida JJ (1998) Mechanisms of phytoremediation: biochemical and ecological interactions between plants and bacteria. Environ Rev 6:65–79

Siciliano S, Fortin N, Himoc A, Wisse G, Labelle S, Beaumier D, Ouellette D, Roy R, Whyte LG, Banks MK, Schwab P, Lee K, Greer CW (2001) Selection of specific endophytic bacterial genotypes by plants in response to soil contamination. Appl Environ Microbiol 67:2469–2475

Singer AC, Crowley DE, Thompson IP (2003) Secondary plant metabolites in phytoremediation and biotransformation. Trends Biotechnol 21:123–130

Stockholm convention 2004

Strobel G, Daisy B, Castillo U, Harper J (2004) Natural products from endophytic microorganisms. J Nat Prod 62:257–268

Taghavi S, Barac T, Greenberg B, Borremans B, Vangronsveld J, van der Lelie D (2005) Horizontal gene transfer to endogenous endophytic bacteria from poplar improves phytoremediation and development of poplar trees. Appl Environ Microbiol 71: 8500–8505

Tang X, Qiao J, Chen C, Chen L, Yu C, Shen C, Chen Y (2013) Bacterial communities of Polychlorinated biphenyls polluted soil around an E- waste recycling workshop. Soil Sed Contam 22:562–573. doi:10.1080/15320383.2013.75026

Taranenko NI, Hurt R, Zhou J, Isola NR, Huang H, Lee SH, Chen CH (2002) Laser desorption mass spectrometry for microbial DNA analysis. J Microbiol Methods 48:101–110

Teng Y, Shen Y, Luo Y, Sun X, Sun M, Fu D, Li Z, Christie P (2011) Influence of Rhizobium meliloti on phytoremediation of polycyclic aromatic hydrocarbons by alfalfa in an aged contaminated soil. J Hazard Mater 186:1271–1276

Toussaint JP, Pham T, Barriault D, Sylvestre M (2012) Plant exudates promote PCB degradation by a Rhodococcal rhizobacteria. Appl Microbiol Biotechnol 95:1589–1603

Trapp S, Zambrano KC, Kusk KO, Karlson U, Karlson U (2000) A Phytotoxicity test using transpiration of willow. Arch Environ Contam Toxicol 39:154–160

Trapp S, Kohler A, Larsen LC, Zambrano KC, Karlson U (2001) Phytotoxicity of fresh and weathered diesel and gasoline to willow and poplar trees. J Soil Sed 1:71–76

Uhlik O, Strejcek M, Junkova P, Sanda M, Hroudova M, Vlcek MM, Macek M (2011) Matrix assisted laser desorption ionization (MALDI)- Time of flight mass spectrometry and MALDI biotyper based identification from contaminated horseradish rhizosphere soil. Appl Environ Microbiol 77(19):6858–6866

Uhlik O, Wald J, Strejcek M, Musilova L, Ridi J, Hroudova M, Vlcek C, Cardenas E, Mackova M, Macek T (2012) Identification of bacteria utilizing biphenyl, benzoate, and naphthalene in long term contaminated soil. PLoS One 7(7):e40653. doi:10.1371/journal.pone.0040653

Uhlik O, Musilova L, Ridl J, Hroudova RM, Vlcek C, Koubek J, Holeckova M, Mackova M, Macek T (2013) Plant secondary metabolites- induced shifts in bacterial community structure and degradative ability in contaminated soil. Appl Micobiol Biotechnol 97(20): 9245–9256

Van Loon LC, Geraats BPJ, Linthorst HJM (2006) Ethylene as a modulator of disease resistance in plants. Trends Plant Sci 11:184–189

Wang G, Gentry T, Grass G, Josephson K, Rensing C, Pepper IL (2004) Real-time PCR quantification of a green fluorescent protein-labelled, genetically engineered Pseudomonas putida strain during 2-chlorobenzoate degradation in soil. FEMS Microbiol Lett 233:307–314

Watanabe K, Teramoto M, Futamata H, Harayama S (1998) Molecular detection, isolation, and physiological characterization of functionally dominant phenol-degrading bacteria in activated sludge. Appl Environ Microbiol 64:4396–4402

Watts JEM, Wu Q, Schreier SB, May HD, Sowers KR (2001) Comparative analysis of polychlorinated biphenyl-dechlorinating communities in enrichment cultures using three different molecular screening techniques. Environ Microbiol 3:710–719

Weyens N, Taghavi S, Barac T, van der Lelie D, Boulet J, Artois T, Carleer R, Vangronsveld J (2009) Bacteria associated with oak and ash on a TCE contaminated site: characterization of isolates with potential to avoid evapotranspiration of TCE. Environ Sci Pollut Res 16:830–843

Weyens N, Monchy S, Vangronsveld J, Taghavi S, van der Lelie D (2010a) Role of plant–microbe partnerships to deal with environmental challenges. In: Timmis KN (ed) Handbook of hydrocarbon and lipid microbiology. Springer, Berlin, pp 3357–3359

Weyens N, Monchy S, Vangronsveld J, Taghavi S, van der Lelie D (2010b) Plant–microbe partnerships. In: Timmis KN (ed) Handbook of hydrocarbon and lipid microbiology. Springer, Berlin, pp 2545–2574

Wolff MS, Toniolo PG, Lee EW, Rivera M, Dlin NB (1993) Blood levels of organochlorine residues and risk of breast cancer. J Natl Cancer Inst 85: 648–652

Yousaf S, Afzal M, Reichenauer TG, Brady CL, Sessitsch A (2011) Hydrocarbon degradation, plant colonization and gene expression of alkane degradation genes by endophytic Enterobacter ludwigii strains. Environ Pollut 159:2675–2683

Harnessing Plant-Microbe Interactions for Enhanced Protection Against Phytopathogens

5

Sandhya Mishra, Akanksha Singh, Chetan Keswani, Amrita Saxena, B.K. Sarma, and H.B. Singh

Abstract

Beneficial plant-microbe interactions have utmost importance for enhancing plant growth, improving soil structure, and managing plant diseases. Not surprisingly, such mutual interactions, where plants provide nourishment to rhizospheric microbes and in return microbes help in facilitating plant growth and stress amelioration, actually lay the foundation of sustainable agriculture. To cope with the major challenge of pathogen attack, beneficial rhizospheric microbes have proven their efficacy by induced systemic resistance (ISR). Therefore, such microbes are increasingly used in the form of biofertilizers and biopesticides. Moreover, such plant-microbe interactions elicit a range of defense-responsive activities in order to combat the pathogen challenge. The main microbes-mediated defense strategies adopted by plants include activation of antioxidant status of the plant by reprogramming defense-related enzymes, modulation of quorum sensing phenomenon, and activation of phenylpropanoid pathway leading to phenolics production, lignin deposition, and transgenerational defense response. In this chapter, we highlight the relevance of beneficial interactions between plant and microbes in enhancing plants' innate immune system against pathogen attack. This review provides a better understanding of the recent advances and major outcome of positive plant-microbe interactions and linking their relevance to plant defense response.

Introduction

Plants are the basis of life on earth that provide carbon source for all non-photosynthetic organisms. However, ~300,000 plant species are attacked by a huge number of detrimental organisms including pathogens and insects (Pieterse and Dicke 2007). Such biotic factors tremendously

S. Mishra • A. Singh • C. Keswani • A. Saxena
B.K. Sarma • H.B. Singh (✉)
Department of Mycology and Plant Pathology, Institute of Agricultural Sciences, Banaras Hindu University, Varanasi 221005, India
e-mail: hbs1@rediffmail.com

N.K. Arora (ed.), *Plant Microbes Symbiosis: Applied Facets*,
DOI 10.1007/978-81-322-2068-8_5, © Springer India 2015

influence the plant growth and limit total agricultural production. Hence, sustainable approach of enhancing plant growth and managing plant diseases is being integrated to achieve higher crop yield. Plants in nature interact with wide range of beneficial and detrimental microorganisms providing baseline for linking aboveground and belowground community members (Van der Putten et al. 2001; Pineda et al. 2010).

The successful establishment of plant-microbe interaction depends on the ability of roots to interact with microbes as rhizosphere is the main zone where such interactions take place. The rhizospheric interaction could directly affect the plant growth by exerting either beneficial, neutral, or detrimental effects. Beneficial plant-microbe interactions require involvement of plant growth-promoting rhizobacteria (PGPR), endophytes, and mycorrhizal fungi that enhance plant growth by improving nutritional status for plants and helping the plants to combat abiotic and biotic stresses (Harrison 2005; Berendsen et al. 2012). In contrary, the detrimental interaction is imparted by pathogenic microorganisms resulting into various kinds of plant diseases. In the crowd of mixed population of beneficial and detrimental microbes residing in the rhizospheric region, the selection of potent biocontrol microbes is necessary to eliminate the pathogens and to combat the challenge imposed by pathogens. Moreover, plants also develop strategies spontaneously to recognize biotic and abiotic interactions and further translate the signal into defense response (Pei et al. 2000; Jones and Dangl 2006; Dicke and Hilker 2003).

Interestingly, plant-microbe interaction is regulated through signal-transduction pathways allowing plants to prioritize defense responses following stress conditions. It is well known that plants' response to pathogen attack is regulated by jasmonic acid (JA) and salicylic acid (SA)-dependent pathways effective against necrotrophic and biotrophic pathogens, respectively (Pieterse et al. 2012). The induced resistance imparted by these pathways represents two distinct responses: systemic acquired resistance (SAR) and ISR (van Loon et al. 1998). SAR is mediated by SA which is frequently produced following pathogen infection (Park and Kloepper 2000; Jeun et al. 2004). In contrary, JA- and ethylene (ET)-dependent ISRs are activated by beneficial microbes proven to produce antimicrobial compounds, siderophores, O-antigen of lipopolysaccharides (LPS) and salicylate. Moreover, ISR leads to the expression of pathogenesis-related (PR) proteins such as PR-1, PR-2, chitinases, and some peroxidases (POxs) (van Wees et al. 2000; Silva et al. 2004; Jetiyanon 2007). Considering the importance of ISR, recently, noteworthy consideration is being given to exploiting the beneficial soil microbes for enhancing plants' immunity against pathogen attack. Moreover, microbe-mediated suppression of plant diseases provides eco-friendly and sustainable approach of plant disease management. It is believed that such beneficial microbes can enhance the plant's innate immunity level against the invading pathogens by inducing an array of defense responses that include enhancement in antioxidant status of the plant by reprogramming defense-related enzymes, modulation of quorum sensing activities, and activation of phenylpropanoid pathway leading to phenolics production and lignin deposition.

Microbe-Mediated Antioxidants Status in the Host

Successful pathogen infection in plants results into oxidative burst that lead to production of reactive oxygen species (ROS) causing oxidative destruction of the cell (Asada 1999; Dat et al. 2000). These ROS are formed as a result of excitation of O_2 to singlet form of oxygen or via formation of a superoxide radical (O_2^-), hydrogen peroxide (H_2O_2), or a hydroxyl radical (HO^-). The enhanced production of ROS after pathogen infection acts as cellular indicator of stress conditions and is triggered by the activity of NADPH oxidases (Cazale et al. 1999; Pei et al. 2000). However, recently, it has been investigated that apart from NADPH oxidases, the other sources such as amine oxidases and cell wall-bound POxs also participate in the formation of ROS during programmed cell death and pathogen defense (Dat et al. 2000; Grant and Loake 2000). It is interesting to note that though the elevated level of ROS imposes threat condition to the cells, it also acts as a signal for commencement of stress

Table 5.1 Major ROS scavengers in plants

Scavengers	Localization	Target ROS
Superoxide dismutase	Chl, Cyt, Mit, Per, Apo	O_2^-
Ascorbate peroxidase	Chl, Cyt, Mit, Per, Apo	H_2O_2
Catalase	Per	H_2O_2
Glutathione peroxidase	Cyt	H_2O_2, ROOH
Peroxidases	CW, Cyt, Vac	H_2O_2
Thioredoxin peroxidase	Chl, Cyt, Mit	H_2O_2
Ascorbic acid	Chl, Cyt, Mit, Per, Apo	H_2O_2, O_2^-
Glutathione	Chl, Cyt, Mit, Per, Apo	H_2O_2
α-Tocopherol	Membranes	ROOH, O_2^1
Carotenoids	Chl	O_2^1

Apo apoplast, *Chl* chloroplast, *CW* cell wall, *Cyt* cytosol, *Mit* mitochondria, *Per* peroxisome, *Vac* vacuole

response and defense pathways (Desikan et al. 2001). The steady-state level of ROS in plant cells should always be under control as their over accumulation eventually result in cell death as a consequence from various kinds of oxidative processes, viz., lipid peroxidation, protein oxidation, nucleic acid damage, and enzyme inactivation.

In order to cope with the detrimental effects of elevated level of ROS, plants are bestowed with efficient antioxidant enzymes that contribute to ROS-scavenging mechanisms of plants. Major ROS scavengers in plants are reported to be superoxide dismutase (SOD), ascorbate peroxidase (APX), and catalase (CAT) (Mittler 2002) (Table 5.1). To maintain the balance of superoxide radicals in cells, the activities of SOD and APX or CAT enzymes play crucial role as this balance avert formation of hydroxyl radical via the metal-dependent Haber–Weiss or the Fenton reactions (Asada and Takahashi 1987; Bowler et al. 1991). Moreover, the key pathways of ROS scavengers in plants include SOD present in more or less all cellular compartments. The major threat to plant system is instigated by various kinds of plant pathogens which upon infection lead to rapid formation of ROS. Plants generally increase their tolerance against invading pathogens by elevating activities of antioxidant enzymes.

The major defense strategy adopted by plants is phenylpropanoid pathway catalyzing transformation of L-phenylalanine into trans-cinnamic acid (Dixon and Paiva 1995) which play an important role in the biosynthesis of phenolics having strong antimicrobial properties (Nicholson and Hammerschmidt 1992). Interestingly, activation of the phenylpropanoid pathway also leads to deposition of lignin and induction of antioxidant enzymes including SOD and peroxidase (POx) (Silva et al. 2004; Singhai et al. 2011). It is interesting to note that beneficial soil microflora in rhizosphere provide supportive environment for the plants by augmenting the antioxidant status. Such microbe-mediated ISR protects the plants from various plant pathogens. Earlier reports have revealed significant role of many microbes in enhancing antioxidant enzymes in plants contributing to their resistance against pathogens (Table 5.2). Previously, rhizobacterial strains and *Serratia marcescens* were reported to enhance activities of POx, phenylalanine ammonia lyase (PAL), polyphenol oxidase (PPO), and lipoxygenase (LOX) in betelvine and tomato after pathogen attack by *Phytophthora nicotianae* and *Pseudomonas syringae*, respectively. Modulation of such antioxidant status in plants treated with beneficial rhizobacteria enhanced disease resistance against pathogen attack (Diallo et al. 2011). Similar observation was recorded by Singhai et al. (2011) where pseudomonad strains were found to increase the level of POx and PAL in potato leading to tolerance against potato scab disease caused by *Streptomyces scabies*. Similarly, in another interesting study by Jain et al. (2012), the microbial consortium enhanced tolerance in pea plants against *Sclerotinia sclerotiorum* by inducing the level of PAL, POx, PPO, and SOD. Most recently, Singh et al. (2013) showed elevation of SOD and PO in *Sclerotium rolfsii* challenged chickpea plants treated with triple consortium of *Pseudomonas*, *Trichoderma*, and *Rhizobium*. Taken together, these reports clearly validate the importance of beneficial rhizosphere microbes in imparting tolerance to the plants against diverse range of pathogens by modulating their innate antioxidant status as shown in Fig. 5.1.

Table 5.2 Examples of microbe-mediated defense response in plants against pathogen infection

Host plant	Microbes	Antioxidant enhanced	Pathogen	References
Chickpea	*Pseudomonas*, *Trichoderma*, *Rhizobium*	Superoxide dismutase, peroxidase	*Sclerotium rolfsii*	Singh et al. (2013)
Pea	*Pseudomonas*, *Trichoderma harzianum*, *Bacillus subtilis*	Phenylalanine ammonia lyase, peroxidase, polyphenol oxidase, superoxide dismutase	*Sclerotinia sclerotiorum*	Jain et al. (2012)
Potato	Pseudomonad strains	Peroxidase, phenylalanine ammonia lyase	*Streptomyces scabies*	Singhai et al. (2011)
Betelvine	*Serratia marcescens*	Phenylalanine ammonia lyase, peroxidase, polyphenol oxidase	*Phytophthora nicotianae*	Lavania et al. (2006)
Tomato	Rhizobacterial strains	Lipoxygenase, phenylalanine ammonia lyase, peroxidase	*P. syringae* pv. *tomato*	Silva et al. (2004)

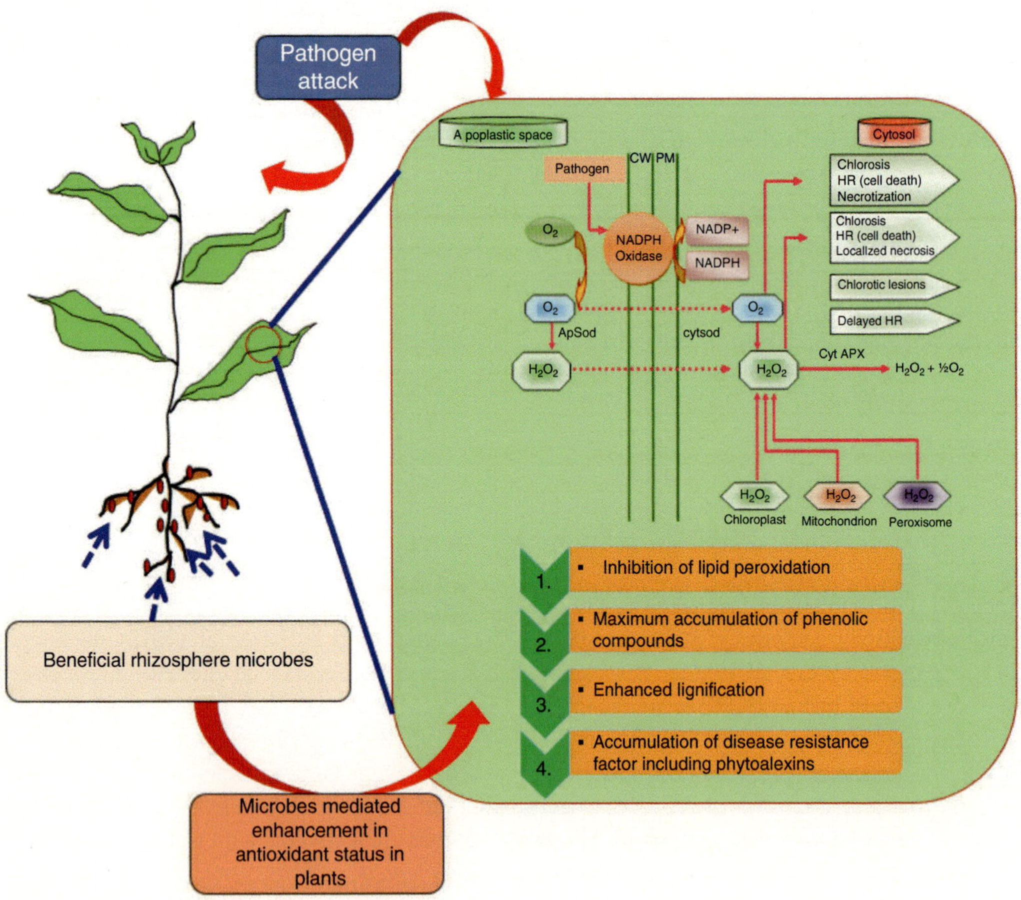

Fig. 5.1 Schematic diagram showing beneficial rhizosphere microbes mediated enhanced antioxidant activity contributing to their defense strategy against pathogen attack

Microbe-Mediated Activation of Phenylpropanoid Pathway

Soil is the locale of numerous microorganisms and can aptly be referred to as the mine of microorganisms. The rhizospheric regions of the plants inhabit most of the microbial communities in its vicinity, be it having beneficial or deleterious effects on the plant. Beneficial microorganisms may include PGPR and mycorrhizal fungi that induce systemic defense response in the host to preclude it from the chronic impairment of phytopathogens comprising of bacteria, fungi, viruses, and nematodes. To avoid the perplexity in the use of different terms for denoting the beneficial microbes, two new terms were proposed by Bashan and Holguin (1998): "biocontrol plant growth-promoting bacteria (PGPB)" that suppress plant diseases by enhancing the plant defense responses and "PGPB" that specifically play eminent role in augmenting plant growth. The upgradation of the conventionally used "PGPR" term coined by Kloepper et al. (1980) to "PGPB" would thereby include the useful microbes that do not inhabit the rhizospheric region of the soil yet enhance the vigor of plants against the phytopathogens.

Diverse interactions prevail between the microbial commune and plant roots, for instance, the symbiotic associations by mycorrhizal fungi that aid in the uptake of water and minerals (Harrison 2005), and the nodule-inhabiting *Rhizobium* bacteria that fix the atmospheric nitrogen for the plant (Spaink 2000). Several other types of beneficial microbes like PGPR and fungi are reported to suppress plant diseases (Van Loon et al. 1998; Harman et al. 2004; Kloepper et al. 2004) or insect herbivory (Van Oosten et al. 2008) by enhancing the defense response of the plant thereby resulting in overall increment in the plant growth parameters. The plethora of benefits endowed to the plants by the microbes can be attributed to either direct effect through mycoparasitism of soil-borne pathogens or indirectly by eliciting the plant defense mechanisms thereby fortifying the plant immune system against the invading pathogens (Van Loon et al. 1998; Pozo and Azcon-Aguilar 2007).

Several microbial determinants are related with the elicitation of defense responses in plant system. Most commonly studied and well associated as the inducers of host immune response conferred by rhizobacteria are the microbe-associated molecular patterns (MAMPs). Also, LPS and flagellin found at the cell surfaces of rhizobacteria are found to be potent in inducing defense response of the associated host against phytopathogens (Bakker et al. 2007). Another important mechanism employed by beneficial microbes in disease suppression is to create competition for iron by forming low molecular weight iron chelators known as siderophores (Meziane et al. 2005). Important chemicals that act as elicitors for the defense response are the variously secreted antibiotics by the rhizobacteria and fungi (Ran et al. 2005); surfactin, a lipoprotein secreted by *Bacillus subtilis* (Ongena et al. 2007); biosurfactant massetolide A from *P. fluorescens* (Tran et al. 2007); N-alkylated benzylamine (Ongena et al. 2005); N-acyl-L-homoserine lactone (Schuhegger et al. 2006); and volatile organic compound 2, 3-butanediol by *Bacillus* spp. (Ryu et al. 2004). Fungal proteins like endochitinase secreted by *Trichoderma* spp. have been shown to enhance plant defense-related proteins (Harman and Shoresh 2007; Keswani et al. 2014). A small protein SM1 produced by strains of *Trichoderma virens* can induce terpenoid phytoalexin biosynthesis and POx activity as studied in cotton plants and other systems as well (Djonovic et al. 2006, 2007). Another class of proteins associated with the immune system of the plants is the products of avirulence like (Avr) genes, produced not only by the phytopathogens but also by the beneficial microbes. They generally function as pathogen-specific elicitors of the hypersensitive responses in plants containing the corresponding resistance (R) gene (Woo et al. 2006). Small secondary metabolites produced by *Trichoderma* species have also been reported to possess defense response induction activity mainly via eliciting the expression of PR proteins on application to plants causing both local and systemic disease suppression (Vinale et al. 2008; Keswani et al. 2014).

Plant defense response gets activated by the metabolism of phenyl propanoid (PP) pathway in which PAL catalyzes the first important step of the general PP metabolism. Further, the pathway leads to the synthesis of other important compounds having indispensible role in providing plant defense including cell wall strengthening and repair (lignin and suberin), antimicrobial activity (pterocarpan, isoflavonoid phytoalexins), and signaling SA (Hummerschmidt 1999). Beneficial microbes like *Trichoderma* spp. have been reported to trigger the terpenoid phytoalexin defense compounds in cotton seedlings thereby controlling *Rhizoctonia solani* infestation apart from the conventional mycoparasitic mode exhibited by the species of the genus (Howell et al. 2000). Also, biocontrol of *Pythium ultimum* on *Arabidopsis* seedlings by *T. harzianum* strain T22 reported the elevated expression level of *NPR1* gene which is the main gene involved in disease resistance (Shoresh et al. 2010). Recently, the concept of using two or three compatible beneficial microbes has been proposed for better management of disease. It has been shown that significant enhancement of PP activity is recorded in chickpea plants treated with the triple consortium developed using *Trichoderma*, *Bacillus*, and *Rhizobium* spp. when challenged with *S. rolfsii* (Singh et al. 2013). Similar results were previously reported by Jain et al. (2012) where the consortium of beneficial microbes (*Trichoderma* spp., *Pseudomonas* spp., *Bacillus* spp.) were reported for the increment of various defense-related enzymes in pea plants against *S. sclerotinia* challenge. Another class of beneficial fungi *Piriformospora indica* has also been reported to cause elevation of defense response in plants against soil-borne pathogens (Serfling et al. 2007; Stein et al. 2008) showing its mechanism to be related with the upregulation of JA-mediated pathway.

A complex cross talk subsists between the plant and the beneficial microbes in response to the pathogenic microbes. The advancement in the research related to the responses triggered in plants against the pathogens by the beneficial microbes demonstrates the involvement of specific MAMPs analogous to the MAMPs of the pathogen highlighting the extensive coordination prevailing between different defense pathways involved.

Microbe-Mediated-Induced Lignification

Tertiary structure of lignin results from the polymerization of polyphenols and free radicals p-coniferyl, coumaryl, and sinapyl alcohols within the plant cell wall. This polymerization also results in the formation of covalent cross links with polysaccharide and protein moieties framing a tremendously resistant cell wall towards mechanical and enzymatic disruption against various classes of plant pathogens, insects, and herbivores (Bernards and Lewis 1998; Sederoff et al. 1999; Davin and Lewis 2000; Hatfield and Vermerris as 2001; Boerjan et al. 2003). Thus, lignin serves as physical defense shield in plant defense. Modifications in lignin composition, content, and distribution affect the strength of the shield which ultimately influences the agro-industrial pertinence of the plant material (Lewis and Yamamoto 1990).

Lignification obstructs phytopathogen growth on plant surface by a pentagonal approach (Ride 1978) as mentioned below:

1. Lignin deposition shields the plant tissue surface from enzymatic hydrolysis and mechanical penetration of phytopathogens by intensifying compressive forces between lignin layers preventing cellular penetration of phytopathogens.
2. Lignification of walls hinders the mobility of water and electrolytes between plant cells and phytopathogens facilitating pathogen killing by starvation.
3. Chemical modification of cell wall components is an effective strategy for disguising the pathogen enzymes. Coniferyl alcohol and ferulic acid covalently bound to cell wall glycoproteins and esterification of cell wall polysaccharides with cinnamic acid derivatives evidently reduces the cell wall damage

due to unavailability of substrates (Friend 1976; Whitmore 1978).

4. Generation of free radicals and low molecular weight phenolic precursors produced during polymerization of lignin may directly inactivate pathogens' membranes, enzymes, toxins, and elicitors.
5. Fungal walls contain chitin, cellulose, and hydroxyproline-rich proteins which can serve as matrices for lignin polymerization. Consequently, the hyphal tips become lignified and lose plasticity necessary for growth and penetration (Gottlieb and Pelczar 1951).

Among various mechanisms of plant defense, lignification is a strong structural defense strategy employed by plants to prevent pathogen penetration, and this relationship of lignification and disease resistance in plants is clearly witnessed in various studies. Comparatively, rapid lignin accumulation and deposition is observed in resistant cultivars than susceptible varieties (Vance et al. 1976; Yates et al. 1997; Durrant and Dong 2004). Not much is known about the role of rhizospheric microbes in strengthening plant's cell wall towards various biotic stresses. A recent study aimed to determine the efficacy of a triple microbial consortium of fluorescent *Pseudomonas* PHU094, *Trichoderma* THU0816, and *Rhizobium* RL091 strain on physiological defense responses in chickpea against the collar rot pathogen *S. rolfsii*. The result clearly illustrates the profound variation of lignin deposition in chickpea infected with *S. rolfsii*, which is attributed to different combinations of plant beneficial microbes to trigger lignification process. Interestingly, on treatment with triple consortium, uniform and maximum lignin deposition in the intrafascicular cambial cells was clearly observed and the phloem cells also displayed an enhanced lignification in sclerenchyma cap. Thus, claiming that beneficial rhizospheric microbes when employed in synergistic consortium can enhance the physical strength and durability of the cell wall towards cell wall-degrading phytopathogen (Singh et al. 2013).

In another study, alterations in phenolic metabolism and lignin deposition were analyzed in the roots of tomato plants after elicitation with *Fusarium* mycelium extract (FME), *Fusarium* culture filtrate (FCF), chitosan (CHT), and *Trichoderma* mycelium extract (TME). Maximum lignin synthesis was observed in plants treated with FME followed by CHT. Lignin deposition in the root cell walls increased to 5.7 times within 24 h after elicitation with FME. Similarly, CHT increased lignin deposition to almost five times, 24 h after elicitation. Thus, it was concluded that cell wall strengthening by lignin deposition was preceded by elicitation of lignin synthesizing enzymes revealing its essential role in defense response of tomato plants in response to various elicitors including one derived from *Fusarium oxysporum* f. sp. *lycopersici*, the causal organism of *Fusarium* wilt of tomato (Mandal and Mitra 2007).

Microbe-Mediated Quorum Sensing in Pathogen Management

The phenomenon of quorum sensing (QS) depicts the bacterial cell-cell communication and is generally cell density dependent. This cell-cell communication network is mediated by signal molecules (also called autoinducers), for example, oligopeptides and N-acylhomoserine lactones (AHL) in Gram-positive and Gram-negative bacteria, respectively. The QS plays a significant role in biofilm formation and in determining virulence factor in pathogenic bacterial species (Gram et al. 2002). In contrary, the phenomenon of antiquorum sensing (anti-QS) also exists where autoinducers interrupt with QS and thereby reducing the pathogenicity in several bacterial species (Truchado et al. 2012). Since QS contribute significantly in regulating virulence factor in many plant pathogens, anti-QS could be of great interest for decreasing pathogenic behavior and in developing innovative approach of disease control (Alvarez et al. 2012).

Considering the importance of anti-QS, presently, this phenomenon is getting noteworthy attention in plant disease management using bacterial biosensors and indicators. Interestingly, the cloning of *aiiA* gene (from *Bacillus* sp.) into transgenic potato and tobacco for enhancing disease resistance against *Erwinia carotovora* is perhaps the very first example (Dong et al. 2001). Expression of this gene resulted in the production of AHL-lactonase in transgenic plants that interrupt with the QS systems arresting the virulence factor of *E. carotovora* leading to reduced disease incidence. Likewise, another example is the generation of transgenic tobacco lines using *expI* (*E. carotovora* AHL gene) which, after expression in plants, trap the pathogen in premature stage when it is unable to cause infection (Mäe et al. 2001). It is interesting to note that the autoinducer molecule AHL also contributes to the production of antimicrobial compounds in nonpathogenic *Pseudomonas chlororaphis* which has been successfully employed in suppressing plant diseases (Pierson et al. 1998a, b). Molina et al. (2003) evaluated the biocontrol potential of AHL-degrading *Bacillus* sp. A24 and genetically engineered AHL-degrading strain P3/pME6863 against soft rot in potato caused by *E. carotovora*. Recently, the novel approach of disrupting QS using structural analogs stimulating AHL-degradative microflora has been investigated by some researchers (Crépin et al. 2012a, b). In conclusion, these studies altogether clearly validate that the QS inhibition of phytopathogenic bacteria could be successfully employed for plant disease management. Hence, more focused researches are needed towards this approach.

Microbe-Mediated Nutrient Uptake and Defense

Fifty years back, the drive required for feeding the surplus population gave birth to the much talked about "Green Revolution" in India, leading to tremendous increase in food production. The unremitting use of fertilizers and pesticides has undoubtedly contributed to the increment in the food production but had led to the slow death of the soil microflora and fertility as well. In order to resolve the burning issue of deposition of toxic residues leading to biomagnifications in the food web, urgent need of using eco-friendly alternatives is required. Biopesticides comprising of microbial inoculants have emerged as a silver line for the current scenario with multifaceted benefits like safer approach both to the environment and to human kind, more targeted activity, low dose effectiveness, easily decomposable, natural propagation, and multiplication along with fortifying the plants' immune system (Berg 2009).

A constant conversation exists between the plants and the microbes in its vicinity. The plant-microbe interaction persists owing to the beneficial mutualism between the two partners which upshot various remuneration to plants as well as microbial commune. Positive effect on the growth and health of plants, enhanced stress tolerance, induction of disease reduction, biodiversity enrichment with ability to foster nutrient availability and uptake are the consequence of the interface between plant and microbes (Lugtenberg et al. 2002; Morrissey et al. 2004). These interactions are specific in terms of the host colonization due to the specific secondary metabolism and morphology (Berg and Smalla 2009). However, plant growth promotion and disease reduction or control have been the most noteworthy consequence of this interaction. Growth promotion in plants can be mediated through direct mechanisms by the microbes and also through indirect means through their antagonistic properties thereby reducing disease incidence allowing the healthy proliferation of the plants.

The competence of the microbes to colonize plant habitats is one of the crucial requirements for an effective plant-microbe interaction (Kamilova et al. 2005). Recognition, adherence, invasion (in case of endophytes and pathogens), colonization, and growth are the essential steps required for successful colonization apart from the various strategies employed by the microbial commune to establish the interaction. However, the initiation is executed by the plant itself by releasing signals recognized by the microbes which reciprocate the signals to initiate the colonization (Bais et al. 2006). Generally, motile organisms are preferred to participate and react in this cross talk (Lugtenberg et al. 2002).

Basically, three main types of interactions are related to the increment of plant growth mediated by beneficial microbes (Fig. 5.2). The most commonly known type of interaction is the

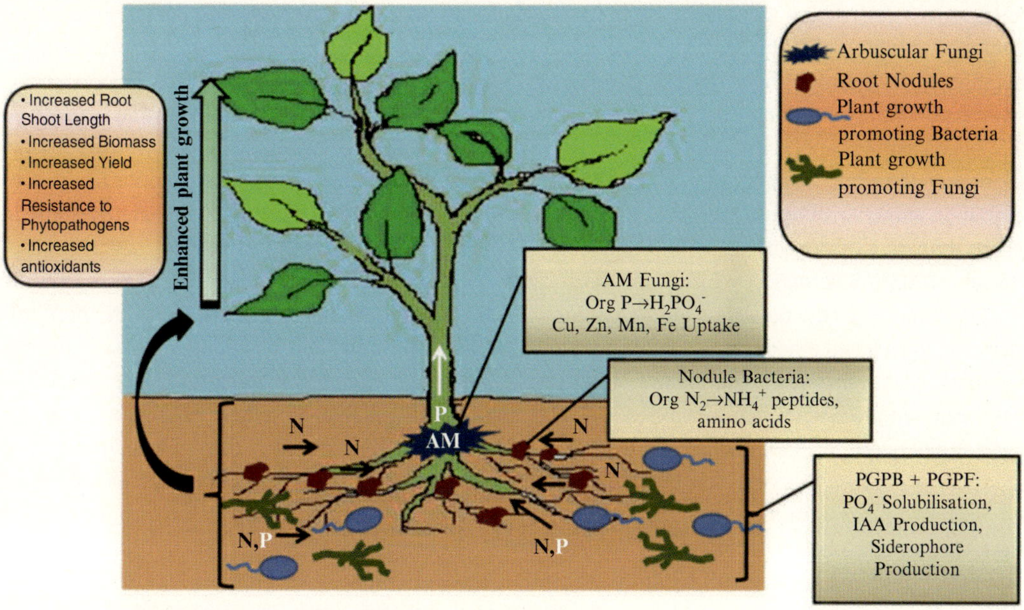

Fig. 5.2 An overview of microbe-mediated nutrient uptake in plants

symbiosis by *Rhizobium* species in fixing atmospheric nitrogen into ammonia in specific organs called nodules, found in leguminous plants (Van Rhijn and Vanderleyden 1995). Other important interaction is that of higher plants, commonly terrestrial flowering plants, with arbuscular mycorrhiza (AM) fungi which facilitate the absorption and translocation of phosphate from the soil (Harrison 1999). Lastly, the outcome of the numerous beneficial microbes that aid in mineralization of organic matter, thereby producing available nitrogen and phosphorous forms along with numerous micronutrients, provides the platform for the third type of interaction prevailing between plants and microbes (Hayatsu et al. 2008).

Phytohormones like indole-3-acetic acid (IAA), ET, cytokinins, and gibberellins are crucial for plant growth. Plants can obtain these hormones by either synthesizing themselves or by obtaining these from the microbes which can even alter the hormonal balance of the plant thereby causing alterations in the growth of the plants. Plant-associated bacteria have been shown to decrease the endogenous ACC levels thereby leading to increased root growth (Glick 2005). ACC deaminase-producing bacteria have also been reported to provide abiotic and biotic stress tolerance to plants thereby protecting them from the unfavorable conditions (Saleem et al. 2007). Seed treatment with auxin-producing *Pseudomonas fluorescens* has been reported to show stimulation of root growth, due to the production of nine times more tryptophan in root exudates of radish plants (Kamilova et al. 2006). This could be directly related to the growth promotion ability as the root growth-promoting hormone auxin is generally found in the root exudate which is synthesized from the amino acid tryptophan. Certain microbes like *B. subtilis*, *B. amyloliquefaciens*, and *Enterobacter cloacae* are known to enhance plant growth by releasing volatiles like 2, 3-butanediol and acetoin (Ryu et al. 2003). Increased photosynthetic efficiency and chlorophyll content were recorded in *A. thaliana* on treatment with *B. subtilis* GB03 which could be possibly related with the modulation of glucose and abscisic acid signaling (Zhang et al. 2008). Apart from auxin-producing bacteria, beneficial fungi like *Trichoderma* have been shown to be responsible for plant growth increment via auxin signaling (Contreras-Cornejo et al. 2009). Nitrogen fixation by the symbiotically associated *Rhizobium* species in the root nodules of leguminous plants or by free living bacteria like *Azospirillum*, *Burkholderia*, and *Stenotrophomonas* (Dobbelare et al. 2003) has been a major example for nutrient acquisition to plants by microbes. Bacterial indirect contribution to plant growth by liberating phosphorous from organic compounds such as phytates play an important role in providing the necessary phosphorous required for proper growth of the plant (Unno et al. 2005). Another important nutrient, sulfate, is also made available to plants through oxidation by bacteria (Banerjee and Yesmin 2002). Siderophores have also been an important mode for the uptake of important micro-elements like Fe and other poorly soluble inorganic nutrients. Also, relation of siderophore production has been reported with the antagonistic activity against pathogens. *P. fluorescens* CHA0 has been reported to produce gluconic acid which acidifies the surrounding environment and thereby solubilizes the mineral phosphate in the soil creating a nutrient-limited condition for the plant pathogens (De Werra et al. 2009). Fungal biocontrol agent *Trichoderma* and mycorrhizal fungus *P. indica* have been reported to produce siderophore as a mechanism to check the growth of pathogens by creating a competitive environment for the availability of iron (Shoresh et al. 2010). Phosphate absorption to plants by AM has been well reported (Smith et al. 2011; Balakrishnan and Subramanian 2012). Also, increment in Cu, Zn, B, Mn, and Fe uptake in plants has been attributed to AM (Lambert et al. 1980; Clark et al. 1999; Liu et al. 2000).

Other important microbes that have been well studied for their mode of action and regulation in promoting plant growth and antagonism are members of the genera *Azospirillum* (Cassan and Garcia 2008), *Serratia* (De Vleeschauwer and Hofte 2007), *Stenotrophomonas* (Ryan et al. 2009), and *Streptomyces* (Schrey and Tarkka 2008) along with fungal genera *Ampelomyces*, *Coniothyrium*, and *Trichoderma* (Harman et al. 2004).

Microbe-Mediated Transgenerational Defense

Plant-recruited beneficial microbes can also prime the plants for enhanced defense responses, and the effect of priming could be passed on to the next generations as well. Priming of plants leads to enhanced perception of MAMPs, recognition of effector molecules secreted by pathogen, and recruitment of beneficial rhizospheric microbes (Conrath 2011). Recent understanding in the subject reflects that similar to animals, epigenetic inheritance in plants also takes place, and the epigenetic modifications of the chromatin as well as DNA methylation in plants could be well preserved in several subsequent generations (Pieterse 2012). Being sessile organisms, plants communicate with their offspring through this mechanism to "inform" the offsprings about the potential threats in their environment. Since, plants encounter potential biotic and abiotic threats from the environment at one or the other point of their life span, they are adapted to such mechanisms through the process of evolution for passing on the information to the next generation. Slaughter et al. (2012) demonstrated that the descendants of *A. thaliana* plants primed with an avirulent strain of *P. syringae* pv. *tomato* (PstavrRpt2) showed enhanced and rapid accumulation of defense-related gene transcripts associated with the SA signaling pathway. Further, the descendents also showed enhanced disease resistance against a virulent isolate of *P. syringae* and another oomycete pathogen *Hyaloperonospora arabidopsidis*. Interestingly, the progeny of transgenerationally primed plants when treated again with the priming agent displayed an even stronger primed phenotype. Recent evidences suggest that SA-mediated systemic resistance in plants also require chromatin remodeling and DNA methylation (Luna et al. 2012). Luna and Ton (2012) showed that transgenerationally acquired resistance was sustained through one stress-free generation, confirming epigenetic basis of the phenomenon. Failure of *non expressor of PR 1* (*npr1*) mutants to sustain transgenerationally acquired resistance in the progenies further signifies the SA-inducible pathway in this phenomenon and the central role of *NPR1*. Further, transgenerationally acquired systemic resistance was also demonstrated against biotrophic pathogens. A study on progenies obtained from diseased *Arabidopsis* also resulted in enhanced resistance towards the downy mildew pathogen (Luna and Ton 2012). Histone deacetylase 6 (HDA6) is a well-studied histone deacetylase that has a prominent role in the silencing of genes. It was reported that HDA6 has also a significant role to play in the process of DNA methylation on its direct target locus. Thus, elucidation of the functions of HDA6 provided some very important clues of epigenetic regulation in plants (Kim et al. 2012). All these findings suggest the importance of transgenerationally acquired systemic resistance in plants and their potential role in managing plant diseases.

Future Prospects

The major challenge in the form of plant pathogens imposed to plant growth in natural and agricultural ecosystems urges for exploiting beneficial plant-microbe interactions. It is a well-known fact that microbial approach for plant disease management is necessary for maintaining the sustainability in agroecosystems. Therefore, it has been an emerging topic and gained considerable attention by many researchers. Though, various facets of plant-microbe interactions have long been studied, there is still a long way to go for achieving greater knowledge. The better understanding of plants' rhizosphere components is necessary to know the cross talk between plants and microbes. Linking this information to stress conditions would certainly provide a clue about developing a favorable and friendly environment for plant growth. Moreover, molecular approaches of studying the regulatory components of plant-microbe interactions can provide better understanding of improving such relationships by their manipulation.

Acknowledgments HBS and BKS are grateful to the Department of Biotechnology, Govt. of India, for providing financial support (BT/PR5990/AGR/5/587/2012). SM is thankful to UGC for awarding Dr. D.S. Kothari Postdoctoral Fellowship.

References

Alvarez MV, Moreira MR, Ponce A (2012) Antiquorum sensing and antimicrobial activity of natural agents with potential use in food. J Food Saf 32:379–387

Asada K (1999) The water–water cycle in chloroplasts: scavenging of active oxygen and dissipation of excess photons. Annu Rev Plant Physiol Plant Mol Biol 50:601–639

Asada K, Takahashi M (1987) Production and scavenging of active oxygen in photosynthesis. In: Kyle DJ, Osmond CB, Arntzen CJ (eds) Photoinhibition. Elsevier, Amsterdam, pp 227–287

Bais HP, Weir TL, Perry LG, Gilroy S, Vivanco JM (2006) The role of root exudates in rhizosphere interactions with plants and other organisms. Annu Rev Plant Biol 57:234–266

Bakker PAHM, Pieterse CMJ, Van Loon LC (2007) Induced systemic resistance by fluorescent pseudomonas spp. Phytopathology 97:239–243

Balakrishnan N, Subramanian KS (2012) Mycorrhizal symbiosis and bioavailability of micronutrients in maize grain. Maydica 57:129–138

Banerjee M, Yesmin L (2002) Sulfur-oxidizing plant growth promoting rhizobacteria for enhanced canola performance. US Patent 07491535

Bashan Y, Holguin G (1998) Proposal for the division of plant growth-promoting rhizobacteria into two classifications: biocontrol-PGPB (plant growth-promoting bacteria) and PGPB. Soil Biol Biochem 30:1225–1228

Berendsen RL, Pieterse CMJ, Bakker PAHM (2012) The rhizosphere microbiome and plant health. Trends Plant Sci 17:478–486

Berg G (2009) Plant–microbe interactions promoting plant growth and health: perspectives for controlled use of microorganisms in agriculture. Appl Microbiol Biotechnol 84:11–18

Berg G, Smalla K (2009) Plant species and soil type cooperatively shape the structure and function of microbial communities in the rhizosphere. FEMS Microbiol Ecol 68:1–13

Bernards MA, Lewis NG (1998) The macromolecular aromatic domain in suberized tissue: a changing paradigm. Phytochemistry 47:915–933

Boerjan W, Ralph J, Baucher M (2003) Lignin biosynthesis. Annu Rev Plant Biol 54:519–546

Bowler C, Slooten L, Vandenbranden S, De Rycke R, Botterman J, Sybesma C, Van Montagu M, Inzé D (1991) Manganese superoxide dismutase can reduce cellular damage mediated by oxygen radicals in transgenic plants. EMBO J 10:1723–1732

Cassán FD, García de Salamone I (2008) *Azospirillum* sp.: cell physiology, plant interactions and agronomic research in Argentina. Asociación Argentina de Microbiología, Argentina, p 266

Cazale AC, Droillard MJ, Wilson C, Heberle-Bors E, Barbier-Brygoo H, Laurière C (1999) MAP kinase activation by hypo-osmotic stress of tobacco cell suspensions: towards the oxidative burst response? Plant J 19:297–307

Clark RB, Zobel RW, Zeto SK (1999) Effects of mycorrhizal fungus isolates on mineral acquisition by *Panicum virgatum* acidic soils. Mycorrhiza 9: 167–176

Conrath U (2011) Molecular aspects of defence priming. Trends Plant Sci 16(10):524–531

Contreras-Cornejo HA, Macias-Rodriguez L, Cortes-Penagos C, Lopez-Bucio J (2009) *Trichoderma virens*, a plant beneficial fungus, enhances biomass production and promotes lateral root growth through an auxin-dependent mechanism in *Arabidopsis*. Plant Physiol 149:1579–1592

Crépin A, Barbey C, Beury-Cirou A, Hélias V, Taupin L, Reverchon S, Nasser W, Faure D, Dufour A, Orange N, Feuilloley M, Heurlier K, Burini JF, Latour X (2012a) Quorum sensing signaling molecules produced by reference and emerging soft-rot bacteria (*Dickeya* and *Pectobacterium* spp.). PLoS One 7(4):e35176

Crépin A, Barbey C, Cirou A, Tanniéres M, Orange N, Orange N, Feuilloley M, Dessaux Y, Burini JF, Faure D, Latour X (2012b) Biological control of pathogen communication in the rhizosphere: a novel approach applied to potato soft rot due to *Pectobacterium atrosepticum*. Plant Soil 358:27–37

Dat J, Vandenabeele S, Vranová E, Van Montagu M, Inzé D, Van Breusegem F (2000) Dual action of the active oxygen species during plant stress responses. Cell Mol Life Sci 57:779–795

Davin LB, Lewis NG (2000) Dirigent proteins and dirigent sites explain the mystery of specificity of radical precursor coupling in lignan and lignin biosynthesis. Plant Physiol 123:453–462

De Vleeschauwer D, Höfte M (2007) Using *Serratia plymuthica* to control fungal pathogens of plant. CAB Rev 2:46

De Werra P, Péchy-Tarr M, Keel C, Maurhofer M (2009) Role of gluconic acid production in the regulation of biocontrol traits of *Pseudomonas fluorescens* CHA0. Appl Environ Microbiol 75:4162–4174

Desikan R, A-HMackerness S, Hancock JT, Neill SJ (2001) Regulation of the *Arabidopsis* transcriptosome by oxidative stress. Plant Physiol 127:159–172

Diallo S, Crépin A, Barbey C, Orange N, Burini JF, Latour X (2011) Mechanisms and recent advances in biological control mediated through the potato rhizosphere. FEMS Microbiol Ecol 75:351–364

Dicke M, Hilker M (2003) Induced plant defences: from molecular biology to evolutionary ecology. Basic Appl Ecol 4:3–14

Dixon RA, Paiva N (1995) Stress induced phenylpropanoid metabolism. Plant Cell 7:1085–1097

Djonovic S, Pozo MJ, Dangott LJ, Howell CR, Kenerley CM (2006) Sm1, a proteinaceous elicitor secreted by the biocontrol fungus *Trichoderma virens* induces plant defense responses and systemic resistance. Mol Plant-Microbe Interact 8:838–853

Djonovic S, Vargas WA, Kolomiets MV, Horndeski M, Wiest A, Kenerley CM (2007) A proteinaceous elicitor Sm1 from the beneficial fungus *Trichoderma virens* is required for induced systemic resistance in maize. Plant Physiol 145:875–889

Dobbelare S, Vanderleydern J, Okon Y (2003) Plant-growth promoting effects of diazotrophs in the rhizosphere. Crit Rev Plant Sci 22:107–149

Dong YH, Wang LH, Xu JL, Zhang HB, Zhang XF, Zhang LH (2001) Quenching quorum-sensing-dependent bacterial infection by an N-acyl homoserine lactonase. Nature 411:813–817

Durrant WE, Dong X (2004) Systemic acquired resistance. Annu Rev Phytopathol 42:185–209

Friend J (1976) Lignification in infected tissue. In: Friend J, Threfall DR (eds) Biochemical aspects of plant-parasite relationships. Academic, London, pp 291–303

Glick BR (2005) Modulation of plant ethylene levels by the bacterial enzyme ACC deaminase. FEMS Microbiol Lett 252:1–7

Gottlieb S, Pelczar MJ (1951) Microbiological aspects of lignin degradation. Bacteriol Rev 15:55–76

Gram L, Grossart H, Schlingloff A, Kiørboe T (2002) Possible quorum sensing in marine snow bacteria: production of acylated homoserine lactones by roseobacter strains isolated from marine snow. Appl Environ Microbiol 8(68):4111–4116

Grant JJ, Loake GJ (2000) Role of reactive oxygen intermediates and cognate redox signaling in disease resistance. Plant Physiol 124:21–29

Harman G, Shoresh M (2007) The mechanisms and applications of opportunistic plant symbionts. In: Vurro M, Gressel J (eds) Novel biotechnologies for biocontrol agent enhancement and management. Springer, Amsterdam, pp 131–155

Harman GE, Howell CR, Viterbo A, Chet I, Lorito M (2004) *Trichoderma* species – opportunistic, avirulent plant symbionts. Nat Rev Microbiol 2:43–56

Harrison MJ (1999) Molecular and cellular aspects of the arbuscular mycorrhizal symbiosis. Annu Rev Plant Physiol 50:361–389

Harrison MJ (2005) Signaling in the arbuscular mycorrhizal symbiosis. Annu Rev Microbiol 59:19–42

Hatfield R, Vermerris W (2001) Lignin formation in plants. The dilemma of linkage specificity. Plant Physiol 126:1351–1357

Hayatsu M, Tago K, Saito M (2008) Various players in the nitrogen cycle: diversity and functions of the microorganisms involved in nitrification and denitrification. Soil Sci Plant Nutr 54:33–45

Howell CR, Hanson LE, Stipanovic RD, Puckhaber LS (2000) Induction of terpenoid synthesis in cotton roots and control of *Rhizoctonia solani* by seed treatment with *Trichoderma virens*. Phytopathology 90:248–252

Hummerschmidt R (1999) Phytoalexins: what have we learned after 60 years? Annu Rev Phytopathol 37:285–306

Jain A, Singh S, Sarma BK, Singh HB (2012) Microbial consortium mediated reprogramming of defence network in pea to enhance tolerance against *Sclerotinia sclerotiorum*. J Appl Microbiol 112:537–550

Jetiyanon K (2007) Defensive-related enzyme response in plants treated with a mixture of *Bacillus* strains (IN937a and IN937b) against different pathogens. Biol Control 42:178–185

Jeun YC, Park KS, Kim CH, Fowler WD, Kloepper JW (2004) Cytological observations of cucumber plants during induced resistance elicited by rhizobacteria. Biol Control 29:34–42

Jones JDG, Dangl JL (2006) The plant immune system. Nature 444:323–329

Kamilova F, Validov S, Azarova T, Mulders I, Lugtenberg B (2005) Enrichment for enhanced competitive plant root tip colonizers selects for a new class of biocontrol bacteria. Environ Microbiol 7:1809–1817

Kamilova F, Kravchenko LV, Shaposhnikov AI, Makarova N, Lugtenberg BJJ (2006) Effects of the tomato pathogen *Fusarium oxysporum* f. sp. *radicis-lycopersici* and of the biocontrol bacterium *Pseudomonas fluorescens* WCS365on the composition of organic acids and sugars in tomato root exudate. Mol Plant-Microbe Interact 19:1121–1126

Keswani C, Mishra S, Sarma BK, Singh SP, Singh HB (2014) Unraveling the efficient applications of secondary metabolites of various *Trichoderma* spp. Appl Microbiol Biotechnol 98:533–544

Kim JM, To TK, Seki M (2012) An epigenetic integrator: new insights into genome regulation, environmental stress responses and developmental controls by histone deacetylase 6. Plant Cell Physiol 53(5): 794–800

Kloepper JW, Leong J, Teintze M, Schroth MN (1980) Enhanced plant growth by siderophores produced by plant growth-promoting rhizobacteria. Nature 286:885–886

Kloepper JW, Ryu CM, Zhang SA (2004) Induced systemic resistance and promotion of plant growth by *Bacillus* spp. Phytopathology 94:1259–1266

Lambert DH, Cole H, Baker DE (1980) The role of boron in plant response to mycorrhizal infection. Plant Soil 57:431–438

Lavania M, Chauhan PS, Chauhan SVS, Singh HB, Nautiyal CS (2006) Induction of plant defense enzymes and phenolics by treatments with plant growth promoting rhizobacteria *Serratia marcescens* NBRI 1213. Curr Microbiol 52:363–368

Lewis NG, Yamamoto E (1990) Lignin: occurrence, biogenesis and biodegradation. Annu Rev Plant Physiol Plant Mol Biol 41:455–496

Liu A, Hamel C, Hamilton RI, Ma BL, Smith DL (2000) Acquisition of Cu, Zn, Mn and Fe by mycorrhizal maize (*Zea mays* L.) growth in soil at different P and micronutrient levels. Mycorrhiza 9:331–336

Lugtenberg BJJ, Chin-A-Woeng TFC, Bloemberg GV (2002) Microbe– plant interactions: principles and mechanisms. Antonie Van Leeuwenhoek 81:373–383

Luna E, Ton J (2012) The epigenetic machinery controlling transgenerational systemic acquired resistance. Plant Signal Behav 7:615–618

Luna E, Bruce TJA, Roberts MR, Flors V, Ton J (2012) Next-generation systemic acquired resistance. Plant Physiol 158:844–853

Mäe A, Montesano M, Koiv V, Palva ET (2001) Transgenic plants producing the bacterial pheromone N-acyl-homoserine lactone exhibit enhanced resistance to the bacterial phytopathogen *Erwinia carotovora*. Mol Plant-Microbe Interact 14:1035–1042

Mandal S, Mitra A (2007) Reinforcement of cell wall in roots of *Lycopersicon esculentum* through induction of phenolic compounds and lignin by elicitors. Physiol Mol Plant Pathol 71:201–209

Meziane H, Van der Sluis I, Van Loon LC, Ho¨fte M, Bakker PAHM (2005) Determinants of *Pseudomonas putida* WCS358 involved in inducing systemic resistance in plants. Mol Plant Pathol 6:177–185

Mittler R (2002) Oxidative stress, antioxidants and stress tolerance. Trends Plant Sci 7:405–410

Molina L, Constantinescu F, Michel L, Reimmann C, Duffy B, Defago G (2003) Degradation of pathogen quorum-sensing molecules by soil bacteria: a preventive and curative biological control mechanism. FEMS Microbiol Ecol 45:71–81

Morrissey JP, Dow JM, Mark L, O'Gara F (2004) Are microbes at the root of a solution to world food production? EMBO Rep 5:922–926

Nicholson RL, Hammerschmidt R (1992) Phenolic compounds and their role in disease resistance. Annu Rev Phytopathol 30:369–389

Ongena M, Jourdan E, Scha¨fer M, Kech C, Budzikiewicz H, Luxen A, Thonart P (2005) Isolation of an N-alkylated benzylamine derivative from *Pseudomonas putida* BTP1 as elicitor of induced systemic resistance in bean. Mol Plant Microbe Interact 18:562–569

Ongena M, Jourdan E, Adam A, Paquot M, Brans A, Joris B, Arpigny JL, Thonart P (2007) Surfactin and fengycin lipopeptides of *Bacillus subtilis* as elicitors of induced systemic resistance in plants. Environ Microbiol 9:1084–1090

Park KS, Kloepper JW (2000) Activation of PR-1a promoter by rhizobacteria which induce systemic resistance in tobacco against *Pseudomonas syringae* pv. *tabaci*. Biol Control 18:2–9

Pei ZM, Murata Y, Benning G, Thomine S, Klüsener B, Allen GJ, Grill E, Schroeder JI (2000) Calcium channels activated by hydrogen peroxide mediate abscisic acid signaling in guard cells. Nature 406:731–734

Pierson EA, Wood D, Cannon JAW, Blachere FM, Pierson LS (1998a) Interpopulation signaling via N-acyl-homoserine lactones among bacteria in the wheat rhizosphere. Mol Plant-Microbe Interact 11: 1078–1084

Pierson LS, Wood DW, Pierson EA (1998b) Homoserine lactone-mediated gene regulation in plant-associated bacteria. Annu Rev Phytopathol 36:207–225

Pieterse CMJ (2012) Prime time for transgenerational defense. Plant Physiol 158:545

Pieterse CM, Dicke M (2007) Plant interactions with microbes and insects: from molecular mechanisms to ecology. Trends Plant Sci 12:564–569

Pieterse CMJ, van der Does D, Zamioudis C, Leon-Reyes A, van Wees SCM (2012) Hormonal modulation of plant immunity. Annu Rev Cell Dev Biol 28:489–521

Pineda A, Zheng SJ, van Loon JJA, Pieterse CMJ, Dicke M (2010) Helping plants to deal with insects: the role of beneficial soil-borne microbes. Trends Plant Sci 15:507–514

Pozo MJ, Azcon-Aguilar C (2007) Unraveling mycorrhiza-induced resistance. Curr Opin Plant Biol 10:393–398

Ran LX, Li ZN, Wu GJ, Van Loon LC, Bakker PAHM (2005) Induction of systemic resistance against bacterial wilt in *Eucalyptus urophylla* by fluorescent *Pseudomonas* spp. Eur J Plant Pathol 113:59–70

Ride JP (1978) The role of cell wall alterations in resistance to fungi. Ann Appl Biol 89:302–306

Ryan RP, Monchy S, Cardinale M, Taghavi S, Crossman L, Avison MB, Berg G, van der Lelie D, Dow JM (2009) Versatility and adaptation of bacteria from the genus *Stenotrophomonas*. Nat Microbiol Rev 7: 514–525

Ryu CM, Farag MA, Hu CH, Reddy MS, Wie HX, Paré PW, Kloepper JW (2003) Bacterial volatiles promote growth of *Arabidopsis*. Proc Natl Acad Sci 100: 4927–4932

Ryu CM, Farag MA, Hu CH, Reddy MS, Kloepper JW, Pare' PW (2004) Bacterial volatiles induce systemic resistance in *Arabidopsis*. Plant Physiol 134: 1017–1026

Saleem M, Arshad M, Hussain S, Bhatti AS (2007) Perspective of plant growth promoting rhizobacteria (PGPR) containing ACC deaminase in stress agriculture. J Ind Microbiol Biotechnol 34:635–648

Schrey SD, Tarkka MT (2008) Friends and foes: *Streptomycetes* as modulators of plant disease and symbiosis. Antonie Van Leeuwenhoek 94:11–19

Schuhegger R, Ihring A, Gantner S, Bahnweg G, Knappe C, Hartmann A, Langebartels C (2006) Induction of systemic resistance in tomato by N-acyl-L-homoserine lactone-producing rhizosphere bacteria. Plant Cell Environ 29:909–918

Sederoff RR, MacKay JJ, Ralph J, Hatfield RD (1999) Unexpected variation in lignin. Curr Opin Plant Biol 2:145–152

Serfling A, Wirsel SGR, Lind V, Deising HB (2007) Performance of the biocontrol fungus *Piriformospora indica* on wheat under greenhouse and field conditions. Phytopathology 97:523–531

Shoresh M, Harman GE, Mastouri F (2010) Induced systemic resistance and plant responses to fungal biocontrol agents. Annu Rev Phytopathol 48:21–43

Silva HSA, Romeiro RDS, Macagnan D, Halfeld-Vieira BDA, Pereira MCB, Mounteer A (2004) Rhizobacterial induction of systemic resistance in tomato plants nonspecific protection and increase in enzyme activities. Biol Control 29:288–295

Singh A, Sarma BK, Upadhyay RS, Singh HB (2013) Compatible rhizosphere microbes mediated alleviation of biotic stress in chickpea through enhanced antioxidant and phenylpropanoid activities. Microbiol Res 168:33–40

Singhai PK, Sarma BK, Srivastava JS (2011) Biological management of common scab of potato through *Pseudomonas* species and vermicompost. Biol Control 57:150–157

Slaughter A, Daniel X, Flors V, Luna E, Hohn B, Mauch-Mani B (2012) Descendants of primed Arabidopsis plants exhibit resistance to biotic stress. Plant Physiol 158:835–843

Smith SE, Jakobsen I, Gronlund M, Smith FA (2011) Roles of arbuscular mycorrhizas in plant phosphorus nutrition: Interactions between pathways of phosphorus uptake in arbuscular mycorrhizal roots have important implications for understanding and manipulating plant phosphorus acquisition. Plant Physiol 156: 1050–1057

Spaink HP (2000) Root nodulation and infection factors produced by rhizobial bacteria. Annu Rev Microbiol 54:257–288

Stein E, Molitor A, Kogel KH, Waller F (2008) Systemic resistance in *Arabidopsis* conferred by the mycorrhizal fungus *Piriformospora indica* requires jasmonic acid signaling and the cytoplasmic function of NPR1. Plant Cell Physiol 49:1747–1751

Tran H, Ficke A, Asiimwe T, Ho¨fte M, Raaijmakers JM (2007) Role of the cyclic lipopeptide massetolide A in biological control of *Phytophthora infestans* and in colonization of tomato plants by *Pseudomonas fluorescens*. New Phytol 175:731–742

Truchado P, Tomás-Barberán F, Larrosa M, Allende A (2012) Food phytochemicals act as quorum sensing inhibitors reducing production and/or degrading autoinducers of *Yersinia enterocolítica* and *Erwinia carotovora*. Food Control 24:78–85

Unno Y, Okubo K, Wasaki J, Shinano T, Osaki M (2005) Plant growth promotion abilities and microscale bacterial dynamics in the rhizosphere of lupin analysed by phytate utilization ability. Environ Microbiol 7: 396–404

Van der Putten WH, Vet LM, Harvey JA, Wäckers FL (2001) Linking above- and belowground multitrophic interactions of plants, herbivores, pathogens, and their antagonists. Trends Ecol Evol 16:547–554

Van Loon LC, Bakker PAHM, Pieterse CMJ (1998) Systemic resistance induced by rhizosphere bacteria. Annu Rev Phytopathol 36:453–483

Van Oosten VR, Bodenhausen N, Reymond P, Van Pelt JA, Van Loon LC, Dicke M, Pieterse CMJ (2008) Differential effectiveness of microbially induced resistance against herbivorous insects in *Arabidopsis*. Mol Plant-Microbe Interact 21:919–930

van Rhijn P, Vanderleyden J (1995) The *Rhizobium*-plant symbiosis. Microbiol Rev 59:124–142

van Wees SCM, de Swart EAM, van Pelt JA, van Loon LC, Pieterse CMJ (2000) Enhancement of induced disease resistance by simultaneous activation of salicylate – and jasmonate-dependent defense pathways in *Arabidopsis thaliana*. Proc Natl Acad Sci U S A 97:8711–8716

Vance CP, Anderson JO, Sherwood RT (1976) Soluble and cell wall peroxidases in reed canary grass in relation to disease resistance and localized lignin formation. Plant Physiol 57:920–922

Vinale F, Sivasithamparam K, Ghisalberti EL, Marra R, Barbetti MJ, Li H, Woo SL, Lorito M (2008) A novel role for *Trichoderma* secondary metabolites in the interactions with plants. Physiol Mol Plant Pathol 72:80–86

Whitmore FW (1978) Lignin-carbohydrate complex formed in isolated cell walls of callus. Phytochemistry 17:421–425

Woo SL, Scala F, Ruocco M, Lorito M (2006) The molecular biology of the interactions between *Trichoderma* spp., phytopathogenic fungi and plants. Phytopathology 96:181–185

Yates IE, Bacon CW, Hinton DM (1997) Effects of endophytic infection by *Fusarium* moniliforme on corn growth and cellular morphology. Plant Dis 81:723–728

Zhang H, Xie X, Kim MS, Kornyeyev DA, Holaday S, Par'e PW (2008) Soil bacteria augment *Arabidopsis* photosynthesis by decreasing glucose sensing and abscisic acid levels *in planta*. Plant J 56:264–273

Rhizospheric Plant-Microbe Interactions: Key Factors to Soil Fertility and Plant Nutrition

6

Brijesh Kumar Yadav, Mohd. Sayeed Akhtar, and Jitendra Panwar

Abstract

Plant roots radiate a wide range of potentially valuable small molecular weight compounds into the rhizosphere which play a key role in the chemical, physical, and biological interaction between roots of the plants and the rhizosphere. The microorganisms present in the rhizosphere react with the numerous metabolites released by plant roots by positive, negative, and neutral ways, and these interactions may influence the plant growth and development, change nutrient dynamics, and also alter the plants susceptibility towards diseases and abiotic stresses. The root produces chemical signals that attract the bacteria and other microbes towards it. Beside this, positive interactions also include growth regulator mimics that support the plant growth and the cross-species signaling with other rhizospheric microorganisms. Plant-microbe interactions can influence the plant growth by providing nutrients and increased biotic and abiotic stress tolerance. Most of the agricultural soils have large amounts of inorganic and organic phosphorus (P), but it is present in immobilized form so is usually unavailable to plants. One of the major reasons why P is not readily available to plants is because of the high reactivity of P with some metal complexes. In this regard, the soil inoculants such as fungi, plant growth-promoting rhizobacteria (PGPR), and mycorrhizal fungi play a significant role in the solubilization of inorganic phosphate and mineralization of organic phosphates into easily available form to plants. Similarly, nitrogen (N) fixers provide available N to the plants. N is a key limiting factor in any ecosystem. For treating heavy metal-contaminated tailings and soils, bioremediation is one

B.K. Yadav (✉)
Punjab Agricultural University Regional Research Station, Bathinda 151001, India
e-mail: bkyadav74@yahoo.co.in

M.S. Akhtar (✉)
Institute of Tropical Agriculture, Universiti Putra Malaysia, UPM, 43400, Serdang, Selangor, Malaysia
e-mail: sayeedbot@yahoo.co.in

J. Panwar (✉)
Centre for Biotechnology, Department of Biological Sciences, Birla Institute of Technology and Science, Pilani 333031, India
e-mail: drjitendrapanwar@yahoo.co.in

N.K. Arora (ed.), *Plant Microbes Symbiosis: Applied Facets*,
DOI 10.1007/978-81-322-2068-8_6, © Springer India 2015

of the cost-effective methods and is emerging as the potential tool for removal of these contaminants from the soil or water. Bioremediation is a versatile process that could be applied in situ or ex situ manner. A wide variety of microorganisms such as bacteria, fungi, yeasts, and algae are being used in bioremediation processes, and some of these have already been employed as biosorbents of heavy metals. Various technologies such as phytoremediation, bioventing, bioleaching, land farming, bioreactor, composting, bioaugmentation, rhizofiltration, and biostimulation are nowadays used for the bioremediation of contaminants from the soil. The aim of this chapter is to focus on the plant-microbe interactions responsible for the maintenance of soil fertility, plant nutrition, and also the remediation of contaminated soil for sustainable agricultural system.

Introduction

Plants and their associated microorganisms play an important role in the formation or modification of soil (Pate and Verboom 2009). The soil formation takes place from the weathering of rocks and minerals and has distinct properties based on their origin, climate, and vegetation. Soil carbon is predominantly derived from plants, directly or indirectly, and while weathering may be due to physical and chemical influences. Most of the weathering processes involve primarily the plant roots, and the microbial activity may depend on the root-derived carbon (Beerling and Berner 2005). Next to water and temperature, nutrients are most crucial environmental factors desired for the development of the terrestrial plants. Rhizosphere is a zone of intense microbial activity, and the microorganisms present in the rhizosphere react with the various metabolites released by plant roots (Akhtar and Siddiqui 2010). Thus, the microorganisms and their products interact with plant roots in a variety of ways such as positive, negative, and neutral (Kuzyakov and Xu 2013). These interactions can influence the plant growth, change nutrient dynamics, and also alter the plants susceptibility towards the disease, abiotic stress, and resistance to heavy metals (Morgan et al. 2005).

The root systems support the aboveground part of the plant. In addition, the soil needs to maintain an appropriate pH, provide the protection from toxic substances and pathogens, and also contain suitable water levels. Besides this, all the essential mineral elements desired by plants are obtained from the soil. Marschner (1995) has recognized about 17 essential elements required for plant growth and reproduction. Amongst all the required essential elements, 14 elements are primarily acquired from the soil solution including the six macronutrients (N, P, K, Ca, Mg, and S) and 8 micronutrients (Fe, Cu, Zn, Mn, Mo, B, Cl, and Ni). Additionally, the plants also accumulate nonessential or toxic elements such as Cd, Pb, and Na from the soil solution.

Most of the essential elements are taken up in the ionic form from the soil solution by plants (White 2003). It is a well-known fact that the plant growth may be limited by the availability of essential elements or by the presence of toxic elements (Morgan et al. 2005). The interactions between microorganisms and the plant roots in rhizosphere assist the plants to acquire essential mineral nutrients and thwart the accumulation of toxic elements. Thus, it would be obvious that the structures of rhizospheric microbial communities are distinct from the bulk soil (Marilley and Aragno 1999), but vary in between plant species and over geographical time scale (Smalla et al. 2001). The different root zones in the same plant can hold distinct microbial communities reflecting qualitative and quantitative differences in root exudation (Yang and Crowley 2000). Moreover, the structure of rhizospheric microbial communities could also be influenced by root infection by pathogenic microorganisms, which promote greater microbial community variability compared to healthy roots (Yang et al. 2001).

The Rhizosphere

Hiltner (1904) recognized rhizosphere as the volume of soil within immediate vicinity of the roots, which is predominantly affected by the activity of plants. The rhizosphere differs from the surrounding soils in most of the physicochemical and biological factors and with extensive microbial population both in numbers and diversity (Phillips et al. 2003). The number of microorganisms present in per gram of soil is much larger in the rhizosphere as compared to bulk soil. This increased microbial activity in the vicinity of the roots could be attributed to root exudates, sloughed senescent root cells, and mucigel described as rhizodeposition (Mukerji et al. 1998). Thus, the rhizosphere is the region in which the materials released from the root and root metabolic activities directly affect microbial density (Table 6.1). The roots continuously release volatile, soluble, and particulate materials through the process of rhizodeposition, and the growth of rhizospheric microorganisms on these materials turn over all the cellular activities and also release the nutrients in the form which could be utilized by plants.

The rhizosphere encompasses not only the region of nutrient uptake by the plant roots but also extends into the soil by the action of their byproducts (Van der Putten et al. 2001). This infusion of organic substrates into the rhizosphere by plants explains very clearly that the biomass and microbial activity are always greater in rhizosphere compared to bulk soil (Bardgett et al. 1998). Root tip is the site of root growth, usually characterized by rapidly dividing cells having the root exudates. The root exudates and sloughed root cells provide carbon to

Table 6.1 Various root zones in the soil

Root zones	Functions
Rhizosphere	Region around the plant root where materials released from the root modify microbial populations and their activities
Endorhizosphere	Regions of the various cell layers of the root itself where microorganisms also colonize
Ectorhizosphere	An area surrounding the root and containing root hairs, plant, and bacterial mucilage
Rhizoplane	Root surface that can be colonized by microorganisms
Mycorrhizosphere	The ectorhizosphere extends a substantial distance from the root with the development of mycorrhizal fungal associations. Materials released from the fungus increase the microbial populations and their activities around the fungal hyphae
Spermosphere	The region around the germinating seed
Rhizodeposition	Release of materials from the roots

rhizospheric microorganisms, which in turn mobilize N and P in soil for the plants.

Chemical Compounds Produced by Plant Roots in the Rhizosphere

The general concept about the plant is that an aerial part such as stem and leaves contains greater biomass than roots, but this is actually a misleading impression. For many plants the biomass ratio of root and shoot is greater than the shoot and root biomass ratio. The materials released by plants include a wide variety of organic compounds (Table 6.2). The nature of these compounds depends upon various environment factors like temperature, moisture content, fertilizer dosages, herb, and plant age. The fine hairs of the roots are critical parts of the root system. They release various root exudate products into the environment due to their metabolic activity, and additionally a variety of gaseous metabolites also flow from the roots in the process of rhizodeposition.

Table 6.2 Different compounds released by plant roots in the process of rhizodeposition

Compounds	Exudates components
Amino compounds	Asparagine, α- alanine, glutamine, aspartic acid, leucine/isoleucine, serine, glycine, cystine/cysteine, methionine, phenylalanine, tyrosine, threonine, lysine, proline, tryptophan, β- alanine, arginine, homoserine, cystathionine
Fatty acids and sterols	Palmitic, stearic, oleic, linoleic, linolenic acids, cholesterol, campesterol, stigmasterol, sitosterol
Growth factors	Biotin, thiamine, niacin, pantothenate, choline, inositol, pyridoxine, N-methyl nicotinic acid
Nucleotides, flavonines, and enzymes	Flavonine, adenine, guanine, uridine/cytidine, phosphatase, invertase, amylase, protease, polygalacturonase
Organic acids	Tartaric, oxalic, citric, malic, propanoic, butyric, succinic, fumaric, glycolic, valeric, malonic
Sugars	Glucose, fructose, sucrose, maltose, galactose, rhamnose, ribose, xylose, arabinose, raffinose, oligosaccharide
Miscellaneous compounds	Auxins, scopoletin, fluorescent substances, hydrocyanic acid, glycosides, saponin (glucosides), organic phosphorus compounds, nematode-cyst or egg-hatching factors, nematodes attractants, fungal mycelium growth stimulants and inhibitors, zoospore attractants

Microbial Miscellanies in the Rhizosphere

The rhizosphere is the "cloud" of microbes surrounding the plant roots and is vital for the plant growth and survival. Plant roots construct novel environments for microbes due to change in increased levels of nutrients and intense microbial population (Giri et al. 2005). The microbial density is always high along the root away from the plant tip because respiration of the root is responsible for the change in the environment of the rhizosphere microbes. However, the microbial community developed in the changed environment will also face additional challenges like the availability of nutrients and edaphic and other biotic and abiotic factors which might be limiting for both the plant and their associated rhizospheric microbes.

The plant has an increasing demand for inorganic unavailable nutrients not present in a sufficient rate in the soil. In this regard, the rhizospheric microorganisms make a major contribution to overcome this demand (Table 6.3). The filamentous fungi also play a unique role in the nutrient uptake available to the plant due to their extensive hyphal network. They can derive the carbon from the plant and other limiting nutrients such as N and P from outside the root zone. Some free-living N_2-fixing bacteria like *Azotobacter*, *Azospirillum*, and *Azoarcus* present in the nitrogen-free or low nitrogen input environment play a significant role in the nitrogen fixation and also make the availability of nutrients to plant. Besides this, the rhizospheric community is not only enriched in bacterial and fungal populations but also has protozoans and nematodes. These patrons feed on the nutrient-rich bacteria and fungi and lead to more rapid turnover of the microbe populations in the rhizosphere which could be responsible for the acceleration in the release of nutrients for plant.

Plant-Microbe Interaction and Nutrient Availability

Continuous application of chemical fertilizers for enhancing soil fertility and crop productivity resulted in unforeseen harmful environmental effects such as leaching of nitrate into ground water, surface runoff of N and P, and eutrophication of aquatic ecosystems (Tilman 1998; Gyaneshwar et al. 2002). Besides this, it also negatively impacted the complex system of the biogeochemical cycles (Perrott et al. 1992; Steinshamn et al. 2004). All these events suggested that long-term application of chemical fertilizers not only reduced the soil fertility but also reduced the microbial population in the various agroclimatic conditions.

Despite the various negative effects on the environment, the use of total amount of fertilizers has increased globally to fulfill the food demand

Table 6.3 Microbial density of major groups of microorganisms present in the rhizospheric and non-rhizospheric soils

Microorganisms	Rhizosphere (microbes/g dry soil)	Non-rhizosphere (microbes/g dry soil)	R:S ratio
Algae	5×10^3	27×10^3	0.2
Actinomycetes	46×10^6	7×10^6	7.0
Bacteria	1200×10^6	53×10^6	23.0
Fungi	12×10^5	1×10^5	12.0

of the growing world population through intensive agriculture (Vitousek et al. 1997; Frink et al. 1999). In the last five decades, the rate of NPK fertilizer application has increased tremendously. Thus, there is a challenge to increase the agricultural productivity by minimizing the use of harmful chemical fertilizers. In this regard, the use of microbial inoculants such as PGPR and mycorrhizal fungi play a crucial role to minimize the demand of chemical fertilizers. Thus, the microbial inoculants are considered as promising components of agro-environmental integrated system because these inoculants possess the capacity to promote plant growth, enhance nutrient availability and uptake, and support plants' health (Barea et al. 1998; Dobbelaere et al. 2001; Hodge et al. 2001; Bonfante 2003; Vessey 2003; Kloepper et al. 2004; Han and Lee 2005; Adesemoye et al. 2008; Akhtar et al. 2011).

The arbuscular mycorrhizal (AM) fungi increase the plant growth by water and nutrient uptake (Ames et al. 1983; Akhtar and Siddiqui 2008a; Akhtar and Panwar 2011). The AM fungi have a high affinity towards P-uptake mechanism that enhances P nutrition in plants. The AM fungi could also scavenge the available P through their extraradical hyphae (Bianciotto and Bonfante 2002; Akhtar and Siddiqui 2008b; Akhtar et al. 2011). Apart from this beneficial association between the AM fungi and plant roots, there are few demerits as well. The AM fungi could not be cultured in vitro, due to their obligate nature, limiting the knowledge about the genetic basis of P solubilization and rhizosphere competence (Amijee et al. 1989; Koide 1991). Moreover, a high concentration of soil phosphate, above 100 ppm, could lead to the reduction in hyphal growth and chlamydospore production (Koide 1991). These limitations directly affect the P uptake and also cause the reduction of the benefits to plants (Stewart et al. 2005).

Similarly, the use of PGPR increases seed germination rate, root growth, yield, leaf area, chlorophyll content, nutrient uptake, protein content, hydraulic activity, tolerance to abiotic stress, and shoot and root weight, biocontrol and also delay the senescence (Raaijmakers et al. 1997; Bashan et al. 2004; Mantelin and Touraine 2004; Siddiqui et al. 2007; Bakker et al. 2007; Yang et al. 2009; Akhtar and Siddiqui, 2010). Other beneficial effects of PGPR include enhancing P availability (Rodriguez and Fraga 1999; Akhtar and Siddiqui 2008a; Akhtar and Panwar 2011; Yadav et al. 2012); fixing atmospheric nitrogen (Bashan et al. 2004; Gupta et al. 2012); sequestering iron for plants by production of siderophores (Raaijmakers et al. 1997; Bakker et al. 2007; Akhtar and Siddiqui, 2009), producing plant hormones such as gibberellins, cytokinins, and auxins (Gutierrez-Manero et al. 2001); and synthesizing 1-amino cyclopropane-1-carboxylate (ACC) deaminase (Glick et al. 2007a, b).

Availability of Nitrogen

Nitrogen is an essential element for plant growth and development, and the complex nitrogen cycle has a great impact on soil fertility (Jetten 2008). This cycle is conquered by four major steps, nitrogen fixation, nitrification, denitrification, and nitrogen mineralization (Ogunseitan 2005). It has been reported by earlier researchers that the microbial inoculants have significant roles in nitrogen cycling and utilization of nitrogenous fertilizers in the plant-soil system (Briones et al. 2003; Adesemoye et al. 2009) (Fig. 6.1). The uptake of nitrogen by leguminous (Elsheikh and Elzidany 1997; Akhtar and Siddiqui, 2006; Gupta et al. 2012) and nonlegume plants have been reviewed by many researchers (Kennedy et al. 1997; Dobbelaere et al. 2001;

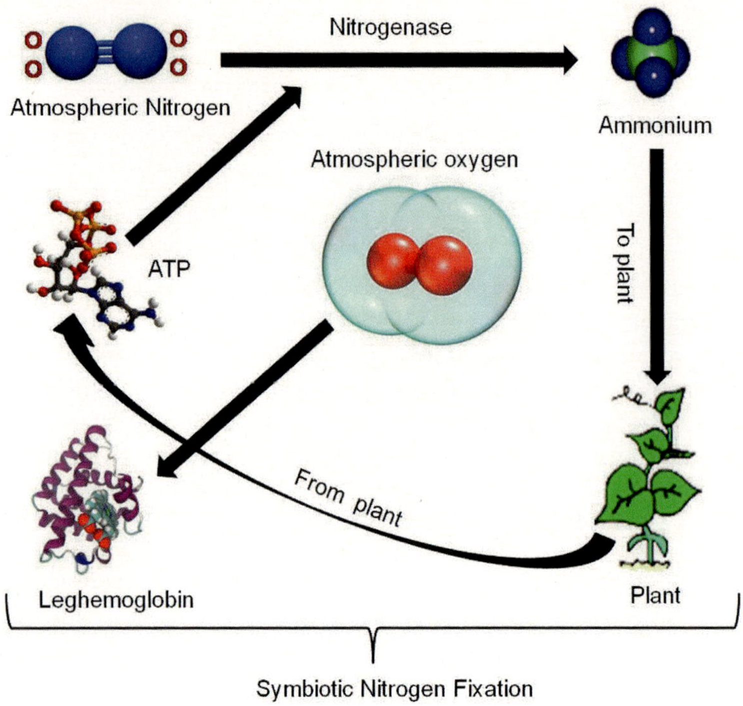

Fig. 6.1 Schematic representation of symbiotic nitrogen fixation in the leguminous plants

Vessey 2003; Egamberdiyeva and Hoflich 2004; Hernandez and Chailloux 2004; Wu et al. 2005; Shaharoona et al. 2008).

Wu et al. (2005) conducted greenhouse experiments on maize utilizing *Glomus mosseae* and *Glomus intraradices* with or without free-living nitrogen fixer, *Azotobacter chroococcum* and found that the co-inoculant increased the plant growth and NPK uptake and improved the soil properties in a much better way. Shaharoona et al. (2008) reported that inoculation of *Pseudomonas fluorescens* (strain ACC50) and *P. fluorescens* biotype F (strain ACC73) increased efficiency at all tested NPK fertilizer levels in wheat under pot and field trials. Amir et al. (2005) found that the inoculation of PGPR enhanced uptake of N and P in oil palm seedlings in nursery. Similarly, Aseri et al. (2008) conducted a field experiment to assess the effectiveness of PGPR (*A. chroococcum*) and AM fungi (*G. mosseae*) on the growth, nutrient uptake, and biomass production of pomegranate in individual or combined inoculations. The results showed that dual inoculation of PGPR and AM fungi led to higher biomass accumulation and uptake of N, P, K, Ca, and Mg. The result of the study thus confirmed that inoculation with mixed strains was more consistent than inoculation of single strain. It is well reported that the uptake of NPK and micronutrients are significantly enhanced in *Azospirillum* spp. inoculated plants under the greenhouse and field conditions. Thus, it would be very crucial to find out the factors behind the successful plant root colonization in *Azospirillum* and other PGPR which is responsible for the increase nutrient uptake.

Nitrogenase is the enzyme responsible for N_2-fixation. It has two components: I (a $\alpha2\beta2$ tetramer encoded by *nifD* and *nifK* genes) and II (a homodimer encoded by *nifH* gene). These two components are conserved in structure, function, and amino acid sequence throughout the diazotrophs. These genes are commonly reported to regulate lateral root development and long-distance movement of nitrogen (de Zamaroczy et al. 1989; Ueda et al. 1995; Minerdi et al. 2001). The nitrogenase enzyme complex has been credited for the capacity of PGPR to convert nitrogen into ammonia in a free state. Egener et al. (1999)

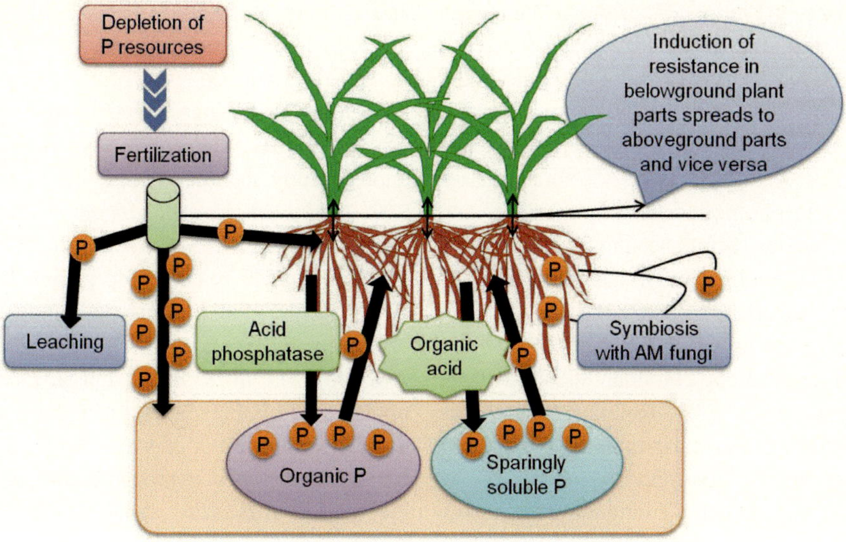

Fig. 6.2 Mechanism of soil P solubilization/mineralization and immobilization

studied root-associated GUS and *nifH* expression to monitor the establishment of N_2-fixing bacteria (*Azoarcus* sp.) on rice roots and successfully localized the expression of bacterial genes of interest in the host plant. They have found that the presence of combined nitrogen such as ammonia has a strong impact on the expression of *nif* gene in most diazotrophs. Similarly, Vande Broek et al. (1993) estimated the qualitative and quantitative associativeness in *nifH* expression in *A. brasilense* on wheat roots through *gusA* fusion plasmid system.

Availability of Phosphorus

Phosphorus is another growth-limiting nutrient generally present in the immobilized organic and inorganic form in the soil. Phosphorus is not readily available to plant due to its high reactivity with some metal complexes such as Fe, Al, and Ca leading to precipitation in the soil or may be present in very low concentration (usually in micromolar amount) (Igual et al. 2001; Gyaneshwar et al. 2002). It is also an important and well-known fact that when the P fertilizers are added to soils, they may not be absorbed or utilized by plants because of their sparingly soluble nature, and thus less amount of P would be available for the growth of agricultural crops (Gyaneshwar et al. 2002). Thus, the farmer may have to add large amount of fertilizers into the fields which later cause environmental problems (Ohno et al. 2005). The inoculants of PGPR and AM fungi play a significant role in the solubilization of inorganic phosphate and mineralization of organic phosphates (Mahmood et al. 2001; Tarafdar and Marschner 1994a; Tawaraya et al. 2006). Moreover, there are many evidences which are related to inorganic phosphate (Pi) transporter and its expression in the external hyphae of AM fungi, which is important in the uptake of P and transfer from the AM fungi to plants (Tarafdar and Marschner 1994b; Harrison and van Buuren 1995).

The mechanism for the solubilization of P was reported by Gharu and Tarafdar (2004) and Chen et al. (2006) (Fig. 6.2). The organic P usually accounting for 30–65 % of total P in soils must be converted to inorganic or low molecular weight organic acids before they could be assimilated by plants. The different forms of organic P in soils are inositol phosphatases, phosphoesters, phosphodiesters, and phosphotriesters. A large part of the organic P is present in the form of inositol phosphatases (phytate) (Rodriguez and Fraga 1999; Zimmermann 2003). Phosphatases refer to any enzyme that can hydrolyze phosphate esters and anhydrides including phosphoprotein phosphatases, phosphodiesterases, diadenosine tetraphosphatases, exonucleases, 5′-nucleotidases phytases, phosphomonoesterases, alkaline, and acid phosphatases (Zimmermann 2003).

The role of phosphatases in the mobilization of phosphorus, originating from the organic soil sources, by AM fungi and PGPR has been reviewed by several researchers (Tarafdar and Marschner 1994b; Idriss et al. 2002; Rodriguez and Fraga 1999). Moreover, molecular tools have been also used to elucidate plant-microbe interactions in phosphorus metabolism (Rodriguez et al. 2000; Chen et al. 2006). Minder et al. (1998) indicated that the genetic control system of phosphate uptake is based on the phosphate regulatory protein *PhoB*, which is mediated by the transmembrane sensor protein *PhoR*. They have suggested that phosphorylated *PhoB* acts as a transcriptional activator to the *Pho* box in the promoter region of genes belonging to the *Pho* regulon and concluded that the product of the *PhoB* gene regulates the cellular response to environmental phosphate limitation. A study on *Bradyrhizobium japonicum* and soybean concluded that *PhoB* is responsible only for phosphate-limited growth, not for symbiotic nitrogen fixation (Minder et al. 1998). Subsequently, Ruiz-Lozano and Bonfante (1999) have investigated the role of *Burkholderia* sp. and AM fungi in P metabolism and found that through shunting off mechanism, the phosphorus is transferred from fungus to the plant. Ruiz-Lozano and Bonfante (1999) also characterized an operon *Pst*-like system in *Burkholderia* similar to *E. coli* phosphorus uptake by *Gigaspora margarita*. By the possession of a DNA region with nitrogenase-coding genes (*nif* operon), the *Burkholderia* sp. could also affect nitrogen uptake. The approaches are promising to elaborate the role of the interaction of bacteria and AM fungi in the nutrient uptake (Akhtar 2011).

Availability of Other Nutrients

Microbial inoculants have shown their influence towards the uptake of other nutrients besides N and P (Peix et al. 2001; Khan 2005; Wu et al. 2005; Adesemoye et al. 2008). Khan (2005) observed that inoculation of *Pseudomonas* and *Acinetobacter* strains resulted in enhanced uptake of Fe, Zn, Mg, Ca, and K by crop plants. In another study, inoculation of chickpea and barley with strains of *Mesorhizobium mediterraneum* significantly increased the K, Ca, and Mg in addition to P and N in both crop plants (Peix et al. 2001). Kohler et al. (2008) have demonstrated the effects of PGPR (*Pseudomonas mendocina*) and AM fungi (*G. intraradices* and *G. mosseae*) on uptake of N, P, Fe, Ca, and Mn in lettuce under three different levels of water stress. Sheng and He (2006) reported improved uptake of K by the inoculation of *Bacillus edaphicus* and suggested that the production of organic acids (citric, oxalic, tartaric, succinic, and α-ketogluconic) by this strain leads to chelation of metals and mobilization of K from K-containing minerals. Similarly, Giri and Mukerji (2004) reported a significant increase in Mg concentrations in the seedling of *Sesbania aegyptiaca* and *Sesbania grandiflora* by the application of *Glomus macrocarpum*, compared to non-mycorrhizal seedlings in saline soil. Liu et al. (2000) reported an increase in acquisition of Fe, Zn, Cu, and Mn by mycorrhizal fungi in maize. Moreover, sulfur uptake has been achieved through sulfur oxidization (Banerjee et al. 2006) and iron uptake through siderophore-producing bacteria (Bakker et al. 2007). Biswas et al. (2000) reported a significant increase in Fe uptake in lowland rice through inoculation of *Rhizobium leguminosarum* bv. *trifolii* and suggested that the increase in uptake of Fe, P, and K was associated with higher nitrogen rates.

Plant Interactions for Remediation of Contaminated Soils

Soil is an important habitat for thousands of organisms including a variety of fungi, actinobacteria, algae, protozoa, and different types of bacteria. These microorganisms in association with soil particles or soil organic matter in the rhizosphere are essential for the plant. Plant-microbe interactions are now being intensively investigated for decontamination and remediation processes. With the discovery of a number of soil microorganisms that are capable of degrading xenobiotic chemicals including herbicides, pesticides, solvents, and other organic compounds, microbial degradation might provide a reasonable and effective means of disposing toxic chemical wastes. Due to the sensitivity and the sequestration ability of the microbial communities towards the heavy metals, microbes have been used for bioremediation of sites contaminated with them (Hallberg and Johnson 2005; Kao et al. 2006; Umrania 2006). Although microbial communities in metal-polluted bulk soils have been studied, there are a limited number of studies on the composition of microbial community in the plant rhizosphere growing in soils highly polluted with heavy metals (Dell'Amico et al. 2005).

Phytoremediation

Phytoremediation is a kind of bioremediation technique that uses plants to manage or remediate polluted soils. It is an emerging and cost-effective technology. It could be defined as "the elimination, attenuation, or transformation of polluting or contaminating substances by plants into their less toxic forms" (Vidali 2001; Kavamura and Esposito 2008). It can be used as in situ or ex situ technology. Soils could be contaminated with thousands of contaminants varying in their composition and concentration through inadequate residue disposal, accidental wastes, and inappropriate use (Knaebel et al. 1994). These contaminants include nitrates, phosphates, and perchlorates (Nozawa-Inoue et al. 2005); explosives such as hexahydro-1,3,5-trinitro-1,3,5-triazine and octahydro-1,3,5,7-tetranitro-1,3,5,7-tetrazocine (Kitts et al. 1994); monoaromatic hydrocarbons like benzene, toluene, ethylbenzene, and xylene (Rooney-Varga et al. 1999); polycyclic aromatic hydrocarbons (Wang et al. 1990); herbicides such as diuron, linuron, and chlorotoluron (Fantroussi et al. 1999); and heavy metals (Glick 2003).

In case of soil remediation, several factors such as soil characteristics, type, and concentration of contaminants should be considered (Boopathy 2000; Sheoran et al. 2008). The remediation of the harmful contaminants from the soil by plants is an emerging alternative to restore the contaminated sites (Singh et al. 2003; Paquin et al. 2004; Vassilev et al. 2004; Shah and Nongkynrih 2007; Padmavathiamma and Li 2007; Rajkumar and Freitas 2008; Lone et al. 2008; Akhtar et al. 2013). Phytoremediation can be classified according to the method and the nature of contaminants (Lasat 2002; Eapen et al. 2003; Newman and Reynolds 2004; January et al. 2008). The various methods used for the phytoremediation are discussed below (Fig. 6.3).

Phytoextraction

Plants can absorb the concentrated metals in their aboveground parts which can then be harvested. Brennan and Shelley (1999) found that plants have the capability to extract large concentrations of heavy metals into their roots, translocate them to the stem, and produce a large quantity of plant biomass.

Phytodegradation

It is also known as phytotransformation. In this process, plants degrade organic pollutants directly via their enzymatic activities. Some enzymes break down and convert ammunition wastes, others degrade chlorinated solvents such as trichloroethylene, and others degrade herbicides.

Phytovolatilization

Phytovolatilization refers to the uptake and transpiration of contaminants, primarily organic compounds, by plants. The contaminant, present in the soil solution, is taken up and modified by the plant and is released to the atmosphere through the plant leaves by evaporation or vaporization processes.

Phytostimulation

It refers to stimulation of rhizospheric microorganisms capable of degrading the contaminants by the growing roots releasing exudates/nutrients such as carbon sources. This method is useful in removing organic contaminants, such as pesticides, aromatics, and polynuclear aromatic hydrocarbons from soil and sediments (Ukiwe et al. 2013).

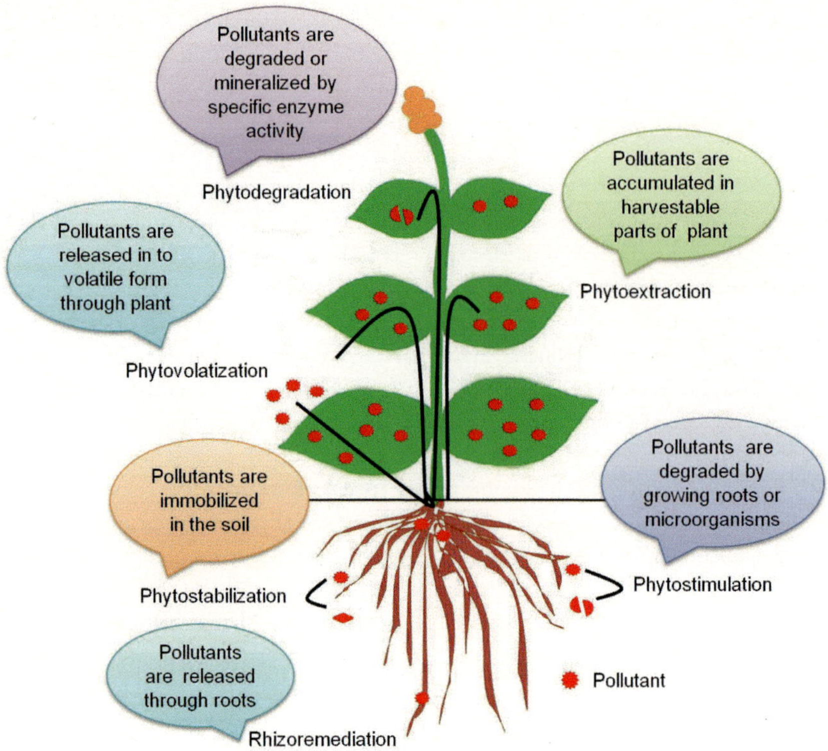

Fig. 6.3 Schematic diagram of different approaches of phytoremediation

Phytostabilization

In this method, the use of plant roots may limit the contaminant by reducing its mobility or leaching in the soil. Plants decrease the amount of water percolating through the soil matrix, which may act as a barrier and reduce the leaching of the contaminant. Phytostabilization can occur through sorption, precipitation, complexation, or metal valence reduction. It is helpful in the treatment of contaminated land areas affected by mining activities (Raskin and Ensley 2000). Phytostabilization is commonly used to treat the metal (arsenic, cadmium, chromium, copper, and zinc) contaminants (Kunito et al. 2001).

Rhizofiltration

It could be used for metals such as Pb, Cd, Cu, Ni, Zn, and Cr which are retained within the roots. It is useful for both terrestrial and aquatic plants for in situ or ex situ purposes. In this method, the contaminants do not translocate to the shoots. The terrestrial plants are more favored for rhizofiltration due to their fibrous and elongated root systems. The main limitation of this method is to adjust to the pH at regular intervals.

Rhizoremediation

In this method, microorganisms are utilized in combination with the plants (Jing et al. 2007). Generally organic pollutants with high hydrophobicity (hence, unable to be absorbed by the plant) are remediated by this method. Microbes play major role in this method. Plants mainly provide the microbes with nutrients and growth factors to proliferate (Siciliano and Germida 1998; Chaudhry et al. 2005).

Rhizoremediation of Organic Contaminants by PGPR

Initially, the PGPR were used for plant growth promotion and biocontrol of plant diseases, but now they are also being used for rhizoremediation of organic contaminants (Narasimhan et al. 2003; Huang et al. 2004, 2005). In contrast to inorganic compounds, microorganisms can degrade and mineralize organic compounds in association with plants (Saleh et al. 2004). Brazil et al. (1995) reported that the bacteria are capable of degrading a certain kind of organic pollutants such as polychlorinated biphenyls isolated from various locations and studied the encoding genes involved in this pathway. Rhizoremediation of various organic pollutants is known when a particular crop plant in combination with their known or unknown microbes are used as reviewed in detail by Kuiper et al. (2004).

Rhizoremediation of Metals by PGPR

A wide range of plants have been tested for their ability to take up high levels of metals by roots form soil and translocate these metals into the leaves and shoots. The use of PGPR as adjuncts in metal phytoremediation can significantly facilitate the growth of plants in the presence of high levels of metals (Zhuang et al. 2007; Glick 2010).

Rhizoremediation by Endophytic Microorganisms

Endophytic bacteria could be defined as bacteria colonizing the internal tissues of plants without causing infection or negative effects on their host (Lodewyckx et al. 2002). With the exception of seed endophytes, the primary site where endophytes gain entry into plants is via the roots (Pan et al. 1997; Germaine et al. 2004). The endophytes either reside inside the plants in specific plant tissues like root cortex and xylem or colonize the plant systematically by transport through the vascular system or apoplast (Mahaffee et al. 1997; Quadt-Hallmann et al. 1997). Endophytic bacteria have been isolated from a variety of healthy plant species ranging from herbaceous crop plants (Lodewyckx et al. 2002; Malinowski et al. 2004; Mastretta et al. 2009), different grass species (Zinniel et al. 2002; Dalton et al. 2004), to woody tree species (Cankar et al. 2005; Moore et al. 2006; Taghavi et al. 2009).

Pseudomonas, *Burkholderia*, and *Enterobacter* are amongst the most common genera of cultivable endophytes (Mastretta et al. 2006). In addition to their beneficial effects on plant growth, endophytes have also been used in phytoremediation (Weyens et al. 2009). Idris et al. (2004) investigated the endophytes from *Thlaspi goesingense*, a hyperaccumulator of Ni in both cultivation-dependent and cultivation-independent techniques. They concluded that the endophytes used in the cultivation-independent techniques have the potential to tolerate higher concentration of Ni than rhizospheric bacteria. This kind of technique is very promising in the phytoremediation of heavy metals, but the actual mechanisms is not well understood, and its application in the phytoremediation of heavy metal is expensive and very complicated (Weyens et al. 2009).

Mycorrhizoremediation

In this advanced approach, symbiotic AM fungi could be used for phytoremediation (Huang et al. 2004; Khan 2006). AM fungi have the potential to efficiently explore the soil volume (Meharg and Cairney 2000). Mycorrhizal association exhibits substantial resistance against the toxic metals (Leyval et al. 1997; Meharg and Cairney 2000) and organic compounds such as m-toluate and petroleum polycyclic aromatic hydrocarbons (Sarand et al. 1998, 1999; Leyval and Binet 1998).

In addition to their protective behavior, mycorrhizae may contribute to resistance of plant-microbial associations through enhanced degradation of organic pollutants in the mycorrhizosphere and lowering the bioavailable concentration of heavy metals in soil (Meharg and Cairney 2000). It is evident from the reports of the earlier researchers that the AM fungi have the potential to increase the uptake of various heavy metals in plants (Liao et al. 2003; Whitfield et al. 2004; Liu et al. 2005; Leung et al. 2006). However, some other studies showed there is no effect of AM fungi or even decreased concentrations in plant tissues (Trotta et al. 2006; Wu et al. 2007). On the basis of contrasting results, it is very difficult to evaluate the potential of mycorrhizal fungi in the uptake of heavy metals in field experiments (Liu et al. 2005; Leung et al. 2006; Wu et al. 2007; Wenzel 2009).

Bioremediation by Microbes

In this process, microorganisms are used for degradation or removal of contaminants from the soil. This method of degradation or removal of environmental contaminant through microbial activity is cost-effective and environmentally safe compared to other physicochemical methodologies used for bioremediation (Akhtar et al. 2013). It could be used both as in situ and ex situ methods. However, there are many loopholes and black holes in this technology such as the use of inappropriate microbial system for removal of metal toxicity, low or inactive microbial populations, or presence of complex pollutant mixtures. The rate and extent of biodegradation depend upon many factors which have been summarized in tabular form (Table 6.4).

For endurance under metal-stressed environment, PGPR have evolved several mechanisms by which they can immobilize, mobilize, or transform metals rendering them inactive so as to tolerate the heavy metal ions. These mechanisms include: (1) exclusion, the metal ions are kept away from the target sites; (2) extrusion, the metals are pushed out of the cell through chromosomal/plasmid mediated events; (3) accommodation, metals form complex with the metal binding proteins or other cell components; (4) bio-transformation, toxic metal is reduced to less toxic forms; and (5) methylation and demethylation. Thus, in general, the immobilization and mobilization are the two main techniques used for the bioremediation of metals by microbes.

Table 6.4 Some major factors affecting the bioremediation process

Factors	Affect
Microbial	Growth until critical biomass is reached, mutation and horizontal gene transfer, enzyme induction, enrichment of the capable microbial populations, and production of toxic metabolites
Environmental	Depletion of preferential substrates, lack of nutrients, and inhibitory environmental conditions
Substrate	Too low concentration of contaminants, chemical structure of contaminants, toxicity of contaminants, and solubility of contaminants
Biological aerobic vs. anaerobic process	Oxidation/reduction potential, availability of electron acceptors, and microbial population present in the site
Growth substrate vs. co-metabolism	Type of contaminants, concentration, alternate carbon source present, and microbial interaction such as competition, succession, and predation
Physicochemical bioavailability of pollutants	Equilibrium sorption, irreversible sorption, and incorporation into humic matters
Mass transfer limitations	Oxygen diffusion and solubility, diffusion of nutrients, and solubility and miscibility in water

Immobilization Techniques

Immobilization is a technique used to reduce the mobility of contaminants by altering the physical or chemical characteristics of the contaminant. This remediation approach can utilize microorganisms to immobilize metal contaminants. It is usually accomplished by physically restricting contact between the contaminant or by chemically altering the contaminant (Evanko and Dzombak 1997, Mulligan et al. 2001; Akhtar et al. 2013). Chemical reagents and bacterial reagents assist with the immobilization of metal contaminants. Most sites contaminated with metals use the solidification and stabilization approach to immobilize metals. Solidification treatment involves mixing or injecting chemical agents to the contaminated soil. The prominent mechanism by which metals are immobilized is by precipitation of hydroxides. The chemical composition of the site, the amount of water present, and the temperatures are all factors important to the successful use of the solidification/stabilization mechanisms (Evanko and Dzombak 1997; Wuana and Okieimen 2011). The stabilization and solidification technique is achieved by mixing the contaminated material with appropriate amounts of stabilizer material and water. The mixture forms a solidified matrix with the waste. The stabilization and solidification techniques can occur both in situ and ex situ. In situ is preferred for volatile or semi-volatile organics. The in situ process is useful for treating surface or shallow contamination.

Mobilization

Microorganisms can mobilize metals through autotrophic and heterotrophic leaching, chelation by microbial metabolites and siderophores, methylation, and redox transformations. Heterotrophic leaching is when microorganisms can acidify their environment by proton efflux thus leading to the acidification resulting in the release of free metal cations. Autotrophic leaching is when acidophilic bacteria retrieve CO_2 and obtain energy from the oxidation of the ferrous iron or reduced sulfate compounds, which causes solubilization of metals (Agrawal 2005). Siderophores are specific iron-chelating ligands and are able to bind to other metals, such as magnesium, manganese, chromium, and gallium, and radionuclide, such as plutonium (Akhtar et al. 2013). Methylation involves methyl groups that are enzymatically transferred to a metal, forming a number of different metalloids. Redox transformations can allow microorganisms to mobilize metals, metalloids, and organometallic compounds by reduction and oxidation processes. There are various metal-mobilization techniques that can also occur in nature (Gadd 2004).

Conclusion

The rhizospheric microorganisms can influence plant growth, nutrition availability, disease susceptibility, resistance towards heavy metals, and various abiotic stresses. Plant growth attributes could be limited by the unavailability of essential elements or the presence of toxic elements. The interactions between plant roots and microorganisms present in the rhizosphere assist them to acquire essential mineral nutrients from the soil and prevent the accumulation of toxic elements. Amongst various rhizospheric microorganisms, free-living and symbiotic nitrogen-fixing bacteria contribute a lot to meet this demand. In the presence of N-free or low nitrogen content, the rhizospheric bacteria accomplish associative nitrogen fixation and thus provide essential nutrients to plants. Moreover, the phosphate-solubilizing bacteria have the capacity to convert inorganic unavailable P form to soluble forms available to plants.

Of the various microorganisms present in the soil, the rhizospheric bacterial community has the potential to increase the plant growth and minimize the disease severity and also is useful in the degradation or removal of toxic elements from water, soil, sludge, and process-waste stream through bioremediation. These technologies could be broadly classified as ex situ and in situ. The ex situ technologies are applied for the physical removal of the contaminated materials for treatment process, while in situ techniques for the treatment of contaminated materials in place. Plant-microbe interactions can thus be applied for diverse aspects by the development of sustainable technologies for enhancement of crop yield, suppression of phytopathogens, degradation of pollutants, and remediation of contaminated sites.

References

Adesemoye AO, Torbert HA, Kloepper JW (2008) Enhanced plant nutrient use efficiency with PGPR and AMF in an integrated nutrient management system. Can J Microbiol 54:876–886

Adesemoye AO, Torbert HA, Kloepper JW (2009) Plant growth promoting rhizobacteria allow reduced application rates of chemical fertilizers. Microb Ecol 58:921–929

Agrawal SK (2005) Advanced environmental biotechnology. APH Publishing Corporation, Darya Ganj, New Delhi

Akhtar MS (2011) In: Akhtar MS (ed) Biocontrol of root-rot disease complex of chickpea by arbuscular mycorrhizal fungi and other phosphate solubilizing microorganisms. LAMBERT Academic Publishing GmbH and Co. KG, Dudweiler Landstrasse

Akhtar MS, Panwar J (2011) Arbuscular mycorrhizal fungi and opportunistic fungi: efficient root symbionts for the management of plant parasitic nematodes. Adv Sci Eng Med 3:165–175

Akhtar MS, Siddiqui ZA (2006) Effects of phosphate solubilizing microorganisms on the growth and root-rot disease complex of chickpea. Mycol Phytopathol 40:246–254

Akhtar MS, Siddiqui ZA (2008a) Arbuscular mycorrhizal fungi as potential bioprotectants against plant pathogens. In: Siddiqui ZA, Akhtar MS, Futai K (eds)

Mycorrhizae: sustainable agriculture and forestry. Springer, Dordrecht, pp 61–98

Akhtar MS, Siddiqui ZA (2008b) Biocontrol of a root-rot disease complex of chickpea by *Glomus intraradices*, *Rhizobium* sp. and *Pseudomonas straita*. Crop Prot 27:410–417

Akhtar MS, Siddiqui ZA (2009) Use of plant growth promoting rhizobacteria for the biocontrol of root-rot disease complex of chickpea. Australas Plant Pathol 38:44–50.

Akhtar MS, Siddiqui ZA (2010) Role of plant growth promoting rhizobacteria in biocontrol of plant diseases and sustainable agriculture. In: Maheshwari DK (ed) Plant growth and health promoting bacteria. Microbiology monographs 18. Springer, Berlin, pp 157–196

Akhtar MS, Siddiqui ZA, Wiemken A (2011) Arbuscular mycorrhizal fungi and *Rhizobium* to control plant fungal diseases. In: Lichtfouse E (ed) Alternative farming systems, biotechnology, drought stress and ecological fertilisation. Sustainable agriculture reviews 6. Springer, Dordrecht, pp 263–292

Akhtar MS, Chali B, Azam T (2013) Bioremediation of arsenic and lead by plants and microbes from contaminated soil. Res Plant Sci 1:68–73

Ames RN, Reid CPP, Porterf LK, Cambardella C (1983) Hyphal uptake and transport of nitrogen from two ^{15}N-labelled sources by *Glomus mosseae*, a vesicular-arbuscular mycorrhizal fungus. New Phytol 95:381–396

Amijee F, Tinker PB, Stribley DP (1989) The development of endomycorrhizal root systems. VII. A detailed study of effects of soil phosphorus on colonization. New Phytol 111:435–446

Amir HG, Shamsuddin ZH, Halimi MS, Marziah M, Ramlan MF (2005) Enhancement in nutrient accumulation and growth of oil palm seedlings caused by PGPR under field nursery conditions. Commun Soil Sci Plant Anal 36:2059–2066

Aseri GK, Jain N, Panwar J, Rao AV, Meghwal PR (2008) Biofertilizers improve plant growth, fruit yield, nutrition, metabolism and rhizosphere enzyme activities of pomegranate (*Punica granatum* L.) in Indian Thar Desert. Sci Hortic 117:130–135.

Bakker PAHM, Pieterse CMJ, Van Loon LC (2007) Induced systemic resistance by fluorescent *Pseudomonas* spp. Phytopathology 97:239–243

Banerjee MR, Yesmin L, Vessey JK (2006) Plant growth-promoting rhizobacteria as biofertilizers and biopesticide. In: Rai MK (ed) Handbook of microbial biofertilizers. Food Products Press, New York, pp 137–181

Bardgett RD, Wardle DA, Yeates GW (1998) Linking above-ground and below-ground food webs: how plant responses to foliar herbivory influences soil organisms. Soil Biol Biochem 30:1867–1978

Barea JM, Andrade G, Bianciotto V, Dowling D, Lohrke S, Bonfante P, O'Gara F, Azcon-Aguilar C (1998) Impact on arbuscular mycorrhiza formation of *Pseudomonas* strains used as inoculants for biocontrol of soil-borne fungal plant pathogens. Appl Environ Microbiol 64:2304–2307

Bashan Y, Holguin G, de-Bashan LE (2004) *Azospirillum-plant* relationships: physiological, molecular, agricultural, and environmental advances (1997-2003). Can J Microbiol 50:521–577

Beerling DJ, Berner RA (2005) Feedbacks and the coevolution of plants and atmospheric CO_2. Proc Natl Acad Sci U S A 102:1302–1305

Bianciotto V, Bonfante P (2002) Arbuscular mycorrhizal fungi: a specialized niche for rhizospheric and endocellular bacteria. Antonie Van Leeuwenhoek 81:365–371

Biswas JC, Ladha JK, Dazzo FB (2000) *Rhizobia* inoculation improves nutrient uptake and growth of lowland rice. Soil Sci Soc Am J 64:1644–1650

Bonfante P (2003) Plants, mycorrhizal fungi, and endobacteria: a dialog among cells and genomes. Biol Bull 204:215–220

Boopathy R (2000) Factors limiting bioremediation technologies. Bioresour Technol 74:63–67

Brazil GM, Kenefick L, Callanan M, Haro A, de Lorenzo V, Dowling DN, O'Gara F (1995) Construction of a rhizosphere pseudomonad with potential to degrade polychlorinated biphenyls and detection of *bph* gene expression in the rhizosphere. Appl Environ Microbiol 61:1946–1952

Brennan MA, Shelley ML (1999) A model of the up takes translocation and accumulation of lead (Pb) by maize for the purpose of phytoextraction. Ecol Eng 12:271–297

Briones AM Jr, Okabe S, Umemiya Y, Ramsing NB, Reichardt W, Okuyama H (2003) Ammonia-oxidizing bacteria on root biofilms and their possible contribution to N use efficiency of different rice cultivars. Plant Soil 250:335–348

Cankar K, Kraigher H, Ravnikar M, Rupnik M (2005) Bacterial endophytes from seeds of Norway spruce (*Picea abies* L. Karts). FEMS Microbiol Lett 244:341–345

Chaudhry Q, Blom-Zandstra M, Gupta S, Joner E (2005) Utilizing the synergy between plants and rhizosphere microorganisms to enhance breakdown of organic pollutants in the environment. Environ Sci Pollut Res Int 12:34–48

Chen YP, Rekha PD, Arun AB, Shen FT, Lai WA, Young CC (2006) Phosphate solubilizing bacteria from subtropical soil and their tricalcium phosphate solubilizing abilities. Appl Soil Ecol 34:33–41

Dalton DA, Kramer S, Azios N, Fusaro S, Cahill E, Kennedy C (2004) Endophytic nitrogen fixation in dune grasses (*Ammophila arenaria* and *Elymus mollis*) from Oregon. FEMS Microbiol Ecol 49: 469–479

de Zamaroczy M, Delorme F, Elmerich C (1989) Regulation of transcription and promoter mapping of the structural genes for nitrogenase (*nifHDK*) of *Azospirillum brasilense* Sp7. Mol Gen Genet 220:88–94

Dell'Amico E, Cavalca L, Andreoni V (2005) Analysis of rhizobacterial communities in perennial Graminaceae from polluted water meadow soil, and screening of metal-resistant, potentially plant growth-promoting bacteria. FEMS Microbiol Ecol 52:153–162

Dobbelaere S, Croonenborghs A, Thys A, Ptacek D, Vanderleyden J, Dutto P, Labandera-Gonzalez C, Caballero-Mellado J, Anguirre JF, Kapulnik Y, Brener S, Burdman S, Kadouri D, Sarig S, Okon Y (2001) Response of agronomically important crops to inoculation with *Azospirillum*. Aust J Plant Physiol 28:871–879

Eapen S, Suseelan KN, Tivarekar S, Kotwal SA, Mitra R (2003) Potential for rhizofiltration of uranium using hairy root cultures of *Brassica juncea* and *Chenopodium amaranticolor*. Environ Res 91:127–133

Egamberdiyeva D, Höflich G (2004) Effect of plant growth-promoting bacteria on growth and nutrient uptake of cotton and pea in a semi-arid region of Uzbekistan. J Arid Environ 56:293–301

Egener T, Hurek T, Reinhold-Hurek B (1999) Endophytic expression of *nif* genes of *Azoarcus* sp. strain BH72 in rice roots. Mol Plant-Microbe Interact 12:813–819

Elsheikh EAE, Elzidany AA (1997) Effects of *Rhizobium* inoculation, organic and chemical fertilizers on yield and physical properties of faba bean seeds. Plant Foods Hum Nutr 51:137–144

Evanko CR, Dzombak DA (1997) Remediation of metals-contaminated soil and groundwater. Environ Sci 412:1–45

Fantroussi S, Verschuere L, Verstraete W, Top EM (1999) Effect of phenylurea herbicides on soil microbial communities estimated by analysis of 16S rRNA gene fingerprints and community-level physiological profiles. Appl Environ Microbiol 65:982–988

Frink CR, Waggoner PE, Ausubel JH (1999) Nitrogen fertilizer: retrospect and prospect. Proc Natl Acad Sci U S A 96:1175–1180

Gadd GM (2004) Microbial influence on metal mobility and application for bioremediation. Geoderma 122:109–119

Germaine K, Keogh E, Garcia-Cabellos G, Borremans B, van der Lelie D, Barac T, Oeyen L, Vangronsveld J, Moore FP, Moore ERB, Campbel CD, Ryan D, Dowling DN (2004) Colonization of poplar trees by gfp expressing endophytes. FEMS Microbiol Ecol 48:109–118

Gharu A, Tarafadar JC (2004) Influence of organic acids on mobilization of inorganic and organic phosphorus in soil. J Indian Soc Soil Sci 52:248–253

Giri B, Mukerji KG (2004) Mycorrhizal inoculant alleviates salt stress in *Sesbania aegyptiaca* and *Sesbania grandiflora* under field conditions: evidence for reduced sodium and improved magnesium uptake. Mycorrhiza 14:307–312

Giri B, Giang PH, Kumari RAP, Oelmuller R, Varma A (2005) Mycorrhizosphere: strategies and function. In: Buscot F, Varma A (eds) Microorganisms in soil: roles in genesis and function. Soil biology, vol 3. Springer, Berlin

Glick BR (2003) Phytoremediation: synergistic use of plants and bacteria to clean up the environment. Biotechnol Adv 21:383–393

Glick BR (2010) Using soil bacteria to facilitate phytoremediation. Biotechnol Adv 28:367–374

Glick BR, Cheng Z, Czarny J, Duan J (2007a) Promotion of plant growth by ACC deaminase-producing soil bacteria. Eur J Plant Pathol 119:329–339

Glick BR, Todorovic B, Czarny J, Cheng Z, Duan J, Mc Conkey B (2007b) Promotion of plant growth by bacterial ACC deaminase. Crit Rev Plant Sci 26:227–242

Gupta G, Panwar J, Akhtar MS, Jha PN (2012) Endophytic Nitrogen-fixing bacteria as biofertilizer. In: Lichtfouse E (ed) Sustainable agriculture reviews 11. Springer, Dordrecht, pp 183–221

Gutierrez-Manero FJ, Ramos-Solano B, Probanza A, Mehouachi J, Tadeo FR, Talon M (2001) The plant-growth promoting rhizobacteria *Bacillus pumilus* and *Bacillus licheniformis* produce high amounts of physiologically active gibberellins. Physiol Plant 111:206–211

Gyaneshwar P, Kumar GN, Parekh LJ, Poole PS (2002) Role of soil microorganisms in improving P nutrition of plants. Plant Soil 245:83–93

Hallberg KB, Johnson DB (2005) Microbiology of a wetland ecosystem constructed to remediate mine drainage from a heavy metal mine. Sci Total Environ 338:53–66

Han HS, Lee KD (2005) Phosphate and potassium solubilizing bacteria effect on mineral uptake, soil availability, and growth of egg plant. Res J Agric Biol Sci 1:176–180

Harrison MJ, van Buuren ML (1995) A phosphate transporter from the mycorrhizal fungus *Glomus versiforme*. Nature 378:626–629

Hernandez MI, Chailloux M (2004) Las micorrizas arbusculares y las bacterias rizosfericas como alternativa a la nutricion mineral del tomate. Cultivos Tropicales 25:5–12

Hiltner L (1904) Uber neue erfahrungen und probleme auf dem gebiete der bodenbakteriologie. Arbeiten der DLG 98:59–78

Hodge A, Campbell CD, Fitter AH (2001) An arbuscular mycorrhizal fungus accelerates decomposition and acquires nitrogen directly from organic material. Nature 413:297–299

Huang XD, El-Alawi Y, Penrose DM, Glick BR, Greenberg BM (2004) A multi-process phytoremediation system for removal of polycyclic aromatic hydrocarbons from contaminated soils. Environ Pollut 130:465–476

Huang XD, El-Alawi Y, Gurska J, Glick BR, Greenberg BM (2005) A multi-process phytoremediation system for decontamination of persistent total petroleum hydrocarbons (TPHs) from soils. Microchem J 81:139–147

Idris R, Trifonova R, Puschenreiter M, Wenzel WW, Sessitsch A (2004) Bacterial communities associated with flowering plants of the Ni hyperaccumulator *Thlaspi goesingense*. Appl Environ Microbiol 70:2667–2677

Idriss EE, Makarewicz O, Farouk A, Rosner K, Greiner R, Bochow H, Richter T, Borriss R (2002) Extracellular phytase activity of *Bacillus amyloliquefaciens* FZB45 contributes to its plant-growth promoting effect. Microbiology 148:2097–2109

Igual JM, Valverde A, Cervantes E, Velazquez E (2001) Phosphate solubilizing bacteria as inoculants for agriculture: use of updated molecular techniques in their studies. Agronomie 21:561–568

January MC, Cutright TJ, Van Keulen H, Wei R (2008) Hydroponic phytoremediation of Cd, Cr, Ni, As, and Fe: can *Helianthus annuus* hyper accumulate multiple heavy metals. Chemosphere 70:531–537

Jetten MSM (2008) The microbial nitrogen cycle. Environ Microbiol 10:2903–2909

Jing Y, He Z, Yang X (2007) Role of soil rhizobacteria in phytoremediation of heavy metal contaminated soils. J Zhejiang Univ Sci B 8:192–207

Kao PH, Huang CC, Hseu ZY (2006) Response of microbial activities to heavy metals in a neutral loamy soil treated with biosolid. Chemosphere 64:63–70

Kavamura NV, Esposito E (2008) Biotechnological strategies applied to the decontamination of soils polluted with heavy metals. Biotechnol Adv 28:61–69

Kennedy IR, Pereg-Gerk LL, Wood C, Deaker R, Gilchrist K, Katupitiya S (1997) Biological nitrogen fixation in non-leguminous field crops: facilitating the evolution of an effective association between *Azospirillum* and wheat. Plant Soil 194:65–79

Khan AG (2005) Role of soil microbes in the rhizospheres of plants growing on trace metal contaminated soils in phytoremediation. J Trace Elem Med Biol 18:355–364

Khan AG (2006) Mycorrhizoremediation-an enhanced form of phytoremediation. J Zhejiang Univ Sci B 7:503–514

Kitts CL, Cunningham DP, Unkefer PJ (1994) Isolation of three hexahydro-1, 3, 5-trinitro-1, 3, 5-triazine-degrading species of the family Enterobacteriaceae from nitramine explosive-contaminated soil. Appl Environ Microbiol 60:4608–4611

Kloepper JW, Ryu CM, Zhang S (2004) Induced systemic resistance and promotion of plant growth by *Bacillus* spp. Phytopathology 94:1259–1266

Knaebel DB, Federle TW, Mc Avoy DC, Vestal JR (1994) Effect of mineral and organic soil constituents on microbial mineralization of organic compounds in a natural soil. Appl Environ Microbiol 60:4500–4508

Kohler J, Hernandez JA, Caravaca F, Roldan A (2008) Plant growth promoting rhizobacteria and arbuscular mycorrhizae fungi modify alleviation biochemical mechanisms in water-stressed plants. Funct Plant Biol 35:141–151

Koide RT (1991) Tansley review No. 29: nutrient supply, nutrient demand, and plant response to mycorrhizal infection. New Phytol 117:365–386

Kuiper I, Lagendijk EL, Bloemberg GV, Lugtenberg BJ (2004) Rhizoremediation: a beneficial plant-microbe interaction. Mol Plant Microbe Interact 17:6–15

Kunito T, Saeki K, Oyaizu K, Mutsumoto S (2001) Characterization of copper resistant bacterial communities in copper contaminated soils. Eur J Soil Biol 37:95–102

Kuzyakov Y, Xu X (2013) Competition between roots and microorganisms for nitrogen: mechanisms and ecological relevance. New Phytol 198:656–669

Lasat MM (2002) Phytoextraction of toxic metals: a review of biological mechanisms. J Environ Qual 31:109–120

Leung HM, Ye ZH, Wong MH (2006) Interactions of mycorrhizal fungi with *Pteris vittata* (As hyperaccumulator) in As-contaminated soils. Environ Pollut 139:1–8

Leyval C, Binet P (1998) Effect of polyaromatic hydrocarbons in soil on arbuscular mycorrhizal plants. J Environ Qual 27:402–407

Leyval C, Turnau K, Haselwandter K (1997) Effect of heavy metal pollution on mycorrhizal colonization and function: physiological, ecological and applied aspects. Mycorrhiza 7:139–153

Liao JP, Lin XG, Cao ZH, Shi YQ, Wong MH (2003) Interactions between arbuscular mycorrhizae and heavy metals under sand culture experiment. Chemosphere 50:847–853

Liu A, Hamel C, Hamilton RI, Ma BL, Smith DL (2000) Acquisition of Cu, Zn, Mn and Fe by mycorrhizal maize (*Zea mays* L.) grown in soil at different P and micronutrient levels. Mycorrhiza 9:331–336

Liu Y, Zhu YG, Chen BD, Christie P, Li XL (2005) Influence of the arbuscular mycorrhizal fungus *Glomus mosseae* on uptake of arsenate by the As hyperaccumulator fern *Pteris vittata* L. Mycorrhiza 15:187–192

Lodewyckx C, Mergeay M, Vangronsveld J, Clijsters H, van der Lelie D (2002) Isolation, characterization, and identification of bacteria associated with the zinc hyperaccumulator *Thlaspi caerulescens* subsp. *calaminaria*. Int J Phytoremed 4:101–115

Lone MI, He Z, Stoffella PJ, Yang X (2008) Phytoremediation of heavy metal polluted soils and water: progresses and perspectives. J Zhejiang Univ Sci B 9:210–220

Mahaffee WF, Kloepper JW, Van Vuurde JWL, Van der Wolf JM, Van den Brink M (1997) Endophytic colonization of *Phaseolus vulgaris* by *Pseudomonas fluorescens* strain 89B-27 and *Enterobacter asburiae* strain JM22. In: Ryder MH, Stephens PM, Bowen GD (eds) Improving plant productivity in rhizosphere bacteria. CSIRO, Melbourne

Mahmood S, Finlay RD, Erland S, Wallander H (2001) Solubilisation and colonisation of wood ash by ectomycorrhizal fungi isolated from a wood ash fertilized spruce forest. FEMS Microbiol Ecol 35:151–161

Malinowski DP, Zuo H, Belesky DP, Alloush GA (2004) Evidence for copper binding by extracellular root exudates of tall fescue but not perennial ryegrass infected with *Neotyphodium* spp. endophytes. Plant Soil 267:1–12

Mantelin S, Touraine B (2004) Plant growth-promoting bacteria and nitrate availability: impacts on root development and nitrate uptake. J Exp Bot 55:27–34

Marilley L, Aragno M (1999) Phylogenetic diversity of bacterial communities differing in degree of proximity of *Lolium perenne* and *Trifolium repens* roots. Appl Soil Ecol 13:127–136

Marschner H (1995) Mineral nutrition of higher plants, 2nd edn. Academic, London

Mastretta C, Barac T, Vangronsveld J, Newman L, Taghavi S, van der Lelie D (2006) Endophytic bacteria and their potential application to improve the phytoremediation of contaminated environments. Biotechnol Genet Eng Rev 23:175–207

Mastretta C, Taghavi S, van der Lelie D, Mengoni A, Galardi F, Gonnelli C, Barac T, Boulet J, Weyens N, Vangronsveld J (2009) Endophytic bacteria from seeds of *Nicotiana tabacum* can reduce cadmium phytotoxicity. Int J Phytoremed 11:251–267

Meharg AA, Cairney JWG (2000) Ectomycorrhizas-extending the capabilities of rhizosphere remediation. Soil Biol Biochem 32:1475–1484

Minder AC, Narberhaus F, Hans-Martin F, Hennecke H (1998) The *Bradyrhizobium japonicum* phoB gene is required for phosphate limited growth but not for symbiotic nitrogen fixation. FEMS Microbiol Lett 161:47–52

Minerdi D, Fani R, Gallo R, Boarino A, Bonfante P (2001) Nitrogen fixation genes in an endosymbiotic *Burkholderia* strain. Appl Environ Microbiol 67:725–732

Moore FP, Barac T, Borremans B, Oeyen L, Vangronsveld J, van der Lelie D, Campbell CD, Moore ERB (2006) Endophytic bacterial diversity in poplar trees growing on a BTEX-contaminated site: the characterization of isolates with potential to enhance phytoremediation. Syst Appl Microbiol 29:539–556

Morgan JAW, Bending GD, White PJ (2005) Biological costs and benefits to plant–microbe interactions in the rhizosphere. J Exp Bot 56:1729–1739

Mukerji KG, Mandeep A, Varma A (1998) Mycorrhizosphere microorganisms: screening and evolution. In: Varma A (ed) Mycorrhizal manual. Springer, Berlin, pp 85–97

Mulligan CN, Yong RN, Gibbs BF (2001) Remediation technologies for metal-contaminated soils and groundwater: an evaluation. Eng Geol 60:193–207

Narasimhan K, Basheer C, Bajic VB, Swarup S (2003) Enhancement of plant-microbe interactions using a rhizosphere metabolomics driven approach and its application in the removal of polychlorinated biphenyls. Plant Physiol 132:146–153

Newman LA, Reynolds CM (2004) Phytodegradation of organic compounds. Curr Opin Biotechnol 15:225–230

Nozawa-Inoue M, Scow KM, Rolston DE (2005) Reduction of perchlorate and nitrate by microbial communities in vadose soil. Appl Environ Microbiol 71:3928–3934

Ogunseitan O (2005) Microbial diversity: form and function in prokaryotes. Blackwell Science Ltd., Malden, p 142

Ohno T, Griffin TS, Liebman M, Porter GA (2005) Chemical characterization of soil phosphorus and organic matter in different cropping systems in Maine, USA. Agric Ecosyst Environ 105:625–634

Padmavathiamma PK, Li LY (2007) Phytoremediation technology: hyper-accumulation metals in plants. Water Air Soil Pollut 184:105–126

Pan MJ, Rademan S, Kuner K, Hastings JW (1997) Ultra structural studies on the colonization of banana tissue and *Fusarium oxysporum* f. sp. *cubense* race 4 by the endophytic bacterium *Burkholderia cepacia*. J Phytopathol 145:479–486

Paquin DG, Campbell S, Li QX (2004) Phytoremediation in subtropical Hawaii-a review of over 100 plant species. Remed J 14:127–139.

Pate JS, Verboom WH (2009) Contemporary biogenic formation of clay pavements by eucalypts: further support for the phytotarium concept. Ann Bot 103:673–685

Peix A, Rivas-Boyero AA, Mateos PF, Rodriguez-Barrueco C, Martinez-Molina E, Velazquez E (2001) Growth promotion of chickpea and barley by a phosphate solubilizing strain of *Mesorhizobium mediterraneum* under growth chamber conditions. Soil Biol Biochem 33:103–110

Perrott KW, Sarathchandra SU, Dow BW (1992) Seasonal and fertilizer effects on the organic cycle and microbial biomass in a hill country soil under pasture. Aust J Soil Res 30:383–394

Phillips DA, Ferris H, Cook DR, Strong DR (2003) Molecular control points in rhizosphere food webs. Ecology 84:816–826

Quadt-Hallmann A, Kloepper JW, Benhamou N (1997) Bacterial endophytes in cotton: mechanisms entering the plant. Can J Microbiol 43:577–582

Raaijmakers JM, Weller DM, Thomashow LS (1997) Frequency of antibiotic-producing *Pseudomonas* spp. in natural environments. Appl Environ Microbiol 63:881–887

Rajkumar M, Freitas H (2008) Influence of metal resistant-plant growth-promoting bacteria on the growth of *Ricinus communis* in soil contaminated with heavy metals. Chemosphere 71:834–842

Raskin I, Ensley D (2000) Phytoremediation of toxic metals: using plants to clean up the environment. Wiley, New York

Rodriguez H, Fraga R (1999) Phosphate solubilizing bacteria and their role in plant growth promotion. Biotech Adv 17:319–339

Rodriguez H, Rossolini GM, Gonzalez T, Li J, Glick BR (2000) Isolation of a gene from *Burkholderia cepacia* IS-16 encoding a protein that facilitates phosphatase activity. Curr Microbiol 40:362–366

Rooney-Varga JN, Anderson RT, Fraga JL, Ringelberg D, Lovley DR (1999) Microbial communities associated with anaerobic benzene degradation in a petroleum contaminated aquifer. Appl Environ Microbiol 65:3056–3063

Ruiz-Lozano JM, Bonfante P (1999) Identification of putative P transporter operon in the genome of a *Burkholderia* strain living inside the arbuscular mycorrhizal fungus *Gigaspora margarita*. J Bacteriol 181:4106–4109

Saleh S, Huang XD, Greenberg BM, Glick BR (2004) Phytoremediation of persistent organic contaminants in the environment. In: Singh A, Ward O (eds) Soil biology: applied bioremediation and phytoremediation. Springer, Berlin, pp 115–134

Sarand I, Timonen S, Nurmiaho-Lassila EL, Koivila T, Haahtela K, Romantschuk M (1998) Microbial biofilms and catabolic plasmid harbouring degradative fluorescent pseudomonads in Scots pine ectomycorrhizospheres developed on petroleum contaminated soil. FEMS Microbiol Ecol 27:115–126

Sarand I, Timonen S, Koivula T, Peltola R, Haahtela K, Sen R, Romantschuk M (1999) Tolerance and biodegradation of m-toluate by Scots pine, a mycorrhizal fungus and fluorescent pseudomonads individually and under associative conditions. J Appl Microbiol 86:817–826

Shah K, Nongkynrih JM (2007) Metal hyperaccumulation and bioremediation. Biol Plant 51:618–634

Shaharoona B, Naveed M, Arshad M, Zahir ZA (2008) Fertilizer dependent efficiency of *Pseudomonads* for improving growth, yield, and nutrient use efficiency of wheat (*Triticum aestivum* L.). Appl Microbiol Biotechnol 79:147–155

Sheng XF, He LY (2006) Solubilization of potassium-bearing minerals by a wild-type strain of *Bacillus edaphicus* and its mutants and increased potassium uptake by wheat. Can J Microbiol 52:66–72

Sheoran V, Sheoran AS, Poonam P (2008) Remediation techniques for contaminated soils. Environ Eng Manag J 7:379–387

Siciliano SD, Germida JJ (1998) Mechanisms of phytoremediation: biochemical and ecological interactions between plants and bacteria. Environ Rev 6:65–79

Siddiqui ZA, Baghel G, Akhtar MS (2007) Biocontrol of *Meloidogyne javanica* by *Rhizobium* and plant growth-promoting rhizobacteria on lentil. World J Microbiol Biotechnol 23:435–441

Singh OV, Labana S, Pandey G, Budhiraja R, Jain RK (2003) Phytoremediation: an overview of metallic ion decontamination from soil. Appl Microbiol Biotechnol 61:405–412

Smalla K, Wieland G, Buchner A, Zock A, Parzy J, Kaiser S, Roskot N, Heuer H, Berg G (2001) Bulk and rhizosphere soil bacterial communities studied by denaturing gradient gel electrophoresis: plant-dependent enrichment and seasonal shifts revealed. Appl Environ Microbiol 67:4742–4751

Steinshamn H, Thuen E, Bleken MA, Brenoe UT, Ekerholt G, Yri C (2004) Utilization of nitrogen (N) and phosphorus (P) in an organic dairy farming system in Norway. Agric Ecosyst Environ 104:509–522

Stewart LI, Hamel C, Hogue R, Moutoglis P (2005) Response of strawberry to inoculation with arbuscular mycorrhizal fungi under very high soil phosphorus conditions. Mycorrhiza 15:612–619

Taghavi S, Garafola C, Monchy S, Newman L, Hoffman A, Weyens N, Barac T, Vangronsveld J, van der Lelie D (2009) Genome survey and characterization of endophytic bacteria exhibiting a beneficial effect on growth and development of poplar trees. Appl Environ Microbiol 75:748–757

Tarafdar JC, Marschner H (1994a) Efficiency of VAM hyphal in utilization of organic phosphorus by wheat plants. Soil Sci Plant Nutr 40:593–600

Tarafdar JC, Marschner H (1994b) Phosphatase activity in the rhizosphere and hyphosphere of VA mycorrhizal wheat supplied with inorganic and organic phosphorous. Soil Biol Biochem 26:387–395

Tawaraya K, Naito M, Wagatsuma T (2006) Solubilization of insoluble inorganic phosphate by hyphal exudates of arbuscular mycorrhizal fungi. J Plant Nutr 29:657–665

Tilman D (1998) The greening of the green revolution. Nature 396:211–212

Trotta A, Falaschi P, Cornara L, Minganti V, Fusconi A, Drava G (2006) Arbuscular mycorrhizae increase the arsenic translocation factor in the As hyper accumulating fern *Pteris vittata* L. Chemosphere 65:74–81

Ueda T, Suga Y, Yahiro N, Matsuguchi T (1995) Remarkable N_2-fixing bacterial diversity detected in rice roots by molecular evolutionary analysis of *nifH* gene sequences. J Bacteriol 177:1414–1417

Ukiwe LN, Egereonu UU, Njoku PC, Nwoko CIA, Allinor JI (2013) Polycyclic aromatic hydrocarbons degradation techniques: a review. Int J Chem 5:43–55

Umrania VV (2006) Bioremediation of toxic heavy metals using acidothermophilic autotrophes. Bioresour Technol 97:1237–1242.

Van der Putten WH, Vet LEM, Harvey JA, Wackers FL (2001) Linking above-ground and below-ground multitrophic interactions of plants, herbivores, pathogens, and their antagonists. Trends Ecol Evol 16:547–554

Vande Broek A, Michiels J, Van Gool A, Vanderleyden J (1993) Spatial-temporal colonization patterns of *Azospirillum brasilense* on the wheat root surface and expression of the bacterial nifH gene during association. Mol Plant Microbe Interact 6:592–600

Vassilev A, Schwitzguebel JP, Thewys T, van Der Lelie D, Vangronsveld J (2004) The use of plants for remediation of metal contaminated soils. Sci World J 4:9–34

Vessey JK (2003) Plant growth promoting rhizobacteria as biofertilizers. Plant Soil 255:571–586

Vidali M (2001) Bioremediation. An overview. Pure Appl Chem 73:1163–1172

Vitousek PM, Aber JD, Howarth RW, Likens GE, Matson PA, Schindler DW, Schlesinger WH, Tilman DG (1997) Technical report: human alteration of the global nitrogen cycle: sources and consequences. Ecol Appl 7:737–750

Wang X, Yu X, Bartha R (1990) Effect of bioremediation on polycyclic aromatic hydrocarbon residues in soil. Environ Sci Technol 24:1086–1089

Wenzel WW (2009) Rhizosphere processes and management in plant-assisted bioremediation (phytoremediation) of soils. Plant Soil 321:385–408

Weyens N, van der Lelie D, Taghavi S, Vangronsveld J (2009) Phytoremediation: plant-endophyte partnerships take the challenge. Curr Opin Biotechnol 20:248–254

White PJ (2003) Ion transport. In: Thomas B, Murphy DJ, Murray BG (eds) Encyclopaedia of applied plant sciences. Academic Press, London, pp 625–634

Whitfield L, Richards AJ, Rimmer DL (2004) Effects of mycorrhizal colonization on *Thymus polytrichus* from heavy-metal contaminated sites in northern England. Mycorrhiza 14:47–54

Wu SC, Cao ZH, Li ZG, Cheung KC, Wong MH (2005) Effects of biofertilizer containing N-fixer, P and K solubilizers and AM fungi on maize growth: a greenhouse trial. Geoderma 125:155–166

Wu FY, Ye ZH, Wu SC, Wong MH (2007) Metal accumulation and arbuscular mycorrhizal status in metallicolous and nonmetallicolous populations of *Pteris vittata* L. and *Sedum alfredii* Hance. Planta 226:1363–1378

Wuana RA, Okieimen FE (2011) Heavy metals in contaminated soils: a review of sources, chemistry, risks and best available strategies for remediation. ISRN Ecol 2011:402647

Yadav RS, Meena SC, Patel SI, Patel KI, Akhtar MS, Yadav BK, Panwar J (2012) Bioavailability of soil P for plant nutrition. In: Lichtfouse E (ed) Farming for food and water security, sustainable agriculture reviews 10. Springer, Dordrecht, pp 177–200

Yang CH, Crowley DE (2000) Rhizosphere microbial community structure in relation to root location and plant iron nutritional status. Appl Environ Microbiol 66:345–351

Yang CH, Crowley DE, Menge JA (2001) 16S rDNA finger printing of rhizosphere bacterial communities associated with healthy and Phytophthora infected avocado roots. FEMS Microbiol Ecol 35:129–136

Yang J, Kloepper JW, Ryu CM (2009) Rhizosphere bacteria help plants tolerate abiotic stress. Trends Plant Sci 14:1–4

Zhuang X, Chen J, Shim H, Bai Z (2007) New advances in plant growth-promoting rhizobacteria for bioremediation. Environ Int 33:406–413

Zimmermann P (2003) Root-secreted phosphomonoesterases mobilizing phosphorus from the rhizosphere: a molecular physiological study in *Solanum tuberosum*. Ph.D. Thesis, Swiss Federal Institute of Technology, Zurich, Switzerland

Zinniel DK, Lambrecht P, Harris NB, Feng Z, Kuczmarski D, Higley P, Ishimaru CA, Arunakumari A, Barletta RG, Vidaver AK (2002) Isolation and characterization of endophytic colonizing bacteria from agronomic crops and prairie plants. Appl Environ Microbiol 68:2198–2208

Plant-Endophyte Interaction and Its Unrelenting Contribution Towards Plant Health

7

Sarangthem Indira Devi and P. Momota

Abstract

Microbes are ubiquitous in nature; likewise, they form a part of flora in the plant species either actively colonizing or latently dormant. Beneficial plant-microbe interactions that promote plant health and development in a broad spectrum of area have been a subject of substantial study. Beneficial microbial associations like mutualists and symbiotic associates in plants show considerable promise due to the ability to provide ample benefits not only to the host plant but also a great prospect to mankind. Endophytes represent a hot spot area of current research, considering their 360 degrees of contribution and a wide range of applications in agricultural science, as potent plant growth promoters and emerging as effective biocontrol agents against various severe plant pathogens, limiting the use of chemical pesticides and other hazardous chemicals. Endophytes also share their contribution to modern medicine, industrial perspectives, nanoscience, forest management, and bioremediation. The endophytes may produce a plethora of substances or compounds with potential therapeutic applications covering plants to humans. There is increase in studies of endophytes for improving yield of food crops and the sustainable production of nonfood crops for biomass and biofuel production. The ability to tolerate both biotic and abiotic stresses is also an added advantage over the other competent microbial flora of its surrounding habitat. Indeed, endophytes are a golden area of researching and exploiting the potent functional properties that aid in understanding the depth of plant-microbe interactions and their contribution to sustainable agriculture, being the focus of concern in the present food crisis scenario.

S. Indira Devi (✉) • P. Momota
Microbial Resources Division, Institute of Bioresources & Sustainable Development (IBSD) (An autonomous research institute, Department of Biotechnology, Govt. of India), Takyelpat, Imphal 795001, India
e-mail: sidevi1@yahoo.co.in

N.K. Arora (ed.), *Plant Microbes Symbiosis: Applied Facets*,
DOI 10.1007/978-81-322-2068-8_7,

Introduction

Research on the bacteria residing in the internal tissues of plants began in the 1870s by Pasteur and others (Hollis 1949). The German botanist, Anton de Bary, known as the father of plant pathology, coined the term endophyte in 1886 giving a description to microorganisms that colonize the internal tissues of stems and leaves (Wilson 1995). The presence of endophytes was first recorded by Vogl in 1898 revealing a mycelium residing in the grass seed of *Lolium temulentum* (Wilson 1996). Later on the term endophyte was used to denote those microorganisms that colonize the plant tissues without causing any symptom throughout their existence in the host. Defining endophyte is simple yet complex, and the term updates with inputs on the current findings; however, endophyte is named by the location and not just with the relationship or interaction it shares with the infected host or the numerous functions it imparts, benefitting the host plant. Microorganisms that gain access to the internal tissues of a plant are termed endophytes, but true endophytes are the ones that colonize internally without causing any disease symptoms throughout their short to long life span in the host. Endophytes with the potential to cause symptoms could be categorized under groups linking to opportunistic, latent pathogens and latent saprophytes. The potential importance of endophytic fungi became clear in 1975, when Charles Bacon discovered endophytes of pasture grasses in the family *Clavicipitaceae* which were toxic to cattle (Bacon et al. 1997). Beneficial endophytic microorganisms comprise especially of fungi and bacteria that colonize the internal plant tissues without causing damage to their host (Petrini 1991).

Endophytes are categorized into three functional groups mainly, plant growth promoters, biocontrol (Bashan and Holguin 1998), and plant stress homeo-regulating microorganisms (Cassan et al. 2009) that can directly or indirectly facilitate the plant growth in optimal, biotic, or abiotic stress conditions. The colonization of plant tissues by endophytes involves several steps including host recognition, spore germination, penetration of the epidermis, and tissue colonization (Petrini 1991, 1996). Studies on microorganisms from plant species are recently becoming more frequent since these microorganisms have widely been explored and investigated for mechanism of biological control, production of novel metabolites with pharmacological properties/importance, and as a potential source for bioremediation. Endophytes intrude phytopathogens and hence have been used as biocontrol agents against microbial plant pathogens, insects, nematodes, and plant pests. Endophytes can promote overall plant growth under biotic (pathogenic microbes, insects, nematode) and abiotic (drought, temperature, pH, salinity, flood, wound, cut or injured part, nutrient deprived) stress conditions. Attributes have been credited to their ability to produce novel bioactive compounds in order to survive in the host environment. One of the considerable roles of endophytic microbes is to initiate the biological degradation of dead or dying host plant, which is necessary for nutrient cycling (Strobel et al. 2002).

Further, endophytic fungi may also produce metabolites with a thermoprotective role, e.g., plants of some volcanic areas in the USA were found to be colonized by the endophytic fungus *Curvularia* species. Plants grown from surface-sterilized seeds in sterile soil that had been incubated with the organism survived constant soil temperature of 50 °C, while the non-symbiotic plants died (Redman et al. 2002). The study of endophytes is expected to become an important component of microbial biology in the future. In forest ecosystems, endophytes contribute by effectively increasing phenotypic plasticity of their long-lived tree hosts under variable or deleterious environmental conditions. Various nanoparticles were reported to be synthesized by a countable number of microorganisms including endophytes. Recently, endophytic actinomycetes have been recognized as potential organisms for biosynthesis of nanoparticles like gold and silver. Nanoparticles from an endophyte may again be used for curing dreaded diseases in the near future with advancement in endophyte research. Therefore, endophytes will play a great role in the future with production and knowledge of more efficient, productive, and cost-effective compounds including nanomedicine for the disease control of humans, animals, and plants. To date, only a few plants have been extensively investigated for their endophytic biodiversity and their potential to produce bioactive secondary metabolites. Various culture-dependent and culture-independent approaches have been adopted to gain maximum possible endophytic isolates. Culture-independent approaches focus on molecular techniques such as quantitative polymerase chain reaction in real time, denaturing gradient gel electrophoresis (DGGE), multivariate analysis to microbial ecology data, clone library construction analysis, and the upcoming high-throughput methodologies such as microarrays and pyrosequencing. Molecular approaches for the isolation and characterization of bacterial endophytes and plant-associated bacterial communities have been reviewed recently by Franks et al. (2006).

The relationship of an endophyte with single or multiple plant hosts can be described in terms of host specificity, host recurrence, and host selectivity or host preference (Zhou and Hyde 2001; Cohen 2006). Consequently the opportunity to find new endophytic microorganisms with potential beneficial properties among a diversity of plants in different ecosystems is intriguing and can lead to identification of novel compounds for drug discovery and developmental strategies in the treatment of catastrophic diseases in humans, animals, and plants.

Ecological Colonization and Biodiversity

Endophytes are sheltered from environmental stress and microbial competition by the host plant, and they seem to be ubiquitous in plant tissues, having been isolated from various parts of plants like flowers, fruits, leaves, stem, root, and seeds in various plant species (Kobayashi and Palumbo 2000). Plants growing in areas of great biodiversity usually have the prospect of harboring endophytes with great biodiversity (Strobel and Daisy 2003). The mode of colonization is systemic with bacterial colonies and biofilms residing latently in intercellular spaces, inside the vascular tissue or within cells (Ulrich et al. 2008). It seems obvious that endophytes are rich and reliable sources of genetic diversity and novel undescribed species. Endophytic fungi that live inside the tissue of living plants are an underexplored group of microorganisms. Dreyfuss and Chapela (1994) estimated that there may be at least one million species of endophytic fungi alone! Specifically, the bacteria enter tissues via germinating radicals, secondary roots, or stomata or as a result of foliar damage. An important aspect of colonization is the ability to compete with indigenous microorganisms already present in the soil and rhizosphere of the developing plant. Our understanding of the factors involved in these interactions has been hindered by our inability to culture and characterize diverse members of the rhizosphere community and to determine how that community varies with plant species, plant age, location on the root, and soil properties. Molecular studies on endophytes have revealed a large richness of species. Endophytic bacterial diversity was estimated in Mexican husk tomato plant roots by amplified rDNA restriction analysis and sequence homology comparison of the 16S rDNA genes. The predominant genera were *Stenotrophomonas* (21.9 %), *Microbacterium* (17.1 %), *Burkholderia* (14.3 %), *Bacillus* (14.3 %), and *Pseudomonas* (10.5 %). The endophytic bacterial diversity within the roots of Mexican husk tomato plants is a subset of the rhizosphere bacterial population, dominated by a few genera (Marquez-Santa Cruz et al. 2010).

The diversity of cultivable bacterial endophyte is exhibited not only in the variety of plant species colonized but also in the many taxa involved, with most being members of common soil bacterial genera such as *Enterobacter*, *Pseudomonas*, *Burkholderia*, *Bacillus*, and *Azospirillum* (Lodewyckx et al. 2002). A recent study (Magnani et al. 2010) describes the diversity of endophytes found in Brazilian sugarcane, reporting the endophytic bacteria in sugarcane to be more diverse than previously reported and also demonstrate niche specificity. Tropical and temperate

Table 7.1 Some natural active products obtained from endophytes

Endophyte	Host	Active metabolites	Target pathogens/ activity	References
Acremonium zeae	Maize	Pyrrocidines A, B	*Aspergillus flavus*, *Fusarium verticillioides*	Wicklow et al. (2005)
Pseudomonas viridiflava	Grass	Ecomycins B and C	*Cryptococcus neoformans*, *C. albicans*	Miller et al. (1998)
Cytonaema sp.	*Quercus* sp. 103	Cytonic acids A and D	*Cytomegalovirus*	Guo et al. (2000)
Muscodor albus	Tropical tree	Tetrahydrofuran, 2-methyl furan, 2-butanone, aciphyllene	*Stachybotrys chartarum*	Atmosukarto et al. (2005)
Streptomyces griseus	*Kandelia candel*	Antimicrobial, p-amino acetophenonic acids	Antimicrobial	Guan et al. (2005)
Streptomyces NRRL 30562	*Kennedia nigriscans*	Munumbicins, munumbicin D	*Bacillus anthracis*, *M. tuberculosis*, *Plasmodium falciparum*	Castillo et al. (2002)
Periconia sp.	*Taxus cuspidata*	Fusicoccane diterpenes	*Bacillus subtilis*, *Staphylococcus aureus*, *Klebsiella pneumoniae*, *Salmonella typhimurium*	Kim et al. (2004)
S. marcescens	*Rhyncholacis penicillata*	Oocydin A	*Pythium ultimum*, *Phytophthora parasitica*, *Phytophthora cinnamomi*, *Phytophthora citrophora*	Strobel et al. (1999), (2004)
Acremonium zeae	Maize	Pyrrocidines A, B	*Aspergillus flavus*, *Fusarium verticillioides*	Wicklow et al. (2005)
Streptomyces NRRL 30566	*Grevillea pteridifolia*	Kakadumycins	*Bacillus anthracis*, *M. tuberculosis*, *Plasmodium falciparum*	Castillo et al. (2003)
Verticillium sp.	*Rehmannia glutinosa*	Massariphenone, ergosterol peroxide	*Pyricularia oryzae* P-2b	You et al. (2009)
Paenibacillus polymyxa	Wheat, lodge pine, green beans, *Arabidopsis thaliana*, canola	Fusaricidin A–D	*Corynebacterium fimi*, *Escherichia coli*, *Leptosphaeria maculans*	Beck et al. (2003), Li et al. (2007), and Beatty and Jensen (2002)
Phomopsis cassiae	*Cassia spectabilis*	Cadinane sesquiterpenes	*Cladosporium sphaerospermum*, *Cladosporium cladosporioides*	Silva et al. (2006)
Ampelomyces sp.	*Urospermum picroides*	3-O-Methylalaternin, altersolanol A	*Staphylococcus aureus*, *S. epidermidis*, *Enterococcus faecalis*	Aly et al. (2008)
Cryptosporiopsis quercina	*Tripterygium wilfordii*	Cryptocin	*Pyricularia oryzae*	Li et al. (2007)
Pestalotiopsis microspora	*Terminalia morobensis*	Pestacin, isopestacin	Antioxidant	Strobel (2002)

The mechanism by which endophytes colonized and localized their host plant and the process through which they confer resistance to attack by broad phytopathogens are a promising area to explore.

As Biocontrol Agents

The reduction in food productivity worldwide is an alarming issue in the present scenario. One of the major causes is diseases of plants that hinder crop production. Therefore, management of diseases by means of harmless bioagents is important. The adaptation of colonizing the internal host tissue has made endophytes more valuable for agriculture, as a tool to improve crop performance and most importantly encourage organic farming, when maximum percent of the agro market is flooded with chemical-based food crops. Many endophytes have mutualistic relationships with their host plant; they obtain nutrients and in turn provide protection to the host from biotic and abiotic stresses. Controlling pests, insects, and phytopathogenic microbes by using anti-ecological chemical compounds like insecticides, fungicides, and herbicides results in the elimination of important species of insects and microbes that control other pests and harmful microbes, performing a crucial role in the environment. Beneficial endophytes are also getting affected, and hence a revolution is essential for preserving such beneficial microbes. The role of endophytes as upcoming biocontrol agents is an area to focus so as to utilize the alternative biosource agents in reducing the use of chemical hazards in agriculture and promoting an eco-friendly substitute.

Against Phytopathogens

Natural products from endophytic microbes have been observed to inhibit or kill a wide variety of harmful disease-causing agents, but not limited to phytopathogens, as well as bacteria, fungi, viruses, and protozoans. Endophytes colonize an ecological niche similar to that of the phytopathogens and hence can be utilized as agents for the control of pathogenic microbes. Endophytic strains of *Microbispora rosea* subsp. *rosea* and *Streptomyces olivochromogenes* were found to be effective in suppressing the club root of Chinese cabbage caused by *Plasmodiophora brassicae* (Lee et al. 2008). Cryptocin, a unique tetramic acid, is also produced by *C. quercina* (Li et al. 2000). This unusual compound possesses potent activity against *Pyricularia oryzae* as well as a number of other plant-pathogenic fungi. An endophytic strain of *Actinoplanes missouriensis* isolated from lupin roots was effective in suppressing the root rot of lupin caused by *Plectosporium tabacinum* (El-Tarabily 2003). Cytosporone B and C were isolated from a mangrove endophytic fungus, *Phomopsis* sp., inhibiting the growth of *C. albicans* and *Fusarium oxysporum* (Huang et al. 2007). Chlorinated metabolites such as (−) mycorrhizin A and (+) cryptosporiopsin isolated from endophytic *Pezicula* strains were reported as strongly fungicidal, herbicidal, and antibacterial agents (Huang et al. 2007). Benzophenone derivatives, "Pestalachlorides A" (C21H21Cl2NO5) and "B" (C20H18Cl2O5), from the plant endophytic fungus *Pestalotiopsis adusta*, display significant antifungal activity against three plant pathogenic fungi, *Fusarium culmorum*, *Gibberella zeae*, and *Verticillium albo-atrum* (Li et al. 2008). Two new bioactive metabolites, "ethyl 2, 4-dihydroxy- 5, 6-dimethylbenzoate" and "phomopsilactone," were isolated from an endophytic fungus *Phomopsis cassiae* from *Cassia spectabilis* and displayed strong antifungal activity against two phytopathogenic fungi, *Cladosporium cladosporioides* and *Cladosporium sphaerospermum* (Silva et al. 2005). There are reported compounds such as amphisin, 2,4-diacetylphloroglucinol (DAPG), hydrogen cyanide, oomycin A, phenazine, pyoluteorin, pyrrolnitrin, tensin, tropolone, and cyclic lipopeptides produced by pseudomonads (Raaijmakers et al. 2002) and oligomycin A, kanosamine, zwittermicin A, and xanthobaccin produced by *Bacillus*, *Streptomyces*, and *Stenotrophomonas* spp., with antimicrobial activities. It is important to note that many strains produce a pallet of secondary antimicrobial metabolites and conditions favoring one compound may not favor another (Duffy and Défago

1999); thus, the varied arsenal of biocontrol strains may enable antagonists to perform their ultimate objective of pathogen suppression under the widest range of environmental conditions. For example, in *Pseudomonas fluorescens* CHA0 biosynthesis of diacetylphloroglucinol (DAPG) is stimulated, and pyoluteorin is repressed in the presence of glucose as a carbon source. As glucose is depleted, however, pyoluteorin becomes the more abundant antimicrobial compound produced (Duffy and Défago 1999).

Against Insect Pests

Recent studies showed that some plant lectins were toxic to sap-sucking insect pests. For example, the lectin from snowdrop (*Galanthus nivalis* agglutinin) is toxic to plant hoppers by binding to gut epithelium and then passing into the hemolymph of a pest (Fitches et al. 2001). The mannose-binding *Pinellia ternata* agglutinin (PTA) gene was also found with significant insecticidal activities against Homopterans (Zhang et al. 2003). Insect bioassays showed that transgenic plant expressing PTA lectin gene significantly inhibited aphid growth and exhibited considerable insecticidal activities against rice brown plant hopper. Population of aphids was smaller in plants infected with recombinant endophyte, *Chaetomium globosum* YY-11 (with antifungal activities) with PTA gene cloned into YY-11 mediated by *Agrobacterium tumefaciens* than in wild-type endophytic fungi-treated plant and controls. It significantly inhibited the growth and reproduction of *Myzus persicae* (green aphid) through endophytes as expressing systems for delivery of insecticidal proteins in plants and enhances resistance of plants to insect pathogens (Qi et al. 2011). Alkaloid produced by endophytes can significantly reduce herbivores (Clay and Schardl 2002). For instance, some endophytes can produce toxic alkaloids such as *Neotyphodium coenophialum*, which exist inside tall fescue *Schedonorus phoenix* (Scop.) *Holub* and greatly limit animal feeding and reproductive performance. In banana, naturally occurring endophytic *F. oxysporum* antagonized nematode *Radopholus similis* in vitro through the production of nematode-antagonistic metabolites (Athman et al. 2006), and inoculation of the endophytes into tissue-cultured plants resulted in improved plant growth and reduced nematode densities.

Against Weeds

Endophytes especially *Neotyphodium* are studied for the ability to protect grasses from other destructive weeds and invading pathogens. Much work has been done on perennial ryegrass and several species of fescue to understand better the mechanism of its defense. Ohio State University demonstrated that fungal endophytes can provide the necessary arsenal for the grass plants to compete more effectively against common weed species. Further exploration on the competitive ability of endophyte-containing grasses can substantially contribute to reduction in weed invasion. Endophyte-containing tall fescue and perennial ryegrass are often better and able to compete against common weed species. Endophyte-containing grasses have greater seed survival, germination, and establishment than cultivars that lack endophytes. The presence of endophyte in perennial ryegrass resulted in a significant decrease in white clover. Endophyte-containing perennial ryegrass maintained greater cover under severe insect stress, whereas those lacking endophytes were more heavily invaded by weeds (Grewal and Richmond 2010).

Role in Detoxification and Degradation of Pathogen Virulence Factors

Another mechanism of biological control is the detoxification of pathogen virulence factors, for example, certain biocontrol agents are able to detoxify albicidin toxin produced by *Xanthomonas albilineans* (Walker et al. 1988; Zhang and Birch 1997). The detoxification mechanisms include production of a protein by an endophyte such as *Klebsiella oxytoca* and *Alcaligenes denitrificans* that reversibly binds with the toxin, (Walker et al. 1988), as well as an irreversible detoxification of albicidin, mediated by an esterase that occurs in *Pantoea dispersa*

(Zhang and Birch 1997). Several microorganisms, including strains of *Burkholderia cepacia* and *Ralstonia solanacearum*, can also hydrolyze fusaric acid, a phytotoxin produced by various *Fusarium* species (Toyoda et al. 1988).

Endophytes and Siderophores

Iron is one among the essential elements required for the sustenance of life. Its availability in the environment is limited; therefore, the scarcity of bioavailable iron in soil habitats and on plant surfaces foments a furious competition (Loper and Henkels 1997). There are certain groups of microbes that help in accumulating iron from the surrounding environment by producing iron-chelating compounds known as siderophores. Various bacterial siderophores differ in their abilities to sequester iron; in general, they deprive pathogenic fungi of this essential element since the fungal siderophores have lower affinity (Loper and Henkels 1999). Therefore, endophytes with high siderophore activity actively colonize the host and protect it from invading pathogens depriving them of the essential micronutrient. Other benefits of using endophytes may include increased nutrient uptake (Malinowski et al. 2000), probably faster than the competing flora so as to overcome and resist the activity of invading pathogens.

Lytic Enzymes and Disease Suppression Mechanisms

The enzymatic activity of enzymes such as chitinase, protease, and glucanase as a defense mechanism is known and studied for applied purposes. Chitinase produced by *Serratia plymuthica* C48 inhibited spore germination and germ tube elongation of *Botrytis cinerea*. The ability to produce extracellular chitinase is considered crucial for *Serratia marcescens* to act as an antagonist against *Sclerotium rolfsii* (Ordentlich et al. 1998). It has also been demonstrated that extracellular chitinase and laminarinase synthesized by *Pseudomonas stutzeri* digest and lyse mycelia of *Fusarium solani* (Lim et al. 1991). The β-1,3-glucanase synthesized by *Paenibacillus* sp. strain 300 and *Streptomyces* sp. strain 385 lyses fungal cell walls of *F. oxysporum* f. sp. *cucumerinum* (Singh et al. 1999). Similar to siderophores and antibiotics, regulation of lytic enzyme production (proteases and chitinases in particular) involves the GacA/GacS (Sacherer et al. 1994) or GrrA/GrrS regulatory systems (Ovadis et al. 2004) and colony phase variation (Lugtenberg et al. 2001).

Host Resistance Induction Mechanism

Certain bacteria trigger a phenomenon known as induced systemic resistance (ISR) phenotypically similar to systemic acquired resistance (SAR). SAR develops when plants successfully activate their defense mechanisms in response to primary infection by a pathogen, notably when the latter induces a hypersensitive reaction through which it becomes limited in a local necrotic lesion of brown desiccated tissue (Van Loon et al. 1998). Several bacterial traits (i.e., flagellation, production of siderophores and lipopolysaccharides) have been proposed to trigger ISR (Van Loon et al. 1998), but there is no compelling evidence for an overall ISR signal produced by bacteria (Haas et al. 2002). ISR is effective against different types of pathogens but differs from SAR in that the inducing bacterium does not cause visible symptoms on the host plant (Van Loon et al. 1998). Bacterial endophytes and their role in ISR have been reviewed by Kloepper and Ryu (2006).

Prospect of Endophyte in Plant Growth Promotion

There are numbers of mechanisms by which bacteria promote plant growth and health. The production of plant growth phytohormones like auxins, cytokinins, gibberellins, and ethylene by an endophyte has been studied. Endophytes are known to produce phytohormones that play an important role in promoting root elongation and plant growth (Li 2005). Nejad and Johnson (2000) colonized an endophyte into seeds and succeeded to improve seed germination rate and seedling length of rape and tomato, but also significantly reduced virus disease symptoms caused by *Catheter wither*. Endophytic actinomycetes have been reported to produce plant growth-promoting compounds such as IAA and siderophores in vitro (De Oliveira et al. 2010).

Plant Growth Hormones and Ability to Fix Nitrogen

Endophytes can promote plant growth without requiring the use of chemical fertilizers. The mechanism and efficiency of the growth promotion effect depends strongly on the species or endophytic isolate, the host plant, and the prevailing environmental condition from where the host is obtained. Plant-associated microorganisms such as fungi and bacteria evolved biochemical pathways, resulting in the production of plant growth hormones reportedly auxins, absicins, ethylene, gibberellins, and kinetins (Goodman et al. 1986). Volatile substances such as 2,3-butanediol and acetoin produced by bacteria seem to be newly discovered mechanisms responsible for plant growth promotion (Ryu et al. 2003). Plant growth promotion has also been caused by endophytic yeast due to production of auxins, as in the case of *Williopsis saturnus* in maize roots (Nassar et al. 2005). *Piriformospora indica* is another root endophyte that has been shown to promote the growth of both food crops (e.g., maize) and shrubs (*Artemisia annua* L.) with increases up to 50 % of their fresh biomass (Varma et al. 1999). It has also been shown to promote the formation of adventitious roots in cuttings of *Pelargonium* and *Poinsettia* (Drüge et al. 2007). Endophytes produce adenine ribosides that stimulate growth and mitigate browning of pine tissues (Pirttilä et al. 2004). The nitrogen transfer improved host uptake of phosphorus conferred by the endophytic fungus *Cladorrhinum foecundissimum* in cotton roots and increased plant height by 50 % compared with control plants (Gasoni and Stegman de Gurfinkel 1997). Many studies demonstrated that plants infected with endophyte obtain growth promotion (Ait Barka et al. 2002) resistance to drought, stress (Swarthout et al. 2009), and tolerance to unsuitable soil conditions (Malinowski et al. 2004). The enhancement of plant growth may be influenced by compounds like phytohormones produced by fungal endophytes, *Colletotrichum* sp., and endophytic fungus in *A. annua* produces substances like indole acetic acid (IAA) to regulate plant processes (Lu et al. 2000). Rhizobia are very well-known endophytes able to fix nitrogen in the roots of legume plants. The rhizobium-legume system not only provides nitrogen to the legume crop but also enhances the fertility of soil where they are grown. In fact legumes because of their endophytic partner are considered as the most important biological system to enhance the nitrogen content of the soil and hence preferred in crop rotation throughout the globe. Rhizobia are also known as very good phytohormone (especially indole acetic acid) producers which result in root hair proliferation and growth enhancement. Therefore the impact of plant growth is correlated with the production of phytohormones with the supplement of favorable growth conditions.

Stress Tolerance Mechanisms of Endophytes

The fact that plants are continuously exposed to a wide range of environmental stress conditions (comprising both biotic and abiotic factors) is unavoidable, and the need to protect plants from such extreme conditions is vital to preserve the plant life and its yield as it is a serious threat to agriculture. The mutual relation between the host plant and the endophyte significantly improves tolerance to stress factors including drought, high soil salinity, extreme soil pH, insect pests, and nematodes, along with increased persistence and vigor, and in turn the plant provides the symbiont with nutrients, protection, and reliable and efficient dissemination (Schardl et al. 2004). It has become apparent that some plants are unable to tolerate habitat-imposed abiotic and biotic stresses in the absence of fungal endophytes (Redman et al. 2002). Therefore biotechnological intervention on implementing endophytes for improving plant immunity and health is indeed an important and essential area of study.

Abiotic Stress Tolerance

The study on endophytes for tolerance to extreme temperatures, salinity, pH, flood and drought, and also enhanced immunity of the host plant is now being carried out. The primary cause of crop loss worldwide, reducing average yields for most major crop plants by more than 50 %, is due to abiotic stresses (Bray et al. 2000). Abiotic stress leads to series of morphological, physiological, biochemical, and molecular changes that adversely affect plant growth and productivity (Wang et al. 2001). There seems to be a vital interaction between the endophyte and host, and the resultant chemistry induces plant protection. Secondary compounds produced by endophytic fungi may be responsible for the resistance to biological and abiotic stresses and improvement in the host competitiveness. Secondary metabolites play an important role in resisting drought. Loliterm B, for instance, can alter osmotic potential to reduce the harmful effect of drought (Yuan et al. 2010). Interestingly, Hasegawa et al. (2005) discovered that tissue-cultured seedlings of mountain laurel harboring endophytic *Streptomyces padanus* became highly tolerant to drought. In tolerant seedlings, callose deposition and lignification of cell walls as well as osmotic pressure of cells were accelerated, suggesting that all these events could be closely associated with enhanced drought tolerance. *P. indica* in salt stress conditions increased the tolerance of salt-sensitive barley (*Hordeum vulgare*) cultivar to severe salt stress. *P. indica*-colonized plants contained higher ascorbate concentrations in roots compared with non-colonized plants, while the ratio of ascorbate vs. dehydroascorbate was not significantly altered, but catalase, ascorbate peroxidase, glutathione reductase, dehydroascorbate reductase, and monodehydroascorbate reductase activities were increased (Lamabam et al. 2011). Plant responses to abiotic stresses are complex, involving signal reception and transduction followed by genetic and physiological responses. It is perceived that all plants are capable of perceiving and responding to stress (Bartels and Sunkar 2005). Likewise endophytes with growth potential under stress conditions can be exploited for use in agricultural wastelands for encouraging plantation.

Conclusion

The time is right to explore these nature-friendly microbes and update their various applications. Can endophytes give a point solution to the alarming food crisis too? In terms of agricultural prospect, it's high time to give more emphasis on technology transfer to fields rather than to confine in laboratories or greenhouse study alone. There are interesting twists and challenges in the study of endophytes, the most important being the isolation process, especially the surface sterilization techniques and identification of true absolute endophytes, ignoring the rest of the microbes. In case of thin plant parts, where removal of the outer tissue portion is not possible, chances of non-endophytic microbial growth may be encouraged following the growth of injured cells, after exposure to favorable conditions. Alternatively, some endophyte growth may also be hindered in such cases due to the penetration of chemicals in the tissues. Hence, proper measures and precautions for isolation are important criteria based on plant type and tissues in particular. Schematic representation of endophyte isolation and in vitro evaluation of different biochemical enzyme assays and biocontrol activity against plant pathogens is shown in Fig. 7.1.

One of the frequently sought issues is that will a single potential endophyte actively overcome the existing problems by acting as a biocontrol agent or as a source to novel metabolites or impart protection from various abiotic stresses? Adding to this the potential isolate with positive response when inoculated into its own plant or other plants in field may not yield similar results as were visible in lab conditions. Understanding plant-microbe interactions will surmount the critical process that an endophyte delivers for host benefit. Another view is, though we may get information on the total endophytic microbiome

Fig. 7.1 Isolation and in vitro evaluation of different biochemical enzyme assays and biocontrol activity of endophytes

of the host by approaching much advanced high-throughput techniques of culture-independent methods for isolation, how far will we be successful in harboring the most potent microbes, for application is not clear at present. The microorganisms isolated with culture dependence are for sure adaptable to any changes in the environment and hence grow easily without complexity rather than microbes obtained by culture-independent methods. These fragile isolates may find it hard to grow in laboratory and hence difficult to take to the fields as inoculants. Endophytes can readily be termed as "probiotics of plant origin". The survival chain of endophyte involves several animals and human consumers too. Hence, another question arises on endophytes existence in connection to the human microbiome and its overall contribution to human health. Encouring endophyte research and exploitation is essential to answer all the queries. Therefore, discovering the wild and much unexplored territories of multifunctional endophytes and their interaction with the plant host is undoubtedly essential to know more and gain more for a better tomorrow.

Acknowledgment We thank the Department of Biotechnology, DBT, Govt. of India, for assisting financial support to carry out the study.

References

Ait Barka E, Gognies S, Nowa J, Audran JC, Belarbi A (2002) Inhibitory effect of endophyte bacteria on *Botrytis cinerea* and its influence to promote the grapevine growth. Biol Control 24:135–142

Aly AH, Edrada-Ebel R, Wray V, Müller WEG, Kozytska S, Hentschel U, Proksch P, Ebel R (2008) Bioactive metabolites from the endophytic fungus *Ampelomyces* sp. isolated from the medicinal plant *Urospermum picroides*. Phytochemistry 69:1716–1725

Athman SY, Dubois T, Coyne D, Gold C, Labuschagne N, Viljoen A (2006) Effect of endophytic *Fusarium similis* to tissue culture banana plants. J Nematol 38:455–460

Atmosukarto I, Castillo U, Hess WM, Sears J, Strobel G (2005) Isolation and characterization of Muscodor albus I-41.3s, a volatile antibiotic producing fungus. Plant Sci 169:854–861

Bacon CW, Porter JK, Robbins JD, Luttrell ES (1997) *Epichloe typhina* from toxic tall fescue grasses. Appl Environ Microbiol 34:76–81

Baker D, Mocek U, Garr C (2000) Natural products vs. combinatorials: a case study. In: Wrigley SK, Hayes MA, Thomas R, Chrystal EJT, Nicholson N (eds) Biodiversity: new leads for pharmaceutical and agrochemical industries. The Royal Society of Chemistry, Cambridge

Bartels D, Sunkar R (2005) Drought and salt tolerance in plants. Crit Rev Plant Sci 24:23–58

Bashan Y, Holguin G (1998) Proposal for the division of plant growth promoting rhizobacteria into two classifications: biocontrol-PGPB (plant growth-promoting bacteria) and PGPB. Soil Biol Biochem 30:1225–1228

Beatty PH, Jensen SE (2002) *Paenibacillus polymyxa* produces fusaricidin-type antifungal antibiotics active against *Leptosphaeria maculans*, the causative agent of blackleg disease of canola. Can J Microbiol 48:159–169

Beck HC, Hansen AM, Lauritsen FR (2003) Novel pyrazine metabolites found in polymyxin biosynthesis by *Paenibacillus polymyxa*. FEMS Microbiol Lett 220:67–73

Bray EA, Bailey-Serres J, Weretilnyk E (2000) Responses to abiotic stresses. In: Gruissem W, Buchannan B, Jones R (eds) Biochemistry and molecular biology of plants. American Society of Plant Physiologists, Rockville, pp 1158–1249

Cafeu MC, Silva GH, Teles HL, Da Bolzani VS, Araujo AR, Young MCM, Pfenning LH (2005) Antifungal compounds of *Xylaria* sp., an endophytic fungus isolated from *Palicourea marcgravii* (Rubiaceae). Quim Nova 28:991–995

Cassan F, Perrig D, Sgroy V, Masciarelli O, Penna C, Luna V (2009) *Azospirillum brasilense* Az39 and *Bradyrhizobium japonicum* E109, inoculated singly or in combination, promote seed germination and early seedling growth in corn (*Zea mays L.*) and soybean (*Glycine max L.*). Eur J Soil Biol 45:28–35

Castillo UF, Strobel GA, Ford EJ, Hess WM, Porter H, Jensen JB, Albert H, Robison R, Condron MAM, Teplow DB, Stevens D, Yaver D (2002) Munumbicins, wide-spectrum antibiotics produced by *Streptomyces* NRRL30562, endophyte on *Kennedia nigriscans*. Microbiology 148:2675–2685

Castillo U, Harper JK, Strobel G, Sears J, Alesi K, Ford E, Lin J, Hunter M, Maranta M, Ge H, Yaver D, Jensen JB, Porter H, Robison R, Millar D, Hess WM, Condron M, Teplow D (2003) Kakadumycins, novel antibiotics from *Streptomyces* sp. NRRL 30566, an endophyte of *Grevillea pteridifolia*. FEMS Lett 224:183–190

Clay K, Schardl CL (2002) Evolutionary origins and ecological consequences of endophyte symbiosis with grasses. Am Nat 160:S99–S127

Cohen SD (2006) Host selectivity and genetic variation of *Discula umbrinella* isolates from two oak species: analyses of intergenic spacer region sequences of ribosomal DNA. Microb Ecol 52:463–469

De Oliveira MF, da Silva MG, Van der Sand ST (2010) Anti-phytopathogen potential of endophytic actinobacteria isolated from tomato plants (*Lycopersicon esculentum*) in southern Brazil, and characterization of *Streptomyces* sp. R18(6), a potential biocontrol agent. Res Microbiol 161:565–572

Dreyfuss MM, Chapela IH (1994) Potential of fungi in the discovery of novel, low-molecular weight pharmaceuticals. In: Gullo VP (ed) The discovery of natural products with therapeutic potential. Butter-worth-Heinemann, London, pp 49–80

Drüge U, Baltruschat H, Franken P (2007) *Piriformospora indica* promotes adventitious roots formation in cuttings. Sci Hortic 112:422–426

Duffy BK, Défago G (1999) Environmental factors modulating antibiotic and siderophore biosynthesis by *Pseudomonas fluorescens* biocontrol strains. Appl Environ Microbiol 65:2429–2438

El-Tarabily KA (2003) An endophytic chitinase-producing isolate of *Actinoplanes missouriensis*, with potential for biological control of root rot of lupin caused by *Plectosporium tabacinum*. Aust J Bot 51:257–266

Elvira-Recuenco M, Van Vuurde JWL (2000) Natural incidence of endophytic bacteria in pea cultivars under field conditions. Can J Microbiol 46:1036–1041

Firakova S, Sturdikova M, Muckova M (2007) Bio-active secondary metabolites produced by microorganisms associated with plants. Biologia 62:251–257

Fitches ESD, Woodhouse JP, Edwards JA, Gatehouse (2001) *In vitro* and *in vivo* binding of snowdrop (*Galanthus nivalis* agglutinin; GNA) and jackbean (*Canavalia ensiformis*; Con A) lectins within tomato moth (*Lacanobia oleracea*) larvae; mechanisms of insecticidal action. J Insect Physiol 47:777–787

Franks A, Ryan R, Abbas A, Mark L, O'Gara F (2006) Molecular tools for studying plant growth-promoting rhizobacteria (PGPR). In: Cooper JE, Rao JR (eds) The molecular approaches to soil, rhizosphere and plant microorganisms. CABI Publishing, Wallingford, pp 116–131

Gasoni L, Stegman de Gurfinkel B (1997) The endophyte *Cladorrhinum foecundissimum* in cotton roots: phosphorus uptake and host growth. Mycol Res 101:867–870

Goodman R, Kiraly Z, Wood KR (1986) The biochemistry and physiology of plant disease. University of Missouri Press, Columbia, pp 374–375

Grewal P, Richmond D (2010) Benefits of endophytic grasses by infecting grasses, endophytes provide enhanced resistance to insects and weeds. https://www.landcareresearch.co.nz/_data/assets/pdf_file/0014/20651/wtsnew53.pdf. Accessed 24 Dec 2014

Guan S, Grabley S, Groth I, Lin W, Christner A, Guo D, Sattler I (2005) Structure determination of germacrane-type sesquiterpene alcohols from an endophyte *Streptomyces griseus* subsp. Magn Reson Chem 43:1028–1031

Guo LD, Hyde KD, Liew ECY (2000) Identification of endophytic fungi from *Livistona chinensis* based on morphology and rDNA sequences. New Phytol 147:617–630

Guo B, Wang Y, Sun X, Tang K (2008) Bio-active natural products from endophytes: a review. Appl Biochem Microbiol 44:136–142

Haas D, Keel C, Reimmann C (2002) Signal transduction in plant-beneficial rhizobacteria with biocontrol properties. Anton Leeuw 81:385–395

Hasegawa S, Meguro A, Toyoda K, Nishimura T, Kunoh H (2005) Drought tolerance of tissue-cultured seedlings of mountain laurel (*Kalmia latifolia* L.) induced by an endophytic actinomycete. II. Acceleration of callose accumulation and lignification. Actinomycetologica 19:13–17

Hollis JP (1949) Bacteria in healthy potato tissue. Phytopathol 41:350–367

Huang WY, Cai YZ, Xing J, Corke H, Sun M (2007) A potential antioxidant resource: endophytic fungi from medicinal plants. Econ Bot 61:14–30

James EK, Gyaneshwar P, Mathan N, Barraquio WL, Reddy PM, Iannetta PP, Olivares FL, Ladha JK (2005) Infection and colonization of rice seedlings by the plant growth-promoting bacterium *Herbaspirillum seropedicae* Z67. Mol Plant Microbe Interact 15: 894–906

Kim S, Shin DS, Lee T, Oh KB (2004) Periconicins, two new fusicoccane diterpenes produced by an endophytic fungus *Periconia* sp. with antibacterial activity. J Nat Prod 67:448–450

Kloepper JW, Ryu CM (2006) Bacterial endophytes as elicitors of induced systemic resistance. In: Schulz B, Boyle C, Sieber TN (eds) Microbial root endophytes. Springer, Berlin, pp 33–52

Kobayashi DY, Palumbo JD (2000) Bacterial endophytes and their effects on plants and uses in agriculture. In: Bacon CW, White JF (eds) Microbial endophytes. Dekker, New York, pp 199–236

Kusari S, Zuhlke S, Spiteller M (2009) An endophytic fungus from *Camptotheca acuminata* that produces camptothecin and analogues. J Nat Prod 72:2–7

Lamabam PS, Sarvajeet SG, Narendra T (2011) Unraveling the role of fungal symbionts in plant abiotic stress tolerance. Plant Signal Behav 6: 175–191

Larrainzar E, O'Gara F, Morrissey JP (2005) Applications of autofluorescent proteins for in situ studies in microbial ecology. Ann Rev Microbiol 59:257–277

Lee SO, Choi GJ, Choi YH, Jang KS, Park DJ, Kim CJ, Kim JC (2008) Isolation and characterization of endophytic actinomycetes from Chinese cabbage roots as antagonists to *Plasmodiophora brassicae*. J Microbiol Biotechnol 18:1741–1746

Li WK (2005) Endophytes and natural medicines. Chin J Nat Med 3:193–199

Li JY, Strobel G, Harper J, Lobkovsky E, Clardy J (2000) Cryptocin, a potent tetramic acid antimycotic from the endophytic fungus *Cryptosporiopsis cf. quercina*. Org Lett 23:767–770

Li WC, Zhou J, Guo SY, Guo LD (2007) Endophytic fungi associated with lichens in Baihua mountain of Beijing, China. Fungal Divers 25:69–80

Li JH, Wang ET, Chen WF, Chen WX (2008) Genetic diversity and potential for promotion of plant growth detected in nodule endophytic bacteria of soybean grown in Heilongjiang province of China. Soil Biol Biochem 40:238–246

Lim HS, Kim YS, Kim SD (1991) *Pseudomonas stutzeri* YPL-1 genetic transformation and antifungal mechanism against *Fusarium solani*, an agent of plant root rot. Appl Environ Microbiol 57:510–516

Lodewyckx C, Vangronsveld J, Porteous F, Moore ERB, Taghavi S, Mezgeay M, Van der Lelie D (2002) Endophytic bacteria and their potential applications. Crit Rev Plant Sci 21:583–606

Loper JE, Henkels MD (1997) Availability of iron to *Pseudomonas fluorescens* in rhizosphere and bulk soil evaluated with an ice nucleation reporter gene. Appl Environ Microbiol 63:99–105

Loper JE, Henkels MD (1999) Utilization of heterologous siderophores enhances levels of iron available to *Pseudomonas putida* in the rhizosphere. Appl Environ Microbiol 65:5357–5363

Lu H, Zou WX, Meng JC, Hu J, Tan RX (2000) New bioactive metabolites produced by *Colletotrichum* sp., an endophytic fungus in *Artemisia annua*. Plant Sci 151:67–73

Lugtenberg BJJ, Dekkers L, Bloemberg GV (2001) Molecular determinants of rhizosphere colonization by *Pseudomonas*. Annu Rev Phytopathol 39:461–490

Magnani GS, Didonet CM, Cruz LM, Picheth CF, Pedrosa FO, Souza EM (2010) Diversity of endophytic bacteria in Brazilian sugarcane. Genet Mol Res 9:250–258

Malinowski DP, Alloush GA, Belesky DP (2000) Leaf endophyte *Neotyphodium coenophialum* modifies mineral uptake in tall fescue. Plant Soil 227: 115–126

Malinowski DP, Zuo H, Belesky DP, Alloush GA (2004) Evidence for copper binding by extracellular root exudates of tall fescue but not perennial ryegrass infected with *Neotyphodium* spp. endophytes. Plant Soil 267:1–12

Marquez-Santacruz HA, Hernandez-Leon R, Orozco-Mosqueda MC, Velazquez-Sepulveda I, Santoyo G (2010) Diversity of bacterial endophytes in roots of Mexican husk tomato plants (*Physalis ixocarpa*) and their detection in the rhizosphere. Genet Mol Res 9:2372–2380

Miche L, Battistoni F, Gemmer S, Belghazi M, Reinhold-Hurek B (2006) Up regulation of jasmonate-inducible defense proteins and differential colonization of roots of *Oryza sativa* cultivars with the endophyte *Azoarcus sp*. Mol Plant Microbe Interact 19:502–511

Miller RV, Miller CM, Garton-Kinney D, Redgrave B, Sears J, Condron M, Teplow D, Strobel GA (1998) Ecomycins, unique antimycotics from *Pseudomonas viridiflava*. J Appl Microbiol 84:937–944

Mittermeier RA, Meyers N, Gil PR, Mittermeier CG (1999) Hotspots: Earth's biologically richest and most endangered ecoregions. Toppan Printing Co., Japan, p 392

Nassar HA, El-Tarabily KA, Sivasithamparam K (2005) Promotion of plant growth by an auxin-producing isolates of the yeast *Williopsis saturnus* endophytic in maize (*Zea mays*) roots. J Biol Fertil Soils 42:2

Nejad P, Johnson (2000) Endophytic bacteria induce growth promotion and wilt disease suppression in oilseed rape and tomato. Biol Control 18:208–215

Ordentlich A, Elad Y, Chet I (1998) The role of chitinase of *Serratia marcescens* in the biocontrol of *Sclerotium rolfsii*. Phytopathology 78:84–88

Ovadis M, Liu X, Gavriel S, Ismailov Z, Chet I, Chernin L (2004) The global regulator genes from biocontrol strain *Serratia plymuthica* IC1270: cloning, sequencing, and functional studies. J Bacteriol 186:4986–4993

Petrini O (1991) Fungal endophytes of tree leaves. In: Andreas JH, Hirano SC (eds) Microbial ecology of the leaves. Springer, New York, pp 179–187

Petrini O (1996) Ecological and physiological aspect of host specificity in endophytic fungi. In: Redlin SC, Carris LM (eds) Endophytic fungi in grasses and woody plants. APS Press, St. Paul, pp 87–100

Pirttilä AM, Joensuu P, Pospiech H, Jalonen J, Hohtola A (2004) Bud endophytes of Scots pine produce adenine derivatives and other compounds that affect morphology and mitigate browning of callus cultures. Physiol Plant 121:305–312

Qi G, Lan N, Ma X, Yu Z, Zhao X (2011) Controlling *Myzus persicae* with recombinant endophytic fungi *Chaetomium globosum* expressing Pinellia ternate agglutinin. J Appl Microbiol 110:1314–1322

Qin JC, Zhang YM, Gao JM, Bai MS, Yang SX, Laatsch H, Zhang AL (2009) Bioactive metabolites produced by *Chaetomium globosum*, an endophytic fungus isolated from *Ginkgo biloba*. Bioorg Med Chem Lett 19:1572–1574

Raaijmakers JM, Vlami M, De Souza JT (2002) Antibiotic production by bacterial biocontrol agents. Anton Leeuw 81:537–547

Redman RS, Sheehan KB, Stout RG, Rodriguez RJ, Henson JM (2002) Thermotolerance generated by plant/fungal symbiosis. Science 298:1581

Ryu C, Farag MA, Hu C, Reddy MS, Wei H, Pare PW, Kloepper JW (2003) Bacterial volatiles promote growth in *Arabidopsis*. Proc Natl Acad Sci U S A 100:4927–4932

Sacherer P, Défago G, Haas D (1994) Extracellular protease and phospholipase C are controlled by the global regulatory gene gacA in the biocontrol strain *Pseudomonas fluorescens* CHA0. FEMS Microbiol Lett 116:155–160

Schardl CL, Leuchtmann A, Spiering MJ (2004) Symbioses of grasses with seedborne fungal endophytes. Annu Rev Plant Biol 55:315–340

Silva GH, Teles HL, Trevisan HC, Bolzani VS, Young MCM, Pfenning LH, Eberlin MN, Haddad R, Costa-Neto CM, Araujo AR (2005) New bioactive metabolites produced by *Phomopsis cassiae*, an endophytic fungus in *Cassia spectabilis*. J Braz Chem Soc 16:1463–1466

Silva GH, Teles HL, Zanardi LM, Marx Young MC, Eberlin MN, Hadad R, Pfenning LH, Costa-Neto CM, Castro-Gamboa I, Bolzani YS, Araújo AR (2006) Cadinane sesquiterpenoids of *Phomopsis cassiae*, an endophytic fungus associated with *Cassia spectabilis* (Leguminosae). Phytochemistry 67:1964–1969

Singh PP, Shin YC, Park CS, Chung YR (1999) Biological control of *Fusarium* wilt of cucumber by chitinolytic bacteria. Phytopathology 89:92–99

Strobel GA (2002) Rainforest endophytes and bioactive products. Crit Rev Biotechnol 22:315–333

Strobel G, Daisy B (2003) Bioprospecting for microbial endophytes and their natural Products. Microbiol Mol Biol Rev 67:491–502

Strobel GA, Sugawara F, Koshino H, Li JY, Harper J, Hess WM (1999) Oocydin A, a chlorinated macrocyclic lactone with potent anti-oomycete activity from *Serratia marcescens*. Microbiology 145:3557–3564

Strobel GA, Ford E, Worapong J, Harper JK, Arif AM, Grant DM, Fung P, Chau RMW (2002) Isopestacin, a unique isobenzofuranone from *Pestalotiopsis microspora* possessing antifungal and antioxidant properties. Phytochemistry 60:179–183

Strobel G, Daisy B, Castillo U, Harper (2004) Natural products from endophytic microorganisms. J Nat Prod 67:257–268

Swarthout D, Harper E, Judd S, Gontheir D, Shyne R, Stowe T, Bultman T (2009) Measures of leaf-level water-use efficiency in drought stressed endophyte infected and non-infected tall fescue grasses. Eviron Exp Bot 66:88–93

Taghavi S, Van der Lelie D, Hoffman A, Zhang Y, Walla MD, Vangronsveld J, Newman L, Monchy S (2010) Genome sequence of the plant growth promoting endophytic bacterium *Enterobacter* sp. 638. PLoS Genet 6:e1000943. doi:10.1371/journal.pgen.1000943

Tan RX, Zou WX (2001) Endophytes: a rich source of functional metabolites. Nat Prod Rep 18:448–459

Tombolini R, Jansson JK (1998) Monitoring of GFP-tagged bacterial cells in methods in molecular biology. In: LaRossa RA (ed) Bioluminescence methods and protocols. Humana Press Inc., Totowa, pp 285–298

Toyoda H, Hashimoto H, Utsumi R, Kobayashi H, Ouchi S (1988) Detoxification of fusaric acid by a fusaric acid-resistant mutant of *Pseudomonas solanacearum* and its application to biological control of *Fusarium* wilt of tomato. Phytopathology 8:1307–1311

Ulrich K, Stauber T, Ewald D (2008) *Paenibacillus* a predominant endophytic bacterium colonising tissue cultures of woody plants. Plant Cell Tissue Organ Cult 93:347–351

Van Loon LC, Bakker PAHM, Pieterse CMJ (1998) Systemic resistance induced by rhizosphere bacteria. Annu Rev Phytopathol 36:453–483

Varma A, Verma S, Sudah SN, Franken P (1999) *Piriformospora indica*, a cultivable plant growth-promoting root endophyte. Appl Environ Microbiol 65:2741–2744

Walker MJ, Birc RG, Pemberton JM (1988) Cloning and characterization of an albicidin resistance gene from *Klebsiella oxytoca*. Mol Microbiol 2:443–454

Wang WX, Vinocur B, Shoseyov O, Altman A (2001) Biotechnology of plant osmotic stress tolerance: Physiological and molecular considerations. Acta Hortic 560:285–292

Wicklow DT, Roth S, Deyrup ST, Gloer JB (2005) A protective endophyte of maize: *Acremonium zeae* antibiotics inhibitory to *Aspergillus flavus* and *Fusarium verticillioides*. Mycol Res 109:610–618

Wilson D (1995) Endophyte: the evolution of a term, and clarification of its use and definition. Oikos 73:274–276

Wilson AD (1996) Resources and testing of endophyte-infected germplasm I national grass repository collections. In: Redlin SC, Carris LM (eds) Endophyte fungi in grasses and woody plants. APS Press, Minnesota, pp 179–195

You FT, Han JZ, Wu BK, Huang LP, Qin (2009) Antifungal secondary metabolites from endophytic *Verticillium* sp. Syst Ecol 37:162–165

Yuan Z, Zhang C, Lin FL (2010) Role of diverse non-systemic fungal endophytes in plant performance and response to stress: progress and approaches. J Plant Growth Regul 29:116–126. doi:10.1007/s00344-009-9112-9

Zhang LH, Birch RG (1997) The gene for albicidin detoxification from *Pantoea dispersa* encodes an esterase and attenuates pathogenicity of *Xanthomonas albilineans* to sugarcane. Proc Natl Acad Sci 94:9984–9989

Zhang HY, Wu XJ, Tang KX, Wang XD, Sun XF, Zhou KD (2003) A primary study of transferring the Pinellia ternata agglutinin (pta) gene into rice and expression. Acta Genet 30:1013–1019

Zhang L, An R, Wang J, Sun N, Zhang S, Hu J, Kuai J (2005) Exploring novel bioactive compounds from marine microbes. Curr Opin Microbiol 8:276–281

Zhou D, Hyde KD (2001) Host-specificity, host-exclusivity, and host-recurrence in saprobic fungi. Mycol Res 105:1449–1457

Abiotic and Biotic Plant Stress-Tolerant and Beneficial Secondary Metabolites Produced by Endophytic *Bacillus* Species

8

Charles W. Bacon, Edwin R. Palencia, and Dorothy M. Hinton

Abstract

Knowledge of endophytic bacteria and their potential for protecting crops has targeted the endophytic species of *Bacillus* as a valued microorganism not only for disease protection but also for inducing plant defense mechanisms. *Bacillus* species and their endophytic strains are also used for a wide range of antibiotics that inhibit pathogens directly impacting cellular structures or at the molecular and physiological levels. The endophytic species and strains produce fungal inhibitory compounds that belong to three broad families of lipopeptides, and these include the bacillomycins, fengycins, and surfactins. Bacilli also produce the ribosomally synthesized antimicrobial peptides, bacteriocins, which have been implemented in plant protection schemes to control fungal and bacterial diseases. Others have yet to be identified. These compounds form the basis of intense activity ranging from acute toxicity to serving as signal transduction systems for specific cellular functions, organelle formation, and responses to environmental changes and challenges. This review addresses some evidence of endophytic bacillus impacts on alleviating plant stresses, both abiotic and biotic with suggestions on future studies necessary for specific mechanisms that may assist in increasing their performance as biocontrol agents.

Introduction

The objective of sustainable agriculture is to reduce the industrial applications for food production and replace it with techniques that use fewer chemicals, designed not to harm the environment, ensuring future generations safe and continued use of the agricultural environment. Biological control is often touted as being a system insuring most of the benefits anticipated from sustainable agriculture. There are several

C.W. Bacon (✉) • D.M. Hinton
USDA, ARS, Russell Research Center, Athens, GA 30605, USA
e-mail: Charles.bacon@ars.usda.gov

E.R. Palencia
National Peanut Research Laboratory, Dawson, GA 39842, USA

N.K. Arora (ed.), *Plant Microbes Symbiosis: Applied Facets*,
DOI 10.1007/978-81-322-2068-8_8, © Springer India 2015

biocontrol systems being used for specific aspects of agricultural management, and there are many organisms that are being tested along and in combination as biocontrol agents. Leading the list are several genera of bacteria with many life habits. There are species of bacteria that utilize the rhizosphere, the soil, the outside foliage, and the inside, referred to as endophytes. Only a few of these are obligate plant dwellers but nevertheless make useful allies for biocontrol of several plant pests. Others are somewhat obligate and tend to be found on plants naturally. One such group includes the microbial endophytes.

Microbial endophytes include both fungi and bacteria that dwell within plant tissue, and these two groups consist of a large diversity of organisms. This group of microbes was recognized by Galippe (1887) as being distinct from pathogenic species of microbes although he used the term "endophyte" to refer to internal infections of a healthy plant by microorganisms which was used earlier to describe infections with pathogenic species by De Bary (1866) some two plus decades earlier. The term is presently used to refer to microorganisms that dwell within tissues of plants forming usually lifelong associations producing very little harm, and in most instances mutualistic expressions are observed. There are both fungal and bacterial species that form this type of association. Many plants colonized by species in this group are often nonculturable requiring specific molecular techniques for their identification, and even then, identifications often fail. Such obligate strains or species apparently have evolved with the wild ancestors within the plant group. During isolation from related free-living species, they have become distinct and highly dependent on plants for specific growth requirements and vice versa. The native plant hosts show no signs of conflict during the association and often fail to live without the infecting organisms. Scientists are aware of the positive value of endophytic associations, and microbial endophytes are being moved into modern and highly hybridized agricultural cultivars.

Current studies indicate that all plants as their natural communities consist of several endophytic organisms, and as will be discussed later, this might be a rationale for inconsistencies of biocontrol performance when an endophyte is managed in crop plants under controlled agricultural habitats. Microbial endophytes offer several advantages over non-endophytic biocontrol species. One, the major advantage that this group offers is an easy procedure for infections into target plants since studies show that they either self-infect through broken roots or infect during seed germination hence not requiring several field applications during crop development. Another advantage is internal growth in the host, colonizing the tissue as the plant grows. We include in the desirable list of bacterial endophytes those that do not invade the plant cells, especially the xylem, but remain intercellular but whose habitat might be localized in the roots and/or the above ground foliage of the plant.

Due to the endophytic habit and genetic diversity, bacterial endophytes have the potential for a variety of uses such as phytoremediation, biological control of pests, and agronomic enhancements of crop plants' growth and maturity. A recent thrust area is the use of bacterial endophytes to reduce biotic and abiotic stresses. Bacterial endophytes are largely being exploited for disease and pest controls primarily because of their ability to produce a variety of secondary metabolites many of which are antibiotics. However, the knowledge of the multitude of compounds and their specific attributes is only suggestive since very few of these compounds have been demonstrated in situ. Decades of studies designed to describe the efficient use of a biocontrol organism have not developed into a successful strategy. We have not defined all the necessary requirements and interactions that are reflective of many unknown mechanisms that may be operative.

Of the many species of bacteria that have been isolated from plants as endophytes, some species offer more promise due to their unique characteristic. The genus *Bacillus* consists of several spore-forming species with additional features of being natural soil dwellers with the genetic potential of undergoing cell differentiation, forming spores under stress, but returns to the vegetative state when conditions are favorable. Signal transduction is considered the basic mechanism through which the vegetative and spore states are initiated (Fuqua and Greenberg 2002). *Bacillus* species are often isolated as endophytes, and since they produce very resistant dormant endospores, storage and keeping qualities are not an issue; hence there are numerous convenient commercial procedures for preserving and extending their shelf life. Indeed, viable species of *Bacillus* have been recovered from viable spores after years of storages from most environments (Kennedy et al. 1994; Potts 1994), and there is one report of the recovery of a viable *Bacillus* relatively isolated from a 25-million-year-old amber sample (Cano and Borucki 1995).

In general, *Bacillus* species are generally recognized as safe and the organisms and their products are allowed in human food industry, used as additives in livestock and poultry feed. There are many species of *Bacillus*, both patented and non-patented, for use as biocontrol agents (Ongena et al. 2007; Ongena and Jacques 2007; Zhou et al. 2008; Raaijmaker et al. 2010; Li et al. 2012; Bacon and Hinton 2014). Considerable discussions will be presented on *Bacillus mojavensis*, a plant-friendly endophytic species that offers several advantages for studies with a strong affinity for a variety of hosts. Strains of this species are desert isolates and exhibit considerable heterogeneity (Olubajo and Bacon 2008). This should translate into a wide range of uses in crop protection such as reduced maize and wheat infections from *Fusarium* species, with the added bonus of mycotoxin reduction produced from fungi of the *Fusarium* species, e.g., deoxynivalenol and the fumonisins.

Due to the large diversity of organisms, and the proposed mechanisms of action that are either contradictory or not observational, we review in this chapter species of the genus *Bacillus* that are considered as endophytic. In some instances and for comparisons, we will detail research from non-endophytic species although endophytic strains of that species have been reported.

Bioactive Substances and Effects on Plants

The foundation for the successful use of bacterial endophytes depends on the antimicrobial and growth-enhancing compounds usually first demonstrated in laboratory. However, demonstration that a *Bacillus* species produce one or several inhibitory substances isolated from culture does not always follow the in planta production. Table 8.1 depicts a list of biologically active compounds that have been isolated from bacilli. The biologically active compounds may be divided into those produced by non-ribosomal synthesis such as the cyclic lipopeptides and to compounds such as the polyketides that are produced by genes or in gene clusters. However, we should indicate first the species of bacteria that have proven useful as biocontrol agents. These species are not necessarily entirely endophytic, although most have strains that have been demonstrated as being endophytic. These species include *B. subtilis*, *B. amyloliquefaciens*, *B. licheniformis*, *B. pasteurii*, *B. cereus*, *B. pumilus*, *B. mojavensis*, *B. mycoides*, and *B. sphaericus*.

Fungicidal and bactericidal compounds have been isolated from most endophytic species, and there are almost similar compounds found throughout the genus, regardless of the species, suggesting an evolutionary strategy based on the importance of the compounds for ecological success within niche diversities. The surfactins apparently occur genus wide, which suggests an important role in the competitive abilities of the species. Others are variations of the surfactin group, although within a species there apparently are strain differences.

Such structural diversity offers more promise for strains within the genus as these variants might prove valuable in selecting strains for specific environments or crop plants with specific problems. Several of the species identified above have been subsequently transferred to a newly erected or related genus, such as *Paenibacillus*, although in this particular genus similar compounds are produced indicating again the importance of these compounds. Demonstration that the *Bacillus*

Table 8.1 Bioactive compounds by *Bacillus* species

Metabolite	Source	Biological activity	References
Bacteriocins	*B. subtilis*	Antagonism against the phytopathogenic fungi, e.g., *Agrobacterium tumefaciens* and *Alternaria solani*	Hammami et al. (2009, 2012), Araújo et al. (2005), and Ashwini and Srividya (2013)
	B. licheniformis	Growth inhibition of fungal pathogens	He et al. (2006)
	B. amyloliquefaciens	Bacterial antagonism to *Ralstonia solanacearum*	Hu et al. (2010)
	B. thuringiensis	Antibacterial and potential plant growth promoter	Lee et al. (2009) and Gray et al. (2006)
	B. clausii	Bactericidal activity against *Agrobacterium tumefaciens*	Mouloud et al. (2013)
Chitinases	*B. amyloliquefaciens*	Fungal growth inhibition	Wang et al. (2002)
	B. licheniformis	Antifungal activity	Gomaa (2012)
	B. thuringiensis	Biocontrol of phytopathogenic fungi	Reyes-Ramirez et al. (2004)
	B. pumilus	Fungal growth inhibition	Shali et al. (2010)
	B. subtilis	Fungal growth suppressor	Ashwini and Srividya (2013)
	B. cereus	Conidial germination inhibitor of *Botrytis elliptica*	Huang et al. (2005)
Plant growth promoters	*B. pumilus*	Indoleacetic acid synthesis for plant growth promotion	Murugappan et al. (2013)
	B. subtilis	Indoleacetic acid and abscisic acid	Araújo et al. (2005)
	B. amyloliquefaciens	Auxin biosynthesis to promote plant growth	Talboys et al. (2014)
Siderophores	*B. pumilus*	Iron sequestration and plant pathogen inhibition	Murugappan et al. (2013)
	B. subtilis	Iron sequestration and plant growth promoter	Yu et al. (2011)
	B. mojavensis	Iron sequestration	Vendan et al. (2010) and Bacon and Hinton (2011a)
Stilbene	*Bacillus* sp.	Antimicrobial activity to phytopathogenic bacteria and fungi	Kumar et al. (2012b)
Phytase	*B. amyloliquefaciens*	Plant growth stimulator under phosphate limitation	Idriss et al. (2002)
Zwittermicin A	*B. cereus*	High antimicrobial activity against oomycetes	Silo-Suh et al. (1998)

species produce one or several inhibitory substances isolated from culture has not been demonstrated *in planta*. Indeed, very little work exists to indicate that such biologically active compounds are produced *in planta*. In addition, the species discussed in this review are not necessarily reported as being endophytic since the techniques of their isolation would not allow for such a distinction.

Lipopeptides

Lipopeptides make up a group of biosurfactants that are closely related metabolites, which include the surfactins, iturins, and the fengycins (plipastatin). This class of metabolites is not produced by ribosomes, rather by a group of peptide synthetases. The lipopeptides, in addition to producing antimicrobial activity, also act as immune stimulators of the host plant. The antibiotic activity is accomplished by destroying or interfering with cell membranes' properties affecting the pathogen's membrane. However, the surfactins and other lipopeptides are ineffective in penetrating phytosterols, and plant cell membranes consist of phytosterols (Ongena and Jacques 2007). The phytosteroidal composition of host plant cells is chemically sufficient to provide protection from these substances indicating some level of selectivity over pathogens. Thus, both sensitivity and specificity reside in the nature of both the pathogen's cell wall and the host cell wall, and perhaps some pathogens might have cell walls that are resistant to these biosurfactants as well.

Of the many compounds that are indicated above, the cyclic lipopeptides are perhaps the most suitable for in planta crop protection (Ongena and Jacques 2007; Romero et al. 2007; Raaijmakers et al. 2010). For example, the physical characteristics of surfactins increase their benefits as biocontrol agents. This group of lipopeptides is stable over a wide range of conditions. They are active from 15 °C up to 100 °C and are structurally sound over a range of pH values from 3 to 11 (Cameotra and Makkar 1998; Makkar and Cameotra 1998; Joshi et al. 2008). These compounds are tolerant to most plant and microbial enzymes such as the pronases and proteinases and somewhat stable in organic solvents. Due to these properties, the lipopeptides have a wide tolerance of environmental conditions. *Bacillus* species and strains producing a combination of these compounds offer more promise as biocontrol agents for crops as compared to others as lipopeptides are associated with a variety of positive biocontrol activities including biofilm formation, signaling, and fungal antagonisms (Bais et al. 2004; Kearns et al. 2005; Ongena and Jacques 2007).

The wide spectrum of activity is based on the uniqueness of the chemical structure of the lipopeptides. The endophytic habit should allow the complete expression of each chemical group produced under endophytic conditions. However, such control expressions potentially can be negated by *Bacillus* species that are soilborne or are non-endophytes. That is, the soil environment can impose constraints on the production and functional activity of such antibiotics as soil is subjected to variations, some of which exceed the limits indicated above, particularly the stability. However, in the in planta situation, the environment is consistent, and generally conditions are amendable to maximum production of single or multiple lipopeptides some of which have not been identified (Snook et al. 2009).

In addition to the variety of surfactins produced by *B. mojavensis*, the organism shares all the benefits afforded by most *Bacillus* species as discussed. Further, the desert strains are all plant friendly, producing symptomless infection in all tissues except the xylem, and the infection is maintained during the season (Bacon and Hinton 2002, 2011b). This species is an osmophile tolerating seawater and capable of anaerobic growth (Folmsbee et al. 2004).

Most of the strains of this species originated from the deserts and have been studied relative to their protection of maize from species of *Fusarium* (Bacon and Hinton 2002, 2011a) and for their ability to reduce the accumulation of the fumonisin mycotoxins (Bacon et al. 2001). It was determined that most strains of the desert group produced the surfactins (Snook et al. 2009). The variation of the acyl chain length of each isoform varies. Thirty-three strains isolated from the deserts around the globe produced surfactins with carbon length ranging from 11 to 17, with the majority producing a carbon length of 15, which is the most active form of the surfactins (Snook et al. 2009; Bacon et al. 2013). The surfactins produced by each strain vary resulting in the accumulation of several isoforms by each strain, and the combination of several isoforms may be synergistic, extending the overall spectrum of activity. The surfactins are inhibitory to virus, bacteria, fungi, and insects. However, in vitro fermentation capacity of biosurfactants by strains of *B. mojavensis* is rather poor, which limits its use for technological exploitations for surfactin production.

Bacteriocins

Bacteriocins are a heterologous group of small bacterial peptides produced by ribosomal synthesis with antimicrobial activities against Gram-positive and Gram-negative bacteria and fungal pathogens (Jack et al. 1995; Benitez et al. 2010). Many of the bacteriocins of the genus *Bacillus* display high specificity of antimicrobial activities against related species; however, recent evidence suggests that some of these peptides target a wide range of unrelated microorganisms (Abriouel et al. 2011). Compared to bacteriocins produced by lactic acid bacteria, another Gram-positive group, those produced by the bacilli, has not attracted the curiosity of plant scientists. The initial in vitro screenings indicated that *Bacillus* antimicrobial peptides exhibited limited antimicrobial activities and were usually restricted to members of the same species. In the last decade, new evidences have revealed a wide array of proteinaceous bactericides by the genus *Bacillus* with remarkable efficiency and broad antimicrobial spectrum against important plant pathogens (Emmert and Handelsman 1999; Kim and Chung 2004; Balhara et al. 2011).

The characterization of *Bacillus* bacteriocins revealed several traits distinctive of these peptide antibiotics. A number of bacteriocins belong to the group of lantibiotics, a group of heat-stable precursor peptides with characteristic polycyclic thioether amino acids lanthionine and methyllantionine that undergo posttranscriptional modifications (Jack et al. 1995; Willey and van der Donk 2007).

Several bacteriocins from plant-associated bacteria of the genus *Bacillus* have been identified as potential candidates to control important plant pathogenic microorganisms. The strain *B. subtilis* 14B, isolated from the rhizosphere of healthy almond plants from Turkey, is capable of producing a monomeric protein (termed Bac 14B) with a molecular weight of 21 kDa. Bac 14B is a bacteriocin with high antimicrobial activity against *Agrobacterium tumefaciens*, the causal agent of crown gall disease (Hammami et al. 2009). Interestingly, Bac 14B displayed a broader antibacterial spectrum against *Agrobacterium* spp. in comparison to two commercially available biocontrol products for this disease: *Agrobacterium rhizogenes* K84 and *Agrobacterium radiobacter* strain K1026. An endophytic *B. amyloliquefaciens* isolated from mangrove (*Rhizophora mangle*) leaves synthesized a class II bacteriocin with a molecular weight of approximately 10 kDa. This newly characterized bacteriocin displayed inhibition against *Ralstonia solanacearum*, the causal agent of *Capsicum* bacterial wilt (Hu et al. 2010). Although other *B. amyloliquefaciens* bacteriocins are known as fungal inhibitors (Yoshida et al. 2001; Yu et al. 2002), this bacteriocin remarkably displayed antibacterial activity against 2 g of negative plant pathogens *Xanthomonas campestris* and *R. solanacearum*.

Bacteriocins are known for their antimicrobial activity against a wide array of microorganisms. However, it has been suggested that a bacteriocin, thuricin 17, obtained from the plant growth-promoting rhizobacterium *Bacillus thuringiensis* might be involved in the enhanced plant growth in soybean (Gray et al. 2006). Current evidence has also shown that thuricin 17, when applied to either leaves or roots of soybean plants, directly increased plant growth (Lee et al. 2009). While more evidence is needed on the role of thuricin 17 in planta growth promotion, these findings opened a new line of investigation to understand the plant-bacterium associations leading to plant growth promotion by the rhizobacterium, *B. thuringiensis*. Finally, a role for this group of compounds has been suggested in the autoinduction and activation of gene clusters as well as involvement in quorum sensing (Kleerebezem et al. 1997).

Polyketides and Other Defensive Metabolites

Other non-ribosomal antimicrobial secondary metabolites produced by the bacilli include the polyketides. Examples of such metabolites are bacillomycin D, bacillaene, and anticapsin. Most of these are antifungal, antiviral, and antimycoplasmal and a few are broad-spectrum antibactericidal. Other secondary metabolites that exhibit remarkably strong antagonisms to fungi include difficidin, mersacidin, subtilin, subtilosin A, sublancin, and TasA. These compounds are also not produced by ribosome, but are produced by polyketide synthases. Their range of activity also includes toxicity to Gram-negative bacteria, viruses, filamentous fungi, and as presented below bacillibactins that function as siderophores. The major uses of the polyketides are associated with crop protection since their spectrum of activity is restricted to microbial antagonisms.

Bacillus species are a potent source of bioactive secondary metabolites (see Table 8.1) with a variety of functions, including antibiotic activity, enzyme inhibitors, mineral scavengers, and surfactants (Vining 1990; Sansinenea and Ortiz 2011; Makkar and Cameotra 2002). While a great percentage of research has been dedicated to study antimicrobial peptides, current studies suggest that endophytic bacilli and their metabolites might be able to modulate a wide array of plant responses to biotic and abiotic stresses that enable the host to survive under a variety of harsh natural conditions. Ecologically, these observations have been demonstrated on a variety of wild populations with the assumption that they are also operative in many crop plants. This assumption removes the involvement of the host in endophytic relationships, which, however, is probably not the case always.

The endophytes of *B. mojavensis* secrete siderophores as well as do other *Bacillus* species (de Silva et al. 1996; Vendan et al. 2010; Bacon and Hinton 2011a). Siderophores are low molecular weight iron chelators that increase abiotic plant resistance to deficiency of iron, zinc, copper, lead, cadmium, chromium, mercury, and nickel. Included in this group are pyoverdine, pyochelin, and several organic acids. Siderophores enhance either the solubility or uptake of these metals, some of which are important micronutrients for plants and microorganisms. While these substances are present in the soils, they are usually combined with other compounds and not in soluble forms. For example, although iron is among the most abundant micronutrients in agricultural soils, the low solubility of the ferric ion in soil and marine sediments limits iron acquisition by plants. Current data show that siderophore-mediated iron sequestration by endophytic *Bacillus* may confer a competitive advantage to the plant host with respect to pathogen inhibition (Murugappan et al. 2013). In addition to mobilizing forms of micronutrients, microbial siderophores also immobilize some metals preventing them from accumulating to toxic amounts in host plants. Immobilization may be accomplished either through a chemical reduction from one form to a less soluble or mobile form.

It is not certain but future research will determine to what extent siderophore production promotes plant growth, but presently the effects on reducing plant disease should be interpreted as indirect. In greenhouse experiments, *B. subtilis* CAS15 strain isolated from the rhizosphere of pepper fields secreted the catecholic siderophore bacillibactin (2,3-dihydroxybenzoate-glycine-threonine trimeric ester) (Yu et al. 2011). In these studies, the authors demonstrated that increasing the iron concentration resulted in significant decrease in bacillibactin by *B. subtilis*. The plant growth-promoting bacterium, *B. cereus*, isolated as an endophyte from *Sophora alopecuroides* in China, in addition to other compounds, secretes siderophores, suggesting that these molecules might be involved in plant growth promotion (Zhao et al. 2011).

Stilbene phytoalexins are also important bioactive compounds secreted by *Bacillus* sp. that have been associated with increased plant disease resistance (Dercks and Creasy 1989; Paul et al. 1998). Phytoalexins are low molecular weight antibiotics synthesized by plants and bacteria that promptly accumulate in response to biotic or abiotic stress at the infection sites (van Etten et al. 1994; Hammerschmidt and Dann 1999). The stilbene 3,4,5-trihydroxystilbene, known as resveratrol, and 3,5-dihydroxy-4-isopropylstilbene were isolated from ethyl acetate extracts from *Bacillus* strains that showed high antifungal activity against several medical and agriculturally important crop pathogens such as *Ralstonia solani* (Kumar et al. 2012b). Current data show that stilbenes produced by *Bacillus* sp. act synergistically with the antibiotic fluoroquinolone, an important cephalosporin used to control Gram-negative and anaerobic bacteria (Kumar et al. 2012a). In addition, stilbenes are potentially useful for controlling important plant pathogens such as oomycetes (Kumar and Nambisan 2014), and resveratrol, a stilbene-type phenol, is an important antioxidant used in human medicine for cancer prevention as well as a powerful antioxidant that protects plants from environmental stresses.

Many endophytic bacilli produce plant hormones such as gibberellins, auxin, and cytokinins (Ortiz-Castro et al. 2008; Malfanova et al. 2011; Murugappan et al. 2013). To evaluate the role of cytokinins by the plant-associated bacterium *Bacillus megaterium*, in vitro experiments were performed using *Arabidopsis thaliana* (Ortiz-Castro et al. 2008). Compared to the *A. thaliana* ecotype, and in the presence of the cytokinin-producing *B. megaterium*, *A. thaliana* mutants lacking cytokinin receptors showed reduced plant development. These studies showed that *B. megaterium* cytokinins are an important factor in *A. thaliana* growth. A plant-associated bacterium, *B. pumilus*, synthesizes indoleacetic acid (IAA), the most naturally occurring phytohormone in the auxin class (Murugappan et al. 2013). Similar to cytokinins, IAA produced by plant-associated *Bacillus* was associated with promotion of root growth and increased nodulation in *Phaseolus vulgaris* by *Rhizobium etli* (Srinivasan et al. 1996).

B. cereus and *B. thuringiensis* produce zwittermicin A, a linear aminopolyol antibiotic with broad-spectrum antimicrobial activity against oomycetes and Gram-negative and Gram-positive bacteria (Silo-Suh et al. 1998; Zhou et al. 2008). Structurally, zwittermicin A is unique as it is a combination of a polyketide, a peptide that includes moieties consisting of a D-amino acid, an ethanolamine group, and a glycolyl group. Further, this antibiotic is produced by both polyketide and non-ribosomal biosynthetic pathways (Milner et al. 1996). Studies on the production of zwittermicin A by *B. cereus* revealed that the acetyltransferase gene to catalyze the acetylation of zwittermicin A belongs to a new class of acetyltransferase with self-resistance properties for *B. cereus* providing a novel method of disease protection (Zhou et al. 2008).

In Planta Interactions and Functions of Bioactive Compounds

Knowledge of the actions of one or more mechanisms is required to predict control of target species. This will lead to the dependable and efficient use of endophytic bacteria as biocontrol agents. Very little is known about the specific mechanisms involved, including in planta production, accumulation, and regulation of responsible metabolites. Questions arise, such as can these be regulated under current crop production practices and if not how to modify production practices to achieve the desired result? Very few studies are designed to test if in fact metabolites isolated in culture are produced in planta.

There are several proposed mechanisms for the action of biocontrol bacteria, and these include antimicrobial activity mediated by one or more secondary metabolites such as the production of antibiotics, allelochemicals, nutrient stimulators, niche exclusion, and systemic acquired resistance and the production of volatile secondary metabolites that can be interactive in soils (Ryu et al. 2004). Many have been implicated in the elaboration of specific morphological structuring compounds such as colony surface attachments allowing for multicellular aggregation and attachment to cell surfaces, particularly biofilms. Of importance is their role in mediating cell signaling and quorum sensing. Included in these mechanisms are also those defensive hydrolytic proteins such as hydrolases, chitinases, and lipases (Idriss et al. 2002; Huang et al. 2005; Gomaa 2012). Others involve the detoxification and degradation of various pathogen virulence factors. Certain secondary metabolites produced by bacterial endophytes can also induce systemic acquired resistance.

Prolonged planting of a crop over time or even rotational plantings will result in a group of resistant organisms that dominate within these locations. The result is a group of resistant pathogens to both host natural defenses and possibly to any biocontrol agent. However, there is emerging evidence that some bacteria also are effective modifiers of their host natural defensive compounds such that the host metabolites, often not effective to resistant pathogens, are modified by their endophytic partner into compounds that are highly effective to control pathogens as well as other competing endophytes within specific crops. Such chemical modifications can be viewed as a control mechanism at the population level since they are competing with one or more pathogenic species naturally associated with crop plantings. For example, it has been observed that *B. mojavensis* chemically modifies a group of maize phytoanticipins, the hydroxamic acids (benzoxazolinones), into the toxic compound 2-amino-3*H*-phenoxazin-3-one (APO) (Bacon et al. 2007), that are toxic to *Fusarium verticillioides* and other hydroxamic acid-resistant pathogens (Fig. 8.1). The hydroxamic acid in spite of being short lived is not an effective inhibitor to the usual maize endophytic pathogenic fungi, the *Fusarium* species and *Diplodia*, which over time can and do dominate fields under constant cultivation of maize. However, these maize pathogens are very susceptible to the bacterial modified compounds, APO, which have the potential for controlling populations of pathogens (Glenn et al. 2001, 2002; Bacon et al. 2007). In a more detailed study, Glenn et al. (2003) have used this as an explanation for the dominance of *F. verticillioides* and other maize pathogens that dominate field under constant maize cultivation. Further, there is some evidence for host genetic modifications indicating a two-point control system (Glenn et al. 2003). Of course, there are converse examples as well. A target pathogen can and does produce a variety of substances and mechanisms that are capable of controlling or nullifying any activity of a biocontrol agent. The wide-scale production of fusaric acid, an antibiotic substance, by *Fusarium* species is one example (Bacon et al. 1996, 2004). Fusaric acid is highly toxic to most bacteria providing some measure of control over a biocontrol bacterium. Thus, there are mutual inhibiting properties that have to be considered, resulting in the degree of success in the field that is only partial from a few biocontrol bacterial endophytes.

Since most of the researches dealing with the production of biologically active compounds by bacterial endophytes, such as antibiotics, have been demonstrated under careful laboratory conditions, very little have been established for in planta production. Experiments designed to test their regulations with or without interaction from the plant physiological processes have not been determined. There is, however, considerable information for increased plant performance by bacterial endophytes when cultured under conditions using a single endophytic microbe co-inoculated with a single pathogen. However, the bacterial endophytes are in competition with all soil and foliage microbes as well as the target pathogen or pest. Clearly, many enhanced effects observed by endophytes and usually reported are not produced along with multiple competing organisms. The point here is if bacterial endophytes are to be tested under laboratory conditions, such multiple infections by naturally occurring organisms should be used in standard procedures which is not the case at present. There are, however, inocula consisting of multiple species or strains used as biocontrol agents.

Abiotic and Biotic Stress-Tolerant Applications

Plants are subjected to a variety of environmental stresses such as soil pH, drought, hot and cold temperatures, air pollution, soil salinity, heavy metal contamination, and soil and foliage parasitism. In an agricultural environment, all these factors affect plant functions, which corresponds to yield reduction to less than optimal. And if excessive, death of the crop is assured. Conceivably each of these abiotic and biotic stresses can be controlled by separate mechanisms based on adaptations of each crop species, its genetic potential to withstand stresses, and the importance of each stress on the physiology of the plant, that is, whether the mechanism affected is a primary process such as photosynthesis or a minor process as production of secondary root development. Distinct from the basic physiology of the plant, there are microorganisms that allow tolerance or increase the tolerance against stresses in the hosts (Warren 1998; Mayak et al. 2004; Vinocur and Altman 2005; Zaidi et al. 2006; Hardoim et al. 2008;

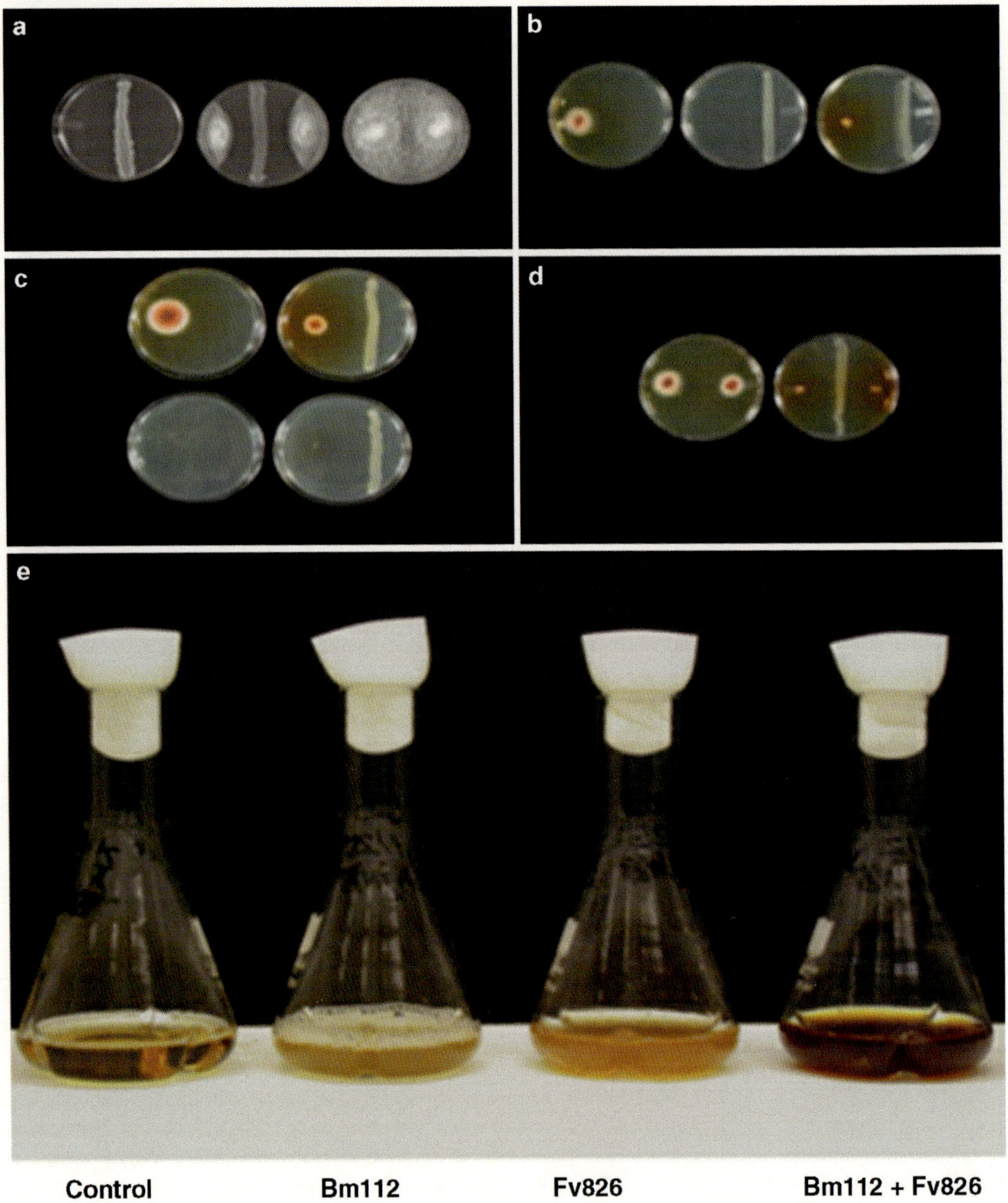

Fig. 8.1 Culture of *Bacillus mojavensis* RRC102 and *Fusarium verticillioides* MRC826 on nutrient agar amended with benzoxazolinones, specifically benzoxazolin-2(3*H*)-one (BOA). (**a**) The zone of inhibition produced by the toxic metabolite transformed from BOA by the bacterium, center plate. (**b**) Culture of *Fusarium verticillioides* cultured with and without *B. mojavensis* on nutrient agar and on BOA-amended nutrient agar showing expression of red pigment, identified as 2-amino-3*H*-phenoxazin-3-one (APO), and severely inhibited on plate far right. (**c**) *Top row*, production of APO, the red pigment, by *F. verticillioides* (RRC 826) with and without the bacterium *B. mojavensis* (RRC 112); bottom row, agar plates showing no growth with and without one of the benzoxazolinone-sensitive genetic strains of *F. verticillioides* with and without the bacterium. (**d**) Production of the red pigment, APO, achieved in the culture in the presence of *F. verticillioides* RRC826 along (*left*) and with the bacterium (*right*). (**e**) Liquid culture of *B. mojavensis* (Bm112) and *F. verticillioides* (Fv826) on nutrient broth amended with BOA resulting in the production of APO during the culture of both organisms (Bm112 and Fv826) (Bacon et al. 2007)

Ryan et al. 2008; Vardharajula et al. 2011; Luo et al. 2012; Yandigeri et al. 2012). A major objective is to increase agricultural production in the land located in unfavorable environments. Traditionally this research is being done by classical genetic and crop breeding techniques. *Bacillus* species have been shown to promote growth in the presence of abiotic stress such as drought. Bacterial endophytes have the potential to induce stress resistance in crops grown under adverse conditions. Endophytes can be used as phytoremediators of soil contaminated with several heavy metals. This approach is novel and recent. The mechanism of action is complex involving several metabolites such as organic acids, along with siderophores resulting in metal detoxification, nutrient acquisition, and often for relief from biotic and abiotic stresses. The process also results in decontamination of heavy metals from plant tissues (Wani et al. 2007; Luo et al. 2012; Rajkumar et al. 2012). While highly desirable, most of this work has not been attempted and is highly theoretical. There is an expected host interaction along with several of its endophytes that will allow the host to grow under environmental extremes (Sziderics et al. 2007; Schalk et al. 2008; Sgroy et al. 2009).

Mixed bacterial populations are also used to provide an array of enhanced effects in a complex environment. Thus, plant hormone-producing bacteria are used along with antibiotic or antagonistic bacteria. Other combinations involve nitrogen-fixing bacteria as well. Apparently, it is also important to determine whether the addition of a single or mixed bacterial population will be beneficial or synergistic for the host. Sturz and Christie (1995) indicated that inhibitory bacteria when added as mixed cultures actually stimulated plant growth. Detailed discussions on the problems and benefits on the uses of mixed cultures of biocontrol bacteria in crop protection are discussed by Stuz et al. (2000).

Conclusion

Bacterial endophytes offer a highly desirable approach for control of plant pathogens. Most of these are self-infecting, co-habitating with the host, and invading the host tissue as the host grows. In spite of the accumulation of antibiotic substances by biocontrol bacterial endophytes, demonstration of their effects under field conditions is inconsistent or lacking. This may be due to unfavorable environmental conditions for the combination being tested; host genetic effect might also compound the results, or competing microbes other than the target microbe might be interfering with colonization or decomposing the inhibitory metabolites. Suitable ecological niches are expected to play a role in biocontrol parameters, and this should occur without doing harm to the host plant. As there are several endophytic species isolated from a host, there is the concept of microniche that should be examined for a functional role in crop production. Acceptance by the host plant could also influence the results, so it is important that only plant-friendly strains be used.

Environmental conditions, especially in the soil, may impose additional stresses on the host, reducing its tolerance to endophytic colonization. Disease recognition and plant induction mechanisms may be required for success with very little to do with specific inhibitory compounds. The mode of action might be based on an induced systemic resistance response. Signaling pathways are expected to be interactive in this regard, resulting in either a positive or no response from the plant host to a disease. The knowledge on basic contribution of a host is lacking although there is an interaction (Smith and Goodman 1999; Smith et al. 1999). The utilization of specific nutrients by the plants or incorporating these in the soil might stimulate the association to produce more of the bioactive compounds. However, to date there are no indications that soil additions will increase positive responses from an endophyte-infected host.

Consistent and effective biocontrol of pathogens will depend on the complex nature of events resulting from the compatible response resulting from plant infection and colonization of tissues by the endophytes. This is followed by the initiation of a cascade of several mechanisms, any one of which if blocked might prevent one or all desired mechanisms necessary for biocontrol.

Overcoming the challenges induced by invading pathogens is monumental; however, successful uses of biocontrol bacterial endophytes are dependent on continued fundamental studies of host-bacterial-pests interactions.

References

Abriouel H, Franz CMAP, Omar NB, Gálvez A (2011) Diversity and applications of *Bacillus* bacteriocins. FEMS Microbiol Rev 35(1):201–232

Araújo FF, Henning AA, Hungria M (2005) Phytohormones and antibiotics produced by *Bacillus subtilis* and their effects on seed pathogenic fungi and on soybean root development. World J Microbiol Biotechnol 21(8–9):1639–1645

Ashwini N, Srividya S (2013) Potentiality of *Bacillus subtilis* as biocontrol agent for management of anthracnose disease of chilli caused by *Colletotrichum gloeosporioides* OGC1. Biotech 3:1–10

Bacon CW, Hinton DM (2002) Endophytic and biological control potential of *Bacillus mojavensis* and related species. Biol Control 23:274–284

Bacon CW, Hinton DM (2011a) In planta reduction of maize seedling stalk lesions by the bacterial endophyte *Bacillus mojavensis*. Can J Microbiol 57:485–492

Bacon CW, Hinton DM (2011b) *Bacillus mojavensis*: its endophytic nature, the surfactins, and their role in the plant response to infection by *Fusarium verticillioides*. In: Maheshwari DK (ed) Bacteria in agrobiology: plant growth responses. Springer, Berlin, pp 21–39

Bacon CW, Hinton DM (2014) Microbial endophytes: future challenges. In: Gange AC, Verma VC (eds) Advances in endophytic research. Springer, New York, pp 441–451

Bacon CW, Porter JK, Norred WP, Leslie JF (1996) Production of fusaric acid by *Fusarium* species. Appl Environ Microbiol 62:4039–4043

Bacon CW, Yates IE, Hinton DM, Meredith F (2001) Biological control of *Fusarium moniliforme* in maize. Environ Heal Persp 109:325–332

Bacon CW, Hinton DM, Porter JK, Glenn AE, Kuldau GA (2004) Fusaric acid, a *Fusarium verticillioides* metabolite, antagonistic to the endophytic biocontrol bacterium *Bacillus mojavensis*. Can J Bot 82:878–885

Bacon CW, Hinton DM, Glenn AE, Macias FA, Marin D (2007) Interaction of *Bacillus mojavensis* and *Fusarium verticillioides* with a benzoxazolinone (BOA) and its transformation products, APO. J Chem Ecol 33:1885–1897

Bacon CW, Hinton DM, Mitchell TR, Snook ME, Olubajo BA (2013) Characterization of endophytic strains of *Bacillus mojavensis* and their production of surfactin isomers. Biol Control 62:1–9

Bais HP, Fall R, Vivanco JM (2004) Biocontrol of *Bacillus subtilis* against infection of *Arabidopsis* roots by *Pseudomonas syringae* is facilitated by biofilm formation and surfactin production. Plant Physiol 134:307–319

Balhara M, Ruhil S, Dhankhar S, Chhillar AK (2011) Bioactive compounds hold up- *Bacillus amyloliquefaciens* as a potent bio-control agent. Nat Prod J 1:8

Benitez L, Velho R, Lisboa M, Costa Medina L, Brandelli A (2010) Isolation and characterization of antifungal peptides produced by *Bacillus amyloliquefaciens* LBM5006. J Microbiol 48(6):791–797

Cameotra S, Makkar R (1998) Synthesis of biosurfactants in extreme conditions. Appl Microbiol Biotechnol 50:520–529

Cano RJ, Borucki MK (1995) Revival and identification of bacterial spores in 25- to 40-million-year-old Dominican amber. Science 268:160–1064

De Bary A (1866) Mophololgie und physiologie pilze, flechten, und myxomyceten. Hofmeister's Handbook of Physiol Bot Leipzig 2:1

de Silva DM, Askwith CC, Kaplan J (1996) Molecular mechanisms of iron uptake in eukaryotes. Physiol Rev 76(1):31–47

Dercks W, Creasy LL (1989) The significance of stilbene phytoalexins in the *Plasmopara viticola*-grapevine interaction. Physiol Mol Plant Pathol 34(3):189–202

Emmert EA, Handelsman J (1999) Biocontrol of plant disease: a (gram-) positive perspective. FEMS Microbiol Lett 171(1):1–9

Folmsbee MJ, McInerney MJ, Nagle DP (2004) Anaerobic growth of *Bacillus mojavensis* and *Bacillus subtilis* requires deoxyribonucleosides or DNA. Appl Environ Microbiol 70:5252–5257

Fuqua C, Greenberg EP (2002) Listening in on bacteria: acyl-homoserine lactone signalling. Nat Rev 3:685–695

Galippe MLV (1887) Note sur la presence de micro-organismes dan les tissus vegetaux and filiales et associees. Comptes Rendus Hebdomodaires des Seances et Memoires de la Societe de Biologie et des ses Filales et Associees 39:410–416

Glenn AE, Hinton DM, Yates IE, Bacon CW (2001) Detoxification of corn antimicrobial compounds as the basis for isolating *Fusarium verticillioides* and some other *Fusarium* species from corn. Appl Environ Microbiol 67:2973–2981

Glenn AE, Gold SE, Bacon CW (2002) *Fdb1* and *Fdb2*, *Fusarium verticillioides* loci necessary for detoxification of preformed antimicrobials from corn. Mol Plant Microbe Interact 15:91–101

Glenn AE, Meredith FI, Morrison WHI, Bacon CW (2003) Identification of intermediate and branch metabolites resulting from biotransformation of 2-benzoxazolinone by *Fusarium verticillioides*. Appl Environ Microbiol 69:3165–3169

Gomaa E (2012) Chitinase production by *Bacillus thuringiensis* and *Bacillus licheniformis*: their potential in antifungal biocontrol. J Microbiol 50(1):103–111

Gray EJ, Lee KD, Souleimanov AM, Di Falco MR, Zhou X, Ly A, Charles TC, Driscoll BT, Smith DL (2006) A novel bacteriocin, thuricin 17, produced by plant growth promoting rhizobacteria strain *Bacillus thuringiensis* NEB17: isolation and classification. J Appl Microbiol 100(3):545–554

Hammami I, Rhouma A, Jaouadi B, Rebai A, Nesme X (2009) Optimization and biochemical characterization of a bacteriocin from a newly isolated *Bacillus subtilis* strain 14B for biocontrol of Agrobacterium spp. strains. Lett Appl Microbiol 48(2):253–260

Hammami I, Jaouadi B, Bacha A, Rebai A, Bejar S, Nesme X, Rhouma A (2012) *Bacillus subtilis* bacteriocin Bac 14B with a broad inhibitory spectrum: purification, amino acid sequence analysis, and physicochemical characterization. Biotechnol Bioproc E 17(1):41–49

Hammerschmidt R, Dann EK (1999) The role of phytoalexins in plant protection. Novartis Found Sympos 223:188–190

Hardoim PR, Overbeek V, Leo S, Elsas DJV (2008) Trend Microbiol 16:463–471

He L, Chen W, Liu Y (2006) Production and partial characterization of bacteriocin-like peptides by *Bacillus licheniformis* ZJU12. Microbiol Res 161(4):321–326

Hu HQ, Li XS, He H (2010) Characterization of an antimicrobial material from a newly isolated *Bacillus amyloliquefaciens* from mangrove for biocontrol of Capsicum bacterial wilt. Biol Control 54(3):359–365

Huang CJ, Wang TK, Chung SC, Chen CY (2005) Identification of an antifungal chitinase from a potential biocontrol agent, *Bacillus cereus* 28-9. J Biochem Mol Biol 38(1):82–88

Idriss EE, Makarewicz O, Farouk A, Rosner K, Greiner R, Bochow H, Richter T, Borriss R (2002) Extracellular phytase activity of *Bacillus amyloliquefaciens* FZB45 contributes to its plant-growth-promoting effect. Microbiology 148(7):2097–2109

Jack RW, Tagg JR, Ray B (1995) Bacteriocins of gram-positive bacteria. Microbiol Rev 59(2):171–200

Joshi C, Bharucha JHA, Yadav S, Nerukar A, Desai AJ (2008) Biosurfactant production using molasses and whey under thermophilic conditions. Bioresh Technol 99:195–199

Kearns DB, Chu F, Rudner R, Losick R (2005) A master regulator for biofilm formation by *Bacillus subtilis*. Mol Microbiol 55:739–749

Kennedy MJ, Reader SL, Swierczynski LM (1994) Preservation records of micro-organisms: evidence of the tenacity of life. Microbiology 140:2513–2529

Kim PI, Chung KC (2004) Production of an antifungal protein for control of *Colletotrichum lagenarium* by *Bacillus amyloliquefaciens* MET0908. FEMS Microbiol Lett 234(1):177–183

Kleerebezem M, Quadri LE, Kuiper P, de Vos WM (1997) Quorum sensing by peptide pheromones and two component signal-transduction systems in Gram-positive bacteria. Mol Microbiol 24:895–904

Kumar SN, Nambisan B (2014) Antifungal activity of diketopiperazines and stilbenes against plant pathogenic fungi in vitro. Appl Biochem Biotechnol 172(2):741–754

Kumar SN, Siji JV, Nambisan B, Mohandas C (2012a) Activity and synergistic interactions of stilbenes and antibiotic combinations against bacteria in vitro. World J Microbiol Biotechnol 28(11):3143–3150

Kumar SN, Siji JV, Rajasekharan KN, Nambisan B, Mohandas C (2012b) Bioactive stilbenes from a *Bacillus* sp. N strain associated with a novel rhabditid entomopathogenic nematode. Lett Appl Microbiol 54(5):410–417

Lee K, Gray E, Mabood F, Jung W-J, Charles T, Clark SD, Ly A, Souleimanov A, Zhou X, Smith D (2009) The class IId bacteriocin thuricin-17 increases plant growth. Planta 229(4):747–755

Li H, Wang X, Han M, Zhao Z, Wang Q, Tang Q, Liu C, Kemp B, Gu Y, Shuang J (2012) Endophytic *Bacillus subtilis* ZZ120 and its potential application in control of replant diseases. Afr J Biotechnol 11:231–242

Luo S, XuT CL, Rao C, Xiao X (2012) Endophyte-assisted promotion of biomass production and metal-uptake of energy crop sweet sorghum by plant-growth-promoting endophytic *Bacillus* sp. SLS18. Appl Microbiol Biotechnol 93:1745–1753

Makkar RS, Cameotra SS (1998) Production of biosurfactants at mesophilic and thermophilic conditions by a strain of *Bacillus subtilis*. J Ind Microbiol Biotechnol 20:48–52

Makkar RS, Cameotra SS (2002) An update on the use of unconventional substrates for biosurfactant production and their new applications. Appl Microbiol Biotechnol 58:428–434

Malfanova N, Kamilova F, Validov S, Shcherbakov A, Chebotar V, Tikhonovich I, Lugtenberg B (2011) Characterization of *Bacillus subtilis* HC8, a novel plant-beneficial endophytic strain from giant hogweed. Micro Biotechnol 4(4):523–532

Mayak S, Tirosh T, Glick BR (2004) Plant growth-promoting bacteria that confer resistance in tomato to salt stress. Plant Physiol Biochem 42:565–572

Milner JL, Stohl EA, Handelsman J (1996) Zwittermicin A resistance gene from *Bacillus cereus*. J Bact 178(14):4266–4272

Mouloud G, Daoud H, Bassem J, Laribi Atef I, Hani B (2013) New bacteriocin from *Bacillus clausii* strainGM17: purification, characterization, and biological activity. Appl Biochem Biotechnol 171(8):2186–2200

Murugappan RM, Begum SB, Roobia RR (2013) Symbiotic influence of endophytic *Bacillus pumilus* on growth promotion and probiotic potential of the medicinal plant *Ocimum sanctum*. Symbiosis 60(2):91–99

Olubajo BA, Bacon CW (2008) Electrotransformation of *Bacillus mojavensis* with fluorescent protein markers. J Microbiol Met 74:102–105

Ongena M, Jacques P (2007) *Bacillus* lipopeptides: versatile weapons for plant disease biocontrol. Trends Microbiol 16:115–125

Ongena J, Emmanuel J, Adam A, Paquot M, Brans A, Joris B (2007) Surfactin and fengycin lipopeptides of *Bacillus subtilis* as elicitors of induced systemic resistance in plants. Environ Microbiol 9:1084–1090

Ortiz-Castro R, Valencia-Cantero E, Lopez-Bucio J (2008) Plant growth promotion by *Bacillus megaterium* involves cytokinin signaling. Plant Signal Behav 3(4):263–265

Paul B, Chereyathmanjiyil A, Masih I, Chapuis L, Benoît A (1998) Biological control of *Botrytis cinerea* causing grey mould disease of grapevine and elicitation of stilbene phytoalexin (resveratrol) by a soil bacterium. FEMS Microbiol Lett 165(1):65–70

Potts M (1994) Desiccation tolerance of prokaryotes. Microbiol Rev 58:755–805

Raaijmaker JM, Bruijn I, Nybroe O, Ongena M (2010) Natural function of lipopeptides from *Bacillus* and *Pseudomonas*: more than surfactants and antibiotics. FEMS Microbiol Rev 34:1037–1062

Rajkumar M, Sandhya S, Prasad MNV, Fretas H (2012) Perspectives of plant-associated microbes in heavy phytoremediation. Biotechnol Adv 30:1562–1574

Reyes-Ramírez A, Escudero-Abarca BI, Aguilar-Uscanga G, Hayward-Jones PM, Barboza-Corona JE (2004) Antifungal activity of *Bacillus thuringiensis* chitinase and its potential for the biocontrol of phytopathogenic fungi in soybean seeds. J Food Sci 69(5):M131–M134

Romero DA, Vicente A, Rakotoaly R, Dufour S, Veening J, Arrebola A (2007) The iturin and fengycin families of lipopeptides are key factor in antagonism of *Bacillus subtilis* toward *Podosphaera fusca*. Mol Plant Microbe Interact 20:430–440

Ryan RP, Germaine K, Franks A, Ryan DJ, Dowling DN (2008) Bacterial endophytes: recent developments and applications. FEMS Microbiol Lett 278:1–9

Ryu CM, Farag MA, Hu CH, Reddy MS, Kloepper JW, Pare PW (2004) Bacterial volatiles induce systemic resistance in Arabidopsis. Plant Physiol 134:1017–1026

Sansinenea E, Ortiz A (2011) Secondary metabolites of soil *Bacillus* spp. Biotechnol Lett 33(8):1523–1538

Schalk IJ, Hannauer M, Braud A (2008) Effects of inoculation of biosurfactant-producing *Bacillus* sp J119 on plant growth and cadmium uptake in a cadmium-amended soil. J Hazard Mater 155:17–22

Sgroy V, Cassan F, Masciarelli O (2009) Isolation and characterization of endophytic plant growth promoting (PGPB) or stress homeostasis-regulating bacteria assocaited to the halophyte *Prosopis strombulifera*. Appl Microbiol 85:371–381

Shali A, Ghasemi S, Ahmadian G, Ranjbar G, Dehestani A, Khalesi N, Motallebi E, Vahed M (2010) *Bacillus pumilus* SG2 chitinases induced and regulated by chitin, show inhibitory activity against *Fusarium graminearum* and *Bipolaris sorokiniana*. Phytoparasitica 38(2):141–147

Silo-Suh LA, Stabb EV, Raffel SJ, Handelsman J (1998) Target range of zwittermicin A, an aminopolyol antibiotic from *Bacillus cereus*. Curr Microbiol 37(1):6–11

Smith KP, Goodman RM (1999) Host variation for interactions with beneficial plant-associated microbes. Annu Rev Plant Physiol Plant Mol Biol 37:473–491

Smith KP, Handelsman J, Goodman RM (1999) Genetic basis in plant for interactions with disease-suppressive bacteria. Proc Natl Acad Sci U S A 96:4786–4790

Snook ME, Mitchell T, Hinton DM, Bacon CW (2009) Isolation and characterization of Leu7-surfactin from the endophytic bacterium *Bacillus mojavensis* RRC 101, a biocontrol agent for *Fusarium verticillioides*. Agric Food Chem 57:4287–4295

Srinivasan M, Holl FB, Petersen DJ (1996) Influence of indoleacetic-acid-producing Bacillus isolates on the nodulation of *Phaseolus vulgaris* by *Rhizobium etli* under gnotobiotic conditions. Can J Microbiol 42(10):1006–1014

Sturz AV, Christie BR (1995) Endophytic bacterial systems governing red clover growth and development. Ann Appl Biol 126:285–290

Sturz AV, Christie BR, Nowaki J (2000) Bacterial endophytes: potential role in developing sustainable systems of crop production. Crit Rev Plant Sci 19:1–30

Sziderics AH, Rasche F, Trognitz F, Sessitsch A, Wilhelm E (2007) Bacterial endophytes contribute to abiotic stress adaptation in pepper plants (*Capsicum annuum* L). Can J Microbiol 53:1195–1202

Talboys P, Owen D, Healey J, Withers P, Jones D (2014) Auxin secretion by *Bacillus amyloliquefaciens* FZB42 both stimulates root exudation and limits phosphorus uptake in *Triticum aestivum*. BMC Plant Biol 14(1):51

Van Etten HD, Mansfield JW, Bailey JA, Farmer EE (1994) Two classes of plant antibiotics: phytoalexins versus "phytoanticipins". Plant Cell Online 6(9):1191–1192

Vardharajula S, Ali SZ, Grover M, Reddy G, Bandi V (2011) Drought-tolerant plant growth promoting *Bacillus* spp.: effect on growth, osmolytes, and antioxidant status of maize under drought stress. J Plant Interact 6:1–14

Vendan R, Yu Y, Lee S, Rhee Y (2010) Diversity of endophytic bacteria in ginseng and their potential for plant growth promotion. J Microbiol 48(5):559–565

Vining LC (1990) Functions of secondary metabolites. Ann Rev Microbiol 44(1):395–427

Vinocur B, Altman A (2005) Recent advances in engineering plant tolerance to abiotic stress: achievements and limitations. Curr Opin Biotechnol 16:123–132

Wang SL, Shih IL, Liang TW, Wang CH (2002) Purification and characterization of two antifungal chitinases extracellularly produced by *Bacillus amyloliquefaciens* V656 in a shrimp and crab shell powder medium. J Agric Food Chem 50(8):2241–2248

Wani PA, Kahn MS, Zaidi A (2007) Chromium reduction, plant growth promoting potentials, and metal solubilization by *Bacillus* sp. isolated from alluvial soil. Curr Microbiol 54:237–243

Warren GF (1998) Spectacular increases in crop yields in the twentieth century. Weed Technol 12:752–760

Willey JM, Van der Donk WA (2007) Lantibiotics: peptides of diverse structure and function. Ann Rev Microbiol 61:477–501

Yandigeri MS, Meena KK, Singh D, Malviya N, Singh DP, Solanki MK, Yadav AK, Arora DK (2012) Drought-tolerant endophytic actinobacteria promote growth of wheat (*Triticum aestivum*) under water stress conditions. Plant Growth Regul 68:411–420

Yoshida S, Hiradate S, Tsukamoto T, Hatakeda K, Shirata A (2001) Antimicrobial activity of culture filtrate of *Bacillus amyloliquefaciens* RC-2 isolated from Mulberry leaves. Phytopathology 91(2):181–187

Yu GY, Sinclair JB, Hartman GL, Bertagnolli BL (2002) Production of iturin A by Bacillus amyloliquefaciens suppressing *Rhizoctonia solani*. Soil Biol Biochem 34(7):955–963

Yu X, Ai C, Xin L, Zhou G (2011) The siderophore-producing bacterium, *Bacillus subtilis* CAS15, has a biocontrol effect on *Fusarium* wilt and promotes the growth of pepper. Eur J Soil Biol 47(2):138–145

Zaidi S, Usmani S, Singh BR, Musarrat J (2006) Significance of *Bacillus subtilis* strain SJ-101 as a bioinoculant for concurrent plant growth portion and nickel accumulation in *Brassica juncea*. Chemosphere 64:991–997

Zhao L, Xu Y, Sun R, Deng Z, Yang W, Wei G (2011) Identification and characterization of the endophytic plant growth prompter *Bacillus cereus* strain mq23 isolated from Sophora alopecuroides root nodules. Braz J Microbiol 42(2):567–575

Zhou Y, Choi YL, Sun M, Yu Z (2008) Novel roles of *Bacillus thuringiensis* to control plant diseases. Appl Microbiol Biotechnol 80(4):563–572

Antifungal Compounds from Pseudomonads and the Study of Their Molecular Features for Disease Suppression Against Soil Borne Pathogens

9

Urja Pandya and Meenu Saraf

Abstract

The identification and biological and molecular characterization of *Pseudomonas*, a very versatile microbe with biocontrol potential, is of great interest for the modern and eco-compatible agriculture. The mechanisms of biocontrol by *Pseudomonas* include antibiotic production, siderophore production, and production of fungal cell wall-lysing enzymes and induced systemic resistance. Current genome analyses of biocontrol traits will likely lead to the development of novel tools for effective management of indigenous and inoculated pseudomonads as biocontrol agents and a better exploitation of their properties for sustainable agriculture. The chapter summarizes and discusses various studies of pseudomonads from the plant rhizosphere and their use for exploring disease management in integrated disease management (IDM) and the study of genome sequences of *Pseudomonas* spp. for sustainable application.

Introduction

Food security is the one of the most pressing challenges faced currently by humanity. The world population is predicted to reach nine billion by 2050, requiring food production to increase by 50 % against a backdrop of climate change and the loss of arable land. Despite agricultural advances plant pathogens still account for losses of 10–16 % in the global harvest. The need for new control strategies for this and other plant diseases is therefore urgent. At present, the major strategies against damages caused by plant pathogens are chemical pesticides or resistant plant cultivars. However, there are major limitations in using both strategies (Borriss 2011): (1) agrochemicals do not prevent all kinds of diseases, and toxic residues can accumulate in the soil and food chain. Therefore, the application of many agrochemicals was banned or restricted, because of the environmental and health risks to humankind and soil fertility, and (2) resistance of genetically resistant cultivars is often broken by the pathogen within a few years and frequently accompanied by a reduction in yield. Typically, there is a lack of acceptance among the public for genetically modified (GM) crops.

U. Pandya • M. Saraf (✉)
Department of Microbiology and Biotechnology, University School of Sciences, Gujarat University, Ahmedabad 380 009, Gujarat, India
e-mail: sarafmeenu@gmail.com

N.K. Arora (ed.), *Plant Microbes Symbiosis: Applied Facets*,
DOI 10.1007/978-81-322-2068-8_9, © Springer India 2015

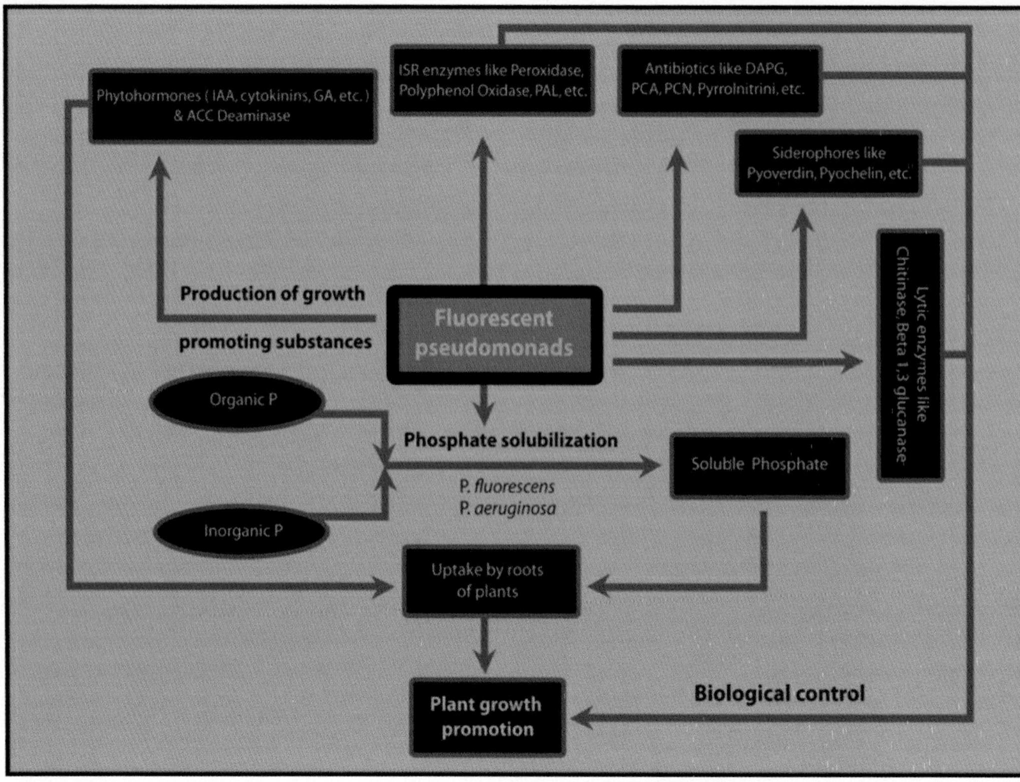

Fig. 9.1 Direct and indirect mechanisms involved in plant growth promotion by fluorescent pseudomonads (Adapted and modified from Khan et al. (2010))

IDM, which combines biological, cultural, physical, and chemical control strategies in a holistic way rather than using a single component strategy, proved to be more effective and sustainable. In practice and in the majority of cropping systems today, emphasis is still being placed on a single technology. The concept of IDM, where diseases are managed by integrating a range of control methods and practices, is becoming more popular among farmers and researchers. The success and sustainability of IDM strategy, especially with resource-poor farmers, greatly depend on their involvement in helping generate locally specific techniques and solutions suitable for their particular farming systems and integrating control components that are ecologically sound and readily available to them (Khoury and Makkouk 2010). Biological control of plant diseases has been considered a viable alternative method to manage diseases. Biological control is the inhibition of growth, infection, or reproduction of one organism using another organism. Biocontrol is environmentally safe and in some case studies is the only option available to protect plants against pathogens (Heydari and Pessarakli 2010). Fravel (2005) reported 14 bacteria and 12 fungi registered with the United States Environmental Protection Agency for the control of plant diseases. Among the 14 registered bacterial biocontrol agents, six are based on *Bacillus*, five on *Pseudomonas*, two on *Agrobacterium*, and one on *Streptomyces*.

Pseudomonas spp. are particularly suited as biocontrol agents because they can use diverse root exudates as nutrient source; they are abundantly present in natural soils, especially in the rhizosphere; they have a high growth rate; they can be directly plant growth promoting; and they have the ability to control diseases by a variety of mechanisms as shown in Fig. 9.1. They are also the most extensively studied group of bacterial biocontrol agents, since *Pseudomonas* bacteria are amenable to mutation and modification using genetic tools (Hofte and Altier 2010).

The chapter summarizes and discusses various studies of pseudomonads from the plant rhizosphere and their use for exploring disease management in IDM for sustainable agriculture.

Ecological Distribution of Pseudomonads in Plants

Pseudomonas is one of the most ubiquitous bacterial genera in the world, and different species have been isolated from very diverse ecological niches. Only the representatives of rRNA group I were included in the genus *Pseudomonas*, and up to now, 128 species have been validly described for this genus, including the fluorescent pseudomonads that have the capacity to produce fluorescent pyoverdine-type siderophores under low-iron conditions. Most species are saprophytes that are commonly found in water and soil; 23 species are pathogenic to plants, including *P. syringae* with 36 pathovars affecting different plants. In addition, 16 species are associated with diseases in humans and animals (Peix et al. 2009). Some species, such as *P. aeruginosa*, are ubiquitous and can be associated with both plants and animals. Some of the saprophytic species have interesting characteristics and are used in biotechnological applications to improve plant growth and plant health, but also in water and soil bioremediation. *P. fluorescens* biovar III has been reported as the dominant group of bacteria among fluorescent pseudomonads associated with the rhizosphere of rice, the most important food crop of the world. *P. fluorescens* biovar V constituted more than 75 % of fluorescent pseudomonads in at least some areas of the world and represented as the predominant group in Australian soils and wheat rhizosphere. Nearly 90 % of fluorescent pseudomonads in Colombia were reported to be *P. putida*. Thus, vast microbial diversity occurs among fluorescent pseudomonads in different geographic origins, and this group of bacteria exhibits an array of mechanisms that mediate biocontrol of pathogens and growth promotion of crops (Pathama et al. 2011).

Pseudomonas spp. diversity is also found in many plant rhizosphere and phyllosphere environments (van den Broek et al. 2005; Hirano and Upper 2000). Many researchers reported culture-independent analysis for examination of *Pseudomonas* assemblage in the endosphere by using genus-specific primers (Garbeva et al. 2001; Reiter et al. 2002; Sessitsch et al. 2002). Root tissue analysis indicated that the dominant pseudomonad species in flax was *P. putida* and in tomato, *P. fluorescens*. Fluorescent *Pseudomonas* were reported as competitive in the rhizoplane of maize, as there was an increase in the heterotrophic and total gram-negative communities present inside the rhizoplane. Thirty percent of isolates were identified as *P. fluorescens* and 50 % were identified as *P. putida* (Chaudhary et al. 2009).

Pseudomonads as Potential PGPR by Direct Mechanisms Under Field Conditions

Rhizosphere microorganisms may provide a front line defense against pathogen attack and are ideal for use as biocontrol agents. Some bacteria are associated with roots of crop plants and exert beneficial effects on their hosts and are referred to as plant growth-promoting rhizobacteria (PGPR) (Lugtenberg and Kamilova 2009). PGPR are free-living soil bacteria that can either directly or indirectly facilitate rooting and growth of plants. Direct mechanisms include nitrogen fixation, siderophore production, phosphate solubilization, and synthesis of phytohormones. PGPR indirectly enhance plant growth via suppression of phytopathogens by a variety of mechanisms. These include the ability to produce siderophores that chelate iron, making it unavailable to pathogens; the ability to synthesize antifungal metabolites such as antibiotics, fungal cell wall-lysing enzymes, or production of volatiles such as hydrogen cyanide, which suppresses the growth of fungal pathogens; the ability to successfully compete with pathogens for nutrients or specific niches on the root; and the ability to induce systemic resistance (ISR) (Saraf et al. 2010). Zabihi et al. (2011) studied that *P. putida* 108 was the most effective strain-enhancing wheat P uptake and grain yield under greenhouse (96 and 58 %) and field (80 and 37 %) conditions, respectively. Hence, although *Pseudomonas* sp. could be a suitable replacement for high P fertilization, however, the optimum wheat yield resulted when the bioinoculants are combined with 50 % (25 kg/ha P_2O_5) P fertilization. This finding has great agricultural and environmental implications. The presence of *P. fluorescens* inoculants in the combination of microbial fertilizer plays an effective role in stimulating the yield and growth traits of chickpea (Rokhzadi et al. 2008). Five *P. putida*, one *P. graminis*, and one *P. fluorescens* strains showed significant increases in fresh and dry masses of sugarcane plantlets in tissue culture under gnotobiotic conditions (Mehnaz et al. 2009). Husen et al. (2011) evaluated the effects of eight promising isolates of ACC deaminase-producing pseudomonads on soybean growth and yield under field soil conditions. Field performance of five *P. fluorescens* isolates (Pf-173, Pf-193, Pf-547, Pf-551, and Pf-572) for biocontrol and plant growth-promoting activities were evaluated in French bean (var. Pusa contender). Germinated French bean seedlings exhibited enhanced root and shoot lengths in all the treatments. The combination of the five isolates proved to be the best, with maximum disease reduction (93.33 %) and enhancement of various plant growth parameters. Among individual isolates, Pf-551 had the best performance for most of the parameters assessed (Negi et al. 2011). Wahyudi et al. (2011) examined 14 isolates from the soybean rhizosphere as pseudomonads spp. that showed phosphate-solubilizing activity, siderophore production, IAA production, and promoted enhancement of root length, shoot length, or a number of lateral roots of soybean. Plant growth attributes produced by fluorescent pseudomonads that promote plant growth are listed in Table 9.1.

Naz and Bano (2010) reported three species of *Pseudomonas*, viz., *P. mendocina* Khsr2, *P. stutzeri* Khsr3, and *P. putida* Khsr4, for their phosphate solubilization and phytohormone production (indole-3-acetic acid, gibberellic acid, *trans*-zeatin riboside, and abscisic acid) in culture media and were found to be efficient in stimulating root/shoot length and dry weight and proline contents of *Zea mays* L. (advance germplasm line: Islamabad Gold) seedlings grown under normal and NaCl stress (20 dS m^{-1}). The strain *P. stutzeri* Khsr3 appears to be a potential candidate as a bioinoculant for saline fields. Mishra et al. (2011) observed that inoculation with *Pseudomonas* sp. strain PGERs17 exhibits more vigorous vegetative growth with increase in nodulation (55.0 %), leghaemoglobin content (27.7 %), physiologically available iron (24.1 %), total iron (90.7 %), chlorophyll content (43.9 %), P uptake (37.1 %), and N uptake (27.1 %) over mono-inoculation of *R. leguminosarum*-PR1 60 days after sowing (DAS) of lentil.

Pseudomonads as Biocontrol Agents Against Soil borne Pathogens

Pseudomonads are effective against a number of Soil borne pathogens through their both direct and indirect mechanisms. Antibiosis involving secretion of antibiotics, siderophore, and fungal cell wall-lysing enzymes is a direct mechanism of biocontrol, whereas induction of disease resistance in plants is an indirect mechanism of plant growth promotion.

Antibiosis

Phloroglucinol

Phloroglucinols are phenolic compounds by nature and bacterial and plant metabolites with broad-spectrum antiviral, antibacterial, antifungal, antihelminthic, and phytotoxic properties. The compound 2,4-diacetylphloroglucinol (2,4-DAPG), which is produced by certain plant-associated fluorescent *Pseudomonas* species, is of particular significance to agriculture because of its activity against a variety of root and seedling-borne pathogens of plants (Pathama et al. 2011).

Phenazines

Phenazines (Phz) are N-containing heterocyclic pigments produced by various pseudomonad species. Phz also play an important role of microbial

Table 9.1 Plant growth-promoting characteristics of fluorescent pseudomonads

Pigments, enzymes, and hormones	*Pseudomonas* species	References
Siderophore		
Pyoverdine	*Pseudomonas* sp.	Tank and Saraf (2009)
	P. fluorescens 3551	Loper (1988)
	P. fluorescens CHAO	
Pyochelin	*P. putida* WCS358	Maurhofer et al. (1994)
	P. aeruginosa 7NSK2	
Pseudomonine	*P. stutzeri*	Van Wees et al. (1977)
	P. fluorescens WCS374	Buysens et al. (1996)
	P. aeruginosa NJ-15	Lewis et al. (2000)
	P. monteilii FPB59, FPB63	Mercado-Blanco et al. (2001)
	P. plecoglossicida FPB31	Bano and Musarrat (2003)
		Ravindra Naik et al. (2008)
		Ravindra Naik et al. (2008)
Phosphate		
Solubilization	*P. mosselii* FP13	Jha et al. (2009)
Phosphatase	*P. fluorescens* CHAO	De Werra et al. (2009)
	P. putida MSC1 and *P. pseudoalcaligens* MSC4	Patel et al. (2011)
Denitrifying enzymes	*P. putida* GR 12-2	Patten and Glick (2002)
	P. mosselii FP13	Ravindra Naik et al. (2008)
	P. aeruginosa FP 10	Ayyadurai et al. (2006)
Indole-3-acetic acid	*P. fluorescens* Psd	Upadhyay and Srivastava (2010)
	P. aeruginosa PS1	Ahemad and Khan (2011a)
	P. putida PS9	Ahemad and Khan (2011b)
ACC deaminase	*P. fluorescens*	Shaharoona et al. (2008)
	P. jessenii	Rajkumar and Freitas (2008)
	P. aeruginosa and *P. fluorescens*	Tank and Saraf (2010)

competition in rhizosphere, including survival and competence. Phenazine-1-carboxylic acid (PCA) has been reported from fluorescent pseudomonads such as *P. fluorescens*, *P. chlororaphis*, and *P. aeruginosa* (Quan et al. 2010).

Pyrrolnitrin

Pyrrolnitrin (PRN) (3-chloro-4-(20-nitro-30-chlorophenyl) pyrrole) has a broad-spectrum antifungal activity as first described by Arima et al. (1964). PRN is a secondary metabolite derived from L-tryptophan during metabolism. It is produced by fluorescent pseudomonads such as *P. fluorescens* (Kirner et al. 1998) and *P. aureofaciens* (Elander et al. 1968). Other variants of PRN such as isopyrrolnitrin, oxypyrrolnitrin, and monodichloropyrrolnitrin have lower antifungal activity. PRN is active against a wide range of deuteromycete, ascomycete, and basidiomycete fungi.

Cyclic Lipopeptides (CLP)

Production of different kinds of CLP is common among fluorescent *Pseudomonas* spp. All CLPs have either 9 or 11 amino acids in the peptide ring with a C10 fatty acid attached at one of the amino acids (Nielsen et al. 2002). Its synthesis is non-ribosomal and catalyzed by large peptide synthetase complexes (Marahiel et al. 1997). CLP is involved in the promotion of bacterial swarming (Givskov et al. 1998; Lindum et al. 1998), with antimicrobial (Takesako et al. 1993; Gerard et al. 1997; Vollenbroich et al. 1997) and biosurfactant properties (Rosenberg and Ron 1999). Strains of *P. fluorescens*, DR54, 96.578, and DSS73, produce three different CLPs, viz., viscosinamide (Nielsen et al. 2002), tensin (Henriksen et al. 2000), and amphisin (Sorensen et al. 2001). Apart from the antifungal action of viscosinamide, it is also involved in the primary metabolism and cell proliferation and strongly binds to the producing cells of the strain DR54 (Nielsen et al. 1999). Tensin and amphisin produced are released into the surrounding medium and suppress the ingress of the pathogen (Nielsen et al. 2000). Amphisin is a new member of a group of dual-functioning compounds like tensin, viscosin, and viscosinamide that have both biosurfactant and antifungal properties.

Amphisin is produced during the stationary phase. *amsY* gene codes for the synthesis of amphisin synthetase, controlled by a two-component regulatory system GacA/GacS (Koch et al. 2002). The ability of *P. fluorescens* strain DSS73 to control *P. ultimum* and *R. solani* arises from amphisin-dependent surface translocation and growth by which the bacterium inhibits *P. ultimum* and *R. solani* (Andersen et al. 2003). The synergistic effect of surface motility and the synthesis of antifungal compounds could efficiently check and terminate the growth of pathogen and prevent the plants from infection by the pathogens. Massetolides are cyclic lipopeptide antibiotics produced by various *Pseudomonas* strains. Three genes were found to be involved in the massetolide A biosynthesis in *P. fluorescens* strain SS101. Massetolide A production is essential in the swarming motility of *P. fluorescens* SS101 and plays an important role in biofilm formation (de Bruijn et al. 2008).

Siderophore: Competition for Iron

Pyoverdine

Pyoverdine has a great affinity for ferric iron than the pyochelin. These compounds are water-soluble pigments that fluoresce yellow green under ultraviolet light. They all are structurally similar being comprised of a chromophore moiety, which confers on the bacterium its fluorescent nature, to which is bound a peptide chain of variable composition and a dicarboxylic acid (Glick et al. 1999).

Pyochelin

When plant-associated pseudomonads such as biocontrol strains of *P. fluorescens* and *P. putida* are deprived of iron, they produce two different types of siderophore: one is pyochelins and pyoverdines. Pyochelins are small, phenolate siderophores derived from salicylic acid and cysteine. Two pyochelin molecules bind with one molecule of ferric iron with a relatively low efficiency (Cox et al. 1981; Crosa 1989).

Salicylic Acid

Some *Pseudomonas* spp. can produce a nonfluorescent siderophore called pyochelin, consisting of a salicylic-substituted cysteinyl peptide. Salicylic acid is considered to be not only an intermediate in pyochelin biosynthesis but also an endogenous siderophore for some *Pseudomonas* spp. Salicylic acid is produced under iron limitation conditions (Leeman et al. 1996).

HCN Production

HCN is an effective inhibitor of cytochrome c oxidase (Knowles 1976) and other metalloenzymes. Its production by PGPR is implicated in biological control of black root rot of tobacco (Voisard et al. 1989), root rot of tomato, and *Pythium* damping-off of cucumbers (Ramette et al. 2003). Cyanogenesis by *Pseudomonas* spp. is the product of an HCN synthase, encoded by three biosynthetic genes (hcnABC) (Laville et al. 1998). These three genes are important for HCN biosynthesis from glycine, the primary precursor (Blumer and Haas 2000).

Fungal Cell Wall-Degrading Enzymes

Antagonistic effects on phytopathogens can also be mediated by extracellular lytic enzymes, and many biocontrol pseudomonads produce such enzymes like chitinase, β-1,3-glucanase, protease, and cellulase (Chernin and Chet 2002). Chitinolysis plays an important role in the biological control of plant diseases and has been substantiated with increased disease control by chitin-supplemented application of chitinolytic biocontrol agents. Chitinase and β-1,3-glucanase are the key enzymes associated with the decomposition of the fungal cell wall of Soil borne fungal pathogens. It has also been demonstrated that extracellular chitinase and laminarinase synthesized by *P. stutzeri* digest and lyse mycelium of *Fusarium solani* (Kumar et al. 2010). Chitinase and cellulase produced by pseudomonad strains showed mycelial growth inhibition of different fungal pathogens on PDA medium (Sindhu and Dadarwal 2001). Production of extracellular β-1,3-glucanases, chitinases, and proteinase increases significantly when *Pseudomonas* spp. are grown in media supplemented with either autoclaved mycelium or isolated purified host fungal cell walls (Viswanathan and Samiyappan 2000). Such induction of chitinases and their antifungal activity together with the fact that chitin, β-1,3-glucan, and protein are the main structural components of most fungal cell walls suggested that hydrolytic enzymes produced by some fluorescent pseudomonads play an important role in destruction of plant pathogens. Saraf et al. (2008) studied that *Pseudomonas* spp. (M1P3) showed maximum chitinase activity (40 μg/ml N-acetyl-glucosamine production) and also induced maximum cellulase production. Arora et al. (2008) reported fluorescent *Pseudomonas* (PGC 1 and PGC 2) for their antifungal potential against *R. solani* and *P. capsici*. The results of this study indicated the role of chitinase and β -1,3-glucanase in the inhibition of *R. solani*; however, antifungal metabolites of nonenzymatic nature were responsible for inhibition of *P. capsici*. Aeron et al. (2011) reported that *P. fluorescens* PS1 caused morphological alteration in mycelia of *S. sclerotiorum* as evidenced by hyphal perforation and fragmented lysis by extracellular enzymes chitinase (0.69 units/ml) and β-1,3-glucanase (1.98 units/ml).

Induced Systemic Resistance

Colonization of the rhizosphere by certain bacteria such as nonpathogenic pseudomonads can trigger a systemic resistance in plants by activating the plants' defense mechanisms phenotypically similar to pathogen-induced systemic acquired resistance (SAR), and this form of induced resistance is referred as rhizobacteria-mediated ISR. Fluorescent pseudomonads are also involved in the induction of plant resistance against insects and nematodes, and specific metabolites produced by these bacteria have been involved in the elicitation of defense reactions of the host plants. Similar to classical induced resistance, PGPR-mediated resistance is known to produce broad-spectrum resistance, which is correlated with increased amounts of pathogenesis-related (PR) proteins, peroxidase, chitinases, and β-1,3-glucanase in plant tissue (Pathama et al. 2011). A salicylate-overproducing recombinant of *P. fluorescens* strain P3 causes enhanced protection to tobacco against tobacco necrosis virus compared with the wild-type P3, which indicates that salicylate might also stimulate defense. In another strain of *Pseudomonas*, a combination of pyocyanin and pyochelin seems to be most effective for inducing resistance in tomato (Audenaert et al. 2002). Rajkumar et al. (2008) studied fluorescent pseudomonads (SE21 and RD 41) and resistance inducers (chitin and salicylic acid) for plant growth promotion and biological control of damping-off of pepper caused by *Rhizoctonia solani*. *P. fluorescens* strains CHA0 and Pf1 were investigated for their biocontrol efficacy against banana bunchy top virus (BBTV) in banana (*Musa* spp.) alone and in combination with chitin under glasshouse and field conditions. Bioformulation of *P. fluorescens* strain CHA0 with chitin was effective in reducing the banana bunchy top disease (BBTD) incidence in banana under glasshouse and field conditions. In addition to disease control, the bioformulation increased the economic yield significantly compared to the untreated control. Increased accumulation of oxidative enzymes, peroxidase (PO), polyphenol oxidase (PPO), phenylalanine ammonia lyase (PAL), pathogenesis-related (PR) proteins, chitinase, β-1,3-glucanase, and phenolics was observed in CHA0 bioformulation amended with chitin-treated plants challenged with BBTV under glasshouse conditions (Kavino et al. 2008).

Biosynthesis of Antifungal Metabolites and Their Genome Sequence Features

There are many strains of pseudomonads reported for their biofertilizer and biocontrol potential, but only few strains are known for their whole genome sequences with their metabolic pathway. Three *P. fluorescens* strains (biocontrol agents SWB25 and Pf-5 and the soil inhabitant Pfo-1) were compared, and results showed that they share only 61 % of their genes. The study revealed that these strains are a species complex rather than a single species (Silby et al. 2009). The antibiotic-productive gene loci are highly conserved and most intensively studies found that they correlated with DAPG-producing strains (Keel et al. 1996; Frapolli et al. 2007). *P. fluorescens* F113 complete genome sequence contained a large number of unique protein-coding genes that encode traits potentially involved in the rhizocompetence of this strain, and the features include protein-coding genes required for denitrification, diterpenoid catabolism, motility and chemotaxis, protein secretion, and production of antimicrobial compounds and insect toxins (Renodo-Nieto et al. 2013).

Nowadays, scientists are concentrating on the biotechnology of *P. fluorescens* for effective application on agricultural crops (Wiyono et al. 2008). Loper et al. (2007) reported that *P. fluorescens* Pf-5 produced three novel metabolite synthesis gene clusters that were identified in the genome attribute to biological control mechanisms. In general practices, sequenced strains present only a fraction of pseudomonad species diversity, and much of the group's metabolic and genetic diversity remains unexplored till today. Comparative genome analyses of ten strains within the *Pseudomonas fluorescens* group including seven new genome sequences were reported by Loper et al. (2012). They discovered genes for traits that were not known previously, including genes for the biosynthesis of the siderophores achromobactin and pseudomonine and the antibiotic 2-hexyl-5-propyl-alkylresorcinol; novel bacteriocins; type II, III, and VI secretion systems; and insect toxins. The genome of the nitrogen-fixing bacterium *Azotobacter vinelandii*, classified in the *Pseudomonadaceae*, has also been sequenced (Setubal et al. 2009). Cyanogenesis by *Pseudomonas* spp. is the product of an HCN synthase, encoded by three biosynthetic genes (hcnABC) (Laville et al. 1998). These three genes are sufficient for HCN biosynthesis from glycine, the primary precursor (Blumer and Haas 2000). Conservation of *phl* genes for biosynthesis of DAPG among ecologically and geographically diverse antagonistic pseudomonads further supports the global importance of DAPG production in biocontrol (Duffy et al. 2004). Biosynthesis of allelochemicals by *Pseudomonas* biocontrol potential strains is closely regulated by molecules produced by the organism itself and by external environmental factors such as nutritional components, soil chemical and physical properties, host plant genotype, and nonpathogenic soil bacteria, because these factors can determine the ability of particular strains to suppress disease. Identification and characterization at molecular levels would facilitate the targeted application of strains into environments that are more favorable for effective and consistent biocontrol activity (Saraf et al. 2014). Four biosynthetic genes (prnABCD) are highly conserved among strains of *P. fluorescens* that produce pyrrolnitrin (Hammer et al. 1997, 1999; Kirner et al. 1998). Expression of prnABCD in heterologous bacteria resulted in pyrrolnitrin production. Results concluded that these genes are sufficient for biosynthesis of the antibiotics from primary precursors (Hammer et al. 1997).

Role of Antifungal Metabolites in Disease Management Under Field Conditions

Indigenous DAPG-producing populations have been identified as the driving force behind the development of natural disease suppressiveness in certain soils under long-term monoculture. Meynet et al. (2011) reported that DAPG-producing *P. fluorescens* F113 enhanced *Azospirillum brasilense* Sp245-Rif traits involved in root colonization (cell motility, biofilm formation, and poly-β-hydroxybutyrate production) and phytostimulation (auxin production) of wheat. Pyoluteorin-producing bacteria *P. putida* strain NH-50 significantly reduced disease severity on both sugarcane varieties, irrespective of fungal inoculation, i.e., either inoculated through stem or

Table 9.2 Commercially available biocontrol products

Commercial product	Name of strain	Crop	Manufacturer
Bio-Save 10, 11, 100, 110, 1000™	*P. syringae ESC-100*	Pome fruit (Bio-Save 100) and citrus (Bio-Save 1000)	Eco Science Corp, Produce Systems Div., Orlando
Blight Ban A506	*P. fluorescens* A 506	Almond, apple, apricot, blueberry, cherry, peach, pear, potato, strawberry, tomato	Plant Health Technologies, USA
Conquer™	*P. fluorescens*	Mushrooms	Mauri Foods, Australia
Victus™	*P. fluorescens*	Mushrooms	Mauri Foods, Australia
Biomonas	*P. fluorescens*	Cotton, cereals, pulses, vegetables, oilseeds, fruit plants, and floriculture	Biotech International Limited, New Delhi
BIOWIN	*P. fluorescens*	Paddy, turmeric, banana, sugarcane, spices, oilseeds, pulses, cotton, and vegetables	Esvin Advanced Technologies Limited, Tamil Nadu
Phasal Rakshak powder	*P. fluorescens* 1 % W.P.	Wheat, chickpea, pigeon-pea, paddy, groundnut, mustard, sunflower, rapeseed, vegetables, fruits, citrus, apple, mango and all other crops	International Panaacea Limited, New Delhi
Bactvipe liquid	*P. fluorescens* 2 % A.S.		

Modified from Nakkeeran et al. (2005)

through soil. This strain also possessed other plant growth characteristics and can be used as a biopesticide for sugarcane crop (Hassan et al. 2011).

Siddiqui et al. (2003) have clearly proved the protective nature of HCN produced by *P. aeruginosa* against *Meloidogyne javanica*, the root-knot nematode of tomato. Siderophores such as pyoverdine and pyochelin have been documented well for the suppression of *Pythium*-induced damping-off disease of tomato (Buysens et al. 1996). Recently, Shen et al. (2014) reported that 100-fold dilution of *P. fluorescens* CZ powder could be used as application under greenhouse and field conditions which inhibited virus infection by 59.2 % on *N. tabacum* cv. NC89 plants in the greenhouse. In field trials, 100-fold dilution of CZ suppressed TMV infection by 58.2 % and 47.6 % in 2010 and 2011, respectively. D-gluconic acid was the most significant antifungal agent produced by *Pseudomonas* spp. strain AN5 for the biocontrol of take all of wheat roots (Kaur et al. 2006). These results interpreted that diverse compounds can be associated for disease suppression. Botelho and Selecaode (2001) concluded that phenazine production was associated with the "sudden death syndrome" caused by *F. solani* infecting soybean. Simultaneous production of DAPG/HCN from *Pseudomonas* sp. LBUM300 showed great potential for controlling the bacterial canker of tomato caused by *Clavibacter michiganensis* (Lanteigne et al. 2012). *Pseudomonas* strain GRP3 is known to produce rhamnolipids and siderophore and has been shown to reduce the occurrence of pre- and postemergence damping-off in chili and tomato (Sharma et al. 2007). Le et al. (2011) studied that phenazine- and lipopeptide-producing *Pseudomonas* spp. inhibited the hyphal growth of *S. rolfsii* and significantly reduced the stem rot disease of groundnut in greenhouse and field experiments. Kishore et al. (2006) also found that the biocontrol agent *P. aeruginosa* GSE 18 reduced the preemergence of groundnut rotting by 60 % in *A. niger*-infested potting mixture. Bhatia et al. (2008) reported increased seed germination, growth promotion, and suppression of charcoal rot due to *M. phaseolina* with fluorescent pseudomonads. Table 9.2 lists some of the commercially available biocontrol products obtained from pseudomonads in the market around the globe. The products are based on the success of pseudomonads as biocontrol agents because of their diverse mechanisms working under variable conditions and for different phytopathogens.

Conclusion

Pseudomonas spp. are prolific producers of many secondary metabolites, and the availability of the genome features for several strains now opens new insight for the discovery of novel natural products for commercialization. The forthcoming developments by functional genomics approach using sequencing and proteome technology will help us to understand the performance of biocontrol strains from fluorescent and nonfluorescent pseudomonad species and overcome the current bottlenecks restricting their commercial use. The major challenges to be resolved prior to widespread commercial exploitation of biocontrol strains lie in the ability to predict more confidently the behavior of such strains in the field. Fortunately, as interest in these organisms grows, more information on the genetics, physiology, and ecology of metabolite production is becoming available. Such data are of immense importance for the selection of wild-type strains with desirable traits from nature and to provide a more rational framework for the choice of strains for use in microbial consortial inoculants. Recombinant DNA methods have enabled genetic manipulation of metabolite production with promising results. The use of metagenomics in parallel with established or novel molecular approaches such as genome mining will undoubtedly lead to a fresh look on the biology and practical applications.

Acknowledgment We gratefully acknowledge financial support by the Department of Science and Technology (DST), New Delhi, under Women Scientist Scheme (WOS-A).

References

Aeron A, Dubey RC, Maheshwari DK, Pandey P, Bajpai VK, Kang SC (2011) Multifarious activity of bioformulated *Pseudomonas fluorescens* PS1 and biocontrol of *Sclerotinia sclerotiorum* in Indian rapeseed (*Brassica campestris* L.). Eur J Plant Pathol 131:81–93

Ahemad M, Khan MS (2011a) Assessment of plant growth promoting activities of rhizobacterium *P. putida* under insecticide stress. Microbiol J 1:54–64

Ahemad M, Khan MS (2011b) Phosphate solubilizing and plant-growth-promoting *Pseudomonas aeruginosa* PS1 improves greengram performance in quizalafop-p-ethyl and Clodinafop amended soil. Arch Environ Contam Toxicol 58:361–372

Andersen JB, Koch B, Nielsen TH, Sorensen D, Hansen M, Nybroe O, Christophersen C, Sorensen J, Molin S, Givskov M (2003) Surface motility in *Pseudomonas* sp. DSS73 is required for efficient biological containment of the root-pathogenic microfungi *Rhizoctonia solani* and *Pythium ultimum*. Microbiology 149:37–46

Arima K, Imanaka H, Kousaka M, Fukuda A, Tamura G (1964) Pyrrolnitrin, a new antibiotic substance produced by *Pseudomonas*. Agric Biol Chem 28:575–576

Arora NK, Khare E, Oh JH, Kang SC, Maheshwari DK (2008) Diverse mechanisms adopted by fluorescent *Pseudomonas* PGC 2 during the inhibition of *Rhizoctonia solani* and *Phytophthora capsici*. World J Microbiol Biotechnol 24:581–585

Audenaert K, Pattery T, Cornelis P, Hofte M (2002) Induction of systemic resistance to *Botrytis cinerea* in tomato by *Pseudomonas aeruginosa* 7NSK2: role of salicylic acid, pyochelin, and pyocyanin. Mol Plant-Microbe Interact 15:1147–1156

Ayyadurai N, Ravindra Naik P, Sreehari Rao M, Sunish Kumar R, Samrat SK, Manohar M, Sakthivel N (2006) Isolation and characterization of a novel banana rhizosphere bacterium as fungal antagonist and microbial adjuvant in micropropagation of banana. J Appl Microbiol 100:926–937

Bano N, Musarrat J (2003) Characterization of a new *Pseudomonas aeruginosa* strain NJ-15 as a potential biocontrol agent. Curr Microbiol 46:324–328

Bhatia S, Maheshwari DK, Dubey RC, Arora DS, Bajpai VK, Kang SC (2008) Beneficial effects of fluorescent pseudomonads on seed germination, growth promotion, and suppression of charcoal rot in groundnut (*Arachis hypogaea* L.). J Microbiol Biotechnol 18:1578–1583

Blumer C, Haas D (2000) Mechanism, regulation, and ecological role of bacterial cyanide biosynthesis. Arch Microbiol 173:170–177

Borriss R (2011) Use of plant-associated *Bacillus* strains as biofertilizers and biocontrol agents in agriculture. In: Maheshwari DK (ed) Bacteria in agrobiology: plant growth responses. Springer, Berlin, pp 41–76

Botelho GR, Selecaode (2001) *Pseudomonas fluorescents* para controle biologic da podrado vermelha da Raiz causada for *R. solani*. PhD thesis, Biotechnologia Vegetal, UFRJ, Rio de Janeiro, p 108

Buysens S, Heungens K, Poppe J, Hofte M (1996) Involvement of pyochelin and pyoverdine in suppression of *Pythium*- induced damping off of tomato by *Pseudomonas aeruginosa* 7NSK2. Appl Environ Microbiol 62:865–871

Chaudhary DK, Prakash A, Wray V, Johri BN (2009) Insights of the fluorescent pseudomonads in plant growth regulation. Curr Sci 97(2):170–179

Chernin L, Chet I (2002) Microbial enzymes in the biocontrol of plant pathogens and pests. In: Burns RG, Dick RP (eds) Enzymes in the environment: activity, ecology and applications. Marcel Dekker, New York, pp 171–226

Cox CD, Rinehart KL, Moore ML, Cook JC (1981) Pyochelin: novel structure of an iron chelating growth promoter for *P. aeruginosa*. Proc Natl Acad Sci U S A 78:4256–4260

Crosa JH (1989) Genetics and molecular biology of siderophore mediated iron transport in bacteria. Microbiol Rev 53(4):517–530

de Bruijn I, de Kock MJD, de Waard P, van Beek TA, Raaijmakers JM (2008) Massetolide a biosynthesis in *Pseudomonas fluorescens*. J Bacteriol 190:2777–2789

De Werra P, Tarr MP, Keel C, Mauehofer M (2009) Role of gluconic acid production in the regulation of biocontrol traits of *P. fluorescens* CHAO. Appl Environ Microbiol 75:4162–4174

Duffy BK, Keel C, Defago G (2004) Potential role of pathogen signaling in multitrophic plant–microbe interactions involved in disease protection. Appl Environ Microbiol 70:1836–1842

Elander RP, Mabe JA, Hamill RH, Gorman M (1968) Metabolism of tryptophans by *Pseudomonas aureofaciens* VI production of pyrrolnitrin by selected *Pseudomonas* species. Appl Microbiol 16:753–758

Frapolli M, De'fago G, Moenne-Loccoz Y (2007) Multilocus sequence analysis of biocontrol fluorescent *Pseudomonas* spp. producing the antifungal compound 2,4-diacetylphloroglucinol. Environ Microbiol 9:1939–1955

Fravel DR (2005) Commercialization and implementation of biocontrol. Ann Rev Phytopathol 43:337–359

Garbeva P, van Overbeek LS, van Vuurde JWL, van Elsas JD (2001) Analysis of endophytic bacterial communities of potato by plating and denaturing gradient gel electrophoresis (DGGE) of 16S rRNA based PCR fragments. Microb Ecol 41:369–383

Gerard JR, Lloyd T, Barsby P, Haden M, Kelly T, Andersen RJ (1997) Massetolides A-H, antimycobacterial cyclic depsipeptides produced by two pseudomonads isolated from marine habitats. J Nat Prod 60:223–229

Givskov M, Östling J, Eberl L, Lindum P, Christensen AB, Christiansen G, Molin S, Kjelleberg S (1998) Two separate regulatory systems participate in control of swarming motility of *Serratia liquefaciens* MG1. J Bacteriol 180:742–745

Glick BR, Patten CL, Holguin G, Penrose DM (1999) Biochemical and genetic mechanisms used by plant growth-promoting bacteria. Imperial College Press, London, pp 1008–1015

Hammer PE, Hill DS, Lam ST, Van Pe'e KH, Ligon JM (1997) Four genes from *Pseudomonas fluorescens* that encode the biosynthesis of pyrrolnitrin. Appl Environ Microbiol 63:2147–2154

Hammer PE, Burd W, Hill DS, Ligon JM, van Pe'e K (1999) Conservation of the pyrrolnitrin biosynthetic gene cluster among six pyrrolnitrin-producing strains. FEMS Microbiol Lett 180:39–44

Hassan MN, Afghan HS, Hafeez FY (2011) Biological control of red rot in sugarcane by native pyoluteorin-producing *Pseudomonas putida* strain NH-50 under field conditions and its potential modes of action. Pest Manag Sci 67:1147–1154. doi:10.1002/ps.2165

Henriksen A, Anthoni U, Nielsen TH, Sorensen J, Christophersen C, Gajhede M (2000) Cyclic lipoundecapeptide Tensin from *P. fluorescens* strain 96. Acta Crystallogr C 56:113–115

Heydari A, Pessarakli M (2010) A review on biological control of fungal plant pathogens using microbial antagonists. J Biol Sci 10:273–290

Hirano SS, Upper CD (2000) Bacteria in the leaf ecosystem with emphasis on *Pseudomonas syringae*-a pathogen, ice nucleus, and epiphyte. Microbiol Mol Biol Rev 64:624–653

Hofte M, Altier N (2010) Fluorescent pseudomonads as biocontrol agents for sustainable agricultural systems. Res J Microbiol 161:464–471

Husen E, Wahyudi AT, Suwanto A, Giyanto (2011) Growth enhancement and disease reduction of soybean by1-Aminocyclopropane-1-Carboxylate deaminase-producing *Pseudomonas*. Am J Appl Sci 8:1073–1080

Jha BK, Pragash MG, Cletus J, Raman G, Sakhtivel N (2009) Simultaneous phosphate solubilization potential and antifungal activity of new fluorescent pseudomonad strains, *Pseudomonas aeruginosa*, *P. plecoglossicida* and *P. mosselii*. World J Microbiol Biotechnol 25:573–581

Kaur R, Macleod J, Foley W, Nayudu M (2006) Gluconic acid: an antifungal agent produced by *Pseudomonas* species in biological control of take all. Phytochemistry 67:595–604

Kavino M, Harish S, Kumar N, Saravanakumar, Samiyappan R (2008) Induction of systemic resistance in banana (*Musa* spp.) against banana bunchy top virus (BBTV) by combining chitin with root-colonizing *Pseudomonas fluorescens* strain CHA0. Eur J Plant Pathol 120:353–362

Keel C, Weller DM, Natsch A, De'fago G, Cook RJ, Thomashow LS (1996) Conservation of the 2,4-diacetylphloroglucinol biosynthesis locus among fluorescent *Pseudomonas* strains from diverse geographic locations. Appl Environ Microbiol 62:552–563

Khan MS, Zaidi A, Ahemad M, Oves M, Wani PA (2010) Plant growth promotion by phosphate solubilizing fungi – current perspective. Arch Agron Soil Sci 56:73–98

Khoury WE, Makkouk K (2010) Integrated plant disease management in developing countries. J Plant Pathol 92:35–42

Kirner S, Hammer PE, Hill DS, Altmann A, Fischer I, Weislo LJ, Lanahan M, van Pee KH, Ligon JM (1998) Functions encoded by pyrrolnitrin biosynthetic genes from *Pseudomonas fluorescens*. J Bacteriol 180:1939–1943

Kishore GK, Pande S, Podile AR (2006) *Pseudomonas aeruginosa* GSE 18 inhibits the cell wall degrading enzymes of *Aspergillus niger* and activates defense-related enzymes of groundnut in control of collar rot disease. Aust Plant Pathol 35:259–263

Knowles CJ (1976) Microorganisms and cyanide. Bacteriol Rev 40:652–680

Koch B, Nielsen TH, Sorensen D, Andersen JB, Christophersen C, Molin S, Givskov M, Sorensen J, Nybroe O (2002) Lipopeptide production in *Pseudomonas* sp. strain DSS73 is regulated by components of sugar beet exudates via the Gac two component regulatory system. Appl Environ Microbiol 68:4509–4516

Kumar H, Bajpai VK, Dubey RC, Maheshwari DK, Kang SC (2010) Wilt disease management and enhancement of growth and yield of *Cajanus cajan* (L) var. Manak by bacterial combinations amended with chemical fertilizer. Crop Prot 29:591–598

Lanteigne C, Gadkar VJ, Wallon T, Novinscak A, Filion M (2012) Production of DAPG and HCN by *Pseudomonas* sp. LBUM300 contributes to the biological control of bacterial canker of tomato. Phytopathology 102(10):967–973

Laville J, Blumer C, Von Schroetter C, Gaia V, Defago G, Keel C, Haas D (1998) Characterization of the hcnABC gene cluster encoding hydrogen cyanide synthase and anaerobic regulation by ANR in the strictly aerobic biocontrol agent *P. fluorescens* CHA0. J Bacteriol 180:3187–3196

Le CN, Kruijit M, Raaijmakers JM (2011) Involvement of phenazines and lipopeptides in interactions between *Pseudomonas* species and *Sclerotium rolfsii*, causal agent of stem rot disease on groundnut. J Appl Microbiol 112:390–403

Leeman M, den Ouden FM, van Pelt JA, Dirkx FPM, Steijl H, Bakker PAHM, Schippers B (1996) Iron availability affects induction of systemic resistance to *Fusarium* wilt of radish by *P. fluorescens*. Phytopathology 86:149–155

Lewis TA, Cortese MS, Sebat JL, Green TL, Lee CH, Crawford RL (2000) A *Pseudomonas stutzeri* gene cluster encoding biosynthesis of the CCl4-dechlorination agent pyridine-2,6-bis (thiocarboxylic acid). Environ Microbiol 2:407–416

Lindum P, Anthoni U, Christophersen C, Eberl L, Molin S, Givskov M (1998) *N*Acyl- L-homoserine lactone autoinducers control production of an extracellular lipopeptide biosurfactant required for swarming motility of *Serratia liquefaciens* MG1. J Bacteriol 180:6384–6388

Loper JE (1988) Role of fluorescent siderophore production in biological control of *Pythium ultimum* by a *Pseudomonas fluorescent* strain. Phytopathology 78:166–172

Loper JE, Kobayashi DY, Paulsen IT (2007) The genomic sequence of *Pseudomonas fluorescens* Pf-5: insights into biological control. Phytopathology 97:233–238

Loper JE, Hassan KA, Mavrodi DV, Davis EW II, Lim CK et al (2012) Comparative genomics of plant-associated *Pseudomonas* spp.: insights into diversity and inheritance of traits involved in multitrophic interactions. PLoS Genet 8(7):e1002784

Lugtenberg B, Kamilova F (2009) Plant-growth-promoting rhizobacteria. Ann Rev Microbiol 63:541–556

Marahiel MA, Stacelhaus T, Mootz HD (1997) Modular peptide synthetases involved in nonribosomal peptide synthesis. Chem Rev 97:2651–2673

Maurhofer M, Keel C, Haas D, Defago G (1994) Pyoluteorin production by *Pseudomonas fluorescent* strain CHAO is involved in the suppression of *Pythium* damping –off of cress but not cucumber. Eur J Plant Pathol 100:221–232

Mehnaz S, Weselowski B, Aftab F, Zahid S, Lazarovits G, Iqbal J (2009) Isolation, characterization, and effect of fluorescent *pseudomonads* on micropropagated sugarcane. Can J Microbiol 55:1007–1011

Mercado- Blanco J, Van der Drift LMGM, Olsson P, Thomas Oates JE, Van Loon LC, Bakker PAHM (2001) Analysis of the pmsCEAB gene cluster involved in biosynthesis of salicyclic acid and the siderophore pseudomonine in the biocontrol strain *Pseudomonas fluorescens* WCA374. J Bacteriol 183:1909–1920

Meynet CE, Pothier JF, Moënne-Loccoz Y, Prigent-Combaret C (2011) The *Pseudomonas* secondary metabolite 2,4-diacetylphloroglucinol is a signal inducing rhizoplane expression of *Azospirillum* genes involved in plant-growth promotion. Mol Plant Microb Interact 24(2):271–284

Mishra PK, Bisht SC, Ruwari P, Joshi GK, Singh G, Bisht JK, Bhatt JC (2011) Bioassociative effect of cold tolerant *Pseudomonas* spp. and *Rhizobium leguminosarum-PR1* on iron acquisition, nutrient uptake and growth of lentil (*Lens culinaris* L.). Eur J Soil Biol 47:35–43

Nakkeeran S, Fernado WGD, Siddiqui ZA (2005) Plant growth promoting rhizobacteria formulations and its scope in commercialization for the management of pests and diseases. In: Siddiqui ZA (ed) PGPR: biocontrol and biofertilization. Springer, Dordrecht, pp 257–296

Naz I, Bano A (2010) Biochemical, molecular characterization and growth promoting effects of phosphate solubilizing *Pseudomonas* sp. isolated from weeds grown in salt range of Pakistan. Plant Soil 334:199–207

Negi YK, Prabha D, Garg SK, Kumar J (2011) Genetic diversity among cold-tolerant fluorescent pseudomonas isolates from Indian Himalayas and their characterization for biocontrol and plant growth-promoting activities. J Plant Growth Regul 30:128–143

Nielsen TH, Christophersen C, Anthoni U, Sorensen J (1999) Viscosinamide, a new cyclic depsipeptide with surfactant and antifungal properties produced by *Pseudomonas fluorescens* DR54. J Appl Microbiol 86:80–90

Nielsen TH, Thrane C, Christophersen C, Anthoni U, Sorensen J (2000) Structure, production characteristics and fungal antagonism of tensin – a new antifungal cyclic lipopeptide from *P. fluorescens* strain 96.578. J Appl Microbiol 89:992–1001

Nielsen TH, Sorensen D, Tobiasen C, Andersen JB, Christophersen C, Givskov M et al (2002) Antibiotic and biosurfactant properties of cyclic lipopeptides produced by fluorescent *Pseudomonas* spp. from the sugar beet rhizosphere. Appl Environ Microbiol 68:3416–3423

Patel D, Jha CK, Tank N, Saraf M (2011) Growth enhancement of chickpea in saline soils using plant growth promoting rhizobacteria. J Plant Growth Regul 31:53–62. doi:10.1007/s00344-011-9219-7

Pathama J, Kennedy RK, Sakthivel N (2011) Mechanisms of fluorescent pseudomonads that mediate biological control of phytopathogens and plant growth promotion of crop plants. In: Maheshwari DK (ed) Bacteria in agrobiology: plant growth responses. Springer, Berlin, pp 77–105

Patten CL, Glick BR (2002) Role of *Pseudomonas putida* indole acetic acid in development of the host plant root system. Appl Environ Microbiol 68:3795–3801

Peix A, Ramirez-Bahena MH, Velazquez E (2009) Historical evolution and current status of the taxonomy of genus *Pseudomonas*. Infect Genet Evol 9(6):1132–1147. doi:10.1016/j.meegid.2009.08.001

Quan Z, Su J, Jiang H, Huang X, Xu Y (2010) Optimization of phenazine-1-carboxylic acid production by a gacA/qscR-inactivated Pseudomonas sp. M18GQ harboring pME6032Phz using response surface methodology. Appl Microbiol Biotechnol 86(6):1761–1773

Rajkumar M, Freitas H (2008) Influence of metal resistant plant growth promoting bacteria on the growth of *Ricinus communis* in soil contaminated with heavy metals. Chemosphere 71:834–842

Rajkumar M, Lee KJ, Freitas H (2008) Effects of chitin and salicyclic acid on biological control of *Pseudomonas* spp. against damping off of pepper. S Afr J Bot 74:268–273

Ramette A, Frapolli M, De'fago G, Moenne-Loccoz Y (2003) Phylogeny of HCN synthase-encoding hcnBC genes in biocontrol fluorescent pseudomonads and its relationship with host plant species and HCN synthesis ability. Mol Plant Microbe Interact 16:525–535

Ravindra Naik P, Raman G, Badri Narayanan K, Sakthivel N (2008) Assessment of genetic and functional diversity of phosphate solubilizing fluorescent pseudomonads isolated from rhizospheric soil. BMC Microbiol 8:230

Reiter B, Pfeifer U, Schwab H, Sessitsch A (2002) Response of endophytic bacterial communities in potato plants to infection with *Erwinia carotovora* subsp. Atroseptica. Appl Environ Microbiol 68:2261–2268

Renodo-Nieto M, Barret M, Morrissey J, Germaine K, Martinez-Granero F, Barahona E, Navazo A, Sanchez-Contreras M, Moynihan JA, Muriel C, Dowling D, O'Gara F (2013) Genome sequence reveals that *Pseudomonas fluorescens* F113 possesses a large and diverse array of systems for rhizosphere function and host interaction. BMC Genomics 14:54

Rokhzadi A, Asgharzadeh A, Darvish F, Nour-Mohammadi, Majidi E (2008) Influence of plant growth- promoting rhizobacteria on dry matter accumulation and yield of chickpea (*Cicer arietinum* L.) under field conditions. Am-Euras J Agric Environ Sci 3(2):253–257

Rosenberg E, Ron EZ (1999) High- and low-molecular-mass microbial surfactants. Appl Microbiol Biotechnol 52:154–162

Saraf M, Thakker A, Patel BV (2008) Biocontrol activity of different species of *pseudomonas* against phytopathogenic fungi in vivo and in vitro conditions. Int J Biotechnol Biochem 4:217–226

Saraf M, Jha CK, Patel D (2010) The role of ACC deaminase producing PGPR in sustainable agriculture. In: Maheshwari DK (ed) Plant growth and health promoting bacteria, Microbiology monographs 18. Springer, Berlin, pp 365–385

Saraf M, Pandya U, Thakkar A (2014) Role of allelochemicals in plant growth promoting rhizobacteria for biocontrol of phytopathogens. Microbiol Res 169:18–29

Sessitsch A, Reiter B, Pfeifer U, Wilhelm E (2002) Cultivation-independent population analysis of bacterial endophytes in three potato varieties based on eubacterial and *Actinomycetes*-specific PCR of 16S rRNA genes. FEMS Microbiol Ecol 39:23–32

Setubal JC, dos Santos P, Goldman BS, Ertesvag H, Espin G, Rubio LM, Valla S, Alemeida NF, Balasubramanian D, Cromes L, Curatti L, Du Z, Godsy E, Goodner B, Hellner-Burris K, Hernandez JA, Houmiel K, Imperial J, Kennedy C, Larson TJ, Ligon LS, Lu J, Maerk M, Miller NM, Norton S, O'Carroll IP, Paulsen I, Raulfs EC, Roemer R, Rosser J, Segura D, Slater S, Stricklin SL, Studholme DJ, Sun J, Viana CJ, Wallin E, Wang B, Wheeler C, Zhu J, Dean DR, Dixon R, Wood D (2009) Genome sequence of *Azotobacter vinelandii*, an obligate aerobe specialized to support diverse anaerobic metabolic processes. J Bacteriol 191:4534–4545

Shaharoona B, Naveed M, Arshad M, Zahir ZA (2008) Fertilizer-dependent efficiency of *Pseudomonads* for improving growth, yield, and nutrient use efficiency of wheat (*Triticum aestivum* L.). Appl Microbiol Biotechnol 79:147–155

Sharma A, Victor W, Johri B (2007) Rhizosphere *Pseudomonas* sp. strains reduce occurrence of pre- and post-emergence damping-off in chile and tomato in Central Himalayan region. Arch Microbiol 187(4):321–335

Shen L, Wang F, Yang J, Qian Y, Dong X, Zhan H (2014) Control of tobacco mosaic virus by *Pseudomonas fluorescens* CZ powder in greenhouses and the field. Crop Prot 56:87–90

Siddiqui IA, Shaukat SS, Khan GH, Ali NA (2003) Suppression of *Meloidogyne javanica* by *Pseudomonas aeruginosa* IE in tomato: the influence of NaCl, oxygen and iron levels. Soil Biol Biochem 35:1625–1634

Silby MW, Cerdeno-Tarraga AM, Vernikos GS, Giddens SR, Jackson RW, Preston GM, Zhang XX, Moon CD, Gehrig SM, Godfrey SAC, Knight CG, Malone JG, Robinson Z, Spiers AJ, Harris D, Seeger K, Murphy L, Rutter S, Squares R, Quail MA, Saunders E,

Mavromatis K, Brettin TS, Bentley SD, Hothersall J, Stephens E, Thomas CM, Parkhill J, Levy SB, Rainey PB, Thomson NR (2009) Genomic and genetic analyses of diversity and plant interactions of *Pseudomonas fluorescens*. Genome Biol 10(5):R51. doi:10.1186/gb-2009-10-5-r51

Sindhu SS, Dadarwal KR (2001) Chitinolytic and cellulolytic *Pseudomonas* sp. antagonistic to fungal pathogens enhances nodulation by *Mesorhizobium* sp. Cicer in chickpea. Microbiol Res 156:353–358

Sorensen D, Nielsen TH, Christophersen C, Sorensen J, Gajhede M (2001) Cyclic lipoundecapeptide amphisin from *Pseudomonas* sp. strain DSS73. Acta Crystallogr C 57:1123–1124

Takesako K, Kuroda H, Inoue T, Haruna F, Yoshikawa Y, Kato I, Uchida K, Hiratani T, Yamaguchi H (1993) Biological properties of aureobasidin A, a cyclic depsipeptide antifungal antibiotic. J Antibiot 46:1414–1420

Tank N, Saraf M (2009) Enhancement of plant growth and decontamination of nickel spiked soil using PGPR. J Basic Microbiol 49:195–204

Tank N, Saraf M (2010) Salinity-resistant plant growth promoting rhizobacteria ameliorates sodium chloride stress on tomato plants. J Plant Interact 5:51–58

Upadhyay A, Srivasatava S (2010) Evaluation of multiple plant growth promoting traits of an isolate of *Pseudomonas fluorescens* strain Psd. Indian J Exp Biol 48:601–609

van den Broek D, Bloemberg GV, Lugtenberg B (2005) The role of phenotypic variation in rhizosphere *Pseudomonas* bacteria. Environ Microbiol 7:1686–1697

Van Wees SCM, Pieterse CMJ, Trijssenaar A, Van T, Westende YAM, Hartog F, Van Loon LC (1977) Differential induction of systemic resistance in *Arabidopsis* by biocontrol bacteria. Mol Plant-Microbe Interact 10:716–724

Viswanathan R, Samiyappan R (2000) Antifungal activity of chitinases produces by some fluorescent pseudomonas against *Colletotrichum falcatum* causing red rot disease in sugarcane. Microbiol Res 155:1–6

Voisard C, Keel C, Haas D, De'fago G (1989) Cyanide production by *P. fluorescens* helps suppress black root rot of tobacco under gnotobiotic conditions. EMBO J 8(2):351–358

Vollenbroich D, Ozel M, Vater J, Kamp RM, Pauli G (1997) Mechanism of inactivation of enveloped viruses by the biosurfactant surfactin from *B. subtilis*. Biologicals 25(3):289–297

Wahyudi AT, Astiti R, Giyanto (2011) Screening of *Pseudomonas* sp. isolated from rhizosphere of soybean plant as plant growth promoter and biocontrol agent. Am J Agric Biol Sci 6:134–141

Wiyono S, Schulz DF, Wolf GA (2008) Improvement of the formulation and antagonistic activity of *Pseudomonas fluorescens* B5 through selective additives in the pelleting process. Biol Control 46:348–357

Zabihi HR, Savaghebi GR, Khavazi K, Ganjali A, Miransari M (2011) *Pseudomonas* bacteria and phosphorous fertilization, affecting wheat (*Triticum aestivum* L.) yield and P uptake under greenhouse and field conditions. Acta Physiol Plant 33:145–152

Pseudomonads: Plant Growth Promotion and Beyond

10

Govindan Selvakumar, Periyasamy Panneerselvam, Gurupadam Hema Bindu, and Arakalagud Nanjundaiah Ganeshamurthy

Abstract

Members of the genus *Pseudomonas* are a ubiquitous and important component of the soil and rhizospheric ecosystems, where they play multifarious roles such as the recycling of organic matter, promotion of plant growth, alleviation of abiotic stress effects in plants, and degradation of xenobiotic compounds. The versatility and ecological fitness of this genus has been often attributed to its metabolic versatility and its ability to produce antagonistic molecules, thereby gaining a niche advantage in the rhizosphere. This chapter attempts to briefly explore the historical evolution of this genus and explore the various facets of Pseudomonads, which confer upon them a distinct edge over other soil microbes. The utility of Pseudomonads both within and beyond the realm of plant growth promotion, in the agro and related ecosystems, is also discussed in brief.

Introduction

The genus *Pseudomonas* (derived from the Greek words *pseudes* "false" and *monas* "a single unit" or "false unit") comprises one of the most diverse and ecologically fit groups of bacteria on this planet, whose members are collectively referred to by the generic term Pseudomonads. Though at times the term "Pseudomonad" is also used to refer to former members of the genus *Pseudomonas* that have been reclassified into other genera at present, for greater clarity sake we shall be limiting our usage of this term to denote only the existing members of the genus *Pseudomonas*. Taxonomically the genus *Pseudomonas* falls within the γ subclass of Proteobacteria, and its ubiquity can be gauged by its ability to colonize terrestrial, freshwater, and marine environments with relative ease, besides forming intimate associations with higher forms of life. But its marked preference for plants in forming associations is well known. Pseudomonads are able to utilize an array of organic compounds as energy sources (Lessie and Phibbs 1984) and produce diverse secondary metabolites (Raaijmakers et al. 2002; Gross and Loper 2009). Universal distribution and remarkable degrees of physiological adaptations make them one of the most sought after and studied bacterial species.

G. Selvakumar (✉) • P. Panneerselvam
G.H. Bindu • A.N. Ganeshamurthy
Indian Institute of Horticultural Research, Hessaraghatta Lake Post, Bangalore 560089, India
e-mail: gselvakumar@iihr.ernet.in

N.K. Arora (ed.), *Plant Microbes Symbiosis: Applied Facets*, DOI 10.1007/978-81-322-2068-8_10, © Springer India 2015

In the agro-ecosystem, Pseudomonads play roles as diverse as the decomposition of organic matter, promotion of plant growth, protection of plants from diseases, control of insect pests, and nutrient solubilization. Other associated roles that have been attributed to Pseudomonads include their ability to metabolize a variety of xenobiotic compounds (Pak et al. 2000) and the alleviation of abiotic stress effects on plants (Mishra et al. 2011). The genus *Pseudomonas* comprises of more than 100 named species that are divided into lineages, groups, and subgroups based on multilocus sequence analysis (Yamamoto et al. 2000; Guttman et al. 2008; Mulet et al. 2010). The *Pseudomonas fluorescens* group alone encompasses more than 50 named species of plant-associated commensals and has therefore emerged as an agriculturally important group among the Pseudomonads. Plant-associated roles of the genus *Pseudomonas* vary from being the causative agents of several diseases to abiotic stress alleviation in plants. Therefore, considering their ubiquity and metabolic and functional versatility, it can be rightly said that Pseudomonads are the bacterial species of prime importance in the agro-ecosystem. This chapter attempts to trace the reasons for Pseudomonad diversity and their utility in the agro-ecosystem, within and beyond the realm of plant growth promotion.

Pseudomonads: A Historical Perspective

The genus *Pseudomonas* was first described by Migula (1894) and has been a part of the history of bacteriology ever since its inception. Though initially *Pseudomonas* was described as a genus of Gram-negative, rod-shaped, and polar flagellated bacteria with some sporulating species, it was later proved that the so-called spores were actually refractive granules of reserve materials. The physiological diversity of this bacterial genus was unraveled by the pioneering studies of bacteriologists such as Beijerinck (1921) and Winogradsky (1949) during the first half of the twentieth century. The 1923 edition of the Bergey's Manual (Bergey et al. 1923) included a chapter on the genus *Pseudomonas*, where the classification of species was based on phenotypic characteristics. But the initial classification of *Pseudomonas* as Gram-negative, aerobic, non-sporulated rods that are motile by means of polar flagella did not differentiate it sufficiently from other Gram-negative bacteria and led to the dumping of several improperly characterized Gram-negative bacteria. But latter-day observations that incorporated other phenotypic characteristics were able to bring about a degree of clarity on the taxonomical boundaries of this genus. A monumental work in the 1960s, to clarify the taxonomy of the genus, was carried out by Stanier et al. (1966). This exhaustive and highly cited work which reported the nutritional characteristics of 267 strains on 146 different organic compounds plus a wide range of associated characteristics is considered as a classic study on bacterial taxonomy (Spiers et al. 2000). The development of DNA-DNA hybridization methodologies, during the same decade, led to the confirmation and reclassification of several phenotypic data, which ultimately led to the inclusion of the G+C content of bacteria in the Bergey's Manual edition published in 1974, which included 29 species (Doudoroff and Palleroni 1974). During this period another development in the classification of genus *Pseudomonas* by genotypic criteria was made by Palleroni and coworkers, who classified this bacterial group into five rRNA subgroups based on the measurements of RNA-DNA relatedness (Palleroni et al. 1973). These rRNA subdivisions were phylogenetically very distant, and only the representatives included in the rRNA group I were finally included in the genus *Pseudomonas*, while the members of the other subgroups were reclassified as separate genera. Though this approach failed to find a place in the 1974 edition of the Bergey's Manual, it was accommodated in the 1984 edition of the Bergey's Manual, with minor modifications in the number of species (Palleroni 1984).

The 1980s saw the emergence of bacterial taxonomy based on the sequencing of the 16S rRNA gene (Woese et al. 1984). But this scheme of classification was not reflected in the Bergey's Manual of Determinative Bacteriology published in 1994, where species determination was carried solely on the basis of phenotypic characters. This edition included only the classical species of *Pseudomonas*, while the newly described species were included as an additional list. The current edition of Bergey's Manual of Systematic Bacteriology contains an exhaustive review of the criteria used in the modern taxonomy of genus *Pseudomonas* and includes 59 species that were described till the year 2005 (Palleroni 2005). Since then several new species of genus *Pseudomonas* have been described, and their number is continuously increasing. Other novel approaches used in the *Pseudomonas* taxonomy include siderotyping of both fluorescent and nonfluorescent Pseudomonads (Meyer et al. 2002) and fluorescent spectroscopy fingerprinting, by which Tourkya and collaborators were able to discriminate *Pseudomonas* at the genus level from *Burkholderia*, *Xanthomonas*, and *Stenotrophomonas*, with very high sensitivity. At the species level they were able to group *P. chlororaphis*, *P. lundensis*, *P. fragi*, *P. taetrolens*, and *P. stutzeri* separately from *P. putida*, *P. pseudoalcaligenes*, and *P. fluorescens* (Tourkya et al. 2009). It is obvious that ever since its discovery, the genus *Pseudomonas* has comprised a large number of species that have increased in magnitude over time often reaching alarming proportions (Palleroni 2008). But constant developments in bacterial taxonomy have led to the realignments at the generic level, and the genus *Pseudomonas* still retains its identity as a versatile and ubiquitous genus. A list of novel species of Pseudomonads from agro-environments that have been validly published subsequent to the year 2005 is listed in Table 10.1.

Table 10.1 Validly published *Pseudomonas* species originating from the agro-environment since the year 2005

Pseudomonas species	Origin	Reference
P. chengduensis	Landfill leachate	Tao et al. (2014)
P. kunmingensis	Phosphate mine	Xie et al. (2014)
P. punonensis	Straw	Ramos et al. (2013)
P. formosensis	Food-waste compost	Lin et al. (2013)
P. guariconensis	Rhizospheric soil	Toro et al. (2013)
P. sagittaria	Oil-contaminated soil	Lin et al. (2013)
P. zeshuii	Herbicide-contaminated soil	Feng et al. (2012)
P. baetica	Wedge sole	López et al. (2012)
P. bauzanensis	Soil	Zhang et al. (2011)
P. taiwanensis	Soil	Wang et al. (2010)
P. cedrina	Phyllosphere of grasses	Behrendt et al. (2009)
P. brassicacearum subsp. *Neoaurantiaca*	Soil and the rhizosphere of agricultural plants	Ivanova et al. (2009)
P. panipatensis	Oil-contaminated soil	Gupta et al. (2008)
P. lurida	Phyllosphere of grasses	Behrendt et al. (2007)
P. reinekei	Soil	Cámara et al. (2007)
P. moorei	Soil	Cámara et al. (2007)
P. mohnii	Soil	Cámara et al. (2007)
P. moraviensis	Soil	Tvrzová et al. (2006)
P. segetis	Soil	Tvrzová et al. (2006)
P. panacis	Soil	Park et al. (2006)

Diversity of Pseudomonads

Diversity in living organisms occurs as a result of interplay between ecological and genetic factors. The ultimate cause of diversity is mutation, but the variations generated by mutations and recombination are subject to selection pressures and genetic drift (Bell 1997; Rainey et al. 2000). Pseudomonads are one of the most diverse groups of bacteria on this planet and comprise an array of environmental and clinical species. This genus includes the well-known plant-pathogenic bacterium *Pseudomonas syringae* and the opportunistic human pathogen *Pseudomonas aeruginosa*.

Apart from these, several environmental species that are either beneficial or neutral in their interaction with plants and several opportunistic human/veterinary pathogens are also found. Pseudomonads occur in environments as diverse as distilled water (Favero et al. 1971), ocean water (Khan et al. 2007), dairy products (Dogan and Boor 2003), meat (Ayres 1960), internal tissues of plants (Zinniel et al. 2002), animal tissues (Ramphal et al. 1980), domestic and hospital areas (Whitby and Rampling 1972). Environmental isolates of *Pseudomonas* have been isolated from hot deserts (Bhatnagar and Bhatnagar 2005), cold alpine environments (Meyer et al. 2004), Antarctic soils (Panicker et al. 2002), oil spill sites (Das and Mukherjee 2007), heavy metal-contaminated sites (Roane and Kellogg 1996), acid soils (De Boer and Kowalchuk 2001), and mine spoils (Piotrowska-Seget et al. 2005). This diversity of Pseudomonads is not entirely limited to its phenotype alone, but is also reflected at the genetic level wherein the architecture of the chromosome and extrachromosomal elements are known to contribute to the ecological fitness and diversity of this genus (Spiers et al. 2000). The phenotypic and chemotaxonomic heterogeneity of the genus is reflected in its fairly broad DNA guanine plus cytosine (G+C) content ranging from 58 to 70 mol % (Palleroni 1984). Thus the diversity of Pseudomonads can be rightly ascribed to its metabolic and genetic versatility, both of which confer a level of ecological fitness to this genus that is rarely seen among bacteria. We shall briefly discuss the ecological and genetic basis of diversity among Pseudomonads in the following section.

Ecological Causes of Pseudomonad Diversity

Charles Darwin proposed that "ecological opportunity" is an essential factor and the primary cause of diversity within an organism. This statement is all the more relevant for bacteria which have the remarkable capacity to exploit the smallest of opportunities. The classical observations of Gause, that two distinct test tube populations of *Paramecium* competing for the same niche could not coexist and therefore led to the extinction of one population (Gause 1934), were followed by the niche exclusion principle, which states that no two organisms with similar requirements and characteristics can occupy the same niche. This principle postulates that if an environment lacks spatial or temporal heterogeneity (i.e., is homogeneous) and affords just a single niche, therefore in principle, it will house just a single species (Hardin 1960). Therefore, it is obvious that in a homogenous environment, as a population evolves, the more successful species simply replaces the existing species. This theory found support in the laboratory studies of Rainey and Travisano (1998) who observed that isogenic populations of *P. fluorescens* when grown in identical environments except for the degree of ecological opportunity (spatial structure that was provided by shaking), diverse niche-specialist genotypes, emerged rapidly in the spatially heterogeneous environment but not in the spatially homogeneous environment. Pseudomonad diversity in nature can therefore partly be attributed to the ability of this genus to utilize the ecological opportunity.

Competition is commonly attributed to be the driving force, or engine, of diversification (Abrams 1987). Let us assume an environment where a single bacterial species is grown; as the density of population increases, growth becomes limited due to depletion of resources such as oxygen and nutrients. Therefore, the competition becomes intense, and natural selection would favor a mutant that can gain access to the limiting resources. Hence the most competent genotypes are those that occupy newer niches unlike the one occupied by the dominant genotype, since scarce resources will be most abundant in the newer niches. This selection would continue to favor mutants that show increasing degrees of niche specialization so that eventually there will be minimal niche overlap among genotypes. If one were to apply this principle of diversity to Pseudomonads, it could be concluded that the metabolic diversity among Pseudomonads could have arisen due to the intense competition within and among populations. Other possible ecological causes of Pseudomonad diversity can be attributed to the increased productivity and lowered mortality rates in populations and increased resistance to parasitic agents in heterogeneous environments (Spiers et al. 2000).

Genetic Basis of Pseudomonad Diversity

Considering the ecological, metabolic, and biochemical diversity of this genus, it is quite obvious that diversity extends to the genome sequence level. The complete genomes of many species have now been sequenced, and it has been observed that only 25–35 % of the genomes of the sequenced isolates are composed of core genes that are shared by all members of the genus (Silby et al. 2011). The genomes of *Pseudomonas* spp. are composed of relatively stable core regions that are interspersed with regions that vary among the strains (Silby et al. 2009, 2011). Some regions that are unique to a specific strain are predicted to confer the distinctive characteristics. The combined repertoire of the core and variable regions of a genome reflects the ecological history of the strain and the various environments or selective pressures that it has encountered over evolutionary time (Loper et al. 2012).

The genomes of *Pseudomonas* are usually larger in size with an extensive gene complement. While in other bacterial species, genomes are smaller in pathogenic isolates compared to the environmental isolates, in *Pseudomonas* there seems to be no correlation between the genome size and the lifestyles (pathogenic or saprophytic). Plant-pathogenic strains of *P. syringae*, opportunistic human pathogens of *P. aeruginosa*, and the insect pathogen *P. entomophila* have genome sizes ranging from 5.9 to 6.9 Mb (Loper et al. 2012). This observation finds support in the fact that plant-pathogenic strains of *P. syringae* can exist epiphytically on plant surfaces (O'brien and Lindow 1989), while *P. aeruginosa* is a common environmental habitant (Das and Mukherjee 2007), which indicates that the pathogenic lifestyle is just another mode of existence, rather than the sole mode of survival. The large genome seems to be an adaptation to handle the range of environments that these bacteria encounter.

Mutations are thought to be the foremost reason for the evolution of novel genotypes among bacteria, and they are thought to occur with rapid frequency under stress conditions because of the SOS response and the reduced ability to deal with DNA-damaging free radicals generated by metabolism (Friedberg et al. 1995). But it is to be deciphered as to whether the genome of *Pseudomonas* is more prone to mutation. This argument finds support with the isolation of mutator strains (with elevated mutation rates) of *P. aeruginosa* containing defective mismatch repair genes (Oliver et al. 2000). Since *Pseudomonas* strains occupy diverse environments, it's quite obvious that they are more prone to stress conditions and hence may be more prone to diversity-generating mutations.

The acquisition of genes and accessory genetic elements (plasmids, transposons, integrons, and phages) by lateral gene transfer is a pervasive and important source of innovation for bacterial evolution (Ochman et al. 2000). This is especially intriguing since with the exception of *P. stutzeri*, others are not known to be naturally competent (Spiers et al. 2000). Population genetic studies in *Pseudomonas* have shown that recombination occurs, but this seems to be within an overall pervasive clonal framework as evidenced by the repeated sampling of identical clones from different environments often years apart (Haubold and Rainey 1996; Kiewitz and Tummler 2000). But accessory genetic elements play an important role in the lateral movement and recombination of external DNA. Large plasmids encoding the entire catabolic pathways and associated regulatory genes are common in *Pseudomonas* (Sesma et al. 2000; Thomas 2000). Such plasmids often have host ranges that extend beyond the genus. An interesting observation by Selvakumar et al. (2009a), that the mountain aspect affects the genetic clustering of cold-tolerant Pseudomonads from high-altitude regions, has thrown another dimension to the ecological-genetic interaction determinants of diversity. They observed from the rep-PCR profile that isolates originating from the cooler northern mountain aspects and warmer southern mountain aspects clustered separately and had distinct phosphate solubilization abilities. To conclude, it can be said that large genomes increased the frequencies of mutation and genetic recombination both at the chromosomal and extrachromosomal levels and play a major role in Pseudomonad diversity.

Plant Growth Promotion and Rhizospheric Competence of Pseudomonads

Pseudomonads are the most studied bacteria with reference to their ability to promote plant growth by a myriad of activities, and therefore a non-exhaustive compilation of plant growth promotion by Pseudomonads is listed in Table 10.2. Since several extensive reviews on plant growth promotion by Pseudomonads are available, we shall not dwell at length on the mechanisms of plant growth mechanism, but rather focus on the rhizospheric competence traits that confer upon *Pseudomonas* a pre-eminent position among rhizobacteria.

To achieve effective root colonization, the following traits, viz., motility (de Weger et al. 1987); production of the O-antigen of lipopolysaccharide (LPS) (de Weger et al. 1989); utilization of cellulose (Matthysse and McMahan 1998), thiamine (Simons et al. 1996), amino acids (Simons et al. 1997), and biotin (Streit et al. 1996); and the synthesis of an isoflavonoid-inducible efflux pump (Palumbo et al. 1998), are thought to be

Table 10.2 Plant growth promotion by different *Pseudomonas* species

Pseudomonas species	PGP effect/trait	Reference
Pseudomonas sp.	Seed piece inoculation caused a reduction of soft rot disease caused by *Erwinia carotovora*	Kloepper (1983)
P. fluorescens	Prevention of citrus canker caused by *Xanthomonas campestris* pv. *citri* by siderophore production	Unnamalai and Gnanamanickam (1984)
P. putida W4P63	Suppression of soft rot of *Erwinia carotovora* in tuber siderophore production	Xu and Gross (1986)
P. cepacia B5	Suppression of *R. solanacearum* by 2-ketogluconic acid	Aoki et al. (1991)
P. glumae	Suppression of *R. solanacearum* by induction of systemic resistance	Furuya et al. (1991)
P. fluorescens PF59	Suppression of *R. solanacearum* by siderophore production	Hartmann et al. (1992)
P. fluorescens A506	Reduction in populations of *Erwinia amylovora* in pear flowers due to competition	Wilson and Lindow (1993)
Pseudomonas sp.	Suppression of the tomato wilt pathogen by siderophore production	Jagadeesh et al. (2001)
P. fluorescens	Resists rice bacterial blight caused by *Xanthomonas oryzae* pv *oryzae* by induced systemic resistance (ISR)	Vidhyasekharan et al. (2001)
P. fluorescens PRS9 and *P. fluorescens* GRS1	Production of IAA, siderophores, and phosphate solubilization	Gupta et al. (2005)
Pseudomonas sp.	Phosphate solubilization, production of IAA and siderophores	Wani et al. (2007)
P. fluorescens	Induced systemic resistance and antifungal activity	Saravanakumar et al. (2007)
Pseudomonas sp.	Production of IAA, siderophores, antifungal activity, ammonia, and HCN production	Ahmad et al. (2008)

(continued)

Table 10.2 (continued)

Pseudomonas species	PGP effect/trait	Reference
P. tolaasii ACC23, *P. fluorescens* ACC9	Protects canola plants against the inhibitory effects of cadmium	Dell'Amico et al. (2008)
Pseudomonas spp.	Increased ability to control bacterial and fungal root pathogens of rice	Lawongsa et al. (2008)
Pseudomonas sp.	Cold-tolerant plant growth-promoting bacterium	Mishra et al. (2008)
Pseudomonas sp.	Production of ACC deaminase, IAA, and siderophore	Poonguzhali et al. (2008)
P. jessenii	Production of ACC deaminase, IAA, siderophore, heavy metal solubilization, and phosphate solubilization	Rajkumar and Freitas (2008)
P. aeruginosa	Stimulation of plant growth, reduced Cd uptake in Indian mustard and pumpkin	Sinha and Mukherjee (2008)
P. aeruginosa, *P. fluorescens*	Production of siderophores	Braud et al. (2009)
P. fragi	Solubilizes rock phosphate at low temperatures and promotes wheat seedling growth	Selvakumar et al. (2009a, b)
Pseudomonas sp. NARs9	Cold-tolerant plant growth-promoting bacterium	Mishra et al. (2009)
Pseudomonas sp. A3R3	Production of IAA and siderophores	Ma et al. (2011)
P. lurida	Cold-tolerant phosphate-solubilizing bacterium that promotes wheat seedling growth	Selvakumar et al. (2011)
Pseudomonas sp.	Significant increase in soil enzyme activities, total productivity and nutrient uptake in soybean and wheat	Sharma et al. (2011)
Pseudomonas sp.	Significant increase in plant dry weight, nodule numbers, total chlorophyll content, leghemoglobin, root N, shoot N, root P, shoot P, seed yield, and seed protein in green gram	Ahemad and Khan (2012a)
P. putida	IAA, siderophores, HCN, ammonia exo-polysaccharides, phosphate solubilization	Ahemad and Khan (2012b)
P. fluorescens SS101 (*Pf. SS101*)	Protects *Arabidopsis* from the infection of *P. syringae* pv *tomato* through induction of systemic resistance	Judith et al. (2012)
Pseudomonas sp.	Increases iron uptake in pea when co-inoculated with *Rhizobium*	Mishra et al. (2012)
P. brassicacearum J12	Suppresses *R. solanacearum* by 2,4-diacetylphloroglucinol (2,4-DAPG)	Zhou et al. (2012)
Pseudomonas spp.	Solubilizes rock phosphate at low temperatures and promotes lentil growth	Selvakumar et al. (2013)

essential, and mutants deficient in one or more of these traits have been proven to be poor root colonizers (Lugtenberg et al. 2001). Though motility confers a distinctive edge to an organism, it has been disputed by some workers (Howie et al. 1987). But it has been clearly proved that nonmotile mutants are clearly deficient in root colonization (Simons et al. 1996). Since *Pseudomonas* is known to produce up to nine flagella per cell, it is easy to screen for motile strains in laboratory media. It is further seen that chemotaxis-derived motility rather than random motility plays an important role in the root tip colonization (Lugtenberg et al. 2001). Hence chemotaxis and motility can serve as suitable traits for the selection of efficient root colonization strains. The pili or fimbriae give the bacterium the necessary anchorage to adhere to surfaces or form biofilms (Hahn 1997). The type 4 pili which are so extensively described in pathogenic strains are also thought to play a role in root adherence and biofilm formation (Dörr et al. 1998).

LPS of *Pseudomonas* and of most other Gram-negative bacteria consists of lipid A, core, and O-antigen. The later can consist of many repeating units. LPS of a pure culture is heterogeneous with respect to the number of repeating units in its O-antigen (Goldman and Leive 1980). Mutants devoid or defective in the O-antigenic side chain appear to be impaired in potato root colonization (de Weger et al. 1989). The ability to utilize amino acids which is an integral part of the root exudates confers a distinctive nutritional edge on a bacterial strain (Simons et al. 1997). Since root exudates are comprised mainly of monosaccharides and organic acids (Phillips and Streit 1995), mutants deficient in the utilization of individual crop-specific root exudates are impaired in root colonization, and hence one has to choose sets of beneficial microbe-plant pairs based on individual exudate composition and high bacterial growth rate on the major exudate carbon sources (Lugtenberg et al. 2001). Other rhizospheric competence traits that have been reported for *Pseudomonas* include putrescine utilization (Kuiper et al. 2001), protein secretion (Preston et al. 2001), enhanced outer membrane permeability (Lugtenberg et al. 2001), NADH functionality (Dekkers et al. 1998), and the ability to rapidly colonize root surfaces (O'Toole and Kolter 1998).

A new dimension of Pseudomonad utility in the agro-ecosystem is their ability to alleviate abiotic stress effects in plants. This has gained momentum over the years, and today *Pseudomonas* strains are known to confer tolerance to a variety of abiotic stresses (Grover et al. 2011). Of specific interest in this area is the ability of cold-adapted *Pseudomonas* strains to alleviate the frost injury/cold stress effects on plants (Xu et al. 1998; Mishra et al. 2011). This gains significance since the cropping cycle in several parts of the world is subject to transient cold periods, which are deleterious to plant beneficial microbial processes. Another emerging area is the alleviation of drought stress effects in plants by the action of inoculated osmotolerant *Pseudomonas* strains. A compilation of abiotic stress alleviation effects on various plants by *Pseudomonas* strains is presented in Table 10.3.

Table 10.3 Pseudomonads that alleviate abiotic stress effects in plants

Pseudomonas species	Effects/traits	Reference
P. fluorescens strain MSP-393	Tolerant to high osmolarity by the synthesis of novel proteins that nullify detrimental effects of drought	Paul et al. (2006)
P. fluorescens	Protects green gram plants from salt stress under in vitro conditions	Saravanakumar and Samiyappan (2007)
P. chlororaphis O6	Induced systemic drought and salt tolerance in asparagus plants	Cho et al. (2008)
P. mendocina	Tolerant to drought stress	Kohler et al. (2008)
Pseudomonas spp.	Promotes asparagus growth under water stress	Liddycoat et al. (2009)
Pseudomonas sp. strain AKM-P6	Increased heat tolerance of sorghum seedlings	Ali et al. (2009)
P. putida strain GAP-P45	Increased drought tolerance in sunflower	Sandhya et al. (2009)
P. mendocina	Promotes lettuce growth under elevated CO_2 and water stress conditions	Kohler et al. (2010)
P. putida Rs-198 strain	Protects and promotes cotton seedlings against salt stress	Yao et al. (2010)
Pseudomonas sp.	Confers drought tolerance to maize seedlings	Sandhya et al. (2010)
P. putida	Promotes wheat growth under high temperatures	Ali et al. (2011)
Pseudomonas spp.	Alleviates cold stress effects on wheat	Mishra et al. (2011)
Pseudomonas sp.	Drought tolerant and ACC deaminase producing	Ali et al. (2014)
Pseudomonas sp.	Thermotolerant trehalose synthase-producing strain	Ali et al. (2014)

(continued)

Table 10.3 (continued)

Pseudomonas species	Effects/traits	Reference
P. putida	Tolerates high temperature (45 °C) and salinity stress (1 M NaCl) through involvement of heat stress responsive molecular chaperones and membrane proteins during heat stress	Rangeshwaran et al. (2013)
P. chlororaphis	Confers salt stress tolerance in gladiolus in sodic soils	Damodaran et al. (2014)
Pseudomonas spp.	High freezing survival and low ice nucleation activity	Bisht et al. (2013)
Pseudomonas spp.	Tolerate salinity levels of 1.2 M. (7 % NaCl) and high temperature (50 °C)	Praveen Kumar et al. (2014)
Pseudomonas P59	Tolerates osmotic stress levels up to (−1.2 MPa)	Praveen Kumar et al. (2014)

Table 10.4 Pesticide degradation by Pseudomonads

Pseudomonas species	Functions	Reference
Pseudomonas sp.	Degrades atrazine by N-dealkylation and dehalogenation	Behki and Khan (1986)
P. putida	Utilizes methyl parathion as C-Source and produces the enzyme organophosphorus acid anhydrase	Lakshmi Rani and Lalitha kumari (1994)
Pseudomonas sp.	Enhanced biodegradation of γ-HCH upon 8 days of incubation	Nawab et al. (2003)
Pseudomonas sp.	Complete degradation of phorate was observed within 40 days	Bano and Musarrat (2003)
Pseudomonas spp.	Degradation of chlorpyrifos up to 71–77 % was recorded	Madhuri and Rangaswamy (2009)
P. fluorescens and *P. aeruginosa*	99 % degradation of chlorpyrifos was observed	Lakshmi and Prakash (2009)
Pseudomonas strain SFP 13	Recorded 70.2 % degradation of fenpropathrin	Sarkar et al. (2011)
P. aeruginosa (S1)	Degrades malathion	Tamer et al. (2013)
P. desmolyticum NCIM 2112	Degrades benzoyl benzoate	Rokade and Mali (2013)
Pseudomonas sp. ED1	Degrades endosulfan (250 ppm)	Siva and Rajam (2013)

Other Roles in the Agro-Ecosystem

The presence of organic pollutants such as pesticides and heavy metals above the permissible levels has adverse effects on plant growth and development and often renders the soils unfit for cultivation. Considering the variety of organic pollutants and heavy metals and the extent of cultivable land that are prone to such stresses, it's imperative to identify suitable and sustainable mechanisms of remediation. In such a scenario, *Pseudomonas* strains have the potential for remediation of contaminated environments due to their metabolic versatility (Aislabie and Lloyd-Jones 1995). The vast array of *Pseudomonas* enzyme systems confers upon them the ability to degrade organic pollutants such as pesticides (Foght et al. 2001), while the energy-dependent efflux pump mechanisms help them overcome stress imposed by heavy metals (Silver 1996). A non-exhaustive compilation of the instances of pesticide degradation and heavy metal tolerance by *Pseudomonas* strains is presented in Tables 10.4 and 10.5. Though the utility of Pseudomonads in this arena is well documented, their utility in the remediation of vast tracts of cultivable land is yet to be exploited.

Table 10.5 Heavy metal-tolerant *Pseudomonas* and their utility

Pseudomonas species	Functions/traits	Reference
P. fluorescens NCIMB 11764	Utilization of cyanide as a nitrogenous substrate	Kuntz et al. (1992)
P. fluorescens LB 300	Capable of reducing Cr (VI)	De Leo and Ehrlich (1994)
P. mendocina	Resistant to Cr (VI) up to 1,560 (mg/l)	Bhide et al. (1996)
Pseudomonas sp.	Resistance to mercury	Barbiere et al. (1996)
P. aeruginosa	Biosorption of uranium	Hu and Reeves (1997)
P. putida Spi3	Removal of mercury from electrolysis wastewater	Canstein et al. (1999)
Pseudomonas sp.	Biodegradation of cyanide compounds	Dhillon and Shivaraman (1999)
Pseudomonas sp.	Resistant to Cr (VI) up to 520 (mg/L)	McLean and Beveridge (2001)
P. aeruginosa strain BU2	Copper removal from aqueous solutions	Kazy et al. (2002)
P. fluorescens	Resistant to Cr (VI) up to 208 (mg/L)	Hussein et al. (2004)
P. putida	Resistant to Cr (VI) up to 156 (mg/L)	Hussein et al. (2004)
P. aeruginosa PU21	Biosorption of mercury	Chang and Hong (2004)
Pseudomonas sp.	Lanthanum biosorption	Kazy et al. (2006)
P. putida	Arsenic resistant strain	Rathinasabapathi et al. (2006)
P. brassicacearum	Stimulation of root growth and enhanced nutrient uptake under cadmium stress	Safronova et al.(2006)
P. aeruginosa strain BU2	Removal of uranium from aqueous solution	Kazy et al. (2008)
P. oleovorans	Resistant to Cr (VI) up to 780 (mg/l)	Mistry et al. (2009)
P. plecoglossicida	Resistant to Cr (VI) up to 600 (mg/l)	Poornima et al. (2010)
P. aeruginosa	Biosorption of metals Cr, Zn, and Cu	Nagashetti et al. (2013)
P. stutzeri ASP3	Resistant to arsenic (sodium arsenate up to 4,680.15 mg/l)	Shakya et al. (2012)

Conclusion

The genus *Pseudomonas* is unique and versatile in its metabolic and genome architecture and is therefore ubiquitous in distribution and varied in function. In the agro-ecosystem, though the plant growth promotion and related biological control properties of Pseudomonads are well studied, several other beneficial properties like the degradation of xenobiotic compounds and bioremediation of heavy metal-contaminated soils and environments remain as artifacts of academic interest only and have not seen much utility in terms of commercial exploitation. A recent interest on the abiotic stress alleviation properties of Pseudomonads is bound to generate interest in the emerging scenario of climate change and associated disturbances in the growing environments. As diversity of Pseudomonads continues to expand with time, it can be rightfully said that the attempts to unearth the real diversity of this genus have been clearly disproportionate, especially in the pristine environments such as the alpine and subalpine environments that are treasure troves of these unique bacteria. Since there is no dearth of basic information, the real need of the day is the unearthing of the undiscovered diversity and the viable commercial utilization of Pseudomonads for the welfare of mankind.

References

Abrams PA (1987) Alternative models of character displacement and niche shift 2. Displacement when there is competition for a single resource. Am Nat 130:271–282

Ahemad M, Khan MS (2012a) Alleviation of fungicide-induced phytotoxicity in green gram *Vigna radiata* (L.) Wilczek using fungicide-tolerant and plant growth promoting *Pseudomonas* strain. Saudi J Biol Sci 19:451–459

Ahemad M, Khan MS (2012b) Effect of fungicides on plant growth promoting activities of phosphate solubilizing *Pseudomonas putida* isolated from mustard (*Brassica campestris*) rhizosphere. Chemosphere 86:945–950

Ahmad F, Ahmad I, Khan MS (2008) Screening of free-living rhizospheric bacteria for their multiple plant growth promoting activities. Microbiol Res 163:173–181

Aislabie J, Lloyd-Jones G (1995) A review of bacterial degradation of pesticides. Aust J Soil Res 33:925–942

Ali SZ, Sandhya V, Grover M, Kishore N, Rao LV, Venkateswarlu B (2009) *Pseudomonas* sp. strain AKM-P6 enhances tolerance of sorghum seedlings to elevated temperatures. Biol Fertil Soils 46:45–55

Ali SZ, Sandhya V, Grover M, Rao LV, Venkateswarlu B (2011) Effect of inoculation with a thermotolerant plant growth promoting *Pseudomonas putida* strain AKMP7 on growth of wheat (*Triticum* spp.) under heat stress. J Plant Interact 6:239–246

Ali SZ, Sandhya V, Rao LV (2014) Isolation and characterization of drought-tolerant ACC deaminase and exopolysaccharide-producing fluorescent *Pseudomonas* sp. Ann Microbiol 64(2):493–502

Aoki MK, Uehara K, Koseki K, Tsuji M, Iijima K, Ono T, Samejima (1991) An antimicrobial substance produced by *Pseudomonas cepacia* B5 against the bacterial wilt disease pathogen *Pseudomonas solanacearum*. Agric Biol Chem 55:715–722

Ayres JC (1960) The relationship of organisms of the genus *Pseudomonas* to the spoilage of meat, poultry and eggs. J Appl Bacteriol 23:471–486

Bano N, Musarrat J (2003) Isolation and characterization of phorate degrading soil bacteria of environmental and agronomic significance. Lett Appl Microbiol 36:349–353

Barbiere P, Bestetti G, Reniero D, Galli E (1996) Mercury resistance in aromatic compound degrading *Pseudomonas* strains. FEMS Microb Ecol 20:185–194

Behki RM, Khan US (1986) Degradation of atrazine by *Pseudomonas*: N-dealkylation and dehalogenation of atrazine and its metabolites. J Agric Food Chem 34:746–749

Behrendt U, Ulrich A, Schumann P, Meyer JM, Spröer C (2007) *Pseudomonas lurida* sp. nov., a fluorescent species associated with the phyllosphere of grasses. Int J Syst Evol Microbiol 57:979–985

Behrendt U, Schumann P, Meyer JM, Ulrich A (2009) *Pseudomonas cedrina* subsp. *fulgida* subsp. nov., a fluorescent bacterium isolated from the phyllosphere of grasses. Int J Syst Evol Microbiol 59:1331–1335

Beijerinck MW (1921) Beobachtungen und Betrachtungen uber Wurzelknospen und Nebenwurzeln. Verz Geschr Beijer 2:7–121

Bell G (1997) Selection: the mechanism of evolution. Springer, Berlin

Bergey DH, Harrison FC, Breed RS, Hammer BW, Huntoon FM (1923) Bergey's manual of determinative bacteriology. Williams and Wilkins, Baltimore, pp 1–442

Bhatnagar A, Bhatnagar M (2005) Microbial diversity in desert ecosystems. Curr Sci 1:91–100

Bhide JV, Dhakephalkar PK, Paknikar KM (1996) Microbiological process for the removal of Cr (VI) from chromate-bearing cooling tower effluent. Biotechnol Lett 18:667

Bisht SC, Mishra PK, Joshi GK (2013) Genetic and functional diversity among root-associated psychrotrophic Pseudomonad's isolated from the Himalayan plants. Arch Microbiol 195:605–615

Braud A, Je'ze'quel K, Bazot S, Lebeau T (2009) Enhanced phytoextraction of an agricultural Cr- Hg and Pb-contaminated soil by bioaugmentation with siderophore producing bacteria. Chemosphere 74:280–286

Cámara B, Strömpl C, Verbarg S, Spröer C, Pieper DH, Tindall BJ (2007) *Pseudomonas reinekei* sp. nov., *Pseudomonas moorei* sp. nov. and *Pseudomonas mohnii* sp. nov., novel species capable of degrading chlorosalicylates or isopimaric acid. Int J Syst Evol Microbiol 57:923–931

Canstein HV, Li Y, Timmis KN, Deckwer WD, Timmis KN, Deckwer WD, Wagner-Döbler I (1999) Removal of mercury from chloralkali electrolysis wastewater by a mercury resistant *Pseudomonas putida* strain. Appl Environ Microbiol 65:5279–5284

Chang JS, Hong J (2004) Biosorption of mercury by the inactivated cells of *Pseudomonas aeruginosa* PU21 (Rip64). Biotechnol Bioeng 44:999–1006

Cho SM, Kang BR, Han SH, Anderson AJ, Park JY, Lee YH (2008) 2R, 3R-Butanediol, a bacterial volatile produced by *Pseudomonas chlororaphis* O6, is involved in induction of systemic tolerance to drought in *Arabidopsis thaliana*. Mol Plant Microbe Interact 21:1067–1075

Damodaran T, Rai RB, Jha SK, Kannan R, Pandey BK, Sah V, Sharma DK (2014) Rhizosphere and endophytic bacteria for induction of salt tolerance in gladiolus grown in sodic soils. J Plant Interact 9:577–584

Das K, Mukherjee AK (2007) Crude petroleum-oil biodegradation efficiency of *Bacillus subtilis* and *Pseudomonas aeruginosa* strains isolated from a petroleum-oil contaminated soil from North-East India. Bioresour Technol 98:1339–1345

De Boer W, Kowalchuk GA (2001) Nitrification in acid soils: micro-organisms and mechanisms. Soil Biol Biochem 33:853–866

De Leo PC, Ehrlich HL (1994) Reduction of hexavalent chromium by *Pseudomonas fluorescens* LB 300 in batch and continuous cultures. Appl Microbiol Biotechnol 40:756

De Weger LA, Van der Vlugt CIM, Wijfjes AHM, Bakker PAHM, Schippers B, Lugtenberg BJJ (1987) Flagella of a plant-growth-stimulating *Pseudomonas fluorescens* strain are required for colonisation of potato roots. J Bacteriol 169:2769–2773

De Weger LA, Van Loosdrecht MC, Klaassen HE, Lugtenberg BJJ (1989) Mutational changes in physiochemical cell surface properties of plant-growth-stimulating *Pseudomonas* spp. do not influence the attachment properties of the cells. J Bacteriol 171:2756–2761

Dekkers LC, Bloemendaal CJP, deWeger LA, Wijffelman CA, Spaink HP, Lugtenberg BJJ (1998) A two-component system plays an important role in the root-colonizing ability of *Pseudomonas fluorescens* strain WCS365. Mol Plant Microbe Interact 11:45–56

Dell'Amico E, Cavalca L, Andreoni V (2008) Improvement of *Brassica napus* growth under cadmium stress by cadmium resistant rhizobacteria soil. Biol Biochem 40:74–84

Dhillon JK, Shivaraman N (1999) Biodegradation of cyanide compounds by a *Pseudomonas* species (S1). Can J Microbiol 45:201–208

Dogan B, Boor KJ (2003) Genetic diversity and spoilage potentials among *Pseudomonas* spp. isolated from fluid milk products and dairy processing plants. Appl Environ Microbiol 9:130–138

Dörr J, Hurek T, Reinhold Hurek HB (1998) Type IV pili are involved in plant–microbe and fungus–microbe interactions. Mol Microbiol 30:7–17

Doudoroff M, Palleroni NJ (1974) Genus I: *Pseudomonas* Migula 1984, 237 Nom. cons. Opin. 5, Jud. Comm. 1952, 121. In: Buchanan RE, Gibbons NE (eds) Bergey's manual of determinative bacteriology. Williams and Wilkins, Baltimore, pp 217–243

Favero MS, Carson LA, Bond WW, Petersen NJ (1971) *Pseudomonas aeruginosa*: growth in distilled water from hospitals. Science 173:836–838

Feng Z, Zhang J, Huang X, Zhang J, Chen M, Li S (2012) *Pseudomonas zeshuii* sp. nov., isolated from herbicide-contaminated soil. Int J Syst Evol Microbiol 62:2608–2612

Foght J, April T, Biggar K, Aislabie J (2001) Bioremediation of DDT-contaminated soils: a review. Bioremediat J 5:225–246

Friedberg EC, Walker GC, Siede W (1995) DNA repair and mutagenesis. Academic, New York

Furuya NY, Kushima K, Tsuchiya K (1991) Protection of tomato by pretreatment with *Pseudomonas glumae* from infection with *Pseudomonas solanacearum* and its mechanisms. Ann Phytopathol Soc Jpn 57:363–370

Gause GF (1934) The struggle for existence. Williams & Wilkins, Baltimore

Goldman RC, Leive L (1980) Heterogeneity in antigenic side chain length in lipopolysaccharide from *Escherichia coli 0111* and *Salmonella typhimurium* LT2. Eur J Biochem 107:145–153

Gross H, Loper JE (2009) Genomics of secondary metabolite production by *Pseudomonas* spp. Nat Prod Rep 26:1408–1446

Grover M, Ali SZ, Sandhya V, Rasul A, Venkateswarlu B (2011) Role of microorganisms in adaptation of agriculture crops to abiotic stresses. World J Microbiol Biotechnol 27:1231–1240

Gupta A, Rai V, Bagdwal N, Goel R (2005) *In situ* characterization of mercury resistant growth promoting fluorescent pseudomonads. Microbiol Res 160:385–388

Gupta SK, Kumari R, Prakash O, Lal R (2008) *Pseudomonas panipatensis* sp. nov., isolated from an oil-contaminated site. Int J Syst Evol Microbiol 58:1339–1345

Guttman DS, Morgan RL, Wang PW (2008) The evolution of the Pseudomonads. In: Fatmi M, Collmer A, Iacobellis NS, Mansfield JW, Murillo J, Schaad NW, Ullrich M (eds) *Pseudomonas syringae* pathovars and related pathogens–identification, epidemiology and genomics. Springer, Dordrecht, pp 307–319

Hahn HP (1997) The type-4 pilus is the major virulence-associated adhesin of *Pseudomonas aeruginosa* review. Gene 192:99–108

Hardin G (1960) The competitive exclusion principle. Science 131:1292–1997

Hartman GL, Hong W, Hayward AC (1992) Potential of biological and chemical control of bacterial wilt. In: Hartman GL, Hayward AC (eds) Bacterial wilt. ACIAR, Canberra, pp 322–326

Haubold B, Rainey PB (1996) Genetic and ecotypic structure of a fluorescent *Pseudomonas* population. Mol Ecol 5:747–761

Howie WJ, Cook RJ, Weller DM (1987) Effects of soil matric potential and cell motility on wheat root colonization by fluorescent pseudomonads suppressive to take all. Phytopathology 77:286–292

Hu MZC, Reeves M (1997) Biosorption of uranium by *Pseudomonas aeruginosa* strain CSU immobilized in a novel matrix. Biotechnol Prog 13:60–70

Hussein H, Moawad H, Farag S (2004) Isolation and characterization of *Pseudomonas* resistant to heavy metals contaminants. Arab J Biotechnol 7:13

Ivanova EP, Christen R, Bizet C, Clermont D, Motreff L, Bouchier C, Zhukova NV, Russell J, Elena C, Kiprianova A (2009) *Pseudomonas brassicacearum* subsp. *neoaurantiaca* subsp. nov., orange-pigmented bacteria isolated from soil and the rhizosphere of agricultural plants. Int J Syst Evol Microbiol 59:2476–2481

Jagadeesh KS, Kulkarni JH, Krishnaraj PU (2001) Evaluation of the role of fluorescent siderophore in the biological control of bacterial wilt in tomato using Tn5 mutants of fluorescent *Pseudomonas* sp. Curr Sci 81:882–889

Judith EM, De Vos RCH, Dekkers E, Pineda A, Guillod L, Bouwmeester K, Van Loon JJA, Dicke M, Raaijmakers JM (2012) Metabolic and transcriptomic changes induced in *Arabidopsis* by the rhizobacterium *Pseudomonas fluorescens SS101*. Plant Physiol 160:2173–2188

Kazy SK, Sar P, Sen AK, Singh SP, D'Souza SF (2002) Extracellular polysaccharides of a copper-sensitive and a copper-resistant *Pseudomonas aeruginosa* strain: synthesis, chemical nature and copper binding. World J Microbiol Biotechnol 18:583–588

Kazy SK, Das SK, Sar P (2006) Lanthanum biosorption by a *Pseudomonas* sp: equilibrium studies and chemical characterization. J Ind Microbiol Biotechnol 33:773–783

Kazy SK, Sar P, D'Souza SF (2008) Studies on uranium removal by the extra cellular polysaccharide of a

Pseudomonas aeruginosa strain. Bioremediat J 12:47–57

Khan NH, Ishii Y, Kimata-Kino N, Esaki H, Nishino T, Nishimura M, Kogure K (2007) Isolation of *Pseudomonas aeruginosa* from open ocean and comparison with freshwater, clinical, and animal isolates. Microb Ecol 53:173–186

Kiewitz C, Tummler B (2000) Sequence diversity of *Pseudomonas aeruginosa* impact on population structure and genome evolution. J Bacteriol 182:3125–3135

Kloepper JW (1983) Effect of seed piece inoculation with plant growth-promoting rhizobacteria on populations of *Erwinia carotovora* on potato roots and daughter tubers. Phytopathology 73:217–219

Kohler J, Hernandez JA, Caravaca F, Roldàn A (2008) Plant-growth-promoting rhizobacteria and arbuscular mycorrhizal fungi modify alleviation biochemical mechanisms in water-stressed plants. Funct Plant Biol 35:141–151

Kohler J, Caravaca F, Roldán A (2010) An AM fungus and a PGPR intensify the adverse effects of salinity on the stability of rhizosphere soil aggregates of *Lactuca sativa*. Soil Biol Biochem 42:429–434

Kuiper I, Bloemberg GV, Noreen S, Thomas-Oates JE, Lugtenberg BJ (2001) Increased uptake of putrescine in the rhizosphere inhibits competitive root colonization by *Pseudomonas fluorescens* strain WCS365. Mol Plant Microbe Interact 14:1096–1104

Kuntz DA, Nagappan O, Avalos JS, Delong GT (1992) Utilization of cyanide as a nitrogenous substrate by *Pseudomonas fluorescens* NCIMB 11764. Appl Environ Microbiol 58:2022–2029

Lakshmi CV, Prakash NT (2009) *In situ* bioremediation of chlorpyrifos in cotton fields: possible role of plant-microbe interaction. J Pure Appl Microbiol 3:543–550

Lakshmi Rani N, Lalitha Kumari D (1994) Degradation of methyl parathion by *Pseudomonas putida*. Can J Microbiol 40:1000–1006

Lawongsa P, Boonkerd N, Wongkaew S, O'Gara F, Teaumroong N (2008) Molecular and phenotypic characterization of potential plant growth-promoting *Pseudomonas* from rice and maize rhizospheres. World J Microb Biotechnol 24:1877–1884

Lessie TG, Phibbs PVJ (1984) Alternative pathways of carbohydrate utilization in *Pseudomonads*. Annu Rev Microbiol 38:359–388

Liddycoat SM, Greenberg BM, Wolyn DJ (2009) The effect of plant growth-promoting rhizobacteria on asparagus seedling and germinating seeds subjected to water stress under greenhouse conditions. Can J Microbiol 55:388–394

Lin SY, Hameed A, Liu YC, Hsu YH, Lai WA, Young CC (2013) *Pseudomonas formosensis* sp. nov., a gammaproteobacteria isolated from food-waste compost in Taiwan. Int J Syst Evol Microbiol 63:3168–3174

Loper JE, Hassan KA, Mavrodi DV, Davis EWII, Lim CK (2012) Comparative genomics of plant-associated *Pseudomonas* spp.: insights into diversity and inheritance of traits involved in multitrophic interactions. PLoS Genet 8:e1002784. doi:10.1371/journal.pgen.1002784

López JR, Diéguez AL, Doce A, De la Roca E, De la Herran R, Navas JI, Romalde JL (2012) *Pseudomonas baetica* sp. nov., a fish pathogen isolated from wedge sole, *Dicologlossa cuneata* (Moreau). Int J Syst Evol Microbiol 62:874–882

Lugtenberg BJJ, Dekkers L, Bloemberg GV (2001) Molecular determinants of rhizosphere colonization by *Pseudomonas*. Annu Rev Phytopathol 39:461–490

Ma Y, Rajkumar M, Luo Y, Freitas H (2011) Inoculation of endophytic bacteria on host and non-host plants-effects on plant growth and Ni uptake. J Hazard Mater 195:230–237

Madhuri RJ, Rangaswamy V (2009) Biodegradation of selected insecticides by *Bacillus* and *Pseudomonas* sps in ground nut fields. Toxicol Int 16:127–132

Matthysse AG, McMahan S (1998) Root colonization by *Agrobacterium tumefaciens* is reduced in *cel, attB, attD*, and *attR* mutants. Appl Environ Microbiol 64:2341–2345

McLean J, Beveridge TJ (2001) Chromate reduction by a Pseudomonad isolated from a site contaminated with chromated copper arsenate. Appl Environ Microbiol 67:1076

Meyer JM, Geoffroy VA, Baida N, Gardan L, Izard D, Lemanceau P, Achouak W, Palleroni N (2002) Siderophore typing, a powerful tool for the identification of fluorescent and non-fluorescent *Pseudomonas*. Appl Environ Microbiol 68:2745–2753

Meyer AF, Lipson DA, Martin AP, Schadt CW, Schmidt SK (2004) Molecular and metabolic characterization of cold-tolerant alpine soil *Pseudomonas sensu* stricto. Appl Environ Microbiol 70:483–489

Migula W (1894) U¨ber ein neues System der Bakterien. Arb Bakteriol Inst Karlsruhe 1:235–238

Mishra PK, Mishra S, Selvakumar G, Bisht SC, Kundu S, Bisht JK, Gupta HS (2008) Characterization of a psychrotrophic plant growth promoting *Pseudomonas* PGERS 17 (MTCC 9000) isolated from North Western Indian Himalayas. Ann Microbiol 58:1–8

Mishra PK, Mishra S, Bisht SC, Selvakumar G, Kundu S, Bisht JK, Gupta HS (2009) Isolation, molecular characterization and growth-promotion activities of a cold tolerant bacterium *Pseudomonas* sp. NARs9 (MTCC9002) from the Indian Himalayas. Biol Res 42:305–313

Mishra PK, Bisht SC, Ruwari P, Selvakumar G, Bisht JK, Bhatt JC (2011) Alleviation of cold stress effects in wheat (*Triticum aestivum* L.) seedlings by application of psychrotolerant pseudomonads from N.W. Himalayas. Arch Microbiol 193:497–513

Mishra PK, Mishra S, Bisht SC, Selvakumar G, Bisht JK, Gupta HS (2012) Co-inoculation of *Rhizobium leguminosarum* PR-1 with a cold tolerant *Pseudomonas* sp improves iron acquisition, nutrient uptake and growth of field pea. J Plant Nutr 35:243–256

Mistry K, Desai C, Patel K (2009) Reduction of chromium (VI) by bacterial strain KK15 isolated from contaminated soil. J Cell Tissue Res 9:1821

Mulet M, Lalucat J, García-Valdés E (2010) DNA sequence-analysis of the *Pseudomonas* species. Environ Microbiol 12:1513–1530

Nagashetti V, Mahadevaraju GK, Muralidhar TS, Javed A, Trivedi D, Bhusa KP (2013) Biosorption of heavy metals from soil by *Pseudomonas aeruginosa*. Int J Innov Technol Explor Eng 2:22–24

Nawab A, Aleem A, Malik A (2003) Determination of organochlorine pesticides in agricultural soil with special reference to γ–HCH degradation by *Pseudomonas* strains. Bioresour Technol 88:41–46

O'brien RD, Lindow SE (1989) Effect of plant species and environmental conditions on epiphytic population sizes of *Pseudomonas syringae* and other bacteria. Phytopathology 79:619–627

O'Toole GA, Kolter R (1998) Initiation of biofilm formation in *Pseudomonas fluorescens* WCS365 proceeds via multiple, convergent signalling pathways: a genetic analysis. Mol Microbiol 28:449–461

Ochman H, Lawrence JG, Groisman EA (2000) Lateral gene transfer and the nature of bacterial innovation. Nature 405:299–344

Oliver A, Canton R, Campo P, Baquero F, Blazquez J (2000) High frequency of hyper mutable *Pseudomonas aeruginosa* in cystic fibrosis lung infection. Science 288:1251–1253

Pak JW, Knoke KL, Noguera DR, Fox BG, Chambliss GH (2000) Transformation of 2, 4, 6-trinitrotoluene by purified xenobiotic reductase B from *Pseudomonas fluorescens* I-C. Appl Environ Microbiol 66:4742–4750

Palleroni NJ (1984) *Pseudomonas*, Topley and Wilson's microbiology and microbial infections. Hodder Arnold, London

Palleroni NJ (2005) Genus I. *Pseudomonas* Migula. In: Brenner DJ, Krieg NR, Staley JT, Garitty GT (eds) Bergey's manual of systematic bacteriology. Springer, New York, pp 323–379, 2(2)

Palleroni NJ (2008) The road to the taxonomy of *Pseudomonas*. In: Pseudomonas: genomics and molecular biology. Caister Academic Press, Norfolk, pp 1–18

Palleroni NJ, Kunisawa R, Contopoulou R, Doudoroff M (1973) Nucleic acid homologies in the genus *Pseudomonas*. Int J Syst Bacteriol 23:333–339

Palumbo J, Kado C, Phillips DA (1998) An isoflavonoid-inducible efflux pump in *Agrobacterium tumefaciens* is involved in competitive colonization of roots. J Bacteriol 180:3107–3113

Panicker G, Aislabie J, Saul D, Bej A (2002) Cold tolerance of *Pseudomonas* sp. 30/3 isolated from oil-contaminated soil. Antarct Polar Biol 225:5–11

Park YD, Yi H, Baik KS, Seong CN, Bae KS, Moon EY, Chun J (2006) *Pseudomonas segetis* sp. nov., isolated from soil. Int J Syst Evol Microbiol 56:2593–2595

Paul D, Dinesh Kumar N, Nair S (2006) Proteomics of a plant growth-promoting rhizobacterium, *Pseudomonas fluorescens* MSP-393, subjected to salt shock. World J Microbiol Biotechnol 22:369–374

Phillips DA, Streit W (1995) Legume signals to rhizobial symbionts: a new approach for defining rhizosphere colonization. In: Stacey G, Keen NT (eds) Plant-microbe interactions. Chapman & Hall, New York, pp 236–271

Piotrowska-Seget Z, Cycon M, Kozdrój J (2005) Metal-tolerant bacteria occurring in heavily polluted soil and mine spoil. Appl Soil Ecol 28:237–246

Poonguzhali S, Madhaiyan M, Sa T (2008) Isolation and identification of phosphate solubilizing bacteria from Chinese cabbage and their effect on growth and phosphorus utilization of plants. J Microbiol Biotechnol 18:773–777

Poornima K, Karthik L, Swadhini SP, Mythili S, Sathiavelu A (2010) Degradation of chromium by using a novel strains of *Pseudomonas* species. J Microb Biochem Technol 2:95–99

Praveen Kumar G, Mir Hassan Ahmed SK, Desai S, Leo Daniel Amalraj E, Rasul A (2014) In vitro screening for abiotic stress tolerance in potent biocontrol and plant growth promoting strains of *Pseudomonas* and *Bacillus* spp. Int J Bacteriol. doi:10.1155/2014/195946596

Preston GM, Bertrand N, Rainey PB (2001) Type III secretion in plant growth-promoting *Pseudomonas fluorescens* SBW25. Mol Microbiol 41:999–1014

Raaijmakers JM, Vlami M, de Souza JT (2002) Antibiotic production by bacterial biocontrol agents. Antonie Van Leeuwenhoek 81:537–547

Rainey PB, Travisano M (1998) Adaptive radiation in a heterogeneous environment. Nature 394:69–72

Rainey PB, Buckling A, Kassen R, Travisano M (2000) The emergence and maintenance of diversity: insights from experimental bacterial populations. Trends Ecol Evol 15:243–247

Rajkumar M, Freitas H (2008) Effects of inoculation of plant-growth promoting bacteria on Ni uptake by Indian mustard. Bioresour Technol 99:3491–3498

Ramos E, Ramírez-Bahena MH, Valverde A, Velázquez E, Zúñiga D, Velezmoro C, Peix A (2013) *Pseudomonas punonensis* sp. nov., isolated from straw. Int J Syst Evol Microbiol 63:1834–1839

Ramphal R, Small PM, Shands JW Jr, Fischlschweiger W, Small PA Jr (1980) Adherence of *Pseudomonas aeruginosa* to tracheal cells injured by influenza infection or by endotracheal intubation. Infect Immun 27:614–619

Rangeshwaran R, Ashwitha K, Sivakumar G, Jalali SK (2013) Analysis of proteins expressed by an abiotic stress tolerant *Pseudomonas putida* (NBAII-RPF9) isolate under saline and high temperature conditions. Curr Microbiol 67:659–667

Rathinasabapathi B, Raman SB, Kertulis G, Ma L (2006) Arsenic resistant proteobacterium from the phyllosphere of arsenic hyper accumulating fern (*Pteris vittata* (L.) reduces arsenate to arsenite. Can J Microbiol 52:695–700

Roane TM, Kellogg ST (1996) Characterization of bacterial communities in heavy metal contaminated soils. Can J Microbiol 42:593–603

Rokade KB, Mali GV (2013) Optimization of soil parameters for benzyl benzoate degradation by *Pseudomonas desmolyticum* NCIM 2112. Res J Agric Sci 1(5):1–9

Safronova VI, Stepanok VV, Engqvist GL, Alekseyev YV, Belimov AA (2006) Root-associated bacteria containing 1-aminocyclopropane-1-carboxylate deaminase improve growth and nutrient uptake by pea genotypes cultivated in cadmium supplemented soil. Biol Fertil Soils 42:267–272

Sandhya V, Grover M, Reddy G, Venkateswarlu B (2009) Alleviation of drought stress effects in sunflower seedlings by the exopolysaccharides producing *Pseudomonas putida* strain GAP-P45. Biol Fertil Soils 46:17–26

Sandhya V, Ali SZ, Grover M, Reddy G, Venkateswarlu B (2010) Effect of plant growth promoting *Pseudomonas* spp. on compatible solutes, antioxidant status and plant growth of maize under drought stress. Plant Growth Regul 62:21–30

Saravanakumar D, Samiyappan R (2007) ACC deaminase from *Pseudomonas fluorescens* mediated saline resistance in groundnut (*Arachis hypogea*) plants. J Appl Microbiol 102:1283–1292

Saravanakumar D, Vijayakumar C, Kumar BN, Samiyappan R (2007) PGPR-induced defense responses in the tea plant against blister blight disease. Crop Prot 26:556–565

Sarkar S, Sathesh Kumar A, Jayanthi R, Premkumar R (2011) Biodegradation of Fenpropathrin by *Pseudomonas* sp. isolated from tea rhizosphere microflora. Adv Biotechnol 10:6–9

Selvakumar G, Joshi P, Mishra PK, Bisht JK, Gupta HS (2009a) Mountain aspect influences the genetic glustering of psychrotolerant phosphate solubilizing Pseudomonads in the Uttarakhand Himalayas. Curr Microbiol 59:432–438

Selvakumar G, Kundu S, Joshi P, Nazim S, Gupta AD, Mishra PK, Gupta HS (2009b) Phosphate solubilization and growth promotion by *Pseudomonas fragi* CS11RH1 (MTCC 8984) a psychrotolerant bacterium isolated from a high altitude Himalayan rhizosphere. Biologia 64:239–245

Selvakumar G, Joshi P, Suyal P, Mishra PK, Joshi GK, Bisht JK, Bhatt JC, Gupta HS (2011) *Pseudomonas lurida* M2RH3 (MTCC 9245), a psychrotolerant bacterium from the Uttarakhand Himalayas, solubilizes phosphate and promotes wheat seedling growth. World J Microbiol Biotechnol 27:1129–1135

Selvakumar G, Joshi P, Suyal P, Mishra PK, Joshi GK, Venugopalan R, Bisht JK, Bhatt JC, Gupta HS (2013) Rock phosphate solubilization by psychrotolerant *Pseudomonas* spp. and their effect on lentil growth and nutrient uptake under polyhouse conditions. Ann Microbiol 63:1353–1362

Sesma A, Sundin GW, Murillo J (2000) Phylogeny of the replication regions of pPT23A-like plasmids from *Pseudomonas syringae*. Microbiology 146:2375–2384

Shakya S, Pradhan B, Smith L, Tuladhar S, Shrestha J (2012) Isolation and characterization of aerobic culturable arsenic-resistant bacteria from surface water and groundwater of Rautahat District, Nepal. J Environ Manag 95:S250–S255

Sharma SK, Johri BN, Ramesh A, Joshi OP, Prasad SVS (2011) Selection of plant growth-promoting *Pseudomonas* spp. that enhanced productivity of soybean-wheat cropping system in central India. J Microbiol Biotechnol 21:1127–1142

Silby MW, Cerden˜o-Tarraga AM, Vernikos GS, Giddens SR, Jackson RW (2009) Genomic and genetic analyses of diversity and plant interactions of *Pseudomonas fluorescens*. Genome Biol 10:R51

Silby MW, Winstanley C, Godfrey SAC, Levy SB, Jackson RW (2011) *Pseudomonas* genomes: diverse and adaptable. FEMS Microbiol Rev 35:652–680

Silver S (1996) Bacterial resistances to toxic metal ions – a review. Gene 179:9–19

Simons M, Van der Bij AJ, Brand I, de Weger LA, Wijffelman CA, Lugtenberg BJJ (1996) Gnotobiotic system for studying rhizosphere colonisation by plant growth-promoting *Pseudomonas* bacteria. Mol Plant Microbe Interact 9:600–607

Simons M, Permentier HP, De Weger LA, Wijffelman CA, Lugtenberg BJJ (1997) Amino acid synthesis is necessary for tomato root colonization by *Pseudomonas fluorescens* strain WCS365. Mol Plant Microbe Interact 10:102–106

Sinha S, Mukherjee SK (2008) Cadmium-induced siderophore production by a high Cd-resistant bacterial strain relieved Cd toxicity in plants through root colonization. Curr Microbiol 56:55–60

Siva GV, Rajam A (2013) Degradation of endosulfan using *Pseudomonas* sp. ED1 isolated from pesticide contaminated soil. J Acad India Res 2:170

Spiers AJ, Buckling A, Rainey P (2000) The causes of *Pseudomonas* diversity. Microbiology 146: 2345–2350

Stanier RY, Palleroni NJ, Doudoroff M (1966) The aerobic Pseudomonads: a taxonomic study. J Gen Microbiol 43:159–271

Streit WR, Joseph CM, Phillips DA (1996) Biotin and other water-soluble vitamins are key growth factors for alfalfa root colonization by *Rhizobium meliloti* 1021. Mol Plant Microbe Interact 9:330–338

Tamer MAT, Medhat AH, Naggar EL (2013) Malathion degradation by soil isolated bacteria and detection of degradation products by GC-MS. Int J Environ Sci 3:1467–1476

Tao Y, Zhou Y, He X, Hu X, Li D (2014) *Pseudomonas chengduensis* sp. nov., isolated from landfill leachate. Int J Syst Evol Microbiol 64:95–100

Thomas CM (2000) The horizontal gene pool. Harwood Academic Publishers, Amsterdam

Toro M, Ramírez-Bahena MH, Cuesta MJ, Velázquez E, Peix A (2013) *Pseudomonas guariconensis* sp. nov., isolated from rhizospheric soil. Int J Syst Evol Microbiol 63:4413–4420

Tourkya B, Boubellouta T, Dufour E, Leriche F (2009) Fluorescence spectroscopy as a promising tool for a polyphasic approach to Pseudomonad taxonomy. Curr Microbiol 58:39–46

Tvrzová L, Schumann P, Spröer C, Sedláček I, Páčová Z, Šedo O, Zdráhal Z, Steffen M, Lang E (2006) *Pseudomonas moraviensis* sp. nov. and *Pseudomonas vranovensis* sp. nov., soil bacteria isolated on nitroaromatic compounds, and emended description of *Pseudomonas asplenii*. Int J Syst Evol Microbiol 56:2657–2663

Unnamalai N, Gnanamanickam SS (1984) *Pseudomonas fluorescens* is an antagonist to *Xanthomonas citri* (Hasse) dye, the incitant of citrus canker. Curr Sci 53:703–704

Vidhyasekaran P, Kamala N, Ramanathan A, Rajappan K, Paranidharan V, Velazhahan R (2001) Induction of systemic resistance by *Pseudomonas fluorescens* Pf1 against *Xanthomonas oryzae* pv. oryzae in rice leaves. Phytoparasitica 29:155–166

Wang LT, Tai CJ, Wu YC, Chen YB, Lee FL, Wang SL (2010) *Pseudomonas taiwanensis* sp. nov., isolated from soil. Int J Syst Evol Microbiol 60:2094–2098

Wani PA, Khan MS, Zaidi A (2007) Synergistic effects of the inoculation with nitrogen fixing and phosphate solubilizing rhizobacteria on the performance of field grown chickpea. J Plant Nutr Soil Sci 170:283–287

Whitby JL, Rampling A (1972) *Pseudomonas aeruginosa* contamination in domestic and hospital environment. Lancet 1:15–17

Wilson M, Lindow SE (1993) Effect of phenotypic plasticity on epiphytic survival and colonization by *Pseudomonas syringae*. Appl Environ Microbiol 59:410–416

Winogradsky S (1949) Microbiologie du Sol. Problèmes et Méthodes: Cinquante Ans de Recherches. Masson et Cie, Paris, p 863

Woese CR, Stackebrandt E, Weisburg WG, Paster BJ, Madigan MT, Fowler VJ, Hahn CM, Blanz P, Gupta R, Nealson KH, Fox GE (1984) The phylogeny of purple bacteria: the alpha subdivision. Syst Appl Microbiol 5:315–326

Xie F, Ma H, Quan S, Liu D, Chen G, Chao Y, Qian S (2014) *Pseudomonas kunmingensis* sp. nov., an exopolysaccharide-producing bacterium isolated from a phosphate mine. Int J Syst Evol Microbiol 64:559–564

Xu GW, Gross DC (1986) Field evaluations of the interactions among fluorescent pseudomonads, *Erwinia carotovora* and potato yields. Phytopathology 76:423–430

Xu H, Griffith M, Patten CL, Glick BR (1998) Isolation and characterization of an antifreeze protein with ice-nucleation activity from the plant growth promoting rhizobacterium *Pseudomonas putida* GR12-2. Can J Microbiol 44:64–73

Yamamoto S, Kasai H, Arnold DL, Jackson RW, Vivian A, Harayama S (2000) Phylogeny of the genus *Pseudomonas*: intrageneric structure reconstructed from the nucleotide sequences of *gyrB* and *rpoD* genes. Microbiology 146:2385–2394

Yao K, Wu Z, Zheng Y, Kaleem I, Li C (2010) Growth promotion and protection against salt stress *by Pseudomonas putida* Rs-198 on cotton. Eur J Soil Biol 46:49–54

Zhang DC, Hong CL, Zhou YG, Schinner F, Margesin R (2011) *Pseudomonas bauzanensis* sp. nov., isolated from soil. Int J Syst Evol Microbiol 61:2333–2337

Zhou T, Chen D, Li C, Sun Q, Li L, Liu F, Shen Q, Shen B (2012) Isolation and characterization of *Pseudomonas brassicacearum* J12 as an antagonist against *Ralstonia solanacearum* and identification of its antimicrobial components. Microbiol Res 167:388–394

Zinniel DK, Lambrecht P, Harris BN (2002) Isolation and characterization of endophytic colonizing bacteria from agronomic crops and prairie plants. Appl Environ Microbiol 68:2198–2208

Perspectives of Rhizobial Inoculation for Sustainable Crop Production

11

Muhammad Naveed, Ijaz Mehboob,
M. Baqir Hussain, and Zahir Ahmad Zahir

Abstract

The need for sustainable agricultural practices is revitalizing the worldwide interest in biological nitrogen fixation (BNF) and rhizobia-legume symbiosis, particularly in organic production system. Rhizobia-legume symbiosis is a complex process involving a set of plant and bacterial genes leading to the formation and development of root nodules on legume roots where rhizobia fix nitrogen from the atmosphere. About 60 % of the Earth's available nitrogen is fixed via BNF. The success of the symbiotic process depends on the competitiveness, specificity, compatibility, and effectiveness of rhizobia with its host legume under variable soil and environmental conditions. In general, the interaction of rhizobia with nonlegumes has been neglected because of the lack of ability to respond to the rhizobial lipochitin Nod factors, which are the essential signal molecules that trigger legume nodulation. Work on rhizobial interaction with nonlegumes, over the years, has been done progressively; it has been demonstrated that rhizobia can associate with roots of nonlegumes also and can promote their growth by using one or more of the direct or indirect mechanisms of actions. The beneficial association/symbiosis of rhizobia can be negatively affected by many rhizosphere processes and other environmental stresses, resulting in important ecological, economic, and nutritional losses. The application of environment-friendly agricultural techniques, based on the use of plant growth-promoting rhizobacteria (PGPR) and/or plant growth regulators (PGRs), can increase the efficiency of the symbiotic/associative processes. The application of appropriate

M. Naveed • M.B. Hussain • Z.A. Zahir (✉)
Institute of Soil and Environmental Sciences,
University of Agriculture, Faisalabad, Pakistan
e-mail: zazahir@yahoo.com

I. Mehboob
District Fertility lab, Kasur, Pakistan

N.K. Arora (ed.), *Plant Microbes Symbiosis: Applied Facets*,
DOI 10.1007/978-81-322-2068-8_11, © Springer India 2015

rhizobia together with PGPR/PGRs is considered an effective and environment-friendly approach which may profoundly increase the crop yield by different mechanisms of actions under variable conditions. Plant growth promotion of legumes/nonlegumes by rhizobia and their practical application as selected commercially available rhizobial inoculants are the major focus of this chapter.

Introduction

Most of the natural nitrogen inputs for life on Earth are fulfilled by biological reduction of molecular dinitrogen to ammonium. Such fixed nitrogen is acquired by the plants of the legume family through a symbiotic interaction with gram-negative soil bacteria called rhizobia. In the symbiotic interaction with rhizobia, legumes develop nodules which involve signal exchange between the host and the bacterium. Often ineffective nodulation in the legumes is experienced because of the competition of introduced rhizobial strains with the native rhizobial population and/or environmental stresses. In other words, the process of symbiotic nitrogen fixation is dependent on the efficiency and competitiveness of the rhizobial strain for nodulation against indigenous soil bacteria and environmental factors. Investigations have established that the use of compatible, efficient, and effective PGPR in combination with rhizobia could not only suppress/eliminate the effect of other rhizosphere strains and adverse environmental factors but also improve the performance of root-nodulating bacteria and hence nodulation, growth, and yield of the inoculated legumes. Also, the role of rhizobia as root colonizer of many nonlegumes is now well recognized. It has been established that rhizobia could express several characteristics like other PGPR. Rhizobia can increase the availability of plant nutrients and can produce various kinds of chemical compounds, namely, phytohormones, lipo-chitooligosaccharides (LCOs), lumichrome, siderophores, hydrogen cyanide (HCN), exopolysaccharides (EPS), and enzymes, which could improve plant growth and yield directly or indirectly. Rhizobial potential as biological control agents against many pathogens of nonlegumes has also been extensively documented. Rhizobia have also been reported as growth modifier in nonlegumes under stress conditions including salinity, temperature, drought, and heavy metal stress. Exploitation of rhizobia in combination with PGPR also constitutes an interesting alternative to improve their performance and association with crop plants under normal and stress conditions. Even more, it has been considered a promising avenue to optimize plant growth via managing interactions between PGPR, arbuscular mycorrhizal (AM) fungi, and the *Rhizobium*-legume symbiosis. Conclusively, this chapter presents a discussion on the *Rhizobium*-legume symbiosis and the mechanisms of action of rhizobia as PGPR to improve growth and yield of nonlegumes. The interactive effects of rhizobia with other PGPR and PGRs under normal and stress conditions to enhance crop production have been discussed. Finally, rhizobia-based commercially available inoculants have been highlighted.

Rhizobia-Legume Symbiosis: An Overview

Rhizobia-legume symbiosis is a highly coordinated effort between root-nodulating bacteria and the legume in soil for exchange of molecular dialogue (Parniske and Downie 2003). A leguminous plant initiates this symbiotic relationship with the release of certain compounds called flavonoids into the rhizosphere. In response to plant-released flavonoids, the special receptor (nodD) of the only specific rhizobia become activated. In turn, the nodD of rhizobia produces LCOs to sense and to recognize the chemicals released by the host legume roots (Russelle 2008). The LCOs in turn activate nodulins (a set of plant genes). The activated nodulins then cause induction of various biochemical and morphological changes as a result of which cell division in the root cortex occurs (Oldroyd and Downie 2008) and a nodule meristem is thus formed within the root. Resultantly, the growth of root hair is redirected to curl and deform to entrap the attached rhizosphere rhizobia. The entrapped rhizobia start infections as a result of which plant produces a tube called infection thread. Through the infection thread, rhizobia enter into the root cortex and reach the nodule primordia. Then a plant-derived peribacteroid membrane quickly develops around the rhizobia after which the unit is called symbiosome. Within the symbiosome the bacteroids utilize the nitrogenase enzyme to catalyze the conversion of atmospheric N_2 to NH_3 (ammonia), a process called BNF. The bacteroids need large amounts of energy to support their nitrogen-fixing capacity which is provided by the host legume in the form of sugar produced through photosynthesis. It is estimated that the *Rhizobium*-legume symbiosis requires about 10 kg of carbohydrates (sugar) for each kg of N_2 fixed (Schulze 2004; Hubbell and Kidder 2009). Hence, in addition to having good health, the legume host must have enough water and other nutrients under proper sunlight. Likewise, fully effective nitrogen fixation requires rhizobial strains that are highly effective nitrogen fixers. Generally, nodules providing nitrogen to plants formed by effective rhizobia are usually large, often pink and red inside, whereas nodules formed by ineffective rhizobia tend to be white or light green inside, small, often quite numerous, and scattered throughout the root system. The ammonia formed as a result of *Rhizobium*-legume symbiosis is used to form amino acids which move out of the nodule via a compound provided by the plant, to other parts of the plant where it is used mainly to produce proteins.

Mechanisms of Plant Growth Promotion by Rhizobia

Rhizobium spp. could have an impact on plant growth and development in two different ways: directly or indirectly. The direct growth promotion of the plant by *Rhizobium* spp. generally entails providing the plant with compounds called phytohormones, regulating ethylene, solubilizing insoluble phosphates, and facilitating the uptake of nutrients from the environment. Rhizobia can benefit plant growth indirectly by decreasing or preventing some of the deleterious effects of phytopathogenic organisms through induced systemic resistance (ISR), parasitism, antibiosis, competition for nutrients, and production of metabolites suppressive to deleterious rhizobacteria. These mechanisms are discussed as under:

Direct Mechanisms of Growth Promotion

Rhizobia cause promotion in plant growth directly by regulating plant ethylene levels; producing plant hormones; providing useful compounds like lumichrome, riboflavin, and LCOs; sequestering iron by releasing siderophores; and solubilizing insoluble phosphorus via releasing organic acids and enzymes (Mehboob et al. 2009).

Ethylene, a plant hormone, generally shows enhancement in root initiation and growth at its low level, but its higher levels can lead to suppression in root elongation (Shaharoona et al. 2006a). Usually increased ethylene level damages/retards growth of plants facing a variety of environmental stresses like heavy metals, pathogens, drought, and salinity. The enzyme ACC deaminase has been known to hydrolyze ACC (the immediate biosynthetic precursor of ethylene in plants) into ammonia and α-ketobutyrate which can be used by the bacterium as a source of nitrogen and carbon. Several rhizobial strains investigated to produce ACC deaminase enzyme including *Rhizobium japonicum*, *Rhizobium leguminosarum*, *Rhizobium hedysari*, *Rhizobium gallicum*, *Bradyrhizobium japonicum*, *Bradyrhizobium elkanii*, *Mesorhizobium loti*, and *Sinorhizobium meliloti* could stimulate the growth of plants by regulating their ethylene levels upon inoculation (Ma et al. 2004; Uchiumi et al. 2004; Nukui et al. 2006; Hafeez et al. 2008; Duan et al. 2009).

Plant hormones are not nutrients but are chemicals that in small amounts promote and influence the growth and development. In general, it is accepted that there are five major classes of plant hormones, namely, abscisic acid, auxins, cytokinins, ethylene, and gibberellins. Other plant hormones and growth regulators such as brassinosteroids, salicylic acid, jasmonates, plant peptide hormones, polyamines, nitric oxide, strigolactones, and karrikins are not easily grouped into these classes; they exist naturally or are synthesized by other organisms that interrupt the physiological processes within plants. Each hormone has positive and inhibitory functions. It is now well established that there are two sources of phytohormones naturally available for plants: endogenous production by the plant tissues and exogenous production by associated microorganisms. Auxins, abscisic acid, cytokinins, gibberellins, ethylene, and nitric oxide are the most studied plant hormones released by *Rhizobium* spp. during their interaction with plants (Khalid et al. 2006; Perrine-Walker et al. 2007). *Azorhizobium caulinodans*, *B. japonicum*, *M. loti*, *R. leguminosarum*, and *S. meliloti* are the examples of studied strains promoting plant growth via producing various phytohormones (Naidu et al. 2004; Dazzo et al. 2005; Mishra et al. 2006; Boiero et al. 2007; Chandra et al. 2007; Pandey and Maheshwari 2007; Afzal and Bano 2008; Vargas et al. 2009; Chi et al. 2010).

Lumichrome is a signaling compound that can affect plant growth in both natural and agricultural ecosystems (Beveridge et al. 2003). Rhizosphere accumulation of lumichrome has been reported to act as developmental signal (Kanu and Dakora 2009). Lumichrome, a compound that is produced by rhizobia, acts as an enhancer of plant growth prior to the onset of nitrogen fixation (Cooper 2007). Lumichrome is a flavin identified from *Rhizobium* spp. which enhanced nodule number when applied to lotus (*Lotus japonicus*) (Gouws et al. 2012). The stimulation of seedling development in cereal, by lumichrome, has also been recognized (Matiru and Dakora 2005). Also, it is known that lumichrome could orchestrate plant development. It has been identified from culture filtrate of *S. meliloti* cells (Dakora 2003). Moreover, rhizobial inoculation of plants has been suggested to alleviate the effect of water stress through the production of lumichrome (Phillips et al. 1999).

Riboflavin is a vitamin that enzymatically or photochemically changes into lumichrome which in turn stimulates plant growth via motivating root respiration (Phillips et al. 1999; Dakora et al. 2002). It is a necessary component of the flavin coenzymes FMN and FAD. It has been reported that riboflavin could also act as a signal of development upon accumulation in the rhizosphere (Kanu and Dakora 2009). *Rhizobium*, *S. meliloti*, and *R. leguminosarum* bv. *viciae* have been recognized to produce riboflavin as growth promoter of plants (Yang et al. 2002; Dakora et al. 2002).

LCOs commonly known as Nod factors are bacterium-to-plant signal molecules produced by rhizobia during rhizobia-legume symbiosis (Spaink 1992). These have been shown to induce rapid and transient alkalinization of plant cells and to restore or resume cell division and embryo development in various plants (De Jong et al. 1993; Staehelin et al. 1994; Baier et al. 1999; Dyachok et al. 2000). Seed germination, root mass, and length enhancement could also result via rhizosphere application of LCOs at low concentration (10^{-7} to 10^{-9} M). The production of photosynthate and grain yield could also be increased via leaf spray of LCOs in micromolar concentration, i.e., 10^{-6} to 10^{-10} (Miransari and Smith 2009). *Rhizobium* spp. are also known for the induction in expression of flavonoid genes, cell division, and embryogenesis via releasing LCOs (Smith et al. 2002).

Iron, a plant nutrient needed for chlorophyll synthesis, is often present in a highly insoluble form in soil. Deficiency of iron causes initial yellow or white areas between the veins of young leaves leading to spots of dead leaf tissues. Siderophore, a low molecular weight compound, is produced to sequester Fe^{3+} as well as to mobilize it in soil. Sequestration and transportation of iron into plant cell via siderophores by rhizobia is assumed to serve as a source of plant-available iron under iron-deficient environment. Several *Rhizobium* spp. have been known to bind insoluble Fe^{3+} and reduce it to soluble and plant-preferred form, i.e., Fe^{2+} (Antoun et al. 1998; Carson et al. 2000; Arora et al. 2001).

Phosphorus (P) is a major and key nutrient limiting plant growth (Podile and Kishore 2006).

A large portion of applied and soil phosphorus becomes unavailable to plants because of its immobilization in either organic or inorganic form due to low or high pH of soil. At low pH, P is fixed in soil by free oxides and hydroxides of Al and Fe, while at high pH it is fixed by Ca. Generally, rhizosphere bacteria convert the fixed/insoluble phosphates into soluble form through the process of acidification, chelation, and exchange reactions and production of gluconic acid (Rodriguez et al. 2004; Chung et al. 2005; Hameeda et al. 2008). *Rhizobium* spp. have the ability to mobilize the bound phosphorus either by releasing enzymes, phosphatases, or organic acid, namely, citric acid, glutamic acid, carbonic acid, and gluconic acid (Abd-Alla 1994; Alikhani et al. 2006; Jayasinghearachchi and Seneviratne 2006; Afzal and Bano 2008). *B. japonicum*, *R. leguminosarum*, *S. meliloti*, and *Mesorhizobium mediterraneum* are rhizobia that have been investigated to solubilize or mineralize phosphate in the rhizosphere of plants (Egamberdieva et al. 2004; Hara and de Oliveira 2004; Puente et al. 2004; Rodriguez et al. 2006; Fernandez et al. 2007; Afzal and Bano 2008).

Eighteen elements have been declared as essential for plants, each of which has at least one specially defined role in plant growth. *Rhizobium* spp. could contribute significantly in maintaining adequate nutrient uptake from soils by reducing the accumulation of nutrients in agricultural soils. Several studies can be cited about the relations with *Rhizobium* spp. and enhancement of nutrient uptake. For example, Khokhar and Qureshi (1998), Biswas et al. (2000), and Yanni et al. (2001) have reported significantly increased uptake of N, P, K, Ca, Mg, Zn, Na, Mo, and Fe by rice due to inoculation with *Rhizobium* spp. Similarly, Hafeez et al. (2004) demonstrated increased uptake of K and Ca by cotton due to *R. leguminosarum* bv. *trifolii* and enhanced uptake of N, K, Na, Zn, Fe, and Cu by wheat due to *Rhizobium* inoculation (Amara and Dahdoh 1997).

Conclusively, growth and yield of leguminous crops could be triggered by rhizobial inoculation directly via producing various metabolites/substances such as plant hormones, ACC deaminase enzymes, LCOs, siderophores, lumichrome, riboflavin, etc., and/or by fixing/solubilizing/increasing the uptake of mineral nutrients.

Indirect Mechanisms of Growth Promotion

Rhizobia also exhibit several indirect mechanisms of plant growth promotion which involve antibiosis, parasitism, competition for nutrients, and induction of systemic resistance.

Antibiosis is one of the effective mechanisms of rhizobia to prevent the proliferation of phytopathogens via synthesizing antimicrobial compounds known as antibiotics. Many rhizobial species have been reported to secrete antibiotics and cell wall-degrading enzymes that can inhibit/kill the phytopathogen (Siddiqui and Mahmood 2001; Shaukat and Siddqui 2003; Bardin et al. 2004; Chandra et al. 2007; Mabrouk and Belhadj 2010).

Parasitism is a type of symbiosis in which one organism (called parasite) benefits at the expense of another organism usually of different species (called host). The association may also lead to the injury or death of the host. The mechanism of parasitism has been used successfully by *Rhizobium* spp., namely, *R. leguminosarum*, *S. meliloti*, and *B. japonicum*, against fungal pathogens belonging to genera *Macrophomina*, *Rhizoctonia*, and *Fusarium* (Siddiqui et al. 2000; Ozkoc and Deliveli 2001).

A contest in which a winner is selected from among two or more parties is called competition. It is believed that a phytopathogen could be displaced through competition mechanism between rhizobia and pathogen. The best understood example of competition is that for iron. Siderophores are secreted by the rhizobial complex with available iron in the rhizosphere to make it less available to competing phytopathogens. More clearly, in iron competition, the biocontrol agent produces siderophores that sequester iron in the rhizosphere and make it less available to certain deleterious rhizosphere microorganisms. The latter cannot obtain sufficient iron for growth and thus are outcompeted. In other words, plant growth and development could be triggered by biocontrol agents indirectly through starvation by producing siderophore. *Rhizobium* spp. are able to produce siderophores and can inhibit a widely occurring plant pathogen, *Macrophomina phaseolina* (Arora et al. 2001). Various *Rhizobium* species, namely, *R. leguminosarum* bv. *viciae*, *R. leguminosarum* bv. *trifolii*, *R. leguminosarum* bv. *phaseoli*, *R. meliloti*, and *R. tropici*, are able to produce siderophores in iron-deficient environment to chelate it (Carson et al. 2000; Arora et al. 2001).

To make the host plant more resistant to pathogens via inducing a change in its susceptibility is termed as ISR. Rhizobia are able to control pathogens by induction of plant defense mechanisms (Abdelaziz et al. 1996). Several studies have demonstrated the improvements in defense mechanisms of plant via release of biostimulatory agents by rhizobia (Peng et al. 2002; Singh et al. 2006). *Rhizobium etli*, *R. leguminosarum* bv. *phaseoli*, and *R. leguminosarum* bv. *trifolii* have been reported to provide bioprotection during pathogenic stresses via induction of enhanced systemic resistance in potato and rice plants (Reitz et al. 2000; Mishra et al. 2006).

In conclusion, rhizobia can indirectly promote the growth of legumes by inhibiting the phytopathogens via producing antibiotics and fungal cell wall-degrading enzymes and HCN or through starvation via siderophore production.

Synergistic Interaction of Rhizobia with Beneficial Microorganisms: Improving Biomass and Nodulation

A synergistic interaction is an interaction in which two or more elements/parties work together in a system to produce an effect greater than the sum of their individual effects. Figure 11.1 illustrates various strategies to improve rhizobia-legume

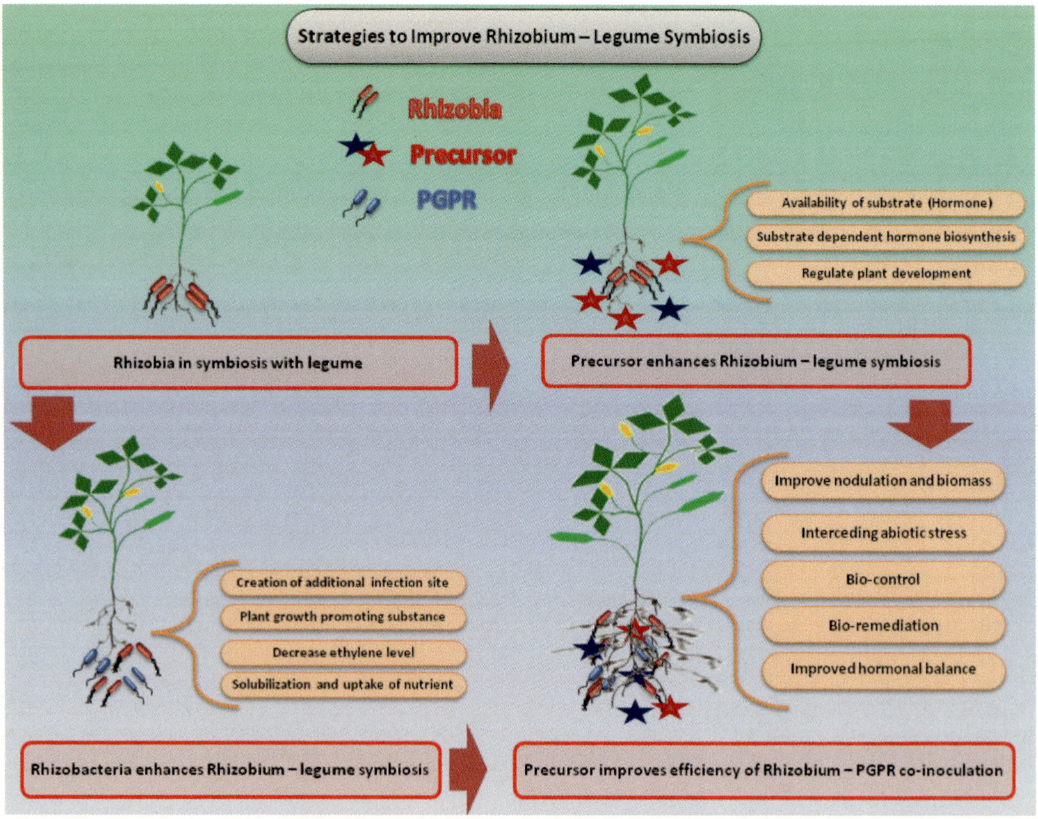

Fig. 11.1 Schematic view of different strategies to improve *Rhizobium*-legume symbiosis

symbiosis. As microbial inoculant formulations containing single strain have often shown inconsistent results, emphasis has been given on co-inoculations in which a co-inoculant interacts synergistically to improve the performance of other beneficial microorganisms (Bashan and de-Bashan 2005). Table 11.1 depicts examples of the synergistic impact of co-inoculation of rhizobia with other PGPR. Synergistic effects of rhizobia when used in combination with beneficial microorganisms have been focused in detail below.

Plant Growth-Promoting Rhizobacteria

Studies have recognized the improvement in plant biomass and nodulation by co-inoculation of rhizobia with PGPR. However, the ability of co-inoculant PGPR depends on the inoculant strain's inherent potential and effectiveness/specificity to certain plant species, cultivar, and genotype, composition of root exudates of host plant, cell density of applied inocula/optimal inoculation dose, temperature variation, and interaction of applied inocula with rhizosphere microflora predominant in the particular crop (Medina et al. 2003; El-Sawy et al. 2006; Remans et al. 2008; Mishra et al. 2009; Fox et al. 2011). Some PGPR from a range of genera improve plant growth and yield when used in combination with rhizobia. Some PGPR from various genera, which improved plant biomass and nodulation when used in combination with rhizobia, are the major focus in the following review:

Azospirillum, a free-living N_2 fixer and a PGPR, has been recognized for its ability to improve nodulation and biomass of legumes as a co-inoculant of rhizobia (Roseline et al. 2008). Improved rhizobial infection, nodule formation,

Table 11.1 Rhizobial inoculation in combination with rhizobacteria or arbuscular mycorrhizal fungi to improve nodulation and biomass of legumes

Rhizobia	Co-inoculants	Host plant	Proposed mechanism(s)/ plant response	References
Rhizobium spp. strain Mg6	PGPR strains A1 and A2	*Phaseolus vulgaris*	ACC deaminase activity	Aamir et al. (2013)
Rhizobium leguminosarum bv. *viciae*	Arbuscular mycorrhizal fungi	*Vicia faba*	Mobilization of P, Fe, K, and other minerals	Abd-Alla et al. (2014)
Rhizobium sp.	PGPR, P-enriched compost	*Cicer arietinum*	Improved growth and nodulation through ACC deaminase activity	Shahzad et al. (2014)
B. japonicum strains MN-S and TAL-102	AM fungi, *Glomus intraradices*	*V. radiata*	Significant improvement in plant biomass and N contents	Yasmeen et al. (2012)
Rhizobium sp. strain PK20	*Pseudomonas* sp. M9	–	ACC deaminase activity of *Pseudomonas* sp. M9	Ahmad et al. (2011, 2013)
R. leguminosarum-PR1	*Pseudomonas* sp. strain NARs1	*Lens culinaris*	Increased growth and nutrient uptake	Mishra et al. (2011)
R. leguminosarum	*Pseudomonas* spp.	*V. faba*	P solubilization, phytohormone and siderophore production	Saidi et al. (2013)
Mesorhizobium sp. BHURC03	*Pseudomonas aeruginosa* BHUPSB02	*C. arietinum*	Enhanced P and Fe uptake, nodulation, and production of IAA	Verma et al. (2013)
R. leguminosarum	PGPR, enriched compost	*L. culinaris*	ACC deaminase activity of PGPR and symbiotic competence of rhizobia	Iqbal et al. (2012)
Rhizobium sp.	*Pseudomonas* sp. LG or *Bacillus* sp. Bx	*P. vulgaris*	P solubilization, IAA, ammonia, and siderophore production	Stajković et al. (2011)

and N_2 fixation activity as a result of the combined inoculation with *Azospirillum* and *Rhizobium* via the enhanced production of plant hormones and siderophores and enhancement of total mineral uptake have been claimed as a mechanism for growth and biomass enhancement of various legumes (Bashan 1998; Wani et al. 2007a; Dardanelli et al. 2008; Cassan et al. 2009). Numerous studies have demonstrated the enhancement of nodulation and growth of a wide array of legumes because of the positive interaction between rhizobia and bacteria of the genus *Azospirillum*, namely, in white clover (Tchebotar et al. 1998), faba bean and chickpea (Rodelas et al. 1999; Wani et al. 2007b), pigeon pea (Deanand et al. 2002), and soybean (Cassan et al. 2009). Species of *Azospirillum*, viz., *A. lipoferum*, *A. brasilense*, *A. amazonense*, *A. halopraeferens*, and *A. irakense*, have been identified so far as co-inoculant of rhizobia (Tarrand et al. 1978; Magalhaes et al. 1983; Reinhold et al. 1987; Khammas et al. 1989).

Bacillus, one of the most familiar soil bacterial groups, has been found to be an effective PGPR. The potential of *Bacillus* together with rhizobia as a promoter of nodulation and biomass of various legumes via improving rhizobial colonization, nodule occupancy, choice of changing nodulation competition between strains, number of infection sites, production of phytohormones and siderophores, biocontrol of diseases, and efficient uptake of nutrients has been investigated (Vessey and Buss 2002; Yuming et al. 2003; Elkoca et al. 2008; Rajendran et al. 2008; Mishra et al. 2009; Medeot et al. 2010; Tsigie et al. 2012; Atieno et al. 2012). Enhanced root and shoot number, length, surface area, nodulation, nodule number, dry weight, and plant height and branches per plant are the reported stimulatory effects of co-inoculation of some of the *Bacillus* strains with rhizobia on various legumes (Guinazu et al. 2010; Stajkovic et al. 2011; Singh et al. 2011).

Another most studied plant PGPR is *Pseudomonas* which belongs to Gram-negative, rod-shaped, nonspore-forming bacteria. Species of *Pseudomonas* including *P. fluorescens* 2137, BHUPSB06, and P-93; *Pseudomonas* sp. strain PGERs 17, FM7d, MRS 13, MRS 16, CDB 35, BWB 21, LG, CPS63, and MPS; *P. striata*; *P. putida*; *P. trivialis* 3Re27; *P. maltophilia*; and *P. jessenii* PS06 have been found to have a role in improving nodulation and biomass of various legumes when used in combination with rhizobia. Root/shoot length, surface area, nitrogenase activity, N_2 fixation, total N content, and plant dry weight are the features of plants which were improved by *Pseudomonas* when used in combination with rhizobia (Deanand et al. 2002; Zaidi et al. 2003; Garcia et al. 2004; Tilak et al. 2006; Malik and Sindhu 2008; Stajkovic et al. 2011; Mishra et al. 2012; Verma et al. 2012).

Azotobacter, free-living, aerobic, N_2-fixing bacteria, is another PGPR which has shown increases in nodulation, plant biomass, and yield of various legumes, namely, lentil, urdbean, soybean, clover, chickpea, and peanut, as a result of its dual inoculation with *Rhizobium* (Chandra and Pareek 2002; Qureshi et al. 2009; Dashadi et al. 2011; Akhtar et al. 2012).

Also, a PGPR, namely, *Serratia*, has proven its ability to invigorate nodulation and biomass of legumes in combination with rhizobia. Increase in root proliferation, infection sites, number and mass of nodule in various legumes are the demonstrated activities of *Serratia* when used in combination with rhizobial strains (Bai et al. 2002; Radwan et al. 2005; Badawi et al. 2011; Zahir et al. 2011).

Increased nodule occupancy and grain yield of various legumes has been reported due to another PGPR, *Enterobacter*, when used in combination with *Rhizobium* (Gupta et al. 1998; Mirza et al. 2007; Mishra et al. 2009).

Arbuscular Mycorrhizal (AM) Fungi

Another promising avenue to optimize plant growth is through managing interactions between plant growth-promoting bacteria (PGPB), AM fungi, and the *Rhizobium*-legume symbiosis. Perhaps fortuitously, the mycorrhizal fungi that are known to form associations with more than 80 % of plant species often enhance nutrient and water uptake. PGPB are able to increase AM fungal development by affecting root colonization as well as by enhancing plant N and P uptake (Richardson et al. 2009). Production of EPS by PGPB significantly enhances the attachment of bacteria to mycorrhizal roots and AM fungal structures that influence the movement of bacteria in the rhizosphere (Bianciotto et al. 2001). Soil microbes are able to enhance the amounts of root exudates resulting in the activation of AM hyphae and hence higher rate of root colonization (Barea et al. 2005). Meyer and Linderman (1986) reported higher nodulation and shoot dry weight in clovers as a result of dual inoculation with AM fungi and PGPB as compared to inoculation with mycorrhizae or PGPB alone, whereas Azcon (1993) reported maximum growth and nutrient uptake as a result of the addition of PGPB cell-free culture to the mycorrhizal and nodulated legume *Hedysarum coronarium*. Silveira et al. (1995) revealed increased nodulation, root growth, and nitrogen and phosphorus uptake by beans as a result of mixed inoculation of fluorescent *Pseudomonas* and *Glomus etunicatum*. Vivas et al. (2003) reported enhanced shoot biomass from 18 % (at 13.6 mg Cd kg^{-1} soil) to 35 % (at 85.1 mg Cd kg^{-1} soil) and N, P, Zn, and Ni contents and transfer of reduced Cd from soil to plants by up to 37.5 % as a result of combined use of indigenous AM fungi plus indigenous PGPB in an experiment with clover grown in Cd-contaminated soil.

However, further research is needed for detailed illustration of interactions between the host plant, AM fungi ,and nodular bacteria by using different molecular techniques so as to enhance the ecosystem productivity.

Rhizobium spp. Containing ACC Deaminase for Improving Nodulation and Biomass

Ethylene production in the infected roots enhances as a result of *Rhizobium* invasion into legume roots during the nodulation process because of which the inhibition of nodulation in various legumes (viz., alfalfa, mung bean, *Pisum sativum*, and *Medicago sativa*) has been reported (Lee and LaRue 1992; Caba et al. 1998). Though enhancement of nodulation in various legumes by lowering ethylene levels via chemicals like aminooxyacetic acid (AOA), L-α-(aminoethoxyvinyl)-glycine (AVG), and rhizobitoxine has been established, it has also been documented (Nukui et al. 2000; Tamimi and Timko 2003) that there are rhizobial strains having an enzyme ACC deaminase which could reduce this stress by decreasing the level of ethylene in the host plant (Ma et al. 2003b). Hence, isolation and introduction of efficient rhizobial cultures with ACC deaminase have been stressed by Elsheikh and Ibrahim (1999) in order to improve the nodulation status, seed quality, and legume productivity.

Strains of *R. leguminosarum* having ACC deaminase action were more effective for pea nodulation (Ma et al. 2003b). Likewise, *S. meliloti* containing ACC deaminase gene (*acdS*) derived from *R. leguminosarum* also indicated increased capability to nodulate alfalfa (Ma et al. 2004). Uchiumi and colleagues (2004) showed that the inactivation of the gene *acdS* in *M. loti* resulted in a decreased number of nodules on *L. japonicus*, in comparison with the number of nodules formed by the wild-type strain.

Brígido et al. (2013) evaluated the symbiotic performance of two *Mesorhizobium ciceri* strains, transformed with an exogenous 1-aminocyclopropane-1-carboxylate deaminase gene (*acdS*), in chickpea plants under salinity stress. The EE-7 (salt-sensitive) and G-55 (salt-tolerant) *M. ciceri* strains were transformed with an *acdS* gene present on plasmid *pRKACC*. They recorded a significantly higher shoot dry weight in plants inoculated with the *acdS*-transformed salt-sensitive strain compared with the plants inoculated with the native strain in the presence of salt. The negative effects of salt stress were reduced when using *acdS*-transformed strains in comparison with the wild-type strains. They further demonstrated that in expressing the exogenous *acdS* gene, the salt-sensitive strain was able to induce nodules to the same extent as the salt-tolerant strain. Similarly, Nascimento et al. (2012) described that rhizobial strains expressing the enzyme ACC deaminase displayed an augmented symbiotic performance as a consequence

Table 11.3 (continued)

Rhizobium spp.	Host plant	Proposed mechanism(s)/plant response	Reference
Temperature stress			
Mesorhizobium sp.	Not described	Increased transcriptional induction of chaperone genes	Alexandre and Oliveira (2011)
Bradyrhizobia strains	*Glycine max*	Efficient nitrogen fixation at high temperatures	Rahmani et al. (2009)
Rhizobium sp.	*Prosopis juliflora*	Improved symbiosis and nitrogen fixation	Kulkarni and Nautiyal (2000)
Sinorhizobium sp.	*Acacia senegal*	Competent colonization and nodulation ability	Raësaënen and Lindstroëm (1999)
Rhizobium sp.	*P. vulgaris*	Increased uptake of methionine and protein synthesis	Michiels et al. (1994)
Heavy metal stress			
B. japonicum E109	*Glycine max*	Reduced symbiosis due to As toxicity	Talano et al. (2013)
Rhizobium, *Sinorhizobium* spp.	*Lathyrus sativus*, *Lens culinaris*, *Medicago marina*, *M. truncatula*, *M. minima*	Phytoremediation of Cd-contaminated soil	Guefrachi et al. (2013)
Cupriavidus necator	*Leucaena leucocephala*, *Mimosa pudica*, *Mimosa caesalpiniaefolia*	Bioremediation of Zn-, Cu-, Pb-, and Cd-contaminated soils	Ferreira et al. (2013)
S. meliloti	*M. sativa*	Promotes the bioaccumulation of Zn in nodulated roots	Zribi et al. (2013)
Rhizobium spp.	*L. culinaris*	Decreased uptake of Ni in contaminated soil and improved plant growth	Wani and Khan (2013)
Mesorhizobium spp.	*Anthyllis vulneraria*	Decreased Zn concentration in both roots and shoots	Soussou et al. (2013)
Bradyrhizobium sp.	*V. radiata*	Survive under Zn stress and developed efficient symbiosis	Pongsilp and Boonkerd (2011)
Azorhizobium caulinodans	Not described	Avoidance of excessive Cd through absorption with lipopolysaccharides and cadmium extrusion	Zhengwei et al. (2005)
Biotic stress			
Rhizobia	*C. arietinum*	Suppression of *Rhizoctonia solani*	Hemissi et al. (2013)
R. leguminosarum strain RhOF4	*V. faba*	Reduction in the impact of cyanotoxin biohazard	Lahrouni et al. (2013)
Mesorhizobium loti	*Brassica campestris*	Suppression of white rot disease/*Sclerotinia sclerotiorum*	Chandra et al. (2007)
R. leguminosarum bv. *phaseoli* strain RRE6, R. *leguminosarum* bv. *trifolii* strain ANU843	*Oryza sativa*	Suppression of *Rhizoctonia solani*	Mishra et al. (2006)
R. etli strain *G12*	Potato	Suppression of cyst nematode (*Globodera pallida*)	Reitz et al. (2000)
R. meliloti	Sunflower, okra	Suppressed *Macrophomina phaseolina*, *Rhizoctonia solani*, *Fusarium solani*	Ehteshamul-Haque and Ghaffar (1993), Arora et al. (2001)
R. leguminosarum bv. *viciae*	Sugar beet	Suppressed pythium/damping-off disease	Bardin et al. (2004)

probably produce antioxidants, osmolytes, and stress proteins to survive under harsh environments. The ability of these beneficial bacteria to tolerate severe water-deficit conditions can be utilized to ameliorate drought impact on plants (Goyal et al. 1986). They can induce stress tolerance in plants by producing physical and chemical changes (Yang et al. 2009). Different mechanisms for stress amelioration and plant growth promotion by rhizobia have been observed including production of chaperones and sugars (Berjak 2006), synthesis of stress enzyme ACC deaminase (Zahir et al. 2009), EPS production (Alami et al. 2000), production of low molecular weight organic compounds like trehalose (Zahran 1999), phosphate solubilization (Zaidi et al. 2009), improved nutrient availability (Hussain et al. 2009), production of siderophores (Arora et al. 2001), phytohormone production (Khalid et al. 2006), and enhanced root respiration (Volpin and Phillips 1998) by influencing plant physiology.

Rhizobia can relieve the effect of water-deficit stress by altering root morphology, carbon assimilation rate, transpiration rate, and leaf stomatal conductance in inoculated non-nodulating crops (Chi et al. 2005). A significant improvement in root/shoot length and nutrient and water use efficiency of sunflower seedlings was recorded due to *Rhizobium* species (strain YAS34) inoculation under drought by Alami et al. (2000). They suggested that the improvement was due to the EPS-producing ability of the strain YAS34. Sandhya et al. (2009) recorded substantial capability of rhizobia for improving water-holding capacity of the rhizosphere by developing microaggregates and biofilm around the plant roots to improve nutrient uptake and water flow. In the same way, Kaci et al. (2005) isolated the rhizobial strain KYGT207 from the drying soil and measured improvement in wheat growth under water-deficit conditions due to better rhizosphere soil aggregation. Similarly, Belimov et al. (2009) recorded improvement in drought-abiding ability of potato by the inoculation of PGPB containing ACC deaminase in a pot trial. Zahir et al. (2008) measured improved water use efficiency in plants inoculated with ACC deaminase producing PGPB due to increase in root length. Microbial chaperones have also been reported to improve plant growth under desiccated conditions by Castiglioni et al. (2008). They inoculated chaperones producing bacteria to maize and observed water-deficit stress alleviation in maize. Forchetti et al. (2010) isolated PGPB from sunflower and measured their ability to produce salicylic acid under drought. They also observed phosphate solubilization and oxidase, catalase, and jasmonic acid production ability of these PGPB. They concluded that bacteria having capability to produce antioxidants and osmoprotectants can be exploited as biofertilizers to sunflower under water-limited conditions.

Inoculation with drought-tolerant PGPR to crops can be a tool to induce drought tolerance in cereals leading to sustainable production goals. Further, rhizobia having EPS, sugars, and protective protein-producing ability and antioxidant activity can help plants to survive under stress and improve the efficiency of nutrient (Tilak 2004; Walley et al. 2007) and water use (Miller et al. 2003; Lenssen et al. 2007).

Salinity Stress

Certain rhizobial isolates are capable to survive under saline conditions, and their distribution in the rhizosphere varies with their tolerance ability (Tate 1995; Naz et al. 2009; Sgroy et al. 2009). However, salinity tolerance may vary due to the growth habit of rhizobial strains. *Rhizobium* species grow fast, they are more salt tolerant, whereas *Bradyrhizobium* species are comparatively less salt tolerant due to slow growth rate (Elsheikh and Wood 1995). The strains from genus *Sinorhizobium* are more salinity tolerant than the strains belonging to *Rhizobium* and/or *Mesorhizobium* (Rejili et al. 2012). On the other hand, it has been reported by Amarger et al. (1997) that salinity tolerance is a strain-specific property rather than species specific. Moreover, salinity tolerance limit of rhizobia has been reported up to 100 mM NaCl and higher levels (Yelton et al. 1983; Zahir et al.

2010b; Laranjo and Oliveira 2011). Salinity tolerance in rhizobia is a complex phenomenon, and many regulatory mechanisms get started to safeguard the cell from damage (Zahran 1999). One of them includes intracellular accumulation of osmolytes (low molecular weight solutes) like trehalose, polyamines, glycine betaine, glutamate, and ionic potassium. *S. meliloti* has been reported to accumulate glycine betaine under stress through uptake or synthesis or both (Boscari et al. 2002).

Several studies have indicated improvement in crop growth and productivity with the inoculation of salinity-tolerant rhizobia which also minimizes the production costs and environmental hazards (Hartmond et al. 1987; Dixon et al. 1993). In a study conducted by Bano and Fatima (2009), alleviation of salinity stress in maize was observed due to the combined inoculation of *Rhizobium* with *Pseudomonas*. They recorded increases in the uptake of Ca, P, and K in maize; however, Na accumulation remained unaffected. Inoculation also increased the proline and protein content in maize but decreased leakage of electrolytes. Similarly, Hafeez et al. (1987) recorded significant improvement in growth and development of *Vigna radiata* due to inoculation with salt-tolerant rhizobia under salinity. Similarly, a higher number of nodules and an increase in nodule dry weight of mung bean were observed by Bouhmouch et al. (2005) due to *Rhizobium* strain RP119 inoculation at 340 mM NaCl salinity. Rhizobial isolates with indoleacetic acid (IAA)-producing ability have also demonstrated improvement in host plants under osmotic stress conditions (Malhotra and Srivastava 2006; Spaepen et al. 2007). Comparison between two strains of *S. meliloti*, 1021 and RD64, for inducing salinity tolerance at 0.3 M NaCl salinity revealed that the *M. truncatula* plants inoculated with RD64 showed lower expression of ethylene-related genes and improved nitrogen fixation and shoot dry weight and showed less symptoms of senescence over 1,021 inoculated plants (Bianco and Defez 2009). Zahir et al. (2010b) inoculated lentil with *Rhizobium phaseoli* strain N20 which was efficient in auxin production and was salt tolerant. They observed increase in the growth and yield of the crop under salinity due to inoculation, whether the precursor of auxin (L-tryptophan) was present or not. However, the combined application of *Rhizobium* N20 and L-tryptophan gave the highest increases in plant height, nodule numbers, plant biomass, grain yield, and grain nitrogen up to 28, 71, 61, 65, and 22 %, respectively, over uninoculated control treatment. After a series of experiments, Abdel-Salam et al. (2011) suggested that *Rhizobium* species have great potential for inducing salinity tolerance and increasing production/nodulation of legumes in arid and semiarid areas of the world. Moreover, Soliman et al. (2012) revealed that the combined inoculation of *Sinorhizobium terangae* and AM fungi mitigated the impact of salinity in *Acacia saligna* in sandy loam soil. They further explained that the combined inoculation increased plant vigor and growth rate at higher salinity; improved nodulation and chlorophyll, carbohydrate, and proline contents; increased N, P, K, and Ca contents; and reduced the Na uptake/accumulation.

Conclusively, rhizobial isolates capable to survive high salinity and having plant growth-promoting characteristics can be a better option to safeguard the crop plants from hazardous impacts of salinity. Moreover, the production of osmolytes, phytohormones, and stress-relieving enzymes would be helpful in rescuing the plants' defense system under salinity stress.

Temperature Stress

Rhizobia are highly sensitive to high temperature (Meghvansi 2006) because it hampers their growth, limits survival, and reduces root colonization capacity (Munevar and Wollum 1981). However, optimal temperature for their growth ranges from 25 to 31 °C (Harwani 2006), whereas most of the rhizobia cannot grow at 37 °C (Graham 1992). But high temperature tolerance ability varies from species to species and even from strain to strain (Laranjo and Oliveira 2011). Several studies have indicated the ability of rhizobia to survive at high temperatures due to the production of heat-shock proteins (Vriezen et al. 2007) and accumulation of compatible solutes or protectants (Santos and da Costa 2002). Therefore, researchers tried to find out that the rhizobial isolates, which can survive and tolerate high temperatures, could be helpful to plants under heat stress upon inoculation. Rahmani et al. (2009) inoculated soybean with thermotolerant *Bradyrhizobium* strain and recorded significant increases in nitrogen fixation, nitrogen contents, shoot dry weight, and grain yield. They suggested that the isolates from geographically hot regions are the potential candidates to be used as inoculants under heat stress. Zahran (1999) has also recognized the selection of rhizobia on the basis of stress tolerance as a good criterion. Bianco and Defez (2009) inoculated heat-tolerant *S. meliloti* strain 1021 to *M. truncatula* and observed reduced senescence, more nitrogen fixation, less reduction in shoot dry weight, and reduced expression of ethylene signaling genes. However, the use of compatible rhizobial strains along with intrinsically compatible cultivars suited to the local environment and rhizobia would be more appropriate (Shamseldin 2007). For this reason, Shamseldin (2007) conducted an experiment in Egyptian soil using indigenous rhizobia and common bean cultivar to measure the competitive ability of *R. etli* (strains EBRI 1 and EBRI 26) to occupy nodules as compared to gus-labeled *R. tropici* CIAT 899G. Results from the study described 78 and 62.5 % more nodule occupancy by EBRI 1 and EBRI 26, respectively, over strain CIAT 899G at 35 °C. Similarly, Kulkarni and Nautiyal (1999) inoculated *Prosopis juliflora* with five rhizobial strains NBRI12, NBRI133, NBRI329, NBRI330, and NBRI332 (tolerant to high temperature stress) and observed 30–89 % improvement in dry weight with respect to uninoculated control.

Thus it can be assumed that the rhizobial strains efficient in their survivability under high temperature, competent in root/nodule colonization, and belonging to the indigenous environment could be suitable inoculants in ameliorating the impact of high temperature stress in plants.

Heavy Metal Stress

Unlike organic pollutants, metals cannot be degraded/mineralized, but they can be transformed from toxic to nontoxic form. Therefore, microorganisms have evolved intrinsic ability to reduce or detoxify heavy metals through various mechanisms. These mechanisms may include biosorption, bioaugmentation, biostimulation, metal reduction, and metal ion efflux from the cell (Outten et al. 2000). For this reason, using heavy metal detoxifying or reducing microbes is an economical and preferred approach for reducing stress impact on plant growth and bioremediation technologies. Keeping this in view, Gupta et al. (2004) reported PGPB as heavy metal stress-reducing agents. Similarly, enhanced phytoremediation efficiency has been reported by the inoculation of metal-resistant and nitrogen-fixing bacteria (Khan et al. 2009). Bacteria have also the capability to adsorb or desorb toxic heavy metals like Zn and Ni to reduce their toxic effect on plants (Mamaril et al. 1997). *R. leguminosarum* bv. *viciae* isolated from the root nodules of faba bean has been reported as an efficient biosorbent for Cd^{2+} and Co^{2+}. The strain was highly tolerant to heavy metal stress in descending order, i.e., Ba^{2+}, 80 ppm; Zn^{2+}, 70 ppm; Co^{2+}, 50 ppm; Al^{3+}, 40 ppm; Ni^{2+}, 30 ppm; and Cd^{2+}, 10 ppm (Abd-Alla et al. 2012). Increased dry matter and improved symbiosis in peas were observed due to the inoculation of *Rhizobium* in Cd-contaminated soils (Ausili et al. 2010). Panigrahi and Randhawa (2010) alleviated the impact of As toxicity in alfalfa through the inoculation of *S. meliloti* strain 1021, and they observed higher concentration of As in the shoots of inoculated plants. Similarly, Dary et al. (2010) reported 4–10 times increase in metal accumulation and more biomass and nitrogen content in *Lupinus luteus* plants inoculated with *Bradyrhizobium* sp. in a multimetal-contaminated soil. They suggested that the inoculation improved phytostabilization potential of *L. luteus* in multi-metal-contaminated soil. Fan et al. (2011) inoculated *Medicago lupulina* with Cu^{2+}-tolerant *S. meliloti* CCNWSX0020 and recorded increase in biomass up to 46 and 78 %, respectively, in a growth medium containing 0 and 100 mg Cu^{2+} kg^{-1}. Inoculation with *S. meliloti* also increased the Cu^{2+} contents in plant tissues up to 39 % where Cu^{2+} was supplied at 100 mg kg^{-1}. Finally, they suggested that the symbiotic relationship between *S. meliloti* CCNWSX0020 and *M. lupulina* can be utilized for the bioremediation of Cu^{2+}-contaminated soil. Rhizobia in effective symbiosis improve the soil fertility with nitrogen fixation and act as bioremediator for As in contaminated soils (Mandal et al. 2008). In an experiment conducted by Wani et al. (2008a), lentil was inoculated with Zn-tolerant *Rhizobium* RL9 under Zn contamination at 4,890 mg kg^{-1}. They measured increases in the grain protein, grain yield, nodule number, nodular dry mass, leghemoglobin, and plant dry matter up to 8, 10, 15, 27, 30, and 150 %, respectively. In another study, Wani et al. (2008b) demonstrated that the Cr contamination in soil at 136 mg kg^{-1} caused toxicity to chickpea plant, but the inoculation with Cr-reducing *Mesorhizobium* strain RC3 improved the grain protein, seed yield, dry matter, and nodule number up to 16, 36, 71, and 86 %, respectively, as compared to uninoculated control.

Biotic Stress

Plants suffer from several bacterial, fungal, or viral diseases during their life cycle where PGPB may help crop plants to overcome the situation through antibiosis, antagonism, competition, and ISR. Rhizobia have been reported as effective biocontrol agents in crop plants. They produce secondary metabolites in the form of antibiotics (Robleto et al. 1998), chitinases (Kacem et al. 2009), HCN (Ahemad and Khan 2009), and β-1-3-glucanase and proteases (Compant et al. 2005). Rhizobia produce bacteriocins and proteinaceous compounds to kill closely related pathogenic bacteria (Hafeez et al. 2005). Although the bacteriocins are narrow-spectrum antibacterial compounds, they are very effective against bacterial plant pathogens (Cladera–Olivera et al. 2006). Kacem et al. (2009) recorded the highest potential of *Rhizobium* species strains ORN 24 and ORN 83 for controlling the olive knot disease causing agent *Pseudomonas savastanoi* by the production of bacteriocins. Similarly, *R. leguminosarum* bv. *trifolii* has been reported to produce high-spectrum antibiotic peptide trifolitoxin against plant pathogens (Triplett et al. 1994; Breil et al. 1996). Certain rhizobia are efficient siderophore producers and their affinity to sequester Fe^{3+} varies tremendously (Ahemad and Khan 2010). Siderophore-producing ability makes rhizobia competent inhabitants in the rhizosphere where they outcompete the pathogenic microorganisms. Arora et al. (2001) reported that iron-deprived pathogens (*M. phaseolina*) die due to iron deficiency in the presence of siderophore-producing rhizobia. In the same way, Deshwal et al. (2003) evaluated the siderophore-producing *Bradyrhizobium* strains AHR 2, AHR 5, and AHR 6 against *M. phaseolina* which caused charcoal rot disease in peanut. The isolates inhibited the growth of *M. phaseolina* and reduced its population in peanut rhizosphere. Vargas et al. (2009) showed varied antagonistic potential of *R. leguminosarum* bv. *trifolii* against fungus *Verticillium* species, and they suggested that the variation was due to the difference in the type of siderophores produced by rhizobia. In addition to an increased iron availability in the rhizosphere due to siderophore activity, plants and microbes can also be saved from Al^{3+} toxicity. *R. leguminosarum* bv. *viciae* alleviates Al^{3+} toxicity by the production of hydroxamate siderophore vicibactin (Rogers et al. 2001). However, chitinase produced by rhizobia degrades the cell wall of pathogenic fungi. Rhizobia belonging to *R. leguminosarum*, *B. japonicum*, and *S. meliloti* have been reported to kill pathogenic fungi (*Rhizoctonia*, *Macrophomina*, and *Fusarium*) by dissolving their cell walls (Ehteshamul–Haque and Ghaffar 1993; Ozkoc and Deliveli 2001). In the same way, Hossain and Martensson (2008) also observed dissolution of fungal mycelium at the initial growth stage in the presence of chitinase-producing rhizobia. Moreover, rhizobia have been documented as elicitors or inducers of resistance against disease/pathogens in crop plants (Elbadry et al. 2006). They co-inoculated

R. leguminosarum bv. *viciae* FBG 05 with *P. fluorescens* FB 11 and recorded significant reduction in the incidence of bean yellow mosaic potyvirus (BYMV) disease in faba bean. Plants have multicomponent defense systems including structural barriers, hypersensitive response, and chemical weapons which are activated at varied extent with the inoculation of PGPB (Dixon et al. 1994). Rhizobial inoculation in chickpea activated the defense against *Fusarium oxysporum* which ultimately leads to the accumulation of phytoalexins and stimulated chalcone synthase, ammonia lyase, and isoflavone reductase enzymes (Arfaoui et al. 2007).

Rhizobia as Plant Growth-Promoting Bacteria (*Rhizobium*: Nonlegumes Association)

Rhizobia have been reported as symbionts of legumes where they fix atmospheric nitrogen and synthesize certain PGRs for the improved growth and productivity of plant. However, few researchers have demonstrated their ability to live in association with nonlegumes where they could be beneficial and harmful or have no effect on plant growth (Prayitno et al. 1999; Mehboob et al. 2011). But those who benefit plant growth while living in the rhizosphere/rhizoplane of the plant are termed as PGPR (Vessey 2003). However, these PGPR account up to 2–5 % of total bacterial population in the rhizosphere (Kloepper and Schroth 1978). For mutual association, both the partners contribute to have that relationship; therefore, plants exude substances (40 % of the plant dry matter) including amino acids, vitamins, organic acids, enzymes, nucleotides, sugars, and other plant metabolites in the rhizosphere (Rovera 1965; Raaijmakers et al. 2009). Associated microbes use these exudates as nutrient source and perform plant growth-promoting activities. Naher et al. (2009) observed a significantly higher population of *Rhizobium* species in the rhizosphere of rice where exudates produced by the rice roots included sucrose, arabinose, mannose, and xylose. Hence, it can be said that inoculated rhizobia/PGPR need to be associated with the plant roots with substantial population size so that they can improve plant growth by improving nutrient availability and rescuing hormonal and osmotic gradients by the production of osmolytes, PGRs, and stress-regulating enzymes. Induction of stress tolerance in plants, suppression of pathogenic fungi and bacteria in the rhizosphere by the production of chitinase and antibiotics, and competition for iron due to the production of siderophores are also the attributes of PGPB. Improvement in germination, root architecture, seedling vigor, shoot growth, stomatal conductance, chlorophyll contents, leaf area, photosynthesis, protein contents, grain yield, harvest index, and tolerance to abiotic stresses has been observed in nonlegumes due to the inoculation of rhizobia (Hilali et al. 2001; Hafeez et al. 2004; Hossain and Mårtensson 2008), whereas certain researchers have recorded plant growth inhibition in nonlegumes due to the overproduction of IAA (Antoun et al. 1998), HCN production (O'Sullivan and O'Gara 1992), auxin, and nitrates (Perrine-Walker et al. 2007) and bacteriocin-like compound production (Perrine-Walker et al. 2009) by inoculated rhizobia. The association between rhizobia and nonlegumes varies with respect to ecological conditions, soil type, indigenous microflora, cultural conditions, inoculated bacterial species, and plant variety (Depret et al. 2004; Mehboob et al. 2008; Hussain et al. 2009).

Rhizobia have associative relationship with nonlegumes, but that association can be loose or strong. They can associate to non-nodulating plants in three different ways depending on the place of the plant body. If they live in the rhizosphere, inside plant tissues, or upper plant parts, then their association is termed as rhizosphere, endophytic, or phyllospheric, respectively (Weyens et al. 2009; Mehboob et al. 2012). Colonization and survival potential of rhizobia with the roots of nonlegumes are comparable to their colonization capability in host legumes (Bhattacharjee et al. 2008). Many scientists have reported rhizobial colonization ability in different nonlegume crops including wheat (Afzal and Bano 2008), rice (Senthilkumar et al. 2009), maize (Cassan et al. 2009), cotton (Hafeez et al.

2004), potato (Sturz and Kimpinski 1999), tobacco (Ji et al. 2010), and lettuce (Noel et al. 1996). Nodule-like structures (hypertrophies) have been reported on the roots of nonlegumes due to rhizobia, but there was no evidence for atmospheric nitrogen fixation (Naidu et al. 2004). Shimshick and Hebert (1979) calculated 8×10^9 number of binding sites of rhizobia per gram fresh root (gfr) of wheat and rice at 22 °C by using the Langmuir adsorption isotherm. Similarly, Chabot et al. (1996) observed 4.1 and 3.7 CFU/gfr on maize and lettuce roots after 4 and 5 weeks of seeding. Rhizobial colonization of the root cortex, root meristem, and xylem of wheat was demonstrated by Sabry et al. (1997). In the same way, Chi et al. (2005) recorded the colonization of rhizobia from the rhizoplane to the stem base, leaf sheath, and leaves of rice plants.

Rhizobia are remarkable colonizers of non-nodulating plant roots, and during their stay in the rhizosphere, they utilize root exudates as carbon source and perform different plant growth-promoting activities to benefit plant growth. Hence, it can be concluded that the rhizobia and plants have an association where they contribute their parts to continue that beneficial mutual association.

Precursors (Phytohormone): Inoculum (Rhizobial) Interaction for Enhancing Crop Production

It is a well-established fact that rhizobial inoculation of legumes/nonlegumes (with suitable and efficient strains of rhizobia) significantly increases the yield of crops. Use of biological inoculants based on rhizobia/PGPR appears to be a promising supplement to chemicals, and biological inoculants are used worldwide for improving the growth and yield of inoculated legumes/cereals. Although inoculation of crop plants with efficient rhizobial/rhizobacterial strains significantly improves the growth and yield of legumes/nonlegumes and various researchers have reported 5–20 % increase in yield of different crops by these inoculants, there are certain limitations in this technology, especially the inconsistent results in field conditions due to a number of soil and environmental conditions. To improve the effectiveness of the inoculum and enhance reproducibility, Frankenberger and Poth (1987) developed the idea of precursor-inoculum interaction based upon the hypothesis that inoculation of PGPB in the presence of specific physiological precursor of a PGR is often more effective in promoting plant growth than inoculation alone. Several published studies have demonstrated the success of precursor-inoculum interactions on nonlegumes; however, scanty information of this approach is available on legumes. This area needs intensive research in depth and detail.

Tryptophan is considered to be the most suitable precursor of auxins in plant species and soil microorganisms (Khalid et al. 2006; Sridevi and Mallaiah 2008). Upon their addition to soil, an increase in auxin contents and its influence on plant growth and yield has been observed and correlated to microbial-derived auxins (Frankenberger and Arshad 1995; Zahir et al. 2005, 2010a, b).

Frankenberger and Poth (1987) conducted greenhouse experiments to study the effect of an auxin-producing ectomycorrhizae, *Pisolithus tinctorius*, and L-TRP on the growth of Douglas fir. Results revealed that the growth of Douglas fir was stimulated by *P. tinctorius* inoculation in supplementation with lower concentrations of L-TRP (10^{-8} to 10^{-6} M) most probably due to the interaction of soil indigenous microbiota with L-TRP. Hussain et al. (1995) used precursor-inoculum interaction for improving lentil yields. They observed that *Rhizobium* inoculation alone increased the grain yield by 21 % over uninoculated control; however, when combined with L-TRP (1.7 mg kg^{-1}), the *Rhizobium* inoculation further increased the grain yield by 31 % over the control. However, they further reported that auxin production by rhizobial inoculation increased by increasing L-TRP levels and incubation period. Anjum et al. (2009) studied precursor-inoculum interactions and reported that although rhizobia produced auxin without L-TRP supplementation, its addition to the medium increased auxin biosynthesis many folds under axenic conditions. They further reported that rhizobial inoculation increased mung bean seedling growth and correlated to microbial-derived auxins in a pouch experiment under axenic conditions. Khalid et al. (2004) conducted trials on auxin biosynthesis in rhizosphere vs. non-rhizosphere soil and reported higher auxin production with selected PGPB in the rhizosphere soils compared to uninoculated soil. They concluded that inoculation with suitable strains and/or amendment with L-TRP could promote auxin biosynthesis in the rhizosphere soils ultimately resulting in better plant/crop growth and development.

Similarly, Zahir et al. (2010a) studied precursor-inoculum interactions and reported that although the separate application of rhizobia and/or L-TRP improved the plant growth and yield, their combined effect was more pronounced in improving growth and yield of mung bean (*V. radiata* L.) under field conditions. *Rhizobium* inoculation supplemented with L-TRP (10^{-4} M) gave the most promising results and significantly increased the nodule number and biomass plant^{-1}, grain yield, and 1,000-grain weight significantly as compared to uninoculated control. In another study, Zahir and co-workers (2010b) evaluated the performance of a rhizobial isolate N20 in the presence and absence of L-TRP for improving the growth and yield of mung bean under saline conditions. Upon their combined application to seed/soil, an increased influence on plant growth and yield was observed and correlated to microbial-derived auxins produced in the rhizosphere that reduced the adverse effects of salinity. They further concluded that supplementing *Rhizobium* inoculation with L-TRP could be a useful approach for improving the growth and yield of mung bean. Very recently, Qureshi et al. (2013) evaluated the potential of *Rhizobium* to enhance growth and fodder yields of maize in the presence and absence of L-TRP. The interaction of L-TRP and rhizobial species (Cp3, Br3, and Lt2) increased the fresh fodder and dry matter yield than their separate application. Likewise, Qureshi and his co-workers (2012) assessed the co-inoculation effect of *Rhizobium* and P-solubilizing *Bacillus* sp. in the presence of L-TRP. They observed that co-inoculation improved the pod and straw yield considerably, but the effect was more pronounced with L-TRP. They finally concluded that precursor-inoculum interaction is an efficient approach; however, it should be studied more comprehensively in different ecological zones to sustain the crop yield.

Overall, the published literature revealed the effectiveness of precursor-inoculum interactions than their separate application for improving the growth and yield of legume/nonlegume crops. However, further research is needed to unlock several horizons like physiochemical transformation of PGRs in the soil, screening efficient and inexpensive precursors of PGRs, and agronomic practices which could enhance the stability and bioavailability of PGRs in plant root zone and to develop an effective approach based upon precursor-inoculum interaction for the prosperity of agriculture industry.

Rhizobium-Based Commercially Available Inoculants

The application of PGPB for improving crop production is becoming an emerging technology owing to their environment-friendly traits. To take advantage of the demonstrated beneficial effects of various soil microbial groups in increasing plant growth and yields, many types of microbial inoculants (biofertilizers) have been in use for a long time. Biofertilizers are biological preparations containing live or latent cells of microorganisms or their metabolites, which when inoculated to seed, soil, or roots of seedlings promote plant growth and enhance harvestable yield. Biofertilizers, generally marketed, contain microbes capable of N_2 fixation, phosphate solubilization/mineralization, phytohormone production, and biocontrol activities. Rhizobial strains that are often used (singly or in mixture) include a range of genera: *Allorhizobium*, *Azorhizobium*, *Bradyrhizobium*, *Mesorhizobium*, *Rhizobium*, and *Sinorhizobium*. Species of other genera are also used but somewhat less frequently. Rhizobia-based inoculants have been used to improve legume nodulation and yield of legumes/nonlegumes, while bacteria belonging to *Azotobacter* and *Azospirillum* have been applied

Table 11.4 Commercially available *Rhizobium*-based biofertilizers

Bacteria	Product	Company
Rhizobia	VAULT® HP plus INTEGRAL®	Becker Underwood Corporate, USA
Delftia acidovorans and *Bradyrhizobium*	BioBoost	BrettYoung Seeds Ltd., Canada
Rhizobium sp.	SeedQuest®	Soygro (Pty) Ltd., South Africa
Rhizobium sp.	Legumefix	Legume Technology Ltd., UK
Bacillus subtilis and *Bradyrhizobium Japonicum*	HiStick N/T, Turbo-N	Becker Underwood Corporate, USA
B. subtilis and *B. japonicum*	Patrol N/T	United Agri Products (UAP) Inc., Canada
Burkholderia cepacia type Wisconsin	Deny	Market VI LLC, Vern Illum 6613 Naskins Shawnee KS 66216, USA
Rhizobium spp.	*Fasloon Ka Jarasimi Teeka*	AARI, Faisalabad, Pakistan
Rhizobium spp.	BioPower	NIBGE, Faisalabad, Pakistan
Rhizobium spp.	Biozote	NARC, Islamabad, Pakistan
Rhizobium spp. and PGPR	Rhizogold	ISES, UAF, Faisalabad, Pakistan

to enhance cereal growth (as biofertilizers), and mainly *Bacillus* and *Pseudomonas* have been applied for biocontrol (as biopesticides) of plant diseases (Fravel 2005; Bravo et al. 2011). Table 11.4 shows some selected commercially available rhizobial inoculants with their producers/trade name. The development of techniques for the large-scale production of pure inoculants, with high infectivity potential, is the main issue to be tackled in order to allow a wide use of biofertilizers. The key aspects in rhizobial inoculation (biofertilizers) technology are the use of a proper formulation of inoculant preparations, the selection of an adequate carrier, and the design of correct delivery methods.

Conclusions and Future Perspectives

The conventional approach to increase agricultural productivity through massive inputs of chemical fertilizers/pesticides is not sustainable because of high costs and concerns about global warming, environmental pollution, and safety. The search for microbes that improve soil health and enhance plant nutrition has continued to attract attention due to the increasing cost of fertilizers and their negative environmental effects. For a more comprehensive development and utilization of rhizobial inoculants, there are several issues/recommendations which need to be taken care of in future research.

- The repertoire of rhizobial effects and functions mentioned in the review clearly suggests that they would be promising to increase crop yield, remove contaminants, inhibit pathogens, and produce fixed nitrogen or novel substances if we manage effective communities to favor plant colonization by rhizobial bacteria.
- It also demonstrates the possibility of reducing chemical fertilizers with biofertilizers, which in turn will help in mitigating the major global problem of environmental pollution and also fetch premium in the agriculture market.
- In the regions where legumes have no response to inoculation with rhizobia, co-inoculant PGPR may be an alternative for the enhancement of crop production.
- Co-inoculation with other PGPR is an efficient strategy that would have better prospects to act synergistically with rhizobia for better productivity of legumes compared to simple inoculation under sustainable agriculture system.
- Moreover, the commercial inoculant products are beginning to emerge that could prove useful in a commercial agriculture setting.
- Exploitation of multifunctional microbial inoculants seems feasible to enable plants to behave in a more competitive way and to

survive when established under stress conditions/environments.

- Comprehensive studies on the analysis of nodulation kinetics are needed which could enable to investigate the role of specific substances as signals or molecular components involved in the nodulation process.
- PGPB may communicate with the plant by means of various mechanisms to augment their growth and interfere with each other when used in consortium. However, actual mechanisms involved during consortium interactions still remain to be elucidated to harness the real potential of multi-strain inoculants.
- Studies on characterization of the interaction between sets of PGPR and rhizobia under various ecological soil environments and the changes in metabolic activities caused by co-inoculants affecting plant performance are also needed to be researched for achieving commercially realistic and effective co-inoculants.

Acknowledgement We are thankful to Mr. Muhammad Usman Jamshaid for his help in drawing Fig. 11.1.

References

Aamir M, Aslam A, Khan MY, Jamshaid MU, Ahmad M, Asghar HN, Zahir ZA (2013) Co-inoculation with rhizobium and plant growth promoting rhizobacteria (PGPR) for inducing salinity tolerance in mung bean under field condition of semi-arid climate. Asian J Agric Biol 1:17–22

Abd-Alla MH (1994) Use of organic phosphorous by *Rhizobium leguminosarum* bv. *viciae* phosphatases. Biol Fertil Soils 8:216–218

Abd-Alla MH, Morsy FM, El-Enany AE, Ohyama T (2012) Isolation and characterization of a heavy-metal- resistant isolate of *Rhizobium leguminosarum* bv. *viciae* potentially applicable for biosorption of Cd^{+2} and Co^{+2}. Int Biodeter Biodegr 67:48–55

Abd-Alla MH, El-Enany AWE, Nafady NA, Khalaf DM, Morsy FM (2014) Synergistic interaction of *Rhizobium leguminosarum* bv. *viciae* and arbuscular mycorrhizal fungi as a plant growth promoting biofertilizers for faba bean (*Vicia faba* L.) in alkaline soil. Microbiol Res 169:49–58

Abdelaziz RA, Radwansamir MA, Abdel-Kader M, Barakat MA (1996) Biocontrol of faba bean root-rot using VA mycorrhizae and its effect on biological nitrogen fixation. Egypt J Microbiol 31:273–286

Abdel-Salam MS, Ibrahim SA, Abd-El-Halim MM, Badawy FM, Abo-Aba SEM (2011) Phenotypic characterization of indigenous Egyptian Rhizobial strains for abiotic stresses performance. J Am Sci 7:168–174

Abolhasani M, Lakzian A, Tajabadipour A, Haghnia G (2010) The study salt and drought tolerance of *Sinorhizobium* bacteria to the adaptation to alkaline condition. Aust J Basic Appl Sci 4:882–886

Afzal A, Bano A (2008) Rhizobium and phosphate solubilizing bacteria improve the yield and phosphorus uptake in wheat (*Triticum aestivum*). Int J Agric Biol 10:85–88

Ahemad M, Khan MS (2009) Effect of insecticide-tolerant and plant growth-promoting Mesorhizobium on the performance of chickpea grown in insecticide stressed alluvial soils. J Crop Sci Biotech 12:213–222

Ahemad M, Khan MS (2010) Comparative toxicity of selected insecticides to pea plants and growth promotion. Crop Prot. doi:10.1016/j.cropro.2010.01.005

Ahmad M, Zahir ZA, Asghar HN, Asghar M (2011) Inducing salt tolerance in mung bean through co-inoculation with rhizobia and plant-growth-promoting rhizobacteria containing 1-aminocyclopropane-1-carboxylate-deaminase. Can J Microbiol 57:578–589

Ahmad M, Zahir ZA, Khalid M, Nazli F, Arshad M (2013) Efficacy of *Rhizobium* and *Pseudomonas* strains to improve physiology, ionic balance and quality of mung bean under salt-affected conditions on farmer's fields. Plant Physiol Biochem 63:170–176

Akhtar N, Qureshi MA, Iqbal A, Ahmad MJ, Khan KH (2012) Influence of *Azotobacter* and IAA on symbiotic performance of *Rhizobium* and yield parameters of lentil. J Agric Res 50:361–372

Alami Y, Achouak W, Marol C, Heulin T (2000) Rhizosphere soil aggregation and plant growth promotion of sunflowers by an exopolysaccharides producing *Rhizobium* sp. strain isolated from sunflower roots. Appl Environ Microbiol 66:3393–3398

Alexandre A, Oliveira S (2011) Most heat-tolerant rhizobia show high induction of major chaperone genes upon stress. FEMS Microbiol Ecol 75:28–36

Alikhani HA, Saleh-Rastin N, Antoun H (2006) Phosphate solubilization activity of rhizobia native to Iranian soils. Plant Soil 287:35–41

Amara MAT, Dahdoh MSA (1997) Effect of inoculation with plant growth promoting rhizobacteria (PGPR) on yield and uptake of nutrients by wheat grown on sandy soil. Egypt J Soil Sci 37:467–484

Amarger N, Macheret V, Laguerre G (1997) *Rhizobium gallicum* sp. nov. and *Rhizobium giardinii* sp. nov., from *Phaseolus vulgaris* nodules. Int J Syst Bacteriol 47:996–1006

Anjum MA, Zahir ZA, Arshad M, Ashraf M, Shehzad SM, Arif MS, Naveed M (2009) Substrate dependent microbial biosynthesis of auxins and their effect on growth and yield of mung bean (*Vigna radiata* L.). In: Annual meetings of Soil Science Society of America, Abstract published on line at Soil Sci. Soc. America website, http://soils.org. Accessed on Aug 2009

Antoun H, Beauchamp CJ, Goussard N, Chabot R, Lalande R (1998) Potential of *Rhizobium* and *Bradyrhizobium* species as plant growth promoting rhizobacteria on non-legumes: effect on radishes (*Raphanus sativus* L.). Plant Soil 204:57–67

Arfaoui A, El Hadrami A, Mabrouk Y, Sifi B, Boudabous A, El Hadrami I, Daayf F, Cherif M (2007) Treatment of chickpea with *Rhizobium* isolates enhances the expression of phenylpropanoid defense-related genes in response to infection by *Fusarium oxysporum* f. sp. *ciceris*. Plant Physiol Biochem 45:470–479

Arora NK, Kang SC, Maheshwari DK (2001) Isolation of siderophore producing strains of *Rhizobium meliloti* and their biocontrol potential against *Macrophomina phaseolina* that causes charcoal rot of groundnut. Curr Sci 81:673–677

Atieno M, Herrmann L, Okalebo R, Lesueur D (2012) Efficiency of different formulations of *Bradyrhizobium japonicum* and effect of co- inoculation of *Bacillus subtilis* with two different strains of *Bradyrhizobium japonicum*. World J Microbiol Biotechnol 28:2541–2550

Ausili P, Borisov A, Lindblad P, Martensson A (2010) Cadmium affects the interaction between peas and root nodule bacteria. Acta Agric Scand Sect B Plant Soil Sci 52:8–17

Azcon R (1993) Growth and nutrition of nodulated mycorrhizal and non-mycorrhizal *Hedysarum coronarium* as a results of treatments with fractions from a plant growth-promoting rhizobacteria. Soil Biol Biochem 25;1037–1042

Badawi FSF, Biomy AMM, Desoky AH (2011) Peanut plant growth and yield as influenced by co-inoculation with *Bradyrhizobium* and some rhizo-microorganisms under sandy loam soil conditions. Ann Agric Sci 56:17–25

Bai Y, Pan B, Charles TC, Smith DL (2002) Co-inoculation dose and root zone temperature for plant growth promoting rhizobacteria on soybean [*Glycine max* (L.) Merr] grown in soil-less media. Soil Biol Biochem 34:1953–1957

Baier R, Schiene K, Kohring B, Flaschen E, Niehaus K (1999) Alfalfa and tobacco cells react differentially to chitin oligosaccharides and *Sinorhizobium meliloti* nodulation factors. Planta 210:157–164

Bano A, Fatima M (2009) Salt tolerance in *Zea mays* (L) following inoculation with *Rhizobium* and *Pseudomonas*. Biol Fertil Soils 45:405–413

Bano A, Batool R, Dazzo F (2010) Adaptation of chickpea to desiccation stress is enhanced by symbiotic rhizobia. Symbiosis 50:129–133

Bardin SD, Huang HC, Pinto J, Amundsen EJ, Erickson RS (2004) Biological control of Pythium damping-off of pea and sugar beet by *Rhizobium leguminosarum* bv. *viceae*. Can J Bot 82:291–296

Barea JM, Pozo MJ, Azcon A, Azcon-Aguilar C (2005) Microbial co-operation in the rhizosphere. J Exp Bot 56:1761–1778

Bashan Y (1998) Inoculants of plant growth promoting bacteria for use in agriculture. Biotechnol Adv 16:729–770

Bashan Y, de-Bashan LE (2005) Bacterial plant growth-promotion. In: Hillel D (ed) Encyclopedia of soils in the environment, vol 1. Elsevier, Oxford, pp 103–115

Belimov AA, Dodd IC, Safronova VI, Davies WJ (2009) ACC deaminase-containing rhizobacteria improve vegetative development and yield of potato plants grown under water-limited conditions. Asp Appl Biol 98:163–169

Berjak P (2006) Unifying perspectives of some mechanisms basic to desiccation tolerance across life forms. Seed Sci Res 16:1–15

Beveridge CA, Gresshoff PM, Rameau C, Turnbull CGN (2003) Additional signaling compounds are required to orchestrate plant development. J Plant Growth Regul 22:15–24

Bhattacharjee RB, Singh A, Mukhopadhyay SN (2008) Use of nitrogen-fixing bacteria as biofertilizer for non-legumes: prospects and challenges. Appl Microbiol Biotechnol 80:199–209

Bianciotto V, Andreotti S, Balestrini R, Bonfante P, Perotto S (2001) Extracellular polysaccharides are involved in the attachment of *Azospirillum brasilense* and *Rhizobium leguminosarum* to arbuscular mycorrhizal structures. Eur J Histochem 45:39–49

Bianco C, Defez R (2009) *Medicago truncatula* improves salt tolerance when nodulated by an indole-3-acetic acid over-producing *Sinorhizobium meliloti* strain. J Exp Bot 60:3097–3107

Biswas JC, Ladha JK, Dazzo FB (2000) Rhizobial inoculation improves nutrient uptake and growth of lowland rice. Soil Sci Soc Am J 64:1644–1650

Boiero L, Perrig D, Masciarelli O, Penna C, Cassan F, Luna V (2007) Phytohormone production by three strains of *Bradyrhizobium japonicum* and possible physiological and technological implications. Appl Microbiol Biotechnol 74:874–880

Boscari A, Mandon K, Dupont L, Poggi M-C, Le Rudulier D (2002) BetS is a major glycine betaine/proline betaine transporter required for early osmotic adjustment in *Sinorhizobium meliloti*. J Bacteriol 184:2654–2663

Bouhmouch I, Souad-Mouhsine B, Brhada F, Aurag J (2005) Influence of host cultivars and Rhizobium species on the growth and symbiotic performance of *Phaseolus vulgaris* under salt stress. J Plant Physiol 162:1103–1113

Bravo A, Likitvivatanavong S, Gill SS, Soberón M (2011) *Bacillus thuringiensis*: a story of a successful bioinsecticide. Insect Biochem Mol Biol 41:423–431

Breil BT, Bornemanand J, Triplett EW (1996) A newly discovered gene, tfuA, involved in the production of the ribosomally synthesized peptide antibiotic trifolitoxin. J Bacteriol 178:4150–4156

Brígido C, Nascimento FX, Duan J, Glick BR, Oliveira S (2013) Expression of an exogenous 1-aminocyclopropane-1-carboxylate deaminase gene in *Mesorhizobium* spp. reduces the negative effects of salt stress in chickpea. FEMS Microbiol Lett 349:46–53

Caba JM, Recalde L, Ligero F (1998) Nitrate-induced ethylene biosynthesis and the control of nodulation in alfalfa. Plant Cell Environ 21:87–93

Carson KC, Meyer JM, Dillworth MJ (2000) Hydroxamate siderophores of root nodule bacteria. Soil Biol Biochem 32:11–21

Cassan F, Perrig D, Sgroy V, Masciarelli O, Penna C, Luna V (2009) *Azospirillum brasilense* Az39 and *Bradyrhizobium japonicum* E109, inoculated singly or in combination, promote seed germination and early seedling growth in corn (*Zea mays* L.) and soybean (*Glycine max* L.). Eur J Soil Biol 45:28–35

Castiglioni P, Warner D, Benson RJ, Anstrom DC, Harrison J, Stoecker M, Abad M, Kumar G, Salvador S, Ordine RD, Navarro S, Back S, Fernandes M, Targolli J, Dasgupta S, Bonin C, Luethy MH, Heard JE (2008) Bacterial RNA chaperones confer abiotic stress tolerance to plants and improved grain yield in maize under water-limited conditions. Plant Physiol 147:446–455

Chabot R, Antoun H, Kloepper JW, Beauchamp CJ (1996) Root colonization of maize and lettuce by bioluminescent *Rhizobium leguminosarum* biovar *phaseoli*. Appl Environ Microbiol 62:2767–2772

Chandra R, Pareek RP (2002) Effect of rhizobacteria in urdbean and lentil. Indian J Pulse Res 15:152–155

Chandra S, Choure K, Dubey RC, Maheshwari DK (2007) Rhizosphere competent *Mesorhizobium loti* MP6 induces root hair curling, inhibits *Sclerotinia sclerotiorum* and enhances growth of Indian mustard (*Brassica campestris*). Braz J Microbiol 38:128–130

Chi F, Shen S, Cheng H, Jing Y, Yanni YG, Dazzo FB (2005) Ascending migration of endophytic rhizobia, from roots to leaves, inside rice plants and assessment of benefits to rice growth physiology. Appl Environ Microbiol 71:7271–7278

Chi F, Yang P, Han F, Jing Y, Shen S (2010) Proteomic analysis of rice seedlings infected by *Sinorhizobium meliloti* 1021. Proteomics 10:1861–1874

Chung TP, Wu PC, Juang RS (2005) Use of microporous hollow fibers for improved biodegradation of high-strength phenol solutions. J Membr Sci 258:55–63

Cladera-Olivera F, Caron GR, Motta AS, Souto AA, Brandelli A (2006) Bacteriocin-like substance inhibits potato soft rot caused by *Erwinia carotovora*. Can J Microbiol 52:533–539

Compant S, Duffy B, Nowak J, Clement C, Barka EA (2005) Use of plant growth-promoting bacteria for biocontrol of plant diseases: principles, mechanisms of action, and future prospects. Appl Environ Microbiol 71:4951–4959

Cooper J (2007) Early interactions between legumes and rhizobia: disclosing complexity in a molecular dialogue. J Appl Microbiol 103:1355–1365

Dakora FD (2003) Defining new roles for plant and rhizobial molecules in sole and mixed plant cultures involving symbiotic legumes. New Phytol 58:39–49

Dakora FD, Matiru V, King M, Phillips DA (2002) Plant growth promotion in legumes and cereals by lumichrome, a rhizobial signal metabolite. In: Finan TM, O'Brian MR, Layzell DB, Vessey K, Newton WE (eds) Nitrogen fixation: global perspectives. CABI Publishing, Wallingford, pp 321–322

Dardanelli MS, de Cordoba FJF, Espuny MR, Carvajal MAR, Diaz MES, Serrano AMG, OkonY MM (2008) Effect of *Azospirillum brasilense* coinoculated with *Rhizobium* on *Phaseolus vulgaris* flavonoids and Nod factor production under salt stress. Soil Biol Biochem 40:2713–2721

Dary M, Chamber-Pérez M, Palomares A, Pajuelo E (2010) "In situ" phytostabilisation of heavy metal polluted soils using *Lupinus luteus* inoculated with metal resistant plant-growth promoting rhizobacteria. J Hazard Mater 177:323–330

Dashadi M, Khosravi H, Moezzi A, Nadian H, Heidari M, Radjabi R (2011) Co-inoculation of *Rhizobium* and *Azotobacter* on growth indices of faba bean under water stress in the green house condition. Adv Stud Biol 3:373–385

Dazzo FB, Yanni YG, Rizk R, Zidan M, Abu-Bakr, Gomaa M, Squartini A, Jing Y, Chi F, Shen S-H (2005) Recent studies on the Rhizobium cereal association. In: Wang YP, Lin M, Tian ZX, Elmerich C, Newton WE (eds) Biological nitrogen fixation, sustainable agriculture and the environment. Proceedings of the 14th international nitrogen fixation congress. Springer, Dordrecht, pp 379–380

De Jong AJ, Heidstra R, Spaink HP, Hartog MV, Meijer EA, Hendrisk T, Schiavo FL, Terzi M, Bisseling T, van Kammen A, de Vries SC (1993) A plant somatic embryo mutant is rescued by rhizobial lipooligosaccharides. Plant Cell 5:615–620

Deanand BJ, Patil AB, Kulkaarni JH, Algawadi AR (2002) Effect of plant growth promoting rhizobacteria on growth and yield of pigeon pea (*Cajanus cajan* L.) by application of plant-growth promoting rhizobacteria. Microbiol Res 159:371–394

Depret G, Houot S, Allard MR, Breuil MC, Nouaim R, Laguerre G (2004) Long-term effects of crop management on *Rhizobium leguminosarum* biovar *viciae* populations. FEMS Microbiol Ecol. doi:10.1016/j.femsec.2004.07.009

Deshwal VK, Dubey RC, Maheshwari DK (2003) Isolation of plant growth-promoting strains of *Bradyrhizobium*(*Arachis*) sp. with biocontrol potential against *Macrophomina phaseolina* causing charcoal rot of peanut. Curr Sci 84:443–448

Dixon RK, Garg VK, Rao MV (1993) Inoculation of *Leucaena* and *Prosopis* seedlings with *Glomus* and *Rhizobium* species in saline soil: rhizosphere relations and seedlings growth. Arid Soil Res Rehabil 7:133–144

Dixon RA, Harrison MJ, Lamb CJ (1994) Early events in the activation of plant defense responses. Annu Rev Plant Physiol Plant Mol Biol 32:479–501

Duan J, Muller KM, Charles TC, Vesely S, Glick BR (2009) 1-Aminocyclopropane-1carboxylate (ACC) deaminase gene in *Rhizobium* from Southern Saskatchewan. Microb Ecol 57:423–436

Dyachok JV, Tobin AE, Price NPJ, von Arnold S (2000) Rhizobial Nod factors stimulated somatic embryo development in *Picea abies*. Plant Cell Rep 19:290–297

Echeverria M, Sannazzaro AI, Ruiz OA, Menéndez AB (2013) Modulatory effects of *Mesorhizobium tianshanense* and *Glomus intraradices* on plant proline and polyamine levels during early plant response of Lotus tenuis to salinity. Plant Soil 364:69–79

Egamberdieva D, Juraeva D, Poberejskaya S, Myachina O, Teryuhova P, Seydalieva L, Aliev A (2004) Improvement of wheat and cotton growth and nutrient uptake by phosphate solubilizing bacteria. In: Mckimmon JS (ed) The 26th southern conservation tillage conference for sustainable agriculture. North Carolina State University, Raleigh, pp 58–66

Egamberdieva D, Berg G, Lindström K, Räsänen LA (2013) Alleviation of salt stress of symbiotic *Galega officinalis* L. (goat's rue) by co-inoculation of *Rhizobium* with root-colonizing *Pseudomonas*. Plant Soil 369:453–465

Ehteshamul-Haque S, Ghaffar A (1993) Use of rhizobia in the control of root rot diseases of sunflower, okra, soybean and mung bean. J Phytopathol 138:157–163

El-Akhal MR, Rincon A, Coba de la Pena T, Lucas MM, El Mourabit N, Barrijal S, Pueyo JJ (2013) Effects of salt stress and rhizobial inoculation on growth and nitrogen fixation of three peanut cultivars. Plant Biol 15:415–421

Elbadry M, Taha RM, Eldougdoug KA, Gamal-Eldin H (2006) Induction of systemic resistance in faba bean (*Vicia faba* L.) to bean yellow mosaic potyvirus (BYMV) via seed bacterization with plant growth promoting rhizobacteria. J Plant Dis Prot 113:247–251

Elboutahiri N, Thami-Alami I, Udupa SM (2010) Phenotypic and genetic diversity in *Sinorhizobium meliloti* and *S. medicae* from drought and salt affected regions of Morocco. BMC Microbiol 10:15

Elkoca E, Kantar F, Sahin F (2008) Influence of nitrogen fixing and phosphorus solubilizing bacteria on the nodulation, plant growth and yield of chickpea. J Plant Nutr 31:157–171

El-Sawy WA, Mekhemar GAA, Kandil BAA (2006) Comparative assessment of growth and yield responses of two peanut genotypes to inoculation with *Bradyrhizobium* conjugated with Cyanobacteria or rhizobacteria. Minufiya J Agric Res 31:1031–1049

Elsheikh EAE, Ibrahim KA (1999) The effect of *Bradyrhizobium* inoculation on yield and seed quality of guar (*Cyamopsis tetragonoloba* L.). Food Chem 65:183–187

Elsheikh EAE, Wood M (1995) Nodulation and nitrogen fixation by soybean inoculated with salt-tolerant rhizobia or salt-sensitive bradyrhizobia in saline soil. Soil Biol Biochem 27:657–661

Esfahani MN, Mostajeran A, Emtiazi G (2010) The effect of drought stress on nitrogenase and antioxidant enzymes activities in nodules formed from symbiosis of chickpea with two strains of *Mesorhizobium ciceri*. World Appl Sci J 10:621–626

Faghire M, Mohamed F, Taoufiq K, Fghire R, Bargaz A, Mandri B, Oufdou K, Laury A, Drevon JJ, Ghoulam C (2013) Genotypic variation of nodules' enzymatic activities in symbiotic nitrogen fixation among common bean (*Phaseolus vulgaris* L.) genotypes grown under salinity constraint. Symbiosis 60:115–122

Fan L, Ma Z, Liang J, Li H, Wang E, Wei G (2011) Characterization of a copper-resistant symbiotic bacterium isolated from *Medicago lupulina* growing in mine tailings. Bioresour Technol 102:703–709

Fernandez LA, Zalba P, Gomez MA, Sagardoy MA (2007) Phosphate-solubilization activity of bacterial strains in soil and their effect on soybean growth under greenhouse conditions. Biol Fertil Soils 43:805–809

Ferreira PAA, Lopes G, Bomfeti CA, de Oliveira Longatti SM, de Sousa Soares CRF, Guilherme LRG, de Souza Moreira FM (2013) Leguminous plants nodulated by selected strains of *Cupriavidus necator* grow in heavy metal contaminated soils amended with calcium silicate. World J Microbiol Biotechnol 29:2055–2066

Forchetti G, Masciarelli O, Izaguirre MJ, Alemano S, Alvarez D, Abdala G (2010) Endophytic bacteria improve seedling growth of sunflower under water stress, produce salicylic acid, and inhibit growth of pathogenic fungi. Curr Microbiol 61:485–493

Fox SL, O'Hara GW, Bräu L (2011) Enhanced nodulation and symbiotic effectiveness of *Medicago truncatula* when co-inoculated with *Pseudomonas fluorescens* WSM3457 and *Ensifer* (Sinorhizobium) *medicae* WSM419. Plant Soil 348:245–254

Frankenberger WT Jr, Arshad M (1995) Phytohormones in soil: microbial production and function. Marcel Dekker, New York, p 503

Frankenberger WT Jr, Poth M (1987) Biosynthesis of indole-3-acetic acid by the Pine ectomycorrhizal fungus *Pisolithus tinctorius*. Appl Environ Microbiol 53:2908–2913

Franzini VI, Azcon R, Mendes FL, Aroca R (2013) Different interaction among *Glomus* and *Rhizobium* species on *Phaseolus vulgaris* and *Zea mays* plant growth, physiology and symbiotic development under moderate drought stress conditions. Plant Growth Regul 70:265–273

Fravel DR (2005) Commercialization and implementation of biocontrol. Annu Rev Plant Physiol Plant Mol Biol 43:337–359

Garcia JAL, Probanza A, Ramos B, Barriuso J, Mañero FJG (2004) Effects of inoculation with plant growth promoting rhizobacteria (PGPR) and *Sinorhizobium fredii* on biological nitrogen fixation, nodulation and growth of *Glycine max* cv. *Osumi*. Plant Soil 267:143–153

Gouws LM, Botes E, Wiese AJ, Trenkamp S, Torres-Jerez I, Tang Y, Hills PN, Usadel B, Lloyd JR, Fernie AR (2012) The plant growth promoting substance, lumichrome, mimics starch, and ethylene-associated symbiotic responses in lotus and tomato roots. Front Plant Sci. doi:10.3389/fpls.2012.00120

Goyal V, Chetal S, Nainawatee HS (1986) Alterations in *Rhizobium trifolii* catalase under water stress. Folia Microbiol 31:164–166

Graham PH (1992) Stress tolerance in *Rhizobium* and *Bradyrhizobium*, and nodulation under adverse soil conditions. Can J Microbiol 38:475–484

Guefrachi I, Rejili M, Mahdhi M, Mars M (2013) Assessing genotypic diversity and symbiotic efficiency of five rhizobial legume interactions under cadmium stress for soil phytoremediation. Int J Phytoremediation 15:938–951

Guinazu LB, Andres JA, Del Papa MF, Pistorio M, Rosas SB (2010) Response of alfalfa (*Medicago sativa* L.) to single and mixed inoculation with phosphate-solubilizing bacteria and *Sinorhizobium meliloti*. Biol Fertil Soils 46:185–190

Gupta A, Saxena AK, Gopal M, Tilak KVBR (1998) Effect of plant growth promoting rhizobacteria on competitive ability of introduced *Bradyrhizobium* sp. (Vigna) for nodulation. Microbiol Res 153:3–117

Gupta DK, Rai UN, Sinha S, Tripathi RD, Nautiyal BD, Rai P, Inouhe M (2004) Role of *Rhizobium*(CA-1) inoculation in increasing growth and metal accumulation in *Cicer arietinum* L. grown under fly-ash stress condition. Bull Environ Contam Toxicol 73:424–431

Hafeez FY, Idris M, Malik KA, Asad S, Malik IA (1987) Screening of chickpeas (*Cicer arietinum* L.) and lentils (*Lens culinaris* Medic) for their nitrogen fixation potential. In: llahi L, Hussain F (eds) Modern trends of plant science research in Pakistan. Jadoon Printing Press, Peshawar, pp 140–144

Hafeez FY, Safdar ME, Chaudhry AU, Malik KA (2004) Rhizobial inoculation improves seedling emergence, nutrient uptake and growth of cotton. Aust J Exp Agric 44:617–622

Hafeez FY, Naeem FI, Naeem R, Zaidi AH, Malik KA (2005) Symbiotic effectiveness and bacteriocin production by *Rhizobium leguminosarum* bv. *viciae* isolated from agriculture soils in Faisalabad. Environ Exp Bot 54:142–147

Hafeez FY, Hassan Z, Naeem F, Basher A, Kiran A, Khan SA, Malik KA (2008) *Rhizobium leguminosarum* bv. *viciae* strain LC-31: analysis of novel bacteriocin and ACC-deaminase gene(s). In: Dakora FD, Chimphango SBM, Valentine AJ, Elmerich C, Newton WE (eds) Biological nitrogen fixation: towards poverty alleviation through sustainable agriculture. Springer, Dordrecht, pp 247–248

Hameeda B, Harini G, Rupela O, Wani S, Reddy G (2008) Growth promotion of maize by phosphate-solubilizing bacteria isolated from composts and macrofauna. Microbiol Res 163:234–242

Hara F, de Oliveira LA (2004) Physiological and ecological characteristics of rhizobia isolated deriving of acid and alic soils of Presidente Figueiredo. Acta Amazon 34:343–357

Hartmond U, Schaesberg NV, Graham JH, Syversten JP (1987) Salinity and flooding stress effects on mycorrhizal and non-mycorrhizal citrus rootstock seedlings. Plant Soil 104:37–43

Harwani D (2006) Biodiversity and efficiency of *Bradyrhizobium* strains are arbuscular mycorrhizoal fungi of soybean cultivars grown in Haroti region of Rajasthan. PhD thesis, Maharshi Dayanand Saraswati University, Ajmer

Hemissi I, Mabrouk Y, Mejri S, Saidi M, Sifi B (2013) Enhanced defence responses of chickpea plants against *Rhizoctonia solani* by pre-inoculation with rhizobia. J Phytopathol 161:412–418

Hilali A, Prevost D, Broughton WJ, Antoun A (2001) Effects de I'inoculation avec des souches de *Rhizobium leguminosarum* bv. *trifolii* sur la croissance du bl'e dans deux sols du Marco. Can J Microbiol 47:590–593

Hossain MS, Mårtensson A (2008) Potential use of *Rhizobium* spp. to improve fitness of non-nitrogen-fixing plants. Acta Agric Scand Sect B Soil Plant Sci 58:352–358

Hubbell DH, Kidder G (2009) Biological nitrogen fixation. Univ Fla IFAS Ext Publ SL16: 1–4

Hussain I, Arshad M, Zahir ZA, Javed M, Hussain A (1995) Substrate dependent microbial production of auxins and their influence on growth and nodulation of lentil. Pak J Agric Sci 32:2–3

Hussain MB, Mehboob I, Zahir ZA, Naveed M, Asghar HN (2009) Potential of *Rhizobium* spp. for improving growth and yield of rice (*Oryza sativa* L.). Soil Environ 28:49–55

Hussain MB, Zahir ZA, Asghar HN (2014a) Can catalase and EPS producing rhizobia ameliorate drought in wheat. Int J Agric Biol 16:3–13

Hussain MB, Zahir ZA, Asghar HN, Mahmood S (2014b) Scrutinizing rhizobia to rescue maize growth under reduced water conditions. Soil Sci Soc Am J. doi:10.2136/sssaj2013.07.0315 (Published online)

Iqbal MA, Khalid M, Shahzad SM, Ahmad M, Soleman N, Akhtar N (2012) Integrated use of *Rhizobium leguminosarum*, plant growth promoting rhizobacteria and enriched compost for improving growth, nodulation and yield of lentil (*Lens culinaris* Medik.). Chil J Agric Res 72:104–110

Jayasinghearachchi HS, Seneviratne G (2006) Fungal solubilization of rock phosphate is enhanced by forming fungal-rhizobial bio films. Soil Biol Biochem 38:405–408

Jha B, Gontia I, Hartmann A (2012) The roots of the halophyte *Salicornia brachiata* are a source of new halotolerant diazotrophic bacteria with plant growth-promoting potential. Plant Soil 356:265–277

Ji K, Chi F, Yang MF, Shen SH, Jing YX, Dazzo FB, Cheng HP (2010) Movement of rhizobia inside tobacco and lifestyle alternation from endophytes to free-living rhizobia on leaves. J Microbiol Biotechnol 20:238–244

Kacem M, Kazouz F, Merabet C, Rezki M, de Lajudie P, Bekki A (2009) Antimicrobial activities of *Rhizobium* sp. strains against *Pseudomonas savastanoi*, the agent responsible for the olive knot disease in Algeria. Grasas Aceites 60:139–146

Kaci Y, Heyraud A, Barakat M, Heulin T (2005) Isolation and identification of an EPS-producing *Rhizobium*

Nukui N, Minamisawa K, Ayabe SI, Aoki T (2006) Expression of the 1-aminocyclopropane-1 carboxylic acid deaminase gene requires symbiotic nitrogen-fixing regulator gene nifA2 in *Mesorhizobium loti* MAFF303099. Appl Environ Microbiol 72:4964–4969

O'Sullivan DJ, O'Gara F (1992) Traits of *fluorescent Pseudomonas* spp. involved in suppression of plant root pathogens. Microbiol Rev 56:662–676

Öğütçü H, Kasimoğlu C, Elkoca E (2010) Effects of rhizobium strains isolated from wild chickpeas on the growth and symbiotic performance of chickpeas (*Cicer arietinum* L.) under salt stress. Turk J Agric For 34:361–371

Oldroyd GE, Downie JA (2008) Coordinating nodule morphogenesis with rhizobial infection in legumes. Annu Rev Plant Biol 59:519–546

Oufdou K, Benidire L, Lyubenova L, Daoui K, Fatemi ZEA, Schröder P (2014) Enzymes of the glutathione-ascorbate cycle in leaves and roots of rhizobia-inoculated faba bean plants (*Vicia faba* L.) under salinity stress. Eur J Soil Biol 60:98–103

Outten FW, Outten CE, O'Halloran T (2000) Metalloregulatory systems at the interface between bacterial metal homeostasis and resistance. In: Storz G, Hengge Aronis R (eds) Bacterial stress responses. ASM Press, Washington, DC, pp 29–42

Ozkoc I, Deliveli MH (2001) In vitro inhibition of the mycelial growth of some root rot fungi by *Rhizobium leguminosarum* biovar *phaseoli* isolates. Turk J Biol 25:435–445

Pandey P, Maheshwari D (2007) Bioformulation of *Burkholderia* sp. MSSP with a multispecies consortium for growth promotion of *Cajanus cajan*. Can J Microbiol 53:213–222

Panigrahi DP, Randhawa GS (2010) A novel method to alleviate arsenic toxicity in alfalfa plants using a deletion mutant strain of Sinorhizobium meliloti. Plant Soil 336:459–467

Parniske M, Downie JA (2003) Plant biology: locks, keys and symbioses. Nature 425:569–570

Peng S, Biswas JC, Ladha JK, Gyaneshwar P, Chen Y (2002) Influence of rhizobial inoculation on photosynthesis and grain yield of rice. Argon J 94:925–929

Perrine-Walker FM, Gartner E, Hocart CH, Becker A, Rolfe G (2007) Rhizobium-inhibited rice growth inhibition caused by nitric oxide accumulation. Mol Plant Microbe Interact 20:283–292

Perrine-Walker FM, Hynes MF, Rolfe BG, Hocart CH (2009) Strain competition and agar affect the interaction of rhizobia with rice. Can J Microbiol 55:1217–1223

Phillips DA, Joseph CM, Yang GP, Martinez-Romero E, Sanborn JR, Volpin H (1999) Identification of lumichrome as a *Sinorhizobium* enhancer of alfalfa root respiration and shoot growth. Proc Natl Acad Sci U S A 96:12275–12280

Podile AR, Kishore GK (2006) Plant growth-promoting rhizobacteria. In: Gnanamanickam SS (ed) Plant-associated bacteria. Springer, Dordrecht, pp 195–230

Pongsilp N, Boonkerd N (2011) Transposon mutant of *Vigna radiata*-nodulating *Bradyrhizobium* sp. impaired in both resistance to stress conditions and symbiotic performance. Afr J Microbiol Res 5:4303–4309

Prayitno J, Mathesius U (2010) Differential regulation of the nodulation zone by silver Ions, L-α – (2-Amino-ethoxyvinyl)-glycine, and the skl mutation in *Medicago truncatula*. Hayati J Biosci 17:15–20

Prayitno J, Stefaniak J, McIver J, Weinman JJ, Dazzo FB, Ladha JK, Barraquio W, Yanni YG, Rolfe BG (1999) Interactions of rice seedlings with bacteria isolated from rice roots. Aust J Plant Physiol 26:521–535

Puente ME, Bashan Y, Li CY, Lebsky GK (2004) Microbial populations and activities in the rhizoplane of rock-weathering desert plants. 1. Root colonization and weathering of igneous rock. Plant Soil 6:629–642

Qureshi MA, Ahmad MJ, Naveed M, Iqbal A, Akhtar N, Niazi KH (2009) Co-inoculation with *Mesorhizobium ciceri* and *Azotobacter chroococcum* for improving growth, nodulation and yield of chickpea (*Cicer arietinum* L.). Soil Environ 28:124–129

Qureshi MA, Iqbal A, Akhtar N, Shakir MA, Khan A (2012) Co-inoculation of phosphate solubilizing bacteria and rhizobia in the presence of L-tryptophan for the promotion of mash bean (*Vigna mungo* L.). Soil Environ 31:47–54

Qureshi MA, Shahzad H, Imran Z, Mushtaq M, Akhtar N, Ali MA, Mujeeb F (2013) Potential of rhizobium species to enhance growth and fodder yield of maize in the presence and absence of tryptophan. J Anim Plant Sci 23(5):1448–1454

Raaijmakers JM, Paulitz TC, Steinberg C, Alabouvette C, Moënne-Loccoz Y (2009) The rhizosphere: a playground and battlefield for soil-borne pathogens and beneficial microorganisms. Plant Soil 321:341–361

Radwan SS, Dashti N, El-Nemr I (2005) Enhancing the growth of *Vicia faba* plants by microbial inoculation to improve their phytoremediation potential for oily desert areas. Int J Phytoremediation 7:19–32

Raësaënen LA, Lindstroëm K (1999) The effect of heat stress on the symbiotic interaction between *Sinorhizobium* sp. and *Acacia Senegal*. FEMS Microbiol Ecol 28:63–74

Rahmani H, Saleh-Rastin N, Khavazi K, Asgharzadeh A, Fewer D, Kiani S, Lindstrom K (2009) Selection of thermotolerant Bradyrhizobial strains for nodulation of soybean (*Glycine max* L.) in semi-arid regions of Iran. World J Microbiol Biotechnol 25:591–600

Rajendran G, Sing F, Desai AJ, Archana G (2008) Enhanced growth and nodulation of pigeon pea by co-inoculation of *Bacillus* strains with *Rhizobium* spp. Bioresour Technol 99:544–550

Reinhold B, Hurek T, Fendrik I, Pot B, Gillis M, Kersters K, Thielemans S, De Ley J (1987) *Azospirillum halopraeferens* sp. nov., a nitrogen-fixing organism associated with roots of Kallar grass (*Leptochloa fusca* (L.) Kunth). Int J Syst Bacteriol 37:43–51

Reitz M, Rudolph K, Schroder I, Hoffmann-Hergarten S, Hallmann J, Sikora RA (2000) Lipopolysaccharides of *Rhizobium etli* strain G12 act in potato roots as an inducing agent of systemic resistance to infection by the cyst nematode *Globodera pallida*. Appl Environ Microbiol 66:3515–3518

Rejili M, Mahdhi M, Fterich A, Dhaoui S, Guefrachi I, Abdeddayem R, Mars M (2012) Symbiotic nitrogen fixation of wild legumes in Tunisia: soil fertility dynamics, field nodulation and nodules effectiveness. Agric Ecosyst Environ 157:69–69

Remans R, Ramaekers L, Schelkens S, Hernandez G, Garcia A, Reyes JL, Mendez N, Toscano V, Mulling M, Galvez L, Vanderleyden J (2008) Effect of *Rhizobium – Azospirillum* coinoculation on nitrogen fixation and yield of two contrasting *Phaseolus vulgaris* L. genotypes cultivated across different environments in Cuba. Plant Soil 312:25–37

Richardson AE, Barea JM, McNeil AM, Prigent-Combaret C (2009) Acquisition of phosphorus and nitrogen in the rhizosphere and plant growth promotion by microorganisms. Plant Soil 321:305–339

Robleto EA, Kmiecik K, Oplinger ES, Nienhuis J, Triplett EW (1998) Trifolitoxin production increases nodulation competitiveness of *Rhizobium etli* CE3 under agricultural conditions. Appl Environ Microbiol 64:2630–2633

Rodelas B, Gonzales Lopez J, Martinez Toledo MV, Pozo C, Salmeron V (1999) Influence of *Rhizobium/Azotobacter* and *Rhizobium/Azospirillum* combined inoculation on mineral composition of faba bean (*Vicia faba* L.). Biol Fertil Soils 29:165–169

Rodriguez H, Fraga R, Gonzalez T, Bashan Y (2006) Genetics of phosphate solubilization and its potential applications for improving plant growth promoting bacteria. Plant Soil 287:15–21

Rodrıguez E, Nuero O, Guillén F, Martınez A, Martınez M (2004) Degradation of phenolic and non-phenolic aromatic pollutants by four *Pleurotus* species: the role of laccase and versatile peroxidase. Soil Biol Biochem 36:909–916

Rogers N, Carson K, Glenn A, Dilworth M, Hughes M, Poole R (2001) Alleviation of aluminum toxicity to *Rhizobium leguminosarum* bv. *viciae* by the hydroxamate siderophore vicibactin. Biometals 14:59–66

Roseline R, Lara R, Sarah S, German H, Aurelio G, Jorge R, Nancy M, Vidalina T, Miguel M, Lazaro G, Jos V (2008) Effect of *Rhizobium – Azospirillum* coinoculation on nitrogen fixation and yield of two contrasting *Phaseolus vulgaris* L. genotypes cultivated across different environments in Cuba. Plant Soil 312:25–37

Rovera AD (1965) Plant root exudates and their influence upon soil microorganisms. In: Baker KF, Synder WC (eds) Ecology of soil borne plant pathogens – prelude to biological control. John Murray, London, pp 170–186

Russelle MP (2008) Biological dinitrogen fixation in agriculture. In: Schepers JS, Raun WR (eds) Nitrogen in agricultural systems, vol 22, 2nd edn, Agronomy monograph. American Society of Agronomy, Madison, pp 281–359

Sabry RS, Saleh SA, Batchelor CA, Jones J, Jotham J, Webster G, Kothari SL, Davey MR, Cocking EC (1997) Endophytic establishment of *Azorhizobium caulinodans* in wheat. Proc R Soc B Biol Sci 264:341–346

Saıdi S, Chebil S, Gtari M, Mhamdi R (2013) Characterization of root-nodule bacteria isolated from Vicia faba and selection of plant growth promoting isolates. World J Microbiol Biotechnol 29:1099–1106

Sandhya V, Ali SZ, Grover M, Reddy G, Venkateswarlu B (2009) Alleviation of drought stress effects in sunflower seedlings by exopolysaccharides producing *Pseudomonas putida* strain P45. Biol Fertil Soils 46:17–26

Santos H, da Costa MS (2002) Compatible solutes of organisms that live in hot saline environments. Environ Microbiol 4:501–509

Sassi-Aydi S, Aydi S, Abdelly C (2012) Inoculation with the native *Rhizobium gallicum* 8a3 improves osmotic stress tolerance in common bean drought-sensitive cultivar. Acta Agric Scand Sect B Soil Plant Sci 62:179–187

Schulze J (2004) How are nitrogen fixation rates regulated in legumes? J Plant Nutr Soil Sci 167:125–137

Senthilkumar M, Madhaiyan M, Sundaram SP, Kannaiyan S (2009) Intercellular colonization and growth promoting effects of *Methylobacterium* sp. with plant-growth regulators on rice (*Oryza sativa* L. Cv CO-43). Microbiol Res 164:92–104

Sgroy V, Cassan F, Masciarelli O, Del Papa MF, Lagares A, Luna V (2009) Isolation and characterization of endophytic plant growth-promoting (PGPB) or stress homeostasis-regulating (PSHB) bacteria associated to the halophyte *Prosopis strombulifera*. Appl Microbiol Biotechnol 85:371–381

Shaharoona B, Arshad M, Zahir ZA (2006a) Effect of plant growth-promoting rhizobacteria containing ACC-deaminase on maize (*Zea mays* L.) growth under axenic conditions and on nodulation in mung bean (*Vigna radiata* L.). Lett Appl Microbiol 42:155–159

Shaharoona B, Bibi R, Arshad M, Zahir ZA, Zia-ul-Hassan (2006b) 1-aminocylopropane-1-carboxylate (ACC)-deaminase rhizobacteria extenuates ACC-induced classical triple response in etiolated pea seedlings. Pak J Bot 38:1491–1499

Shahzad SM, Khalid A, Arif MS, Riaz M, Ashraf M, Iqbal Z, Yasmeen T (2014) Co-inoculation integrated with P-enriched compost improved nodulation and growth of Chickpea (*Cicer arietinum* L.) under irrigated and rainfed farming systems. Biol Fertil Soils 50:1–12

Shamseldin A (2007) Use of DNA marker to select well-adapted Phaseolus-symbionts strains under acid conditions and high temperature. Biotechnol Lett 29:37–44

Shaukat SS, Siddqui IA (2003) The influence of mineral and carbon sources on biological control of charcoal rot fungus, *Macrophomina phaseolina* by fluorescent pseudomonads in tomato. Lett Appl Microbiol 36:392–398

Shimshick EJ, Hebert RR (1979) Binding characteristics of N_2-fixing bacteria to cereal roots. Appl Environ Microbiol 38:447–453

Siddiqui ZA, Mahmood I (2001) Effects of rhizobacteria and root symbionts on the reproduction of *Meloidogyne javanica* and growth of chickpea. Bioresour Technol 79:41–45

Siddiqui IA, Ehteshamul-Haque S, Zaki MJ, Ghaffar A (2000) Effect of urea on the efficacy of *Bradyrhizobium* sp. and *Trichoderma harzianum* in the control of root infecting fungi in mungbean and sunflower. Sarhad J Agric 16:403–406

Silveira APD, Freitas SS, Silva LRC, Lombardi MLCO, Cardoso EJBN (1995) Interactions between arbuscular mycorrhizas and plant-growth promoting rhizobacteria in beans. Rev Bras Ciênc Solo 19:205–211

Singh RK, Mishra RPN, Jaiswal HK, Kumar V, Pandev SP, Rao SB, Annapurna K (2006) Isolation and identification of natural endophytic rhizobia from rice (*Oryza sativa* L.) through rDNA PCR-RFLP and sequence analysis. Curr Microbiol 52:345–349

Singh G, Sekhon HS, Sharma P (2011) Effect of irrigation and biofertilizer on water use, nodulation, growth and yield of chickpea (*Cicer arietinum* L). Arch Agron Soil Sci 57:715–726

Smith DL, Prithiviraj B, Zhang F (2002) Rhizobial signals and control of plant growth. In: Finan TM, O'Brian MR, Layzell DB, Vessey JK, Newton W (eds) Nitrogen fixation: global perspectives. CABI, Wallingford, pp 327–330

Soliman AS, Shanan NT, Massoud ON, Swelim DM (2012) Improving salinity tolerance of *Acacia saligna* (Labill.) plant by arbuscular mycorrhizal fungi and *Rhizobium* inoculation. Afr J Biotechnol 11:1259–1266

Soussou S, Mahieu S, Brunel B, Escarré J, Lebrun M, Banni M, Boussetta H, Cleyet-Marel JC (2013) Zinc accumulation patterns in four *Anthyllis vulneraria* subspecies supplemented with mineral nitrogen or grown in the presence of their symbiotic bacteria. Plant Soil 371:423–434

Spaepen S, Vanderleyden J, Remans R (2007) Indole-3-acetic acid in microbial and microorganism-plant signaling. FEMS Microbiol Rev 31:425–448

Spaink HP (1992) Rhizobial lipo-oligosaccharide: answers and questions. Plant Mol Biol 20:977–986

Sridevi M, Mallaiah KV (2008) Production of indole-3-acetic acid by *Rhizobium* isolates from *Sesbania* species. Plant Sci Res 1:13–16

Staehelin C, Granado J, Muller J, Wiemeken A, Mellor RB, Felix G, Regenass M, Broughton WJ, Boller T (1994) Perception of *Rhizobium* nodulation factor by tomato cells and inactivation by root chitinases. Proc Natl Acad Sci U S A 91:2196–2200

Stajkovic O, Delic D, Josic D, Kuzmanovic D, Rasulic N, Knezevic-Vukcevic J (2011) Improvement of common bean growth by co-inoculation with Rhizobium and plant growth promoting bacteria. Rom Biotechnol Lett 16:5919–5926

Sturz AV, Kimpinski J (1999) The influence of fosthiazate and aldicarb on populations of plant growth promoting bacteria and bacterial feeding nematodes in the potato root zone. Plant Pathol 48:26–32

Suárez R, Wong A, Ramírez M, Barraza A, Orozco MC, Cevallos MA, Lara M, Hernández G, Iturriaga G (2008) Improvement of drought tolerance and grain yield in common bean by overexpressing trehalose-6-phosphate synthase in rhizobia. Mol Plant Microbe Interact 21:958–966

Talano MA, Cejas RB, González PS, Agostini E (2013) Arsenic effect on the model crop symbiosis *Bradyrhizobium*-soybean. Plant Physiol Biochem 63:8–14

Tamimi SM, Timko MP (2003) Effects of ethylene and inhibitors of ethylene synthesis and action on nodulation in common bean (*Phaseolus vulgaris* L.). Plant Soil 257:125–131

Tarrand JJ, Krieg NR, Dobereiner J (1978) A taxonomic study of the Azospirillum lipoferum group, with descriptions of a new genus, Azospirillum gen. nov. and two species, Azospirillum lipoferum (Beijerinck) comb. nov. and Azospirillum brasilense sp. nov. Can J Microbiol 24:967–980

Tate RL (1995) Soil microbiology (symbiotic nitrogen fixation). Wiley, New York, pp 307–333

Tchebotar VK, Kang UG, Asis CAJ, Akao S (1998) The use of GUS-reporter gene to study the effect of *Azospirillum* – *Rhizobium* coinoculation on nodulation of white clover. Biol Fertil Soils 27:349–352

Tilak KVBR (2004) Response of *Sesbania* green manuring and mung bean residue incorporation on microbial activities for sustainability of a rice-wheat cropping system. Int J Agric Rural Dev Trop Subtrop 105:189–196

Tilak KVBR, Ranganayaki N, Manoharachari C (2006) Synergistic effects of plant-growth promoting-rhizobacteria and *Rhizobium* on nodulation and nitrogen fixation by pigeon pea (*Cajanus cajan*). Eur J Soil Sci 57:67–71

Triplett EW, Breil BT, Splitter GA (1994) Expression of tfx and sensitivity to the rhizobial antipeptide trifolitoxin in a taxonomically distinct group of α-proteobacteria including the animal pathogen Brucella abortus. Appl Environ Microbiol 60:4163–4166

Tsigie A, Tilak KVBR, Anil KS (2012) Field response of legumes to inoculation with plant growth promoting rhizobacteria. Biol Fertil Soils 47:971–974

Uchiumi T, Oowada T, Itakura M, Mitsui H, Nukui N, Dawadi P, Kaneko T, Tabata S, Yokoyama T, Tejima T, Saeki K, Oomori H, Hayashi M, Maekawa T, Sriprang R, Murooka Y, Tajima S, Simomura K, Nomura M, Suzuki A, Shimoda S, Sioya K, Abe M, Minamisawa K (2004) Expression islands clustered on symbiosis island of *Mesorhizobium loti* genome. J Bacteriol 186:2439–2448

Vargas LK, Lisboa BB, Schlindwein G, Granada CE, Giongo A, Beneduzi A, Passaglia LMP (2009) Occurrence of plant growth-promoting traits in clover-nodulating rhizobia strains isolated from different soils in Rio Grande do Sul state. R Bras Ci Solo 33:1227–1235

Verma JP, Yadav J, Tiwari KN (2012) Enhancement of nodulation and yield of chickpea by coinoculation of indigenous *Mesorhizobium* spp. and plant growth-promoting rhizobacteria in eastern Uttar Pradesh. Commun Soil Sci Plant Anal 43:605–621

Verma JP, Yadav J, Tiwari KN, Kumar A (2013) Effect of indigenous *Mesorhizobium* spp. and plant growth promoting rhizobacteria on yields and nutrients uptake of chickpea (*Cicer arietinum* L.) under sustainable agriculture. Ecol Eng 51:282–286

Vessey JK (2003) Plant growth promoting rhizobacteria as biofertilizers. Plant Soil 255:571–586

Vessey JK, Buss TJ (2002) *Bacillus cereus* UW85 inoculation effects on growth, nodulation and N-accumulation in grain legumes, controlled environment studies. Can J Plant Sci 82:282–290

Vivas A, Marulanda A, Ruiz-Lozana JM, Barea JM, Azcon R (2003) Influence of a *Bacillus* sp. on physiological activities of two *arbuscular mycorrhizal fungi* and on plant responses to PEG induced drought stress. Mycorrhiza 13:249–256

Volpin H, Phillips DA (1998) Respiratory elicitors from *Rhizobium meliloti* affect intact alfalfa roots. Plant Physiol 116:777–783

Vriezen JAC, De Bruijn FJ, Nusslein K (2007) Responses of rhizobia to desiccation in relation to osmotic stress, oxygen and temperature. Appl Environ Microbiol 73:3451–3459

Walley FL, Clayton GW, Miller PR, Carr PM, Lafond GP (2007) Nitrogen economy of pulse crop production in the northern Great Plains. Agron J 99:1710–1718

Wani PA, Khan MS (2013) Nickel detoxification and plant growth promotion by multi metal resistant plant growth promoting *Rhizobium* species RL9. Bull Environ Contam Toxicol 91:117–124

Wani PA, Khan MS, Zaidi A (2007a) Co-inoculation of nitrogen-fixing and phosphate-solubilizing bacteria to promote growth, yield and nutrient uptake in chickpea. Acta Agron Hung 55:315–323

Wani PA, Khan MS, Zaidi A (2007b) Synergistic effects of the inoculation with nitrogen-fixing and phosphate-solubilizing rhizobacteria on the performance of field-grown chickpea. J Plant Nutr Soil Sci 170:283–287

Wani PA, Khan MS, Zaidi A (2008a) Effects of heavy metal toxicity on growth, symbiosis, seed yield and metal uptake in pea grown in metal amended soil. Bull Environ Contam Toxicol 81:152–158

Wani PA, Khan MS, Zaidi A (2008b) Effect of metal-tolerant plant growth-promoting Rhizobium on the performance of pea grown in metal-amended soil. Arch Environ Contam Toxicol 55:33–42

Weyens N, Van Der Lelie D, Taghavi S, Newman L, Vangronsveld J (2009) Exploiting plant-microbe partnerships to improve biomass production and remediation. Trends Biotechnol 27:591–598

Yang G, Bhuvaneswari TV, Joseph CM, King MD, Phillips DA (2002) Role for riboflavin in the *Sinorhizobium*-alfalfa association. Mol Plant-Microbe Interact 15:456–462

Yang J, Kloepper JW, Ryu CM (2009) Rhizosphere bacteria help plants tolerate abiotic stress. Trends Plant Sci 14:1–4

Yanni YG, Rizk RY, Abd El-Fattah FK, Squartini A, Corich V, Giacomini A, De Bruijn F, Rademaker J, Maya-Flores J, Ostrom P, Vega-Hernandez M, Hollingsworth RI, Martinez-Molina E, Mateos P, Velazquez E, Wopereis J, Triplett E, Umali-Gracia M, Anarna JA, Rolfe BG, Ladha JK, Hill J, Mujoo R, Ng PK, Dazzo FB (2001) The beneficial plant growth-promoting association of *Rhizobium leguminosarum* bv. *trifolii* with rice roots. Aust J Plant Physiol 28:845–870

Yasmeen T, Hameed S, Tariq M, Ali S (2012) Significance of arbuscular mycorrhizal and bacterial symbionts in a tripartite association with *Vigna radiata*. Acta Physiol Plant 34:1519–1528

Yelton MM, Yang SS, Edie SA, Lim ST (1983) Characterization of an effective salt-tolerant, fast growing strain of *Rhizobium japonicum*. J Gen Microbiol 129:1537–1547

Yuming B, Xiaomin Z, Smith DL (2003) Enhanced soybean plant growth resulting from coinoculation of *Bacillus* strains with *Bradyrhizobium japonicum*. Crop Sci 43:1774–1778

Zahir ZA, Asghar HN, Akhtar MJ, Arshad M (2005) Precursor (L-tryptophan)-inoculum (*Azotobacter*) interactions for improving yields and nitrogen uptake of maize. J Plant Nutr 28:805–817

Zahir ZA, Munir A, Asghar HN, Arshad M, Shaharoona B (2008) Effectiveness of rhizobacteria containing ACC-deaminase for growth promotion of peas (*Pisum sativum*) under drought conditions. J Microbiol Biotechnol 18:958–963

Zahir ZA, Ghani U, Naveed M, Nadeem SM, Asghar HN (2009) Comparative effectiveness of *Pseudomonas* and *Serratia* sp. containing ACC-deaminase for improving growth and yield of wheat (*Triticum aestivum L.*) under salt-stressed conditions. Arch Microbiol 191:415–424

Zahir ZA, Shah MK, Naveed M, Akhter MJ (2010a) Substrate-dependent auxin production by *Rhizobium phaseoli* improves the growth and yield of *Vigna radiata* L. under salt stress conditions. J Microbiol Biotechnol 20:1288–1294

Zahir ZA, Yasin HM, Naveed M, Anjum MA, Khalid M (2010b) L-Tryptophan application enhances the effectiveness of *rhizobium* inoculation for improving growth and yield of mung bean (*Vigna radiata* L.). Pak J Bot 42:1771–1780

Zahir ZA, Zafar-ul-Hye M, Sajjad S, Naveed M (2011) Comparative effectiveness of *Pseudomonas* and *Serratia* sp. containing ACC-deaminase for coinoculation with *Rhizobium leguminosarum* to improve growth, nodulation, and yield of lentil. Biol Fertil Soils 47:457–465

Zahran HH (1999) *Rhizobium*–legume symbiosis and nitrogen fixation under severe conditions and in an arid climate. Microbiol Mol Biol Rev 63:968–989

Zaidi A, Khan MS, Amil M (2003) Interactive effect of rhizotrophic microorganisms on yield and nutrient uptake of chickpea (*Cicer arietinum* L.). Eur J Agron 19:15–21

Zaidi A, Khan MS, Ahemad M, Oves M (2009) Plant growth promotion by phosphate solubilizing bacteria. Acta Microbiol Immunol Hung 56:283–284

Zhengwei Z, Fang W, Lee HY, Yang Z (2005) Responses of *Azorhizobium caulinodans* to cadmium stress. FEMS Microbiol Ecol 54:455–461

Zribi K, Nouairi I, Slama I, Talbi-Zribi O, Mhadhbi H (2013) *Medicago sativa – Sinorhizobium meliloti* symbiosis promotes the bioaccumulation of Zinc in nodulated roots. Int J Phytoremediation 17:49-55. doi: 10.1080/15226514.2013.828

Legume-Rhizobia Symbiosis Under Stress

12

Kapudeep Karmakar, Anjul Rana, Asmita Rajwar, Manvika Sahgal, and Bhavdish N. Johri

Abstract

Legume-rhizobia symbioses are of practical importance in provision of sustained food supply. Currently, it provides approximately 45 % of N used in agriculture. This quantity will have to be augmented as the world's population increases and natural resources used in production of fertilizer-N diminish. The major constraints for food security are soil salinity, sodicity, nutrient deficiency, and toxicity which are driven by indiscriminate use of agrochemicals including chemical fertilizers and pesticides, utilization of excess water resource coupled with climate changes, and periodic droughts. In this scenario, legume-rhizobia symbioses are nonpolluting, cost-effective ways to enhance high N_2-fixing potential under stress conditions. Several symbiotic systems that are tolerant in extreme conditions of salinity, alkalinity, acidity, drought, toxic doses of fertilizer, and metal toxicity have been identified including rhizobial symbiosis in arid region. Extreme environmental conditions adversely affect rhizobia-legume interactions wherein rhizobial partner utilizes endogenous or exogenous osmolytes and secretes specific proteins to alleviate the problem of aridity, salinity, and toxicity. Thus, change in rhizobial population can be an indicator of soil fertility. Hence, osmoadapted rhizobial strains can be used as biofertilizers for salt-sensitive crops in saline soils. In this context, interaction among rhizobia, plant growth-promoting rhizobacteria (PGPR), and mycorrhiza are also important. Here, we give an account of relevance of biological nitrogen fixation (BNF) in sustained food supply, effects of extreme conditions on legume-rhizobia symbiosis, as well as interaction

K. Karmakar • A. Rana • A. Rajwar • M. Sahgal (✉)
Department of Microbiology, G.B. Pant University of Agriculture & Technology, Pantnagar 263 145, Uttarakhand, India
e-mail: sahgal.manvika@gmail.com

B.N. Johri
Department of Biotechnology, Barkatullah University, Bhopal 462 026, Madhya Pradesh, India

N.K. Arora (ed.), *Plant Microbes Symbiosis: Applied Facets*,
DOI 10.1007/978-81-322-2068-8_12, © Springer India 2015

of rhizobia with belowground microbial diversity including mycorrhiza and adaptive strategies of rhizobia under condition of stress. We also discuss about models wherein osmotolerant rhizobia can be used as biofertilizers and sustain green revolution to evergreen revolution.

Introduction

The major challenge for agricultural industry is to provide sustained food supply for nearly eight to nine billion people, when 40 % of the world's agricultural land is affected by salinity-, drought-, sodicity-, and toxicity-related problems. The intensive agricultural practices like cultivation of short duration crops which require large amount of water, excessive application of fertilizers, and indiscriminate use of pesticides and herbicides that played a key role in bringing about green revolution are major causes of soil quality degradation. The harsh soil conditions adversely affect legume-rhizobia symbiosis. The importance of legume-rhizobia symbiosis in agricultural and forestry ecosystems cannot be underestimated. It fixes about 45×10^6 metric tons of nitrogen annually and provides nearly half of all the nitrogen used in agriculture (Brockwell et al. 1995). To maintain sustained food supply, the quantity of nitrogen needed for agriculture is projected to double by year 2030 (Tilman 1999). BNF also sequesters carbon. Biologically fixed nitrogen at current estimate, 45×10^6 metric tons of nitrogen per year, is equivalent to sequestering an additional 770 to 990×10^6 metric tons of carbon year^{-1} (Vance 2001; Abberton 2010). Thus, legume-rhizobia symbiosis that is environmentally prudent in N_2 fixation will benefit both agriculture and the environment. Reactive oxygen species (ROS) can adversely affect the survival of rhizobial cells during desiccation, when the water content drops below 40 %. Arid regions are characterized by high temperature, low relative humidity, high evaporation, and scanty rainfall, all leading to salinity. Drought can also lead to hyperosmotic stress, accumulation of salts and solute, impairment of metabolism when water activity has reduced, and formation of ROS.

Agrochemical Stress

Natural resources are not sufficient to fulfill food requirement of the growing population. Hence, various agrochemicals are being used to enhance crop production and protection. Agrochemicals like pesticides are very important in crop protection from various pests. The excessive use of pesticides over the years has, however, caused environmental problems such as accumulation of agrochemicals in soil and groundwater. Out of total agrochemicals applied to a crop, only 15 % reach the target. Their fate in soil system depends upon factors such as soil sorption, leaching, volatilization, surface runoff, and their uptake by plant system. Agrochemicals present in soil affect micro- and macroflora including bacteria, actinomycetes, fungi, nematodes, arthropods, crustaceans, earthworms, and leguminous plants. These soil flora play an important role in the degradation of organic matter as well as in nitrification, nitrogen fixation, and release of various nutrients from soil (Pandey and Singh 2004; Seghers et al. 2003). Thus, the use of agrochemicals has become a major threat to beneficial soil microbes and affects the sustainability of agricultural crops. Agrochemicals also interact with microorganisms that reside in the rhizosphere like rhizobia and reduce the number of available root sites for infection (Anderson et al. 2004). Furthermore, agrochemicals block the signaling mechanisms between rhizobia and leguminous plants and inhibit the interaction between them (Fox et al. 2007).

Rhizobia: Diversity and Occurrence in Soil

Rhizobial population in soil is often diverse because of biotic as well as abiotic factors. Genotypes of host plants and their distribution affect the occurrence and distribution of rhizobial species in soil. For example, *Phaseolus vulgaris* and *Mimosa affinis* differ in their nodulation specificity for legumes. Whereas, *P. vulgaris* is rather nonspecific for its microsymbiotic nodulating partner and is nodulated by six rhizobial species, viz., *Rhizobium etli*, *R. giardinii*, *R. gallicum*, *R. tropici*, *R. leguminosarum* bv. *phaseoli*, and *Bradyrhizobium* spp.; on the other hand, *M. affinis* is rather specific and is nodulated with *R. etli* alone (Wang et al. 1999). Genistoid legumes (brooms) in Canary Islands, Morocco, and Spain are nodulated by *Bradyrhizobium* strains that belong to four distinct evolutionary lineages that include *B. japonicum*, *B. canariense*, and two unnamed genospecies (Vineusa et al. 2005). Abiotic factors like pH, rainfall, soil type, and temperature affect the diversity of rhizobial species, while soil types may also affect the composition of rhizobial community. This is very well exemplified by a legume wherein a host growing in different geographical locations is nodulated by different rhizobial species/genera. For example, *Glycine max* (common name soybean) is usually nodulated by *B. japonicum* globally; however, in the Xinjiang region of China, *Mesorhizobium tianshanense* and *Sinorhizobium fredii* have been recovered from this legume. In another example, *R. leguminosarum* bv. *viciae* and bv. *trifolii* have been reported in bean nodules growing in Leon, France, whereas in addition to these two rhizobia, *R. etli*, *R. gallicum*, and *S. fredii* have been reported in beans growing in Andalucia region (Velázquez et al. 2001). Traditionally two rhizobial species (*Mesorhizobium ciceri* and *Mesorhizobium mediterranean*) have been reported from nodules of *Cicer arietinum*, but *Ensifer meliloti* (formerly *Sinorhizobium meliloti*) was reported from nodules of this legume growing in Tunisia under water-deficient conditions (Romdhane et al. 2009). Similarly, *E. meliloti* has also been identified in *C. arietinum* plants growing in Almora and Terai region of Uttarakhand Himalayas (Rajwar et al. 2013). The presumption that rhizobial diversity is affected by soil type is strengthened by identification of different rhizobial species from Caragana plant growing in three ecoregions of China varying in soil types. *Mesorhizobium* genospecies I, II, IV, VI, and VII were reported from Caragana plants growing in sandy soils of Mongolia. *M. temperatum*, *M. tianshanense*, *M. septentrionale*, *M.* genospecies III, *R. yanglingense*, and *Rhizobium* sp. IV were reported in those growing in saline/alkaline soils and *M. plurifarium*, *M.* genospecies V and VII, and *Rhizobium* sp. IV in fertile/forest soils of Northwestern Yunnan region (Lu et al. 2009).

Stress Responses in Plant and Rhizobia

There are stress responses in both plant and rhizobia. The following sections discuss stress responses and adaptive mechanisms in plants and rhizobia separately and how various osmolytes help to form an effective symbiosis and maintain good productivity under stress.

Stress Responses in Plants

Drought

Drought is an environmental condition that produces water stress in plant (Kozlowski and Pallardy 1997). The primary mechanism which plants adopt to grow in drought is regulation of water loss. Water loss in plants is regulated either by morphological modifications or physiological adaptations. Morphological modifications include thickening of cuticle and development of deep root systems. Physiological adaptations include leaf rolling, responsive stomata, metabolism at low leaf water status, and phytohormone secretion (Turner et al. 2000). Thus, plants with water potential less than that of soil survive (Tiaz and Zeiger 2002).

Salinity

Soil salinity leads to accumulation of high concentration of sodium (Na^+), chloride (Cl^-), and sulfate (SO_4^{2-}) ions in plant cells, inactivating enzymes and inhibiting protein synthesis. This reduces photosynthesis and in turn plant growth (Shannon 1997). Salinity-adaptive responses in plants include accumulation of compatible solutes that raise the osmotic pressure in cytoplasm,

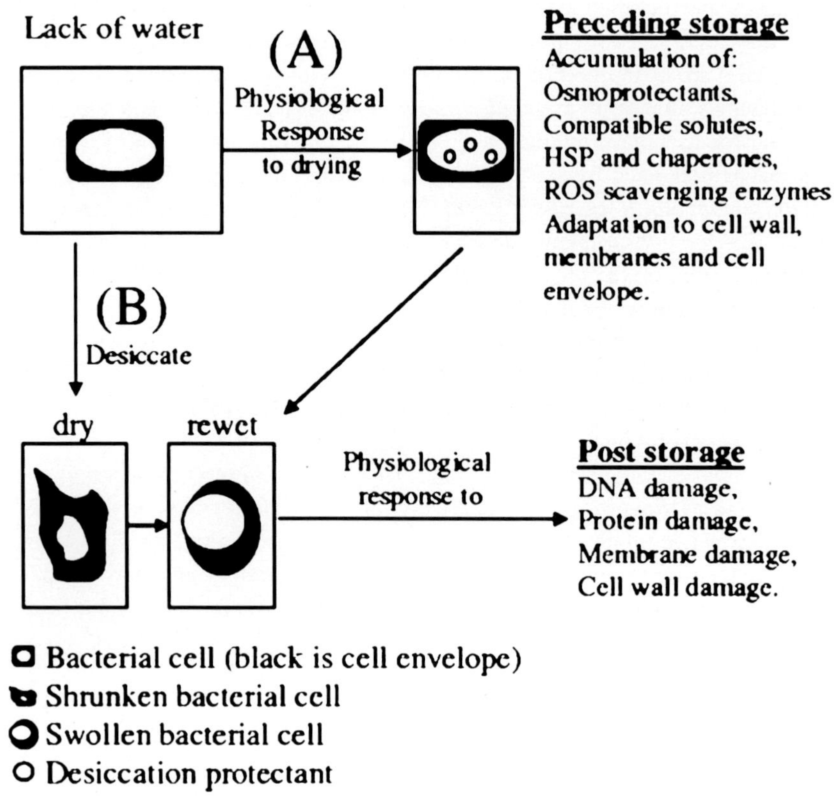

Fig. 12.1 Pathway highlighting response of rhizobia under desiccation and salt stress

stabilize protein and membrane (McNeil et al. 1999), and increase in sodium ion uptake (Hasegawa et al. 2000). The plant can synthesize various compatible solutes, i.e., betaines tertiary sulfonium compounds, polyols, sugars (e.g., trehalose), and amino acids (e.g., proline).

Stress Responses in Rhizobia

Rhizobial species vary in their intrinsic osmotolerance which is measured in terms of their ability to grow in varying concentration of NaCl. *B. japonicum*, *R. etli*, and *R. leguminosarum* (all three biovars) are salt sensitive with their growth inhibited completely at 100 mM NaCl (Boncompagni et al. 1999); *Mesorhizobium huakuii*, *R. tropici* IIB, and *S. fredii* are moderately sensitive with their growth inhibited at 200 mM NaCl, whereas *S. meliloti* and *Agrobacterium tumefaciens* are highly salt tolerant with the ability to grow at 300 mM NaCl (Bernard et al. 1986). *Rhizobium* spp. from nodules of *Acacia*, *Hedysarum*, *Leucaena*, and *Prosopis* plants can tolerate up to 500 mM NaCl (Zhang et al. 1991). Rhizobia survive in stress due to expression of NaCl-responsive loci leading to the accumulation of osmoprotectants, enhanced production of exopolysaccharides, ROS-scavenging enzymes, and heat shock proteins and chaperons (Vriezen et al. 2007). Vriezen and coworkers have proposed two hypothetical pathways for rhizobial response under desiccation and salt stress (Fig. 12.1).

Compatible Solutes

Rhizobial cells are affected by changes in osmolarity of their surroundings. They are plasmolyzed in saline and arid habitat and adapt to changes in osmolarity by accumulation of compatible solutes or osmolytes that confer protection against the deleterious effects of the low water activity, maintain appropriate cell volume, and protect intracellular macromolecules (Le Rudulier et al. 1984). The uptake of compatible solutes is preferred over synthesis. Compatible solutes are of various types: amino acids and related structures (glycine betaine, proline betaine, ectoine, glutamate, and dimethylsulfoniopropionate), carbohydrates (sucrose, trehalose, maltose, and cellobiose), alcohols (glycerol, mannitol), amino acids (pipecolate), and potassium. The type and amount of compatible solutes that accumulate in rhizobia depend on the level of osmotic stress and growth phase of culture (Smith et al. 1994). Osmolytes are either synthesized de novo or from precursors acquired from external environments (Galinski 1995). Osmolytes such as trehalose, sucrose, potassium, and glutamate are synthesized de novo (Miller and Wood 1996; Welsh 2000). In *S. meliloti*, glutamate is accumulated at low and high salt concentrations; but at higher levels besides glutamate, N-acetyl glutaminyl glutamine amide (NAGGN) is also accumulated (Smith et al. 1994). Varying types of compatible solute accumulate within rhizobial cell viz-á-viz growth phases under stress. Glutamate and NAGGN are accumulated at late exponential phase and trehalose at stationary phase (Talibart et al. 1997).

Compatible solute glycine betaine is synthesized from choline through its oxidation. Choline is widespread in soil and rhizosphere, as its derivative phosphatidylcholine is a component of eukaryotic membrane (Boncompagni et al. 1999). Glycine betaine confers osmotolerance to cells grown in inhibitory concentration of salts as well as during lag and early exponential phases. β-dimethylsulfoniopropionate and α-pipecolic acids are accumulated in *Sinorhizobium* spp. (Pichereau et al. 1998; Gouffi et al. 2000).

Role of Exopolysaccharides

Exopolysaccharides (EPS) are involved in maintaining successful symbiotic interaction (Broughton et al. 2000). EPS are involved in several steps of nodule development like invasion, release of bacteria from infection thread, and differentiation of bacteroid (Fraysse et al. 2003). In stress, EPS play a major role in the survival of microorganisms (Rinaudo 2004). They help in two ways: exclusion of toxic compounds, such as chloride (Cl^-) (Lloret et al. 1998), and their ability to form a diffusion barrier to oxygen (Potts 1994), thus leading to higher survival of rhizobia. There are two types of EPS: EPS I (succinoglycan) and EPS II (galactoglycan). EPS I is further divided into low-molecular weight (LMW) and high-molecular weight (HMW) types. EPS I and II are helpful in mitigating the harmful effect of ROS (Lehman and Long 2013), which is enhanced under drought stress (Jebara et al. 2010).

Role of ROS-Scavenging Enzymes

ROS-scavenging enzymes are accumulated during desiccation-induced salinity stress in interacting partners of successful legume-rhizobia symbiosis. Leguminous plants are able to detoxify ROS by two mechanisms: nonenzymatic and enzymatic. Nonenzymatic is through accumulation of flavanones, carotenoids, and ascorbic acid, whereas enzymatic involves superoxide dismutases, peroxidases, and catalases. In bacteroids, peroxidases are not produced; hence, an alternative ascorbate-glutathione (ASC-GSH) cycle enzymes become operative (Jebara et al. 2010). It has earlier been reported that peat-based rhizobial formulations show increased accumulation of manganese-based superoxide dismutase (Feng et al. 2002).

Regulation of Stress Responses in Rhizobia

Water-stressed soils are low in nitrogen and phosphorous. The production of EPS I under nitrogen-limiting conditions involves three genes: *exo*Q, *exo*T, and *exo*P (Gonzalez et al. 1998; Skorupska et al. 2006). ExoQ (product of *exo*Q gene) is responsible for the production of HMW succinoglycan and exoT for LMW succinoglycan, whereas ExoP1 and ExoP2 (the product of *exo*P genes) block the polymerization of HMW succinoglycan and hence promote production of LMW succinoglycan. The production of ExoP is regulated by a two-component signal transduction system in which ExoS and ChvI act as sensor kinase (Cheng and Walker 1998) and SyrM and SyrA as response regulator (Mulligan and Long 1989). Thus, it can be concluded that the expression of genes involved in EPSI production is upregulated during salt stress (Ruberg et al. 2003). During osmotic stress induced by the abundance of solutes out of two, the production of HMW succinoglycan is more than LMW succinoglycan (Breedveld et al. 1990). Under drought stress, available phosphorous in the soil is low (0.1–10 μM). This stimulates the production of EPS II (Mendrygal and Gonzalez 2000; Schmeisser et al. 2009). The production of EPS I is regulated by proteins of the phosphotransferase (PTS) system (Pinedo et al. 2008). The genes involved are hpr (for substrate), hprK (for phosphorylation), and EIIA (manX) (for catabolite repression) and are located immediately downstream of the exoS gene (Pinedo and Gage 2009). Phosphorylation/dephosphorylation of its substrate (HPr) is controlled by hprK genes through succinate-mediated catabolite repression mechanism. A phosphorylated form of HPr activates EIIA and enhances catabolite repression. Out of three genes of PTS system, hpr gene is responsible for the production of EPS I and EIIA for accumulation of EPS I. The strains that have mutated EIIA and hpr genes form an effective nodulation on *Medicago* spp. (Pinedo et al. 2008).

The synthesis of EPS II is regulated by two proteins: WggR (ExpG) and MucR (Baumgarth et al. 2005). WggR, a product of the gene located in the galactoglucan synthesis gene cluster, positively affects the synthesis of EPS II under phosphorous limitation (Becker et al. 1998), whereas MucR represses galactoglucan synthesis (Mendrygal and Gonzalez 2000). Under P starvation, WggR and phosphorylated PhoB bind to the promoter regions and activate the transcription of a few other *wga*, *wgd*, and *wge* genes (Bahlawne et al. 2008). PhoB activates transcription of *wggR*.

Bacterial motility is also an important factor for the establishment of symbiosis under stress conditions. The regulatory systems controlling the production of EPS also control the motility of rhizobia (Bahlawane et al. 2008). The motility genes are regulated along with chemotaxis and flagellar genes and placed into class I, II, and III. Class I genes comprise the *visN* and *visR* genes. Class I genes control Class II genes, which comprise the flagellar assembly and motor genes. Class III genes include flagellin and chemotaxis genes (Sourjik et al. 2000). The regulatory system ExoR/ExoS/ChvI, which is involved in the regulation of EPS I synthesis, affects motility genes. It was reported that rhizobial strains with mutations in *exoR* and *exoS* lose their ability to swim and swarm as the expressions of all three classes of flagellar biosynthesis genes were downregulated (Yao et al. 2004).

Under oxygenic stress, four genes *katA*, *katB*, *katC*, and *sodA* are expressed. The gene *katA* is induced by H_2O_2, while *katC* is induced by heat, salt, ethanol, and paraquat. Thus *katC* is important for alleviation of ROS. In leguminous plants during osmotic stress, induced by low water activity or ionic radical, an NaCl-responsive locus becomes active. It shuts down photosynthetic apparatus, and hence ROS production stops (Vriezen et al. 2007). *S. meliloti* can survive desiccation as it has a genetic mechanism induced by NaCl. The NaCl-responsive locus is induced mainly due to osmotic stress either by reduction of water activity or accumulation of ionic radicals. Two NaCl-responsive genes *asnO* and *ngg* increase the ability of rhizobial strains to grow in high NaCl concentrations. The product of these genes (AsnO and Ngg) is involved in the production of a compatible solute NAGGN by a two-step regulatory mechanism where Ngg (N-acetyl glutaminyl glutamine synthetase) converts glutamine to a dipeptide N-acetyl glutaminyl glutamine (NAGG), and AsnO transfers the amide nitrogen from glutamine to N-acetyl glutaminyl glutamine to produce NAGGN (Sagot et al. 2010). *S. meliloti* strain are not able to grow optimally at higher NaCl concentration if anyone of the two genes of NaCl-responsive locus is not expressed (Vriezen et al. 2013).

Legume-Rhizobium Symbiosis Under Stress

Structural and Biochemical Changes Under Water Stress

Drought- and high salinity-related water stress has negative effect on the nodulation ability of leguminous plants (Rao and Sharma, 1995). If at all nodulation occurs, there is a reduction in nodule size, increase in rate of nodule senescence, and thus BNF is less (Babber et al. 2000). Structural changes in nodules of leguminous plants grown in salinity-affected soils have been observed. These are lobed nucleus with modified chromatin distribution, endoplasmic reticulum, microtubules, mitochondria, and plastids (Serraj et al. 1995). Several metabolic changes are also observed during stress-induced senescence in nodules. These are increased concentration of sucrose due to inhibition of enzyme sucrose synthase (Ben Salah et al. 2009; Sanchez et al. 2011), decrease in malate concentration leading to reduced bacteroid respiration (Prell and Poole. 2006; Marino et al. 2007), and decrease in ascorbic acid content that has a free radical-quenching ability. Effective utilization of enzyme phosphoenolpyruvate (PEP) carboxylase maintains structural integrity of nodule in stress (Drevon et al. 1998; Galvez et al. 2005). Drought leads to decrease in leghemoglobin content (Gordon et al. 1997). Although an antioxidant enzyme catalase is present, but peroxidase is absent; thus ascorbate-glutathione cycle enzymes serve as an alternative peroxidase-scavenging system in nodules (Jebara et al. 2010).

Sheokand et al. (1995) observed that catalase activity in nodules of *C. arietinum* and *Cajanus cajan* growing in the presence of 100 mM NaCl concentration decreases to 44 %. Vadez and Sinclair (2002) reported higher accumulation of ureides in drought-sensitive lines of soybean as compared to tolerant lines and that high accumulation of ureide in nodule decreases nitrogen fixation (King and Prucell 2005). Lodwig et al. (2003) suggested that a complex amino acid cycling occurs in nodules. Plants provide amino acids to the bacteroids enabling them to shut down their ammonium assimilation through BNF. Some non-nitrogenous signals are also involved in the inhibition of BNF under drought. These are oxidative signals created by ROS. ROS can affect the expression of sucrose synthase at transcriptional and posttranslational levels (Marino et al. 2008). It is now recognized that redox signals play a crucial role in establishing and maintaining symbiosis between rhizobia and legumes (Chang et al. 2009). The natural stress hormone like abscisic acid (ABA) concentration was increased fivefold in soybean nodule under drought stress (Clement et al. 2008). Although ABA is not involved in the decline of sucrose synthase, though it might affect the leghemoglobin-facilitated O_2 diffusion and BNF (Gonzalez et al. 2001).

Proteomics: A Tool for Understanding Legume-Rhizobia Symbiosis Under Stress

Several genes that are present in micro symbionts during normal conditions but expressed under conditions of stress have been identified as early as 1998, when techniques for proteome analysis were still not being frequently used (Zurayk et al. 1998). It has already been revealed that rhizobia adapt to stress at physiological (by producing osmolytes and EPS) and enzymatic level (ROS-scavenging enzymes). The genes for their production are present, but expressed under specific condition. Proteome analysis may give an insight into key plant and microbial proteins expressed during the interaction of legume with rhizobia under stress only. Since gene analysis reveals little about their expression, proteomics have become an important tool for studying the biology of legume-rhizobia symbiosis in real time. Djordjevic (2004) has characterized proteins specifically produced in nodules when *S. meliloti* nodulates *Melilotus truncatula* and *Melilotus alba* under stress. Proteins identified are chaperonic, those involved in detoxification of ROS and osmoregulation. Chaperonin proteins, GroEL and GroES, and Hsp are upregulated under stress (Yeh et al. 2002). ABC transporter proteins that are responsible for sodium ion (Na^+) transport are downregulated under stress and hence prevent cell rupturing (Shamseldin et al. 2006). During early colonization of rhizobia, cell receptors (like Myc factors) recognize stigmasterols released by plants and induce calcium spiking within cell which is responsible for activation of calmodulin-like proteins within the nucleus that will further activate the nodulation-specific genes of microsymbiont (reviewed by Rajwar et al. 2013). Under drought stress, these calmodulin-like proteins CaML2 and CaML4 get downregulated, thereby inhibiting nodulation (Estibaliz et al. 2007).

Integrating proteomics, genomics, transcriptomics, and metabolomics studies with bioinformatics and statistical tools may help identify genes and proteins which govern legume-rhizobia interactions under stress. These studies will help design crop plants that are better suited to interact with microbes. It is proposed that better understanding of beneficial and detrimental interactions between the plants and microbes may offer unprecedented opportunities to increase crop productivity.

S. meliloti-Legume Symbiosis Under Stress: A Case Study

M. ciceri are natural symbionts of *C. arietinum* (Nour et al. 1994). But *Sinorhizobium* spp. have been reported in the same legume growing under drought stress in Tunisia (Romdhane et al. 2009), Almora, and Terai region of Uttarakhand, India (Rajwar et al. 2013). This indicates that *Sinorhizobium* spp. have stress adaptation mechanism. In a study carried out in our laboratory at the Department of Microbiology, G.B. P. U. A & T., Pantnagar, the amount of EPS produced by *S. meliloti* strain MTCC 11403 increased from 146 to 234 μg ml^{-1} under drought stress. We also found that the same strain when subjected to water stress colonize *C. arietinum* rhizosphere (7.65×10^6 cfu/g of root) more than the rhizosphere of normal watered plant (1.25×10^5 cfu/g of root) (Karmakar, unpublished). In *S. meliloti* strains, EPS I of both types: HMW and LMW are produced. During osmotic stress induced by abundance of solutes, HMW succinoglycan is produced more than LMW succinoglycan (Breedveld et al. 1990). Thus, it can be concluded that expression of genes involved in EPSI production is upregulated during salt stress (Ruberg et al. 2003). Under drought stress, available phosphorous in the soil is low (0.1–10 μM). This stimulates the production of EPS II (Mendrygal and Gonzalez 2000; Schmeisser et al. 2009). This may be the reason for *Sinorhizobium* to occupy the nodule under drought stress. The mechanism of EPS production in *Sinorhizobium* was elucidated by Skorupska et al. (2006). They reported that two regulatory systems, one with three genes, *exo*Q, *exo*T, and *exo*P, and another with ExoS and ChvI, are involved. Similarly under phosphorous-limiting condition, PhoB and ExpG get upregulated, which collectively enhance the production of EPS II. The study of Davies and Walker (2007) revealed that EPS-producing genes (*exo*P, *exo*Q, and *exo*D) are assisting in nodulation because strains mutated in any one of the EPS-producing genes were unable to nodulate *Medicago*.

Legume-Rhizobium Symbiosis Under Agrochemical Stress

Pesticides are agrochemicals that protect agricultural crops from damaging influences such as weeds, insects, or phytopathogens. While pesticides are used for crop protection, they ultimately reach soil system and influence microbes, other soil communities, as well as plants. A few pesticides can affect rhizobial communities in soil and legume-rhizobia symbiosis (reviewed by Ahemad and Khan 2013). Pesticides affect legume-rhizobia symbiosis in several ways. These affects are summarised in Table 12.1. One of them is reduction of root biomass because fewer sites are available for rhizobial infection: this adversely affects the supply of carbohydrates to the existing nodules with resultant decrease in the potentiality of rhizobia for infection. As a consequence, meristematic activity decreases that is otherwise essential for nodule development (Anderson et al. 2004).

Other is blocking of molecular communication between flavonoids of plants and nod proteins of rhizobia thus preventing nodulation. Rhizobia and pesticides compete for common binding site(s) on plant roots. For example, glyphosate [N-(phosphonomethyl) glycine] is a nonselective herbicide and competes for binding site(s) on plant roots and moves to the plants through phloem. Glyphosate acts upon shikimate acid pathway; blocks synthesis of aromatic amino acids and other phenolic compounds such as cinnamic acids (Canal et al. 1987), flavones (Ishikura et al. 1986), and flavonoids (Laanset 1987); and eventually decreases plant growth. The action of glyphosate decreases synthesis of starch (Greiger and Bestman 1990), causes imbalance in hormone

Table 12.1 Commonly used pesticides and their effects on rhizobia

SN	Name of pesticide	Effect of pesticides
1.	Metribuzin (H), Glyphosate (H), Thiamethoxam (I)	Decrease production of indole-3-acetic acid and siderophore
2.	Thiaram (F)	Decrease population of *Rhizobium* and *Mesorhizobium*
3.	Mancozeb (F)	Reduce the growth by alteration in the membrane composition
4.	Chlorpyrifos (one), Lindane(I)	Inhibit the growth of *Rhizobium japonicum*
5.	Captan(F)	Inhibit the growth of *Rhizobium*, *Mesorhizobium*, *Sinorhizobium*, *Bradyrhizobium*
6.	2,4-D(H)	Produce deleterious effect by modifying the membrane fluidity
7.	Hexaconazole(F)	Inhibit the dehydrogenase system in *Rhizobium*
8.	Atrazine (H)	Decrease the nodules number

H herbicides, *F* fungicides, *I* insecticides

levels (Lee 1984), reduces synthesis of chlorophyll (Kitchen et al. 1981a) and its precursor δ-aminolevulinic acid (ALA) (Kitchen et al. 1981b), and reduces the activity of nitrogenase enzyme complex (Malik and Tesfai 1985). Herbicides inhibit photosynthesis and indirectly reduce the amount of total nitrogen fixed by legumes. Metribuzin and fluchloralin inhibit transfer of electrons from compound Q to plastoquinone in photosystem II preventing $NADP^+$ reduction that is required for CO_2 fixation (Fedtke 1982). However, the activity of nitrogenase enzyme remains unaffected in the presence of herbicides. Dexon p-(dimethylamino) benzenediazo sodium sulfonate a fungicide commonly used to control "root rot" or "damping off" in legume seedling leads to nodulation failure due to interferences in plant hormones, indole acetic acid, and gibberellic acid. Herbicides negatively affect legume-rhizobia interaction reducing the production of surface polysaccharides (LPS and EPS) of rhizobia.

Effect of Macronutrient-Containing Agrochemicals on Rhizobia-Legume Interaction

Nitrogen is the most critical nutrient for plant growth. Plant yield increases when compatible rhizobia are inoculated along with fertilizers in fields (Elsheikh 1992). Leguminous plants fix nitrogen with the help of nitrogen-fixing bacteria and remove the problem of nitrogen limitation (Rasanen and Lindstrom 2003). Nitrogen concentration in soil also regulates the formation of nodules (Rhoades and Coleman 1999). Phosphorus, on the other hand, is essential for synthesis of nucleic acids, phospholipids and ATP, protein synthesis, and N_2 fixation (Buchanan et al. 2000). Available phosphorous increases the number of nodules formed in legume roots (Rasanen and Lindstrom 2003). Potassium activates numerous enzymes and acts as a counterion and cellular osmoticum (Epstein and Bloom 2005). It affects the growth rate of internodes (de la Gardia and Benlloch 1980). The growth of *R. leguminosarum* bv. *trifolii* and *S. meliloti* is restricted in medium which is low or deficient in potassium. Calcium spiking is necessary for efficient nodulation in the root hair (Wais et al. 2002). The role of calcium redistribution in the infection process is also reported (Downie and Walker 1999; Niebel et al. 1999).

Tripartite Relationship Under Drought/Salinity Stress

Presently arid or semiarid land areas are increasing worldwide because soil is becoming saline on account of intensive irrigation in the agricultural fields and rising temperatures, thus limiting crop production. Although plants have their own mechanisms to cope up with drought stress, they become more tolerant to drought when associated with different soil microorganisms such as rhizobia, arbuscular mycorrhizal fungi (AMF), and other PGPRs. Interaction among these microorganisms enhances plant drought tolerance at physiological and molecular levels through decreased oxidative stress, improved water status, or regulation of aquaporins. There is a correlation between drought tolerance in rhizobia and rhizobia-induced plant drought tolerance. Under drought conditions, PGPRs regulate the levels of stress-related hormones, i.e., abscisic acid and ethylene. The regulation of these hormones is the cause of an enhancement of plant drought tolerance mediated by PGPR. At the same time, AMF in combination with rhizobia or PGPR usually has an accumulative beneficial effect on plant drought tolerance (Aroca and Lozano 2009).

For sustainable crop production, it is necessary to improve salt stress tolerance in crops. Thus, development of salt-tolerant crop plants through plant breeding programs is one approach, but this may be available for a few plant species. Alternatively, seeds and seedlings of leguminous crops can be inoculated with various PGPR(s) to alleviate effects of salt stress in crop plants (Dixon et al. 1993; Tank and Saraf 2010). It has been indicated that PGPR strains produce polysaccharides that bind to sodium ions (Na^+) in root zone and hence alleviate effect of salt stress on plant and enhance microbial growth and activities (Tank and Saraf 2010).

Salinity affects rhizobia-legume symbiosis by direct impairment of early interactions during nodule formation and nodule development (El-Hamdaoui et al. 2003). Moreover, symbiotic N_2 fixation in legumes is a phosphorous-requiring process, and salinity significantly reduces plant nutrient uptake, especially phosphorus, since phosphate ions are precipitated under salt-stressed conditions (Grattan and Grieve 1999). It has been shown that some PGPRs can increase P availability to plants by solubilizing insoluble phosphates; this improves P nutrition through increased BNF as well as availability of other nutrients to plants (Gyaneshwar et al. 2002).

There are reports of unsuccessful nodulation under salt stress. It may be due to failure in the infection process because of the adverse effect of salinity on the establishment of rhizobia in rhizosphere of leguminous plant (Singleton and Bohlool 1984; Elsheikh and Wood 1990). The legume-rhizobia symbiosis and the process of nodule initiation on legumes are sensitive to salt stress (Lauter et al. 1981; Rai and Prasad 1983; Velagaleti and Marsh 1989). Salinity enhances ethylene production which causes nodule senescence (Glick 2004; Glick et al. 2007; Madhaiyan and Poonguzhali. 2007). Under high salinity, most of the available energy is used in osmotic adjustments by the plant (Munns and Termaat 1986) and thus growth is affected. Under macronutrient deficiencies, especially nitrogen and phosphorous, the increase in ethylene levels in roots is observed (Lynch and Brown 1997). Borch et al. (1999) reported that P-deficient roots of common bean produced twice as much ethylene based on root dry weight than P-sufficient roots. The increased ethylene production in roots causes reduction in nodulation under low P conditions caused by salinity. It has been reported that increase in root-shoot biomass ratio under salt stress is an index of salt tolerance. An increased functional root biomass has been shown to strengthen the root Na^+ detoxification capability (Hua et al. 1982). It seems that the increased ethylene production in salt-stressed roots may contribute towards reduction in root growth compared with shoot growth.

PGPRs have also evolved a variety of adaptive mechanisms in order to restore the cell turgor pressure and reduce the osmotic potential between cell and environment, subsequently maintaining cell division and cell elongation under salt stress conditions. Among these mechanisms, one response of PGPR is the transient adjustment in ionic balance (Egamberdiyeva 2007). Sodium and chloride ions are regarded as being biologically aggressive osmolytes (Zhu 2002). High concentrations of these ions in the plant accordingly lead to imbalances in water and ion relations. Giri et al. (2003) reported that accumulation of Na^+ was strongly influenced by storage of other cations, particularly potassium ions. Increased K^+ concentration under saline conditions may help to decrease Na^+ uptake, and this can indirectly maintain the growth of plant (Giri and Mukerji 2004). Co-inoculation of *Rhizobium* and *Pseudomonas* significantly increased the K^+ content of plants under salt stress conditions.

Mycorrhiza is relatively more important to plant growth under dry conditions than when soil moisture is plentiful (Michelsen and Rosendahl 1990). Plant infected by AMF can tolerate and recover more rapidly from soil water deficit than uninfected plants (Henderson and Davies 1990). Root hairs are very efficient in increasing the root surface area, although their effect on uptake may not be proportional because of competition among them. AM develop symbiotic associations with plants, and their hyphae are more efficient than root hairs and provide better transport of minerals from the soil to the root over much larger distances than by diffusion.

Alleviation of the Impact of Stress

Medicago sativa, *Arachis hypogea*, *Cyamopsis tetragonoloba*, and *Melilotus* spp. are major crop species, and *Acacia* and *Prosopis* are important leguminous trees growing in arid regions. The osmoadaptation of both the partners is essential for maintaining successful symbiosis and nitrogen fixation in these regions. Rhizobia can nodulate their host even under stress conditions.

Drought Stress

Jenkins et al. (1987) identified rhizobia that effectively nodulate arid legumes. Wadisirisuk et al. (1989) observed that osmotolerant rhizobia can migrate even under scarce moisture condition. Athar and Johnson (1996) reported that osmotolerant strains of *R. meliloti* (now *S. meliloti*) performed better than those of the non-tolerant alfalfa rhizobia and formed effective symbiotic relationship under drought. Peoples et al. (1998) compared the annual herbage production under stress and reported that *Medicago-based* pastures could fix more than 70 kg of Nha^{-1} $year^{-1}$, making them efficient and sustainable. Straub et al. (1997) reported that seed yield in plants inoculated with proline mutant (pro^-) rhizobia decrease twice than those inoculated with pro^+ strains. Thus, rhizobial strains with ability for increased accumulation of compatible solutes are able to perform better in stressed soil. Another approach to improve legume-rhizobia symbiosis is to isolate rhizobia from wild legumes and inoculate them in cultivated legumes. Zhao et al. (1997), working on this approach, isolated *Rhizobium* strains from *Astragalus* spp. and *Cicer* spp. and successfully inoculated them into *M. sativa* and *P. vulgaris*.

Trees of genera *Acacia* and *Prosopis* are of central importance in rural economy (Roshetko 2001). Hocking (1993) reported that these trees can grow in areas with rainfall as low as 40–50 mm $year^{-1}$. During rainless period, long tap roots of trees reach the groundwater (Deans et al. 1999). Fagg and Stewart (1994) reported the contribution of these trees towards soil stabilization and improvement through nitrogen fixation. Masutha et al. (1997) evaluated their agroforestry potential, nodulation ability, and nitrogen fixation potential. Kadiata et al. (1996) reported that nitrogen fixed by different tree legumes were: *Leucaena* (Indian subabul), 20–74 %; *Albizia*, 28–72 %; and *Cliridia*, 44–84 %. Even under water stress prevailing in these regions, they could fix 43–581 kg of nitrogen per hectare as compared to annual legumes which fix 15–210 kg of nitrogen per hectare (Dakora and Keya 1997).

Salt Stress

Root nodule-colonizing bacteria are more salt tolerant than their legume host and show variation in salt tolerance (Zurayk et al. 1998). The growth of many rhizobial species is inhibited by 100 mM NaCl (Yelton et al. 1983); however, Mohammad et al. (1991) have reported that a few can tolerate NaCl concentration of 300–700 mM. Breedveld et al. (1991) reported that *R. leguminosarum* can tolerate up to 350 mM NaCl. The fast-growing strains of soybean and chickpea rhizobia can also tolerate up to 340 mM NaCl compared to slow-growing strains (Elsheikh and Wood 1995). Mashhady et al. (1998) reported that *S. meliloti* formed successful symbiosis with *M. sativa* under saline condition, i.e., 100 mM

NaCl. Among tree species, *Prosopis articulata*, *P. pallida*, and *P. tamarugo* can grow and fix nitrogen under 300 mM NaCl concentration (Felker et al. 1981). Salt-tolerant Australian *Acacia* (*A. auriculiformis* and *A. stenophylla*) can survive even up to 1.7–1.8 M NaCl (Aswathappa et al. 1987).

Conclusion

In stressed soils, signaling between legumes and rhizobia which is important for development of successful legume-rhizobia interaction is negatively affected. Signaling involves three key steps: (1) flavonoid is released by plants to activate the *nod* D genes of rhizobia, (2) activated nod D gene upregulates nod boxes and form nod factors, (3) and endogenous plant hormones are activated by nod factors and mediate nodule initiation. Individually legume and rhizobia possess mechanisms to overcome stress. But several biotic and abiotic factors influence legume-rhizobia symbiosis under stress. Therefore, for the establishment of successful legume-rhizobia symbiosis, the following strategies are proposed.:

Selection of Competitive Strains of Rhizobia

Rhizobial strains with the ability to colonize rhizosphere competitively (Howieson et al. 1995), catabolize secondary metabolites present in root exudates (Boivin et al. 1991), synthesize osmolytes under stress de novo, and produce high EPS under water and nutrient limitation must be selected by screening natural populations. The performance of selected strains under stress can be further enhanced by genetically modifying the strains for traits like rhizosphere competence, rhizopine catabolism, and toxin production. For example, introduction of NolA (a repressor of *nod* genes) prevents colonization of legume roots by native bacteria (Vlassak and Vanderleyden 1997). Rhizobia with *moc* genes have rhizopine-catabolizing ability which enhances their nodule occupancy (Murphy et al. 1995). Incorporation of toxin-producing genes enhances bactericidal effect of inoculant rhizobia over native populations. EPS-producing genes like *exoQ*, *exoT*, and *exoP* may be overexpressed constitutively, and NaCl-responsive genes transferred to develop effective legume-rhizobia symbiosis in stress environment.

Breeding for Competitive Legumes

Breeding programs should target quantitative trait loci (QTL) controlling morphological traits like nodule number per plant and nodule weight (Neo et al. 1996). Legume traits such as high shoot nitrogen content, nitrate tolerance, antioxidant enzymes, and hormones can be also targeted.

Biofertilizers Augmented with Elicitor Molecules

In the current scenario of mass production of biofertilizers, it is not only the quality that is essential but also its specificity towards alleviation of stressed conditions. Towards this end, newer technologies are being employed using various strategies. In such an attempt, signal molecules characterized from root exudates of leguminous plants are being added to the end product. The molecules being used in this refined technology include flavonoids, sugars, acids, amino acids, amines, and secondary metabolites that function as elicitor or signaling molecules for successful microbe-plant interaction (Skorupska et al. 2010). As discussed earlier, all these signal molecules are usually altered during stress conditions like drought and salinity. Currently, companies like Nitragin are marketing formulations with biosignals that improve the symbiotic relationship, activate mechanisms to resist abiotic stress conditions, and induce pathogen-defensive response in plants (www.nitragin.com.ar).

Thus, selecting competitive rhizobial strains, breeding for competitive legume host, and using elicitor-loaded biofertilizers can be useful in sustaining green revolution to evergreen revolution.

Acknowledgements Our work on rhizobia has been supported through the Centre for Research on Bacteria and Archaea, under All India Coordinated Project on Taxonomy (AICOPTAX) of the Ministry of Environment and Forests, Govt. of India.

References

Abberton MT (2010) Enhancing the role of legumes: potential and obstacles. In: Abberton MT, Conant R, Batello C (eds) Grassland carbon sequestration: management, policy and economics, vol 338. FAO, Rome, pp 177–187

Ahemad M, Khan SM (2013) Pesticides as antagonists of rhizobia and the legume-*Rhizobium* symbiosis: a paradigmatic and mechanistic outlook. Biochem Mol Biol 1:63–75

Anderson A, Baldock JA, Rogers SL, Bellotti W, Gill G (2004) Influence of chlorsulfuron on rhizobial growth, nodule formation, and nitrogen fixation with chickpea. Aust J Agric Res 55:1059–1070

Aroca R, Ruiz-Lozano JM (2009) Induction of plant tolerance to semi-arid environments by beneficial soil microorganisms – a review. In: Lichtfouse E (ed) Climate change, intercropping, pest control and beneficial microorganisms, sustainable agriculture reviews. Springer, Dordrecht, pp 121–135

Aswathappa N, Marcar NE, Thomson LAJ (1987) Salt tolerance of Australian tropical and subtropical acacias. In: Turnbull JW (ed) Australian acacias in developing countries: proceedings of an international workshop held at Forestry Training Centre, Gympie, Qld., Australia, 4–7 August 1986. Brown Prior Anderson, Burwood, Victoria 3125, pp 70–73

Athar M, Johnson DA (1996) Influence of drought on competition between selected *Rhizobium meliloti* strains and naturalized soil rhizobia in alfalfa. Plant Soil 184:231–241

Babber S, Sheokand S, Malik S (2000) Nodule structure and functioning in chickpea (*Cicer arietinum*) as affected by salt stress. Biol Plant 43:269–27

Bahlawane C, McIntosh M, Krol E, Becker A (2008) *Sinorhizobium meliloti* regulator MucR couples exopolysaccharide synthesis and motility. Mol Plant Microbe Interact 21:1498–1509

Baumgarth B, Bartels FW, Anselmetti D, Becker A, Ros R (2005) Detailed studies of the binding mechanism of the *Sinorhizobium meliloti* transcriptional activator ExpG to DNA. Microbiology 151:1–10

Becker B, Kosch K, Parniske M, Müller P (1998) Exopolysaccharide synthesis in *Bradyrhizobium japonicum*: sequence, operon structure and mutational analysis of an *exo* gene cluster. Mol Gen Genet 259:161–171

Ben Salah I, Albacete A, Martinez Andujar C, Haouala R, Labidi N, Zribi F, Martinez V, Perez-Alfocea F, Abdelly C (2009) Response of nitrogen fixation in relation to nodule carbohydrate metabolism in *Medicago ciliaris* lines subjected to salt stress. J Plant Physiol 166:477–488

Bernard T, Pocard JA, Perroud B, Le Rudulier D (1986) Variations in the response of salt-stressed *Rhizobium* strains to betaines. Arch Microbiol 143:359–364

Boivin C, Barran LR, Malpica CA, Rosenberg C (1991) Genetic analysis of a region of the *Rhizobium meliloti* pSym plasmid specifying catabolism of trigonelline, a secondary metabolite present in legumes. J Bacteriol 173:2809–2817

Boncompagni E, Osterås M, Poggi MC, Le Rudulier D (1999) Occurrence of choline and glycine betaine uptake and metabolism in the family rhizobiaceae and their roles in osmoprotection. Appl Environ Microbiol 65:2072–2077

Borch K, Bouma T, Lynch P, Brown KM (1999) Ethylene: a regulator of root architectural responses to soil phosphorus availability. Plant Cell Environ 22:425–431

Breedveld MW, Zevenhuizen LPTM, Zehnder AJB (1990) Osmotically induced oligo- and polysaccharide synthesis by *Rhizobium meliloti* SU-47. J Gen Microbiol 136:2511–2519

Breedveld MW, Zevenhuizen LPTM, Zehnder AJB (1991) Osmotically-regulated trehalose accumulation and cyclic beta-(1,2)-glucan excreted by *Rhizobium leguminosarum* bv. trifolii TA. Arch Microbiol 156:501–506

Brockwell J, Bottomley PJ, Thies JE (1995) Manipulation of rhizobia microflora for improving legume productivity and soil fertility: a critical assessment. Plant Soil 174:143–180

Broughton WJ, Jabbouri S, Perret X (2000) Keys to symbiotic harmony. J Bacteriol 182:5641–5652

Buchanan BB, Gruissem W, Jones RL (eds) (2000) Biochemistry and molecular biology of plants. American Society of Plant Physiologists, Rockville, SanterreA, p 849.

Canal MJ, Tames RS, Fernandez B (1987) Glyphosate increased levels of indole–3-acetic acid in yellow nutsedge leaves correlated with gentisic acid levels. Physiol Plant 71:384–388

Chang C, Damiani I, Puppo A, Frendo P (2009) Redox change during the legume rhizobium symbiosis. Mol Plant 2:370–377

Cheng HP, Walker GC (1998) Succinoglycan production by *Rhizobium meliloti* is regulated through the ExoS-ChvI two-component regulatory system. J Bacteriol 180:20–26

Clement M, Lambert A, Herouart D, Boncompagni E (2008) Identification of new up-regulated genes under drought stress in soybean nodules. Gene 426:15–22

Dakora FD, Keya SO (1997) Contribution of legume nitrogen fixation to sustainable agriculture in Sub-Saharan Africa. Soil Biol Biochem 29:809–817

Davies BW, Walker GC (2007) Identification of novel *Sinorhizobium meliloti* mutants compromised for oxidative stress protection and symbiosis. J Bacteriol 189:2110–2113

De la Gardia MD, Benlloch M (1980) Effects of potassium and gibberellic acid on stem growth of whole sunflower plants. Physiol Plant 49:443–448

Deans JD, Diagne O, Lindley DK, Dione M, Parkinson JA (1999) Nutrient and organic-matter accumulation in *Acacia senegal* fallows over 18 years. For Ecol Manage 124:153–167

Dixon RK, Garg VK, Rao MV (1993) Inoculation of Leucaena and Prosopis seedlings with *Glomus* and *Rhizobium* species in saline soil: rhizosphere relations and seedlings growth. Arid Soil Res Rehabil 7:133–144

Djordjevic MA (2004) *Sinorhizobium meliloti* metabolism in the root nodule: a proteomic perspective. Proteomics 4:1859–1872

Downie JA, Walker SA (1999) Plant responses to nodulation factors. Curr Opin Plant Biol 2:483–489

Drevon JJ, Frangne N, Fleurat-Lessard PH, Ribet J, Vadez V, Serraj R (1998) Is nitrogenase-linked respiration regulated by osmocontractile cells in legume nodules? In: Elmerich C, Kondorosi A, Newton WE (eds) Biological nitrogen fixation for the 21st century. Kluwer Academic, Dordrecht, pp 465–466

Egamberdiyeva D (2007) The effect of plant growth promoting bacteria on growth and nutrient uptake of maize in two different soils. Appl Soil Ecol 36:184–189

El-Hamdaoui A, Redondo-Nieto M, Rivilla R, Bonilla I, Bolanos L (2003) Effects of boron and calcium nutrition on the establishment of the *Rhizobium leguminosarum*–pea (*Pisum sativum*) symbiosis and nodule development under salt stress. Plant Cell Environ 26:1003–1012

Elsheikh EAE (1992) Effect of salinity on growth, nodulation and nitrogen yield of inoculated and nitrogen fertilized chickpea (*Cicer arietinum* L.). Arch Biotechnol 1:17–28

Elsheikh EAE, Wood M (1990) Effect of salinity on growth, nodulation and nitrogen yield of chickpea (Cicer arietinum L.). J Exp Bot 41:1263–1269

Elsheikh EAE, Wood M (1995) Nodulation and N_2 fixation by soybean inoculated with salt-tolerant rhizobia or salt-sensitive Bradyrhizobia in saline soil. Soil Biol Biochem 27:657–661

Epstein E, Bloom AJ (2005) Mineral nutrition of plants: principles and perspectives, 2nd edn. Sinauer Associates, Sunderland, pp 1–400

Estibaliz L, Stefanie W, Wolfram W, Ruben L, Cesar A, Esther MG (2007) *Medicago truncatula* root nodule proteome analysis reveals differential plant and bacteroid responses to drought stress. Plant Physiol 144:1495–1507

Fagg CW, Stewart JL (1994) The value of *Acacia* and *Prosopis* in arid and semi-arid environments. J Arid Environ 27:3–25

Fedtke C (1982) Biochemistry and physiology of herbicide action. Springer, Berlin, pp 1–193

Felker P, Clark PR, Laag AE, Pratt PF (1981) Salinity tolerance of the tree legumes: mesquite (*Prosopis glandulosa* var. Torreyana, *P. velutina* and P. articulata), algarrobo (*P. chilensis*), kiawe (*P. pallida*) and tamarugo (*P. tamarugo*) grown in sand culture on nitrogen-free media. Plant Soil 61:311–317

Feng ZH, Wang TG, Li DD, Fung P, Wilson BC, Liu B, Ali SF, Langenbach R, Hong JS (2002) Cyclooxygenase-2-deficient mice are resistant to 1-methyl-4-phenyl-1-2,3,6-tetrahydropyridine induced damage of dopaminergic neurons in the substantia nigra. Neurosci Lett 329:354–358

Fox JE, Gulledge J, Engelhaupt E, Burow ME, McLachlan JA (2007) Pesticides reduce symbiotic efficiency of nitrogen-fixing rhizobia and host plants. Proc Natl Acad Sci U S A 104:10282–10287

Fraysse N, Couderc F, Poinsot V (2003) Surface polysaccharide involvement in establishing the rhizobium-legume symbiosis. Eur J Biochem 270:1365–1380

Galinski EA (1995) Osmoadaptation in bacteria. Adv Microb Physiol 37:273–328

Galvez L, Gonzalez EM, Arrese-Igor C (2005) Evidence for carbon flux shortage and strong carbon/nitrogen interactions in pea nodules at early stages of water stress. J Exp Bot 56:2551–2561

Giri B, Mukerji KG (2004) Mycorrhizal inoculant alleviates salt stress in *Sesbania aegyptiaca* and *Sesbania grandiflora* under field conditions: evidence for reduced sodium and improved magnesium uptake. Mycorrhiza 14:307–312

Giri B, Kapoor R, Mukerji KG (2003) Influence of arbuscular mycorrhizal fungi and salinity on growth, biomass, and mineral nutrition of *Acacia auriculiformis*. Biol Fertil Soils 38:170–175

Glick BR (2004) Changes in plant growth and development by rhizosphere bacteria that modify plant ethylene levels. Acta Hortic 631:265–273

Glick BR, Cheng Z, Czarny J, Duan J (2007) Promotion of plant growth by ACC deaminase-producing soil bacteria. Eur J Plant Pathol 119:329–339

Gonzalez JE, Semino CE, Wang LX, Castellano-Torres LE, Walker GC (1998) Biosynthetic control of molecular weight in the polymerization of the octasaccharide subunits of succinoglycan, a symbiotically important exopolysaccharide of *Rhizobium meliloti*. Proc Natl Acad Sci U S A 95:13477–13482

Gonzalez EM, Galvez L, Arrese-Igor C (2001) Abscisic acid induces a decline in nitrogen fixation that involves leghaemoglobin, but is independent of sucrose synthase activity. J Exp Bot 52:285–293

Gordon AJ, Minchin FR, Skot L, James CL (1997) Stress-induced declines in soybean N_2 fixation are related to nodule sucrose synthase activity. Plant Physiol 114:937–946

Gouffi K, Bernard T, Blanco C (2000) Osmoprotection by pipecolic acid in *Sinorhizobium meliloti*: specific effects of D and L isomers. Appl Environ Microbiol 66:2358–2364

Grattan SR, Grieve CM (1999) Salinity mineral nutrient relations in horticultural crops. Sci Hortic (Amst) 78:127–157

Greiger DR, Bestman HD (1990) Self limitation of herbicide mobility by phytotoxic action. Weed Sci 38:324–329

Gyaneshwar P, Naresh KG, Parekh LJ (2002) Effect of buffering on the phosphate solubilizing ability of microorganisms. World J Microbiol Biotechnol 14:669–673

Hasegawa PM, Bressan RA, Zhu JK, Bohnert HJ (2000) Plant cellular and molecular responses to high salinity. Annu Rev Plant Physiol Plant Mol Biol 51:463–499

Henderson JC, Davies FT (1990) Drought acclimation and the morphology of mycorrhizal *Rosa hybrida* L. cv. "Frey" is independent of leaf elemental content. New Phytol 115:503–510

Hocking DE (1993) Trees for drylands. International Science Publisher, New York, pp 1–370

Howieson JG, Loi A, Carr SJ (1995) *Biserrula pelecinus* L.- a legume pasture species with potential for acid, duplex soils which is nodulated by unique root-nodule bacteria. Aust J Agric Res 46:997–1009

Hua S, Tsai VY, Lichens GM, Noma AT (1982) Accumulation of amino acids in *Rhizobium* sp. strain WRlOOl in response to NaCl salinity. Appl Environ Microbiol 44:135–140

Ishikura N, Teramoto S, Takeshima Y, Mitsui S (1986) Effects of glyphosate on the shikimate pathway and regulation of phenylalanine ammonia-lyase in *Cryptomeria* and *Perilla* cell suspension cultures. Plant Cell Physiol 27:677–684

Jebara S, Drevon JJ, Jebara M (2010) Modulation of symbiotic efficiency and nodular antioxidant enzyme activities in two *Phaseolus vulgaris* genotypes under salinity. Acta Physiol Plant 32:925–932

Jenkins MB, Virginia RA, Jarrel WM (1987) Rhizobial ecology of the woody legume mesquite *Prosopis glandulosa* in the Sonoran desert. Appl Environ Microbiol 53:36–40

Kadiata BD, Mulongoy K, Isirimah NO (1996) Time course of biological nitrogen fixation, nitrogen absorption and biomass accumulation in three woody legumes. Biol Agric Hortic 13:253–266

King CA, Prucell LC (2005) Inhibition of N_2 fixation in soybean is associated with elevated ureides and amino acids. Plant Physiol 137:1389–1396

Kitchen LM, Witt WW, Rieck LE (1981a) Inhibition of aminolevulinic acid synthesis by glyphosate. Weed Sci 29:571–577

Kitchen LM, Witt WW, Rieck LE (1981b) Inhibition of chlorophyll accumulation by glyphosate. Weed Sci 29:513–516

Kozlowski TT, Pallardy SG (1997) Physiology of woody plants. Academic, San Diego, pp 1–411

Laanset L (1987) Incorporation of exogenous tyrosine and phenylalanine into c-glycosylflavones in glyphosate treated barley seedlings. Eesti NSV Tead Akad Toim Biol 36:204–209

Lauter DJ, Munns DN, Clarkin KL (1981) Salt response of chickpeas influenced by N supply. Agric J 73:961–966

Le Rudulier D, Bernard T, Goas G (1984) Osmoregulation in *Klebsiella pneumoniae*: enhancement of anaerobic growth and nitrogen fixation under stress by proline betaine, γ-butyrobetaine, and other related compounds. Can J Microbiol 30:299–305

Lee T (1984) Release of lateral buds from apical dominance by glyphosate in soybean and pea seedlings. J Plant Growth Regul 3:227–235

Lehman AP, Long SR (2013) Exopolysaccharides from *Sinorhizobium meliloti* can protect against H_2O_2-dependent damage. J Bacteriol 195:5362–5369

Lloret J, Wulff BB, Rubio JM, Downie JA, Bonilla I, Rivilla R (1998) Exopolysaccharide II production is regulated by salt in the halotolerant strain *Rhizobium meliloti* EFB1. Appl Environ Microbiol 64:1024–1028

Lodwig EM, Hosie AHF, Bordes A, Findley K, Allaway D, Karunakaran R, Downie JA, Poole PS (2003) Amino acid cycling drives nitrogen fixation in legume rhizobia symbiosis. Nature 422:722–726

Lu YL, Chen WF, Wang ET, Guan SH, Yan XR, Chen WX (2009) Genetic diversity and biogeography of rhizobia associated with *Caragana* species in three ecological regions of China. Syst Appl Microbiol 32:351–361

Lynch J, Brown KM (1997) Ethylene and plant responses to nutritional stress. Physiol Plant 100:613–619

Madhaiyan M, Poonguzhali SST (2007) Characterization of 1-Aminocyclopropane-1-carboxylate (ACC) deaminase containing *Methylobacterium oryzae* and interactions with auxins and ACC regulation of ethylene in canola (*Brassica campestris*). Planta 226:867–876

Malik MAB, Tesfai K (1985) Pesticidal effect on soybean rhizobia symbiosis. Plant Soil 85:33–41

Marino D, Frendo P, Ladrera R, Zabalza A, Puppo A, Arrese-Igor C, Gonzalez EM (2007) Nitrogen fixation control under drought stress. Localized or systemic? Plant Physiol 143:1968–1974

Marino D, Hohnjec N, Kuster H, Moran JF, Gonzalez EM, Arrese-Igor C (2008) Evidence for transcriptional and post translational regulation of sucrose synthase in pea nodule by cellular redox state. Mol Plant Microbe Interact 21:622–630

Mashhady AS, Salem SH, Barakh FN, Heggo AM (1998) Effect of salinity on survival and symbiotic performance between *Rhizobium meliloti* and *Medicago sativa* in Saudi Arabian soils. Arid Soil Res Rehabil 12:3–14

Masutha TH, Muofhe ML, Dokora FD (1997) Evaluation of N_2 fixation and agroforestry potential in selected tree legumes for sustainable use in South Africa. Soil Biol Biochem 29:993–998

McNeil SD, Nuccio ML, Hanson AD (1999) Betaines and related osmoprotectants. Targets for metabolic engineering of stress resistance. Plant Physiol 120:945–949

Mendrygal KE, Gonzalez JE (2000) Environmental regulation of exopolysaccharide production in *Sinorhizobium meliloti*. J Bacteriol 182:599–606

Michelsen A, Rosendahl S (1990) The effect of VA mycorrhizal fungi, phosphorus and drought stress on the growth of *Acacia nilotica* and *Leucaena leucocephala* seedling. Plant Soil 124:7–13

Miller KJ, Wood JM (1996) Osmoadaptation by rhizosphere bacteria. Annu Rev Microbiol 50:101–136

Mohammad RM, Akhavan-Kharazian M, Campbell WF, Rumbaugh MD (1991) Identification of salt-and drought-tolerant *Rhizobium meliloti* L. strains. Plant Soil 134:271–276

Mulligan JT, Long SR (1989) A family of activator genes regulates expression of *Rhizobium meliloti* nodulation genes. Genetics 122:7–18

Munns R, Termaat A (1986) Whole-plant responses to salinity. Aust J Plant Physiol 13:143–160

Murphy PJ, Wexler W, Grzemski W, Rao JP, Gordon D (1995) Rhizopines – their role in symbiosis and competition. Soil Biol Biochem 27:525–529

Neo HH, Hunt S, Layzell DB (1996) Can genotypes of soybean (*Glycine max*) selected for nitrate tolerance provide good 'models' for studying the mechanism of nitrate inhibition of nitrogenase activity? Physiol Plant 98:653–660

Niebel A, Gressent F, Bono JJ, Ranjeva R, Cullimore J (1999) Recent advances in the study of Nod factor perception and signal transduction. Biochimie 81:669–674

Nour SM, Fernandez MP, Normand P, Cleyet-Maret JC (1994) *Rhizobium ciceri* sp. nov. consisting of strains that nodulate chickpea (*Cicer arietinum* L). Int J Syst Bacteriol 44:511–522

Pandey S, Singh DK (2004) Total bacterial and fungal population after chlorpyrifos and quinalphos treatments in groundnut (*Arachis hypogaea* L.) soils. Chemosphere 55:197–205

Peoples MB, Gault RR, Scammell GJ, Dear BS, Vigona J, Sandral GA, Paul J, Wolf EC, Angus JF (1998) Effect of pasture management on the contributions of fixed N to the N economy of ley-farming systems. Aust J Agric Res 49:459–474

Pichereau V, Pocard JA, Hamelin J, Blanco C, Bernard T (1998) Differential effects of dimethylsulfoniopropionate, dimethylsulfonioacetate, and other S-methylated compounds on the growth of *Sinorhizobium meliloti* at low and high osmolarities. Appl Environ Microbiol 64:1420–1429

Pinedo CA, Gage DJ (2009) HPrK regulates succinate-mediated catabolite repression in the gram-negative symbiont *Sinorhizobium meliloti*. J Bacteriol 191:298–309

Pinedo CA, Bringhurst RM, Gage DJ (2008) *Sinorhizobium meliloti* mutants lacking phosphotransferase system enzyme HPr or EIIA are altered in diverse processes, including carbon metabolism, cobalt requirements and succinoglycan production. J Bacteriol 190:2947–2956

Potts M (1994) Desiccation tolerance of prokaryotes. Microbiol Rev 58:755–805

Prell J, Poole P (2006) Metabolic changes of rhizobia in legume nodules. Trends Microbiol 14:161–168

Rai R, Prasad V (1983) Salinity tolerance of *Rhizobium* mutants: growth and relative efficiency of symbiotic nitrogen fixation. Soil Biol Biochem 15:217–219

Rajwar A, Sahgal M, Johri BN (2013) Legume-rhizobia symbiosis and interactions in agroecosystems. In: Arora NK (ed) Plant microbe symbiosis-fundamentals and advances. Springer, India, pp 233–265

Rao DLN, Sharma PC (1995) Effectiveness of rhizobial strains for chickpea (*Cicer arietinum* L.) under salinity stress and recovery of nodulation on desalinization. Indian J Exp Biol 33:500–504

Rasanen LA, Lindstrom K (2003) Effects of biotic and abiotic constraints on the symbiosis between rhizobia and the tropical leguminous trees *Acacia and Prosopis*. Indian J Exp Biol 41:1142–1159

Rhoades CC, Coleman DC (1999) Nitrogen mineralization and nitrification following land conversion in montane Ecuador. Soil Biol Biochem 31:1347–1354

Rinaudo M (2004) Role of substituents on the properties of some polysaccharides. Biomacromolecules 5:1155–1165

Romdhane SB, Trabelsi M, Aouani ME, de Lajudie P, Mhamdi R (2009) The diversity of rhizobia nodulating Chickpea (*Cicer arietinum*) under water deficiency as a source of more efficient inoculants. Soil Biol Biochem 41:2568–2572

Roshetko JM (2001) Agroforestry species and technologies: a compilation of the highlights and factsheets published by NFTA and FACT Net 1895–1999. A Publication of Winrock International, Morrilton, pp 1–232

Ruberg S, Tian ZX, Krol E, Linke B, Meyer F, Wang YP, Puhler A, Weidner S, Becker A (2003) Construction and validation of a *Sinorhizobium meliloti* whole genome DNA microarray: genome-wide profiling of osmoadaptive gene expression. J Biotechnol 106:255–268

Sagot B, Gaysinski M, Mehiri M, Guigonis JM, Le Rudulier D, Alloing G (2010) Osmotically induced synthesis of the dipeptide N-acetylglutaminylglutamine amide is mediated by a new pathway conserved among bacteria. Proc Natl Acad Sci U S A 107:12652–12657

Sanchez DH, Pieckenstain FL, Escaray F, Erban A, Kraemer U, Udvardi MK, Kopka J (2011) Comparative ionomics and metabolomics in extremophile and glycophytic *Lotus* species under salt stress challenge the metabolic pre-adaptation hypothesis. Plant Cell Environ 34:605–617

Schmeisser C, Liesegang H, Krysciak D, Bakkou N, Le Quéré A, Wollherr A, Heinemeyer I, Morgenstern B, Pommerening-Röser A, Flores M (2009) *Rhizobium* sp. Strain NGR234 possesses a remarkable number of secretion systems. Appl Environ Microbiol 75:4035–4045

Seghers D, Verthe K, Reheul D, Bulcke R, Siciliano SD, Verstraete W, Top EM (2003) Effect of long-term herbicide applications on the bacterial community structure and function in an agricultural soil. FEMS Microbiol Ecol 46:139–146

Serraj R, Fleurat-Lessard P, Jaillard B, Drevon JJ (1995) Structural changes in the inner-cortex cells of soybean root nodules are induced by short-term exposure to high salt or oxygen concentrations. Plant Cell Environ 18:455–62

Shamseldin A, Nyalwidhe J, Werner D (2006) A proteomic approach towards the analysis of salt tolerance in *Rhizobium etli* and *Sinorhizobium meliloti* strains. Curr Microbiol 52:333–339

Shannon MC (1997) Adaptation of plants to salinity. Adv Agron 60:75–120

Sheokand S, Dhandi S, Swaraj K (1995) Studies on nodule functioning and hydrogen peroxide scavenging enzymes under salt stress in chick pea nodule. Plant Physiol Biochem 33:561–566

Singleton PW, Bohlool BB (1984) Effect of salinity on nodule formation by soybean. Plant Physiol 74:72–76

Skorupska A, Janczarek M, Marczak M, Mazur A, Król J (2006) Rhizobial exopolysaccharides: genetic control and symbiotic functions. Microb Cell Fact 5:1–19

Skorupska A, Wielbo J, Kidaj D, Marek-Kozaczuk M (2010) Enhancing *Rhizobium*-legume symbiosis using signaling factors. In: Khan MS, Zaidi A, Musarrat J (eds) Microbes for legume improvement. Springer, New York, pp 27–44

Smith LT, Smith GM, D'souza MR, Pocard JA, Le Rudulier D, Madkour MA (1994) Osmoregulation in *Rhizobium meliloti*: mechanism and control by other environmental signals. J Exp Zool 268:162–165

Sourjik V, Muschler P, Scharf B, Schmitt R (2000) VisN and VisR are global regulators of chemotaxis, flagellar, and motility genes in *Sinorhizobium* (*Rhizobium*) *meliloti*. J Bacteriol 182:782–788

Straub PF, Shearer G, Reynolds PHS, Sawyer SA, Kohl D (1997) Effect of disabling bacteroid proline catabolism on the response of soybeans to repeated drought stress. J Exp Bot 48:1299–1307

Talibart R, Jebbar M, Gouffi K, Pichereau V, Gouesbet G, Blanco C, Bernard T, Pocard JA (1997) Transient accumulation of glycine betaine and dynamics of endogenous osmolytes in salt-stressed cultures of *Sinorhizobium meliloti*. Appl Environ Microbiol 63:4657–4663

Tank ND, Saraf MS (2010) Salinity resistant PGPR ameliorates NaCl stress on tomato plants. J Plant Interact 5:51–58

Tiaz L, Zeiger E (2002) Plant physiology. Sinauer Associates, Sunderland, pp 591–621

Tilman D (1999) Global environmental impacts of agricultural expansion: the need for sustainable and efficient practices. Proc Natl Acad Sci U S A 96:5995–6000

Turner NC, Wright GC, Siddique KHM (2000) Adaptation of grain legumes (pulses) to water-limited environments. Adv Agron 71:193–231

Vadez V, Sinclair TR (2002) Sensitivity of N_2 fixation trait in soybean cultivar Jackson to manganese. Crop Sci 42:791–796

Vance CP (2001) Symbiotic nitrogen fixation and phosphorus acquisition. Plant nutrition in a world of declining renewable resources. Plant Physiol 127:390–397

Velagaleti RR, Marsh S (1989) Influence of host cultivars and *Bradyrhizobium* strains on the growth and symbiotic performance of soybean under salt stress. Plant Soil 119:133–138

Velázquez E, Martínez-Romero E, Rodríguez-Navarro DM, Trujillo ME, Daza A, Mateos PE, Martínez-Molina E, van Berkum P (2001) Characterization of rhizobial isolates of *Phaseolus vulgaris* by staircase electrophoresis of low-molecular weight RNA. Appl Environ Microbiol 67:1008–1010

Vineusa P, Léon-Barrios M, Silva C, Willems A, Jabaro-Lorenzo A, Pérez-Galdona R, Werner D, Martínez-Romero E (2005) *Bradyrhizobium canariense* sp. nov., an acid-tolerant endosymbiont that nodulates endemic genistoid legumes (Papilionoideae: Genisteae) from canary Islands, along with *Bradyrhizobium japonicum* bv. *genistearum*, *Bradyrhizobium* genospecies alpha and *Bradyrhizobium* genospecies beta. Int J Syst Evol Microbiol 55:569–575

Vlassak KM, Vanderleyden J (1997) Factors influencing nodule occupancy by inoculant rhizobia. Crit Rev Plant Sci 16:163–229

Vriezen JAC, de Bruijn FJ, Nusslein K (2007) Responses of rhizobia to desiccation in relation to osmotic stress, oxygen, and temperature. Appl Environ Microbiol 73:3451–3459

Vriezen JAC, de Bruijn FJ, Nüsslein KR (2013) Identification and Characterization of a NaCl responsive genetic locus involved in survival during desiccation in *Sinorhizobium meliloti*. Appl Environ Microbiol 79:5693–5700

Wadisirisuk P, Danso SKA, Hardarson G, Bowen GD (1989) Influence of *Bradyrhizobium japonicum* location and movement on nodulation and nitrogen fixation in soybeans. Appl Environ Microbiol 35:1711–1716

Wais RJ, Keating DH, Long SR (2002) Structure function analysis of Nod factor induced calcium spiking in rhizobium-legume symbiosis. Plant Physiol 129:211–224

Wang ET, Rogel A, Santos AG d, Martínez-Romero J, Cevallos MA, Martínez-Romero E (1999) *Rhizobium etli* bv *mimosae*, a novel biovar isolated from *Mimosa affinis*. Int J Syst Bacteriol 49:1479–1491

Welsh DT (2000) Ecological significance of compatible solute accumulation by micro-organisms: from single cells to global climate. FEMS Microbiol Rev 24:263–290

Yao SY, Luo L, Har KJ, Becker A, Rüberg S, Yu GQ, Zhu JB, Cheng HP (2004) *Sinorhizobium meliloti* ExoR and ExoS proteins regulate both succinoglycan and flagellum production. J Bacteriol 186:6042–6049

Yeh KC, Peck MC, Long SR (2002) Luteolin and GroESL modulation in vitro activity of NodD. J Bacteriol 84:525–530

Yelton MM, Yang SS, Edie SA, Lim ST (1983) Characterization of an effective salt-tolerant fast-growing strain of *Rhizobium japonicum*. J Gen Microbiol 129:1537–1547

Zhang XP, Karsisto M, Harper R, Lindstrom K (1991) Diversity of *Rhizobium* bacteria isolated from the root nodules of leguminous trees. Int J Syst Bacteriol 41:104–113

Zhao Z, Williams SE, Schuman GE (1997) Renodulation and characterization of *Rhizobium* isolates from cicer milkvetch (*Astragalus cicer* L.). Biol Fertil Soils 25:169–174

Zhu JK (2002) Salt and drought stress signal transduction in plants. Annu Rev Plant Biol 53:247–274

Zurayk R, Adlan M, Baalbaki R, Saxena MC (1998) Interactive effects of salinity and biological nitrogen fixation on chickpea (*Cicer arietinum* L.) growth. J Agron Crop Sci 180:249–258

Legume Root Exudates: Their Role in Symbiotic Interactions

13

David L. Biate, Annu Kumari, K. Annapurna, Lakkineni Vithal Kumar, D. Ramadoss, Kiran K. Reddy, and Satish Naik

Abstract

Legumes are in constant dynamic interactions with rhizobia and arbuscular mycorrhizal (AM) fungi which mutually benefit the partners. Legumes provide the carbon energy to the microbes, and they in turn provide the nutrients such as nitrogen and phosphorus. The growth of these symbionts largely depends on the secretions of the legume roots which include both high and low molecular weight compounds. These molecules also act as cues in plant-microbe signaling and recognition. A cascade of reactions take place between the legume and the microbe before specific refined symbiotic partnership manifests mutually benefiting both the partners. Here, we provide an overview of the functions of legume root exudates with emphasis on the interactions between legume and rhizobia.

Introduction

The root system, often referred to as the "hidden" part of the plant, not only provides anchorage and facilitates nutrients and water uptake to the plant but also excretes numerous chemical compounds which act as mediators of the plants' various interactions in the rhizosphere. The secreted exudates include sugars, polysaccharides, amino acids, aromatic acids, aliphatic acids, fatty acids, sterols, phenolics, enzymes, proteins, plant growth regulators, and secondary metabolites (Badri et al. 2009). These exudates result in complex biological and chemical interactions in the rhizosphere including beneficial interactions of the plant with rhizobia, mycorrhizal fungi, and plant growth-promoting rhizobacteria (PGPR) or negative or defensive interactions with parasitic plants, pathogenic bacteria or fungi, and invertebrate herbivory. The plant root and its various interactions result in the creation of unique conditions in the rhizosphere which is different compared to the bulk soil (Benfey and Scheres 2000). However, secretions of these compounds come with a significant carbon cost to the plant (Marschner 1995), which could be as much as 30–40 % of fixed carbon in young seedlings (Whipps 1990). Root exudates also help regulate the microbial community in the rhizosphere.

D.L. Biate • A. Kumari • K. Annapurna (✉)
L.V. Kumar • D. Ramadoss • K.K. Reddy • S. Naik
Division of Microbiology, Indian Agricultural Research Institute, New Delhi 110012, India
e-mail: annapurna93@yahoo.co.in

N.K. Arora (ed.), *Plant Microbes Symbiosis: Applied Facets*,
DOI 10.1007/978-81-322-2068-8_13, © Springer India 2015

It has been observed that microbes tend to be more abundantly present in the rhizosphere (Hiltner 1904). The bulk of the organic carbon in the soil is contributed via root exudates of plants (Hutsch et al. 2000; Nguyen 2003). The aim of this review is to provide recent overview of the functions of legume root exudates in the legume-rhizobia and AM symbiosis with special emphasis on flavonoids and strigolactones.

Flavonoids and Strigolactones: Two Important Groups of Legume Root Exudates

Flavonoids are one of the most well-studied compounds in root exudates (Table 13.1). Flavonoids are key signaling compounds in the legume-rhizobia symbiotic interactions (Fisher and Long 1992) and AM symbiosis (Vierheilig et al. 1998; Vierheilig and Piche 2002). They are also known to be involved in the defensive interactions of the plant with pathogens (He et al. 1996; Wasmann and Etten 1996; Enkerli et al. 1998). Flavonoids are phenylpropanoid, biologically active low molecular weight secondary metabolites, mostly synthesized from p-coumaroyl-CoA and malonyl-CoA precursors (Stafford 1990). Flavonoids are a diverse group of compounds containing more than 10,000 members and encompassing flavones, flavonols, flavan-3-ols, flavanones, isoflavonoids, isoflavans, and pterocarpans (Ferrer et al. 2008). Their diversity is derived by the substitution of the flavonoid basal structure by glycosylation, malonylation, methylation, hydroxylation, acylation, prenylation, or polymerization (Winkel 2001). Among the flavonoids, isoflavonoids restricted primarily to the *Leguminosae* have been widely reported to be involved in the legume-rhizobia symbiotic interactions and defense against microbial pathogens. In the legume-rhizobia symbiosis, they act as inducers of a nodulation gene (Ferguson and Mathesius 2003) and also function as a precursor in the biosynthesis of phytoalexin during plant-pathogen/herbivore interactions (Aoki et al. 2000; Dastmalchi and Dhaubhadel 2014). Biosynthesis of isoflavonoids occurs through the general phenylpropanoid pathway. L-phenylalanine, the precursor of isoflavonoids, is first converted to cinnamate in a reaction catalyzed by the enzyme phenylalanine ammonia-lyase (PAL). Hydroxylation of cinnamate at the para position of the phenyl ring by cinnamate 4-hydroxylase then produces p-coumarate. The final step of the general phenylpropanoid pathway is completed with the production of CoA thiol esters by the 4-coumarate-CoA ligase. The first step in the flavonoid pathway is catalyzed by the enzyme chalcone synthase, which is the formation of a chalcone scaffold via the condensation of one CoA thiol ester and three malonyl-CoA molecules. Chalcone reductase then reduces the flavonoid intermediate to form deoxychalcone, which is in turn acted upon to produce the flavonoid core flavonone through cyclization by the enzyme chalcone isomerase. The reaction at this point then enters the legume-specific branch pathway to produce the isoflavone intermediate 2-hydroxyisoflavonone from flavonone by a novel aryl ring migration. The final products in the pathway such as the isoflavone genistein and daidzein are formed either through spontaneous dehydration of 2-hydroxyisoflavonone or through the catalytic reaction of 2-hydroxyisoflavonone dehydratase (Dixon et al. 1995; Wang 2010).

Legume roots also release strigolactones, a small class of carotenoid-derived compounds with only 15 members having been structurally characterized so far (Matusova et al. 2005; Alder et al. 2012). Strigolactones are known as key signaling molecules in the initial stage of the AM symbiosis (Besserer et al. 2006, 2008) and have also been recently reported to play a role in the legume-rhizobia symbiosis as positive regulators of nodulation (Soto et al. 2010; Foo and Davies 2011; Foo et al. 2013). Structurally, strigolactones are composed of four rings (A–D). The A and B rings show variation due to the different side groups, but the C and D rings owing to their role in the essential biological activity are highly conserved (Xi et al. 2010; Cheng et al. 2013). Interestingly, strigolactones and abscisic acid share a common carotenoid biosynthetic pathway but diverge at β-carotene (Matusova et al. 2005; Rani et al.

Table 13.1 Root exudates identified from legumes and their functions

Legume host plant	Rhizobial strain	Compound	Function	Reference
Medicago sativa	*Sinorhizobium meliloti/3654, 1683, 1354*	Luteolin and chrysoeriol	Nod gene and chemotaxis inducer Nematode repellent	Hartwig et al. 1990
Trifolium repens	*Rhizobium leguminosarum* bv. trifolii	Umbelliferone Formononetin, 7,4′-dihydroxyflavone	Inhibition of nod gene Transcriptional activation of nod gene promoters	Djordjevic et al. 1987
Glycine max	*Bradyrhizobium japonicum/9105*	Daidzein Genistein Coumestrol	Induces nod gene, nematode repellent. Induces nod gene	Kosslak et al 1987; Bassam et al. 1988
Vigna unguiculata	*Rhizobium* sp. strain NGR234	Daidzein, genistein Coumestrol 7,4′-dihydroxyflavone	Nod gene and chemotaxis inducer	Dakora 2000
Phaseolus vulgaris	*R. etli* CFN42/4381, 642, 605, 371, 250, 194, 184	Eriodictyol, naringenin Genistein, 7-O-glycoside Delphinidin Petunidin Malvidin Myricetin Quercetin Kaempferol	Induces nod gene	Hungria et al. 1991
Vicia sativa	*R. leguminosarum* bv. Viciae/5057,870, 684, 488, 352, 151, 147	3,5,7,30-Tetrahydroxy-40-methoxyflavone, 7,30-dihydroxy-40-methoxyflavone	Induction of the nod genes	Zaat et al. 1989
Sesbania rostrata	*Azorhizobium caulinodans*/5369	7,40-Dihydroxyflavone	Induction of the nod genes	Messens et al. 1991
Medicago truncatula	*S. meliloti*	7,40-Dihydroxyflavone	Induction of the nod genes	Zhang et al. 2007
Robinia pseudoacacia	*Mesorhizobium* sp.	7,40-Dihydroxyflavone Apigenin Naringenin Chrysoeriol Isoliquiritigenin	Induction of the nod genes	Scheidemann and Wetzel 1997
Cicer arietinum	*R. ciceri*	Naringenin, hesperetin, naringin, 70H coumarin, luteolin	Induction of the nod genes Anti-inducer of the nod ABC genes of *R. ciceri*	Srivastava et al. 1999

2008; Lopez-Raez et al. 2010). Consequently, abscisic acid has been assumed to have a regulatory effect on strigolactone biosynthesis (Lopez-Raez et al. 2010). Key enzymes involved in the synthesis of strigolactones are the carotenoid cleavage dioxygenases 7 and 8 (CCD7 and CCD8) and the iron-binding protein D27. Strigolactone levels in the root are also regulated by the levels of nitrogen and phosphorus in the soil. Depletion of nitrogen and phosphorus in the soil usually results in elevated levels of strigolactones in the roots; however, there is no positive correlation between the regulation of mycorrhizal symbiosis and nodulation by strigolactones with nutrient deficiency (Foo et al. 2013).

Root Exudates in the Legume-Rhizobia Symbiosis

Flavonoids: The Primary Signal Molecule of Legume Root Exudates

In addition to their diverse role in plants such as the regulation of auxin transport (Brown et al. 2001; Wasson et al. 2006), modulating levels of reactive oxygen species (Taylor and Grotewold 2005; Agati et al. 2012), and imparting colors to various tissues including flowers (Davies et al. 2012), flavonoids also play an essential role in the formation of the root nodule during the early stages of the legume-rhizobia symbiosis. Flavonoids act as chemoattractants to compatible rhizobia in the rhizosphere, near the root tips where the greatest amounts are released (Graham 1991). The sites of rhizobial infection, which are the emerging root hair zone, have been shown to have the optimal concentrations of flavonoids (Zuanazzi et al. 1998). The legume-rhizobia interaction leading to the formation of a symbiotic nitrogen-fixing nodule is a very specific interaction. In part, flavonoids are known to play an important role in influencing and determining this specificity. The expression of *Sinorhizobium meliloti nod* genes was inhibited by the soybean (*Glycine max*)-produced daidzein and genistein, while the same isoflavonoids induced the expression of *nod* genes in *Bradyrhizobium japonicum*. However, luteolin was found to induce the *S. meliloti nod* gene expression (Peters et al. 1986). NodD proteins, which are members of the LysR family of transcriptional regulators, are usually needed for the transcription of genes required for the synthesis of Nod factor. Activation of the transcription of Nod factor biosynthesis genes requires both NodD proteins and a flavonoid coinducer. Binding of NodD and an appropriate flavonoid to the *nod* boxes located at the promoter regions induces bending in DNA at the binding site which allows RNA polymerase to initiate gene transcription (Chen et al. 2005; Cooper 2007).

Flavonoids have also been reported to enhance the growth of rhizobia. Flavonoids such as coumestrol and daidzein extracted from soybean roots were shown to promote the growth of *Rhizobium japonicum* USDA 138 and *R. leguminosarum* (D'Arcy-Lameta and Jay 1987). Promotion of rhizobial growth by flavonoids could probably be due to the ability of rhizobia to both convert flavonoids to monocyclic aromatics such as protocatechuic acid and 4-hydroxybenzoate (Rao and Cooper 1994) and utilize the β-ketoadipate pathway to channel these products into the citric acid cycle (Harwood and Parales 1996; Parke 1997: Cooper 2007). In another mechanism employed by rhizobia, biotin released by *Medicago* is perceived by an *S. meliloti* regulatory locus *bioS* resulting in its enhanced growth and root colonization (Streit et al. 1996; Heinz et al. 1999). Further, flavonoids also influence the symbiotic efficiency, metabolism of rhizobia, and regulation of cell division during development of nodules (D'Arcy-Lameta and Jay 1987; Subramanian et al. 2007). Legume root exudates are also known to act as chemoattractants for rhizobia. Besides flavonoids, rhizobia also show positive chemotaxis to other compounds present in root exudates such as simple sugars, amino acids, and carboxylic and hydroxyaromatic acid (Cooper 2007).

Studies in our laboratory have found differential induction of proteins in *Mesorhizobium ciceri* strain DKA-1 when grown with the root exudates of four different genotypes of chickpea. The effect was concentration dependent as well as genotype dependent, clearly indicating the role of the chemical composition of the root exudates (Fig. 13.1). Chemical profiling of a large number of chickpea genotypes would help in identifying the metabolites which are conducive for effective symbiotic interactions as well as develop a diagnostic marker for segregating high and low nodulators and nitrogen fixers.

Strigolactones Involved in the Legume-Rhizobia Symbiosis

While the importance of strigolactones in mycorrhizal symbiosis is well established, their role in the legume-rhizobia symbiosis has only recently gathered interest (Steinkellner et al. 2007).

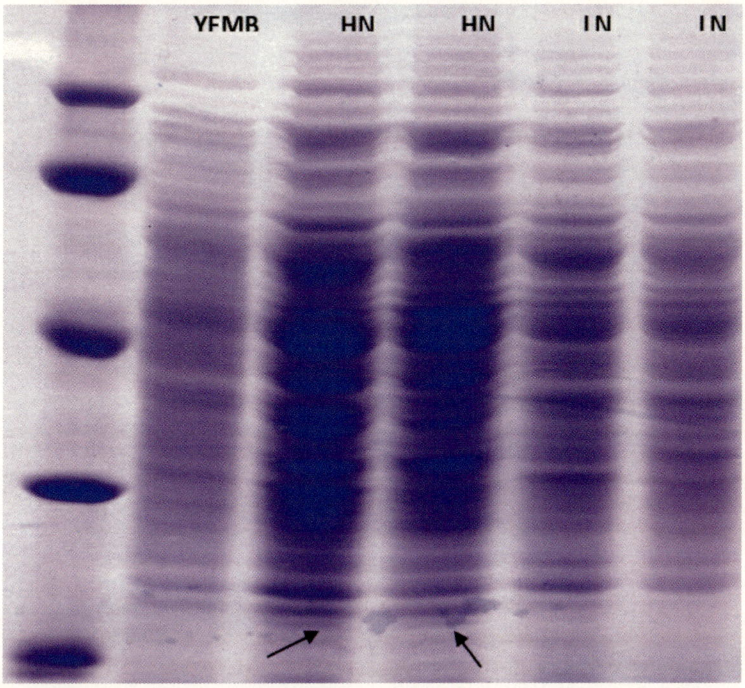

Fig. 13.1 SDS-PAGE profiling of *Mesorhizobium* sp. DKA-1. A 26kD protein was induced in DKA-1 grown with root exudates of high nodulating genotypes (HN). This was absent in DKA-1 grown in YEMB or with root exudates of low nodulating genotype (LN)

Application of synthetic strigolactone analogue Gr24 was associated with increase in the number of nodules formed in *Medicago sativa* and wild-type pea (Soto et al. 2010; Foo and Davies 2011). Contradictory observations were, however, reported in pea (Foo et al. 2013). Flavonoid-induced nodulation-specific rhizosphere signals such as enhanced growth, stimulation of *nod* factor production, or induction of calcium spiking were not detectable when Gr24 was inoculated directly to rhizobial cell cultures (Moscatiello et al. 2010; Soto et al. 2010). Foo et al. (2013) suggested that since strigolactones also play a pivotal role as an endogenous plant hormone in both root and shoot tissue, their role in promoting nodulation could also be as an endogenous plant signal within the root.

Nod Factors Are Induced by Flavonoids

Nod factors are key signal molecules produced by rhizobia in response to flavonoid secretion by legumes. They are lipo-chito-oligosaccharides (LCOs) with an oligomeric backbone of β-1, 4-linked *N*-acetyl-D-glucosaminyl residues, which are N-acylated at the nonreducing terminal residue (Denarie et al. 1996; Kamst et al. 1998). The number of GlcNAc residues present in the chito-oligosaccharide backbone, in the nature of the fatty acyl group, and in the substituents at the nonreducing and/or reducing terminal residues results in variation in the structure of Nod factors (D'Haeze and Holsters 2002). Expressions of nodulation (nod) genes (*nod*, *nol*, and *noe* genes) are required for Nod factor synthesis. Several physical, physiological, and molecular responses are elicited in plants at an active Nod factor concentration of c.10^{-12} mol l^{-1}. These responses include root hair deformation; depolarization of plasma membrane; calcium spiking; alterations in the root hair cytoskeleton; formation of preinfection thread in deformed root hair; nodule primordial cortical cell division; reactive oxygen-generating system inhibition; perturbation of auxin flow in roots (in conjunction with flavonoids); and induction of nodulin gene expression at the preinfection, infection, nodule development, and nodule function stages of the symbiosis (Bartsev et al. 2004; Cooper 2007). The expression of symbiosis-specific gene and subsequent nodule formation is initiated by the binding of specific Nod factors to the LysM receptor kinase (Perret et al. 2000; Antolin–Llovera et al. 2012). The oligosaccharide backbone of Nod factors can undergo various substitutions which influences the binding of Nod factors to the LysM receptor kinases (Perret et al. 2000).

Secreted Proteins Determine Host Specificity/Range

Proteins secreted by rhizobia are known to play important roles in determining host range in the legume-rhizobia interactions. At least three mechanisms of secretions of proteins by rhizobia are known. NodD-dependent NodO, a flavonoid-induced protein secreted by *R. leguminosarum* bv. viciae via a type I secretion system, was the first rhizobial protein whose role in symbiosis was elucidated (de Maagd 1989). NodO has been reported to confer host range extension in *Rhizobium* sp. (Vlassak et al. 1998).

In rhizobia, the type III secretion system is responsible for the secretion of nodulation outer proteins (NOPs). Besides flavonoids and NodD1, the *B. japonicum* type III secretion system also requires NodW (Krause et al. 2002). The functional type III secretion system has so far been reported in *S. fredii* USDA257 (Krishnan et al. 2003), *S. fredii* HH103 (deLyra et al. 2006), *Rhizobium* strain NGR234 (Viprey et al. 1998), *B. japonicum* USDA110 (Krause et al. 2002), *Bradyrhizobium elkanii* (Okazaki et al. 2009), and *Mesorhizobium loti* MAFF303999 (Okazaki et al. 2010). *S. fredii* USDA257 has been reported to secrete several NOPs including nopA, nopB, nopC, nopL, nopP, and nopX via the type III secretion system (Deakin and Broughton 2009). In *Rhizobium* sp. strain NGR234, a promiscuous symbiont capable of nodulating 112 genera of legumes (Pueppke and Broughton 1999) and removal of all secreted or specific NOPs significantly diminishes its nodulation ability on many of its hosts (Ausmees et al. 2004). Several studies have established that proteins secreted by the type III secretion system can have both positive and negative influence on the rhizobial nodulation ability of legume both at the genus and species level (Viprey et al. 1998; Krause et al. 2002; Krishnan et al. 2003; Okazaki et al. 2009, 2010).

Surface Polysaccharides Contribute to Various Stages of Symbiotic Development

Rhizobia produce four different surface polysaccharides – extracellular polysaccharides (EPS), lipopolysaccharides (LPS), K polysaccharides (K antigens, capsular polysaccharides, or KPS), and cyclic glucans. These polysaccharides play an important role at various stages of symbiotic development including the colonization of root, recognition of host, formation of infection thread, and invasion of nodule (Cooper 2007). EPS are essential during the early infection process of the symbiotic development, but at a later stage such as the penetration of the infection thread into the cortical cells or the setting up of the nitrogen-fixing phenotype, LPS are known to play a specific role. Flavonoids are not required for the synthesis of surface polysaccharides, but they act to influence the final structures and symbiotic activity of these compounds either during or after their biosynthesis (Cooper 2007). The role of surface polysaccharides in the establishment of the legume-rhizobia symbiosis has been comprehensively reviewed by Fraysee et al. (2003).

Rhizobial Secretome Also Contains Several Other Important Compounds

Rhizobia also produce several other compounds such as the flavonoid-, NodD1-, NodD2-, and SyrM2-dependent IAA in *Rhizobium* sp. NGR234 (Theunis et al. 2004); pentacyclic triterpenoid lipids, hopanoids, in bradyrhizobia and *Rhizobium* sp. (Kannenberg et al. 1995); a quorum sensor-AHL and a non-AHL quorum sensor, bradyoxetin, in bradyrhizobia (Loh et al. 2002; Mathesius et al. 2003); and a plant growth enhancer, lumichrome, in *S. meliloti* (Phillips et al. 1999; Matiru and Dakora 2005). These compounds are required at different stages of the symbiotic development from root colonization to the formation of a functioning nodule (Cooper 2007).

Transporters Are Necessary for the Secretion of Root Exudates

Many compounds including flavonoids often have to be transported for exudation into the rhizosphere from their site of synthesis in the cytoplasm. Contrary to earlier opinion, secretion of phytochemicals into the rhizosphere is now believed to involve both primary and secondary transporters. Sugiyama et al. (2007), using a specific ABC transporter inhibitor, sodium orthovanadate, showed that secretion of the flavonoid genistein was dependent on ATP which suggests the involvement of ABC transporters in the secretion processes. This report was supported by the work of Badri et al. (2009) using *Arabidopsis* knockout mutants lacking expression of specific ABC transporter proteins. Their study not only provides direct evidence of the ABC transporters in the process of root secretion but also showed that structurally different compounds can be transported by one ABC transporter, or different transporters could transport one compound. ABC transporters comprise one of the largest protein families ubiquitously present in all phyla (Higgins 1992). The largest numbers of ABC transporters (more than 100) are encoded in the plant genome (Kang et al. 2011). Functional ABC transporters are generally made up of two hydrophobic transmembrane domains (TMD) and two cytosolic domains (Martinoia et al. 2002). Initially ABC transporters in plants were thought to be exclusive for detoxification processes (Martinoia et al. 1993), but it is now established that they are also involved in diverse processes as in the response to pathogen, deposition of surface lipid, accumulation of phytate in seeds, transport of flavonoids, and transport of the phytohormones auxin and abscisic acid (Sugiyama et al. 2007; Kang et al. 2011).

Root Exudates in Mycorrhizal Symbiosis

More than 80 % of all terrestrial plants are known to be associated with mycorrhizal fungi. Mycorrhizal associations are broadly divided into two groups – endomycorrhizal (arbuscular mycorrhiza, AM) and ectomycorrhizal associations. The association is mutually beneficial with the fungi helping the plant acquire nutrients such as phosphate from the soil which were otherwise unavailable to the plant and the plant in turn supplying photosynthates to the fungi. AM symbiosis, by far, is one of the most prevalent and the oldest known plant symbioses and is believed to have played an important role in the land colonization of plants. Here, we will be focusing on AM symbiosis due to their similarity with the legume-rhizobia symbiosis. Based on fossil records they are thought to have evolved 400–450 mya (million years ago) (Remy et al. 1994). Approximately 6,000 species of fungi and 200,000 plant species are known to be involved in mycorrhizal association (Wang and Qui 2006; Bonfante and Anca 2009). Hyphae produced by AM fungal soil propagules, asexual spores, or mycorrhizal roots initiate the AM symbiosis when they colonize compatible root.

The establishment of AM symbiosis also involves chemical communication facilitated by root exudates similar to the legume-rhizobia symbiosis. Owing to their presence in root exudates, flavonoids have been suggested to be involved in the signaling in AMF symbiosis. Various workers have reported the role of flavonoids in hyphal growth, hyphal differentiation, and root colonization (Morandi 1996 and Vierheilig et al. 1998). The stimulatory effect of flavonoids on hyphal growth is determined by the structure of flavonoids, and the presence of CO_2 enhances this effect (Becard et al. 1992; Chabot et al. 1992; Poulin et al. 1993; Scervino et al. 2006). Colonization of plant roots by mycorrhizal fungi results in the level of flavonoids (Vierheilig et al. 1998). Interestingly, there seems to exist some genus-species specificity among fungi to alter the pattern of flavonoids in roots during the establishment of AM symbiosis (Morandi et al. 1984; Larose et al. 2002). Generally, the level of flavonoids is known to follow a predictable pattern where the concentrations of flavonoids are usually found to be highest at the later stage of AM symbiosis (Larose et al. 2002). However, the role of flavonoids as signal molecules in AM symbioses is now under a cloud with the recent identification of strigolactones as host recognition signals for AM fungi.

Strigolactones: Evolving Role in AM Symbiosis

In AM symbiosis, exchange of nutrients between the partners occurs through the arbuscules or the highly branched haustoria. The roots of the host plant release specific hyphal branching-promoting compounds, a prerequisite for root colonization in AM fungi (Giovannetti et al.1993; Harrison 2009). Reception of these compounds by the fungi results in the release of signal compounds both by the plant and the fungi called *branching factors* and *Myc factors*, respectively, required for successful colonization (Buee et al. 2000; Kosuta et al. 2003, Maillet et al. 2011) (Fig. 13.2). With the successful isolation and characterization of 5-deoxystrigol, Akiyama and Hayashi (2006) were the first to provide evidence of the probable involvement of strigolactones in promoting hyphal branching in AM fungi. The extensive work of Besserer et al. (2006, 2008, 2009) later corroborated this finding. The germination of spores was reported to be enhanced by the presence of strigolactones in *Glomus intraradices* and *Glomus claroideum.* Strigolactones have also been suggested to be involved in directing the growth of fungi towards roots (Akiyama and Hayashi 2006). Despite the probable existence of other alternative mechanisms, it is now established that the induction of fungal hyphal

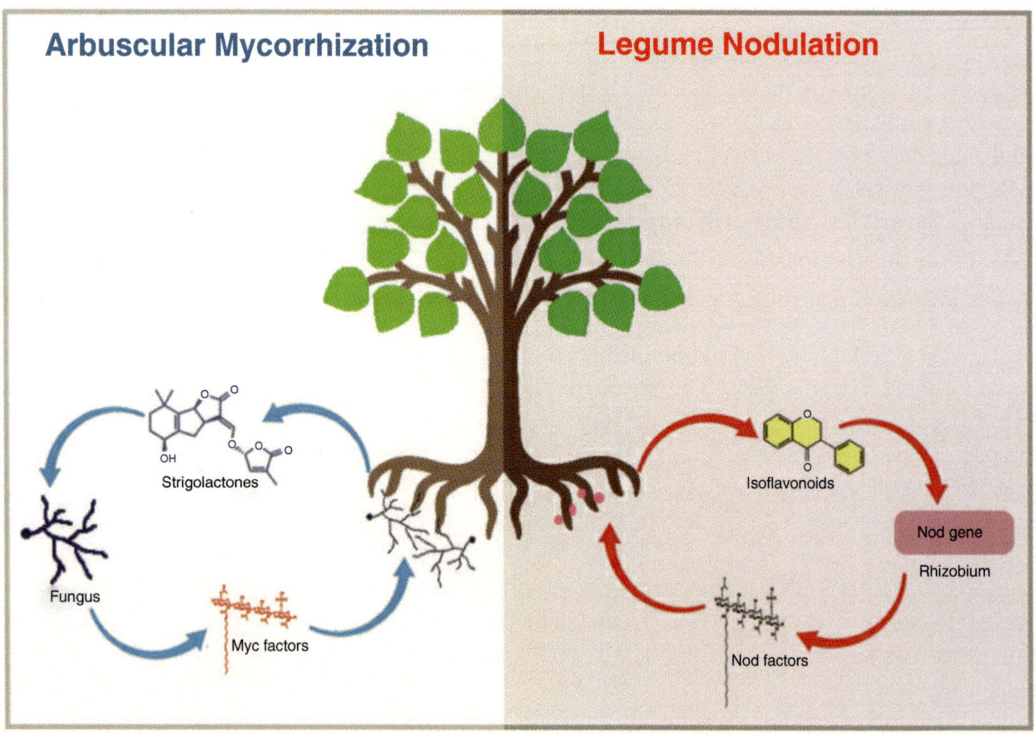

Fig. 13.2 Signaling pathway in the establishment of legume-rhizobia and AM symbioses. Specific isoflavonoid secreted by roots of legume induces the production of a lipo-chito-oligosaccharide nodulation signal (Nod factor) in rhizobia triggering a series of cascading events which ultimately lead to the development of a functional root nodule. Strigolactones are considered as the signal molecules from the host plant in the AM symbiosis

branching by strigolactones is an important step in AM symbiosis (Giovannetti et al. 1994; Gomez-Roldan et al. 2007). The acquisition of nutrients by plants was initially believed to be regulated by strigolactones (Gu et al. 2011; Koltai and Kapultnik 2011; Umehara 2011). However, the recent work of Yoneyama et al. (2012) with the strigolactone-deficient *ccd8* mutant indicates that strigolactones are not essential for nutrient regulation of mycorrhizal symbioses. Foo et al. (2013) also observed that suppression of mycorrhizal colonization could not be altered by strigolactone-deficient *ccd8* mutants and strigolactone response *rms4* mutants. The characterization of two fungal hyphal branching inducers 2-hydroxydodecanoic acid and 2-hydroxytetradecanoic acid in the root exudates of carrots suggests the presence of other branching factors besides strigolactones (Nagahashi and Douds 2011). Concentration as low as 10 nM was sufficient for the strigolactone analogue GR24 to induce nuclear division and enlargement of mitochondria and rapid increase of NADH content, NADH dehydrogenase activity, and ATP content in *Glomus rosea,* suggesting the presence of a sensitive strigolactone perception system in AM fungi (Besserer et al. 2006, 2008; Gutjahr and Parniske et al. 2013).

Conclusions

The plant rhizosphere teems with many microbes each affecting the other and in turn being influenced by plant derived factors. Legume-rhizobia and legume-mycorrhiza interactions are mutually beneficial associations in nature. These associations are facilitated by chemical cues exchanged between the host and the symbionts. In the rhizosphere, which includes plant roots and the surrounding area of soil influenced by the roots, plants exude chemicals to effectively communicate with their neighboring soil organisms. Some of these chemical cues are specific towards a particular organism, and others seem to be present commonly in root exudates of many plants. Rhizobia and mycorrhizae respond to such cues, the former specifically and the latter nonspecifically. However, there are many plant-microbe interactions taking place in the rhizosphere which have neither been studied nor understood well. The chemical nature of the root exudates seems to differentially regulate the microbial community and its activities in the soil. Research should now focus on characterizing and identifying chemical components in root exudates of various plants and especially legumes which have a stimulatory effect on other microbial interactions apart from rhizobia and mycorrhizae.

Acknowledgments The first author is grateful to the Department of Biotechnology, Govt. of India, for the DBT-RA fellowship.

References

Agati G, Azzarello E, Pollastri S, Tattini M (2012) Flavonoids as antioxidants in plants: location and functional significance. Plant Sci 196:67–76

Akiyama K, Hayashi H (2006) Strigolactones: chemical signals in fungal symbionts and parasitic weeds in plant roots. Ann Bot 97:925–931

Alder A, Jamil M, Marzorati M, Bruno M, Vermathen M, Bigler P, Ghisla S, Bouwmeester H, Beyer P, Al-Babili S (2012) The path from beta-carotene to carlactone, a strigolactone-like plant hormone. Science 335: 1348–1351

Antolin-Llovera M, Ried MK, Binder A, Parniske M (2012) Receptor kinase signaling pathways in plant-microbe interactions. Annu Rev Phytopathol 50:451–473

Aoki T, Akashi T, Ayabe S (2000) Flavonoids of leguminous plants: structure, biological activity, and biosynthesis. J Plant Res 113:475–488

Ausmees N, Kobayashi H, Deakin WJ, Marie C, Krishnan HB, Broughton WJ, Perret X (2004) Characterization of NopP, a Type III secreted effector of *Rhizobium* sp. Strain NGR234. J Bacteriol 186:4774–4780

Badri DV, Weirl TL, van der Lelie D, Vivanco JM (2009) Rhizosphere chemical dialogues: plant–microbe interactions. Curr Opin Biotechnol 20:642–650

Bartsev A, Kobayashi H, Broughton WJ (2004) Rhizobial signals convert pathogens to symbionts at the legume interface. In: Gillings M, Holmes A (eds) Plant microbiology. Garland Science/BIOS Scientific Publishers, Abingdon, pp 19–31

Bassam BJ, Djordjevic MA, Redmond JW, Batley M, Rolfe BG (1988) Identification of a nodD dependent locus in the *Rhizobium* strain NGR234 activated by phenolic factors secreted by soybeans and other legumes. Mol Plant-Microbe Interact 1:161–168

Bécard G, Douds DD, Pfeffer PE (1992) Extensive *in vitro* hyphal growth of vesicular-arbuscular mycorrhizal fungi in the presence of CO_2 and flavonols. Appl Environ Microbiol 68:1260–1264

Benfey PN, Scheres B (2000) Root development. Curr Biol 16:813–815

Besserer A, Puech-Pagès V, Kiefer P, Gomez-Roldan V, Jauneau A, Roy S, Portais JC, Roux C, Bécard G, Séjalon-Delmas N (2006) Strigolactones stimulate arbuscular mycorrhizal fungi by activating mitochondria. PLoS Biol 4:226

Besserer A, Bécard G, Jauneau A, Roux C, Séjalon- Delmas N (2008) GR24, a synthetic analog of strigolactones, stimulates the mitosis and growth of the arbuscular mycorrhizal fungus *Gigaspora rosea* by boosting its energy metabolism. Plant Physiol 148:402–413

Besserer A, Becard G, Roux C, Se'jalon-Delmas N (2009) Role of mitochondria in the response of arbuscular mycorrhizal fungi to strigolactones. Plant Signal Behav 4:75–77

Bonfante P, Anca IA (2009) Plants, mycorrhizal fungi and bacteria: a network of interactions. Annu Rev Microbiol 63:363–383

Brown DE, Rashotte AM, Murphy AS, Normanly J, Tague BW, Peer WA, Taiz L, Muday GK (2001) Flavonoids act as negative regulators of auxin transport in vivo in Arabidopsis. Plant Physiol 126:524–535

Buee M, Rossignol M, Jauneau A, Ranjeva R, Becard G (2000) The pre-symbiotic growth of arbuscular mycorrhizal fungi is induced by a branching factor partially purified from plant root exudates. Mol Plant Microbe Interact 13:693–698

Chabot S, Bel-Rhlid R, Chênevert R, Piché Y (1992) Hyphal growth promotion in vitro of the VA mycorrhizal fungus, *Gigaspora margarita* Becker & Hall, by the activity of structurally specific flavonoids compounds under CO_2-enriched conditions. New Phytol 122:461–467

Chen XC, Feng J, Hou BH, Li FQ, Li Q, Hong GF (2005) Modulating DNA bending affects NodD mediated transcriptional control in *Rhizobium leguminosarum*. Nucleic Acids Res 33:2540–2548

Cheng X, Ruyter-Spira C, Bouwmeester H (2013) The interaction between strigolactones and other plant hormones in the regulation of plant development. Front Plant Sci 17:4–199

Cooper JE (2007) Early interactions between legumes and rhizobia: disclosing complexity in a molecular dialogue. J Appl Microbiol 103:1355–1365

D'Arcy-Lameta A, Jay M (1987) Study of soybean and lentil root exudates. III. Influence of soybean isoflavonoids on the growth of rhizobia and some rhizospheric microorganisms. Plant Soil 101:267–72

D'Haeze W, Holsters M (2002) Nod factor structures, responses, and perception during initiation of nodule development. Glycobiology 12(6):79–105

Dakora FD (2000) Commonality of root nodulation signals and nitrogen assimilation in tropical grain legumes belonging to the tribe Phaseoleae. Aust J Plant Physiol 27:885–892

Dastmalchi M, Dhaubhadel S (2014) Soybean seed isoflavonoids: biosynthesis and regulation. Phytochemicals – biosynthesis, function and application. Recent Adv Phytochem 44:1–21

Davies KM, Albert NW, Schwinn KE (2012) From landing lights to mimicry: the molecular regulation of flower colouration and mechanisms for pigmentation patterning. Funct Plant Biol 39:619–638

de Lyra MCCP, Lo'pez-Baena FJ, Madinabeitia N, Vinardell JM, Espuny MR, Cubo MT, Bellogı'n RA, Ruiz-Sainz JE, Ollero FJ (2006) Inactivation of the Sinorhizobium fredii HH103 rhcJ gene abolishes nodulation outer proteins (Nops) secretion and decreases the symbiotic capacity with soybean. Int Microbiol 9:125–133

de Maagd RA, Wijfjes AH, Spaink HP, Ruiz-Sainz JE, Wijffelman CA, Okker RJ, Lugtenberg BJ (1989) NodO, a new nod gene of the *Rhizobium leguminosarum* biovar viciae sym plasmid pRL1JI, encodes a secreted protein. J Bacteriol 171:6764–6770

Deakin WJ, Broughton WJ (2009) Symbiotic use of pathogenic strategies: rhizobial protein secretion systems. Nat Rev Microbiol 7:312–320

Dénarié J, Debellé F, Promé JC (1996) *Rhizobium* lipochitooligosaccharide nodulation factors: Signaling molecules mediating recognition and morphogenesis. Annu Rev Biochem 65:503–535

Dixon RA, Harrison MJ, Paiva NL (1995) The isoflavonoid phytoalexin pathway: from enzymes to genes to transcription factors. Physiol Plant 93:385–392

Djordjevic MA, Redmond JW, Batley M, Rolfe BG (1987) Clovers secrete specific phenolic compounds which either stimulate or repress nod gene expression in *Rhizobium trifolii*. EMBO J 6:1173–1179

Enkerli J, Bhatt G, Covert SF (1998) Maackiain detoxification contributes to the virulence of *Nectria haematococca* MP VI on chickpea. Mol Plant Microbe Interact 11:317–326

Ferguson BJ, Mathesius U (2003) Signaling interactions during nodule development. J Plant Growth Regul 22:47–72

Ferrer JL, Austin MB, Stewart C, Noe JP (2008) Structure and function of enzymes involved in the biosynthesis of phenylpropanoids. Plant Physiol Biochem 46:356–370

Fisher RF, Long SR (1992) *Rhizobium*–plant signal exchange. Nature 387:655–660

Foo E, Davies NW (2011) Strigolactones promote nodulation in pea. Planta 234:1073–1081

Foo E, Yoneyama K, Hugill CJ, Quittenden LJ, Reid JB (2013) Strigolactones and the regulation of pea symbioses in response to nitrate and phosphate deficiency. Mol Plant 6:76–87

Fraysse N, Couderc F, Poinsot V (2003) Surface polysaccharide involvement in establishing the *Rhizobium*–legume symbiosis. Eur J Biochem 270:1365–1380

Giovannetti M, Sbrana C, Avio L, Citernesi AS, Logi C (1993) Differential hyphal morphogenesis in arbuscular mycorrhizal fungi during preinfection stages. New Phytol 125:587–593

Giovannetti M, Sbrana C, Logi C (1994) Early processes involved in host recognition by arbuscular mycorrhizal fungi. New Phytol 127:703–709

Gomez-Roldan V, Roux C, Girard D, Bécard G, Puech V (2007) Strigolactones: promising plant signals. Plant Signal Behav 2:163–164

Graham TL (1991) Flavonoid and isoflavonoid distribution in developing soybean seedling tissues and in seed and root exudates. Plant Physiol 95:594–603

Gu M, Chen A, Dai X, Liu W, Xu G (2011) How does phosphate status influence the development of the arbuscular mycorrhizal symbiosis? Plant Signal Behav 6:1300–1304

Gutjahr C, Parniske M (2013) Cell and developmental biology of Arbuscular Mycorrhiza Symbiosis. Annu Rev Cell Dev Biol 29:593–617

Harrison MJ (2009) Signaling in the arbuscular mycorrhizal symbiosis. Annu Rev Microbiol 59:19–42

Hartwig UA, Maxwell CA, Joseph CM, Phillips DA (1990) Chrysoeriol and Luteolin Released from Alfalfa Seeds Induce nod Genes in *Rhizobium meliloti*. Plant Physiol 92:116–122

Harwood CS, Parales RE (1996) The beta-ketoadipate pathway and the biology of self-identity. Annu Rev Microbiol 50:553–590

He J, Ruan Y, Straney D (1996) Analysis of determinants of binding and transcriptional activation of the pisatin-responsive DNA binding factor of *Nectria haematococca*. Mol Plant Microbe Interact 9:171–179

Heinz EB, Phillips DA, Streit WR (1999) BioS, a biotin-induced, stationary phase, and possible LysR-type regulator in *Sinorhizobium meliloti*. Mol Plant Microbe Interact 12:803–812

Higgins CF (1992) ABC transporters: from microorganisms to man. Annu Rev Cell Biol 8:67–113

Hiltner L (1904) Uber neure erfahrungen und probleme auf dem Gebiet der Boden-bakteriologie und unter besondere Berucksichtigung der grundungung und Bracke. Arb DLG 98:59–78

Hungria M, Joseph CM, Phillips DA (1991) Anthocyanidins and flavonols, major nod gene Inducers from seeds of a black-seeded common bean (*Phaseolus vulgaris L.*). Plant Physiol 97:751–758

Hutsch BW, Augustin J, Merbach W (2000) Plant rhizodeposition an important source for carbon turnover in soils. J Plant Nutr Soil Sci 165:397–407

Kamst E, Spaink HP, Kafetzopoulos D (1998) Biosynthesis and secretion of rhizobial lipochitin-oligosaccharide signal molecules. In: Biswas BB, Das HK (eds) Plant-microbe interactions-subcellular biochemistry. Plenum Press, New York, pp 29–71

Kang J, Parka J, Choia H, Burlab B, Kretzschmarb T, Leea Y, Martinoiaa E (2011) Plant ABC transporters. The Arabidopsis book 9: e0153. http://dx.doi.org/10.1199/tab.0153

Kannenberg EL, Perzl M, Ha¨rtner T (1995) The occurrence of hopanoid lipids in Bradyrhizobium bacteria. FEMS Microbiol Lett 127:255–262

Koltai H, Kapulnik Y (2011) Strigolactones as mediators of plant growth responses to environmental conditions. Plant Signal Behav 6:37–41

Kosslak RM, Bookland R, Barkei J, Paaren HE, Appelbaum ER (1987) Induction of *Bradyrhizobium japonicum* common nod genes by isoflavones isolated from *Glycine max*. Proc Natl Acad Sci U S A 84:7428–7432

Kosuta S, Chabaud M, Lougnon G, Gough C, Denarie J (2003) A diffusible factor from arbuscular mycorrhizal fungi induces symbiosis specific MtENOD11 expression in roots of *Medicago truncatula*. Plant Physiol 131:952–962

Krause A, Doerfel A, Go¨ttfert M (2002) Mutational and transcriptional analysis of the type III secretion system of *Bradyrhizobium japonicum*. Mol Plant Microbe Interact 15:1228–1235

Krishnan HB, Lorio J, Kim WS, Jiang G, Kim KY, DeBoer M, Pueppke SG (2003) Extracellular proteins involved in soybean cultivar-specific nodulation are associated with pilus-like surface appendages and exported by a type III protein secretion system in *Sinorhizobium fredii* USDA257. Mol Plant Microbe Interact 16:617–625

Larose G, Chenevert R, Moutoglis P, Gagne S, Piché Y, Vierheilig H (2002) Flavonoid levels in roots of *Medicago sativa* are modulated by the developmental stage of the symbiosis and the root colonizing arbuscular mycorrhizal fungus. J Plant Physiol 159:1329–1339

Loh J, Carlson RW, York WS, Stacey G (2002) Bradyoxetin, a unique chemical signal involved in symbiotic gene regulation. Proc Natl Acad Sci U S A 99:14446–14451

Lopez-Raez JA, Kohlen W, Charnikhova T, Mulder P, Undas AK, Sergeant MJ (2010) Does abscisic acid affect strigolactone biosynthesis? New Phytol 187:343–354

Maillet F, Poinsot V, Andre O, Puech-Pages V, Haouy A (2011) Fungal lipochitooligosaccharide symbiotic signals in arbuscular mycorrhiza. Nature 469:58

Marschner H (1995) Mineral nutrition of higher plants. Academic, London, p 889

Martinoia E, Grill E, Tommasini R, Kreuz K, Amrhein N (1993) ATP-dependent glutathione S-conjugate export pump in the vacuolar membrane of plants. Nature 364:247–249

Martinoia E, Klein M, Bovet L, Forestier C, Kolukisaoglu Ü, Müller-Rover B, Schulz B (2002) Multifunctionality of plant ABC transporters-more than just detoxifiers. Planta 214:345–355

Mathesius U, Mulders S, Gao MS, Teplitski M, Caetano-Ano'lles G, Rolfe BG, Bauer WD (2003) Extensive and specific responses of a eukaryote to bacterial quorum sensing signals. Proc Natl Acad Sci U S A 100:1444–1449

Matiru VN, Dakora FD (2005) Xylem transport and shoot accumulation of lumichrome, a newly recognized rhizobial signal, alters root respiration, stomatal conductance, leaf transpiration and photosynthetic rates in legumes and cereals. New Phytol 165: 847–855

Matusova R, Rani K, Verstappen FWA, Franssen MCR, Beale MH, Bouwmeester HJ (2005) The strigolactone germination stimulants of the plant-parasitic *Striga* and *Orobanche* spp. are derived from the carotenoid pathway. Plant Physiol 139:920–934

Messens E, Geelen D, van Montagu M, Holsters M (1991) 7,4-Dihydroxyflavanone is the major *Azorhizobium* nod gene-inducing factor present in *Sesbania rostrata* seedling exudate. Mol Plant-Microbe Interact 4:262–267

Morandi D (1996) Occurrence of phytoalexins and phenolic compounds on endomycorrhizal interactions and their potential role in biological control. Plant Soil 185:241–251

Morandi D, Branzanti B, Gianinazzi-Pearson V (1984) Isoflavonoid accumulation in soybean roots infected with vesicular-arbuscular mycorrhizal fungi. Physiol Plant Pathol 24:357–364

Moscatiello R, Squartini A, Mariani P, Navazio L (2010) Flavonoid-induced calcium signaling in *Rhizobium leguminosarum bv. Viciae*. New Phytol 188:814–823

Nagahashi G, Douds DD (2011) The effects of hydroxy fatty acids on the hyphal branching of germinated spores of AM fungi. Fungal Biol 115:351358

Nguyen C (2003) Rhizodeposition of organic C by plants: mechanisms and controls. Agronomie 23:375–396

Okazaki S, Zehner S, Hempel J, Lang K, Gottfert M (2009) Genetic organization and functional analysis of the type III secretion system of *Bradyrhizobium elkanii*. FEMS Microbiol Lett 295:88–95

Okazaki S, Okabe S, Higashi M, Shimoda Y, Sato S, Tabata S (2010) Identification and functional analysis of type III effector proteins in *Mesorhizobium loti*. Mol Plant Microbe Interact 23:223–234

Parke D (1997) Acquisition, reorganization, and merger of genes: novel management of the beta-ketoadipate pathway in *Agrobacterium tumefaciens*. FEMS Microbiol Lett 146:3–12

Perret X, Staehelin C, Broughton WJ (2000) Molecular basis of symbiotic promiscuity. Microbiol Mol Biol Rev 64:180–201

Peters NK, Frost JW, Long SR (1986) A plant flavone, luteolin, induces expression of *Rhizobium meliloti* nodulation genes. Science 233:977–980

Phillips DA, Joseph CM, Yang GP, Martı'nez-Romero E, Sanborn JR, Volpin H (1999) Identification of lumichrome as a *Sinorhizobium* enhancer of alfalfa root respiration and shoot growth. Proc Natl Acad Sci U S A 96:12275–12280

Poulin MJ, Bel-Rhlid R, Piché Y, Chênevert R (1993) Flavonoids released by carrot (*Daucus carota*) seedlings stimulate hyphal development of vesicular-arbuscular mycorrhizal fungi in the presence of optimal CO_2 enrichment. J Chem Ecol 19:2317–2327

Pueppke SG, Broughton WJ (1999) *Rhizobium* sp. strain NGR234 and *R. fredii* USDA257 share exceptionally broad, nested host ranges. Mol Plant Microbe Interact 12(4):293–318

Rani K, Zwanenburg B, Sugimoto Y, Yoneyama K, Bouwmeester HJ (2008) Biosynthetic considerations could assist the structure elucidation of host plant produced rhizosphere signaling compounds (strigolactones) for arbuscular mycorrhizal fungi and parasitic plants. Plant Physiol Biochem 46:617–626

Rao JR, Cooper JE (1994) Rhizobia catabolise nod gene-inducing flavonoids via C-ring fission mechanisms. J Bacteriol 176:5409–5413

Remy W, Taylor TN, Hass H, Kerp H (1994) Four hundred-million-year old vesicular arbuscular mycorrhizae. Proc Natl Acad Sci U S A 91:11841–11843

Scervino JM, Ponce MA, Erra-Bassels R, Vierheilig H, Ocampo JA, Godeas A (2006) Glycosidation of apigenin results in a loss of activity on different growth parameters of arbuscular mycorrhizal fungi from the genus *Glomus* and *Gigaspora*. Soil Biol Biochem 38:2919–2922

Scheidemann P, Wetzel A (1997) Identification and characterization of flavonoids in the root exudate of *Robinia pseudoacacia*. Trees 11:316–321

Soto MJ, Fernandez-Aparicio M, Castellanos-Morales V, Garcia-Garrido JA, Delgado MJ, Vierheilig H (2010) First indications for the involvement of strigolactones on nodule formation in alfalfa (*Medicago sativa*). Soil Biol Biochem 42:383–385

Srivastava P, Sharma PK, Dogra RC (1999) Inducers of nod genes of *Rhizobium ciceri*. Microbiol Res 154:49–55

Stafford HA (1990) Flavonoid metabolism. CRC Press, Boca Raton, p 298

Steinkellner S, Lendzemo V, Langer I, Schweiger P, Khaosaad T, Toussaint JP, Vierheilig H (2007) Flavonoids and strigolactones in root exudates as signals in Symbiotic and Pathogenic Plant-Fungus Interactions. Molecules 12:1290–1306

Streit WR, Joseph CM, Phillips DA (1996) Biotin and other water-soluble vitamins are key growth factors for alfalfa root colonization by *Rhizobium meliloti* 1021. Mol Plant Microbe Interact 9:330–338

Subramanian S, Stacey G, Yu O (2007) Distinct, crucial roles of flavonoids during legume nodulation. Trends Plant Sci 12:282–285

Sugiyama A, Shitan N, Yazaki K (2007) Involvement of a soybean ATP-binding cassette-type transporter in the secretion of genistein, a signal flavonoid in legume-*Rhizobium* symbiosis. Plant Physiol 144: 2000–2008

Taylor LP, Grotewold E (2005) Flavonoids as developmental regulators. Curr Opin Plant Biol 8:317–323

Theunis M, Kobayashi H, Broughton WJ, Prinsen E (2004) Flavonoids, NodD1, NodD2, and nod box NB15 modulate expression of the y4wEFG locus that is required for indole-3-acetic acid synthesis in *Rhizobium* sp. Strain NGR234. Mol Plant Microbe Interact 17:1153–1161

Umehara M (2011) Strigolactone, a key regulator of nutrient allocation in plants. Plant Biotechnol 28:429–437

Vierheilig H, Piche Y (2002) Signalling in arbuscular mycorrhiza: facts and hypotheses. In: Buslig B, Manthey J (eds) Flavonoids in cell function. Kluwer Academic/Plenum Press, New York, pp 23–39

Vierheilig H, Bago B, Albrecht C, Poulin MJ, Piche Y (1998) Flavonoids and arbuscular-mycorrhizal fungi. In: Buslig BS, Manthey JA (eds) Flavonoids in the living system. Plenum Press, New York, pp 9–33

Viprey V, Del Greco A, Golinowski W, Broughton WJ, Perret X (1998) Symbiotic implications of type III protein secretion machinery in *Rhizobium*. Mol Microbiol 28:1381–1389

Vlassak KM, Luyten E, Verreth C, van Rhijn P, Bisseling T, Vanderleyden J (1998) The *Rhizobium* sp. BR816 nodO gene can function as a determinant for nodulation of *Leucaena leucocephala*, *Phaseolus vulgaris*, and *Trifolium repens* by a diversity of *Rhizobium* spp. Mol Plant Microbe Interact 11:383–392

Wang X (2010) Structural studies and mechanisms of isoflavonoid biosynthesis. In: Thompson MJ (ed) Isoflavones biosynthesis, occurrence and health effects. Nova Publishers, New York, pp 239–254

Wang B, Qui YL (2006) Phylogenetic distribution and evolution of mycorrhizas in land plants. Mycorrhiza 16:299–363

Wasmannm CC, Van Etten HD (1996) Transformation-mediated chromosome loss and disruption of a gene for pisatin demethylase decrease the virulence of *Nectria haematococca* on pea. Mol Plant Microbe Interact 9:793–803

Wasson AP, Pellerone FI, Mathesius U (2006) Silencing the flavonoid pathway *in Medicago truncatula* inhibits root nodule formation and prevents auxin transport regulation by rhizobia. Plant Cell 18:1617–1629

Whipps JM (1990) Carbon economy. In: Lynch JM (ed) The rhizosphere. Wiley, Chichester, pp 59–97

Winkel SB (2001) Flavonoid biosynthesis. A colorful model for genetics, biochemistry, cell biology, and biotechnology. Plant Physiol 126:485–493

Xie X, Yoneyama K, Yoneyama K (2010) The strigolactone story. Annu Rev Phytopathol 48:93–117

Yoneyama K, Xie X, Kim HI, Kisugi T, Nomura T, Sekimoto H, Yokota T, Yoneyama K (2012) How do nitrogen and phosphorus deficiencies affect strigolactone production and exudation? Planta 235:1197–1207

Zaat SAJ, Schripsema J, Wijffelman CA, Brussel AAN, Lugtenberg BJJ (1989) Analysis of the major inducers of the *Rhizobium* nodA promoter from *Vicia sativa* root exudate and their activity with different nodD genes. Plant Mol Biol 13:175–188

Zhang J, Subramanian S, Zhang Y, Yu O (2007) Flavone synthases from *Medicago truncatula* are flavanone-2-hydroxylases and are important for nodulation. Plant Physiol 144:741–751

Zuanazzi J, Clergeot PH, Quirion JC, Husson HP, Kondorosi A, Ratet P (1998) Production of *Sinorhizobium meliloti nod* gene activator and repressor flavonoids from *Medicago sativa* roots. Mol Plant Microbe Interact 11:784–794

There are many comparative features of the legume-*Rhizobium* symbiosis with those of the actinorhizal symbiosis regarding physiological resistance to stress conditions and their ability to interact with microbial inoculants able to enhance their tolerance (N_2-fixing symbiont bacteria and beneficial soil microbiota). Thus, this chapter focuses on main resistance mechanisms deployed by these two main plant-bacteria symbioses and their applications or their associations with beneficial telluric microorganisms (mainly PGPRs and mycorrhizae) under soil stressed systems. This chapter also focuses on main applications of plant-bacteria symbioses in contaminated soils and advances in soil bioremediations. Figure 14.1 summarizes mechanisms deployed by N_2-fixing bacteria, PGPRs, and AM fungi to alleviate abiotic stresses.

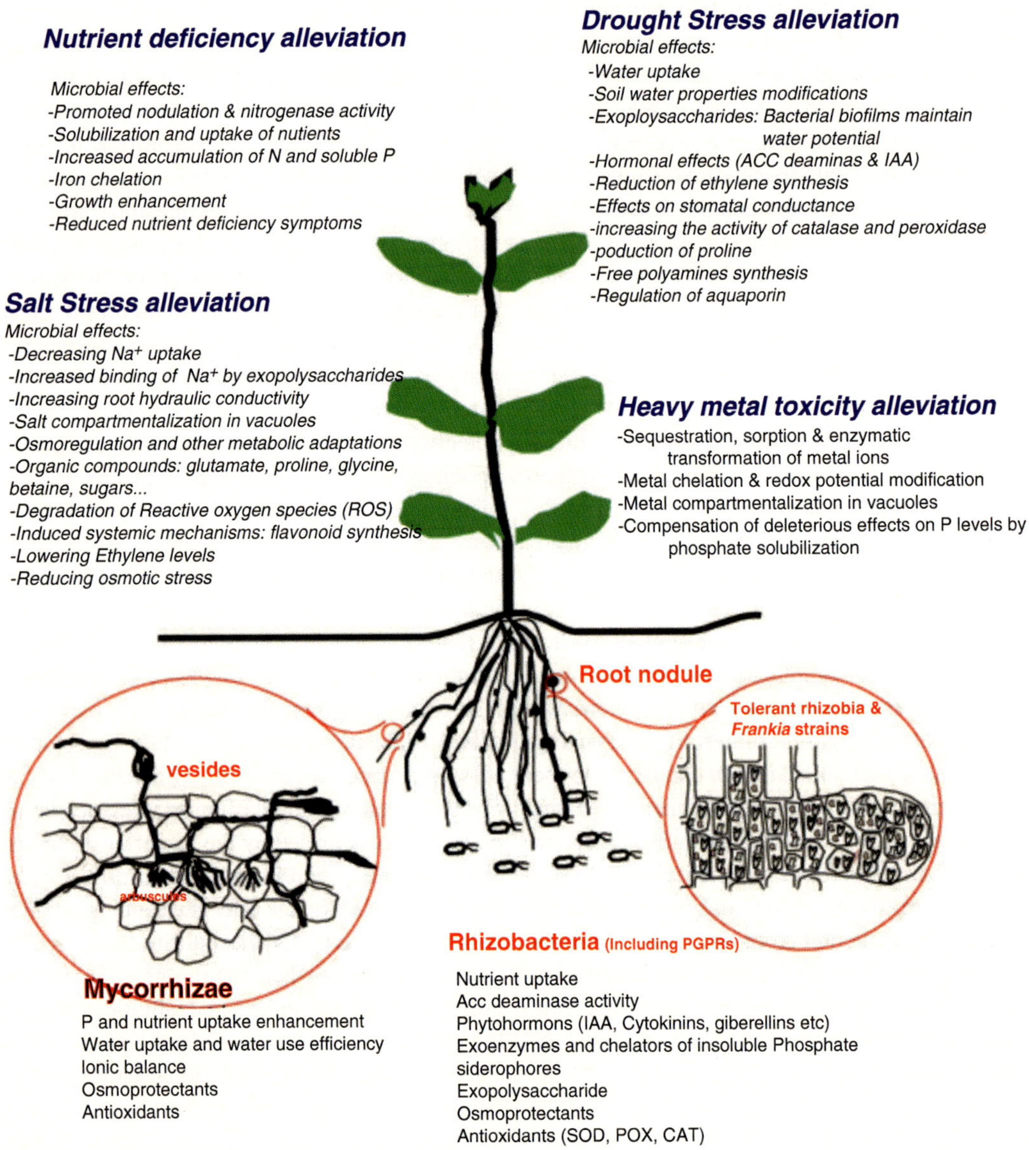

Fig. 14.1 Scheme summarizing the mechanisms deployed by N_2-fixing bacteria, PGPRs, and AM fungi to alleviate abiotic stresses. *CAT* catalase, *IAA* indoleacetic acid, *PGPR* plant growth-promoting rhizobacteria. *POX* peroxidase, *SOD* superoxide dismutase

Application of Plant-Bacteria Symbioses to Alleviate Abiotic Stresses

Tolerance and Growth Promotion Under Nutrient Deficiency

Soil nitrogen status is considered as one of the main factors affecting sustainability of plant growth in a crop or forest ecosystem. N status of soils under N-fixing plants including actinorhizals and legumes is often considered as high. They are very important in agriculture and forestry as they are responsible for a substantial part of the global flux of nitrogen from atmospheric N_2 to fixed mineral forms. Alders are, for example, of major interest for soil nitrogen status as their leaf litter is highly N rich and therefore mineralized much faster than that of other tree species (Claessens et al. 2010). However, in many nutrient-poor areas, colonization of primary successional substrates is largely limited by low levels of nutrients. In most soils where low levels of available nitrogen are a major limitation, N_2-fixing plants which are the primary source of fixed nitrogen (Zahran 1999; Barritt and Facelli 2001) could succeed. Nitrogen fixation and root nodule formation are substantially affected by soil nutrient availability of mineral N (Arnone et al. 1994; Bollman and Vessey 2006) and P (Reddell et al. 1986a; Sanginga et al. 1989; Yang 1995; Kouas et al. 2005; Sadowsky 2005). It is well known that nodulation is stimulated in poor N soils, whereas high levels of mineral forms of nitrogen (nitrate and ammonium) have a negative effect on nodulation and nitrogen fixation activity (Arnone et al. 1994; Baker and Parsons 1997; Wall et al. 2000; Tiffany and Williams 2005; Bollman and Vessey 2006; Shamseldin and Moawad 2010). Nodule development requires P consumption for DNA and plasma membrane synthesis, for cell division, and also for symbiotic N_2 fixation due to the need for large amount of ATP for nitrogenase activity (reduction of N_2). In general, the fixation of atmospheric N_2 in plant-microbe symbioses can negatively be impacted by phosphorus limitation (Reddell et al. 1986a; Tang et al. 2001). Indeed, it was demonstrated that nodulation and growth in some actinorhizal plants are frequently limited by P in early successional environments (Chapin et al. 1991; Uliassi et al. 2000). Also, *Frankia* infective abilities following deglaciation were only showed if P was supplemented (Chapin et al. 1994). Phosphorus has been shown to have a positive effect on nodulation in many actinorhizal and legume plants (Wall et al. 2000; Olivera et al. 2004; Gentili et al. 2006). This was also observed in some soils from Australia, where Reddell et al. (1986a) found that nodulation occurred only when P amendments were performed.

N_2-fixing plants able to grow under nitrogen deficiency are considered as important for revegetation in such soils as these can accumulate nitrogen. Different perennial legumes were studied for their potential use in revegetating Australian soils with phosphorus deficiency (Pang et al. 2010). In addition, efforts have been made to select efficient legume plants for N_2 fixation and biomass and seed yield, particularly under conditions of N and P deficiency (Elisondo Barron et al. 1999; Tang et al. 2004; Remans et al. 2008). The selection of N_2-fixing plants tolerant to N deficiency in combination with tolerant bacterial symbionts seems to be a valuable strategy. Consequently, screening and inoculation of efficient N_2-fixing bacteria in order to provide improvement in plant production equal to or better than that of nitrogen fertilization are aspects that have been investigated by numerous researchers (Sellstedt and Huss-Danell 1986; Romdhane et al. 2008).

Many researches are aimed to select and apply PGPRs for their beneficial effects on plant fitness and productivity. These rhizobacteria are well known to promote plant growth by various direct and indirect mechanisms including phytohormone synthesis and phosphate solubilization and are consequently used as phytostimulators and biofertilizers for agricultural crops. In addition, the use of microbial inoculants could represent an alternative strategy to the use of chemical fertilizers notably NP-based fertilizers. Rhizobia can also be considered as PGPR endophytes as they contribute to plant growth. However, researchers prefer to denote PGPR as free root-colonizing rhizobacteria. Most known PGPRs belong to the genera *Arthrobacter*, *Bacillus*, *Micromonospora*, *Pseudomonas*, and *Streptomyces*. Some of these bacteria were applied for growth stimulation and nodulation in both legume and actinorhizal plants (Solans 2007; Malik and Sindhu 2011). The most important traits of PGPRs in association with N_2-fixing plants are the improvement of nodulation (Knowlton and Dawson 1983; Zhang et al. 1996; Lucas-Garcia et al. 2004) and supply of soluble P by solubilization (Toro et al. 1998; Valverde et al. 2006). The application of PGPRs to N_2-fixing plants often resulted in enhanced legume or actinorhizal plants' growth. Hence, fitness was enhanced in various crops when PGPRs were co-inoculated with rhizobia or with *Frankia* (Knowlton and Dawson 1983; Probanza et al. 1996, 1997; Solans 2007; Wani et al. 2007a; Dardanelli et al. 2008; Egamberdieva et al. 2010; Malik and Sindhu 2011). An example of such application is the single, dual, or triple seed inoculations of common bean (*Phaseolus vulgaris*) with *Rhizobium leguminosarum* bv. *phaseoli*, N_2-fixing *Bacillus subtilis*, and P-solubilizing *Bacillus megaterium* in a highland region in Turkey as indicated by significantly increased nodulation, plant growth, nutrient uptake, and seed yield equal to or higher than nitrogen (N), phosphorus (P), and NP mineral fertilizer application. Moreover, triple inoculations were most efficient for enhancement of macro- and micronutrient uptake resulting in an increase in mineral concentration in the plant and seeds (Elkoca et al. 2010). Similar results were found in *Vicia faba* L. inoculated with four different *Rhizobium*/*Azospirillum* and *Rhizobium*/*Azotobacter* combinations which led to changes in mineral macro- and micronutrients, K, P, Ca, Mg, Fe, B, Mn, Zn, and Cu, compared with plants inoculated with *Rhizobium* alone (Rodelas et al. 1999). Recently, Solans et al. (2011) reported that three root-associated saprophytic actinomycetes (*Streptomyces* sp., *Actinoplanes* sp., and *Micromonospora* sp.) stimulated nodulation by *Frankia* in the actinorhizal plant *Ochetophila trinervis*. These root associated actinobacteria produced phytohormones (indole acetic acid (IAA), gibberellic acid, and zeatin) at levels higher than produced by the symbiotic *Frankia* strain alone. One interesting strain *Micromonospora* MM18 showed direct promotion effect on shoot growth of *Ochetophila trinervis* when inoculated as mycelium as well as cell-free growth medium (Solans 2007). However, in some cases, effects of dual inoculation with *Rhizobium* and PGPR depend on various parameters including legume species, PGPR species, and the nature and concentration of secondary metabolites released by PGPR. Co-inoculation with *Pseudomonas fluorescens* 2137 increased colonization of *Bradyrhizobium japonicum* on soybean roots. Nodule numbers and acetylene reduction increases were observed probably due to release of growth-promoting substances by *Pseudomonas* that stimulate *B. japonicum*, whereas co-inoculation with another strain *P. fluorescens* WCS365 had an opposite effect (Chebotar et al. 2001). Under field conditions, co-inoculation of PGPR, *Serratia* and *P. fluorescens* with *R. leguminosarum*, had positive effects on lentil but had no effect on pea (Chanway et al. 1989). These opposite effects are likely due to PGPR species and/or legume plant species. Elo et al. (2000) isolated a number of bacteria from the humus of spruce and tested their potential beneficial effects on alder (*Alnus incana*), red fescue (*Festuca rubra*), and birch (*Betula pendula*).

Under stress conditions of soil nutrient deficiency, the effectiveness of PGPRs as bioinoculants for N_2-fixing plants has been demonstrated in legume plants. Positive effects were reported in relation to iron, phosphorus, magnesium, calcium, and the nutrient uptake under iron-limiting conditions. Siderophores produced by certain bacteria could contribute to enhanced availability of iron. Inoculation of mung bean (*Vigna radiata*), with the siderophore-producing *Pseudomonas* strain GRP3, resulted in a reduced chlorotic symptoms and enhanced chlorophyll level (Sharma et al. 2003). In general, differences in the responsiveness to PGPR among cultivars of legumes were observed. Thus, PGPRs could be used efficiently for enhancement of tolerance to P deficiency in non-tolerant cultivars. In the

P. vulgaris genotype DOR364, a poor nitrogen fixer under P deficiency, Remans et al. (2008) succeeded to partly alleviate this constrain by using a combination of *Rhizobium etli* and *Azospirillum* inoculation. However, in another *Phaseolus* genotype, the beneficial effect of co-inoculated *R. etli* and *Azospirillum* was shown to be dependent on P nutrition (Remans et al. 2007). These authors showed that co-inoculation resulted in efficient growth enhancement only at high P conditions and combinations failed under low P conditions. It is also worth mentioning that in general, the interplay between *Rhizobium*, PGPR species, and plant host species plays a major role in the success of these combinations under P stress conditions. Under deficient P conditions, while combination of *Rhizobium* and *Azospirillum* was ineffective, the beneficial effect of co-inoculation of *Rhizobium* with *Pseudomonas putida* was strongly expressed (Remans et al. 2007).

In addition, it is well known that N_2-fixing plants can also harbor ecto- or endomycorrhizae forming either tripartite or tetrapartite associations. These kinds of associations were found in both legume plants and actinorhizae (Vance 2001; Yamanaka et al. 2003; Roy et al. 2007; He and Critchley 2008; Gherbi et al. 2008), and their importance for plant growth promotion in poor soils has been documented extensively. Their beneficial effects on plant growth are mainly due to the positive effects of the mycorrhizal fungi through P supply (Barea et al. 2005). In many cases, dual inoculation of actinorhizal plants with arbuscular mycorrhizal fungi (AMF) and *Frankia* resulted in a more enhanced growth than single inoculation with *Frankia* alone (Vasanthakrishna et al. 1994; Oliveira et al. 2005; Elumalai and Raaman 2009; Muthukumar and Udaiyan 2010; Orfanoudakis et al. 2010). In addition, the usefulness of AMF co-inoculation with N_2-fixing bacteria is well established in nutrient-poor soils. Inoculation of mycorrhiza fungi promotes the growth of actinorhizal plants by various mechanisms including increased uptake of phosphorus (Theodorou and Reddell 1991; He and Critchley 2008; Elumalai and Raaman 2009). It was demonstrated that *Elaeagnus* and *Shepherdia* inoculated with soil containing *Frankia* and mycorrhizal fungi under nutrient deficiency had a superior nodulation than those uninoculated (Visser et al. 1991). Theodorou and Reddell (1991) found that symbiotic mycorrhizae could contribute to a successful *Frankia* symbiosis in *Casuarina* plants by increasing N_2 fixation especially in P-deficient soils, in arid climates. However, inoculation experiments showed that mutual interactions and delays between inoculations of the two symbionts could play a major role in the efficiency of tripartite association (Gardner 1986; Theodorou and Reddell 1991). Theodorou and Reddell (1991) observed that *Amanita* sp. formed ectomycorrhizae on *Allocasuarina littoralis* and *Casuarina cunninghamiana* but not on *Casuarina equisetifolia*. It was also shown that the intensity of ectomycorrhizal infection with *Pisolithus tinctorius* and *Laccaria laccata* was higher in *C. equisetifolia* rather than *C. cunninghamiana* suggesting specificity between the plant and mycorrhizae. In addition, under limited supply of mineral N and P, Sempavalan et al. (1995) found a high degree of coordination between *Frankia* and *Glomus* for nodulation and mycorrhizal colonization of roots of *C. equisetifolia*, while Gardner (1986) reported a possible competition between *Frankia* and AMF microsymbionts for infection sites and/or nutrients in AMF-colonized *C. equisetifolia* resulting in negative effect on nodulation. It was thus suggested that efficiency of these tripartite associations in poor soils could be ensured if the moments of inoculations of both symbionts are skillfully monitored (Gardner 1986). All these observations emphasize the importance of knowledge of specific combinations involving mycorrhizae/*Frankia*/actinorhizal plant species.

Alleviation of Drought Stress

Arid or semiarid land areas are increasing worldwide due to climate change and global warming. This condition has deleterious effects on plants in these areas. Drought stress affects plant hormone balance by decreasing the endogenous cytokinin level and increasing the levels of abscisic acid

(ABA) content in the leaves. This effect leads to stomatal closure. Many N_2-fixing species undergo deleterious physiological effects of water deficiency and show a reduction in nitrogen fixation when subjected to drought (Zahran 1999; Sharma et al. 2010). In addition to its deleterious effects on nodules and nitrogenase activity (Sundström and Huss-Danell 1987; Devries et al. 1989; Porcel et al. 2003), drought could affect the survival of rhizobia and the size of the soil population of infective *Frankia* and their infectivity (Mahler and Wollum 1981; Sayed et al. 1997). Drought also alters the soil properties and renders it unsuitable for soil microbial activity and crop growth (Selvakumar et al. 2012).

Actinorhizae are more drought tolerant than other nitrogen-fixing microorganisms. Numerous actinorhizal plants mainly those belonging to the genus *Casuarina* were reported as drought tolerant (Diem and Dommergues 1990). Though they are high water demanders, alders (*Alnus*) are well adapted to water deficits due to their root systems (Claessens et al. 2010). The performance of tolerant *Casuarina* under conditions of drought could be due to their high rate of photosynthesis and decreased stomatal conductance. Thus, researches focused on propagation of *Casuarina* species tolerant to drought via in vitro tissue culture for seedling production or via cuttings of plant parts.

Numerous legume plants are adapted to drought stress conditions. Among these are arid legumes, such as cluster bean (*Cyamopsis tetragonoloba*), dew bean (*Vigna aconitifolia*), and cowpea (*Vigna unguiculata*) (Kumar 2005). It was observed that numerous drought-tolerant species (cultivars) are characterized by a deeper root profiles, a greater root biomass, and a greater root-to-shoot ratio. Those variabilities could be exploited in breeding programs for N_2 fixation drought tolerance. In addition, efforts are made to select more tolerant N_2-fixing plant species to drought by using conventional selection and breeding (El-Lakany 1983; Pimentel et al. 1999; Sinclair et al. 2001; Khan et al. 2010; Asfaw et al. 2012), and efforts are also made to screen more tolerant *Frankia* and rhizobia strains (Sayed et al. 1997; Romdhane et al. 2009; Zahran 2001; Sayed 2011). For instance, salt-tolerant rhizobia that were able to nodulate chickpea (*Cicer arietinum*) were selected for their efficiency in conditions of water deficiency (Romdhane et al. 2009). Mnasri et al. (2007) reported that when inoculated to *P. vulgaris*, a salt-tolerant *Ensifer meliloti* induced more tolerance to drought than a salt-sensitive bacterial strain *Rhizobium tropici*. *Frankia* may be absent in some arid soils, whereas non-leguminous N_2-fixing plants are often nodulated in well-hydrated areas. Thus, there is a need to inoculate host plants with a selected microsymbiont in arid and semiarid areas. Attempts to select more tolerant *Frankia* strains were performed by Sayed et al. (1997) who found that the infectivity of a soil inoculum decreased by storage under dry conditions and pointed out the importance of screening *Frankia* strains for drought tolerance during preparation of inoculum for use in hot climates. Thereafter, inoculation of tolerant bacterial symbionts was also used as a strategy to alleviate drought stress in many N_2-fixing plants. Bacterial aggregation as biofilms in the vicinity of roots was reported as a mechanism that could contribute to maintain convenient water potential (Khan et al. 2011). Bacteria producing extracellular materials were reported for their beneficial effects. Exopolysaccharides (EPS) act as adhering material that help bacteria to attach to soil particles and by binding soil particles together lead to soil aggregate formation. EPS contribute to hold water within soil aggregate and thus lead to ameliorate plant growth under drought conditions. Among EPS-producing bacteria, rhizobia were applied by Hartel and Alexander (1986). They observed a significant correlation between the levels of EPS produced by cowpea *Bradyrhizobium* strains and their drought tolerance.

On the other hand, rhizobacteria and mycorrhizae were also investigated for their ability to help plants to alleviate drought stress. PGPR bacteria could also contribute to drought alleviation in legume plants through hormonal effects that involve IAA and ACC deaminase. Marulanda et al. (2009) used three bacteria: *Pseudomonas sp.*, *P. putida*, and *B. megaterium* with IAA-producing abilities to alleviate drought stress in

Trifolium repens. Alleviation effect could also involve the regulation of ethylene synthesis by ACC-deaminase-producing *Pseudomonas* which showed to be effective in both pot and field trials (Arshad et al. 2008). Indeed, reduction of ethylene synthesis by such bioinoculants effect could result in better growth under drought stress (Zahir et al. 2008). ACC-deaminase-producing rhizobacteria could obviously reduce the negative effects of ethylene on nodulation and N_2 fixation. Dual inoculation involving PGPRs and rhizobia proved their effectiveness in mitigating water stress deficiency in N_2-fixing plants. In bean, Figueiredo et al. (2008) used dual inoculation of *Paenibacillus polymyxa* and *R. tropici* to alleviate drought stress. Plants co-inoculated with *Rhizobium* and *Paenibacillus* exhibited greater nodulation and nitrogen content compared to drought-stressed plants inoculated with only *Rhizobium*. In addition, PGPR induced systemic mechanisms to drought could represent a powerful tool to alleviate such alterations. IST by increasing the activity of catalase and peroxidase in water stress was found in *P. fluorescens* Pf1-treated green gram plants (*V. radiata* L.) (Saravanakumar et al. 2012).

Mycorrhizae could also be used as bioinoculants for drought stress alleviation (Ruiz-Lozano et al. 2001). Several studies have stressed their role in improving legume and actinorhizal plant tolerance to drought (Goicoechea et al. 1998; Ruiz-Lozano et al. 2001; Valdenegro et al. 2001; Zhang et al. 2010). Mycorrhizal inoculation significantly improved both nodulation and nitrogen fixation in these conditions (Azcón et al. 1988; Theodorou and Reddell 1991; Ruiz-Lozano et al. 2001). In soybean, mycorrhizae were reported to have antioxidant activities and to protect legume plants against premature nodule senescence induced by drought stress (Porcel et al. 2003). It was also established that they could reduce the production of proline, a marker of the water stress effect in plants (Goicoechea et al. 1998). In addition, they induce free polyamine synthesis (Goicoechea et al. 1998). This contributes to increased accumulation of nonstructural carbohydrates that could serve as osmoprotectants by lowering the osmotic potential under drought conditions. The regulation of root hydraulic properties by AM symbiosis under drought stress is also correlated with the regulation of aquaporins which are membrane intrinsic proteins that form a pore in cell membranes of living organisms (Aroca et al. 2007). Uehlein et al. (2007) found an AM-induced gene expression of specific aquaporin genes in mycorrhized *Medicago truncatula*. The application of mycorrhizae for drought tolerance in actinorhizal plants was also reported (Zhang et al. 2010). Zhang et al. (2010) found that AM associations improve the drought tolerance of *C. equisetifolia* seedlings by lowering the permeability of plasma membrane and malondialdehyde contents, enhancing concentrations of P nutrition, soluble sugars, soluble proteins, and peroxidase activities. However, efficient dual inoculation involving symbiotic mycorrhizae and rhizobia seems to be variable according to bacterial interactions in a given combination. Ruiz-Lozano et al. (2001) reported that dual inoculation of *Bradyrhizobium* and *Glomus intraradices* was less effective than dual inoculation with *G. mosseae*, in alleviating drought-induced nodule senescence in soybean plants, suggesting the existence of a competitive interaction between the very aggressive *G. intraradices* and *Bradyrhizobium*. The importance of efficient combinations was also discussed by Valdenegro et al. (2001) who used combinations of three mycorrhizae and one PGPR, *Enterobacter* sp., and found that the positive effects of AMF isolates on growth of *Medicago arborea* plants under drought conditions were stimulated by the co-inoculation. This effect was however found to be dependent on the inoculated rhizobial strain and on the AMF isolate suggesting the importance of efficient combinations of AMF, PGPR, and rhizobia in such conditions. Table 14.1 presents some selected examples of drought stress alleviations in legume plants by rhizobia and microbial inoculants (i.e., rhizobacteria or mycorrhizae).

Alleviation of Salt Stress

Salinity in soil is one of the main environmental stress parameters, especially in arid and semiarid

Table 14.1 Selected examples of drought stress alleviation in legume plants by rhizobia and microbial inoculants

Stress	Plant species	Symbiotic bacteria	Microbial inoculant	References
Drought	*Glycine max* (soybean)	*Bradyrhizobium japonicum*	*Glomus mosseae*	Ruiz-Lozano et al. (2001)
	Medicago arborea	*Rhizobium meliloti*	Arbuscular mycorrhizal fungi and *Enterobacter* sp.	Valdenegro et al. (2001)
	Medicago sativa L. cv. Arag6n (Alfalfa)	*R. meliloti*	*Glomus fasciculatum*	Goicoechea et al. (1998)
	Phaseolus vulgaris (common bean)	*Rhizobium tropici*	*Paenibacillus polymyxa*	Figueiredo et al. (2008)
	Vicia faba (faba bean)	*R. leguminosarum* bv. *viciae*	*Azotobacter chroococcum*	Dashadi et al. (2011)

regions, and can severely reduce crop production. Salinization is mainly due to human activities such as excessive tillage or irrigation with saline water. Over 800 million hectares of land throughout the world is salt affected (Rengasamy 2006) with over 70 million ha being farmlands (FAO 2005). The osmotic stress is currently one of the most serious environmental factors in agriculture (Manchanda and Garg 2008). In legume plants, salt toxicity results in reduction in survival and productivity (Shannon 1998; Yuan et al. 2007), whereas actinorhizae, particularly *Casuarina* species, are well known for their relatively high resistance to salt stress conditions (Luard and El-Lakany 1984; Graves and Gallagher 2003; Diagne et al. 2013). Many reports stressed that toxicity is mainly due to plant uptake of salts. Under elevated salt levels, movement of water in response to the water potential gradient results in a water stress. It was stressed that salinity above 50 mM NaCl has deleterious effects on legume plants. It was also reported that nodule formation in N_2-fixing plants is very sensitive to salt stress (Zahran and Sprent 1986; Young et al. 1992). Salinity also affects nitrogenase activity in legumes (Cordovilla et al. 1999) and actinorhizal plants (Sande and Young 1992; Oshone et al. 2013) and consequently reduces N_2 fixation. For instance, in the common bean, high Na^+ and Cl^- concentrations were found in nodules and leaves (Ashraf and Bashir 2003). Nodule formation is even more sensitive than the rhizobia themselves (Zahran 1991). Moreover, in the nodule formation process, it appears that the ability of the rhizobia to infect roots and nodule initiation are more sensitive to salt stress than nodule development (Zahran and Sprent 1986; Craig et al. 1991). Consequently, salinity affects plant performance resulting in reduction in shoot and root growth and legume crop production (Oufdou et al. 2014). In addition, salinity could aggravate this effect through increase in levels of ethylene, a stress hormone that significantly reduces plant growth. Salinity can also indirectly exert deleterious effects by restricting the carbon supply from the host plant as a result of photosynthesis inhibition (Soussi et al. 1999). Other deleterious effects are calcium deficiency, production of reactive oxygen species (ROS), tissue necrosis, and increase in ABA levels that lead to early senescence of older leaves.

Several works emphasized that salt tolerance depends on both plant and bacterial species. For example, most reports agreed that fast-growing rhizobial strains are more tolerant than slow-growing strains (Ghittoni and Bueno 1996). Members of the fast-growing genus *Rhizobium* are generally more tolerant to salts than members of the genus *Bradyrhizobium.* It is also well known that some legumes, e.g., *Sesbania aculeata*, *V. faba*, and *G. max*, are more salt tolerant than other legume species such as *Pisum sativum.* In a given species, tolerance could be variable according to cultivars (Soussi et al. 1999). Resistance to saline conditions is the result of adaptive processes and is mainly due to ability to exclude Na and Cl ions and to prevent the transport of these ions to shoots. They also involve control of ion transport and compartmentation and osmotic solute synthesis and accumulation, which lead to osmotic adjustment. Development of plant symbiont-tolerant systems for cultivation in salt-affected soils is currently considered as necessity. Indeed, several studies are aimed at identifying salt-tolerant genotypes belonging to legumes (Moreno et al. 2000; Sadiki and Rabih 2001) and actinorhizal plants (El-Lakany and Luard 1982; Luard and El-Lakany 1984; Nambiar-Veetil et al. 2011). It is thus well established that N_2-fixing plant symbiosis tolerance to salt stress conditions is variable among species or cultivars. Moreno et al. (2000) found a large genotypic variability in bean cultivars for salt tolerance at the seedling stage and identified some salt-tolerant cultivars with higher performances. Fast-growing salt-tolerant actinorhizal plants showing variable tolerance levels could also be used in salt-affected terrains. Actinorhizal trees such as alders (Graves and Gallagher 2003) and members of the *Casuarinaceae* (*C. equisetifolia* and *C. glauca*) are able to grow under saline conditions (Nambiar-Veetil et al. 2011), whereas *Myrica cerifera* are intolerant to high salinity (Young et al. 1992). Tani and Sasakawa (2006) found that in *C. equisetifolia* plants, which are highly tolerant to salt stress, proline is primarily synthesized as a major compatible solute to adjust the osmotic pressure when Na^+ accumulates in the cells, allowing to maintain cell homeostasis under salt stress conditions. Currently, several members of the *Casuarinaceae* are used in the reclamation of salt-affected lands, and more recent researches focused on tissue culture of salt-tolerant *C. equisetifolia* clones (Liu et al. 2003) and creation of transgenic trees that are more tolerant to salt in tropical regions (Nambiar-Veetil et al. 2011).

Greater performance of N_2-fixing plants under saline conditions seems to be determined mainly by the tolerance of the host plant and its interaction with the bacterial symbionts (Zahran 1991; Tani and Sasakawa 2003). Consequently, inoculation with more tolerant symbiotic bacterial strains with efficient nodulation and N_2 fixation under saline conditions may contribute to the compensation of the salinity-induced plant growth loss (Girgis et al. 1992; Hafeez et al. 1999; Zahran 2001; Tani and Sasakawa 2003). Therefore, the selection of osmotolerant strains of symbiotic bacteria able to efficiently nodulate plants under saline conditions is considered as a great challenge to improve the productivity of N_2-fixing plants. Several salt-tolerant *Rhizobium* and *Frankia* species have been reportedly isolated that could tolerate saline environments (Rai 1983; Hafeez et al. 1999; Zahran 1999; Tani and Sasakawa 2003; Srivastava et al. 2012; Oshone et al. 2013). Among them, some strains of *Rhizobium meliloti* and *R. fredii* (Kassem et al. 1985) and *Sinorhizobium* (*Ensifer*) *arboris* (isolated from *Acacia*) (Zahran et al. 1994) were able to grow at NaCl concentration of more than 300 mM. Similarly, Hafeez et al. (1999) described a *Frankia* strain CcO1 that could tolerate up to 500 mM NaCl (Ec_e of 47 dSm-1), and Tani and Sasakawa (2003) reported nodulation of *C. equisetifolia* with a selected strain of *Frankia* Ceq1 under saline conditions up to 300 mM (28 dSm^{-1}). Accumulation of various solutes was correlated to tolerance of *Rhizobium* (Zahran 1999). Among accumulated compatible solutes are K^+, glutamate, proline, glycine betaine, proline betaine, trehalose, and the dipeptide N-acetylglutaminylglutamine amide. Recently, the role of poly-β-hydroxyl butyrate (PHB) in cell protection of rhizobia during high saline conditions was reported (Arora et al. 2006). Glycine betaine was found to increase more in the salt-tolerant strains of *R. meliloti* than in sensitive strains (Smith et al. 1988). It has also been suggested that rhizobia synthesizing IAA may prevent the deleterious effects of salinity. Bianco and Defez (2009) reported an increased tolerance against salt stress by *M. truncatula* nodulated by a strain of an IAA-overproducing mutant of *Sinorhizobium meliloti*. This effect was accompanied by higher amount of proline and accumulation of antioxidant enzymes in mutant inoculated plants in comparison with plants inoculated with the wild strain. Such inoculation with symbiotic bacteria could help plants to tolerate salt conditions. For instance, in arid saline soils, it was reported that *B. japanicum* S2492 has significantly increased dry weight, plant height, and yield (>35 %) of soybeans (Egamberdiyeva et al. 2004).

The application of other beneficial microbes as inoculants have recently gained interest as beneficial microbes could help plants to mitigate

Table 14.2 Selected examples of salinity stress alleviations in legume plants by combinations of rhizobia and rhizobacteria

Salinity	*Galega officinalis* (goat's rue)	*Rhizobium galegae* sv. *officinalis*	*P. extremorientalis* or *P. trivialis*	Egamberdieva et al. (2013a)
	P. vulgaris (common bean)	*R. tropici* or *R. etli*	*Azospirillum brasilense*	Dardanelli et al. (2008)
	P. vulgaris (common bean)	*R. tropici*	*Chryseobacterium balustinum*	Estevez et al. (2009)
	P. vulgaris (common bean)	*Ensifer fredii*	*C. balustinum*	Estevez et al. (2009)
	Glycine max (soybean)	*B. japonicum*	*Pseudomonas putida*	Egamberdieva et al. (2013b)
	Cicer arietinum (chickpea)	*Rhizobium*	*Azospirillum brasilense*	Hamaoui et al. (2001)

stress conditions. Hamaoui et al. (2001) reported that inoculation with the PGPR *Azospirillum brasilense* significantly enhanced nodulation by native rhizobia in chickpea. This bacterial treatment was shown to significantly enhance growth parameters and to reduce the negative effects on plant growth caused by irrigation with saline water. Dual inoculation including rhizobia and bacterial inoculants could also be used to alleviate salinity stress conditions in N_2-fixing plants (Dardanelli et al. 2008; Estevez et al. 2009; Egamberdieva et al. 2013b). In soybean, Estevez et al. (2009) reported that co-inoculation of the PGPR *Chryseobacterium balustinum* and *Ensifer fredii* led to a significant increase in the number of nodules and root growth under moderate saline conditions. This symbiotic performance was more pronounced than in single inoculation with *E. fredii*. Similar results were found by Egamberdieva et al. (2013b) in soybean co-inoculated by *B. japonicum* and *P. putida*. PGPRs include bacteria with ACC-deaminase activity (Shaharoona et al. 2006; Saravanakumar and Samiyappan 2007; Egamberdieva et al. 2013a). Saravanakumar and Samiyappan (2007) reported that *P. fluorescens* with ACC-deaminase activity led to plant growth promotion and enhanced saline resistance in groundnut plants compared with that inoculated with *Pseudomonas* strains lacking ACC-deaminase activity. PGPR effects could also be mediated by "induced systemic mechanisms" to salt. Indeed, through mechanisms such as osmoregulation and other metabolic adaptations, PGPR could ameliorate the tolerance of plants to saline conditions (Egamberdiyeva and Islam 2008). Phenolic compounds are well known as resistance mechanisms, and their elicitation by PGPRs was reported in many research works as accompanying salinity alleviations in co-inoculated plants (Dardanelli et al. 2008, 2012). Enhanced synthesis of flavonoids (known as phenolic substances involved in *Rhizobium* attraction), following PGPR inoculation, was reported by Dardanelli et al. (2008). They showed that co-inoculation of *A. brasilense* and *R. tropici* strain CIAT899 or *R. etli* ISP42 resulted in enhanced nitrogen fixation concomitant to production of a variety of flavonoid species in *P. vulgaris* cv. *Negro jamapa* under salt stress. This effect is linked to an efficient activation of nod factor genes under such conditions. It was also reported that most of PGPRs could produce EPS able to bind Na^+. Han and Lee (2005) observed that inoculation of EPS-producing strain resulted in a decrease in Na^+ content of soybean grown under saline conditions. The production of osmoprotectants, a kind of organic compounds under inhibitory osmolarities, by either rhizobia or beneficial bacteria that are subsequently taken up by plants, is also reported by several authors. Table 14.2 presents selected examples of salinity stress alleviations in legume plants by combinations of rhizobia and rhizobacteria.

Tripartite association involving ectomycorrhizal and endomycorrhizal fungi was also reported as a powerful tool to alleviate saline stress. Colonization by AMF can help plants to cope with salinity stresses (Ruiz-Lozano 2003) by compensating for its negative effects on nodulation and N_2 fixation (Ruiz-Lozano and Azcón 1993). It was reported for instance that dual inoculation with AM and rhizobia decreased the deleterious effects of sulfate salinity on plant growth and on P and N accumulation in *Lathyrus sativus* (Jin et al. 2010). Zhong et al. (2010) demonstrated that the ectomycorrhizal fungus (EMF) *P. tinctorius* increased *Casuarina* tolerance to salt condition. One mechanism of osmotic stress tolerance induced by AMF colonization was shown by Aroca et al. (2007) in bean (*P. vulgaris*) for which an increase in root hydraulic conductivity of the host plants was observed. Reddell et al. (1986b) observed that dual inoculation of *Frankia* and the EMF *Suillus* excludes salt and toxic metals from metabolic pathways by compartmentalization in vacuoles and cell walls.

Alleviation of Heavy Metal Toxicity in Contaminated Soils

Several soil systems are contaminated by mineral compounds such as heavy metals or by organic compounds of various origins. Many of these compounds originate from industrial and intensive agriculture activities and discharges of untreated or partially treated wastewater. It is well known that heavy metals even in traces are toxic against organisms (Giller et al. 2009; Nonnoi et al. 2012) including humans and result in major health problems. The most common heavy metal contaminants are cadmium (Cd), chromium (Cr), cupper (Cu), mercury (Hg), lead (Pb), nickel (Ni), manganese (Mn), and zinc (Zn). The elevated levels of heavy metals in the environment are excessively absorbed by roots and translocated to shoots, leading to impaired metabolism and reduction in plant growth (Rodríguez-Llorente et al. 2010; Perez-Palacios et al. 2013). For instance, it was reported that in legume plants, Cd contamination could result in a decreased uptake of nutrients, inhibition of various enzyme activities, and induction of oxidative stress (Sandalio et al. 2001) or inhibition of chlorophyll synthesis and photosynthesis (Padmaja et al. 1990). For actinorhizal plants, inhibition of growth, nodulation, and nitrogenase activity by heavy metals such as Cd (Hensley and Carpenter 1987), Cu (Fessenden and Sutherland 1979), Ni (Wheeler et al. 2001), and Zn (Cusato et al. 2007) have been reported. It is currently accepted that the sensitivity to a given heavy metal is variable according to plant species. In *Alnus crispa*, Cu concentration of 60 mg kg^{-1} caused a reduction of nodule dry weight, while nodulation was totally inhibited at 100–150 mg kg^{-1}. However, Lorenc-Pluciñska et al. (2013) described two species of *Alnus*, *A. incana* and *A. glutinosa* tolerant to Cu concentrations higher than 1,500 mg kg^{-1} of soil. Unfortunately, heavy metals cannot be destroyed biologically. The physical-chemical methods to remediate heavy metal pollution are not suitable for practical applications. In addition to their high costs, they led to a large destruction of soil structure and fertility and soil properties. Thus, biological remediation presents the advantage to be of low-cost, effective, eco-friendly, and socially accepted strategy.

Nitrogen-fixing plants could play a key role in the detoxification of heavy metals (Lorenc-Pluciñska et al. 2013; Vivas et al. 2003a, b). Among strategies that should be adopted prior to introduction of N_2-fixing plants in contaminated sites is the selection of tolerant plant species and/or bacterial symbionts. Actinorhizal plants such as alders have been used for the reclamation of sites contaminated with toxic levels of trace elements (Lefrancois et al. 2010; Babu et al. 2013; Lorenc-Pluciñska et al. 2013). Lorenc-Pluciñska et al. (2013) reported that two actinorhizal species *A. incana* and *A. glutinosa* grow well in heavily polluted soil. In spite of accumulation of toxic metals in their roots and nodules and a decreased rate of N_2 fixation, concentrations of Cu, Zn, Cd, and Pb in leaves of both alder species were within normal ranges. In fact, heavy metals are accumulated in roots of these species, and only small proportions are transferred to aboveground parts, including leaves. These species of alders can be recommended for the revegetation of areas polluted with phytotoxic concentrations of trace metals. However, these results were obtained only from short-term pot trials and

Table 14.3 Selected examples of heavy metal bioremediation by legume and actinorhizal plants in association with microbial inoculants

Plant	Symbiont bacterium	Tolerant microbial inoculant species	Heavy metal contamination[a]	Reference
Alnus glutinosa (alder)	Soil as *Frankia* inoculum		Cu	Whitbread-Abrutat (1997)
A. glutinosa (alder)	*Frankia* (Ni tolerant)		Ni	Wheeler et al. (2001)
A. glutinosa and *Alnus incana* (tolerant plants)	*Frankia*		Cu and Pb	Lorenc-Pluciñska et al. (2013)
Alnus firma (alder)		*Bacillus thuringiensis* (endophyte)	Pb, Zn, As, Cd, Cu, and Ni	Babu et al. (2013)
Pisum sativum (pea)	*Rhizobium* (metal tolerant)		Ni, Zn	Wani et al. (2008)
Trifolium (clover)	*Rhizobium*	PGPRs and the mycorrhizal fungus *Glomus mosseae*	Zn	Vivas et al. (2006a)
Lupinus luteus (lupine)	*Bradyrhizobium*	*Pseudomonas* and *Ochrobactrum cytisi*	Cd, Cu, and Pb	Dary et al. (2010)
Pisum sativum (pea)		*Pseudomonas brassicacearum* and *P. marginalis*	Cd	Safronova et al. (2006)
Trifolium pratense (red clover)	*Rhizobium*	*Brevibacillus* sp.	Pb	Vivas et al. (2003a)
Trifolium repens (white clover)	*Rhizobium*	*G. mosseae* (mycorrhizae) and *Brevibacillus* sp.	Cd	Vivas et al. (2003b)
T. repens (white clover)	*Rhizobium trifolii*	*Brevibacillus brevis* and *Glomus mosseae* (AM fungus)	Ni	Vivas et al. (2006b)
Vigna radiata (mung bean)		*P. putida*	Cd, Pb	Tripathi et al. (2005)
V. radiata (mung bean)	*Bradyrhizobium* (metal tolerant)		Ni, Zn	Wani et al. (2007b)

[a]*Cd* cadmium, *Cu* copper, *Mn* manganese, *Ni* nickel, *Pb* lead, *Zn* zinc

require further field validation. *Frankia* and rhizobia strains were also consequently screened for their tolerance to heavy metals (Richards et al. 2002; Carrasco et al. 2005; Dary et al. 2010; Bélanger et al. 2011). Heavy metal-tolerant *Frankia* could allow actinorhizal plants to colonize contaminated terrains, and their spectra of tolerance were reported to vary depending on strains tested (Richards et al. 2002; Bélanger et al. 2011). Richards et al. (2002) investigated the tolerance of 12 *Frankia* strains to a wide variety of heavy metals. While all of the 12 strains were sensitive to low concentrations (<0.5 mM) of Ag^+, AsO_2^-, Cd_2^+, SbO_2^-, and Ni_2^+, most were less sensitive to Pb^{2+} (6–8 mM), CrO_4^{2-} (1.0–1.75 mM), AsO_4^{3-} (>50 mM), and SeO_2^{2-} (1.5–3.5 mM). These authors also found that four strains are resistant to elevated levels of Cu^{2+} (up to 20 mM for strain CN_3). Carrasco et al. (2005) and Dary et al. (2010) have isolated effective rhizobia resistant to heavy metals from polluted soils in Spain. These strains could be exploited for revegetation programs of many other polluted areas.

Other microbial inoculants such as PGPRs or mycorrhizae could act as helper microbia in heavy metal-contaminated areas. Table 14.3 presents selected examples of heavy metal bioremediation by legume and actinorhizal plants in combination with microbial inoculants. PGPRs as bioinoculants proved their utilities to alleviate heavy metal stress conditions and were used as a tool for rhizoremediation in contaminated soils. For instance, elevated levels of heavy metals interfere with P uptake by plants. By using P-solubilizing PGPRs, this effect could be compensated (Gupta et al. 2001). Rhizobacteria have been shown to possess several characteristics that can alter heavy metal bioavailability through acidification of the microenvironment, releasing of chelating substances, and modifying heavy metal redox potential (Abou-Shanab et al. 2003). One of the major traits of these bacteria is the sequestration and accumulation of heavy metals (Richards et al. 2002; Gupta et al. 2004; Lee et al. 2006). It was suggested to be the role of polysaccharides or proteins as potential binding sites for these heavy metals (Richards et al. 2002). Works by Vivas et al. (2003a, b, 2006a) showed the utility of bioinoculants for alleviation of heavy metals as a strategy to grow legumes in contaminated soils. In these studies, bacteria with the ability to survive and colonize the rhizosphere were used in Zn-, Pb-, or Cd-contaminated soils. Clover (*Trifolium*) growing in soil contaminated with Cd inoculated with a Cd-adapted autochthonous PGPR, *Brevibacillus*, resulted in growth-promoting effects and a reduction in Cd transfer from soil to plants (Vivas et al. 2003b). A significant enhancement of nitrogen and phosphorus accumulations and nodule formation concomitant to decrease the amount of Pb absorbed by plants was also reported (Vivas et al. 2003a). Dary et al. (2010) used *Lupinus luteus* and a consortium of metal-resistant PGPR co-inoculated (including *Bradyrhizobium*, *Pseudomonas* sp., and *Ochrobactrum cytisi*) for reclamation of multi-metal-contaminated soil. They observed root accumulation of heavy metals concomitant with increment in lupine growth promotion effect when compared to *Lupinus* inoculated with *Bradyrhizobium* alone. This mixture also succeeded to reduce plant toxicity symptoms and metal accumulation in both shoots and roots. The alleviating effect could also be due to more intimate bacteria-plant relationships such as those ensured by endophytic PGPRs. Babu et al. (2013) reported a significant enhancement of heavy metal phytoremediation by *Alnus firma* with an endophytic strain of *Bacillus thuringiensis*.

Tripartite associations involving mycorrhizae were also used as a tool to alleviate heavy metal stress and for bioremediation (Díaz et al. 1996; Roy et al. 2007; Karimi et al. 2011). Mycorrhizae can help contribute to phytoremediation by their ability to sequester heavy metals through the production of chelates or by absorption. Consequently, less heavy metals are translocated to plant shoots and leaves. AMF were used in association with *C. equisetifolia* to rehabilitate heavy metal-polluted soils (Karimi et al. 2011). Tripartite association established by *Alnus*, mycorrhizae-*Frankia*, was reviewed by Roy et al. (2007). In their work aiming to provide *Trifolium pretense* growth promotion and lead toxicity reduction by *Brevibacillus*, Vivas et al. (2003a) reported that association of bacteria and mycorrhizae could interact synergistically resulting in significant positive effects. They reported that mycorrhization of AM fungi was positively stimulated by the PGPR bacterium resulting in significant growth under Pb contamination. All these associations could provide a basis for selecting nodulating bacteria associated with other microbial inoculants that may have applications in formulating appropriate inocula, effective when used for phytoremediation and revegetation of contaminated soils, but also could be a tenable strategy for improving growth while avoiding high input of chemical fertilizers in such sites.

Conclusions

Enhancing application and management of biologically fixed N could result in very important environmental and economic benefits for the humanity. N_2-fixing plants have an exceptional ability to form symbiotic association with rhizobia and *Frankia*. Hence, they have been used on a broad scale so as to improve the nitrogen status of soils. Symbiosis could represent a key factor in abiotic stress conditions, and many researchers stressed the importance of both host specificity and environmental variations in determining patterns of symbiont efficiencies under such conditions. Therefore, the selection of more tolerant plant/symbionts has been one of the multiple approaches. A valuable selection approach could be performed by targeting multitolerant rhizobia. Recently, Elboutahiri et al. (2010) isolated rhizobia from drought and salt-affected regions of Morocco that were tolerant to salinity, water stress, high temperature, acidity, and heavy metals. However, bacterial strain–plant genotype combination should be considered for selecting the most adapted microbe-plant combinations to stress conditions. On the other hand, the application of microbial inoculants to N_2-fixing plants is one of the most interesting eco-friendly strategies for agriculture and forestry sustainability. Plant inoculation with more tolerant strains of these bacteria could contribute to their tolerance to stress conditions (Figueiredo et al. 2008). In comparison with legume plants, the use and exploitation of the beneficial effects of PGPRs to actinorhizal plants is still a rarely used approach by researchers (Knowlton and Dawson 1983; Probanza et al. 1996; Rojas et al. 2002). Very little information is available on the effectiveness of PGPRs as bioinoculants for actinorhizal plants under stress conditions. This topic needs to be further investigated. Unexplored fluorescent *Pseudomonas* and other PGPR species with beneficial properties (e.g., phytohormone or siderophore production) (Arshad et al. 2008; Figueiredo et al. 2008) could contribute to more tolerance to abiotic stress conditions. There is in fact a need to explore more number of effective combinations of PGPR and *Frankia*. Also, the application of helper bacteria or PGPRs in association with actinorhizal plants for bioremediation of eroded or degraded areas following industrial waste pollution is another aim to be focused on.

In addition, tripartite associations involving mycorrhizae seem to be promising for sustainable agriculture and forestry based on N_2-fixing plants. To help actinorhizal plants to withstand abiotic stresses, mostly used bioinoculants in tripartite associations are mycorrhizae. In researches focusing on tripartite associations, one of the most important technical traits is the selection of the appropriate microbial inoculants (mycorrhizae and PGPRs). In addition, the use of compatible multiple microbial consortia including bacterial symbionts and fungal symbionts acting synergistically, providing various beneficial effects, is also a powerful strategic tool (Requena et al. 1997; Muthukumar and Udaiyan 2010). In general, the use of tripartite and tetrapartite associations is largely studied and applied in both legumes and actinorhizal plants. However, the use of these associations should be approached cautiously in relation to efficient combinations, particularly those involving mutual interactions between the N_2-fixing bacterium and the bioinoculant (rhizobacterium or mycorrhizae) (Requena et al. 1997; Gardner 1986; Ruiz-Lozano et al. 2001; Valdenegro et al. 2001; Remans et al. 2007).

In addition to classical rhizobia and *Frankia* strains, another way is to study and evaluate other endophytic bacteria potentially found in nodules (nodule inhabitants) in experiments aiming to alleviate abiotic stress effects. Recently, an actinobacterium, *Micromonospora*, was found in actinorhizal and legume nodules as holding nitrogenase activity and nifH-like gene sequences demonstrating its potential role in N_2 fixation (Valdès et al. 2005; Trujillo et al. 2006, 2010; Carro et al. 2012). To our knowledge, most works dealing with abiotic stress tolerance are aimed to select only known symbiotic bacteria, i.e., rhizobia and *Frankia*, and very little work has addressed the role of other nodule-colonizing bacteria. Special attention should be drawn on isolating these bacteria, particularly those tolerant to stress conditions. Their use as bioinoculants in combination with rhizobia or *Frankia* strains could help to emerge a new dimension into the application of microbial inoculants to N_2-fixing plants under abiotic stress conditions.

References

Abou-Shanab RA, Angle JS, Delorme TA, Chaney RL, van Berkum P, Moawad H, Ghanem K, Ghozlan HA (2003) Rhizobacterial effects on nickel extraction from soil and uptake by *Alyssum murale*. New Phytol 158:219–224

Arnone JA, Kohls SJ, Baker DD (1994) Nitrate effects on nodulation and nitrogenase activity of actinorhizal *Casuarina* studied in split-root systems. Soil Biol Biochem 26:599–606

Aroca R, Porcel R, Ruiz-Lozano JM (2007) How does arbuscular mycorrhizal symbiosis regulate root hydraulic properties and plasma membrane aquaporins in *Phaseolus* vulgaris under drought, cold or salinity stresses? New Phytol 173:808–816

Arora NK, Singhal V, Maheshwari DK (2006) Salinity-induced accumulation of poly-b- hydroxyl butyrate in rhizobia indicating its role in cell protection. World J Microbiol Biotechnol 22:603–606

Arshad M, Sharoona B, Mahmood T (2008) Inoculation with *Pseudomonas* spp. containing ACC deaminase partially eliminate the effects of drought stress on growth, yield and ripening of pea (*Pisum sativum* L). Pedosphere 18:611–620

Asfaw A, Almekinders CJM, Blair MW, Struik PC (2012) Participatory approach in common bean (*Phaseolus vulgaris* L.) breeding for drought tolerance for southern Ethiopia. Plant Breed 131:125–134

Ashraf M, Bashir A (2003) Salt stress induced changes in some organic metabolites and ionic relations in nodules and other plant parts of two crop legumes differing in salt tolerance. J Flora 198:486–498

Azcón R, El-Atrash F, Barea JM (1988) Influence of mycorrhiza versus soluble phosphate on growth, nodulation, and N_2 fixation ^{15}N in alfalfa under different levels of water potential. Biol Fertil Soils 7:28–31

Babu G, Kim JD, Oh BT (2013) Enhancement of heavy metal phytoremediation by *Alnus firma* with endophytic *Bacillus thuringiensis* GDB-1. J Hazard Mater 250(251):477–483

Baker A, Parsons R (1997) Evidence for N feedback regulation of N_2 fixation in *Alnus glutinosa* L. J Exp Bot 48:67–73

Barea JM, Azcón-Aguilar C (1983) Mycorrhizas and their significance in nodulating nitrogen fixing plants. Adv Agron 36:1–54

Barea JM, Werner D, Azcón-Guilar C, Azcón R (2005) Interactions of arbuscular mycorrhiza and nitrogen-fixing symbiosis in sustainable agriculture. In: Werner D, Newton WE (eds) Nitrogen fixation in agriculture, forestry, ecology and the environment. Springer, Dordrecht, pp 199–222

Barritt AR, Facelli JM (2001) Effects of *Casuarina* pauper litter and grove soil on emergence and growth of understorey species in arid land of South Australia. J Arid Environ 49:569–579

Bélanger PA, Beaudin J, Roy S (2011) High-throughput screening of microbial adaptation to environmental stress. J Microbiol Methods 85:92–97

Bianco C, Defez R (2009) *Medicago truncatula* improves salt tolerance when inoculated by an indole-3-acetic acid-overproducing *Sinorhizobium meliloti* strain. J Exp Bot 60:3097–3107

Bollman MI, Vessey JK (2006) Differential effects of nitrate and ammonium supply on nodule initiation, development and distribution on roots of pea (*Pisum sativum*). Can J Bot 84:893–903

Carrasco JA, Armario P, Pajuelo E, Burgos A, Caviedes MA, Lopez R, Chamber MA, Palomares AJ (2005) Isolation and characterisation of symbiotically effective *Rhizobium* resistant to arsenic and heavy metals after the toxic spill at the Aznalcollar pyrite mine. Soil Biol Biochem 37:1131–1140

Carro L, Sproer C, Alonso P, Trujillo ME (2012) Diversity of *Micromonospora* strains isolated from nitrogen fixing nodules and rhizosphere of *Pisum sativum* analyzed by multilocus sequence analysis. Syst Appl Microbiol 35:73–80

Chaia EE, Wall LG, Huss-Danell K (2010) Life in soil by the actinorhizal root nodule endophyte *Frankia*: a review. Symbiosis 51:201–226

Chanway CP, Hynes RK, Nelson LM (1989) Plant growth-promoting rhizobacteria: effects on growth and nitrogen fixation of lentil (*Lens esculenta* Moench) and pea (*Pisum sativum* L.). Soil Biol Biochem 21:511–517

Chapin DM, Bliss LC, Bledsoe LJ (1991) Environmental regulation of dinitrogen fixation in a high arctic lowland ecosystem. Can J Bot 69:2744–2755

Chapin FS III, Walker RL, Fastie CL, Sharman LC (1994) Mechanisms of primary succession following deglaciation at Glacier Bay, Alaska. Ecol Monogr 64:149–175

Chebotar VKK, Asis CA, Akao S (2001) Production of growth-promoting substances and high colonization ability of rhizobacteria enhance the nitrogen fixation of soybean when coinoculated with *Bradyrhizobium japonicum*. Biol Fertil Soils 34:427–432

Chen WM, Moulin L, Bontemps C, Vandamme P, Béna G, Boivin-Masson C (2003) Legume symbiotic nitrogen fixation by β-proteobacteria is widespread in nature. J Bacteriol 185:7266–7272

Claessens H, Oosterbaan A, Savill P, Rondeux J (2010) A review of the characteristics of black alder (*Alnus glutinosa* (L.) Gaertn.) and their implications for silvicultural practices. Forestry 83:163–175

Cordovilla MP, Ligero F, Lluch C (1999) Effect of salinity on growth, nodulation and nitrogen assimilation in nodules of faba bean (*Vicia faba* L.). Appl Soil Ecol 11:1–7

Craig GF, Atkins CA, Bell DT (1991) Effect of salinity on growth of four strains of *Rhizobium* and their infectivity and effectiveness on two species of *Acacia*. Plant Soil 133:253–262

Cusato MS, Tortosa RD, Valiente L, Barneix AJ, Puelles MM (2007) Effects of Zn^{2+} on nodulation and growth of a South American actinorhizal plant, *Discaria americana* (Rhamnaceae). World J Microbiol Biotechnol 23:771–777

Dardanelli MS, de Cordoba FJF, Espuny MR, Carvajal MAR, Diaz MES, Serrano AMG, Okon Y, Megias M (2008) Effect of *Azospirillum brasilense* coinoculated with *Rhizobium* on *Phaseolus vulgaris* flavonoids and Nod factor production under salt stress. Soil Biol Biochem 40:2713–2721

Dardanelli MS, Fernandez de Cordoba FJ, Estévez J, Contreras R, Cubo MT, Rodriguez-Carvajal MA, Gil-Serrano AM, Lopez-Baena FJ, Bellogin R, Manyani H, Ollero FJ, Megias M (2012) Changes in flavonoids secreted by *Phaseolus vulgaris* roots in the presence of salt and the plant growth-promoting rhizobacterium *Chryseobacterium balustinum*. Appl Soil Ecol 57:31–38

Dary M, Chamber-Pérez MA, Palomares AJ, Pajuelo E (2010) "In situ" phytostabilisation of heavy metal polluted soils using *Lupinus luteus* inoculated with metal resistant plant-growth promoting rhizobacteria. J Hazard Mater 177:323–330

Dashadi M, Khosravi H, Moezzi A, Nadian H, Heidari M, Radjabi R (2011) Co- inoculation of *Rhizobium* and *Azotobacter* on growth indices of faba bean under water stress in the green house condition. Adv Stud Biol 3:373–385

Dawson JO (2008) Ecology of actinorhizal plants. In: Pawlowski K, Newton WE (eds) Nitrogen-fixing actinorhizal symbioses. Springer, Dordrecht, pp 199–234

Devries JD, Bennett JM, Albrecht SL, Boote KJ (1989) Water relations, nitrogenase activity and root development of three grain legumes in response to soil water deficits. Field Crop Res 21:215–226

Diagne N, Diouf D, Svistoonoff S, Kane A, Noba K, Franche C, Bogusz D, Duponnois R (2013) *Casuarina* in Africa: distribution, role and importance of arbuscular mycorrhizal, ectomycorrhizal fungi and *Frankia* on plant development. J Environ Manag 128:204–209

Díaz G, Azcón-Aguilar C, Honrubia M (1996) Influence of arbuscular mycorrhizae on heavy metal (Zn and Pb) uptake and growth of *Lygeum spartum* and *Anthyllis cytisoides*. Plant Soil 180:241–249

Diem HG, Dommergues YR (1990) Current and potential uses and management of *Casuarinaceae* in the tropics and subtopics. In: Schwintzer CR, Tjepkema JD (eds) The biology of *Frankia* and actinorhizal plants. Academic, New York, pp 317–342

Egamberdieva D, Berg G, Lindström K, Räsänen LA (2010) Co-inoculation of *Pseudomonas* spp. with *Rhizobium* improves growth and symbiotic performance of fodder galega *Galega orientalis* Lam. Eur J Soil Biol 46:269–272

Egamberdieva D, Berg G, Lindström K, Räsänen LA (2013a) Alleviation of salt stress of symbiotic *Galega officinalis* L. (Goat's Rue) by co-inoculation of *Rhizobium* with root colonizing *Pseudomonas*. Plant Soil 369:453–465

Egamberdieva D, Jabborova D, Wirth S (2013b) Alleviation of salt stress in legumes by co-inoculation with *Pseudomonas* and *Rhizobium*. In: Arora NK (ed) Plant microbe symbiosis: fundamentals and advances. Springer, New Delhi, pp 291–303

Egamberdiyeva D, Islam KR (2008) Salt tolerant rhizobacteria: plant growth promoting traits and physiological characterization within ecologically stressed environment. In: Ahmad I, Pichtel J, Hayat S (eds) Plant-bacteria interactions: strategies and techniques to promote plant growth. Wiley-VCH Verlag GmbH & Co, Weinheim, pp 257–281

Egamberdiyeva D, Qarshieva D, Davranov K (2004) The use of *Bradyrhizobium japonicum* to enhance growth and yield of soybean varieties in Uzbekistan conditions. J Plant Growth Regul 23:54–57

Elboutahiri N, Thami-Alami I, Udupa SM (2010) Phenotypic and genetic diversity in *Sinorhizobium meliloti* and *S. medicae* from drought and salt affected regions of Morocco. BMC Microbiol 10:15

Elisondo Barron J, Pasini RJ, Davis DW, Stuthman DD, Graham PH (1999) Response to selection for seed yield and nitrogen (N_2) fixation in common bean (*Phaseolus vulgaris* L.). Field Crops Res 62:119–128

Elkoca E, Turan M, Donmez MF (2010) Effects of single, dual and triple inoculations with *Bacillus subtilis*, *Bacillus megaterium* and *Rhizobium leguminosarum* bv. *phaseoli* on nodulation, nutrient uptake, yield and yield parameters of common bean (*Phaseolus vulgaris* l. cv. 'elkoca-05'). J Plant Nut 33:2104–2119

El-Lakany MH (1983) A review of breeding drought resistant *Casuarina* for shelterbelt establishment in arid region with special reference to Egypt. For Ecol Manag 6:129–137

El-Lakany MH, Luard EJ (1982) Comparative salt tolerance of selected *Casuarina* species. Aust For Res 13:11–20

Elo S, Maunuksela L, Salkinoja-Salonen M, Smolander A, Haahtela K (2000) Humus bacteria of Norway spruce stands: plant growth promoting properties and birch, red fescue and alder colonizing capacity. FEMS Microbiol Ecol 31:143–152

Elumalai S, Raaman N (2009) *In vitro* synthesis of *Frankia* and mycorrhiza with *Casuarina equisetifolia* and ultrastructure of root system. Ind J Exp Biol 47:289–297

Estevez J, Dardanelli MS, Megias M, Rodriguez-Navarro DN (2009) Symbiotic performance of common bean and soybean co inoculated with rhizobia and *Chryseobacterium balustinum* Aur9 under moderate saline conditions. Symbiosis 49:29–36

FAO (2005) Global network on integrated soil management for sustainable use of salt-affected soils, Italy. FAO Services, Rome. http://www.fao.org/ag/agl/agll/spush

Fessenden RJ, Sutherland BJ (1979) The effect of excess soil copper on the growth of black spruce and green alder seedling. Bot Gaz 140:S82–S87

Figueiredo MVB, Burity HA, Martinez CR, Chanway CP (2008) Alleviation of drought stress in common bean (*Phaseolus vulgaris* L.) by co-inoculation with *Paenibacillus polymyxa* and *Rhizobium tropici*. Appl Soil Ecol 40:182–188

Freire JRJ (1992) Fixação do nitrogênio pela simbiose rizóbio/leguminosas. In: Cardoso EJN, Tsai SNM, Neves MC (eds) Microbiologia do solo. Sociedade Brasileira de Ciência do Solo, Campinas, pp 121–140

Gardner LC (1986) Mycorrhizae of actinorhizal plants. Mircen J 2:147–160

Gentili F, Wall LG, Huss-Danell K (2006) Effects of phosphorus and nitrogen on nodulation are seen already at the stage of early cortical cell divisions in *Alnus incana*. Ann Bot 98:309–315

Gherbi H, Markman K, Svistoonoff S, Estevan J, Autran D, Giczey G, Auguy F, Péret B, Laplaze L, Franche C, Parniske M, Bogusz D (2008) SymRK defines a common genetic basis for plant root endosymbioses with arbuscular mycorrhiza fungi, rhizobia, and *Frankia* bacteria. Proc Natl Acad Sci U S A 105:4928–4932

Ghittoni NE, Bueno MA (1996) Changes in the cellular content of trehalose in four peanut rhizobia strains cultured under hypersalinity. Symbiosis 20:117–127

Giller KE, Witter E, McGrath SP (2009) Heavy metals and soil microbes. Soil Biol Biochem 41:2031–2037

Girgis MG, Ishac YZ, Diem HG, Dommergues YR (1992) Selection of salt tolerant *Casuarina glauca* and *Frankia*. Acta Oecol 13:443–451

Goicoechea N, Szalai G, Antolín MC, Sánchez-Díaz M, Paldi E (1998) Influence of arbuscular mycorrhizae and *Rhizobium* on free polyamines and proline levels in water-stressed alfalfa. J Plant Physiol 153:706–711

Graves WR, Gallagher JL (2003) Resistance to salinity of *Alnus maritima* from distinct wetlands: symptoms of salt injury, comparison to other shrubs, and effect of inundation. Wetlands 23:394–405

Gupta A, Phung LT, Taylor DE, Silver S (2001) Diversity of silver resistance genes in IncH incompatibility group plasmids. Microbiology 147:3393–3402

Gupta DK, Rai UN, Sinha S, Tripathi RD, Nautiyal BD, Rai P, Inouhe M (2004) Role of *Rhizobium* (CA-1) inoculation in increasing growth and metal accumulation in *Cicer arietinum* L. growing under fly-ash stress condition. Bull Environ Contam Toxicol 73:424–431

Hafeez FY, Sohail Hameed A, Malik KA (1999) *Frankia* and *Rhizobium* strains as inoculum for fast growing trees in saline environment. Pak J Bot 31:173–182

Hamaoui B, Abbadi JM, Burdman S, Rashid A, Sarig S, Okon Y (2001) Effects of inoculation with *Azospirillum brasilense* on chickpeas (*Cicer arietinum* L.) and faba beans (*Vicia faba* L.) under different growth conditions. Agronomie 21:553–560

Han HS, Lee KD (2005) Physiological responses of soybean-inoculation of *Bradyrhizobium japonicum* with PGPR in saline soil conditions. Res J Agric Biol Sci 1:216–221

Hartel PG, Alexander M (1986) Role of extracellular polysaccharide production and clays in the desiccation tolerance of cowpea *Bradyrhizobia*. Soil Sci Soc Am J 50:1193–1198

He XH, Critchley C (2008) *Frankia* nodulation, mycorrhization and interactions between *Frankia* and mycorrhizal fungi in *Casuarina* plants. In: Varma A (ed) Mycorrhizae state of the art, genetics and molecular biology, eco-function, biotechnology, eco-physiology, structure and systematics, 3rd edn. Springer, Berlin/Heidelberg, pp 767–781

Hensley DL, Carpenter PL (1987) The effect of cadmium on growth and acetylene reduction in *Alnus glutinosa*. Hortic Sci 222:69–70

Jin L, Sun XW, Wang XJ, Shen YY, Hou FJ, Chang SH, Wang C (2010) Synergistic interactions of arbuscular mycorrhizal fungi and rhizobia promoted the growth of *Lathyrus sativus* under sulphate salt stress. Symbiosis 50:157–164

Karimi A, Khodaverdiloo H, Sepehri M, Sadaghiani MR (2011) AMF and heavy metal contaminated soils. Afr J Microbiol Res 5:1571–1576

Karthikeyan A, Deeparaj B, Nepolean P (2009) Reforestation in bauxite mine spoils with *Casuarina equisetifolia* frost and beneficial microbes. For Trees Livelihoods 19:153–165

Kassem M, Capellano A, Gounot AM (1985) Effets du chlorure desodium sur la croissance *in vitro*, l'infectivite' etl'efficience de *Rhizobium* meli*l*oti. Mircen J 1:63–75

Khan HR, Paull JG, Siddique KHM, Stoddard FL (2010) Faba bean breeding for drought-affected environments: a physiological and agronomic perspective. Field Crops Res 115:279–286

Khan N, Mishra A, Chauhan PS, Nautiyal CS (2011) Induction of *Paenibacillus lentimorbus* biofilm by sodium alginate and $CaCl_2$ alleviates drought stress in chickpea. Ann Appl Biol 159:372–386

Knowlton S, Dawson JO (1983) Effects of *Pseudomonas cepacia* and cultural factors on the nodulation of *Alnus* roots by *Frankia*. Can J Bot 61:2877–2882

Kouas S, Labidi N, Debez A, Abdelly C (2005) Effect of P on nodule formation and N fixation in bean. Agron Sustain Dev 25:389–393

Kumar D (2005) Breeding for drought resistance. In: Ashraf M, Harris PJC (eds) Abiotic stress: plant resistance through breeding and molecular approaches. Haworth Press, New York, pp 145–175

Lee W, Wood TK, Chen W (2006) Engineering TCE-degrading rhizobacteria for heavy metal accumulation and enhanced TCE degradation. Biotechnol Bioeng 95:399–403

Lefrancois E, Quoreshi A, Khasa D, Fung M, Whyte LG, Roy S, Greer C (2010) Fields performance of adler-*Frankia* symbionts for the reclamation of oil sands sites. Appl Soil Ecol 46:183–191

Liu Y, Zhong C, Bai J, Zhang Y, Chen J (2003) The salt resistance experiment on four clones of *Casuarina equisetifolia* in tissue culture. Guangdong For Sci Technol 19:47–50

Lorenc-Pluciñska G, Walentynowicz M, Alicja N (2013) Capabilities of alders (*Alnus incana* and *A. glutinosa*) to grow in metal-contaminated soil. Ecol Engine 58:214–227

Luard EJ, El-Lakany MH (1984) Effects of *Casuarina and Allocasuarina* species of increasing sodium chloride concentrations in solution culture. Aust J Plant Physiol 11:471–481

Lucas-Garcia JA, Probanza A, Ramos B, Colón-Flores JJ, Gutierrez-Mañéro FJ (2004) Effects of plant growth promoting rhizobacteria (PGPRs) on the biological nitrogen fixation, nodulation and growth of *Lupinus albus* I. cv. Multolupa. Eng Life Sci 4:71–77

Mahler RL, Wollum AG (1981) The influence of soil water potential and soil texture on the survival of *Rhizobium japonicum* and *Rhizobium leguminosarum* isolates in the soil. Soil Sci Soc Am J 45:761–766

Malik D, Sindhu S (2011) Production of indole acetic acid by *Pseudomonas* sp.: effect of coinoculation with *Mesorhizobium* sp. *cicer* on nodulation and plant growth of chickpea *Cicer arietinum*. Physiol Mol Biol Plant 17:25–32

Manchanda G, Garg N (2008) Salinity and its effects on the functional biology of legumes. Acta Physiol Plant 30:595–618

Marulanda A, Barea J-M, Azcón R (2009) Stimulation of plant growth and drought tolerance by native microorganisms (AM fungi and bacteria) from dry environments: mechanisms related to bacterial effectiveness. J Plant Growth Regul 28:115–124

Masson-Boivin C, Giraud E, Perret X, Batut J (2009) Establishing nitrogen-fixing symbiosis with legumes: how many *Rhizobium* recipes? Trends Microbiol 17:458–466

Mnasri B, Aouani ME, Mhamdi R (2007) Nodulation and growth of common bean (*Phaseolus vulgaris*) under water deficiency. Soil Biol Biochem 39:1744–1750

Moreno LS, Maiti RK, Gonzales AN, Star JV, Foroughbakhch R, Gonzales HG (2000) Genotypic variability in bean cultivars (*Phaseolus vulgaris* L.) for resistance to salinity at the seedling stage. Ind Agric 44:1–12

Muthukumar T, Udaiyan K (2010) Growth response and nutrient utilization of *Casuarina equisetifolia* seedlings inoculated with bioinoculants under tropical nursery conditions. New For 40:101–118

Nambiar-Veetil M, Nair DN, Selvakesavan RK, Jayaraj RSC, Roopesh M, Prabhu SJ, Balasubramanian A, Venkatachalam R, Sowmiyarani KS, Sudha S, Bachpai VW, Franche C, Gherbi H, Svistoonoff S, Hocher V, Bogusz D, Vivekanandan R (2011) Development of an in silico gene bank for plant abiotic stresses: towards its utilization for molecular analysis of salt tolerant and susceptible Casuarina equisetifolia clones. In: Zhong C, Pinyopusarerk K, Kalinganire A, Franche C (eds) Improving smallholder livelihoods through improved casuarinas productivity. China Forestry Publishing House, Beijing, pp 144–151

Nonnoi F, Chinnaswamy A, de la Torre VSG, de la Peña TC, Lucas MM, Pueyo JJ (2012) Metal tolerance of rhizobial strains isolated from nodules of herbaceous legumes (*Medicago* spp. and *Trifolium* spp.) growing in mercury-contaminated soils. Appl Soil Ecol 61:49–59

Oliveira RS, Castro PML, Dodd JC, Vosátka M (2005) Synergistic effect of *Glomus intraradices* and *Frankia* spp. on the growth and stress recovery of *Alnus glutinosa* in an alkaline anthropogenic sediment. Chemosphere 60:1462–1470

Olivera M, Tejera N, Iribarne C, Ocaña A, Lluch C (2004) Growth, nitrogen fixation and ammonium assimilation in common bean (*Phaseolus vulgaris*): effect of phosphorus. Physiol Plant 121:498–505

Orfanoudakis M, Wheeler CT, Hooker JE (2010) Both the arbuscular Mycorrhizal fungus *Gigaspora rosea* and *Frankia* increase root system branching and reduce root hair frequency in *Alnus glutinosa*. Mycorrhiza 20:117–126

Oshone R, Mansour SR, Tisa LS (2013) Effect of salt stress on the physiology of *Frankia* sp strain CcI6. J Biosci 38:699–702

Oufdou K, Benidire L, Lyubenova L, Daoui K, Fatemi ZA, Schröder P (2014) Enzymes of the glutathione-ascorbate cycle in leaves and roots of rhizobia-inoculated faba bean plants (*Vicia faba* L.) under salinity stress. Eur J Soil Biol 60:98–103

Padmaja K, Prasad DDK, Prasad ARK (1990) Inhibition of chlorophyll synthesis in *Phaseolus vulgaris* seedlings by cadmium acetate. Photosynthetica 24: 399–405

Pang J, Tibbett M, Denton MD, Lambers H, Siddique KHM, Bolland MDA, Revell CK, Ryan MH (2010) Variation in seedling growth of 11 perennial legumes in response to phosphorus supply. Plant Soil 328:133–143

Pawlowski K, Sirrenberg A (2003) Symbiosis between *Frankia* and actinorhizal plants: root nodules of non-legumes. Ind J Exp Biol 41:1165–1183

Pérez-Palacios P, Agostini E, Ibañez SG, Talano MA, Caviedes MA, Rodriguez-Llorente ID, Pajuelo E (2013) Copper-roots: engineering transgenic tobacco hairy root for extreme Cu accumulation. Poster presentation at VIII Encuentro Latinoamericano y del Caribe de Biotecnología REDBIO, Argentina

Pimentel C, Laffray D, Louguet P (1999) Intrinsic water use efficiency at the pollination stage as a parameter for drought tolerance selection in *Phaseolus vulgaris*. Physiol Plant 106:184–198

Porcel R, Barea JM, Ruiz-Lozano JM (2003) Antioxidant activities in mycorrhizal soybean plants under drought stress and their possible relationship to the process of nodule senescence. New Phytol 157:135–143

Probanza A, Lucas JA, Acero N, Gutierrez Mañero FJ (1996) The influence of native rhizobacteria on European alder (*Alnus glutinosa* (L.) Gaertn.) growth. Plant Soil 182:59–66

Probanza A, Acero N, Ramos B, Gutierrez Mañero FJ (1997) Effects of European alder *Alnus glutinosa* L. Gaertn rhizobacteria on nodular metabolism and root development. Plant Growth Regul 22:145–149

Rai R (1983) The salt tolerance of *Rhizobium leguminosarum* strains and lentil (*Lens esculenta*) genotypes and the effect of salinity on aspects of symbiotic nitrogen fixation. J Agric Sci 100:81–86

Rajwar A, Sahgal M, Johri BN (2013) Legume–rhizobia symbiosis and interactions in agroecosystems. In: Arora NK (ed) Plant microbe symbiosis: fundamentals and advances. Springer, New Delhi, pp 233–265

Reddell P, Bowen GD, Robson AD (1986a) Nodulation of *Casuarina*ceae in relation to host species and soil properties. Aust J Bot 34:435–444

Reddell P, Foster RC, Bowen GD (1986b) The effects of sodium chloride on growth and nitrogen fixation in *Casuarina obesa* Miq. New Phytol 102:397–408

Remans R, Croonenborghs A, Torres Gutierrez R, Michiels J, Vanderleyden J (2007) Effects of plant growth-promoting rhizobacteria on nodulation of *Phaseolus vulgaris* L. are dependent on plant P nutrition. Eur J Plant Pathol 119:341–351

Remans R, Ramaekers L, Schelkens S, Hernandez G, Garcia A, Reyes J, Mendez N, Toscano V, Mulling M, Galvez L, Vanderleyden J (2008) Effect of *Rhizobium-Azospirillum* coinoculation on nitrogen fixation and yield of two contrasting *Phaseolus vulgaris* L. genotypes cultivated across different environments in Cuba. Plant Soil 312:25–37

Rengasamy P (2006) World salinization with emphasis on Australia. J Exp Bot 57:1017–1023

Requena N, Jimenez I, Toro M, Barea JM (1997) Interactions between plant-growth promoting rhizobacteria PGPR, arbuscular mycorrhizal fungi and *Rhizobium* spp. in the rhizosphere of *Anthyllis cytisoides*, a model legume for revegetation in Mediterranean semi-arid ecosystems. New Phytol 136:667–677

Richards JW, Krumholz GD, Chval MS, Tisa LS (2002) Heavy metal resistance patterns of *Frankia* strains. Appl Environ Microbiol 68:923–927

Rodelas B, González-López J, Martínez-Toledo MV, Pozo C, Salmerón V (1999) Influence of *Rhizobium/Azotobacter* and *Rhizobium/Azospirillum* combined inoculation on mineral composition of faba bean (*Vicia faba* L.). Biol Fertil Soils 29:165–169

Rodríguez-Llorente ID, Dary M, Gamane D, El Hamdaoui A, Doukkali B, Lafuente A, Delgadillo J, Caviedes MA, Pajuelo E (2010) Cadmium biosorption properties of the metal resistant *Ochrobactrum cytisi* Azn6.2. Eng Life Sci 10:49–56

Rojas NS, Perry DA, Li CY, Ganio LM (2002) Interactions among soil biology, nutrition, and performance of actinorhizal plant species in the H. J. Andrews Experimental Forest of Oregon. Appl Soil Ecol 19:13–26

Romdhane SB, Aouani ME, Trabelsi M, de Lajudie P, Mhamdi R (2008) Selection of high nitrogen-fixing rhizobia nodulating chickpea (*Cicer arietinum*) for semi-arid Tunisia. J Agron Crop Sci 194:413–420

Romdhane SB, Trabelsi M, Aouani ME, de Lajudie P, Mhamdi R (2009) The diversity of rhizobia nodulating chickpea (*Cicer arietinum*) under water deficiency as a source of more efficient inoculants. Soil Biol Biochem 41:2568–2572

Roy S, Khasa DP, Greer CV (2007) Combining alders, frankiae, and mycorrhizae for the revegetation and remediation of contaminated ecosystems. Can J Bot 85:237–251

Ruiz-Lozano JM (2003) Arbuscular mycorrhizal symbiosis and alleviation of osmotic stress: new perspectives for molecular studies. Mycorrhiza 13:309–317

Ruiz-Lozano JM, Azcón R (1993) Specificity and functional compatibility of VA mycorrhizal endophytes in association with *Bradyrhizobium* strains in *Cicer arietinum*. Symbiosis 15:217–226

Ruiz-Lozano JM, Collados C, Barea JM, Azcón R (2001) Arbuscular mycorrhizal symbiosis can alleviate drought induced nodule senescence in soybean plants. New Phytol 151:493–502

Sadiki M, Rabih K (2001) Selection of chickpea (*Cicer arietinum*) for yield and symbiotic nitrogen fixation ability under salt stress. Agronomie 21:659–666

Sadowsky MJ (2005) Soil stress factors influencing symbiotic nitrogen fixation. In: Werner D, Newton WE (eds) Nitrogen fixation in agriculture, forestry, ecology, and the environment. Springer, Dordrecht, pp 113–141

Sadowsky MJ, Graham PH (1998) Soil biology of the *Rhizobiaceae*. In: Spaink HP, Kondorosi A, Hooykaas PJJ (eds) The rhizobiaceae. Kluwer Academic Publishers, Dordrecht, pp 155–172

Safronova VI, Stepanok VV, Engqvist GL, Alekseyev YV, Belimov AA (2006) Root-associated bacteria containing 1-aminocyclopropane-1-carboxylate deaminase improve growth and nutrient uptake by pea genotypes cultivated in cadmium supplemented soil. Biol Fertil Soils 42:267–272

Sandalio LM, Dalurzo HC, Gomez M, Romero-Puertas MC, del Rio LA (2001) Cadmium-induced changes in the growth and oxidative metabolism of pea plants. J Exp Bot 52:2115–2126

Sande E, Young DR (1992) Effect of sodium chloride on growth and nitrogenase activity in seedlings of *Myrica cerifera* L. New Phytol 120:345–350

Sanginga N, Danso SKA, Bowen GD (1989) Nodulation and growth-response of *Allocasuarina* and *Casuarina* species to phosphorus fertilization. Plant Soil 118:125–132

Saravanakumar D, Samiyappan R (2007) ACC deaminase from *Pseudomonas fluorescens* mediated saline resistance in groundnut (*Arachis hypogea*) plants. J Appl Microbiol 102:1283–1292

Saravanakumar D, Kavino M, Raguchander T, Subbian P, Samiyappan R (2012) Plant growth promoting bacteria enhance water stress resistance in green gram plants. Acta Physiol Plant 33:203–209

Sayed WF (2011) Improving *Casuarina* growth and symbiosis with *Frankia* under different soil environmental conditions-review. Folia Microbiol 56:1–9

Sayed WF, Wheeler CT, Zahran HH, Shoreit AAM (1997) Effect of temperature and soil moisture on the survival and symbiotic effectiveness of *Frankia* spp. Biol Fertil Soils 25:349–353

Sellstedt A, Huss-Danell K (1986) Biomass production and nitrogen utilization by *Alnus incana* when grown on N_2 or NH_4^+ made available at the same rate. Planta 167:387–394

Selvakumar G, Panneerselvam P, Ganeshamurthy AN (2012) Bacterial mediated alleviation of abiotic stress in crops. In: Maheshwari DK (ed) Bacteria in agrobiology: stress management. Springer, Berlin/Heidelberg, pp 205–224

Sempavalan J, Wheeler CT, Hooker JE (1995) Lack of competition between *Frankia* and *Glomus* for infection and colonization of roots of *Casuarina equisetifolia* (L). New Phytol 130:429–436

Shaharoona B, Arshad M, Zahir ZA (2006) Effect of plant growth promoting rhizobacteria containing ACC-deaminase on maize (*Zea mays* L.) growth under axenic conditions and on nodulation in mung bean (*Vigna radiata* L.). Lett Appl Microbiol 42:155–159

Shamseldin A, Moawad H (2010) Inhibition of nitrogenase enzyme and completely suppression of nodulation in common bean (*Phaseolus vulgaris* L.) at high levels of available nitrogen. Am-Eur J Agric Environ Sci 7:75–79

Shannon MC (1998) Adaptation of plants to salinity. Adv Agron 60:75–119

Sharma A, Johria BN, Sharma AK, Glick BR (2003) Plant growth-promoting bacterium *Pseudomonas* sp. strain GRP3 influences iron acquisition in mung bean (*Vigna radiata* L. Wilczek). Soil Biol Biochem 35:887–894

Sharma G, Sharma R, Sharma E (2010) Impact of altitudinal gradients on energetics and efficiencies of N_2-fixation in alder-cardamom agroforestry systems of the eastern Himalayas. Ecol Res 25:1–12

Sinclair TR, Purcell LC, Vadez V, Serraj R (2001) Selection of soybean (*Glycine max*) lines for increased tolerance of N_2 fixation to drying soil. Agronomie 21:653–657

Smith LT, Pocard JA, Bernard T, Le Rudulier D (1988) Osmotic control of glycine betaine biosynthesis and degradation in *Rhizobium meliloti*. J Bacteriol 170:3142–3149

Solans M (2007) *Discaria trinervis-Frankia* symbiosis promotion by saprophytic actinomycetes. J Basic Microbiol 47:243–250

Solans M, Vobis G, Cassán F, Luna V, Wall LG (2011) Production of phytohormones by root-associated saprophytic actinomycetes isolated from the actinorhizal plant *Ochetophila trinervis*. World J Microbiol Biotechnol 27:2195–2202

Soussi M, Lluch C, Ocana A (1999) Comparative study of nitrogen fixation and carbon metabolism in two chick-pea (*Cicer arietinum* L.) cultivars under salt stress. J Exp Bot 50:1701–1708

Srivastava A, Singh SS, Mishra AK (2012) Sodium transport and mechanism(s) of sodium tolerance in *Frankia* strains. J Basic Microbiol 52:1–2

Sundström KR, Huss-Danell K (1987) Effects of water stress on nitrogenase activity in *Alnus incana*. Physiol Plant 70:342–348

Tang C, Hinsinger P, Jaillard B, Rengel Z, Drevon JJ (2001) Effect of phosphorous deficiency on the growth, symbiotic N_2 fixation and proton release by two bean (*Phaseolus vulgaris*) genotypes. Agronomie 21:683–689

Tang C, Drevon JJ, Jaillard B, Souche G, Hinsinger P (2004) Proton release of two genotypes of bean (*Phaseolus vulgaris* L.) as affected by N nutrition and P deficiency. Plant Soil 260:59–68

Tani C, Sasakawa H (2003) Salt tolerance of *Casuarina equisetifolia* and *Frankia* Ceql strain isolated from the root nodules of *C. equisetifolia*. Soil Sci Plant Nutr 49:215–222

Tani C, Sasakawa H (2006) Proline accumulates in *Casuarina equisetifolia* seedlings under salt stress. Soil Sci Plant Nutr 52:21–25

Terpolilli JJ, Hood GA, Poole PS (2012) What determines the efficiency of N_2-fixing *Rhizobium*-legume symbioses? Adv Microb Physiol 60:325–389

Theodorou C, Reddell P (1991) *In vitro* synthesis of ectomycorrhizas on *Casuarinaceae* with range of mycorrhizal fungi. New Phytol 118:279–288

Tiffany ML, Williams RG (2005) Nitrogen inhibits nodulation and reversibly suppress nitrogen fixation in nodules of *Alnus maritimus*. J Am Soc Sci 130:496–499

Toro M, Azcón R, Barea JM (1998) The use of isotopic dilution techniques to evaluate the interactive effects of *Rhizobium* genotype, mycorrhizal fungi, phosphate-solubilizing rhizobacteria and rock phosphate on nitrogen and phosphorus acquisition by *Medicago sativa*. New Phytol 138:265–273

Tripathi M, Munot H, Shouche Y, Meyer JM, Goel R (2005) Isolation and functional characterization of siderophore-producing lead- and cadmium-resistant *Pseudomonas putida* KNP9. Curr Microbiol 50:233–237

Trujillo ME, Kroppenstedt RM, Schumann P, Carro L, Martnez-Molina E (2006) *Micromonospora coriariae* sp nov., isolated from root nodules of *Coriaria myrtifolia*. Int J Syst Evol Microbiol 56:2381–2385

Trujillo ME, Alonso-Vega P, Rodríguez R, Carro L, Cerda E, Alonso P, Martínez-Molina E (2010) The genus *Micromonospora* is widespread in legume root nodules: the example of *Lupinus angustifolius*. ISME J 4:1265–1281

Uehlein N, Fileschi K, Eckert M, Bienert G, Bertl A, Kaldenhoff R (2007) Arbuscular mycorrhizal symbiosis and plant aquaporin expression. Phytochemistry 68:122–129

Uliassi DD, Huss-Danell K, Ruess RW, Doran K (2000) Biomass allocation and nitrogenase activity in *Alnus tenuifolia*: responses to successional soil type and phosphorus availability. Ecoscience 7:73–79

Valdenegro M, Barea JM, Azcón R (2001) Influence of arbuscular-mycorrhizal fungi, *Rhizobium meliloti* strains and PGPR inoculation on the growth of *Medicago arborea* used as model legume for re-vegetation and biological reactivation in a semi-arid Mediterranean area. Plant Growth Regul 34:233–240

Valdés M, Pérez NM, Santo PE D l, Mellado JC, Cabrielas JJP, Normand P, Hirsh M (2005) Non-*Frankia* actinomycetes isolated from surface sterilized roots of *Casuarina equisetifolia* fix nitrogen. Appl Environ Microbiol 71:460–466

Valverde A, Burgos A, Fiscella T, Rivas R, Velazquez E, Rodrìguez-Barrueco C, Cervantes E, Chamber M, Igual JM (2006) Differential effects of coinoculations with *Pseudomonas jessenii* PS06 (a phosphate-solubilizing bacterium) and *Mesorhizobium ciceri*

C-2/2 strains on the growth and seed yield of *chickpea* under greenhouse and field conditions. Plant Soil 287:43–50

Vance CP (2001) Symbiotic nitrogen fixation and phosphorus acquisition: plant nutrition in a world of declining renewable resources. Plant Physiol 127:390–397

Vasanthakrishna M, Bagyaraj JD, Nirmalnath PJ (1994) Responses of *Casuarina equisetifolia* to inoculation with *Glomus fasciculatum* and/or *Frankia*. For Ecol Manag 68:399–402

Visser S, Danielsson RM, Parkinson D (1991) Field performance of *Elaeagnus commutata* and *Shepherdia canadensis* (Elaeagnaceae) inoculated with soil containing *Frankia* and vesicular arbuscular mycorrhizal fungi. Can J Bot 69:1321–1328

Vivas A, Azcòn R, Biro B, Barea JM, Ruiz-Lozano JM (2003a) Influence of bacterial strains isolated from lead-polluted soil and their interactions with arbuscular mycorrhizae on the growth of *Trifolium pratense* L. under lead toxicity. Can J Microbiol 49:577–588

Vivas A, Biro B, Campos E, Barea JM, Azcòn R (2003b) Symbiotic efficiency of autochthonous arbuscular mycorrhizal fungus (*G. mossae*) and *Brevibacillus brevis* isolated from cadmium polluted soil under increasing cadmium levels. Environ Pollut 126:179–189

Vivas A, Barea JM, Biro B, Azcon R (2006a) Effectiveness of autochthonous bacterium and mycorrhizal fungus on *Trifolium* growth, symbiotic development and soil enzymatic activities in Zn contaminated soil. J Appl Microbiol 100:587–598

Vivas A, Biro B, Nemeth T, Barea JM, Azcòn R (2006b) Nickel-tolerant *Brevibacillus brevis* and arbuscular mycorrhizal fungus can reduce metal acquisition and nickel toxicity effects in plant growing in nickel supplemented soil. Soil Biol Biochem 38:2694–2704

Wall LG, Hellsten A, Huss-Danell K (2000) Nitrogen, phosphorous, and the ratio between them affect nodulation in *Alnus incana* and *Trifolium pratense*. Symbiosis 29:91–105

Wani PA, Khan MS, Zaidi A (2007a) Co-inoculation of nitrogen-fixing and phosphate-solubilizing bacteria to promote growth, yield and nutrient uptake in chickpea. Acta Agron Hung 55:315–323

Wani PA, Khan MS, Zaidi A (2007b) Effect of metal tolerant plant growth promoting *Bradyrhizobium* sp. (vigna) on growth, symbiosis, seed yield and metal uptake by green gram plants. Chemosphere 70:36–45

Wani PA, Khan MS, Zaidi A (2008) Effect of metal-tolerant plant growth-promoting *Rhizobium* on the performance of pea grown in metal-amended soil. Arch Environ Contam Toxicol 55:33–42

Wheeler CT, Hughes LT, Oldroyd J, Pulford ID (2001) Effects of nickel on *Frankia* and its symbiosis with *Alnus glutinosa* (L.) Gaertn. Plant Soil 231:81–90

Whitbread-Abrutat PH (1997) The potential of some soil amendments to improve tree growth on metalliferous mine wastes. Plant Soil 192:199–217

Yamanaka T, Li CY, Bormann BT, Okabe H (2003) Tripartite associations in an alder: effects of *Frankia* and *Alpova diplophloeus* on the growth, nitrogen fixation and mineral acquisition of *Alnus tenuifolia*. Plant Soil 254:179–186

Yang Y (1995) The effect of phosphorus on nodule formation and function in the *Casuarina-Frankia* symbiosis. Plant Soil 118:125–132

Yang J, Kloepper JW, Ryu CM (2009) Rhizosphere bacteria help plants tolerate abiotic stress. Trends Plant Sci 14:1–4

Young DR, Sande E, Peters G (1992) Spatial relationships of *Frankia* and *Myrica cerifera* on a Virginia, USA Barrier Island. Symbiosis 12:209–220

Yuan BC, Li ZZ, Liu H, Gao M, Zhang YY (2007) Microbial biomass and activity in salt affected soils under arid conditions. Appl Soil Ecol 35:319–328

Zahir ZA, Munir A, Asghar HN, Shahroona B, Arshad M (2008) Effectiveness of rhizobacteria containing ACC-deaminase for growth promotion of peas (*Pisum sativum*) under drought conditions. J Microbiol Biotechnol 18:958–963

Zahran HH (1991) Conditions for successful *Rhizobium*–legume symbiosis in saline environments. Biol Fertil Soils 12:73–80

Zahran HH (1999) *Rhizobium*–legume symbiosis and nitrogen fixation under severe conditions and in arid climate. Microbiol Mol Biol Rev 12:968–998

Zahran HH (2001) Rhizobia from wild legumes: diversity, taxonomy, ecology, nitrogen fixation and biotechnology. J Biotechnol 91:143–153

Zahran HH, Sprent JI (1986) Effects of sodium chloride and polyethylene glycol on root hair infection and nodulation of *Vicia faba* L. plants by *Rhizobium leguminosarum*. Planta 167:303–309

Zahran HH, Rasanen LA, Karsisto M, Lindstrom K (1994) Alteration of lipopolysaccharide and protein profiles in SDS PAGE of rhizobia by osmotic and heat stress. World J Microbiol 10:100–105

Zhang F, Dashti N, Hynes RK, Smith DL (1996) Plant-growth promoting rhizobacteria and soybean (*Glycine max* [L.] Merr.) nodulation and nitrogen fixation at suboptimal root zone temperatures. Ann Bot 77:453–459

Zhang Y, Zhong CL, Chen Y, Chen Z, Jiang QB, Wu C, Pinyopusarerk K (2010) Improving drought tolerance of *Casuarina equisetifolia* seedlings by arbuscular mycorrhizas under glasshouse conditions. New For 40:261–271

Zhong C, Zhang Y, Chen Y, Jiang Q, Chen Z, Liang J, Pinyopusarerk K, Franche C, Bogusz D (2010) *Casuarina* research and applications in China. Symbiosis 50:107–114

Azospirillum: A Biofertilizer for Every Crop

15

Samina Mehnaz

Abstract

Azospirillum is known for its nitrogen-fixing and phytohormone production ability. It is one of very well-studied plant growth-promoting rhizobacteria, at lab scale to field. None of its species or strain is reported as human or plant pathogen. It is considered as safest bacteria which can be used as a biofertilizer at commercial level for several crops, especially cereals or grasses including wheat and rice which are of economic importance for the whole world. Some of its species are reported for phosphate-solubilizing ability and high salt tolerance. Fifteen of its species have been isolated from variety of hosts and environmental sources; however, a majority have been reported from plants. There are several reviews available on this organism; in this chapter, an overview of this organism covering its plant growth-promoting abilities, used as inoculum in lab and field experiments and used as a commercial biofertilizer for different crops, is provided.

Characteristic Features of Genus *Azospirillum* and Its Species

Azospirillum is a gram-negative curve rod, microaerophilic, non-fermentative, and a chemoorganotroph that belongs to class *Alphaproteobacteria*. This bacterium was first time isolated, reported, and named in 1978 (Tarrand et al. 1978). As the name reflects, it was described as nitrogen-fixing, curved rod-shaped organism. Two species, *A. lipoferum* and *A. brasilense*, were designated at that time. Since then, 15 species of this genus have been described. For a long time, only five species were known, but during the last 10–12 years, ten new species have been described. *Azospirillum* is known to be an organism which has been isolated from most of the parts of the world, mainly from rhizosphere and roots of the members of the family Gramineae including rice, wheat, corn, and sugarcane as most common hosts. Coffee, fruits, and orchids have also been reported as host of this bacterium but less frequently as compared to grasses. Two new species *A. picis* and *A. thiophilum* had been isolated from very different environment, i.e.,

S. Mehnaz (✉)
Department of Biological Sciences, Forman Christian College (A Chartered University), Ferozepur Road, Lahore 54600, Pakistan
e-mail: saminamehnaz@fccollege.edu.pk; saminamehnaz@gmail.com

N.K. Arora (ed.), *Plant Microbes Symbiosis: Applied Facets*,
DOI 10.1007/978-81-322-2068-8_15, © Springer India 2015

road tar and sulfide spring (Lin et al. 2009; Lavrinenko et al. 2010).

Azospirillum was first time described as a nitrogen fixer – ability to convert atmospheric nitrogen into ammonia. Later on, the second most common characteristic feature was phytohormone production, specifically auxins. Among 15 species, *A. palatum* is an exception, lacking the ability to fix nitrogen, as reduction of acetylene (method for the detection of nitrogen fixation) was not detected by it (Zhou et al. 2009). The production of indoles is not reported for *A. picis* and *A. palatum.* The difference in carbon source utilization is helpful for distinguishing among species. Most common carbon sources are sugars and sugar alcohols. Reis et al. (2011) provided a very comprehensive detail on their carbon source utilization.

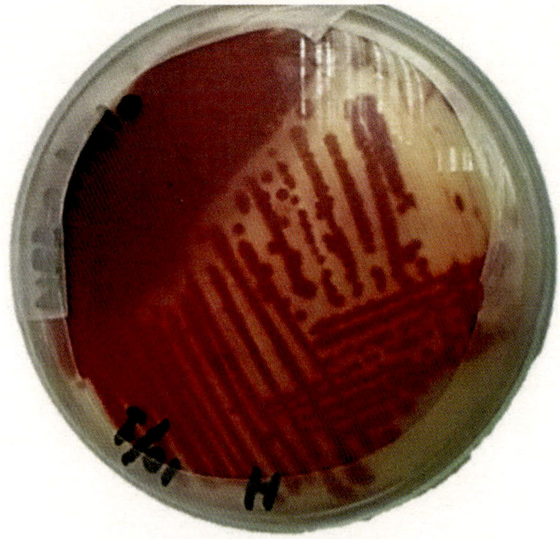

Fig. 15.1 Pure culture of *Azospirillum halopraeferens* NFP9 on LB medium (Mehnaz et al. unpublished results)

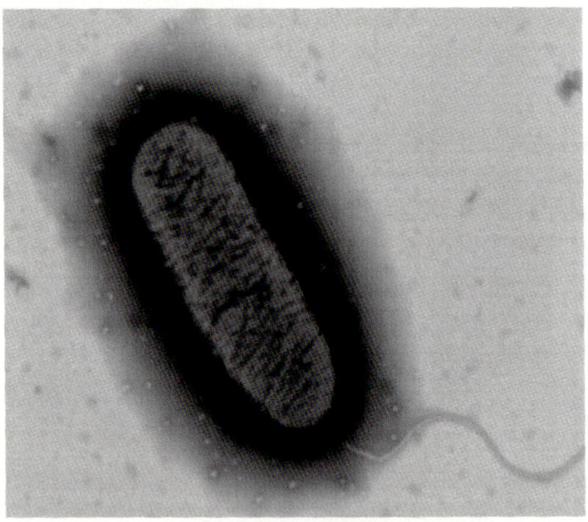

Fig. 15.2 Single cell of *Azospirillum zeae* N7 (Mehnaz et al. 2007a)

During the last 3 years, genomes of *A. lipoferum* 4B (www.genoscope.cns.fr), *A. brasilense* CBG497, *A. brasilense* Sp245, *A. amazonense* Y2, and *Azospirillum* sp. B510 have been sequenced and published (Kaneko et al. 2010; Sant'Anna et al. 2011; Wisniewski-Dye et al. 2012). These sequences provided the insight into genetics of this genus. This information can be used not only to improve the strains but also to explore their potential for purposes other than agriculture, e.g., bioremediation. A comprehensive review about genetic and biochemical traits of *Azospirillum* has been published by Steenhoudt and Vanderleyden (2000). Recently, Reis et al. (2011) discussed about the genetic makeup of *Azospirillum*, mostly based on genome of *Azospirillum* sp. B510. In addition to *nif* gene cluster, *nod* gene has also been identified in *Azospirillum* spp. Genes for indole acetic acid (IAA) production and those involved in chemotaxis signal transduction are also reported. According to information based on *Azospirillum* sp. B510 genome, this genus has nine sets of *rrns* genes which is atypical for the members of α-proteobacteria. Quorum sensing is a very rare feature for this genus. Boyer et al. (2008) reported the N-acyl-L-homoserine lactone (AHL) production in two strains of *A. lipoferum*.

If we look at the published material about this genus, most of the information is based on two species, i.e., *A. brasilense* and *A. lipoferum*. It can be due to worldwide occurrence of these two as compared to the other species. *A. irakense* has not been reported from anywhere after its first isolation. Reports for *A. amazonense* and *A. halopraeferans* are also rare (recently authors' lab has isolated a strain of *A. halopraeferans* from *Kochia indica*; Fig. 15.1, unpublished results). Two of the species, *A. canadense* and *A. zeae*, have been isolated and designated by our group (Mehnaz et al. 2007a, b; Fig. 15.2) and not reported yet from anywhere else. Similarly, the rest of the species have been recently designated;

therefore, commenting on their occurrence is too early to discuss.

Growth-Promoting Abilities of *Azospirillum*

Among plant growth-promoting rhizobacteria (PGPR), *Azospirillum* is one of those bacteria which had never been reported as a pathogen for any living organism. It has been used as an inoculum for several crops due to its beneficial effects. These beneficial effects can be an individual or a combined effect of several mechanisms which exist in this bacterium. *Azospirillum* is known for nitrogen fixation, phytohormone production, phosphate solubilization, and siderophore production. It was never considered as a biocontrol agent although there were always reports available about its anti-pathogenic activity. ACC deaminase activity has not been reported for this genus, and quorum sensing is very rare as mentioned earlier. The growth-promoting attributes of *Azospirillum* are described below in detail.

Nitrogen Fixation

The most important characteristic feature of *Azospirillum* is its ability of biological nitrogen fixation (BNF). Nitrogen fixation is common among bacteria responsible for symbiotic relationship with leguminous plants. These bacteria are known for their endophytic nature and nodule formation and inability to fix nitrogen "in vitro." In contrast, *Azospirillum* is known as a "free-living" nitrogen fixer that does not make nodules. It stays in close association with the plant, in the rhizosphere, and on the root surface.

The most common method to detect the nitrogen fixation in bacteria is acetylene reduction assay. Nitrogenase enzyme responsible for nitrogen fixation has also the ability to convert the ethylene into acetylene. Acetylene reduction assay is based on it. N^{15} dilution method has also been used by several researchers and considered very authentic when it comes to quantifying the contribution of these microorganisms to the plants in terms of nitrogen and minimizing the use of N fertilizers. There is no doubt about the nitrogen-fixing ability of *Azospirillum*, but how much of the fixed nitrogen is transferred to the plants is still a big question. Transfer process is very slow, and a very small part becomes available to the plant. The death and subsequent mineralization of dead nitrogen fixers can release significant amount of fixed nitrogen (Okumura et al. 2013).

Phytohormone Production

Almost 80 % of soil bacteria have the ability to produce phytohormones, and *Azospirillum* is one of them. Phytohormones are divided into five classes, i.e., auxins, gibberellins, cytokinins, abscisic acid, and ethylene. *Azospirillum* is known for the production of all of these except ethylene.

Auxins

Auxins are responsible for division, extension, and differentiation of plant cells and tissues. These compounds increase the rate of xylem and root formation; control vegetative growth, tropism, and florescence; and affect photosynthesis, pigment formation, biosynthesis of several metabolites, and resistance to biotic stress (Bashan and de-Bashan 2010). Among auxins, IAA production is very high in this genus, and it follows more than one pathway.

IAA production by using tryptophan is known since 1979, for *Azospirillum* (Reynders and Vlassak 1979). IAA production depends on composition of medium, pH, and availability of tryptophan as precursor (Ona et al. 2003; Malhotra and Srivastava 2006, 2008). Spaepen et al. (2007) characterized IAA production via indole-3-acetamide (IAM) and indole-3-pyruvate (IPyA) intermediates, in *Azospirillum*. Very low level of pyridoxine and nicotinic acid increased IAA production in *A. brasilense*, indicating that vitamins may also regulate IAA synthesis (Zakharova et al. 2000). IAA is produced during all growth stages of *Azospirillum* and most after the stationary phase (Malhotra and Srivastava 2009). Due to this feature, credit of growth promotion by

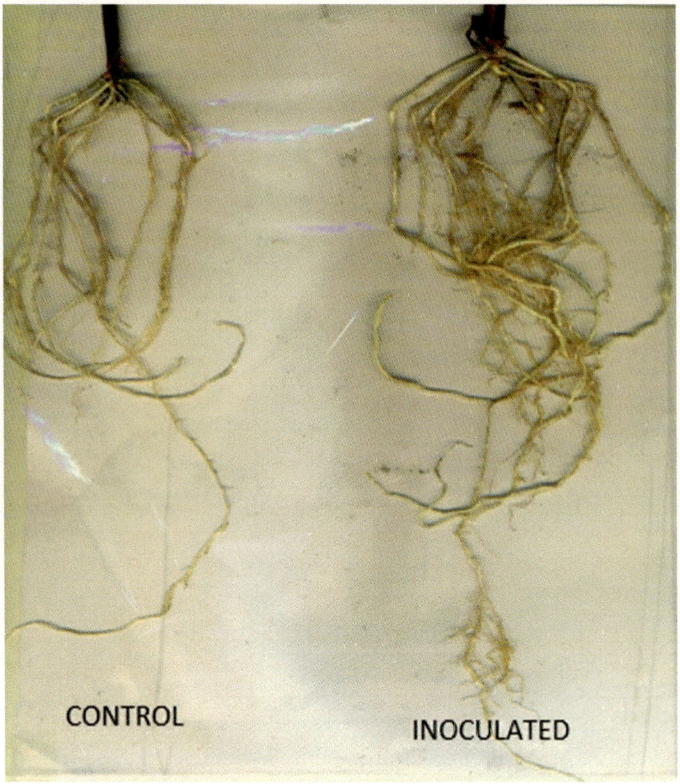

Fig. 15.3 Effect of *Azospirillum* sp. inoculation on root growth of corn plants (Mehnaz et al. unpublished results)

Azospirillum inoculation is given to IAA as effect lasts longer. Several auxins including indole butyric acid, indole lactic acid, indole acetamide, indole acetaldehyde, indole ethanol, indole methanol, and phenyl acetic acid (PAA) have been detected in *Azospirillum* cultures (Hartmann et al. 1983; Crozier et al. 1988; Fallik et al. 1989; Costacurta et al. 1994; Somers et al. 2005).

Plants inoculated with *Azospirillum* strains showed increase in number and length of root hairs, number of lateral roots, diameter and length of lateral and adventitious roots, and root surface area. IAA produced by *Azospirillum* is considered as the major mechanism responsible for increase in root system (Fig. 15.3; Mehnaz et al., unpublished results). Another possibility is that inoculation enhances the endogenous level of IAA in plant. The best way to prove the direct involvement of IAA produced by *Azospirillum* on plant growth is to develop IAA-deficient mutant and use them as inoculant, but due to existence of more than one pathway, it is almost impossible to make a mutant which does not produce IAA (Zakharova et al. 1999; Spaepen et al. 2007).

Gibberellic Acids (GAs)

GAs are responsible for cell division, cell elongation, and effects on different growth stages of plants. GAs are involved in breaking dormancy during seed germination. These compounds induce synthesis of α-amylase which hydrolyzes starch into glucose. *A. brasilense* is known to enhance germination of soybean and wheat seeds (Bacilio et al. 2003; Cassan et al. 2009a). It is reported that improved seed germination coincides with high GA production in *A. brasilense* cultures (Cassan et al. 2009b).

Azospirillum synthesizes and metabolizes GAs in vitro and in planta. The production of different GA compounds and metabolism of exogenously applied GA are reported for different *Azospirillum* species (Reis et al. 2011). Unlike IAA production, the mechanism of GA production is not known. However, dwarfism induced in maize plant by inhibitors of GA biosynthesis was reverted by inoculation with *A. lipoferum* and *A. brasilense* (Lucangeli and Bottini 1997). Recently, Manivannan and Tholkappian (2013) isolated *Azospirillum* strains from tomato rhizosphere and estimated GA production from 3.3 to 9.5 μg/25 ml broth. GA_3 is the main GA identified in *Azospirillum*. Piccoli et al. (1999) reported the effect of water potential on GA_3 production. Water potential and oxygen concentration affect GA_3 production in *A. lipoferum;* production is reduced at high water potential and low O_2 concentration. Lucangeli and Bottini (1997) suggested the involvement of GA_3 produced by *Azospirillum* in growth promotion of maize.

Cytokinins

Cytokinins are derived from N-substituted amino-purines. In plants, these are produced in root tips and germinating seeds. These are transported in shoots to regulate cell division, shoot and root morphogenesis, chloroplast maturation, cell enlargement, leaf expansion, auxiliary bud release, and delaying senescence (Spaepen et al. 2009). Auxin is known to regulate the biosynthesis of cytokinin, and ratio of auxin to cytokinin is crucial during cell division and differentiation of plant tissues. Cytokinin production in defined culture medium has been reported for *Azospirillum* by several researchers (Tien et al. 1979; Horemans et al. 1986; Cacciari et al. 1989; Strzelczyk et al. 1994). Inoculation effect of these strains on plant is based on the balance between both the hormones produced by both the partners (Reis et al. 2011). Some *Bradyrhizobium* strains use cytokinins for nodulation (Giraud et al. 2007). *Azospirillum* being a cytokinin producer might support nodulation when co-inoculated with rhizobia as also reported earlier (Yahalom et al. 1990).

Abscisic Acid (ABA)

The main role of ABA is as a phytohormone produced in response to environmental stress, e.g., decreased soil water potential, heat, and water or salt stress. ABA produced in roots is translocated to leaves and alters osmotic potential of stomata guard cells leading to closing of stomata, preventing further water loss at the time of low water availability (Bartels and Sunkar 2005). ABA production by *A. brasilense* Sp245 was reported by Cohen et al. (2008), but like GA compounds, production mechanism is not known.

Polyamines

Polyamines are low molecular weight organic compounds containing two or more primary amino groups. These compounds are synthesized in cell, but their exact function is not clear. However, if synthesis of these compounds is inhibited, cell growth is stopped or inhibited. Among polyamines, cadaverine is known to promote root growth and control stomata activity in plants (Bashan and de-Bashan 2010). *A. brasilense* Az39, a common inoculant for wheat and maize in Argentina, is known to produce polyamines including spermidine, spermine, putrescine in culture, and cadaverine in growth medium supplemented with precursor L-lysine and in rice plants inoculated with this strain (Thuler et al. 2003; Perrig et al. 2007). Bashan et al. (2004) proposed cadaverine as a contributing factor in whole plant response due to *Azospirillum* inoculation.

Ethylene

Ethylene is known for breaking seed dormancy and inhibition of root elongation if produced in higher amount. Ethylene level is lowered to prevent economic loss in agriculture and can be considered as having potential for promoting plant growth (Glick et al. 1999). 1-Aminocyclopropane-1-carboxylic acid (ACC) is a common precursor of ethylene, and ACC deaminase degrades this precursor. Although ACC deaminase is very common among PGPR, wild strains of *Azospirillum* spp. do not have this enzyme; nevertheless, some strains can produce ethylene (Perrig et al. 2007). However, ACC deaminase gene (*acd* S) was recently detected in *A. lipoferum* 4B, and a homolog of *acd* S gene has been identified in *Azospirillum* B510 genome (Pringent-Combaret et al. 2008; Kaneko et al. 2010).

Nitrogen Oxide (NO)

NO functions as a signal molecule in IAA-induced signaling cascade leading to adventitious root development (Pagnussat et al. 2002, 2003).

The role of NO in lateral root formation and root hair development is demonstrated by Correa-Aragunde et al. (2004) and Lombardo et al. (2006). *Azospirillum* is known to produce NO at low oxygen pressure by denitrification, since a long time ago (Hartmann and Zimmer 1994). Creus et al. (2005) quantified the production of 6.4 nmol NO/g of *A. brasilense* growing under aerobic conditions.

Phosphate Solubilization

Microbes have the ability to solubilize insoluble phosphate by producing enzymes such as phytases and phosphatases or organic acids. Enzymes help to release organic phosphorus, and organic acids release inorganic phosphorous. Around 40 % of culturable bacteria from soil have this ability (Reis et al. 2011). It is not common among species of *Azospirillum*. However, *Azospirillum* spp. can produce different organic acids that assist in phosphorus solubilization depending on sugars in root exudates. There are few reports available about *A. halopraeferans*, *A. lipoferum*, and *A. brasilense* on this. Gluconic acid production and pH reduction were observed in *A. lipoferum* JA4. For *A. brasilense* strains, pH reduction and gluconic acid production were observed when carbon source of the medium was replaced by glucose (Rodriguez et al. 2004). *A. halopraeferans* solubilized inorganic phosphorous (Seshadri et al. 2000). A recent report about phosphate solubilization by *A. brasilense* WS-1 was published by Tahir et al. (2013).

Siderophore Production

Siderophores are low molecular weight iron chelating compounds which help beneficial bacteria to take away the ferric ion and make it unavailable to pathogenic bacteria. It is considered as indirect growth mechanism by which PGPR promote plant growth. Catechol-type siderophore production by *A. lipoferum* and *A. brasilense* was reported earlier by Shah et al. (1992) and Tapia-Hernandez et al. (1990). Recently, *Azospirillum* strains isolated from tomato rhizosphere had shown the production of catechol-type siderophores (Manivannan and Tholkappian 2013). Strains were not identified up to species level in this study. There are several reports for the siderophore production by *A. lipoferum* and *A. brasilense*. Boyer et al. (2008) detected siderophores in two strains TVV3 and B518 of *A. lipoferum*. Pedraza et al. (2007) isolated *A. brasilense* strains from strawberry plants and detected siderophore production by these strains. According to their observation, the amount of siderophore production has a correlation with its area of isolation, i.e., rhizosphere or within the plants. Endophytic strains produced higher amount.

Antimicrobial Activity

Oliveira and Drozdowicz (1987) reported bacteriocin production by *Azospirillum*. Somers et al. (2005) detected the production of Phenyl acetic acid (PAA) – an auxin-like molecule with antimicrobial activity, in culture supernatant, when phenylalanine was added as precursor. Tapia-Hernandez et al. (1990) isolated bacteriocin-producing *Azospirillum* strains that inhibited the growth of several other bacteria. In vitro HCN production by *Azospirillum* strains isolated from Brazil has been reported by Goncalves and de Oliveira (1998). Siderophores also exhibit antimicrobial activity against various bacterial and fungal isolates (Shah et al. 1992).

Azospirillum as a Biofertilizer

Plant Growth Promotion Due to Nitrogen Fixation

The oldest concept about plant growth promotion due to *Azospirillum* inoculation is based on its nitrogen-fixing ability. Although there are doubts about transfer of this fixed nitrogen to the plant, still several reports are available to support the contribution through BNF. A recent review by Okumura et al. (2013) discussed about BNF data for *Azospirillum*. The review is focused on five crops and BNF contribution to these plants. Some of the examples selected from this review are described here. Brasil et al. (2005) inoculated a mixture of *A. brasilense* and *A. lipoferum* to *Brachiaria* plants and noticed higher root, shoot dry weights, and nitrogen content as compared to uninoculated plants. Kennedy et al. (2004) mentioned 26 % increase in corn productivity under field conditions. Saikia et al. (2007) suggested that high yield of *Azospirillum*-inoculated plants is due to higher photosynthesis and stomatal conductance in these plants. Okon and Vanderleyden (1997) reviewed 22-year field inoculation data of *Azospirillum* and concluded that this organism promotes crop productivity under different climates and soil conditions and goes beyond increase in BNF as increase in absorptive surface of root system was also observed. Sala et al. (2007) reported 26 % increase in wheat grain production when inoculated with *Azospirillum*. Diaz-Zorita and Fernandez-Canigia (2009) used liquid-based inoculants of *A. brasilense* for wheat, from 2002 to 2006, at 297 localities in Argentine Pampas region. The increase in productivity was 260 kg/ha and 70 % of grain yield. Several researchers are convinced that a combination of nitrogen fertilizer with *Azospirillum* inoculation is beneficial for wheat plants as inoculation promotes better absorption and utilization of available nitrogen. Dose of fertilizer recommended for combination varies from 15 to 160 kg N/ha (Saubidet et al. 2002; Dalla Santa et al. 2004; Sala et al. 2005).

Many studies showed that the contribution of nitrogen fixation by *Azospirillum* to the plant ranged from 5 to 18 % of total nitrogen increase in plant (Bashan and de-Bashan 2010). However, conclusive proof of the nitrogen contribution is based on ^{15}N dilution technique. Garcia de Salamone et al. (1997) used ^{15}N isotope dilution technique to determine the contribution of BNF to two genotypes of maize, inoculated with a mixture of *Azospirillum* strains. Significant increase in grain yield and total nitrogen content of both genotypes was observed. Saubidet and Barnex (1998) inoculated five cultivars of wheat with most efficient nitrogen-fixing strains of *A. brasilense*. Increase in growth and nitrogen content was observed, but response varied among cultivars. Didonet et al. (2003) inoculated ten lines of *Oryza sativa* with *A. brasilense* and *A. lipoferum* and reported that response depends upon cultivar and strains used. Mehnaz and Lazarovits (2006) reported significant increase in root and shoot dry weight of four corn varieties, inoculated with *A. zeae* (misidentified at that time as *A. lipoferum*); however, response varied with all varieties. Although *Azospirillum* is considered as a nonspecific PGPR with respect to host, there are several studies indicating that response varies among varieties and cultivars of the plants. Rodrigues et al. (2008) inoculated rice plants with *A. amazonense* and observed increased grain dry matter and nitrogen accumulation at maturation. Contribution from nitrogen fixation was up to 27 %. The comprehensive review by Bashan and de-Bashan (2010) based on the analysis of 35-year data suggested that even though there are controversial reports about nitrogen contribution to the plant due to nitrogen fixation by *Azospirillum*, the mechanism cannot be ignored.

Plant Growth Promotion due to Phytohormone Production

When plants are inoculated with bacteria, root system is the first plant part which shows response. Inoculation increases number and length of root hairs, number of lateral roots, diameter and length of lateral roots and adventitious roots, and root surface area. The increased root system can improve plant nutrition and in general is responsible for plant growth promotion. Change in the root system is considered due to the production of auxins, and *Azospirillum* is well known for these compounds. El-Khawas and Adachi (1999) inoculated rice plants growing in hydroponic system, with filter-sterilized supernatants of *Azospirillum* culture. Increase in root dry weight, surface area, lateral roots, root hairs, and root elongation as compared to uninoculated plants was observed. Cell-free supernatant of *A. brasilense* Cd was applied to soybean plants, and increase in root numbers and length was observed (Molla et al. 2001). In contrast, wheat inoculation with *A. brasilense* Sp 245 and Sp7 decreased root length; however, root hair formation increased (Dobbelaere et al. 1999). This type of effect on root system is very common when higher concentrations of IAA are applied exogenously. Remans et al. (2008) proved that response of exogenously applied IAA on bean roots resembled to *Azospirillum* inoculation. Malhotra and Srivastava (2006) engineered *A. brasilense* SM strain and made it to produce threefold higher IAA as compared to wild-type strain. *Sorghum* plants inoculated with this strain showed higher effect on lateral branching and dry weight of roots when compared with plants inoculated with wild type. Although there are several reports of plant growth promotion due to IAA production by *Azospirillum*, considering it as a single mechanism responsible for growth promotion needs more evidences.

Cohen et al. (2009) inoculated *A. lipoferum* in maize plants which were unable to produce ABA and GA, due to application of inhibitors of biosynthetic pathways of these compounds. It causes the effect similar to drought; however, inoculation with *A. lipoferum* reversed the effect which suggested the participation of ABA and GA in alleviating water stress in the presence of the bacteria. So far, several researchers are convinced that water stress alleviation of plants by *Azospirillum* involves the production of ABA, auxins, and GAs (Costacurta and Vanderleyden 1995; Bottini et al. 2004; Cohen et al. 2008).

Plant Growth Promotion Due to Phosphate Solubilization

Among 15 species of *Azospirillum*, *A. halopraeferens* can solubilize insoluble inorganic phosphate in vitro by unknown mechanism as it does not use glucose, and therefore does not produce gluconic acid (Seshadri et al. 2000). Some strains of *A. brasilense* and *A. lipoferum* are capable of producing gluconic acid and can solubilize rock phosphate (Puente et al. 2004; Rodriguez et al. 2004). Chang and Li (1998) isolated *Azospirillum* strains from ectomycorrhizal sporocarp which degraded limestone, marble, and calcium phosphate in vitro. Carrillo et al. (2002) inoculated cardon (cactus) with *A. brasilense* Cd strain and reported enhanced phosphate solubilization and plant growth. Deubel et al. (2000) inoculated pea plants with *Azospirillum* strains and observed a shift in sugars under phosphorous deficiency which increased the capability of *Azospirillum* spp. to mobilize phosphate. Sugars such as glucose are part of pea root exudates grown in phosphorous-deficient soils, thus increasing the capacity of *Azospirillum* spp. to solubilize insoluble $Ca_3(PO_4)_2$. It was observed that under phosphorous deficiency glucose content decreased, while galactose, ribose, xylose, and fructose increased in root exudates.

Plant Growth Promotion by Minimizing the Stresses

The use of *Azospirillum* in addition to other PGPR to alleviate the salt stress and minimize the negative effects of several plant stresses including water stress has been recommended by several researchers (Bacilio et al. 2004; Bashan et al. 2004; Lucy et al. 2004). Salt resistance among *Azospirillum* species increased from *A. amazonense* (lowest) to *A. halopraeferans* (highest), as the latter tolerates over 3 % that is equivalent to seawater salinity (Bashan and de-Bashan 2010). *A. brasilense*-inoculated wheat plants showed a significant increase in shoot elongation, fresh and dry weights, water content, and relative water content in shoots, when grown under 320 mM NaCl and 20 or 30 % polyethylene glycol (PEG) 6000 causing salt or osmotic stress, respectively (Creus et al. 1997). Barassi et al. (2006) used *A. brasilense* Sp245 to inoculate lettuce seeds and observed the effect of salt stress. Higher total fresh and dry weights were recorded as compared to uninoculated plants. Inoculated dried seeds maintained the ability to mitigate the effect of salt on lettuce germination. The same strain alleviated the salt stress in carrot and chickpea as well (Hamaoui et al. 2001; Ayrault 2002). Rueda-Puente et al. (2004) suggested the use of *Klebsiella pneumoniae* and *A. halopraeferens* as auxiliary biofertilizers to improve the growth of *Salicornia bigelovii* (halophyte) in coastal semiarid zones. Later on, the same group reported that yield loss due to stress diminished in *Azospirillum*-inoculated wheat plants, and grains showed 38.4, 22.2, and 125 % higher Mg, K, and Ca, respectively, as compared to uninoculated ones (Creus et al. 2004).

Inoculation with *Azospirillum* improved growth under water stress conditions, which was initially demonstrated in the 1980s (Bashan and Levanony 1990). Okon (1985) stressed in a review about the importance of *Azospirillum* in improving water status of the plants. It suggests that *Azospirillum* inoculation can protect the crops against drought. Sarig et al. (1990) inoculated *Sorghum bicolor* plants to osmotic stress in hydroponic system and observed that due to *A. brasilense* inoculation adverse effects such as reduction in leaf senescence were diminished. Corn plants inoculated with *Azospirillum* showed less effect of water stress on plant growth, as increase in root growth, total aerial biomass, foliar area, and proline content in leaves and roots was observed (Casanovas et al. 2002). *Azospirillum* inoculation alleviated the stress on wheat plants grown under drought conditions (El-Komy et al. 2003). Creus et al. (2004) inoculated wheat seeds with *A. brasilense* Sp245 at flowering stage and subjected the plants to drought. Significantly higher water content, relative water content, water potential, apoplastic water fraction, lower cell wall modulus of elasticity values, reduction in grain yield loss, and significantly higher Mg, K, and Ca in grains were observed in inoculated plants. Moutia et al. (2010) inoculated two cultivars of sugarcane with *Azospirillum* sp. strains under drought and non-drought conditions. Both cultivars showed different results as one showed 15 % increase in shoot height and 75 % increase in dry weight under drought, whereas the other cultivar showed negative results under non-drought conditions. Kasim et al. (2013) used *A. brasilense* NO40 to prime the wheat seeds and grew them under drought conditions. Priming significantly alleviated the deleterious effect of drought stress on wheat. Inoculated plants showed attenuated transcript level suggesting improved homeostatic mechanism due to priming.

Rodriguez-Salazar et al. (2009) used an engineered strain of *A. brasilense* with the ability to produce trehalose (osmotic regulating sugar), more salt tolerant than wild type, to inoculate maize plants grown under stress. Inoculation significantly enhanced survival of these plants under water stress. According to Bashan and de-Bashan (2010), limitations regarding the effect of inoculation under saline stress are valid for osmotic stress.

There are few reports about decreasing the toxic effect of metals on plants due to *Azospirillum* inoculation. Cadmium (Cd) inhibits the growth and nutrient uptake in barley plants. *A. lipoferum-inoculated* barley plants showed decreased toxicity to Cd, enhanced root length and biomass, and significant increase in amount of nutrients absorbed (Belimov and Dietz 2000; Belimov et al. 2004). Lyubun et al. (2006) observed less accumulation of arsenic in *A. brasilense* Sp245-inoculated wheat plants as compared to uninoculated plants. The association of *Azospirillum* changed the speciation, bioavailability, and uptake of arsenic by plants.

There are individual reports about positive effect of *Azospirillum* inoculation on plants growing under different stresses such as pH, toxic substances, high light intensities, humic acid, etc. High pH decreases the microalgal population. *A. brasilense* inoculation in microalgal culture eliminated the negative effect (de-Bashan et al. 2005). High level of tryptophan reduced multiplication of *Chlorella vulgaris*, but co-culturing with *A. brasilense* significantly reduced the inhibition (de-Bashan and Bashan 2008). Sometimes light intensity has negative effect on crops. Wheat inoculation with *A. brasilense* Cd significantly increased quantity of photosynthetic pigments chlorophyll a, b and auxiliary photo-protective pigments violaxanthin, zeaxanthin, antheraxanthin, lutein, neoxanthin, and ß-carotene that help the plant to sustain photosynthesis under unfavorable light conditions (Bashan et al. 2006). Hernandez et al. (2006) used *A. brasilense* to inoculate *Chlorella* spp. that is used for wastewater treatment and observed enhanced growth and metabolism of the microalgae. Bacilio et al. (2003) inoculated wheat seeds with *A. brasilense* or *A. lipoferum* and observed improved seed germination and plant growth in soils amended with compost containing high concentration of humic acid. Both bacterial species survived in high humic acid solutions and modified compost composition "in vitro."

Considering the reports about *Azospirillum* effect on plant growth under different stresses, it is clear that several known and unknown mechanisms are involved in the interaction of this organism with plants. Based on these findings, Cassan et al. (2009b) and Sgroy et al. (2009) proposed a term "plant stress homeo-regulating bacteria (PSHRB)" for all rhizosphere PGPR/PGPB including *Azospirillum* that regulate homeostasis in plants under abiotic stresses.

Miscellaneous Effects of *Azospirillum* on Plants

Azospirillum inoculation in maize plants led to major quantitative and qualitative modification of secondary metabolites especially benzoxazinoids (Walker et al. 2011). *Azospirillum* produces nitric oxide (NO) which participates in metabolic, signaling, defense, and developmental pathways in plants and can also modify root architecture (Molina-Favero et al. 2007, 2008). *Azospirillum* can produce nitrite as part of its normal metabolism which also participates in plant growth promotion (Zimmer et al. 1988). Nitrate reductase (NR) activity of wheat leaves decreased by *Azospirillum* inoculation and increased in nitrate assimilation was also reported (Ferreira et al. 1987). Some of the lectins produced by *Azospirillum* induced changes in mitotic state of growing onion plant cells (Nikitina et al. 2004). Cell surface lectins isolated from *A. brasilense* affected activities of α-glucosidase, ß-glucosidase, and ß-galactosidase in the membrane and apoplast fractions of roots of wheat seedlings (Alenkina et al. 2006). *A. brasilense* Cd inoculation enhanced the proton efflux of wheat root (Bashan 1990). Inoculation of soybean seedlings with the same strain significantly reduced the membrane potential in every root part (Bashan 1991; Bashan and Levanony 1991).

Bashan et al. (1992) proved the effect of *A. brasilense* on root cell membrane of soybean and cowpea plants. Pereyra et al. (2006) showed that fatty acid composition of main phospholipids in roots is affected by *A. brasilense* inoculation. *A. lipoferum* 34H increased the phosphate ion content, resulting in significant increase in root length and fresh and dry weight of shoots of rice plants (Rodriguez and Fraga 1999). During dry storage, seeds deteriorate and germination gets delayed. Lettuce seeds inoculated with *A. brasilense* along or after an osmo-priming treatment could mitigate the negative effect of ageing (Carrozzi 2005).

Azospirillum as a Biocontrol Agent

The role of *Azospirillum* as a biocontrol agent has always been ignored, although reports for its anti-pathogenic activity were available since a long time ago. Mechanisms such as environmental competition, displacement of pathogens, inhibition of seed germination of parasitic weeds, inhibition of fungal growth via the production of toxic substances, and growth enhancement of plants to resist pathogen are suggested for *Azospirillum* to play its role as a biocontrol agent (Bashan and de-Bashan 2010).

As an Antimicrobial Agent

When *A. brasilense* Cd was co-cultured with pathogenic *Staphylococcus* spp., population of pathogenic bacteria got reduced (Holguin and Bashan 1996). Mixed cultures of *A. brasilense* and tomato pathogen *Pseudomonas syringae* were inoculated on tomato seeds. Significant decrease in severity of disease and improved plant growth was observed as *P. syringae* did not survive in the presence of *A. brasilense* (Bashan and de-Bashan 2002). *A. brasilense* reduced the dry weight of mycelium of four soilborne fungal pathogens of cucumber, i.e., *Fusarium* (90–96 %), *Rhizoctonia* (72–94 %), *Pythium* (71–95 %), and *Sclerotinia* (100 %), during an in vitro study (Hassouna et al. 1998). Sudhakar et al. (2000)

applied *A. brasilense* as foliar spray against foliar fungal and bacterial diseases of mulberry. Inoculation reduced fungal pathogens and excelled as a treatment against bacterial blight. Co-inoculation of *Azospirillum* spp. with arbuscular mycorrhizal (AM) suppressed damping off disease in *Capsicum* sp., caused by *Pythium aphanidermatum* (Kavitha et al. 2003). Tortora et al. (2011) isolated *A. brasilense* strains with the ability to produce catechol-type siderophores including salicylic acid. Inoculation of these strains reduced the anthracnose disease caused by *Colletotrichum acutatum* in strawberry plants.

As a Pesticide

Inoculation of *A. brasilense* to *Sorghum* resulted in tenfold decrease of dead heart disease caused by a shoot fly *Atherigona soccata* and also resulted in increase in grain yield (Kishore 1998). Only a few root galls and egg masses per root system, of mung bean plants infected with root-knot nematode, were observed when inoculated with *A. lipoferum* (Khan and Kounsar 2000). The increase in pod yield and reduction in root-knot nematode egg masses and total nematode population was observed in Okra plants, inoculated with *A. brasilense* (Ramakrishnan et al. 1997). A combination of mycorrhizal fungi, phosphate-solubilizing bacteria, and *Azospirillum* reduced the population of *Pratylenchus zeae* (nematode) and increased the cob yield of maize (Babu et al. 1998). *A. lipoferum* inoculation in wheat plants reduced *Heterodera avenae* (nematode) infection (Bansal et al. 1999).

As a Herbicide

Azospirillum spp. inhibited germination of a parasitic weed, *Striga hermonthica* (witch weed), of *Sorghum* fields (Bouillant et al. 1997). Lipophilic compounds produced by *A. brasilense* in culture medium inhibited the germination of *Striga* seeds (Miché et al. 2000). *A. brasilense* inhibited germination and radicle growth of *Orobanche aegyptiaca* (Egyptian broomrape) seeds, a parasitic weed of sunflower (Dadon et al. 2004).

As an Elicitor of Induced Systemic Resistance

The role of rhizobacterial exopolysaccharides (EPS) as elicitor of induced systemic resistance (ISR) was reported by Kyungseok et al. (2008). Recently, Sankari et al. (2011) used different concentrations of EPS produced by *Azospirillum* strains as inoculants of rice plants. EPS-inoculated plants showed better growth and reduction in blast disease caused by *Pyricularia oryzae*. *Azospirillum* sp. B510 induced disease resistance in rice plants against *Xanthomonas oryzae* and fungus *Magnaporthe oryzae*. Resistance was due to a novel mechanism independent of salicylic acid-mediated defense (Yasuda et al. 2009).

Azospirillum as a Commercial Biofertilizer

Azospirillum is capable of colonizing vegetables, cereals, and several other plants and induces beneficial effects as it congregates several plant growth-promoting mechanisms which make it a valuable candidate as a biofertilizer. For more than 40 years, researchers are using *Azospirillum* as an inoculant in lab, pots, and field experiments. Several reviews are available which have covered the 20–35-year data based on growth promotion by different strains of this genus. All of the researchers and reviewers agree on one point that this bacterium is beneficial for plant, and there is not a single report about its harmful effect. However, there is also an agreement that plant growth promotion is not due to any single mechanism; rather, it is a combination of several mechanisms possessed by *Azospirillum*, and in a recent review Bashan and de-Bashan (2010) proposed it as "multiple mechanisms theory." According to this theory "there is no single mechanism involved in promotion of plant growth by *Azospirillum* but a combination of a few or many mechanisms in each case of inoculation." These may vary according to plant species, *Azospirillum* strain, and environmental conditions when interaction occurred. Actually, possession of several mechanisms makes this organism one of the best candidates for bioinoculants.

Bottini R, Cassan F, Piccoli P (2004) Gibberellin production by bacteria and its involvement in plant growth promotion and yield increase. Appl Microbiol Biotechnol 65:497–503

Bouillant ML, Miché L, Ouedraogo O, Alexandre G, Jacoud C, Sallé G, Bally R (1997) Inhibition of *Striga* seed germination associated with sorghum growth promotion by soil bacteria. C R Acad Sci Paris-Sciences de la vie 320:159–162

Boyer M, Bally R, Perrotto S, Chaintreuil C, Wisniewski-Dye F (2008) A quorum quenching approach to identify quorum-sensing regulated functions in *Azospirillum lipoferum*. Res Microbiol 20:72–77

Brasil MS, Baldani VLD, Baldani JI, Souto SM (2005) Effects of inoculation of diazotrophs in grasses Pantanal. Pasturas Tropicales 27:22–33

Cacciari I, Lippi D, Pietrosanti T, Pietrosanti W (1989) Phytohormone-like substances produced by single and mixed diazotrophic cultures of *Azospirillum* spp. and *Arthrobacter*. Plant Soil 115:151–153

Carrillo AE, Li CY, Bashan Y (2002) Increased acidification in the rhizosphere of cactus seedlings induced by *Azospirillum brasilense*. Naturwissenschaften 89:428–432

Carrozzi LE (2005) *Lactuca sativa* (L.) seed priming and *Azospirillum* inoculations a tool for improving germination rate. MSc dissertation, University of Mar del Plata, Argentina

Casanovas EM, Barassi CA, Sueldo RJ (2002) *Azospirillum* inoculation mitigates water stress effects in maize seedlings. Cer Res Commun 30:343–350

Cassan F, Maiale S, Masciarelli O, Vidal A, Luna V, Ruiz O (2009a) Cadaverine production by *Azospirillum brasilense* and its possible role in plant growth promotion and osmotic stress mitigation. Eur J Soil Biol 45:12–19

Cassan F, Perrig D, Sgroy V, Masciarelli O, Penna C, Luna V (2009b) *Azospirillum brasilense* Az39 and *Bradyrhizobium japonicum* E109, inoculated singly or in combination, promote seed germination and early seedling growth in corn (*Zea mays* L.) and soybean (*Glycine max* L.). Eur J Soil Biol 45:28–35

Chang TT, Li CY (1998) Weathering of limestone, marble, and calcium phosphate by ecto-mycorrhizal fungal and associated microorganisms. Taiwan J Sci 13:8590

Cohen AC, Bottini R, Piccoli PN (2008) *Azospirillum brasilense* Sp 245 produces ABA in chemically-defined culture medium and increases ABA content in *Arabidopsis* plants. Plant Growth Regul 54:97–103

Cohen AC, Travaglia CN, Bottini R, Piccoli PN (2009) Participation of abscisic acid and gibberellins produced by endophytic *Azospirillum* in the alleviation of drought effects in maize. Botany 87:455–462

Correa-Aragunde N, Graziano M, Lamattina L (2004) Nitric oxide plays a central role in determining lateral root development in tomato. Planta 218:900–905

Costacurta A, Vanderleyden J (1995) Synthesis of phytohormones by plant-associated bacteria. Crit Rev Microbiol 21:1–18

Costacurta A, Keijers V, Vanderleyden J (1994) Molecular cloning and sequence analysis of an *Azospirillum brasilense* indole-3-pyruvate decarboxylase gene. Mol Gen Genet 243:463–472

Creus CM, Sueldo RJ, Barssi CA (1997) Shoot growth and water status in *Azospirillum* inoculated wheat seedlings grown under osmotic and salt stresses. Plant Physiol Biochem 35:939–944

Creus CM, Suelda RJ, Barassi CA (2004) Water relations and yield in *Azospirillum*-inoculated wheat exposed to drought in the field. Can J Bot 82:273–281

Creus CM, Graziano M, Casanovas EM, Pereyra MA, Simontacchi M, Puntarulo S, Barassi CA, Lamattina L (2005) Nitric oxide is involved in the *Azospirillum brasilense*-induced lateral root formation in tomato. Planta 221:297–303

Crozier A, Arruda P, Jasmin JM, Monteiro AM, Sandberg G (1988) Analysis of indole-3-acetic acid and related indoles in culture medium from *Azospirillum lipoferum* and *Azospirillum brasilense*. Appl Environ Microbiol 54:2833–2837

Dadon T, Bar Nun N, Mayer AM (2004) A factor from *Azospirillum brasilense* inhibits germination and radicle growth of *Orobanche aegyptiaca*. Isr J Plant Sci 52:83–86

Dalla Santa OR, Hernandez RF, Alvarez GLM, Ronzelli JP, Soccol CR (2004) *Azospirillum* sp. inoculation in wheat, barley and oats seeds greenhouse experiments. Braz Arch Biol Tech 47:843–850

de-Bashan LE, Bashan Y (2008) Joint immobilization of plant growth-promoting bacteria and green microalgae in alginate beads as an experimental model for studying plant–bacterium interactions. Appl Environ Microbiol 74:6797–6802

de-Bashan LE, Antoon H, Bashan Y (2005) Cultivation factors and population size control uptake of nitrogen by the microalgae *Chlorella vulgaris* when interacting with the microalgae growth-promoting bacterium *Azospirillum brasilense*. FEMS Microbiol Ecol 54:197–203

Deubel A, Gransee A, Merbach W (2000) Transformation of organic rhizodepositions by rhizosphere bacteria and its influence on the availability of tertiary calcium phosphate. J Plant Nutr Soil Sci 163:387–392

Diaz-Zorita M, Fernandez-Canigia MV (2009) Field performance of a liquid formulation of *Azospirillum brasilense* on dryland wheat productivity. Eur J Soil Biol 45:3–11

Didonet AD, Didonet CCGM, Gomes GF (2003) Evaluation of strains of upland rice inoculated with *Azospirillum lipoferum* Sp59b and *A. brasilense* Sp24. Comunicado Tecnico EMBRAPA, p 69

Dobbelaere S, Croonenborghs A, Thys A, Vande Broek A, Vanderleyden J (1999) Phytostimulatory effect of *Azospirillum brasilense* wild type and mutant strains altered in IAA production on wheat. Plant Soil 212:155–164

El-Khawas H, Adachi K (1999) Identification and quantification of auxins in culture media of *Azospirillum* and *Klebsiella* and their effect on rice roots. Biol Fertil Soils 28:377–381

El-Komy HM, Hamdia MA, El-Baki GKA (2003) Nitrate reductase in wheat plants grown under water stress and inoculated with *Azospirillum* spp. Biol Plant 46:281–287

Fallik E, Okon Y, Epstein E, Goldman A, Fischer M (1989) Identification and quantification of IAA and IBA in *Azospirillum brasilense* inoculated maize roots. Soil Biol Biochem 21:147–153

Ferreira MCB, Fernandes MS, Dobereiner J (1987) Role of *Azospirillum brasilense* nitrate reductase in nitrate assimilation by wheat plants. Biol Fertil Soils 4:47–53

Garcia de Salamone IE, Dobereiner J, Urquiaga S, Boddey RM (1997) Biological nitrogen fixation in *Azospirillum* strain–maize genotype associations as evaluated by the ^{15}N isotope dilution technique. Biol Fertil Soils 23:249–256

Giraud E, Moulin L, Vallenet D, Barbe V, Cytryn E, Avarre JC, Jaubert M, Simon D, Cartieaux F, Prin Y, Bena G, Hannibal L, Fardoux J, Kojadinovic M, Vuillet L, Lajus A, Cruveiller S, Rouy Z, Mangenot S, Segurens B, Dossat C, Franck WL, Chang WS, Saunders E, Bruce D, Richardson P, Normand P, Dreyfus B, Pignol D, Stacey G, Emerich D, Vermeglio A, Medigue C, Sadowsky M (2007) Legumes symbioses: absence of genes in nod in photosynthetic bradyrhizobia. Science 316:1307–1312

Glick BR, Patten CL, Holguin G, Penrose DM (1999) Biochemical and genetic mechanisms used by plant growth promoting bacteria. Imperial College Press, London, pp 125–140

Goncalves AFS, de Oliveira RGB (1998) Cyanide production by Brazilian strains of *Azospirillum*. Rev Microbiol 29:36–39

Hamaoui B, Abbadi JM, Burdman S, Rashid A, Sarig S, Okon Y (2001) Effects of inoculation with *Azospirillum brasilense* on chickpeas (*Cicer arietinum*) and faba beans (*Vicia faba*) under different growth conditions. Agronomie 21:553–560

Hartmann A, Zimmer W (1994) Physiology of *Azospirillum*. In: Okon Y (ed) *Azospirillum*/plant association. CRC Press, Boca Raton, pp 15–39

Hartmann A, Singh M, Klingmuller W (1983) Isolation and characterization of *Azospirillum* mutants excreting high amounts of indole acetic acid. Can J Microbiol 29:916–923

Hassouna MG, El-Saedy MAM, Saleh HMA (1998) Biocontrol of soil-borne plant pathogens attacking cucumber (*Cucumis sativus*) by rhizobacteria in a semiarid environment. Arid Soil Res Rehabil 12:345–357

Hernandez JP, de-Bashan LE, Bashan Y (2006) Starvation enhances phosphorus removal from wastewater by the microalga *Chlorella* spp. co-immobilized with *Azospirillum brasilense*. Enzyme Microb Technol 38:190–198

Holguin G, Bashan Y (1996) Nitrogen-fixing by *Azospirillum brasilense* Cd is promoted when co-cultured with a mangrove rhizosphere bacterium (*Staphylococcus* sp.). Soil Biol Biochem 28:1651–1660

Horemans S, De Koninck K, Neuray J, Hermans R, Vlassak K (1986) Production of plant growth substances by *Azospirillum* sp. and other rhizosphere bacteria. Symbiosis 2:341–346

Kaneko T, Minamisawa K, Isawa T, Nakatsukasa H, Mitsui H, Kawaharada Y, Nakamura Y, Watanabe A, Kawashima K, Ono A, Shimizu Y, Takahashi C, Minami C, Fujishiro T, Kohara M, Katoh M, Nakazaki N, Nakayama S, Yamada M, Tabata S, Sato S (2010) Complete genomic structure of the cultivated rice endophyte *Azospirillum* sp. B510. DNA Res 17(1):37–50

Kasim WA, Osman ME, Omar MN, Abd-Eldeim IA, Bejai S, Meijer J (2013) Control of drought stress in wheat using plant growth promoting bacteria. J Plant Growth Regul 32:122–130

Kavitha K, Meenakumari KS, Sivaprasad P (2003) Effect of dual inoculation of native arbuscular mycorrhizal fungi and *Azospirillum* on suppression of damping off in chilli. Ind Phytopathol 56:112–113

Kennedy IR, Choudhry ATMA, Kecskes ML (2004) Non-symbiotic bacterial diazotrophs in crop-farming systems: can their potential for plant growth promotion be better explored. Soil Biol Biochem 36:1229–1244

Khan MR, Kounsar K (2000) Effect of seed treatment with certain bacteria and fungi on the growth of mung bean and reproduction of *Meloidogyne incognita*. Nematol Mediterr 28:221–226

Kishore P (1998) Response of sorghum variety Pusa chari-121 to carrier based inoculants (*Azotobacter* and *Azospirillum*), fermented residue and shootfly (*Atherigona soccata Rondani*) under field conditions. J Entomol Res 22:101–105

Kyungseok P, Kloepper JW, Ryu CM (2008) Rhizobacterial exopolysaccharides elicit induced resistance on cucumber. J Microbiol Biotechnol 18:1095–1100

Lavrinenko K, Chernousova E, Gridneva E, Dubinina G, Akimov V, Kuever J, Lysenko A, Grabovich M (2010) *Azospirillum thiophilum* sp. nov., a novel diazotrophic bacterium isolated from a sulfide spring. Int J Syst Evol Microbiol 60:2832–2837

Lin SY, Young CC, Hupfer H, Siering C, arun AB, Chen WM, Lai WA, Shen FT, Rekha PD, Yasin AF (2009) *Azospirillum picis* sp. nov., isolated from discarded tar. Int J Syst Evol Microbiol 59:761–765

Lombardo MC, Graziano M, Polacco JC, Lamattina L (2006) Nitric oxide functions as a positive regulator of root hair development. Plant Signal Behav 1:28–33

Lucangeli C, Bottini R (1997) Effects of *Azospirillum* spp. on endogenous gibberellin content and growth of maize (*Zea mays* L.) treated with uniconazole. Symbiosis 23:63–72

Lucy M, Reed E, Glick BR (2004) Application of free living plant growth promoting rhizobacteria. Antonie van Leeuwenhoek Int J Gen Mol Biol 86:1–25

Lyubun YV, Fritzsche A, Chernyshova MP, Dudel EG, Fedorov EE (2006) Arsenic transformation by *Azospirillum brasilense* Sp245 in association with wheat (*Triticum aestivum* L.) roots. Plant Soil 286:219–227

Malhotra M, Srivastava S (2006) Targeted engineering of *Azospirillum brasilense* SM with indole acetamide pathway for indole acetic acid over-expression. Can J Microbiol 52:1078–1084

Malhotra M, Srivastava S (2008) An *ipdc* gene knock out of *Azospirillum brasilense* strain SM and its implications on indole 3-acetic acid biosynthesis and plant growth promotion. Antonie van Leeuwenhoek J Gen 93:425–433

Malhotra M, Srivastava S (2009) Stress responsive indole-3-acetic acid biosynthesis by *Azospirillum brasilense* SM and its ability to modulate plant growth. Eur J Soil Biol 45:73–80

Manivannan M, Tholkappian P (2013) Prevalence of *Azospirillum* isolates in tomato rhizosphere soils of coastal areas of Cuddalore District, Tamil Nadu. Int J Recent Sci Res 4:1610–1613

Mehnaz S, Lazarovits G (2006) Inoculation effects of *Pseudomonas putida, Gluconacetobacter azotocaptans* and *Azospirillum lipoferum* on corn plant growth under green house conditions. Microb Ecol 51:326–335

Mehnaz S, Weselowski B, Lazarovits G (2007a) *Azospirillum canadense* sp. nov., a nitrogen fixing bacterium isolated from corn rhizosphere. Int J Syst Evol Microbiol 57(3):620–624

Mehnaz S, Weselowski B, Lazarovits G (2007b) *Azospirillum zeae* sp. nov., diazotrophic bacteria isolated from rhizosphere soil of *Zea mays*. Int J Syst Evol Microbiol 57(12):2805–2809

Miché L, Bouillant ML, Rohr R, Sallé G, Bally R (2000) Physiological and cytological studies on the inhibition of *Striga* seed germination by the plant growth-promoting bacterium *Azospirillum brasilense*. Eur J Plant Pathol 106:347–351

Molina-Favero C, Creus CM, Lanteri ML, Correa-Aragunde N, Lombardo MC, Barassi CA, Lamattina L (2007) Nitric oxide and plant growth promoting rhizobacteria: common features influencing root growth and development. Adv Bot Res 46:1–33

Molina-Favero C, Creus CM, Simontacchi M, Puntarulo S, Lamattina L (2008) Aerobic nitric oxide production by *Azospirillum brasilense* Sp245 and its influence on root architecture in tomato. Mol Plant Microbe Interact 21:1001–1009

Molla AH, Shamsuddin ZH, Saud HM (2001) Mechanism of root growth and promotion of nodulation in vegetable soybean by *Azospirillum brasilense*. Commun Soil Sci Plant Anal 32:2177–2187

Moutia YJF, Saumtally S, Spaepen S, Vanderleyden J (2010) Plant growth promotion by *Azospirillum* sp. in sugarcane is influenced by genotype and drought stress. Plant Soil 337:233–242

Nikitina VE, Bogomolova NV, Ponomareva EG, Sokolov OI (2004) Effect of azospirilla lectins on germination capacity of seeds. Biol Bull (Moscow) 31:354–357

Okon Y (1985) *Azospirillum* as a potential inoculants for agriculture. Trends Biotechnol 3:223–228

Okon Y, Vanderleyden J (1997) Root-associated *Azospirillum* species can stimulate plants. Appl Environ Microbiol 63:366–370

Okumura RS, Mariano DC, Dallacort R, Nogueira de Albuquerque A, Lobato AKS, Guedes EMS, Neto CFO, Oliveira da Conceicao HE, Alves GAR (2013) *Azospirillum*: a new and efficient alternative to biological nitrogen fixation in grasses. J Food Agric Environ 2(1):1142–1146

Oliveira RGB, Drozdowicz A (1987) Inhibition of bacteriocin producing strains of *Azospirillum lipoferum* by their own bacteriocin. Zentralblatt fur Mikrobiologie 142:387–391

Ona O, Smets I, Gysegom P, Bernaerts K, Impe JV, Prinsen E, Vanderleyden J (2003) The effect of pH on indole-3-acetic acid biosynthesis of *Azospirillum brasilense* Sp7. Symbiosis 35:199–208

Pagnussat GC, Simontacchi M, Puntarulo S, Lamattina L (2002) Nitric oxide is required for root organogenesis. Plant Physiol 129:954–956

Pagnussat GC, Lanteri ML, Lamattina L (2003) Nitric oxide and cyclic GMP are messengers in the indole acetic acid-induced adventitious rooting process. Plant Physiol 132:1241–1248

Pedraza RO, Motok J, Tortora ML, Salazar SM, Dı'az Ricci JC (2007) Natural occurrence of *Azospirillum brasilense* in strawberry plants. Plant Soil 295:169–178

Pereyra MA, Zalazar CA, Barassi CA (2006) Root phospholipids in *Azospirillum* inoculated wheat seedlings exposed to water stress. Plant Physiol Biochem 44:873–879

Perrig D, Boiero ML, Masciarelli OA, Penna C, Ruiz OA, Cassan FD, Luna MV (2007) Plant-growth-promoting compounds produced by two agronomically important strains of *Azospirillum brasilense*, and implications for inoculant formulation. Appl Microbiol Biotechnol 75:1143–1150

Piccoli P, Masciarelli O, Bottini R (1999) Gibberellin production by *Azospirillum lipoferum* cultured in chemically defined medium as affected by oxygen availability and water status. Symbiosis 27:135–145

Prigent-Combaret C, Blaha D, Pothier JF, Vial L, Poirier MA, Wisniewski-Dyé F, Moe¨nne-Loccoz Y (2008) Physical organization and phylogenetic analysis of *acd*R as leucine-responsive regulator of the 1-aminocyclopropane-1-carboxylate deaminase gene *acd*S in phytobeneficial *Azospirillum lipoferum* 4B and other proteobacteria. FEMS Microbiol Ecol 65:202–219

Puente ME, Bashan Y, Li CY, Lebsky VK (2004) Microbial populations and activities in the rhizoplane of rock-weathering desert plants I. Root colonization and weathering of igneous rocks. Plant Biol 6:629–642

Ramakrishnan S, Gunasekaran CR, Vadivelu S (1997) Effect of bio-fertilizers *Azolla* and *Azospirillum* on root-knot nematode, *Meloidogyne incognita* and plant growth of okra. Ind J Nematol 26:127–130

Reis VM, Teixeira KRS, Pedraza RO (2011) What is expected from the genus *Azospirillum* as a plant growth promoting bacteria? In: Maheshwari DK (ed) Bacteria in agrobiology: plant growth responses. Springer, Berlin/Heidelberg, pp 123–138

Remans R, Beebe S, Blair M, Manrique G, Tovar E, Rao I, Croonenborghs A, Torres-Gutierrez R, El-Howeity M, Michiels J, Vanderleyden J (2008) Physiological and genetic analysis of root responsiveness to auxin-producing plant growth-promoting bacteria in common bean (*Phaseolus vulgaris* L.). Plant Soil 302:149–161

Reynders L, Vlassak K (1979) Conversion of tryptophan to indole acetic acid by *Azospirillum brasilense*. Soil Biol Biochem 11:547–548

Rodrigues EP, Rodrigues LS, de Oliveira ALM, Baldani VLD, Teixeira KRD, Urquiaga S, Reis VM (2008) *Azospirillum amazonense* inoculation: effects on growth, yield and N_2-fixation of rice (*Oryza sativa* L.). Plant Soil 302:249–261

Rodriguez H, Fraga R (1999) Phosphate solubilizing bacteria and their role in plant growth promotion. Biotechnol Adv 17:319–339

Rodriguez H, Gonzalez T, Goire I, Bashan Y (2004) Gluconic acid production and phosphate solubilization by the plant growth-promoting bacterium *Azospirillum* spp. Naturwissenschaften 91:552–555

Rodriguez-Salazar J, Suarez R, Caballero-Mellado J, Iturriaga G (2009) Trehalose accumulation in *Azospirillum brasilense* improves drought tolerance and biomass in maize plants. FEMS Microbiol Lett 296:52–59

Rueda-Puente EO, Castellanos T, Troyo-Dieguez E, Diaz de Leon-Avarez JL (2004) Effect of *Klebsiella pneumonia* and *Azospirillum halopraeferens* on the growth and development of two *Salicornia bigelovii* genotypes. Aust J Exp Agric 44:65–74

Saikia SP, Jain V, Khetarpal S, Aravind S (2007) Dinitrogen fixation activity of *Azospirillum brasilense* in maize (*Zea mays*). Curr Sci 93:1296–1300

Sala VMR, Freitas SS, Donzeli VP, Freitas JG, Gallo PB, Silveira APD (2005) Occurrence and effect of diazotrophic bacteria in wheat genotypes. Revista Brasileira de Ciencia do Solo 29:345–352

Sala VMR, Cardoso EJBN, Freitas JG, Silveira APD (2007) Wheat genotypes response to inoculation of diazotrophic bacteria in field conditions. Pesq Agrop Brasileira 42:833–842

Sankari JU, Dinakar S, Sekar C (2011) Dual effect of *Azospirillum* exo-polysaccharides (EPS) on the enhancement of plant growth and biocontrol of blast (*Pyricularia oryzae*) disease in upland rice (var. ASD-19). J Phytol 3(10):16–19

Sant'Anna FH, Almeida LGP, Cecagno R, Reolon LA, Siqueira FM, Machado MRS, Vasconcelos ATR, Schrank IS (2011) Genomics insight into the versatility of the plant growth promoting bacterium *Azospirillum amazonense*. BMC Genomics 12:409

Sarig S, Okon Y, Blum A (1990) Promotion of leaf area development and yield in *Sorghum bicolor* inoculated with *Azospirillum brasilense*. Symbiosis 9:235–245

Saubidet MI, Barneix AJ (1998) Growth stimulation and nitrogen supply to wheat plants inoculated with *Azospirillum brasilense*. J Plant Nutr 21:2565–2577

Saubidet MI, Fatta N, Barneix AJ (2002) The effect of inoculation with *Azospirillum brasilense* on growth and nitrogen utilization by wheat plants. Plant Soil 245:215–222

Seshadri S, Muthukumuramasamy R, Lakshiminarasami C, Ignacimuthu S (2000) Solubilization of inorganic phosphates by *Azospirillum halopraeferans*. Curr Sci 79:565–567

Sgroy V, Cassan F, Masciarelli O, Del Papa MF, Lagares A, Luna V (2009) Isolation and characterization of endophytic plant growth-promoting (PGPB) or stress homeostasis-regulating (PSHB) bacteria associated to the halophyte *Prosopis strombulifera*. Appl Microbiol Biotechnol 85:371–381

Shah S, Karkhanis V, Desai A (1992) Isolation and characterization of siderophore, with antimicrobial activity, from *Azospirillum lipoferum*. Curr Microbiol 25:347–351

Somers E, Ptacek D, Gysegom P, Srinivasan M, Vanderleyden J (2005) *Azospirillum brasilense* produces the auxin like phenylacetic acid by using the key enzyme for indole 3-acetic acid biosynthesis. Appl Environ Microbiol 71:1803–1810

Spaepen S, Vanderleyden J, Remans R (2007) Indole-3-acetic acid in microbial and microorganism-plant signaling. FEMS Microbiol Rev 31:425–448

Spaepen S, Vanderleyden J, Okon Y (2009) Plant growth promoting actions of rhizobacteria. Adv Bot Res 51:283–320

Steenhoudt O, Vanderleyden J (2000) *Azospirillum*, a free living nitrogen fixing bacterium closely associated with grasses: genetic, biochemistry and ecological aspects. FEMS Microbiol Rev 24:487–506

Strzelczyk E, Kampert M, Li CY (1994) Cytokinin-like substances and ethylene production by *Azospirillum* in media with different carbon sources. Microbiol Res 149:55–60

Sudhakar P, Chattopadhyay GN, Gangwar SK, Ghosh JK (2000) Effect of foliar application of *Azotobacter*, *Azospirillum* and *Beijerinckia* on leaf yield and quality of mulberry (*Morus alba*). J Agric Sci 134:227–234

Tahir M, Mirza MS, Zaheer A, Dimitrov MR, Smidt H, Hameed S (2013) Isolation and identification of phosphate solubilizer *Azospirillum*, *Bacillus* and *Enterobacter* strain by 16S rRNA sequence analysis and their effect on growth of wheat (*Triticum aestivum* L.). Aust J Crop Sci 7(9):1284–1292

Tapia-Hernandez A, Mascarua-Esparza M, Caballero-Mellado J (1990) Production of bacteriocins and siderophore-like activity by *Azospirillum brasilense*. Microbios 64:73–83

Tarrand JJ, Kreig NR, Dobereiner J (1978) A taxonomic study of the *Spirillum lipoferum* group with description of a new genus *Azospirillum* gen.nov., and two species, *Azospirillum lipoferum* (Beijerinck) com nov. and *Azospirillum brasilense* sp. nov. Can J Microbiol 24:967–980

Thuler DS, Floh EIS, Handro W, Barbosa HR (2003) Plant growth regulators and amino acids released by *Azospirillum* sp in chemically defined media. Lett Appl Microbiol 37:174–178

Tien TM, Gaskins MH, Hubell DH (1979) Plant growth substances produced by *Azospirillum brasilense* and their effect on the growth of pearl millet (*Pennisetum americanum* L). Appl Environ Microbiol 37:1016–1024

Tortora ML, Diaz-Ricci JC, Pedraza RO (2011) *Azospirillum brasilense* siderophores with antifungal activity against *Colletotrichum acutatum*. Arch Microbiol 193:275–286

Walker V, Bertrand C, Bellvert F, Moenne-Loccoz Y, Bally R (2011) Host plant secondary metabolite profiling shows a complex strain dependent response of maize to plant growth promoting rhizobacteria of the genus *Azospirillum*. New Phytol 189:494–506

Wisniewski-Dye F, Lozano L, Acosta-Cruz E, Borland S, Drogue B, Prigent-Combaret C, Rouy Z, Barbe V, Herrera AM, Gonzalez B, Mavingui P (2012) Genome sequence of *Azospirillum brasilense* CBG497 and comparative analysis of *Azospirillum* core and accessory genomes provide insight into niche adaptation. Genes 3:576–602

Yahalom E, Okon Y, Dovrat A (1990) Possible mode of action of *Azospirillum brasilense* strain Cd on the root morphology and nodule formation in burr medic (*Medicago polymorpha*). Can J Microbiol 36:10–14

Yasuda M, Isawa T, Shinozaki S, Minamisawa K, Nakashita H (2009) Effects of colonization of a bacterial endophyte, *Azospirillum* sp. B510, on disease resistance in rice. Biosci Biotechnol Biochem 73:2595–2599

Zakharova E, Shcherbakov A, Brudnik V, Skripko N, Bulkhin N, Ignatov V (1999) Biosynthesis of indole-3-acetic acid in *Azospirillum brasilense*. Insights from quantum chemistry. Eur J Biochem 259:572–576

Zakharova EA, Iosipenko AD, Ignatov VV (2000) Effect of water soluble vitamins on the production of indole-3-acetic acid by *Azospirillum brasilense*. Microbial Res 155:209–214

Zhou Y, Wei W, Wang X, Xu L, Lai R (2009) *Azospirillum palatum* sp. nov. isolated from forest soil in Zhejiang province, China. J Gen Appl Microbiol 55:1–7

Zimmer W, Roeben K, Bothe H (1988) An alternative explanation for plant growth promotion by bacteria of the genus *Azospirillum*. Planta 176:333–342

Ectomycorrhizal Fungi and Their Applications

16

Antonietta Mello, Elisa Zampieri, and Raffaella Balestrini

Abstract

Ectomycorrhizal (ECM) fungi form association with relatively small number of plants that dominate boreal, temperate, Mediterranean, and some subtropical forest ecosystems. These plant species have been able to acquire metabolic capabilities through symbioses with ECM fungi, thus improving their mineral nutrition and growth in several ecological niches. Mycorrhizal fungi can also play several other important ecological roles, including the protection of plants from abiotic and biotic stresses. Several "targeted" metagenomic projects have been carried out, or are now in progress, in order to identify the fungal communities in soil, including ECM fungi, which are present in various habitats (e.g., forest and truffle-ground soils, etc.). ECM fungi, which are important both because of their economic value as edible fungi (i.e., truffles, boletes) and because of their application in reforestation projects, are the subject of this chapter, in which the recent advances in ECM fungal communities are reviewed, focusing mainly on the applicative aspects related to the use of these fungi.

Introduction

Mycorrhizae are widespread symbiotic interactions between soil fungi and the roots of almost all land plants, including forest trees, grasses, and many crops (Smith and Read 2008). It has been estimated that about 90 % of terrestrial plant species undergo an improvement in mineral nutrient uptake, thanks to root symbiosis with mycorrhizal fungi (Brundrett 2009), which, in turn, provide the fungus with carbon compounds (i.e., sugars). Several mycorrhizal associations exist, and these are identified primarily on the basis of the taxonomic

A. Mello • R. Balestrini (✉)
Istituto per la Protezione Sostenibile delle Piante del CNR, c/o Dipartimento di Scienze della Vita e Biologia dei Sistemi, Università degli Studi di Torino, Viale Mattioli 25, 10125 Torino, Italy
e-mail: r.balestrini@ipp.cnr.it

E. Zampieri
Dipartimento di Scienze Agrarie, Forestali e Alimentari (DISAFA), Università degli Studi di Torino, Largo Paolo Braccini 2 (ex Via Leonardo da Vinci 44), 10095 Grugliasco (Torino), Italy

N.K. Arora (ed.), *Plant Microbes Symbiosis: Applied Facets*,
DOI 10.1007/978-81-322-2068-8_16,

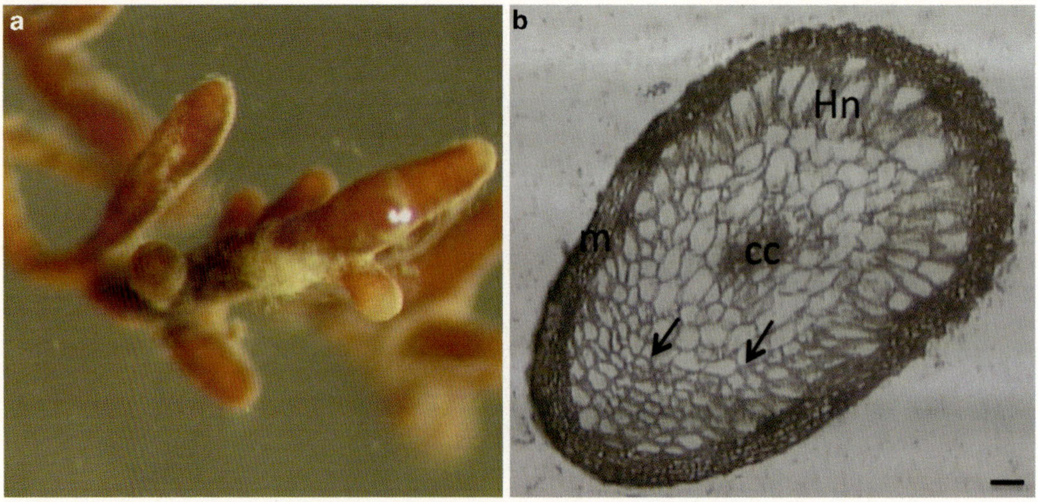

Fig. 16.1 (**a**) *Corylus avellana/Tuber melanosporum* ectomycorrhizae with the typical clavate aspect and the presence of external mycelium. (**b**) Paraffin oblique transverse section of a *C. avellana/T. melanosporum* ectomycorrhiza showing the mantle (m), which consists of several layers of hyphae, and the Hartig net proliferation (Hn and *arrows*). *cc* central cylinder. Bar corresponds to 25 μm

identity of the hosts and the structural features of the symbiotic interfaces (Smith and Read 2008). According to the ability of the mycorrhizal fungus to penetrate the root cells, mycorrhizae can be divided into two main types: endomycorrhizae and ectomycorrhizae (Balestrini et al. 2012; Perotto et al. 2013).

In ECM symbiosis, which is the subject of this review, the fungus forms a hyphal sheath, called mantle, which is made of aggregated hyphae that surround the root surface, and the Hartig net, which is formed by the hyphae that penetrate between the root cells (Balestrini et al. 2012; Fig. 16.1). The mantle is thought to be responsible for the mineral nutrition and water uptake of the symbiotic tissues, while the Hartig net is considered the site in which metabolites are exchanged. A preliminary confirmation of the functional diversity of these two fungal compartments can be found in the study on *Amanita muscaria* ECMs by Nehls et al. (2001) in which the separation of the mantle from the ECM root, using only tweezers, underlined a differential expression for two hexose-regulated fungal genes. A specificity in the mantle and Hartig net transcriptomic profiles, which reflects a functional specificity, has recently been revealed by Hacquard et al. (2013) through a laser microdissection approach, in which the two fungal compartments in truffle ECM were dissected, in combination with microarray gene expression analysis.

ECM fungi occur all over the world, and their host range includes most angiosperm and gymnosperm trees, as well as economically important timber-producing tree species. They play an important ecological role in woodland and forest communities in boreal, temperate, Mediterranean, and some subtropical forest ecosystems (Tedersoo et al. 2010a) that are dominated by hundreds of ECM fungal species (i.e., *Basidio-* and *Ascomycetes*), which form symbiotic associations with the lateral roots of trees and shrubs (Bonfante 2010). ECM symbiosis involves a relatively small number of plants, i.e., those belonging to Betulaceae, Dipterocarpaceae, Fagaceae, Nothofagaceae, Myrtaceae (e.g., *Eucalyptus*), Pinaceae, Salicaceae, and some genera of shrubs (e.g., Cistaceae), which are assisted by ECM fungi in the nutrition and protection against root diseases (Smith and Read 2008). As far as the fungal part is concerned, the majority of ECM fungi belong to *Basidiomycetes*, which can form macroscopic epigeous fruiting bodies that often grow next to tree trunks in woodlands, such as *Boletus edulis* and *Cantharellus cibarius*, while some others belong to *Ascomycetes* and form hypogeous fruiting bodies, such as truffles (Girlanda et al. 2007).

ECM fungi in the soil are highly competitive for nutrient acquisition (Bücking et al. 2012 and reference therein), and their colonization allows plant access to N forms that are not so easily available to non-mycorrhizal plants (Hobbie and Hogberg 2012 and reference therein). In line with these observations, Averill et al. (2014) and the relative commentary by Bradford (2014) have suggested that the presence of greater stores of organic matter in forest soils dominated by ECM fungi, than in those dominated by arbuscular mycorrhizal (AM) fungi, could be due to a reduced nitrogen availability for the free-living microbes that use organic matter. Interestingly, when a tripartite interaction is present, as in the *Alnus* host system (N-fixing bacteria, ECM fungi, *Alnus* roots), the ECM fungal activity seems to be shifted toward a greater capacity for organic P acquisition (Walker et al. 2013). A study about the impact of an ECM fungus (*Hebeloma cylindrosporum*) on the potassium (K^+) nutrition of its host plant (*Pinus pinaster*) has recently been performed, and the involvement of a fungal K^+ transporter has been investigated (Garcia et al. 2014). The results have shown that the K^+ nutrition of mycorrhizal pine plants was significantly improved under potassium-limiting conditions (Garcia et al. 2014).

A positive correlation between ECM symbiosis and plant performance in drought conditions has been reported (Morte et al. 2001; Dunabeitia et al. 2004; Alvarez et al. 2009), although plant response to drought can vary depending on the ECM fungal species (Dosskey et al. 1991; Kennedy and Peay 2007). More recently, a pot experiment, using *Pinus sylvestris* seedlings inoculated with several ECM fungi (*Cenococcum geophilum*, *Paxillus involutus*, *Rhizopogon roseolus*, and *Suillus granulatus*), has shown that only *S. granulatus* has a positive effect on shoot growth. Two different watering regimes (moist *versus* dry) were considered, and it was shown that *S. granulatus* effect on shoot growth was more pronounced under moist conditions (threefold increase) than under dry conditions (twofold increase), thus suggesting that the considered ECM fungus did not provide any additional support during drought stress (Kipfer et al. 2012). Danielsen and Polle (2014) have instead investigated the nutrient status and the physiological responses to drought of young poplar trees in the presence/absence of the ECM fungus *P. involutus*, and they have been shown that root tips from ECM plants have a higher vitality than those from non-mycorrhizal plants. Since this effect is evident in both the colonized and non-colonized tips of ECM plants, the authors have suggested that it could be due to a general improved water supply to the roots of the host plants.

Numerous *in situ* $^{13}CO_2$ pulse-labeling experiments have been conducted on annual crops or grasslands to demonstrate a rapid carbon flux pathway from the host to the roots and from the roots to the rhizosphere (Robin et al. 1990; Nguyen et al. 1999; Johnson et al. 2002; Leake et al. 2006). The studies on carbon allocation in trees using pulse labeling have usually been performed in microcosms or mesocosms (Norton et al. 1990; Ek 1997; Simard et al. 1997), and only a few studies have been conducted in situ with adult trees but have never considered the fruiting bodies of the associated fungi (Högberg et al. 2008; Plain et al. 2009; Subke et al. 2009; Epron et al. 2011). Recently, Le Tacon et al. (2013) have elegantly assessed the allocation of carbon by the host to *Tuber melanosporum* mycorrhizae and ascocarps via an *in situ* $^{13}CO_2$ pulse-labeling experiment performed on a 20-year-old hazel tree in a truffle orchard established in the northeast of France. Almost all of the carbon allocated to the truffle ascocarps came from the host; thus, the hypothesis that it was mainly supplied via saprotrophic pathways was excluded. The development of truffles requires that carbon is stored in the trunk or roots of the host, and the process takes several weeks/months unlike what happens in an ectomycorrhizal member of *Basidiomycotina* producing fruiting bodies over a number of days (Teramoto et al. 2012). This result, as expected, demonstrates that the processes involved in carbon acquisition and ascocarp development are different from those of basidiocarps, since *Tuber* ascocarps take at least 6 months to grow between the production of the primordia and full ascocarp development, unlike *Basidiomycota* sporocarps, which develop over a number of days directly from diploid mycorrhizae.

ECM symbiosis can therefore be considered a crucial component in nutrient cycling in sustainable forest ecosystem, and the current genome sequencing projects on ECM fungi are providing useful information to understand the functional and ecological roles for these fungi (Martin et al. 2008, 2010; Martin and Nehls 2009; Plett and Martin 2011; Martin and Bonito 2012; Balestrini et al. 2013; Marmeisse et al. 2013).

ECM fungi, which are therefore important for both their economic value (i.e., truffles, boletes) and their use in reforestation projects, are the subject of this chapter, in which the recent advances in ECM fungal communities are reviewed, focusing mainly on the applicative aspects related to the use of these fungi.

Research on ECM Fungi: From the Past to the Present

The study of ECM fungal diversity initially focused on the screening and molecular identification of fruiting bodies and, only later, on ECM tips, usually after sorting these in morphotypes (Horton and Bruns 2001; Mello et al. 2006a). As each ECM species is specialized in exploiting specific resources of the soil ecosystem, investigations have focused on the spatial distribution of the extraradical mycelium. Tracking the distribution of a given ECM fungus is considered difficult, since fruiting bodies do not reflect the distribution of ground networks (Dahlberg 2001). *H. cylindrosporum* was the first ECM fungus to be detected in soil (Guidot et al. 2002), and its presence was revealed within 50 cm from the fruiting bodies. Thanks to the use of the β-tubulin gene as a marker, Zampieri et al. (2010) were able to show that *T. magnatum* mycelium is more widespread than could be inferred from the distribution of its fruiting bodies and ECM and were able to identify a new haplotype that had never been described before from fruiting body material. The application of the denaturing gradient gel electrophoresis (DGGE) technique made it possible to discover that *T. melanosporum* is the dominant fungus in an area characterized by scanty vegetation, known as brûlé, which is associated to this fungus, and that *Basidiomycota* ECM fungi decrease within the brûlé, thus indicating a competitive effect of *T. melanosporum* on other ECM fungi (Napoli et al. 2010).

Since each individual within a species has its own functional traits, the next step toward a better understanding of the role of biodiversity in ecosystem functioning will need to consider the intraspecific diversity of mycorrhizal plants and fungi (Johnson et al. 2012).

The possibility of studying (micro)organisms directly in the field (metagenomics or environmental genomics, Chivian et al. 2008) thanks to the introduction of high-throughput sequencing techniques (i.e., 454 pyrosequencing) has given a strong impulse to the development of projects devoted to the study of fungal communities in different environments, including soils. The first studies on fungal diversity, which used a metagenomic approach in combination with high-throughput technology, appeared in 2009 (Buée et al. 2009; Jumpponen and Jones 2009). Buée and colleagues investigated fungal diversity in six different forest soils from a temperate French site using tag-encoded 454 pyrosequencing of the ITS-1 (nuclear ribosomal internal transcribed spacer-1), while Jumpponen and Jones (2009) studied the fungal communities in leaves of *Quercus macrocarpa* of trees located inside and outside a small urban center using the same molecular target. These authors have demonstrated that 454 pyrosequencing can be used successfully to study fungal communities in forest soil and phyllosphere. Starting from these first works, an approach based on 454 GS-FLX pyrosequencing has been widely used to investigate fungal communities in soils and has allowed new information to be provided on ECM fungal communities in several biomes/ecosystems, e.g., Swedish spruce plantations (Wallander et al. 2010); tropical African forests (Tedersoo et al. 2010b); truffle grounds (Mello et al. 2011); transgenic poplar plantations (soils and roots; Danielsen et al. 2012); ECM herb *Bistorta vivipara* roots on the Arctic archipelago of Svalbard (Blaalid et al. 2012, 2014); an urban landscape

(Lothamer et al. 2013); a boreal forest (Clemmensen et al. 2013); boreal and tropical forests (McGuire et al. 2013); truffle grounds, a Mediterranean agro-silvo-pastoral system, serpentine substrates, and a contaminated industrial area (Orgiazzi et al. 2013); an oak-dominated forest in Japan (Toju et al. 2013); and three microsites (decayed wood, mineral soil adjacent to intact logs, control mineral soil) in mature spruce forests in British Columbia (Walker et al. 2014).

Global climate change is supposed to influence soil fungal communities, including ECM fungal communities. It has already been demonstrated that several ecological factors, disturbances (e.g., fire; Kipfer et al. 2011), and management practices (e.g., nitrogen deposition; Peter et al. 2001) affect the composition of ECM fungal communities and can lead to a reduction in the number of ECM fungal species (Koide et al. 2011). Changes in ECM fungal communities have recently been investigated using 454 pyrosequencing. Voříšková et al. (2014) have shown that seasonality and soil depth can influence ECM fungal communities in a temperate oak forest soil. A decrease in ECM fungi abundance has been verified in buried soils in the Siberian tundra (Gittel et al. 2013), where the abiotic conditions (low temperature and anoxia) seem to favor an abundance of bacteria, facultative anaerobic decomposers of soil organic matter (SOM) such as *Actinobacteria*, which increase compared to unburied soils. Hui et al. (2011) have observed, in a boreal coniferous forest site in Southern Finland, that a long-term exposition to Pb contamination can lead to a shift in the composition of the ECM community associated with the dominant pine (*P. sylvestris* L.), as well as an increase in the abundance of the OTUs (operation taxonomic units) assigned to the *Thelephora* genus and a decrease in the frequency of OTUs corresponding to *Pseudotomentella*, *Suillus*, and *Tylospora* in the contaminated zone (Hui et al. 2011).

However, although several factors influence the composition of an ECM community and species richness of ECM fungi, the functional consequences of these shifts on the aboveground communities (e.g., for tree performance), as well as for the soil ecosystems, still require further investigation (Kipfer et al. 2012). An extensive metabolic reprogramming during the colonization between *Laccaria bicolor* and its compatible host *Populus trichocarpa* has recently been demonstrated. However, this extensive metabolic reprogramming is repressed in incompatible interactions where more defensive compounds are produced or retained (Tschaplinski et al. 2014). Moreover, Pena and Polle (2014) have demonstrated, using ^{15}N isotope enrichment, that ECM assemblages provide advantages for inorganic N uptake mainly under environmental constraints with respect to unstressed plants, thus suggesting a stress activation of specific ECM taxa.

In addition to their ecological role, some ECM fungi such as *Tricholoma matsutake*, the so-called pine mushroom, have great value as commercial food. This mushroom is popular in Asia because of its aromatic odor and particular taste, as well as its high nutritional and medicinal value (Ohnuma et al. 2000; Kim et al. 2008; Ding et al. 2010). However, its annual production is limited, and several abiotic factors (e.g., rainfall and temperature) can affect it; moreover, attempts to cultivate it artificially have been unsuccessful. Kim and colleagues (2013) have investigated, through a 454 GS-FLX pyrosequencing platform, the fungal communities in soil where fruiting bodies develop. Several zones have been considered (inside, beneath, and outside the fairy ring zone of *T. matsutake*), with the aim of obtaining information on the fungal communities that could influence the development of the fruiting bodies. Mello et al. (2011) have used a similar approach to verify the fungal populations inside and outside the brûlé (burnt area), in which *T. melanosporum* fruiting bodies (the black truffles) are usually collected. The results show that *Ascomycota*, which was the most dominant phylum in the investigated French truffle ground, are more abundant inside than outside the brûlé, while *Basidiomycota* increase outside the brûlé in agreement with previous results reported by Napoli et al. (2010) using the DDGE technique.

The Application of ECM Fungi to Field Projects

ECM fungi provide trees with several benefits, such as the enhanced ability to absorb water, phosphorus, and nitrogen and protection from soilborne root pathogens such as *Fusarium oxysporum*. Evidence of the positive role of ECM *Pisolithus albus* in enhancing the growth of plants, such as *Acacia spirorbis* and *Eucalyptus globulus*, has been observed by significant increases in shoot and root biomass and mineral nutrition (P, K, and Ca), as well as a limited metal uptake, acting as a protective barrier, in nickel-rich ultramafic topsoils in New Caledonia (Jourand et al. 2014). In the restoration of ultramafic ecosystems degraded by mining activities, it could be convenient to isolate indigenous and stress-adapted beneficial ECM fungi in order to inoculate endemic plants (Jourand et al. 2014). ECM fungi have already been proposed for the ecological restoration of mine sites (in particular chromium and nickel mines) in Australia by Reddell et al. (1999) and in New Caledonia by Perrier et al. (2006) as well as in the post-mining of bauxite in Brazil by Khosla and Reddy (2008). ECM fungi are also expected to play a key role in forest regeneration after major disturbance events such as stand-replacing forest fires. Kipfer et al. (2010) investigated the heat tolerance of ECM fungi of Scots pine. They verified that 60 and 70 °C reduced the mean of the species, but not 45 °C. The composition changed because of heat, but most of the ECM fungi, such as *R. roseolus*, *C. geophilum*, and several unidentified species, survived.

Sustainable soil ecosystem services require the management of the beneficial soil organisms that are considered of economic value and which are available in the market for their application. Reforestation using container-grown seedlings of *P. pinaster* produced in nurseries is a common practice in many countries. Fertilizers are often used in nurseries. However, the use of chemical fertilizers can constitute a threat to the environment, in addition to modifying nutrient availability of the fertilized seedlings, which may not be able to adapt to forest soil conditions when transplanted. Sousa et al. (2012) showed that selected ECM fungi could be used as a beneficial biotechnological tool in the nursery production of *P. pinaster*, without the need of fertilizers. Oliveira et al. (2012) found that inoculation with selected ECM fungi can be an advantageous ecotechnological approach that can be used to improve the nursery production of *P. pinaster*. The inoculation of Chinese pine (*Pinus tabulaeformis* Carr.) seedlings with *Boletus luridus*, under field experimental conditions, has shown to significantly influence bacterial functional diversity in the rhizosphere of *P. tabulaeformis* seedlings, thus highlighting the importance of the application of ECM fungal inoculum in order to promote microbial community diversity of soil in forest restoration projects (Zhang et al. 2010).

The fast growth rate of American chestnut, coupled with its quality timber, makes it a desired species for use in reforestation projects. Bauman et al. (2013) have evaluated various soil preparation methods that promote ECM colonization and American chestnut *Castanea dentata* establishment in coal mine restoration projects. *Quercus ilex* forests play ecological and socioeconomic roles by protecting the environment and providing wood, forage resources, and tourism. Oliveira et al. (2010) considered the management of nursery practices for efficient ECM fungi application in order to establish *Q. ilex* plantations. The ECM inoculants could be rhizosphere soil from forests, spores, or vegetative mycelia, although the soil from forests has the disadvantage of perhaps containing both beneficial (e.g., mycorrhizal fungi) and harmful (pathogenic fungi) microorganisms. The use of spores or vegetative mycelia of ECM fungi seems to be the most convenient and practical technique although the supply of spores is limited by the onset of the rainy season and the maturing of the fruiting bodies. Aggangan et al. (2012) have developed a protocol for the production of quality rooted cuttings for plantation establishment or enrichment planting in the red soil of Caliraya, Philippines, of *Anisoptera thurifera* (Blanco) Blume and *Shorea guiso* (Blanco) Blume, which belong to the *Dipterocarpaceae*, the most important tree family in the tropical forests of southeast Asia, and are considered as endangered species. These authors have shown that ECM mycelia entrapped in alginate beads are effective in promoting the growth and P uptake of *A. thurifera* and *S. guiso*. Immobilized mycelium offers more advantages than non-immobilized inoculum such as longer survival in the soil, easy storage, and greater viability. However, the cultivation in fermenters or in bioreactors for the large-scale production of mycelial inoculum may compromise the quality of the inoculum and thus limit the application of this technique. Forecasting the response of ECM fungi to environmental changes represents an important step in maintaining forest productivity for the future. Jarvis et al. (2013) have analyzed and identified fungal communities from 15 semi-natural Scots pine (*P. sylvestris* L.) forests through ITS sequencing. Their data have demonstrated an important effect of rainfall and soil moisture on community composition at the species level and less influence of temperature on the abundance of ECM exploration types. Valdés et al. (2006) observed that a severe drought had an important effect on both total fine-root biomass and the ECM-root biomass in a tropical pine forest and suggested that forest management practices should consider the effects of drought in reducing the capacity of *Pinus oaxacana* to form ECM.

Food Applications and Toxicity of ECM Fungi

Fungi are the most productive biological sources of primary and secondary metabolites that have long been exploited by the pharmaceutical and food industries. Some of them, such as *Boletus* species, are also an important source of proteins, carbohydrates, fatty acids (mainly linoleic acid), sugars (mainly mannitol and trehalose), and vitamins (tocopherols and ascorbic acid), as well as phenolic acids (Heleno et al. 2011). Since phenolic acids have antioxidative properties, they are currently being exploited by food and pharmaceutical industries. Important antioxidative properties have been found for polysaccharides extracted from *B. edulis*, and these could be employed as ingredients in healthy food to alleviate oxidative stress (Zhang et al. 2011). A lectin with antitumoral properties has also been found, thus opening new perspectives in research, with the aim of developing new drugs for cancer therapies (Bovi et al. 2011). A homologous of the sugar-binding antiviral protein cyanovirin-N (CVN), previously identified in the cyanobacterium *Nostoc ellipsosporum*, has been found by analysis of transcript sequences deriving from a gene expression profiling study conducted in the truffle *Tuber borchii*, and a novel protein family has been described in filamentous fungi and in the fern *Ceratopteris richardii* (Percudani et al. 2005). It has been proposed that these findings provide candidate polypeptides to be tested as antiviral agents (Percudani et al. 2005). Two genes coding for putative lectins belonging to the CVNH (CyanoVirin-N Homolog) family have also been found in *T. melanosporum* genome (http://mycor.nancy.inra.fr/IMGC/TuberGenome/index.html). However, preliminary results have shown that TbCVNH (the CVNH discovered in *T. borchii*) seems to be completely inactive in the antiviral activity test (Koharudin et al. 2008). Mushrooms produce particular aroma compounds that are of interest for industrial applications, such as 1-octen-3-ol, which is often added as flavoring to processed products in order to reintegrate its loss which occurs during the preparation of these food products (Zawirska-Wojtasiak 2004). Long-chain unsaturated fatty acids, such as palmitoleic acid and linoleic acid, show antibacterial activity and are used as antimicrobial food additives (Zheng et al. 2005). To date, more than 200 volatile organic compounds have been described from various truffle species, and the biosynthetic pathways involved in volatile biosynthesis have been traced in the genome of a mushroom that is highly appreciated for its special taste and aroma, the black truffle *T. melanosporum* (Martin et al. 2010). From the ecological point of view, truffle volatiles are used to attract mammals and insects, which are thus able to locate the precious hypogeous fungi, feed on them, and spread their spores. In addition, truffle volatiles diffuse in the soil and mediate complex interactions with microorganisms and plant roots (Splivallo et al. 2011; Mello et al. 2013). Truffles and porcini are greatly appreciated throughout the world, both as fresh fruiting bodies and as ingredients in processed products (Mello 2012).

A qPCR assay has been developed to authenticate and quantify *T. magnatum* and *T. melanosporum* in food matrices that have undergone intensive transformation processes, such as cream and butter (Rizzello et al. 2012). This method is promising in detecting the incorrect labeling of processed products and can therefore be used to protect the consumer and to assess food quality. The rapid identification of mushroom poisoning, which continues to be a public health concern in Europe, the USA, and several other parts of the world, is particularly important. The majority of reported fatal intoxications have been attributed to a few species of the *Amanita* genus and in particular to the death cap *Amanita phalloides*, which can cause a high mortality rate (10–30 % in adults). *A. phalloides* are often mistaken in appearance for nonpoisonous species, and the ingestion of one single mushroom cap may be sufficient to cause death within 2–8 days (Gausterer et al. 2014). Besides the previous publications that have reported the use of conventional and real-time PCR in the cases of suspected mushroom poisoning as an alternative to morphological investigations and as a complementary approach to toxicological analyses, an article on a rapid and sensitive detection of genetic traces from poisonous mushrooms in a variety of matrices, including raw, fried, and digested mushroom homogenates, spiked feces, and clinical samples (vomit, stool), has just been published by Gausterer et al. (2014). Mushrooms belonging to the *Boletus edulis* sensu lato group, a complex of at least four species of ECM fungi in the genus *Boletus* section *Boletus* (Singer 1986), constitute an interesting example of contrasting fungal features. These fruiting bodies are in fact in high demand as mushrooms because of their pleasant flavor and texture, but at the same time, they have been shown to induce allergic symptoms either through inhalation, ingestion, or contact (Helbling et al. 2002). In order to guarantee safe naturally derived food, Mello et al. (2006b) developed specific primers for the unambiguous detection of *B. edulis* sensu stricto, *B. aereus*, *B. pinophilus*, and *B. aestivalis*. In addition, the relationships of *B. violaceofuscus* with the members of *B. edulis* s.l. have been examined. The data that will be obtained from the *B. edulis* genome sequencing project, which is currently in progress, will surely allow our knowledge on its allergenic potential and the presence of allergen orthologues to be improved. Thanks to the availability of the genome sequence, an extremely low allergenic potential and the lack of key mycotoxin biosynthetic enzymes have been found in the black truffle *T. melanosporum* (Martin et al. 2010).

Acknowledgments EZ's fellowship was funded by a grant from the Italian Ministry of Education, University and Research, as part of the FIRB program (Project DEFINE; grant number RBFR128ONN).

References

Aggangan NS, Aggangan JS, Charisse J, Bulan O, Limos CAS (2012) Inoculation of Dipterocarps *Anisoptera thurifera* and *Shorea guiso* with ectomycorrhizal fungi in Philippine red soil. Philipp J Sci 141:229–241

Alvarez M, Huygens D, Olivares E, Saavedra I, Alberdi M, Valenzuela E (2009) Ectomycorrhizal fungi enhance nitrogen and phosphorus nutrition of *Nothofagus dombeyi* under drought conditions by regulating assimilative enzyme activities. Physiol Plant 136:426–436

Averill C, Turner BL, Finzi AC (2014) Mycorrhiza-mediated competition between plants and decomposers drives soil carbon storage. Nature 505:543–545

Balestrini R, Bianciotto V, Bonfante P (2012) Mycorrhizae. In: Huang PM, Li Y, Sumner ME (eds) Handbook of soil sciences: properties and processes, vol 24, 2nd edn. CRC Press, Boca Raton, pp 29–39

Balestrini R, Ghignone S, Sillo F (2013) The contribution of new technologies toward understanding plant–fungus symbioses. In: Arora NK (ed) Plant microbe symbiosis: fundamentals and advances. Springer, New Delhi, pp 201–214

Bauman JM, Keiffer CH, Hiremath S, McCarthy BC (2013) Soil preparation methods promoting ectomycorrhizal colonization and American chestnut *Castanea dentate* establishment in coal mine restoration. J Appl Ecol 50:721–729

Blaalid R, Carlsen T, Kumar S, Halvorsen R, Ugland KI, Fontana G, Kauserud H (2012) Changes in the root-associated fungal communities along a primary succession gradient analysed by 454 pyrosequencing. Mol Ecol 21:1897–1908

Blaalid R, Davey ML, Kauserud H, Carlsen T, Halvorsen R, Høiland K, Eidesen PB (2014) Arctic root-associated fungal community composition reflects environmental filtering. Mol Ecol 23:649–659

Bonfante P (2010) Plant-fungal interactions in mycorrhizas. In: Encyclopedia of Life Sciences (ELS). John Wiley & Sons Ltd, Chichester

Bovi M, Carrizo ME, Capaldi S, Perduca M, Chiarelli LR, Galliano M, Monaco HL (2011) Structure of a lectin with antitumoral properties in king bolete (*Boletus edulis*) mushrooms. Glycobiology 21:1000–1009

Bradford MA (2014) Ecology: good dirt with good friends. Nature 505:486–487

Brundrett MC (2009) Mycorrhizal associations and other means of nutrition of vascular plants: understanding the global diversity of host plants by resolving conflicting information and developing reliable means of diagnosis. Plant Soil 320:37–77

Bücking H, Liepold E, Ambilwade P (2012) The role of the mycorrhizal symbiosis in nutrient uptake of plants and the regulatory mechanisms underlying these transport processes. In: Dhal NK, Sahu SC (eds) Plant science. Intech, Janeza Trdine, p 107

Buée M, Reich M, Murat C, Morin E, Nilsson RH, Uroz S, Martin F (2009) 454 pyrosequencing analyses of forest soil reveal an unexpected high fungal. New Phytol 184:449–456

Chivian D, Brodie EL, Alm EJ, Culley DE, Dehal PS, DeSantis TZ, Gihring TM, Lapidus A, Lin LH, Lowry SR, Moser DP, Richardson PM, Southam G, Wanger G, Pratt LM, Andersen GL, Hazen TC, Brockman FJ, Arkin AP, Onstott TC (2008) Environmental genomics reveals a single-species ecosystem deep within earth. Science 322:275–278

Clemmensen KE, Bahr A, Ovaskainen O, Dahlberg A, Ekblad A, Wallander H, Stenlid J, Finlay RD, Wardle DA, Lindahl BD (2013) Roots and associated fungi drive long-term carbon sequestration in boreal forest. Science 339:1615–1618

Dahlberg A (2001) Community ecology of ectomycorrhizal fungi: an advancing interdisciplinary field. New Phytol 150:555–562

Danielsen L, Polle A (2014) Poplar nutrition under drought as affected by ectomycorrhizal colonization. Environ Exp Bot 108:89–98

Danielsen L, Thürmer A, Meinicke P, Buée M, Morin E, Martin F, Pilate G, Daniel R, Polle A, Reich M (2012) Fungal soil communities in a young transgenic poplar plantation form a rich reservoir for fungal root communities. Ecol Evol 2:1935–1948

Ding X, Tang J, Cao M, Guo CX, Zhang X, Zhong J, Zhang J, Sun Q, Feng S, Yang ZR, Zhao J (2010) Structure elucidation and antioxidant activity of a novel polysaccharide isolated from *Tricholoma matsutake*. Int J Biol Macromol 47:271–275

Dosskey MG, Boersma L, Linderman RG (1991) Role of photosynthate demand of ectomycorrhizas in the response of Douglas fir seedlings to drying soil. New Phytol 117:327–334

Duñabeitia MK, Hormilla S, Garcia-Plazaola JI, Txarterina K, Arteche U, Becerril JM (2004) Differential responses of three fungal species to environmental factors and their role in the mycorrhization of *Pinus radiata* D. Don. Mycorrhiza 14:11–18

Ek H (1997) The influence of nitrogen fertilization on the carbon economy of *Paxillus involutus* in ectomycorrhizal association with *Betula pendula*. New Phytol 135:133–142

Epron D, Ngao J, Dannoura M, Bakker MR, Zeller B, Bazot S, Bosc A, Plain C, Lata JC, Priault P, Barthes L, Loustau D (2011) Seasonal variations of belowground carbon transfer assessed by in situ $^{13}CO_2$ pulse labelling of trees. Biogeosciences 8:1153–1168

Garcia K, Delteil A, Conéjéro G, Becquer A, Plassard C, Sentenac H, Zimmermann S (2014) Potassium nutrition of ectomycorrhizal *Pinus pinaster*: overexpression of the *Hebeloma cylindrosporum* HcTrk1 transporter affects the translocation of both K(+) and phosphorus in the host plant. New Phytol 201:951–960

Gausterer C, Penker M, Krisai-Greilhuber I, Stein C, Stimpfl T (2014) Rapid genetic detection of ingested *Amanita phalloides*. Forensic Sci Int Genet 9:66–71

Girlanda M, Perotto S, Bonfante P (2007) Mycorrhizal fungi: their habitats and nutritional strategies biology of the fungal cell. In: Howard RJ, Gow NAR (eds) The Mycota IV, 2nd edn. Springer, Berlin

Gittel A, Bárta J, Kohoutová I, Mikutta R, Owens S, Gilbert J, Schnecker J, Wild B, Hannisdal B, Maerz J, Lashchinskiy N, Capek P, Santruková H, Gentsch N, Shibistova O, Guggenberger G, Richter A, Torsvik VL, Schleper C, Urich T (2013) Distinct microbial communities associated with buried soils in the Siberian tundra. ISME J 8:841–853

Guidot A, Debaud JC, Marmeisse R (2002) Spatial distribution of the below-ground mycelia of an ectomycorrhizal fungus inferred from specific quantification of its DNA in soil samples. FEMS Microbiol Ecol 42:477–486

Hacquard S, Tisserant E, Brun A, Legué V, Martin F, Kohler A (2013) Laser microdissection and microarray analysis of *Tuber melanosporum* ectomycorrhizas reveal functional heterogeneity between mantle and Hartig net compartments. Environ Microbiol 15:1853–1869

Helbling A, Bonadies N, Brander KA, Pichler WJ (2002) *Boletus edulis*: a digestion-resistant allergen may be relevant for food allergy. Clin Exp Allergy 32:771–775

Heleno SA, Barros L, Sousa MJ, Martins A, Santos-Buelga C, Ferreira ICFR (2011) Targeted metabolites analysis in wild *Boletus species*. LWT Food Sci Technol 44:1343–1348

Hobbie EA, Högberg P (2012) Nitrogen isotopes link mycorrhizal fungi and plants to nitrogen dynamics. New Phytol 196:367–382

Högberg P, Högberg MN, Gottlicher SG, Betson NR, Keel SG, Metcalfe DB, Campbell C, Schindlbacher A, Hurry V, Lundmark T, Linder S, Nasholm T (2008) High temporal resolution tracing of photosynthate carbon from the tree canopy to forest soil microorganisms. New Phytol 177:220–228

Horton TR, Bruns TD (2001) The molecular revolution in ectomycorrhizal ecology: peeking into the black-box. Mol Ecol 10:1855–1871

Hui N, Jumpponen A, Niskanen T, Liimatainen K, Jones KL, Koivula T, Romantschuk M, Strömmer R (2011) ECM fungal community structure, but not diversity, altered in a Pb-contaminated shooting range in a boreal coniferous forest site in Southern Finland. FEMS Microbiol Ecol 76:121–132

Jarvis S, Woodward S, Alexander IJ, Taylor AF (2013) Regional scale gradients of climate and nitrogen deposition drive variation in ectomycorrhizal fungal communities associated with native Scots pine. Glob Chang Biol 19:1688–1696

Johnson D, Leake JR, Ostle N, Ineson P, Read DJ (2002) In situ $^{13}CO_2$ pulse-labelling of upland grassland demonstrates a rapid pathway of carbon flux from arbuscular mycorrhizal mycelia to the soil. New Phytol 153:327–334

Johnson D, Martin F, Cairney JW, Anderson IC (2012) The importance of individuals: intraspecific diversity of mycorrhizal plants and fungi in ecosystems. New Phytol 19:614–628

Jourand P, Hannibal L, Majorel C, Mengant S, Ducousso M, Lebrun M (2014) Ectomycorrhizal *Pisolithus albus* inoculation of *Acacia spirorbis* and *Eucalyptus globulus* grown in ultramafic topsoil enhances plant growth and mineral nutrition while limits metal uptake. J Plant Physiol 171:164–172

Jumpponen A, Jones KL (2009) Massively parallel 454 sequencing indicates hyperdiverse fungal communities in temperate *Quercus macrocarpa* phyllosphere. New Phytol 184:438–448

Kennedy PG, Peay KG (2007) Different soil moisture conditions change the outcome of the ectomycorrhizal symbiosis between *Rhizopogon* species and *Pinus muricata*. Plant Soil 291:155–165

Khosla B, Reddy MS (2008) Response of ectomycorrhizal fungi on the growth and mineral nutrition of *Eucalyptus* seedlings in bauxite mined soil. Am-Eurasian J Agric Environ Sci 3:123–126

Kim JY, Byeon SE, Lee YG, Lee JY, Park J, Hong EK, Cho JY (2008) Immunostimulatory activities of polysaccharides from liquid culture of pine-mushroom *Tricholoma matsutake*. J Microbiol Biotechnol 18:95–103

Kim M, Yoon H, You YH, Kim YE, Woo JR, Seo Y, Lee GM, Kim YJ, Kong WS, Kim JG (2013) Metagenomic analysis of fungal communities inhabiting the fairy ring zone of *Tricholoma matsutake*. J Microbiol Biotechnol 23:1347–1356

Kipfer T, Egli S, Ghazoul J, Moser B, Wohlgemuth T (2010) Susceptibility of ectomycorrhizal fungi to soil heating. Fungal Biol 114:467–472

Kipfer T, Moser B, Egli S, Wohlgemuth T, Ghazoul J (2011) Ectomycorrhiza succession patterns in *Pinus sylvestris* forests after stand-replacing fire in the Central Alps. Oecologia 167:219–228

Kipfer T, Wohlgemuth T, van der Heijden MG, Ghazoul J, Egli S (2012) Growth response of drought-stressed *Pinus sylvestris* seedlings to single- and multi-species inoculation with ectomycorrhizal fungi. PLoS ONE 7:e35275

Koharudin LMI, Viscomi AR, Jee J-G, Ottonello S, Gronenborn AM (2008) The evolutionarily conserved family of Cyanovirin-N homologs: structures and carbohydrate specificity. Structure 16:570–584

Koide RT, Fernandez C, Petprakob K (2011) General principles in the community ecology of ectomycorrhizal fungi. Ann For Sci 68:45–55

Le Tacon F, Zeller B, Plain C, Hossann C, Bréchet C, Robin C (2013) Carbon transfer from the host to *Tuber melanosporum* mycorrhizas and ascocarps followed using a ^{13}C pulse-labeling technique. PLoS ONE 8:e64626

Leake JR, Ostle NJ, Rangel-Castro JI, Johnson D (2006) Carbon fluxes from plants through soil organisms determined by field $^{13}CO_2$ pulse-labelling in an upland grassland. Appl Soil Ecol 33:152–175

Lothamer K, Brown SP, Mattox JD, Jumpponen A (2013) Comparison of root-associated communities of native and non-native ectomycorrhizal hosts in an urban landscape. Mycorrhiza 24:267–280

Marmeisse R, Nehls U, Opik M, Selosse MA, Pringle A (2013) Bridging mycorrhizal genomics, metagenomics and forest ecology. New Phytol 198:343–346

Martin F, Bonito G (2012) Ten years of genomics for ectomycorrhizal fungi: what have we achieved and where are we heading? In: Zambonelli A, Bonito GM (eds) Edible ectomycorrhizal mushrooms, vol 34, Soil biology. Springer, Berlin/Heidelberg, pp 383–401

Martin F, Nehls U (2009) Harnessing ectomycorrhizal genomics for ecological insights. Curr Opin Plant Biol 12:508–515

Martin F, Aerts A, Ahrén D, Brun A, Danchin EG, Duchaussoy F, Gibon J, Kohler A, Lindquist E, Pereda V, Salamov A, Shapiro HJ, Wuyts J, Blaudez D, Buée M, Brokstein P, Canbäck B, Cohen D, Courty PE, Coutinho PM, Delaruelle C, Detter JC, Deveau A, DiFazio S, Duplessis S, Fraissinet-Tachet L, Lucic E, Frey-Klett P, Fourrey C, Feussner I, Gay G, Grimwood J, Hoegger PJ, Jain P, Kilaru S, Labbé J, Lin YC, Legué V, Le Tacon F, Marmeisse R, Melayah D, Montanini B, Muratet M, Nehls U, Niculita-Hirzel H, Oudot-Le Secq MP, Peter M, Quesneville H, Rajashekar B, Reich M, Rouhier N, Schmutz J, Yin T, Chalot M, Henrissat B, Kües U, Lucas S, Van de Peer Y, Podila GK, Polle A, Pukkila PJ, Richardson PM, Rouzé P, Sanders IR, Stajich JE, Tunlid A, Tuskan G, Grigoriev IV (2008) The genome of *Laccaria bicolor* provides insights into mycorrhizal symbiosis. Nature 452:88–92

Martin F, Kohler A, Murat C, Balestrini R, Coutinho PM, Jaillon O, Montanini B, Morin E, Noel B, Percudani R, Porcel B, Rubini A, Amicucci A, Amselem J, Anthouard V, Arcioni S, Artiguenave F, Aury JM, Ballario P, Bolchi A, Brenna A, Brun A, Buée M, Cantarel B, Chevalier G, Couloux A, Da Silva C, Denoeud F, Duplessis S, Ghignone S, Hilselberger B, Iotti M, Marçais B, Mello A, Miranda M, Pacioni G, Quesneville H, Riccioni C, Ruotolo R, Splivallo R, Stocchi V, Tisserant E, Viscomi AR, Zambonelli A, Zampieri E, Henrissat B, Lebrun MH, Paolocci F,

Bonfante P, Ottonello S, Wincker P (2010) Périgord black truffle genome uncovers evolutionary origins and mechanisms of symbiosis. Nature 464:1033–1038

McGuire KL, Allison SD, Fierer N, Treseder KK (2013) Ectomycorrhizal-dominated boreal and tropical forests have distinct fungal communities, but analogous spatial patterns across soil horizons. PLoS ONE 8:e68278

Mello A (2012) State of the art of the research on *Boletus edulis*. In: Zambonelli A, Bonito GM (eds) Edible ectomycorrhizal mushrooms: current knowledge and future prospects, vol 34, Soil biology. Springer, Berlin, p 73

Mello A, Ghignone S, Vizzini A, Sechi C, Ruiu P, Bonfante P (2006a) ITS primers for the identification of marketable Boletes. J Biotechnol 121:318–329

Mello A, Murat C, Bonfante P (2006b) Truffles: much more than a prized and local fungal delicacy. FEMS Microbiol Lett 260:1–8

Mello A, Napoli C, Murat C, Morin E, Marceddu G, Bonfante P (2011) ITS-1 versus ITS-2 pyrosequencing: a comparison of fungal populations in truffle grounds. Mycologia 103:1184–1193

Mello A, Ding GC, Piceno YM, Napoli C, Tom LM, DeSantis TZ, Andersen GL, Smalla K, Bonfante P (2013) Truffle brûlés have an impact on the diversity of soil bacterial communities. PLoS ONE 8:e61945

Morte A, Díaz G, Rodríguez P, Alarcón JJ, Sánchez-Blanco MJ (2001) Growth and water relations in mycorrhizal and nonmycorrhizal *Pinus halepensis* plants in response to drought. Biol Plant 44:263–267

Napoli C, Mello A, Borra A, Vizzini A, Sourzat P, Bonfante P (2010) *Tuber melanosporum*, when dominant, affects fungal dynamics in truffle-grounds. New Phytol 185:237–247

Nehls U, Mikolajewski S, Magel E, Hampp R (2001) Carbohydrate metabolism in ectomycorrhizas: gene expression, monosaccharide transport and metabolic control. New Phytol 150:533–541

Nguyen C, Todorovic C, Robin C, Christophe A, Guckert A (1999) Continuous monitoring of rhizosphere respiration after labelling of plant shoots with $^{14}CO_2$. Plant Soil 212:189–199

Norton JM, Smith JL, Firestone MK (1990) Carbon flow in the rhizosphere of *Ponderosa* pine seedlings. Soil Biol Biochem 22:449–455

Ohnuma N, Amemiya K, Kakuda R, Yaoita Y, Machida K, Kikuchi M (2000) Sterol constituents from two edible mushrooms, *Lentinula edodes* and *Tricholoma matsutake*. Chem Pharm Bull (Tokyo) 48:749–751

Oliveira RS, Franco AR, Vosátka M, Castro PML (2010) Management of nursery practices for efficient ectomycorrhizal fungi application in the production of *Quercus ilex*. Symbiosis 52:125–131

Oliveira RS, Franco AR, Castro PML (2012) Combined use of *Pinus pinaster* plus and inoculation with selected ectomycorrhizal fungi as an ecotechnology to improve plant performance. Ecol Eng 43:95–103

Orgiazzi A, Bianciotto V, Bonfante P, Daghino S, Ghignone S, Lazzari A, Lumini E, Mello A, Napoli C, Perotto S, Vizzini A, Bagella S, Murat C, Girlanda M (2013) 454 pyrosequencing analysis of fungal assemblages from geographically distant, disparate soils reveals spatial patterning and a core mycobiome. Diversity 5:73–98

Pena R, Polle A (2014) Attributing functions to ectomycorrhizal fungal identities in assemblages for nitrogen acquisition under stress. ISME J 8:321–330

Percudani R, Montanini B, Ottonello S (2005) The anti-HIV cyanovirin-N domain is evolutionarily conserved and occurs as a protein module in eukaryotes. Proteins 60:670–678

Perotto S, Angelini P, Bianciotto V, Bonfante P, Girlanda M, Kull T, Mello A, Pecoraro L, Perini C, Persiani AM, Saitta A, Sarrocco S, Vannacci G, Venanzoni R, Venturella G, Selosse MA (2013) Interactions of fungi with other organisms. Plant Biosyst 147:208–218

Perrier N, Amir H, Colin F (2006) Occurrence of mycorrhizal symbioses in the metal-rich lateritic soils of the Koniambo Massif, New Caledonia. Mycorrhiza 16:449–458

Peter M, Ayer F, Egli S (2001) Nitrogen addition in a Norway spruce stand altered macromycete sporocarp production and belowground ectomycorrhizal species composition. New Phytol 149:311–325

Plain C, Gerant D, Maillard P, Dannoura M, Dong Y, Zeller B, Priault P, Parent F, Epron D (2009) Tracing of recently assimilated carbon in respiration at high temporal resolution in the field with a tuneable diode laser absorption spectrometer after in situ $^{13}CO_2$ pulse labelling of 20-year-old beech trees. Tree Physiol 29:1433–1445

Plett JM, Martin F (2011) Blurred boundaries: lifestyle lessons from ectomycorrhizal fungal genomes. Trends Genet 27:14–22

Reddell P, Gordon V, Hopkins MS (1999) Ectomycorrhizas in *Eucalyptus tetrodonta* and *E. miniata* forest communities in tropical Northern Australia and their role in the rehabilitation of these forests following mining. Aust J Bot 47:881–907

Rizzello R, Zampieri E, Vizzini A, Autino A, Cresti M, Bonfante P, Mello A (2012) Authentication of prized white and black truffles in processed products using quantitative real-time PCR. Food Res Int 48:792–797

Robin C, Vaillant V, Vansuyt G, Zinsou C (1990) Assimilate partitioning in *Pachyrhizus erosus* L. during long-day vegetative development. Plant Physiol Biochem 28:343–349

Simard SW, Durall DM, Jones MD (1997) Carbon allocation and carbon transfer between *Betula papyrifera* and *Pseudotsuga menziesii* seedlings using a ^{13}C pulse-labeling method. Plant Soil 191:41–55

Singer R (1986) The agaricales in modern taxonomy, 4th edn. Koeltz Scientific Books, Koenigstein, p 981

Smith SE, Read DJ (2008) Mycorrhizal symbiosis, 3rd edn. Academic, London

Sousa NR, Franco AR, Oliveira RS, Castro PML (2012) Ectomycorrhizal fungi as an alternative to the use of chemical fertilisers in nursery production of *Pinus pinaster*. J Environ Manag 95:S269–S274

Splivallo R, Ottonello S, Mello A, Karlovsky P (2011) Truffle volatiles: from chemical ecology to aroma biosynthesis. New Phytol 189:688–699

Subke JA, Vallack HW, Magnusson T, Keel SG, Metcalfe DB, Högberg P, Ineson P (2009) Short-term dynamics of abiotic and biotic soil $^{13}CO_2$ effluxes after in situ $^{13}CO_2$ pulse labelling of a boreal pine forest. New Phytol 183:349–357

Tedersoo L, May TW, Smith ME (2010a) Ectomycorrhizal lifestyle in fungi: global diversity, distribution, and evolution of phylogenetic lineages. Mycorrhiza 20:217–263

Tedersoo L, Nilsson RH, Abarenkov K, Jairus T, Sadam A, Saar I, Bahram M, Bechem E, Chuyong G, Kõljalg U (2010b) 454 Pyrosequencing and Sanger sequencing of tropical mycorrhizal fungi provide similar results but reveal substantial methodological biases. New Phytol 188:291–301

Teramoto M, Wu B, Hogetsu T (2012) Transfer of ^{14}C-photosynthate to the sporocarp of an ectomycorrhizal fungus *Laccaria amethystina*. Mycorrhiza 22:219–225

Toju H, Yamamoto S, Sato H, Tanabe AS, Gilbert GS, Kadowaki K (2013) Community composition of root-associated fungi in a *Quercus*-dominated temperate forest: "codominance" of mycorrhizal and root-endophytic fungi. Ecol Evol 3:1281–1293

Tschaplinski TJ, Plett JM, Engle N, Deveau A, Cushman K, Martin MZ, Doktycz MJ, Tuskan G, Brun A, Kohler A, Martin FM (2014) *Populus trichocarpa* and *Populus deltoides* exhibit different metabolomic responses to colonization by the symbiotic fungus *Laccaria bicolor*. Mol Plant Microbe Interact 6:546–556

Valdés M, Asbjornsen H, Gómez-Cárdenas M, Juárez M, Vogt KA (2006) Drought effects on fine-root and ectomycorrhizal-root biomass in managed *Pinus oaxacana* Mirov stands in Oaxaca, Mexico. Mycorrhiza 16:117–124

Voříšková J, Brabcová V, Cajthaml T, Baldrian P (2014) Seasonal dynamics of fungal communities in a temperate oak forest soil. New Phytol 201:269–278

Walker JK, Cohen H, Higgins LM, Kennedy PG (2013) Testing the link between community structure and function for ectomycorrhizal fungi involved in a global tripartite symbiosis. New Phytol 202:287–296

Walker JKM, Phillips LA, Jones MD (2014) Ectomycorrhizal fungal hyphae communities vary more along a pH and nitrogen gradient than between decayed wood and mineral soil microsites. Botany 92:453–463

Wallander H, Johansson U, Sterkenburg E, Brandström Durling M, Lindahl BD (2010) Production of ectomycorrhizal mycelium peaks during canopy closure in Norway spruce forests. New Phytol 187:1124–1134

Zampieri E, Murat C, Cagnasso M, Bonfante P, Mello A (2010) Soil analysis reveals the presence of an extended mycelial network in a *Tuber magnatum* truffle-ground. FEMS Microbiol Ecol 71:43–49

Zawirska-Wojtasiak R (2004) Optical purity of (R)-(−)-1-octen-3-ol in the aroma of various species of edible mushrooms. Food Chem 86:113–118

Zhang HH, Tang M, Chen H, Zheng CL (2010) Effects of inoculation with ectomycorrhizal fungi on microbial biomass and bacterial functional diversity in the rhizosphere of *Pinus tabulaeformis* seedlings. Eur J Soil Biol 46:55–61

Zhang A, Xiao N, He P, Sun P (2011) Chemical analysis and antioxidant activity in vitro of polysaccharides extracted from *Boletus edulis*. Int J Biol Macromol 49:1092–1095

Zheng CJ, Yoo JS, Lee TG, Cho HY, Kim YH, Kim WG (2005) Fatty acid synthesis is a target for antibacterial activity of unsaturated fatty acids. FEBS Lett 579:5157–5162

Increasing the Role of Mycorrhizal Symbiosis in Plant-Plant Facilitation Process to Improve the Productivity and Sustainability of Mediterranean Agrosystems

17

S. Wahbi, H. Sanguin, E. Tournier, E. Baudoin, T. Maghraoui, M. Hafidi, Y. Prin, A. Galiana, and R. Duponnois

Abstract

Plant-plant facilitation is an ecological process occurring in most terrestrial ecosystems. Plant-plant facilitation is considered as a positive interaction between both plant partners in which one plant species promotes the growth, survival or reproduction of the neighbouring plant. Recent studies have underlined the role of mycorrhizal fungi, i.e. arbuscular mycorrhizal fungi (AMF) and particularly extraradical hyphae of AMF, in interconnecting plants and consequently their importance in plant-plant facilitation process. Networks of AMF impact soils both physically and biologically and are considered as an important pathway for the transference of nutrients such as nitrogen and phosphorus. In parallel,

S. Wahbi
Laboratoire d'Ecologie et Environnement (Unité associée au CNRST, URAC 32, Unité associée au CNERS). Faculté des Sciences Semlalia, Université Cadi Ayyad, BP. 2390, Marrakech, Morocco

IRD. UMR 113 CIRAD/INRA/IRD/SUP-AGRO/UM2. Laboratoire des Symbioses Tropicales et Méditerranéennes (LSTM), Campus International de Baillarguet, 34398 Montpellier, France

H. Sanguin • E. Tournier • Y. Prin • A. Galiana
CIRAD. UMR 113 CIRAD/INRA/IRD/SUP-AGRO/UM2. Laboratoire des Symbioses Tropicales et Méditerranéennes (LSTM), Campus International de Baillarguet, 34398 Montpellier, France

E. Baudoin • R. Duponnois (✉)
IRD. UMR 113 CIRAD/INRA/IRD/SUP-AGRO/UM2. Laboratoire des Symbioses Tropicales et Méditerranéennes (LSTM), Campus International de Baillarguet, 34398 Montpellier, France
e-mail: robin.duponnois@ird.fr

T. Maghraoui
Laboratoire de Biologie et de Biotechnologie des microorganismes. Faculté des Sciences Semlalia, Université Cadi Ayyad, BP. 2390, Marrakech, Morocco

IRD. UMR 113 CIRAD/INRA/IRD/SUP-AGRO/UM2. Laboratoire des Symbioses Tropicales et Méditerranéennes (LSTM), Campus International de Baillarguet, 34398 Montpellier, France

M. Hafidi
Laboratoire d'Ecologie et Environnement (Unité associée au CNRST, URAC 32, Unité associée au CNERS). Faculté des Sciences Semlalia, Université Cadi Ayyad, BP. 2390, Marrakech, Morocco

N.K. Arora (ed.), *Plant Microbes Symbiosis: Applied Facets*,
DOI 10.1007/978-81-322-2068-8_17, © Springer India 2015

AMF increase plant growth and nutrient uptake and decrease the deleterious effects of pathogens and drought. The aims of this chapter are to describe how the AMF are involved in the plant facilitation process and to assess the main mycorrhizal effects on the plant nutrition (P and N) and plant health. Also the contribution of AMF in cropping systems as well as agricultural strategies which improve AM associations in arid agrosystems will be reviewed and illustrated by experimental results from field studies in Mediterranean environment.

Introduction

In arid and semiarid environments, desertification of land area has increased in size over the last decades due to drastic climatic conditions (long periods of drought, erratic rainfall) combined with human activities (overexploitation of natural resources, grazing, etc.) (Francis and Thornes 1990). Desertification is commonly related to an alteration of the plant cover (species diversity, abundance), but it also impacts the physical, chemical and biological characteristics of soils (soil structure, nutrient bioavailability, microbial activity, etc.) (Garcia et al. 1997; Requena et al. 2001). Indeed, due to the exposure of soils to wind and water erosion, genetic and functional diversity of soil microorganisms is particularly affected by these biotic and abiotic degradations. Among the microbial components particularly susceptible to desertification, the diversity and abundance of mycorrhizal fungi are significantly reduced (Bethlenfalvay and Schüepp 1994). Mycorrhizal fungi play a major role in the dynamic of terrestrial plant ecosystems (Boucher et al. 1982). In addition to their direct impact on the development of the plant mainly by improving the mineral nutrition of the host plant, it has been shown that the structure of the plant cover and its development were closely linked to the functioning of mycorrhizal symbiosis (Odum 1959; van der Heijden et al. 1998). The ability of a plant species to tolerate biotic or abiotic stresses is also highly dependent on the degree of establishment and functioning of the symbiotic relationship between the fungus and the host plant (Odum 1959; Barea et al. 1997). It is also now well demonstrated that mycorrhizal symbiosis affects microbial soil functioning via its influence on some microbial groups involved in the major biogeochemical cycles (N, P and C) (Frey-Klett et al. 2005). Figure 17.1 depicts the various positive impacts of AMF on the plants and ecosystems as a whole.

Since the availability of mineral nutrients and the increase in abiotic and biotic stresses will become important limiting factors for plant production, the integration of key processes from natural ecosystems in agricultural management has been suggested as one strategy to overcome the increasing agricultural problems. In many cases, the non-degraded ecosystems governed by

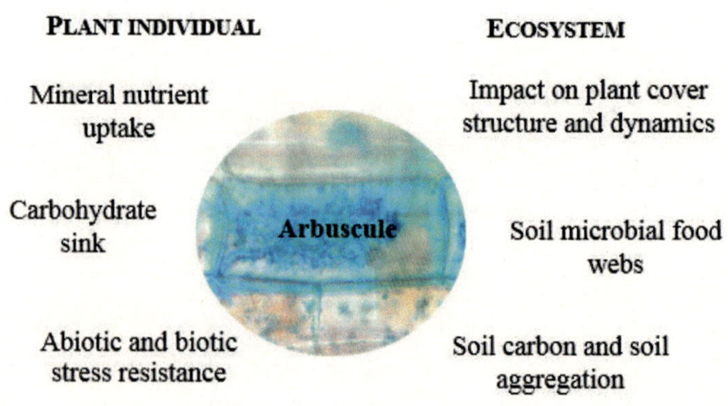

Fig. 17.1 AM symbiosis effects on plant and ecosystem (adapted from Fester and Sawers (2014))

Table 17.1 Key roles of AM symbiosis as an ecosystem service provider (modified from Gianinazzi et al. 2010)

AM function	Ecosystem service	References
Development of a complex, ramifying fungal network	Improvement of soil structure/soil stability/water retention	Bedini et al. (2009)
Root morphology modifications		
Enhancement of plant mineral nutrition and plant water uptake	Plant growth promotion with reduced phosphate fertiliser	Smith and Read (2008)
Secretion by AM fungi of "sticky" proteinaceous substances (glomalin)	Improvement of soil stability and water retention	Rillig et al. (2002)
Increase of crop plant tolerance against abiotic stresses (salinity, drought, etc.)	Improvement of plant resistance to root pathogens (nematodes, pathogenic fungi)	Marulanda et al. (2006)
Plant protection against biotic stresses	Improvement of plant resistance to drought, salinity, heavy metals, pollution, mineral nutrient deficiencies	Hao et al. (2009)
Activation of plant secondary metabolism	Improvement of plant quality (nutritional, medicinal, etc.) for human health	Gianinazzi et al. 2010

these natural processes that usually result from long periods of evolution are highly productive, resistant to pests and retentive of nutrients (Ewel 1999). AMF symbioses show particular potentialities with regard to plant stress resistance and mineral nutrient uptake and have key roles as ecosystem service provider (Gianinazzi et al. 2010) (Table 17.1). The mycorrhizal efficiency has been usually studied as the result of dual interactions between the host plant and the fungal symbiont. But it must be reconsidered as the fungal effect on the ecosystem derived from complex interactions with other microorganisms within the mycorrhizosphere (Bonfante and Anca 2009).

Among natural processes that are involved in the stability and the productivity of non-disturbed ecosystems, plant-plant facilitation could be considered as a powerful tool to sustain the mycorrhizal efficiency in agricultural conditions. Plant-plant facilitation is an ecological process occurring in communities worldwide (Valiente-Banuet et al. 2006) when the presence of one plant enhances the growth, survival and reproduction of its neighbour (Callaway 2007). The maintenance of plant-plant facilitation over time shows that the benefits of this association are not restricted to germination and seedling establishment but are extended to growth and long-term survival (Montesinos-Navarro et al. 2012). Plant facilitation has been experimentally described to be species specific (Castillo et al. 2010). Hence, the benefactor plant species (nurse plant) may promote the development of specific beneficiary (facilitated) species more strongly than others (Valiente-Banuet and Verdu 2007, 2008; Castillo et al. 2010). The main mechanisms involved in the facilitation processes, provided by the nurse plants, are: (i) microclimate improvement beneath their canopies (Gomez-Aparicio et al. 2004); (ii) protection from herbivores (Baraza et al. 2006); and (iii) enhancement of physical, chemical and biological soil properties (Puerta-Pinero et al. 2006). Some studies have suggested that mycorrhizal fungi could be a key ecological component to understand plant-plant facilitation processes (Van der Heijden and Horton 2009; Van der Putten 2009). Mycorrhizal fungi can interconnect plant individuals from different species or genera in natural communities (Newman 1988). These links between plants provide pathways for the transfer of major mineral nutrients such as N (He et al. 2005) and P (Smith et al. 2001).

All these scientific data show that it would be possible to introduce in agricultural practices some "nurse plants" providing mycorrhizal propagules for the benefit of the cultivated plant species (beneficiary plant species).

The Contribution of Mycorrhizal Symbiosis to Plant Nutrition and Plant Health

The Mycorrhizal Symbiosis and the Crop Nutrition

The most important benefit that AMF provide to their host plant is generally attributed to an enhancement of plant P uptake (Graham 2000). AM fungi can play an important role in crop P nutrition, increasing total uptake and sometimes P use efficiency (Koide et al. 2000) leading in some cases to a growth and yield enhancement (Koide et al. 2000). The extraradical hyphae are important to the plants for acquisition of phosphorus (P) and other mineral nutrients (Read and Perez-Moreno 2003). There are examples where no response of crops to colonisation by native AMF was recorded (Ryan et al. 2002). This is generally attributed to a high concentration of available soil P (Sorensen et al. 2005).

Nitrogen (N) is a major nutrient that frequently limits primary productivity in terrestrial ecosystems. AMF could cause changes in N-cycling processes through five general pathways: (i) substrate availability (NO_3), (ii) modification of the abiotic soil environment (improved aeration through macroaggregate formation), (iii) microbial community shift (the influence of the hyphal exudation structuring the hyphosphere microbial communities), (iv) individual host plant level effect and (v) plant community shifts (effect of the plant cover composition). AMF assimilate N predominantly in the form of NH_4^+ (Tanaka and Yano 2006). Moreover, it is largely admitted that AMF can mobilise N from organic sources (Barrett et al. 2011). AMF are also important in the plant acquisition of other nutrients such as zinc (Zn), calcium (Ca), copper (Cu), etc. (Clark and Zeto 2000).

It has been frequently reported that AMF were the main actors in the P nutrition and N uptake enhancement. Other soil microorganisms, e.g. N-fixing bacteria or P-solubilising bacteria, may synergistically interact with AMF and thereby benefit plant development and growth (Puppi et al. 1994). Bacteria associated with the AMF can also assist in the uptake of nutrients from soil. Numerous examples of this are available from bacterial-AM-legume tripartite symbiotic relationships, where diazotrophic bacteria provide fixed N not only for the plant but also for the fungus. Nodulation of legumes by N-fixing bacteria and establishment of AMF often occur simultaneously and synergistically. Genes for N fixation in endosymbiotic *Burkholderia* bacteria have been detected in AM hyphae (Minerdi et al. 2001) suggesting potentialities for improved N supply to mycorrhizal plants through fixation of atmospheric N.

The Mycorrhizal Effects Against Plant Pathogens (Nematodes, Phytoparasitic Plants)

Large quantities of pesticides are usually used to limit the spread of pests that enforce great yield losses. However, these chemical products are detrimental to human health and to the environment, and in addition, they are often partially effective against soilborne diseases. Numerous studies have shown the beneficial effects of AMF in acting against the plant pathogens and in increasing the plant tolerance to these biotic stresses. Most of these researches have been carried out in controlled conditions at early stages of plant growth. In these conditions, it has been demonstrated that AMF exerted antagonistic effects against pathogenic fungi such as *Fusarium*, *Rhizoctonia*, *Verticillium*, *Phytophthora*, *Pythium*, etc. (Whipps 2004) (Table 17.2). The same fungal influences have been recorded on different genera of plant-parasitic nematodes such as *Heterodera*, *Meloidogyne*, *Pratylenchus* and *Radopholus* (Hao et al. 2009) (Table 17.2).

Management of the Soil Mycorrhizal Potential in Mediterranean Cropping System

The adoption of the plant-plant facilitation process for agricultural management has to be adapted to the edaphic conditions encountered in the cultural area. Two main approaches could be followed according to the soil nutrient contents and soil biological diversity: to adapt the process to current agroecosystems (the "reductionist" approach) or to adapt agroecosystems to fit the process in question (the "holistic" approach) (Fester and Sawers 2014) (Fig. 17.2).

The reductionist approach is based on the criteria of the "controlled mycorrhization" practice and is realised by using a selected AM strain adapted to specific conditions and compatible with the target crop, whereas the objectives of the "holistic approach" are to maximise fungal diversity for ensuring AMF-dependant ecosystem services. Hence, the reductionist approach

Table 17.2 Examples of pathogenic fungi and plant-parasitic nematodes controlled by AM fungi (adapted from Gosling et al. 2006)

Pathogen	Diseases	Crop	References
Sclerotium cepivorum	White rot	Onion	Torres-Barragan et al. (1996)
Fusarium oxysporum	Fusarium root rot	Asparagus	Matsubara et al. (2002)
		French bean	Dar et al. (1997)
		Tomato	Datnoff et al. (1995)
Verticillium dahliae	Verticillium wilt	Tomato	Karagiannidis et al. (2002)
		Aubergine	Matsubara et al. (2000)
Rhizoctonia solani	Root and stem rots	Mung bean	Kjoller and Rosendahl (1996)
		Tomato	Berta et al. (2005)
Helicobasidium mompa	Violet root rot	Asparagus	Kasiamdari et al. (2002)
Aphanomyces euteiches	Root rot	Pea	Bodker et al. (2002)
Phytophthora parasitica	Root rots	Tomato	Cordier et al. (1996)
Meloidogyne incognita	Root-knot	Tomato	Talavera et al. (2001)

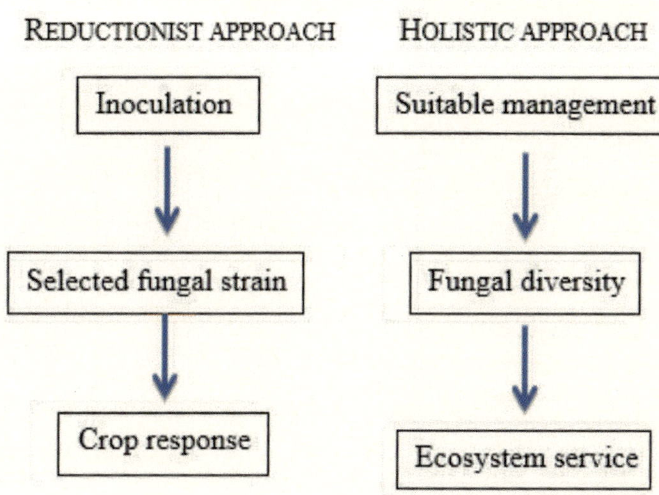

Fig. 17.2 Approaches to promote AM symbiosis for agronomic applications

will be encouraged to promote plant performance under low diversity circumstances. In contrast, the holistic approach will be implemented to conserve or restore native AMF diversity. In order to promote the AM fungal effect on plant growth through the management of the plant-plant facilitation process, some highly mycotrophic plant species (shrubs, tree) could be integrated in intercropping system to play the role of mycorrhizal propagule provider. Moreover and depending on the level of chemical fertility and AMF diversity, these nurse plants could be inoculated by selected AMF strains (reductionist strategy) or not inoculated but highly colonised by native AMF communities. For instance, the use of leguminous shrubs and trees as symbiotic islands under arid conditions could be beneficial for the AM propagule multiplication (Camargo-Ricalde and Dhillion 2003).

Experimental Results

The grain legume faba bean is cultivated worldwide as a protein source for food and feed. In addition, faba bean provides ecosystem services such as renewable inputs of nitrogen (N) into crops and soil resulting from biological N_2 fixation and a diversification of cropping systems (Tables 17.3 and 17.4). In some parts of the world

Table 17.3 Impact of faba bean intercropping on plant height, leaf area index (LAI), biomass accumulation and N uptake of three durum wheat varieties at time of faba bean incorporation (from Tosti and Guidicci (2010))

	Plant height (m)		LAI (m^2 m^{-2})		Biomass (Mg ha^{-1})		N uptake (kg ha^{-1})		N concentration (N)	
	2005	2006	2005	2006	2005	2006	2005	2006	2005	2006
Means of varieties										
Cappelli	0.28	0.30	1.04	1.64	1.22	1.59	18.5	24.7	1.53	1.57
Creso	0.21	0.25	0.82	1.13	0.98	1.23	17.5	20.7	1.80	1.65
Duilio	0.23	0.25	0.74	1.11	0.95	1.14	15.5	18.7	1.62	1.65
Means of techniques										
Intercropping	0.26	0.30	0.70	1.42	0.79	1.35	12.9	22.4	1.66	1.68
Sole crop	0.22	0.22	1.04	1.16	1.30	1.29	21.5	20.3	1.64	1.57
SEM										
Varieties	0.008*	0.015^{ns}	0.051*	0.04**	0.07^{ns}	0.063*	1.19^{ns}	2.28^{ns}	0.118^{ns}	0.100^{ns}
Techniques	0.006**	0.008**	0.040**	0.057*	0.068*	0.058^{ns}	1.12**	1.20^{ns}	0.055^{ns}	0.059^{ns}

Pooled standard error of the means (SEM) and significance of F test for varieties and techniques are reported. The variety × technique interaction was never significant so means of varieties (over techniques) and techniques (over varieties) are reported

ns not significant

**F* significant for $P=0.05$, ***F* significant for $P=0.01$

Table 17.4 Impact of faba bean intercropping on plant height, dry biomass accumulation, plant N uptake, grain N accumulation and nitrogen harvest index (NHI, % of plant allocated to the grain) recorded at maturity in three durum wheat varieties (from Tosti and Guidicci (2010))

	Plant height (m)		Biomass (Mg ha^{-1})		N uptake (kg ha^{-1})		Grain N (kg ha^{-1})		NHI (%)	
	2005	2006	2005	2006	2005	2006	2005	2006	2005	2006
Means of varieties										
Cappelli	1.09	1.06	6.79	8.36	63.0	60.0	45.0	39.8	71.2	66.6
Creso	0.63	0.61	5.18	5.71	57.9	50.8	43.9	40.6	76.1	79.5
Duilio	0.63	0.64	4.81	6.34	51.8	64.9	43.3	52.6	83.0	80.8
Means of techniques										
Intercropping	0.80	0.84	5.50	7.25	63.2	66.4	49.5	51.4	78.6	77.4
Sole crop	0.77	0.71	5.68	6.35	51.9	50.6	38.6	37.3	74.9	73.8
SEM										
Varieties	0.022**	0.030**	0.368*	0.554^{ns}	4.29^{ns}	5.60^{ns}	4.10^{ns}	3.93^{ns}	2.79^{ns}	1.83**
Techniques	0.017^{ns}	0.024**	0.398^{ns}	0.329^{ns}	3.66^{ns}	2.67*	2.65*	1.67*	0.61*	0.56**

Pooled standard error of the means (SEM) and significance of F test for varieties and techniques are reported. The variety × technique interaction was never significant so means of varieties (over techniques) and techniques (over varieties) are reported

ns not significant

* F significant for $P=0.05$, ** F significant for $P=0.01$

(e.g. China), intercropping of faba bean with maize and other cereals is a common practice (Li et al. 2009). In this context, it has been reported that nitrogen sources from N_2 fixation and soil will be uptaken more efficiently (Li et al. 2009; Tosti and Guidicci 2010). Intercropping of faba bean with cereals appeared as an efficient management tool to control weeds and plant diseases (Hauggaard-Nielsen et al. 2008). Planting the cereal with faba bean will ensure earlier canopy closure and soil cover. It has also been shown that legume parasitic weed broomrape can be controlled by intercropping faba bean with cereals (Fernandez-Aparicio et al. 2007). Recently, it has been demonstrated that faba bean, cocultured with wheat in Morocco, could significantly enhance the mycorrhizal soil infectivity which explains all the beneficial effects recorded from previous studies (unpublished data). Hence, these interactions between the leguminous plant species (e.g. faba bean) and the cereal could be interpreted as a plant-plant facilitation interaction where faba bean plays the role of the nurse plant (benefactor plant species) species, whereas the other crop plant (e.g. wheat, maize, etc.) is the beneficiary (facilitated), and the facilitation is mediated by mycorrhizal fungi.

Conclusion

AMF offer an environmentally sustainable approach to increase crop production and health. In order to improve the performances of AM symbiosis on plant growth, the introduction of nurse plants in cultural practices (conventional or alternative agricultural regimes) could be a useful biological tool to sustainably introduce the fungal symbionts into the agricultural areas. However, major research efforts will be necessary to identify some biological parameters that will be discriminant for the choice of the AM fungal management strategy (reductionist vs. holistic strategies). In addition, it will be necessary to refine the techniques for detecting and following AMF strains or communities in the field.

References

Baraza E, Zamora R, Hodar JA (2006) Conditional outcomes in plant-herbivore interactions: neighbors matter. Oikos 113:148–156

Barea JM, Azcon-Aguilar C, Azcon R (1997) Interactions between mycorrhizal fungi and rhizosphere microorganisms within the context of sustainable soil–plant systems. In: Gange AC, Brown VK (eds) Multitrophic interactions in terrestrial systems. Blackwell Science, Cambridge, pp 65–77

Barrett G, Campbell CD, Fitter AH, Hodge A (2011) The arbuscular mycorrhizal fungus *Glomus hoi* can capture and transfer nitrogen from organic patches to its associated host plant at low temperature. App Soil Ecol 48:102–105

Bedini S, Pellegrino E, Avio L, Pellegrini S, Bazzofi P, Argese E, Giovannetti M (2009) Changes in soil aggregation and glomalin-related soil protein content as affected by the arbuscular mycorrhizal fungal species *Glomus mosseae* and *Glomus intraradices*. Soil Biol Biochem 41:1491–1496

Berta G, Sampo S, Gamalero E, Massa N, Lemanceau P (2005) Suppression of Rhizoctonia root-rot of tomato by *Glomus mosseae* BEG12 and *Pseudomonas fluorescens* A6R1 is associated with their effect on the pathogen growth and on the root morphogenesis. Eur J Plant Pathol 111:279–288

Bethlenfalvay GJ, Schuëpp H (1994) Arbuscular mycorrhizas and agrosystem stability. In: Gianinazzi S, Schuëpp H (eds) Impact of arbuscular mycorrhizas on sustainable agriculture and natural ecosystems. Birkhaüser Verlag, Basel, pp 117–131

Bodker L, Kjoller R, Kristensen K, Rosendahl S (2002) Interactions between indigenous arbuscular mycorrhizal fungi and *Aphanomyces euteiches* in field-grown pea. Mycorrhiza 12:7–12

Bonfante P, Anca IA (2009) Plants, mycorrhizal fungi, and bacteria: a network of interactions. Annu Rev Microbiol 63:363–383

Boucher DH, James S, Keeler KH (1982) The ecology of mutualism. Ann Rev Ecol Syst 13:315–347

Callaway RM (2007) Positive interactions and interdependence in plant communities. Springer, Dordrecht, p 415

Camargo-Ricalde SL, Dhillion SS (2003) Endemic Mimosa species can serve as mycorrhizal "resource islands" within semi-arid communities of the Tehuacan-Cuicatlan Valley, Mexico. Mycorrhiza 13:129–136

Castillo JP, Verdu M, Valiente-Banuet A (2010) Neighborhood phylodiversity affects plant performance. Ecology 91:3656–3663

Clark RB, Zeto SK (2000) Mineral acquisition by arbuscular mycorrhizal plants. J Plant Nutr 23:867–902

Cordier C, Gianinazzi S, Gianinazzi-Pearon V (1996) Colonisation patterns of root tissues by *Phytophthora nicotianae* var. *parasitica* related to reduced disease in mycorrhizal tomato. Plant Soil 185:223–232

Dar GH, Zargar MY, Beigh GM (1997) Biocontrol of Fusarium root rot in the common bean (*Phaseolus vulgaris* L.) by using symbiotic *Glomus mosseae* and *Rhizobium leguminosarum*. Microb Ecol 34:74–80

Datnoff LE, Nemec S, Pernezny K (1995) Biological control of *Fusarium* crown and root rot of tomato in Florida using *Trichoderma harzianum* and *Glomus intraradices*. Biol Control 5:427–431

Ewel JJ (1999) Natural systems as models for the design of sustainable systems of land uses. Agrofor Syst 45:1–21

Fernandez-Aparicio M, Sillero JC, Rubiales D (2007) Intercropping with cereals reduces infection by Orobanche crenata in legumes. Crop Prot 26:1166–1172

Fester T, Sawers R (2014) Progress and challenges in agricultural applications of arbuscular mycorrhizal fungi. Crit Rev Plant Sci 30:459–470

Francis DF, Thornes JB (1990) Matorral: erosion and reclamation. In: Albaladejo J, Stocking MA, Diaz E (eds) Soil degradation and rehabilitation in Mediterranean environmental conditions. CSIC, Murcia, pp 87–115

Frey-Klett P, Chavatte M, Clausse ML, Courrier S, Le Roux C, Raaijmakers J, Martinotti MG, Pierrat JC, Garbaye J (2005) Ectomycorrhizal symbiosis affects functional diversity of rhizosphere fluorescent pseudomonads. New Phytol 165:317–328

Garcia C, Hernandez T, Roldan A, Albaladejo L (1997) Biological and biochemical quality of a semiarid soil after induced revegetation. J Environ Qual 26:1116–1122

Gianinazzi S, Gollotte A, Binet MN, Van Tuinen D, Redecker D, Wipf D (2010) Agroecology: the key role of arbuscular mycorrhizas in ecosystem services. Mycorrhiza 20:519–530

Gomez-Aparicio L, Zamora R, Gomez JM, Hodar JA, Castro J, Baraza E (2004) Applying plant facilitation to forest restoration in Mediterranean ecosystems: a meta-analysis of the use of shrubs as nurse plant. Ecol Appl 14:1128–1138

Gosling P, Hodge A, Goodlass G, Bending GD (2006) Arbuscular mycorrhizal fungi and organic farming. Agric Ecosyst Environ 113:17–35

Graham JH (2000) Assessing cost of f arbuscular mycorrhizal symbiosis in agrosystems. In: Podila GK, Douds JDD (eds) Current advances in mycorrhizal research. APS Press, St. Paul, pp 127–140

Hao Z, Fayolle L, Van Tuinen D, Gianinazzi-Pearson V, Gianinazzi S (2009) Mycorrhiza reduce development of nematode vector of Grapevine fanleaf virus in soils and root systems. In: Boudon-Padfieu E (ed) Extended abstract 16th meeting of ICVG, Dijon, pp 100–1001

Hauggaard-Nielsen H, Jørnsgard B, Kinane J, Jensen ES (2008) Grain legume-cereal intercropping: the practical application of diversity, competition and facilitation in arable and organic cropping systems. Renew Agric Food Syst 23:3–12

He XH, Critchley C, Ng H, Bledsoe C (2005) Nodulated N_2-fixing *Casuarina cunninghamiana* is the sink for net N transfer from non-N_2-fixing *Eucalyptus maculata* via an ectomycorrhizal fungus *Pisolithus* sp. using 15NH4+ or 15NO3-Supplied as ammonium nitrate. New Phytol 167:897–912

Karagiannidis N, Bletsos F, Stavropoulos N (2002) Effect of *Verticillium* wilt (*Verticillium dahliae Kleb.*) and mycorrhiza (*Glomus mosseae*) on root colonization, growth and nutrient uptake in tomato and eggplant seedlings. Sci Hortic 94:145–156

Kasiamdari RS, Smith SE, Smith FA, Scott ES (2002) Influence of the mycorrhizal fungus *Glomus coronatum* and soil phosphorus on infection and disease caused by binucleate *Rhizoctonia* and *Rhizoctonia solani* on mung bean (*Vigna radiata*). Plant Soil 238:235–244

Kjoller R, Rosendahl S (1996) The presence of arbuscular mycorrhizal fungus *Glomus intraradices* influences enzymatic activities of the root pathogen *Aphanomyces euteiches* in pea roots. Mycorrhiza 6:487–491

Koide RT, Goff MD, Dickie IA (2000) Component growth efficiencies of mycorrhizal and non mycorrhizal plants. New Phytol 148:163–168

Li YY, Yu CB, Cheng X, Li CJ, Sun JH, Zhang FS, Lambers H, Li L (2009) Intercropping alleviates the inhibitory effect of N fertilisation on nodulation and symbiotic N_2 fixation of faba bean. Plant Soil 323:295–308

Marulanda A, Barea JM, Azcon R (2006) An indigenous drought-tolerant strain of *Glomus intraradices* associated with a native bacterium improves water transport and root development in *Retama sphaerocarpa*. Microb Ecol 52:670–678

Matsubara Y, Kayukawa Y, Yano M, Fukui H (2000) Tolerance of asparagus infected with arbuscular mycorrhizal fungus to violet root rot caused by *Helicobasidium mompa*. J Jpn Soc Hortic Sci 69:552–556

Matsubara Y, Hasegawa N, Fukui (2002) Incidence of *Fusarium* root rot in asparagus seedlings infected with arbuscular mycorrhizal fungus as affected by several soil amendments. J Jpn Soc Hortic Sci 71:370–374

Minerdi D, Fani R, Gallo R, Boarino A, Bonfante P (2001) Nitrogen fixation genes in an endosymbiotic *Burkholderia* strain. Appl Environ Microbiol 67:725–732

Montesinos-Navarro A, Segarra-Moragues JG, Valiente-Banuet A, Verdu M (2012) Plant facilitation occurs between species differing in their associated arbuscular mycorrhizal fungi. New Phytol 196:835–844

Newman EI (1988) Mycorrhizal links between plants: their functioning and ecological significance. Adv Ecol Res 18:243–270

Odum EP (1959) Fundamentals of ecology. Saunders, Philadelphia, p 546

Puerta-Pinero C, Gomez JM, Zamora R (2006) Species-specific effects on topsoil development affect *Quercus ilex* seedling performance. Acta Oecol 29:65–71

Puppi G, Azcon R, Hoeflich G (1994) Management of positive interactions of arbuscular mycorrhizal fungi with essential groups of soil microorganisms. In: Gianinazzi S, Schuepp H (eds) Impact of arbuscular mycorrhizas on sustainable agriculture and natural ecosystems. Birkhauser Verlag, Basel, pp 201–215

Read DJ, Perez-Moreno J (2003) Mycorrhizas and nutrient cycling in ecosystems – a journey towards relevance? New Phytol 157:475–492

Requena N, Perez-Solis E, Azcon-Aguilar C, Jeffries P, Barea JM (2001) Management of indigenous plant–microbe symbioses aids restoration of desertified ecosystems. Appl Environ Microbiol 67:495–498

Rillig MC, Wright SF, Nichols KA, Schmid WF, Torn MS (2002) The role of arbuscular mycorrhizal fungi and glomalin in soil aggregation: comparing effects of five plant species. Plant Soil 238:325–333

Ryan MH, Norton RM, Kirkegaard JA, McCormick KM, Knights SE, Angus JF (2002) Increasing mycorrhizal colonization does not improve growth and nutrition of wheat on vertisols in south-eastern Australia. Aust J Agric Res 53:1173–1181

Smith SE, Read DJ (2008) Mycorrhizal symbiosis, 3rd edn. Academic, London, pp 145–18

Smith SE, Dickson S, Smith FA (2001) Nutrient transfer in arbuscular mycorrhizas: how are fungal and plant processes integrated? Aust J Plant Physiol 28:683–694

Sorensen N, Larsen J, Jakobsen I (2005) Mycorrhiza formation and nutrient concentration in leeks (*Allium porrum*) in relation to previous crop and cover crop management in high P soils. Plant Soil 273:101–114

Talavera M, Itou K, Mizukubo T (2001) Reduction of nematode damage by root colonization with arbuscular mycorrhiza (*Glomus* spp.) in tomato- *Meloidogyne incognita* (Tylenchida, Meloidognidae) and carrot-Pratylenchus *penetrans* (Tylenchida, Pratylenchidae) pathosystems. Appl Entomol Zool 36:387–392

Tanaka Y, Yano K (2006) Nitrogen delivery to maize via mycorrhizal hyphae depends on the form of N supplied. Plant Cell Environ 28:1247–1254

Torres-Barragan A, Zavale-Tamejia E, Gonzales-Chavez C, Ferrera-Cerrato R (1996) The use of arbuscular mycorrhizae to control onion white rot (*Sclerotium cepivorum*) under field conditions. Mycorrhiza 6:253–257

Tosti G, Guidicci M (2010) Durum wheat-faba bean temporary intercropping: effects on nitrogen supply and wheat quality. Eur J Agron 33:157–165

Valiente-Banuet A, Verdu M (2007) Facilitation can increase the phylogenetic diversity of plant communities. Ecol Lett 10:1029–1036

Valiente-Banuet A, Verdu M (2008) Temporal shifts from facilitation to competition occur between closely related taxa. J Ecol 96:489–494

Valiente-Banuet A, Vital Rumebe A, Verdu M, Callaway RM (2006) Modern quaternary plant lineages promote diversity through facilitation of ancient tertiary lineages. PNAS 103:16812–16817

Van der Heijden MGA, Horton TR (2009) Socialism in soil? The importance of mycorrhizal fungal networks for facilitation in natural communities. J Ecol 97:1139–1150

Van der Heijden MGA, Klironomos JN, Ursic M, Poutoglis P, Streitwolf-Engel R, Boller T, Wiemken A, Sanders IR (1998) Mycorrhizal fungal diversity determines plant biodiversity, ecosystem variability and productivity. Nature 396:69–72

Van der Putten WH (2009) A multitrophic perspective on functioning and evolution of facilitation in plant communities. J Ecol 97:1131–1138

Whipps JM (2004) Prospects and limitations for mycorrhizas in biocontrol of root pathogens. Can J Bot 82:1198–1227

Role of Plant Growth-Promoting Rhizobacteria and Their Formulation in Biocontrol of Plant Diseases

18

S.S. Shaikh and R.Z. Sayyed

Abstract

Fungal phytopathogens are a major threat to crop plants. More than 60 fungal phytopathogens are known to cause various plant diseases. These diseases lead to the loss of 30 % crop yield that accounts to about 20,000 crore rupees per year in the global market. Different methods of chemotherapy have been used to control these diseases. However, uncontrolled use of chemical fungicides has severely affected the agroecosystem, leading to the development of pesticide-resistant pests and emergence of new species of pests. Increased dependence on chemicals is inevitably associated with environmental and health hazards. In this regard, plant growth-promoting rhizobacteria (PGPR) have been found as an effective, eco-friendly, and sustainable replacement to the chemical fungicides. The present chapter focuses on antifungal traits such as siderophores, lytic enzymes, hydrogen cyanide, and antibiotics of PGPR, development of effective formulation of PGPR for longer shelf life and competitiveness with the natural environment, obstacles in formulation development, and characteristics of bioformulation. Biotechnological attributes of PGPR with respect to plant growth promotion, disease resistance, and iron nutrition are also discussed.

Introduction

The world population is continuously increasing and it is estimated to reach ~8 billion by 2020. There is a real challenge to feed this huge population, an endeavor that requires agricultural productivity to be increased. Thus, more than ever, obtaining high yields is the main challenge for agriculture. In addition, in recent years, both producers and consumers are increasingly focusing on health and quality of food and on their organoleptic and

S.S. Shaikh • R.Z. Sayyed (✉)
Department of Microbiology, Shri S. I. Patil Arts, G. B. Patel Science and S.T.S.K.V.S. Commerce College, Shahada, Dist., Nandurbar, Maharashtra 425409, India
e-mail: riyaz829@yahoo.co.in; sayyedrz@gmail.com

N.K. Arora (ed.), *Plant Microbes Symbiosis: Applied Facets*,
DOI 10.1007/978-81-322-2068-8_18, © Springer India 2015

nutritional properties. Fungal diseases of crop plants have led to the development of micronutrient-deficient food. Consumption of such staple food has led to micronutrient-deficient anemia, for instance, iron deficiency anemia (IDA). PGPRs are often seen as potential candidates that provide various micronutrients to crop plants by solubilizations, which decrease micronutrient deficiencies. PGPRs have been used as effective bioinoculants, biocontrol agents, mineral solubilizers, etc. In spite of various useful activities of PGPR, they fail to perform these functions under natural field conditions. A formulation of PGPR for sustained release and longer survival may improve their performance under field condition.

Plant Growth-Promoting Rhizobacteria (PGPR)

The term PGPR derived by Kloepper et al. (1980) is a group of plant-beneficial rhizobacteria, potentially useful for stimulating plant growth and increasing crop yields. In the last few years, the number of PGPRs has been found to increase mainly due to their role in the rhizosphere. Various species of bacteria such as *Pseudomonas*, *Azospirillum*, *Azotobacter*, *Klebsiella*, *Enterobacter*, *Alcaligenes*, *Arthrobacter*, *Burkholderia*, *Bacillus*, and *Serratia* have been reported to enhance the plant growth and suppress phytopathogens (Sayyed and Chincholkar 2009; Sayyed and Patel 2011; Sayyed et al. 2012, 2013; Saharan and Nehra 2011).

Bashan and Holguin (1998) proposed the division of PGPR into two classes: biocontrol plant growth-promoting bacteria (biocontrol PGPB) and plant growth-promoting bacteria (PGPB). This classification may include beneficial bacteria that are not rhizosphere bacteria, but this classification system has not been widely accepted. PGPR are known to provide essential nutrients for plant growth or to enhance their availability, to play an important role in pathogen suppression, and to improve the health of soil in the long term (Vessey 2003; Lucy et al. 2004; Lugtenberg and Kamilova 2009), and therefore are potentially important in reducing the use of chemical fertilizers and pesticides.

Biotechnological Attributes of PGPR

Plant Growth Promotion

PGPR increases the plant growth by nitrogen fixation (Kloepper et al. 1989), production of phytohormones (Gutierrez et al. 2001; Patten and Glick 1996), production of siderophore (Sayyed and Chincholkar 2009; Sayyed and Patel 2011), lowering of ethylene concentration (Madhaiyan et al 2007; Chinnadurai et al. 2009), and solubilization of phosphorous and various other minerals (Sayyed et al. 2007b; Sharma et al. 2013a).

Iron Nutrition in Crops

Iron is the fourth most abundant and common element present in the soil. It is an essential element for growth, metabolism, and survival of the majority of cell types on the Earth. It is an essential component of different processes involved in the biosynthetic pathways and in the formation of chlorophyll (Lopez-Millan et al. 2001), but it is rarely found in free form (Höfte 1993). Thus, it remains unavailable for crops even in iron-rich soils (Podile and Kishore 2006). When the availability of iron is inadequate for the growth of crops, leaves become pale green, yellow, or white in color, and eventually brown indicating iron deficiency (Brittenham 1994). Iron deficiency is a common nutritional disorder in many crop plants, resulting in poor yields and iron-deficient staple crops. The application of siderophore-producing PGPR is known to provide iron nutrition and to help in iron uptake by plants (Podile and Kishore 2006). Most microorganisms have developed efficient methods of absorbing iron from the environment and many microorganisms secrete siderophores to mobilize iron (Hider and Kong 2010). Once Fe (III) is mobilized and captured by the secreted siderophore molecule, it is accessible for cellular uptake by the plants. Sharma et al. (2013b) have shown that treatment of rice plants with siderophore-producing PGPRs results in increased iron content of the plants. Treatments with *Pseudomonas putida* B 17 and B 19 almost doubled the iron content of the plants.

Plant Disease Control

Biological control of plant diseases using antagonistic PGPR offers a highly effective, economical, and environment-friendly alternative to the use of synthetic pesticides (Emmert and Handelsman 1999). Plant disease suppression by soil microorganisms is the best possible alternative of reducing the chemical input in agriculture (Compant et al. 2005). Rhizobacteria are effective competitors in the rhizosphere, which can establish and persist on the roots of plants that are agronomically important.

The mode of action of the antagonistic organisms against various soilborne plant pathogenic fungi includes biosynthesis of antibiotics, production of hydrolytic enzymes, synthesis of cell wall lytic enzymes (Kim et al. 2003), production of siderophore, competition for substrates and the sites on the roots, and induced systemic resistance (ISR) (Ramamoorthy et al. 2001; Nakkeeran et al. 2004). Successful bacterial antagonists often show a synergistic combination of mechanisms responsible for a successful antifungal interaction.

Fluorescent pseudomonads and closely related species are important antagonistic bacteria present in the soil. Some of the fluorescent pseudomonads have currently received worldwide attention due to the production of a wide range of antifungal compounds, viz., fluorescent pigments, siderophores, volatile compounds such as hydrocyanic acid (HCN), antibiotics, and lytic enzymes (Sullivan and Gara 1992). Lytic enzymes (chitinase, β-1,3-glucanase, and protease) of PGPR are responsible for the lysis and hyperparasitism of antagonists against deleterious fungal pathogens. Such bacterial strains have been implicated in the plant growth as well as in the inhibition of plant pathogenic fungi and deleterious rhizobacteria showing a significant increase in root colonization. These attributes make PGPR an effective biocontrol agent (Garbeva et al. 2004; Ran et al 2005).

Some PGPRs do not produce metabolites against the pathogens, but still protect the plant from the disease, due to alteration of host defense mechanism accounting for the protection of observed disease. This mechanism includes ISR or systemic acquired resistance (SAR). ISR/SAR is defined as the activation of chemical and physical defense mechanism of the host plant by an inducer that can be a chemical or a microorganism, leading to the control of several pathogens (Kloepper et al. 1992) (Fig. 18.1).

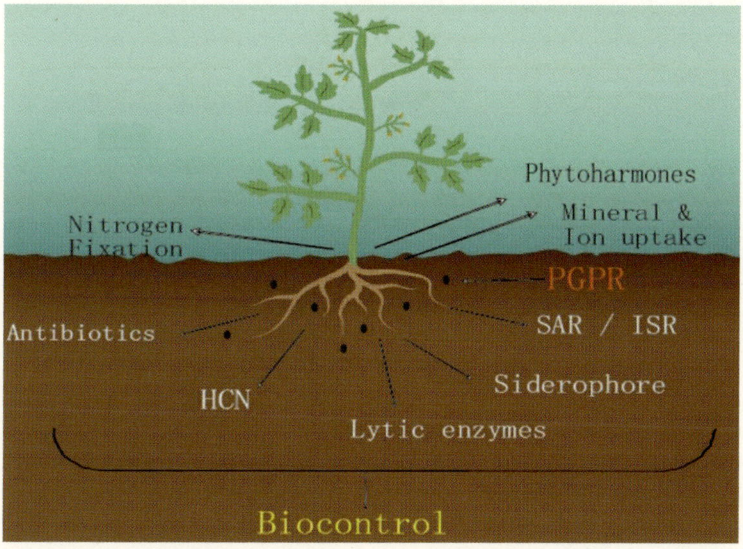

Fig. 18.1 Mechanism of action of PGPR including biocontrol

Antifungal Metabolites of PGPR

Siderophores

A siderophore (Greek word for iron carrier) is a low molecular weight (500–1,000 daltons), high-affinity, ferric, ion-chelating compound secreted by organisms. Under iron stress conditions, some microorganisms produce siderophores that chelate the available iron and prevent the iron nutrition of respective phytopathogen, and thereby restrict the proliferation and root colonization by the phytopathogen (Chaiharn et al. 2009; Miethke and Marahiel 2007; Sayyed and Chincholkar 2009). Siderophore-producing rhizobacteria are also known to impart ISR to plants (Pieterse et al. 2001) and suppressiveness to the soil (Mazzola 2002) and have been implicated in the biocontrol of several plant diseases (Sayyed et al. 2005).

Siderophore-producing PGPRs function as biocontrol agents by depriving the pathogen from iron nutrition (Lemanceau and Albouvette 1993) (Fig. 18.2). Siderophore-producing rhizobacteria rapidly colonize the plant roots of several crops and secrete an array of antifungal metabolites (Hass and Defago 2005). Chaiharn et al. (2009) studied the suppression of rice pathogen by siderophore-producing isolates. Gupta et al. (2002) studied the suppression of *Macrophomina phaseolina* causing charcoal rot of peanut by *Pseudomonas*. Freitas and Pizzinato (1997) have reported the inhibition of *Colletotrichum gossypii* by siderophore-producing rhizobacteria. Sindhu et al. (1997) reviewed the role of PGPR in the inhibition of phytopathogens. Johri et al. (1997) have reported the role of siderophore-producing fluorescent *Pseudomonas* strain RBT 13 that exhibited antagonistic action against several bacterial and fungal plant pathogens. Akhtar and Siddiqui (2009) reported that siderophore-producing pseudomonad significantly reduced the root rot disease in chickpea. However, microorganisms having the ability to produce powerful siderophores can only become ecologically competent biological control agents (BCAs) provided that they exhibit strong root-colonizing potential (Chincholkar et al. 2000).

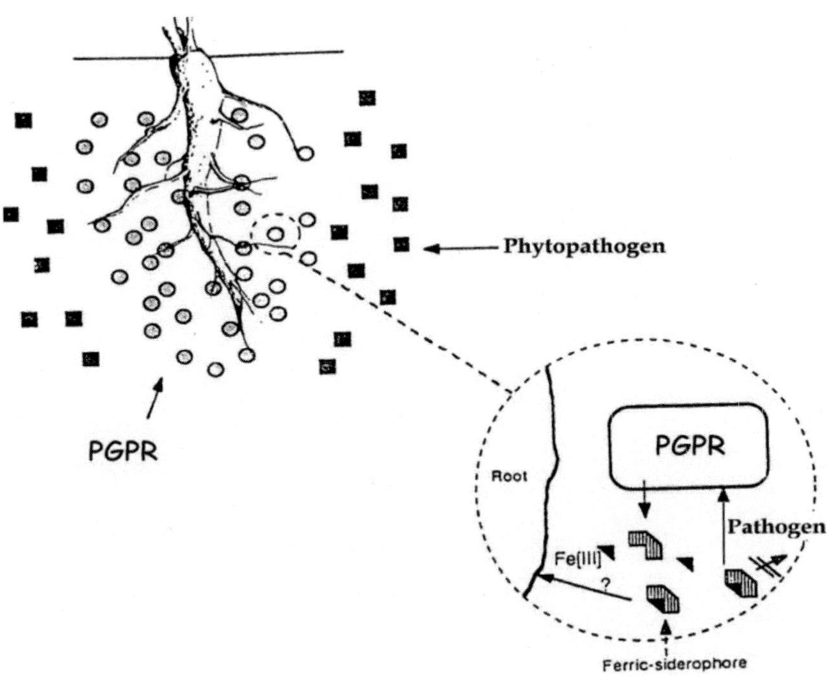

Fig. 18.2 Suppression of phytopathogens by siderophore-producing PGPR

Siderophore-based BCAs are gaining commercial significance because they are safer, their self-replication circumvents the need for repeated application, they do not lead to biomagnification, and the target organisms do not develop pesticide resistance (Sayyed et al. 2005). They also provide iron nutrition to roots, and thereby promote plant growth (Sayyed et al. 2007a).

Lytic Enzymes

Some PGPR strains have been found to produce enzymes that can lyse the cell wall of phytopathogenic fungi. Cell walls of most of the phytopathogenic fungi (except oomycetes) are made up of chitin $(C_8H_{13}O_5N)_n$, an unbranched, long-chain polymer of glucose derivatives, composed of β-1,4-linked units of the amino sugar N-acetyl-D-glucosamine (NAG). Chitinases (EC 3.2.1.14) are hydrolytic enzymes that break down glycosidic bonds in chitin (Kim et al. 2003). Thus, chitinases play a vital role in the biological control of many plant diseases by degrading the chitin polymer in the cell walls of fungal pathogens. It affects fungal growth through the lysis of cell walls, hyphal tips, and germ tubes. Chitinase is an inducible enzyme secreted by many microorganisms in the cultures containing chitin or its oligomers as sole carbon source. Chitinolytic enzymes have been reported from *Streptomyces* (Kim et al. 2003), marine isolates (Park et al. 2000), isolates from alkaline soil (Loni et al. 2011), and rhizosphere (Kishore and Pande 2007; Idris et al. 2007; Kamil et al. 2007). These microorganisms have been used as BCAs because the chitinases are involved in the antagonistic and antifungal actions against many phytopathogenic fungi (Singh et al. 1999; Frankowski et al. 2001; Gohel et al. 2005; Kim et al. 2008; Ramos-Solano et al. 2010). Velusamy and Kim (2011) showed chitinolytic activity of *Enterobacter* species against fungal hyphae. Some biocontrol strains produce enzymes including chitinases, cellulases, β-1,3 glucanases, proteases, and lipases that can lyse a portion of the cell wall of many pathogenic fungi (Glick 2012).

Hydrogen Cyanide

Hydrogen cyanide (HCN) is a secondary metabolite of some bacteria (Schippers et al. 1990). It is a colorless or pale blue liquid or gas with a faint, bitter, almondlike odor. It is a highly toxic, volatile compound that interferes with the cellular respiration by inhibiting the cytochrome oxidase enzyme in the mitochondria. This, in turn, prevents the production of ATP (adenosine-5′-triphosphate), the molecule that transfers energy in the cell. Thus, the cell is deprived of energy, and as a result, its growth is suppressed. It helps in the suppression of aerobic root pathogens (Schippers et al. 1990). The production of HCN requires glycine as a precursor (Castric 1977). Voisard et al. (1989) found that HCN-producing fluorescent *Pseudomonas* strain CHAO has the ability to suppress black root rot of tobacco, while HCN mutant of the wild-type strain of CHAO had lost its ability. HCN production in the rhizosphere has different effects on different plant types; still the role of HCN production in the control of root pathogens is unclear.

The detection of HCN in rhizosphere microbes has been described by many workers (Lorck 1948; Castric 1975; Askeland and Morrison 1983). The HCN produced by Gram-negative *Pseudomonas fluorescens*, *P. aeruginosa*, and *Chromobacterium violaceum* (Askeland and Morrison 1983) is detected by observing the change in color (from yellow to brownish red) of the indicator paper, amended with picric acid solution.

Antibiotics

Some well-known PGPR biocontrol strains such as *Pseudomonas*, *Bacillus*, *Azospirillum*, *Rhizobium*, and *Serratia* species are involved in the production of antibiotics such as phenazine-1-carboxylic acid, 2,4-diacetylphloroglucinol (DAPG), oomycin, pyoluteorin, pyrrolnitrin, kanosamine, zwittermicin A, and pantocin (Fernando et al. 2005). These antibiotics have been reported to be involved in the suppression of different pathogens by inducing fungistasis,

Table 18.1 Role of antifungal metabolites of PGPR in biocontrol of phytopathogens

AFM	Producing PGPR	Target pathogen	Crop
Siderophore	*P. fluorescence*	*Erwinia carotovora*	Potato
		G. graminis	Wheat
		Fusarium glycinia	Wheat
		Sarocladium oryzae	Soybean
	P. putida	*Fusarium* sp. wilt	Radish, cucumber
		Fusarium solani	Beans
		Erwinia carotovora	Potato
	P. cepacia	*F. oxysporum*	Onion
	P. aureofaciens	*G. graminis* var. *tritici*	Wheat
	B. pumilus	*G. graminis* var. *tritici*	Wheat
	Enterobacter aerogenes	*P. cactorum*	Apple
	Bradyrhizobium sp.	*F. solani*	Sunflower
		R. solani	Mung bean
	Rhizobium meliloti	*M. phaseolina*	Groundnut
Chitinase	*Enterobacter* spp.	*Aphanomyces* and *Pythium*	Pea
		Botrytis	Apple
	Streptomycetes	*S. sclerotiorum*	Potato, tomato
	Bacillus thuringiensis	*R. solani*	Cotton
		All phytopathogens	
	Pseudomonas sp.	*Fusarium* sp.	Banana
		M. phaseolina	Maize, payn
HCN	*P. fluorescence*	*Thielaviopsis basicola*	Tobacco, wheat
		G. graminis	
Antibiotics	*P. fluorescens* (DAPG)	*G. graminis tritici*	Take-all disease
		Thielaviopsis basicola	Tobacco
		P. ultimum	Sugar beet
	P. cepacia	*Bipolaris maydis*	Maize
	(Pyrrolnitrin)		
	P. fluorescence	*Sclerotinia homoeocarpa*	Grass
	(Pyoluteorin)	*Pythium* sp.	Cucumber

inhibition of spore germination, lysis of fungal mycelia, or exerting fungicidal effects (Fravel 1988). Cronin et al. (1997) have reported that DAPG produced by *Pseudomonas* species is responsible for suppressing phytopathogens. It was demonstrated that with increasing populations of *Pseudomonas* species producing the antibiotic DAPG, there was a rapid decline in take-all disease in wheat caused by the fungus *Gaeumannomyces graminis* var. *tritici* (de Souza et al. 2003). Several strains of *Bacillus subtilis* produce iturin (Yu et al. 2002; Chitarra et al. 2003) bacillomycin D (Besson and Michel 1992), plipastatins, surfactin, and aminopolyol that exert strong antimicrobial action against *Pythium ultimum*, *Rhizoctonia solani*, *Fusarium oxysporum*, *Sclerotinia sclerotiorum*, and *M. phaseolina* (Constantinescu 2001).

Phenazine, a potent antibiotic produced by *P. fluorescens*, has been used to control take-all disease of wheat caused by *G. graminis*. *Agrobacterium radiobacter*, a first commercially applied BCA for controlling crown gall in dicots, produces agrocin, which specifically inhibits *Agrobacterium tumefaciens* (Sindhu et al. 1997). Table 18.1 shows the role of AFMs of PGPR in the biocontrol of phytopathogens. Genetic analysis of many biocontrol strains of *Pseudomonas* indicated that there is a positive correlation between disease suppression and antibiotic production (Vincent et al. 1991).

Development of Effective Formulation

Though biocontrol with PGPR is an acceptable green approach, the registration of biocontrol agents for commercial availability is very low. Handling these products requires care, but these are not properly handled by the people who use it. This is the reason that there are so many PGPRs that perform very well in controlled conditions but not in field conditions, and hence the commercialization of these products is not easy. Since bioformulation is a living product, utmost care is needed at all the steps beginning from the production till the end use to maintain the microbial load and vigor. In addition, the present-day bioproducts can be further improved to obtain greater efficiency levels. The development of formulations with increased shelf life and broad spectrum of action with consistent performance under field conditions can pave the way for commercialization of the technology at a faster rate.

In general, biofertilizers (rhizobacteria) are living microorganisms. Unlike chemical fertilizers, they themselves are not the source of nutrient, but can help the plants in accessing the nutrients available in its surrounding environment (Table 18.2). The development of a successful inoculant involves several critical elements such as strain selection, selection of a carrier, mass multiplication, formulation of the inoculant, and packaging and marketing. The functions of formulations include stabilizing the microorganism, aiding in its delivery, protecting the microbe at the target zone, and enhancing the functionality of the microbe against the target (if a BCA).

Characteristics of PGPR and Their Formulation

To develop a successful PGPR formulation, rhizobacteria should possess certain characteristics such as high rhizosphere competence; high competitive saprophytic ability; enhanced plant growth; ease for mass multiplication; broad spectrum of action; excellent and reliable control; safe to environment, humans, and animals; compatible with other rhizobacteria; and tolerance to desiccation, heat, oxidizing agents, and UV radiations (Nakkeeran et al. 2005).

To be an ideal formulation, it should have increased shelf life. It should not be phytotoxic to plants. It should be dissolved well in water and should release the bacteria. It should be cost-effective and should give reliable control of plant diseases. It should be compatible with other agrochemicals, and the carriers (for carrying the microbe) must be cheap and readily available for formulation development (Nakkeeran et al. 2005).

Types of Formulations

The formulations are in two forms, dry powder (solid) and liquid suspensions, which can be prepared in several ways.

Dry Powders

Organisms can be formulated into concentrated dry or wet powder for easy storage, transport, and application, which comprises of dust, granules, and briquettes for the preparation of formulations (Jones and Burges 1998). The types of powder preparation are wettable dry powder and moist powder culture. The wettable dry powder is normally spray-dried or lyophilized biomass with practically no free moisture for the growth of bacteria. In moist powder, culture organisms that do not form a mat of biomass can easily be formulated into a moist powder suitable for spraying, seed coating, or direct discharge into the soil (Paau 1998).

Different carrier materials used are peat, turf, talc, lignite, kaolinite, pyrophyllite, zeolite, montmorillonite, alginate, press mud, sawdust, and vermiculite (Kloepper and Schroth 1981; Vidhyasekaran and Muthamilan 1995; Rabindran and Vidhyasekaran 1996; Nakkeeran et al. 2005). Carriers increase the survival rate and shelf life of bacteria by protecting them from desiccation and also provide suitable microenvironment for rapid growth after their release. Rabindran and Vidhyasekaran (1996) used various formulations

Table 18.2 Partial list of commercially available formulations of PGPR as biocontrol agents as practiced in various parts of the world (Adopted from Sayyed et al. 2012)

Trade name	Biocontrol organism	Target pathogen/disease	Crop under application	Formulation	Mode of application
Bio-save 10[5]	*P. syringae*	*Botrytis cinerea*, *Mucor piriformis*, *Geotrichum*, and *Penicillium* sp.	Citrus and pome fruit	Wettable powder	Postharvest Drench, dip, and spray
BlightBan A506[6]	*P. fluorescence*	*Erwinia amylovora*	Almond, cherry, apple, potato, and tomato	Wettable powder	Postharvest Drench, dip, and spray
Conquer[8]¼Victus[26]	*Pseudomonas* sp.	*Pseudomonas tolaasii*	Mushroom	Aqueous biomass Suspension	Spray
Epic[11]	*Bacillus subtilis*	Various fungi	Cotton and legumes	Dry powder 5.5×10^{10} spores/g	Drip and potting mix in row
Mycostop[15]	*Streptomycetes* sp.	*Fusarium*, *Alternaria*, *Pythium*	Field ornamental and vegetable	Powder	Drench, spray, or through ligation
Intercept[14]	*P. cepacia*	*R. solani*, *Fusarium*, and *Pythium* sp.	*R. solani*, *Fusarium*, and *Pythium* sp.	–	–
Galltrol-A[13]	*A. radiobacter*	*A. tumefaciens* crown gall	Fruit, nut, and ornamental nursery plants	Bacterial suspension 1.2×10^{11} cfu	Suspension applied to seeds, seedlings, cuttings, root, stem, and as a soil drench
Kodiak[11], Kodiak HB, Kodiak AT	*B. subtilis*	*R. solani*, *Fusarium*, *Alternaria*, and *Aspergillus* sp.	Cotton and legumes	Dry powder 5.5×10^{10} spores/g	Slurry mixture for seed treatment

of *Pseudomonas* and found them effective in the management of sheath blight disease. Vidhyasekaran and Muthamilan (1995) prepared a formulation of *P. fluorescens* using various carriers such as talc, peat, lignite, vermiculate, and kaolinite to control the chickpea wilt. Kumar and Gupta (2010) studied the lignite-based formulation of *Azotobacter* to evaluate the bio-efficacy in wheat. Dry powder and dry granules are most successfully produced with spore-forming microbes, where drying processes such as spray-dry, freeze-dry, or air-dry are used for this purpose. The performance of microorganisms from the drying process varies with different microbes and with their physiological state. All dry formulations contain dormant microbes, whereas moist powder and granular formulations usually contain metabolically active microbes.

Liquid Suspensions

Solid-based formulations have some problems of survival. To overcome these problems, many formulations of the liquid-based inoculants have been introduced in recent years, which ensure longer survival, no contamination, ease of applicability, and tolerance to adverse environmental conditions. Liquid bioinoculants are special formulations containing not only the desired microorganisms and their nutrients but also the special cell protectants or substances that encourage longer shelf life and tolerance to adverse conditions (Vora et al. 2008). Liquid formulations typically are aqueous, oil-based, or polymer-based products (Jones and Burges 1998). Polysaccharides such as gums, carboxymethyl cellulose, and polyalcohol derivatives are frequently used to alter the fluid properties of liquid formulations (Paau 1988).

Materials used in liquid formulations are synthetic medium, sludge, baker's yeast, effluents, paneer-whey, and vegetable oil (Kremer and Peterson 1982; Hoben et al. 1991; Vendan and Thangaraju 2006; Pandey and Maheshawari 2007; Rebah et al. 2007; Albareda et al. 2008). In an experiment conducted with *Rhizobium* inoculants, liquid inoculant was found equally effective in comparison to the solid carrier-based inoculants (Gupta 2005). Vendan and Thangaraju (2007) developed cyst-based liquid formulation of *Azospirillum* using glycerol as a protective agent. Cyst-based bioinoculant formulations have twin advantages, i.e., longer shelf life and tolerance to adverse or harmful environmental conditions.

Liquid biofertilizer formulation could be considered as a potential strategy for improving the shelf life of biofertilizers. Unlike solid carrier-based biofertilizers, liquid formulation allows to add sufficient amount of nutrient, cell protectants, and inducers responsible for cell/spore/cyst formulation to ensure prolonged shelf life (Brar et al. 2012).

Mass Multiplication

The first requirement for the development of effective formulation is the production of PGPR strain in adequate amount. This requires the selection of proper, cheap, and easily available medium. In many cases, the production of bacterial strains is not easy because of the specific requirement of nutritional and environmental conditions. Mass production is achieved through liquid, semisolid, and solid state fermentation techniques.

The selected medium for mass multiplication should be inexpensive and readily available with appropriate nutrient balance. King's B broth or nutrient broth is used for the mass production of *Pseudomonas* and *Bacillus* species by using liquid fermentation technology (Manjula and Podile 2001; Nakkeeran et al. 2005). Slininger et al. (1996) reported that liquid culture formulation technology has to be optimized for the commercial exploitation of *P. fluorescens* intended for the management of take-all disease of wheat. Solid fermentation medium consisting of inert carriers with food bases was used for mass production of biocontrol agents (Lewis 1991). The solid fermentation technology is most suitable for fungal biocontrol agents.

Formulation Design

The success of formulation to suppress diseases or to enhance plant growth depends on the availability and shelf life of microbes because the shelf life of formulation decides the commercialization of bioinoculants. Formulation should support the viable nature of the product for the increased period of storage. Biocontrol product should have the minimum shelf life of 8–12 months for industrialization (Nakkeeran et al. 2005). Carrier material should not affect the viable nature of the biocontrol agent.

Formulation design also involves the type of formulation in which the choice of carrier is an important task. Many workers have evaluated the superiority of liquid formulations over the solid carrier-based formulations. Singleton et al. (2002) developed liquid formulation of *Rhizobium* by adding various additives in the yeast extract mannitol medium and claimed cell numbers of 1×10^{10} cells/ml in the liquid inoculant. Vendan and Thangaraju (2006) have developed liquid formulation of *Azospirillum* by using various cell protectants such as trehalose, polyvinylpyrrolidone, and glycerol, which resulted in increased shelf life by maintaining population level at 10^8 CFU/ml. Pandey and Maheshwari (2007) evaluated different carriers for the formulation of *Burkholderia* species strain MSSP and found that paneer-whey and wheat bran were efficient carriers in maintaining higher population.

Obstacles in Formulation Development

Formulation is the product of living organism so care needs to be taken to maintain the formulation at all the steps of production and maintenance. Production technology of bioformulations requires proper care and aid of sophisticated equipments to ensure availability of quality products in the market (Arora et al. 2010). During transport and distribution maintaining the temperature of the formulation is an important task. This, in turn, increases the cost of product. The earliest planning in the investigation of bioinoculants must anticipate the ultimate objectives and must meet commercial and scientific requirements (Burges and Jones 1998).

Shelf life of formulation is another obstacle because the storage of bioformulations requires special facilities and skills, which most producers, shopkeepers, and farmers do not possess. Shelf life is a culmination of several factors such as production technology, carrier and packaging material used, and mode and distance of transport. All these are desired to sustain the shelf life (Arora et al. 2010). Shelf life of formulation also depends on the performance of strain in field and laboratory condition. To avoid this problem of reduced survival rate of strain, the preparation should be air-dried and lyophilized (Kosanke et al. 1992, Fravel et al. 1998). Bacteria in the formulation remain inactive, resistant to environmental stresses, and insensitive to contamination and are more compatible with fertilizer application when the water content of the formulation is low (Bashan 1998).

Inconsistent field performance is the major obstacle associated with marketing of formulations. Many PGPRs work well in laboratory conditions, but they are unable to survive in fields, and therefore, this is an obstacle in the industrialization of formulation. These failures have raised concerns about the perspective of the great practical potential offered by microbial releases into soils. A key factor involved in the failure is the rapid decline of the size of the population of active cells. The active cells decline to a level at which the formulation becomes ineffective when introduced into the soil and the desired objective is not achieved. As soil is a heterogeneous system with a mixed biota, the formulation added in soil is subjected to rhizosphere competence, interaction with already adopted soil microflora, and survival in new habitat. A number of factors such as soil composition, temperature, relative humidity, composition of root exudates, presence of recombinant plasmids, as well as the interactions with other soil biota can affect the persistence of a PGPR on the root system making it difficult to predict the behavior of the bacterial strain under natural conditions. Therefore, PGPRs that are effective in the laboratory may not show any significant impact on plants under field conditions (Glick et al. 1999).

Commercial Aspects

With increasing population and public interest, the demand of organically grown food is also increasing. To fulfill the requirement of food, the annual requirement and present production capacity for different biofertilizers not only show a tremendous gap but also indicate the potential for organized production. The requirement for *Rhizobium* cultures to cover the entire cultivation area in India will be around 15,000 tons of carrier-based material, of which the present production is only around 800 tons. Similarly, 400,000 tons of BCA are required to cover the entire rice cultivation area in India. This indicates the need to augment efforts and to plan a phased strategy (Tilak and Saxena 1986).

An ongoing attempt to promote biofertilizer in agriculture through public intervention is expected to increase the commercial market of biofertilizers. To attain production targets of biofertilizers, their production, distribution, and promotion are some of the aspects that need to be focused.

In the recent past, large amount of research has been carried out on the PGPR, which has also led to the development of commercial products. Representative commercially available bacterial biocontrol products are listed in Table 18.2. It has been observed that biocontrol products used to date control possess relatively narrow spectrum of diseases control on a particular host crop. Hence, before approaching for commercial production, a company must asses many factors including demand for the product, potential market size, and existing competing formulation of the bioproducts.

Conclusion

The complex environment of the rhizosphere and its community structure is not fully understood. Hence, we can expect to see new PGPR products becoming available from new emerging strains of PGPR. The success of these products depends on our ability to understand the mechanism of action of PGPR and manage the rhizosphere to enhance survival and competitiveness of these beneficial microorganisms.

In the past few years, there has been an increasing interest in utilizing PGPRs as adjuncts to agricultural practice and environmental cleanup. Moreover, with this interest there have been major efforts worldwide to better understand many of the fundamental mechanisms that PGPRs use to facilitate plant growth and biocontrol. An effective PGPR could be developed for disease control only after understanding its performance under natural environmental conditions and in the presence of other competing microbiota in which it is expected to perform large amount of research has been devoted to PGPR and BCAs (Nielsen et al. 2000; Bloemberg and Lugtenberg 2001; Pieterse et al. 2001; Thrane et al. 2000; Sayyed and Chincholkar 2009; Sayyed and Patel 2011; Sayyed et al. 2012, 2013). In nature, agriculture crops are exposed to diverse environmental conditions and gambling of monsoons, which alter the microclimatic conditions existing around the infection court. A thorough knowledge on the mechanisms and performance related to disease control will help in the selection of promising candidates that suit industries to produce reliable commercial products (Collins and Jacobsen 2003).

Future Prospective

Rhizosphere management will require consideration of soil and crop cultural practices as well as inoculant formulation and delivery. Biotechnology can be applied to further improve strains that have appreciated qualities, namely, formulation ease, stability, and competent root colonization, by creating transgenic strains that combine multiple mechanisms of action. Genetic manipulation of PGPR strains to enhance colonization and effectiveness may involve addition of one or more traits associated with plant growth promotion. The use of PGPR with known functions is of interest as these formulations may increase consistency in the field. They offer the potential to address multiple modes of action, multiple pathogens, and temporal or spatial variability. PGPRs offer an environmentally sustainable approach to increase crop production.

There is no doubt that PGPRs are the potential tools for sustainable agriculture not only in India but also all over the world. The use of biofertilizers (PGPRs) in preference to chemical fertilizers is always welcome taking into consideration its suitability for agriculture. Biofertilizers are always beneficial in terms of soil fertility, ecological health, etc. As the use of biofertilizers, till so far, is grossly inadequate in India, more emphasis needs to be given on their production, consumption, and proper distribution. Selecting the strains with multiple traits for biocontrol will be effective in the field of PGPR.

Designing an effective formulation with the increased shelf life and tolerance to adverse condition is always a challenging task, but the liquid formulation has given an option to this. This has enhanced the shelf life, but has also increased the cost of production. To reduce the production cost of formulation, cheap raw materials need to be used or the shelf life of solid formulations needs to be improved as the commercialization depends on the shelf life of formulation.

Acknowledgement Author RZS is thankful to UGC, New Delhi, for providing financial support in this research project.

References

Akhtar MS, Siddiqui ZA (2009) Use of plant growth promoting rhizobacteria for the biocontrol of root-rot disease complex of chickpea. Aust Plant Pathol 38:44–50

Albareda M, Rodríguez-Navarro DN, Camacho M, Temprano FJ (2008) Alternatives to peat as a carrier for rhizobia inoculants: solid and liquid formulations. Soil Biol Biochem 40:2771–2779

Arora NK, Khare E, Maheshwari DK (2010) Plant growth promoting rhizobacteria: constraints in bioformulation, commercialization, and future strategies. In: Maheshwari DK (ed) Plant growth and health promoting bacteria, microbiology monographs. Springer, Berlin/Heidelberg, pp 97–115

Askeland RA, Morrison SM (1983) Cyanide production by *Pseudomonas fluorescens* and *Pseudomonas aeruginosa*. Appl Environ Microbiol 45:1802–1807

Bashan Y (1998) Inoculants of plant growth promoting bacteria use in agriculture. Biotech Adv 6:729–770

Bashan Y, Holguin G (1998) Proposal for the division of plant growth-promoting rhizobacteria into two classifications: biocontrol-PGPB (plant growth-promoting bacteria) and PGPB. Soil Biol Biochem 30: 1225–1228

Besson F, Michel G (1992) Biosynthesis of bacillomycin D activating enzymes by the use of affinity chromatography. FEBS Lett 308:18–21

Bloemberg GV, Lugtenberg BJJ (2001) Molecular basis of plant growth promotion and biocontrol by rhizobacteria. Curr Opin Plant Biol 4:343–350

Brar SK, Sarma SJ, Chaabouni E (2012) Shelf-life of biofertilizers: an accord between formulation and genetics. J Biofertil Biopestici 3:5

Brittenham GM (1994) New advances in iron metabolism, iron deficiency and iron overload. Curr Opin Hematol 1:549–556

Burges HD, Jones KA (1998) Trends in formulation of microorganisms and future research requirements. In: Burges HD (ed) Formulation of microbial biopesticides- beneficial organisms nematodes and seed treatment. Springer, Dordrecht, pp 311–331

Castric PA (1975) Hydrogen cyanide, a secondary metabolite of *Pseudomonas aeruginosa*. Can J Microbiol 21:613–618

Castric PA (1977) Glycine metabolism by *Pseudomonas aeruginosa*: hydrogen cyanide biosynthesis. J Bacteriol 130:826–831

Chaiharn M, Chunhaleuchanon S, Lumyong S (2009) Screening siderophore producing bacteria as potential biological control agent for fungal rice pathogens in Thailand. World J Microbiol Biotechnol 25(11): 1919–1928

Chincholkar SB, Chaudhari BL, Talegaonkar SK, Kothari RM (2000) Microbial iron chelators: a tool for sustainable agriculture. In: Upadhayay RK, Mukherji KG, Chamola BP (eds) Biocontrol potential and their exploration in crop disease management, vol 1. Kluwer Academic Press, New York, pp 49–70

Chinnadurai C, Balachandar D, Sundaram SP (2009) Characterization of 1-aminocyclopropane-1-carboxylate deaminase producing methylobacteria from phyllosphere of rice and their role in ethylene regulation. World J Microbiol Biotechnol 25:1403–1411

Chitarra GS, Breeuwer P, Nout MJR, Aelst AC, van Rombouts FM, Abee T (2003) An antifungal compound produced by *B. subtilis* YM 10-20 inhibits germination of *Penicillium roqueforti* conidiospores. J Appl Microbiol 94:159–166

Collins DP, Jacobsen RJ (2003) Optimizing a *Bacillus subtilis* isolate for biological control of sugar beet *Cercospora* leaf spot. Biol Cont 26:153–161

Compant S, Duffy B, Nowak J, Clement C, Barka EA (2005) Use of plant growth-promoting bacteria for biocontrol of plant diseases: principles, mechanisms of action and future prospects. Appl Environ Microbiol 71:4951–4959

Constantinescu F (2001) Extraction and identification of antifungal metabolites produced by some *B. subtilis* strains. Analele Institutului de Cercetari Pentru Cereale Protectia Plantelo 31:17–23

Cronin D, Möenne-Loccoz Y, Fenton A, Dunne C, Dowling DN, O'Gara F (1997) Role of 2, 4-diacetylphloroglucinol in the interactions of the biocontrol *pseudomonad* strain F113 with the potato cyst nematode *Globodera rostochiensis*. Appl Environ Microbiol 63:1357–1361

De Souza JT, Weller DM, Raaijmakers JM (2003) Frequency, diversity and activity of 2,4-diacetylphloroglucinol producing fluorescent *Pseudomonas* spp. in Dutch take-all decline soils. Phytopathol 93:54–63

Emmert EA, Handelsman J (1999) Biocontrol of plant disease: a (gram) positive perspective. FEMS Microbiol Lett 171:1–9

Fernando DWG, Nakkeeran S, Zhang Y (2005) Biosynthesis of antibiotics by PGPR and its relation in biocontrol of plant diseases. In: Siddiqui ZA (ed) PGPR: biocontrol and biofertilization. Springer, Dordrecht, pp 67–109

Frankowski J, Lorito M, Scala F, Schmidt R, Berg G, Bahl H (2001) Purification and properties of two chitinolytic enzymes of Serratia plymuthica HRO-C48. Arch Microbiol 176:421–426

Fravel DR (1988) Role of antibiosis in the biocontrol of plant diseases. Annu Rev Phytopathol 26:75–91

Fravel DR, Connick WJ Jr, Jack AL (1998) Formulation of microorganisms to control plant diseases. In: Berges HD (ed) Formulation of microbial biopesticides, beneficial microorganisms, nematodes and seed treatments. Kluwer Academic Press, Dordrecht, pp 187–202

Freitas SDS, Pizzinato MA (1997) Action of rhizobacteria on the *Colletotrichum gossypii* incidence and growth promotion in cotton seedlings. Summa Phytopathol 23:36–41

Garbeva P, van Veen JA, van Elas JD (2004) Assessment of the diversity, and antagonism towards *Rhizoctonia solani* AG3, of *Pseudomonas* species in soil from different agricultural regimes. FEMS Microbiol Ecol 47:51–64

Glick BR (2012) Plant growth-promoting bacteria: mechanisms and applications. Scientifica Article ID 963401, http://dx.doi.org/10.6064/2012/963401

Glick BR, Patten CL, Holguin G, Penrose DM (1999) Biochemical and genetic mechanisms used by plant growth promoting bacteria. Imperial College Press, London

Gohel V, Singh A, Misuria V, Phadnis A, Chhatpar HS (2005) Bioprospecting and antifungal potential of chitinolytic microorganisms. Afr J of Biotech 5(2):54–72

Gupta SC (2005) Evaluation of liquid and carrier based *Rhizobium* inoculants in chickpea. Ind J of Pulses Res 18:40–42

Gupta CP, Dubey RC, Maheshwari DK (2002) Plant growth enhancement and suppression *Macrophomina phaseolina* causing charcoal rot of peanut by fluorescent *Pseudomonas*. Biol Fertil Soils 35:399–405

Gutierrez FJ, Ramos-Solano B, Probanza A, Mehouachi J, Tadeo FR, Talon M (2001) The plant growth-promoting rhizobacteria *Bacillus licheniformis* produce high amounts of physiologically active gibberellins. Physiol Plant 111:206–211

Hass D, Defago G (2005) Biological control of soil borne pathogens by *fluorescent pseudomonads*. Nat Rev Microbiol 3:307–319

Hider RC, Kong X (2010) Chemistry and biology of siderophores. Nat Prod Rep 27:637–657

Hoben HJ, Aung NN, Somasegaran P, Kang UG (1991) Oils as adhesives for seed inoculation and their influence on the survival of *Rhizobium* spp. and *Bradyrhizobium* spp. on inoculated seeds. World J Microbiol Biotechnol 7:324–330

Höfte M (1993) Classes of microbial siderophores. In: Barton LL, Hemming BC (eds) Iron chelation in plants and soil microorganisms. Academic, San Diego, pp 3–26

Idris HA, Labuschagne N, Korsten L (2007) Screening rhizobacteria for biological control of Fusarium root and crown rot of sorghum in Ethiopia. J Biocontrol 40:97–106

Johri BN, Rao CVS, Goel R (1997) Fluorescent *pseudomonads* in plant disease management. In: Dadarwal KR (ed) Biotechnological approaches in soil microorganisms for sustainable crop production. Scientific Publishers, Jodhpur, pp 193–223

Jones KA, Burges HD (1998) Technology of formulation and application. In: Burges HD (ed) Formulation of microbial biopesticides- beneficial organisms nematodes and seed treatment. Springer, Dordrecht, pp 2–27

Kamil Z, Rizk M, Saleh M, Moustafa S (2007) Isolation and identification of rhizosphere soil chitinolytic bacteria and their potential in antifungal biocontrol. Global J Mol Sci 2(2):57–66

Kim KJ, Yang YJ, Kim JG (2003) Purification and characterization of Chitinase from *Streptomyces* sp. M-20. J Biochem Mol Biol 36(2):185–189

Kim YC, Jung H, Kim KY, Park SK (2008) An effective biocontrol bioformulation against Phytophthora blight of pepper using growth mixtures of combined chitinolytic bacteria under different field conditions. Eur J Plant Pathol 120:373–382

Kishore GK, Pande S (2007) Chitin-supplemented foliar application of chitinolytic *Bacillus cereus* reduces severity of Botrytis gray mold disease in chickpea under controlled conditions. Lett Appl Microbiol 44:98–10

Kloepper JW, Schroth MN (1981) Development of a powder formulation of rhizobacteria for inoculation of potato seed pieces. Phytopathol 71:590–592

Kloepper JW, Leong J, Teintze M, Schroth MN (1980) Enhanced plant growth by siderophores produced by plant growth promoting rhizobacteria. Nature 286:885–886

Kloepper JW, Lifshitz R, Zablotowicz RM (1989) Free living bacterial inocula for enhancing crop productivity. Trend Biotechnol 7:39–43

Kloepper JW, Tuzun S, Kuc J (1992) Proposed definitions related to induced disease resistance. Biocontrol Sci Technol 2:349–351

Kosanke JW, Osburn RM, Shuppe GI, Smith RS (1992) Slow rehydration improves the recovery of dried bacterial populations. Can J Microbiol 38:520–525

Kremer RJ, Peterson HL (1982) Effect of inoculant carrier on survival of *Rhizobium* on inoculated seed. Soil Sci 134:117–125

Kumar V, Gupta P (2010) Studies on shelf-life of fly-ash based *Azotobacter Chroococcum* formulation and its bio-efficacy in wheat. Res J Agric Biol Sci 6(3):280–282

Lemanceau P, Albouvette A (1993) Suppression of *Fusarium* wilts by fluorescent *pseudomonas*: mechanism and applications. Biocon Sci Technol 3:219–234

Lewis JA (1991) Formulation and delivery system of biocontrol agents with emphasis on fungi *Beltsville symposia* in agricultural research. In: Keister DL, Cregan PB (eds) The rhizosphere and plant growth. Kluwer Academic Publisher, Dordrecht, pp 279–287

Loni PP, Shyam S, Bajekal (2011) Alkaline chitinase from *Bacillus firmus* SBPL-05 isolated from alkaline-saline environment of Lonar lake. Indian J Fundam Appl Life Sci 1(3):161–165

Lopez-Millan AF, Morales F, Abadia A, Abadia J (2001) Iron deficiency-associated changes in the composition of the leaf apoplastic fluid from field-grown pear (*Pyrus communis*) trees. J Exp Bot 52:1489–1498

Lorck H (1948) Production of hydrocyanic acid by bacteria. Physiol Plant 11:142–146

Lucy M, Reed E, Glick BR (2004) Applications of free living plant growth-promoting rhizobacteria. Antonie Van Leeuwenhoek Int J Gen Mol Microbiol 86:1–25

Lugtenberg B, Kamilova F (2009) Plant-growth-promoting rhizobacteria. Annu Rev Microbiol 63:541–556

Madhaiyan M, Poonguzhali S, Sa T (2007) Characterization of 1-aminocyclopropane-1-carboxylate (ACC) deaminase containing *Methylobacterium oryzae* and interactions with auxins and ACC regulation of ethylene in canola (*Brassica campestris*). Planta 226:867–876

Manjula K, Podile AR (2001) Chitin-supplemented formulations improve biocontrol and plant growth promoting efficiency of *Bacillus subtilis* AF 1. Can J Microbiol 47:618–625

Mazzola M (2002) Mechanism of natural soil suppressiveness to soil borne diseases. Antonie van Leeuwenhock 81:557–564

Miethke M, Marahiel MA (2007) Siderophore-based iron acquisition and pathogen control. Microbiol Mol Biol Rev 71(3):413–451

Nakkeeran S, Kavitha K, Mathiyazhagan S, Fernando WGD, Chandrasekar G, Renukadevi P (2004) Induced systemic resistance and plant growth promotion by *Pseudomonas chlororaphis* strain PA-23 and *Bacillus subtilis* strain CBE4 against rhizome rot of turmeric (*Curcuma longa* L.). Can J Plant Pathol 26:417–418

Nakkeeran S, Fernando DWG, Siddiqui ZA (2005) Plant growth promoting rhizobacteria formulations and its scope. In: Siddiqui ZA (ed) PGPR: biocontrol and biofertilization. Springer, Dordrecht, pp 257–296

Nielsen TH, Thrane C, Christophersen C, Anthoni U, Sorensen J (2000) Structure production characteristic and fungal antagonism of tensin – a new antifungal cyclic lipopeptide from *Pseudomonas fluorescence*. J Appl Microbiol 89:992–1001

Paau AS (1988) Formulation useful in applying beneficial microorganisms to seeds. Trends Biotechnol 6:276–278

Paau AS (1998) Formulation of beneficial organisms applied to soil. In: Burges HD (ed) Formulation of microbial biopesticides- beneficial organisms nematodes and seed treatment. Springer, Dordrecht, pp 236–253

Pandey P, Maheshwari DK (2007) Bioformulation of *Burkholderia* sp. MSSP with multispecies consortium for growth promotion of *Cajanus cajan*. Can J Microbiol 53:213–222

Park SH, Lee JH, Lee HK (2000) Purification and characterization of chitinase from a marine bacterium, *Vibrio* sp. 98CJ11027. J Microbiol 38(4):224–229

Patten CL, Glick BR (1996) Bacterial biosynthesis of indole-3-acetic acid. Can J Microbiol 42:207–220

Pieterse CMJ, Pelt JA, Van Wees SCM, Ton J, L-Kloosterziel KM, Keurentjes JJB, Verhagen BMW (2001) Rhizobacteria mediated induced systemic resistance: triggering, signaling and expression. Eur J Plant Pathol 107:51–61

Podile AR, Kishore GK (2006) Plant growth promoting rhizobacteria. In: Gnanamanickam SS (ed) Plant associated bacteria. Springer, Amsterdam, pp 195–230

Rabindran R, Vidhyasekaran P (1996) Development of formulation of *Pseudomonas fluorescens* PfALR2 for management of rice sheath blight. Crop Prot 15:715–721

Ramamoorthy V, Viswanathan R, Raghuchander T, Prakasam V, Samiyappan R (2001) Induction of systemic resistance by plant growth promoting rhizobacteria in crop plants against pests and diseases. Crop Prot 20:1–11

Ramos Solano B, García JAL, Villaraco AG, Algar E, Cristobal JG, Mañero FJG (2010) Siderophore and chitinase producing isolates from the rhizosphere of *Nicotiana glauca* Graham growth and induce systemic resistance in *Solanum lycopersicum* L. Plant Soil 334:189–197

Ran LX, Liu CI, Wu GJ, van Loon LC, Bakker PAHM (2005) Suppression of bacterial wilt in *Eucalyptus urophylla* by fluorescent *Pseudomonas* spp. in China. Biol Control 32:111–120

Rebah FB, Prevost D, Yezza A, Tyagi RD (2007) Agro-industrial waste materials and wastewater sludge for rhizobial inoculant production: a review. Bioresour Technol 98:3535–3546

Saharan BS, Nehra V (2011) Plant growth promoting rhizobacteria: a critical review life sciences and medicine research. LSMR 21:1–30

Sayyed RZ, Chincholkar SB (2009) Siderophore producing *A. feacalis* more biocontrol potential vis-a'-vis chemical fungicide. Curr Microbiol 58(1):47–51

Sayyed RZ, Patel PR (2011) Biocontrol potential of siderophore producing heavy metal resistant *Alcaligenes* sp. and *Pseudomonas aerogenosa* RZS3. vis-à-vis organophosphorus fungicide. Ind J Microbiol 51(3):266–272

Sayyed RZ, Naphade BS, Chincholkar SB (2005) Ecologically competent rhizobacteria for plant growth promotion and disease management. In: Rai MK, Chikhale NJ, Thakare PV, Wadegaonkar PA, Ramteke AP (eds) Rec trends biotechnol. Scientific Publishers, Jodhpur, pp 1–16

Sayyed RZ, Naphade BS, Chincholklar SB (2007a) Siderophore producing *A. feacalis* promoted the growth of Safed musali and Ashwagandha. J Med Arom Plants 29:1–5

Sayyed RZ, Patel DC, Patel PR (2007b) Plant growth promoting potential of P solubilizing *Pseudomonas* sp. occurring in acidic soil of Jalgaon. Asian J Microbiol Biotechnol Environ Sci 4:925–928

Sayyed RZ, Reddy MS, Kumar KV, Yellareddygari SKR, Deshmukh AM, Patel PR, Gangurde NS (2012) Potential of plant growth-promoting rhizobacteria for sustainable agriculture. In: Maheshwari DK (ed) Bacteria in agrobiology: plant probiotics. Springer, Berlin/Heidelberg, pp 287–313

Sayyed RZ, Chincholkar SB, Reddy MS, Gangurde NS, Patel PR (2013) Siderophore producing PGPR for crop nutrition and phytopathogen suppression. In: Maheshwari DK (ed) Bacteria in agrobiology: disease management. Springer, Berlin/Heidelberg, pp 449–471

Schippers B, Bakker A, Bakker P, van Peer R (1990) Beneficial and deleterious effects of HCN-producing *pseudomonads* on rhizosphere interactions. Plant Soil 129:75–83

Sharma A, Shankhdhar D, Shankhdhar SC (2013a) Enhancing grain iron content of rice by the application of plant growth promoting rhizobacteria. Plant Soil Environ 59(2):89–94

Sharma SB, Sayyed RZ, Trivedi MH, Gobi TA (2013b) Phosphate solubilizing microbes: sustainable approach for managing phosphorus deficiency in agricultural soils. Springer Plus 2:587

Sindhu SS, Suneja S, Dadarwal KR (1997) Plant growth promoting rhizobacteria and their role in crop productivity. In: Dadarwal KR (ed) Biotechnological approaches in soil microorganisms for sustainable crop production. Scientific Publishers, Jodhpur, pp 149–193

Singh PP, Shin YC, Park CS, Chung YR (1999) Biological control of *Fusarium* wilt of cucumber by chitinolytic bacteria. Phytopathol 89:92–99

Singleton P, Keyser H, Sande E (2002) Development and Evaluation of Liquid Inoculants. In: Herridge D (ed) Inoculants and nitrogen fixation of legumes in Vietnam, ACIAR Proceedings Australian Centre for International Agricultural Research, Canberra, Australia 109:52–66

Slininger PJ, Van Cauwenberge JE, Bothast RJ, Weller DM, Thomashow LS, Cook RJ (1996) Effect of growth culture physiological state, metabolites, and formulation on the viability, phytotoxicity, and efficacy of the take-all biocontrol agent *Pseudomonas fluorescens* 2–79 stored encapsulated on wheat seeds. Appl Microbiol Biotechnol 45:391–398

Sullivan DJO, Gara FO (1992) Traits of fluorescent *Pseudomonas* spp. involved in suppression of plant root pathogens. Microbiol Rev 56(4):662–676

Thrane C, Harder NI, Neiendam NM, Sorensen J, Olson S (2000) Visconiamide producing Pseudomonas fluorescence DR 54 exerts a biocontrol effect on *Pythium ultimum* in sugar beet. FEMS Microbiol Ecol 33:139–146

Tilak KVBR, Saxena AK (1986) Biofertilizers – their role in crop production. Fertil Indust 19:187–194

Velusamy P, Kim KY (2011) Chitinolytic activity of *Enterobacter* sp. KB3 antagonistic to *Rhizoctonia solani* and its role in the degradation of living fungal hyphae. Int Res J Microbiol 2(6):206–214

Vendan RT, Thangaraju M (2006) Development and standardization of liquid formulation for *Azospirillum* bioinoculant. Ind J Microbiol 46:379–387

Vendan RT, Thangaraju M (2007) Development and standardization of cyst based liquid formulation of *Azospirillum* bioinoculant. Acta Microbiol Immunol Hung 54:166–177

Vessey JK (2003) Plant growth promoting rhizobacteria as biofertilizers. Plant Soil 255:571–586

Vidhyasekaran P, Muthamilan M (1995) Development of formulations of *Pseudomonas fluorescens* for control of chickpea wilt. Plant Dis 79:782–786

Vincent MN, Harrison LA, Brackin JM, Kovacevich PA, Mukerji P, Weller DM, Pierson EA (1991) Genetic analysis of the antifungal activity of a soilborne *Pseudomonas aureofaciens* strain. Appl Environ Microbiol 57:2928–2934

Voisard C, Keel C, Haas D, Defago G (1989) Cyanide production by Pseudomonas fluorescens helps suppress black root rot of tobacco under gnotobiotic conditions. EMBO J 8:351–358

Vora MS, Shelat HN, Vyas RV (2008) Liquid biofertilizers: a new vistas. In: Vora MS, Shelat HN, Vyas RV (eds) Handbook of biofertilizers and microbial pesticides. Satish Serial Publishing House, New Delhi, pp 87–90

Yu GY, Sinclair JB, Hartman GL, Beragnolli BL (2002) Production of iturin A by *B. amyloliquefaciens* suppressing *R. solani*. Soil Biol Bioche 34:955–963

Effects of Soil Environment on Field Efficacy of Microbial Inoculants

19

Ekta Khare and Naveen K. Arora

Abstract

Although many microorganisms show good performance in specific trials, this is often not translated into consistent, effective plant growth promotion and biocontrol in diverse field situations. The key factors involved in the lack of success are rapid decline in the size of populations of active cells to levels ineffective to achieve the objective and variable production of required metabolites or poor colonization, following the introduction into soil. The physical, chemical, and physicochemical nature of soil and its indigenous microorganisms and predators influence the microbial population both quantitatively and qualitatively. Soil abiotic factors (e.g., texture, pH, temperature, and moisture) exert their (direct) effect on inoculant population dynamics by imposing stresses. On the other side, trophic competitions and antagonistic/synergic and predatory interactions with the resident microbial and fauna populations determine the field efficiency of inoculants. The aim of this review is to throw light on different soil environmental conditions that affect the survival of inoculated microbial strains in the field. A proper characterization of target soils and rhizospheres as habitats for introduced microbes is a key to the development of bioformulations that support beneficial microorganisms in the soil.

Introduction

The increased use of chemical pesticides and fertilizers in agriculture helped the world in achieving self sufficiency in food grain production. However, it has also polluted the environment and caused slow deterioration of soil health. The chemical residues in the food products are also causing harm to human beings and other living organisms. To combat these problems and in

E. Khare
Department of Microbiology, Institute of Biosciences and Biotechnology, Chhatrapati Shahu Ji Maharaj University, Kanpur, India

N.K. Arora (✉)
Department of Environmental Microbiology, Babasaheb Bhimrao Ambedkar University, Lucknow, Uttar Pradesh, India
e-mail: nkarora_net@rediffmail.com

N.K. Arora (ed.), *Plant Microbes Symbiosis: Applied Facets*, DOI 10.1007/978-81-322-2068-8_19, © Springer India 2015 the light of sustainable agriculture, plant growth-promoting microorganisms (PGPMs) as green technology are now being greatly used (Sumathi et al. 2012). Plant growth benefits may be attributed mainly to three mechanisms: (1) PGPMs acting as biofertilizers (such as nitrogen-fixing bacteria and phosphate-solubilizing bacteria) assist plant nutrient uptake by providing fixed nitrogen or other nutrients; (2) phytostimulators (microbes-expressing phytohormones such as *Azospirillum*) can directly promote the growth of plants, usually by producing plant hormones; and (3) biological control agents (such as *Trichoderma*, *Pseudomonas*, and *Bacillus* spp.) protect plants against phytopathogenic organisms (Compant et al. 2005; Trabelsi and Mhamdi 2013). Although PGPMs occur in soil, usually their number is not high enough to compete with other indigenous microorganisms commonly established in the rhizosphere. Therefore, for agronomic utility, inoculation of plants with target microorganisms at a much higher concentration than those normally found in soil is necessary to take advantage of their beneficial properties for plant yield enhancement (Subba Rao 1993).

Formulation/inoculant is an industrial art of converting a promising laboratory-proven microbe to a commercial field product (Bashan 1998). Many bacterial inoculants exert their beneficial effects in laboratory culture, a lower number are successful in a laboratory greenhouse, and a much lower number are functioning under practical conditions, i.e., in a commercial greenhouse or in the field (Lugtenberg and Kamilova 2009). The soil being a heterogeneous, unpredictable environment, the inoculated microbe finds it often difficult to establish a niche for survival among the competitors and predators. Uncertainty about the product performance can lead to poor demand from end users, the farmers. Understanding the reasons for the failures in the field may lead to the development of bioformulations that are effective in natural conditions.

> "...the best type go all over the world, fitting in so perfectly with their background that not even the inhabitants notice they are strangers; in other words they achieve the highest accomplishment possible" (Post 1922)

The "highest accomplishment" for biocontrol and PGP bioformulation lies in "fitting in" to the host environment and being able to tolerate, manipulate, or evade plant defense responses. As climatic conditions, soils, plants, and microorganisms are all variable and/or diverse, there is no general rule for how introductions into soils can be optimized. However, it is clear that since soil generally represents a hostile environment to microbial introductions and since microbial cells in soil are subjected to a range of adverse abiotic and biotic conditions (Fig. 19.1), the success of the application of microbes depends to a large extent on how favorable to its survival and functioning the target environment is or can be made, in terms of either natural or induced ecological selectivity or available protective niches. Abiotic soil factors (e.g., texture, pH, temperature, and moisture) exert their (direct) effect on inoculant population dynamics by imposing stresses of various natures on the cells (van Elsas et al. 1991; Evans et al. 1993). They can also act indirectly by affecting the activity of the indigenous soil microflora. Both abiotic and biotic factors regulate the production of antifungal metabolites under in vitro and in situ conditions (Dwivedi and Johri 2003; Khare et al. 2011). The dynamics of the inoculant population, i.e., the interactions of the inoculants with and response to their environment (predation, antagonism, and death/growth), is responsible for inconsistent performance at the soil microhabitat level (van Veen et al. 1997). Since effectiveness of bioformulation and soil properties is interrelated with one another, the understanding of and recognizing soil properties are important for making sound decisions regarding the use of microbial bioformulations. This review presents a general discussion on the factors that cause the decrease in numbers and activity of PGPMs following their introduction into soil (Fig. 19.2).

Effect of Soil Abiotic Factors

Soil Texture

Soils are made up of four components: minerals, air, water, and organic matter (Brady and Weil 2002). In most soils, minerals represent around

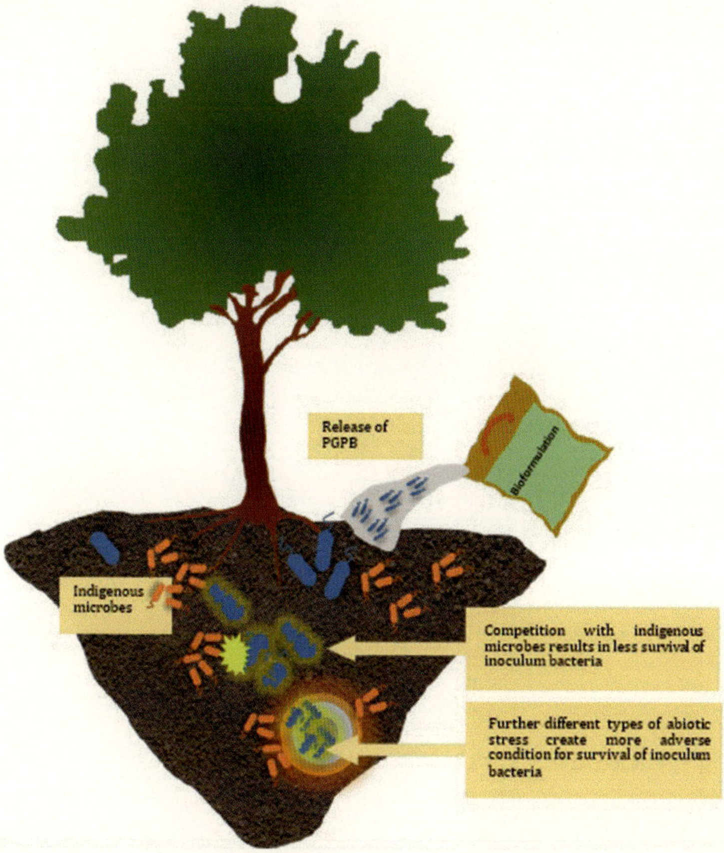

Fig. 19.1 General state of inoculated cells in the hostile rhizosphere ecosystem

45 % of the total volume, water and air about 25 % each, and organic matter from 2 to 5 %. The mineral portion consists of three distinct particle sizes classified as sand, silt, or clay. Sand is the largest size particles that can be considered soil. Sand is largely the mineral quartz, though other minerals are also present. Silt particles are much smaller than sand, but, like sand, silt is mostly quartz. The smallest of all the soil particles is clay. Mineral surfaces in soils are mainly provided by the very small particles (<2 μm) that constitute the soil clay fraction. This fraction is a mixture of phyllosilicates, oxides, and hydroxides and, in some soil types, short-range-order minerals like allophane and imogolite. Traditional concepts were highly focused on the role of phyllosilicates as dominating mineral surfaces for the formation of organo-mineral associations (Theng 1979; Stevenson 1982). Clays are quite different from sand or silt and contain appreciable amounts of plant nutrients (Silver et al. 2000). Clay has large surface area resulting from the platelike shape of the individual particles. The textural designation of a soil is derived from the relative portions of sand, silt, and clay. Another soil characteristic – soil structure – is different from soil texture. Structure refers to the combination or "aggregation" of sand, silt, and clay particles into larger secondary clusters. If you grab a handful of soil, good structure is apparent when the sand, silt, and clay particles are aggregated into granules or crumbs (Nimmo and Perkins 2002).

Both texture and structure determine pore space for air and water circulation, erosion resistance, looseness, ease of tillage, and root penetration (Rieu and Sposito 1991). However, while texture is an innate property of the native soil and does not change with agricultural activities,

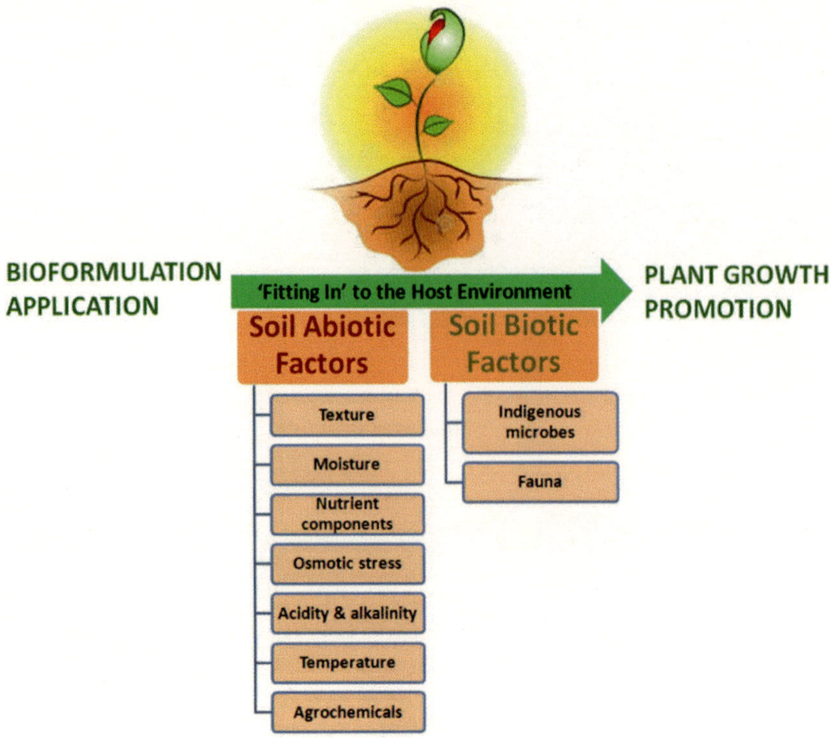

Fig. 19.2 Soil environmental conditions that are necessary to evaluate, for the development of bioformulation to achieve the highest accomplishment possible

structure can be improved or destroyed readily through our choice and timing of farm practices. A well-aggregated soil has increased water entry into the soil, increased aeration, and increased water-holding capacity (Rieu and Sposito 1991). An aggregate formed by physical forces (wetting and drying, freezing, and thawing) becomes stabilized (will remain intact when wet) through microbial processes involving organic matter decomposition and its by-products – chiefly gums, waxes, and other glue-like substances. These by-products cement the soil particles together forming water-stable aggregates. Aggregated soils also prevent glazing or crusting of the soil surface (Boyle et al. 1989). Previous studies have shown that soil types influence the structure of microbial communities, especially bacterial population among soils of different textures (Garbeva et al. 2004; Fang et al. 2005). On the other hand, no significant differences were noticed among average counts of soil fungi due to soil textures (Hamarashid et al. 2010).

The soils in good physical condition have better aeration and moisture content which is essential for optimum microbial activity. Soils of different texture differ in particle size composition and, thus, in pore size distribution. The pores between particles of solid material are just as important to the nature of the soil as are the solids themselves (Miller and Donahue 1995), since it is in the pore space that air and water circulate, roots grow, and most of the microbial population live. A structure of good quality implies a good quality of pore space, with good continuity and stability of pores and a good distribution of the pore size, including both macro- and micropores (Cabeda 1984). Moisture is held in the micropores. Water moves in the macropores, and these tend to be occupied by the air constituting the soil atmosphere. Hence, pore size distribution strongly determines the fate of introduced microorganisms, and differences in the behavior of bacteria released into different textured soils may be related to differences in the protective pore spaces present in these soils. The gaps between sand particles tend to be large favoring aeration but also result in rapid water loss. In contrast, pores between clay particles tend to be small, favoring water retention (Ashman and Puri 2002) but decreased aeration (Miller and Donahue 1995). In this way, soil texture influences most soil processes. In a simplified explanation, it not only determines its moisture and aeration characteristics (Brady and Weil 2002) but ultimately its ability to store and supply organic matter for plant and microbial use. For instance, clays due to their large sorption area as well as their electrostatic nature have a greater tendency to adsorb and retain organic matter into stable aggregates (Ashman and Puri 2002). Colonization of soil particles and aggregates is assumed to be vital to ultimate inoculant survival in soil (Hattori and Hattori 1976).

van Veen et al. (1997) reported higher survival levels of inoculants in fine-textured (clay) than in coarse (sandy) soils. The effects of, for instance, clay surfaces in conferring protection upon bacteria have been convincingly documented by Marshall (1975). These bacterial populations are top-down controlled by the grazing of bacterivorous nematodes and protozoa (Ingham et al. 1986; Moore et al. 1988; Wardle 2002). Nematodes are ubiquitous residents of soils and sediments in most ecosystems. Some soil nematode species feed on bacteria or fungi, while others are omnivorous, predaceous, or parasites of plant roots (Yeates et al. 1993). Soil nematodes range in size from 0.3 to 3 mm and are active in water-filled pores in the soil matrix and in water films that coat soil particles. Due to their small size and inability to migrate over long distances, soil nematodes are closely dependent on the immediate environmental conditions in soil pores to feed and reproduce (Treonis and Wall 2005). Bacterivorous soil protozoa accumulate in "hot spots" of microbial activity as in the humus region or closed to plant roots, attracted by their prey (Griffiths 1990; Hausmann and Hülsmann 1996; Scheu et al. 2005) reaching densities of 10^4–10^7 active protist individuals per gram dry weight forest soil (Adl et al. 2006). Bacterivorous protozoa are more important than nematodes since they have access to bacteria living in small pore spaces unavailable to nematodes (Bamforth 1997). Each microbial species has different optimal conditions, but overall bacterial activity is highest when pore spaces are about 60 % water filled. Soil texture and porosity determine the amount of space available for soil organisms and for the movement of air and water through the soil. Thus, porosity, aeration, and moisture levels are linked (Chapin et al. 2002).

Soil Moisture

The survival and activity of microorganisms may depend on their distribution among microhabitats and changes in soil moisture (Orchard and Cook 1983; Sylvia et al. 2005). It is one of the important factors influencing the microbial population and their activity in soil. In soil, water is held as films on particle surfaces and within soil pores (Stengel and Gelin 2003) through hydrogen bonding (Miller and Donahue 1995). Therefore, the higher the clay or organic matter contents of a soil, the larger the amount of water it can store, but also more tightly it is bond to the large surface area (Brady and Weil 2002). However, if the water is held too tightly, it becomes unavailable to plants and microorganisms, as they can only extract water at certain tensions. "Tension" here refers to the measure of suction required in order to extract water from the soil (Miller and Donahue 1995; Ashman and Puri 2002). Perhaps the most important abiotic factor affecting bacterial dispersion in soil is water (Madsen and Alexander 1982). Understanding bacterial movement in soil is also important for manipulating colonization of plant roots and for optimizing the dispersal of inoculants used in biocontrol strategies (Turnbull et al. 2001). The inability of the introduced organisms to move from the site of inoculation (usually the seed) and establish population in the rhizosphere and on plant roots is one of the main problems (Suslow 1982). The ability to colonize distal parts of the rhizosphere from an initial point of inoculum, especially in the absence of percolating water flow, has been termed "mobility" (Misaghi et al. 1992) and has been implicated as an important trait for biocontrol in the rhizosphere. Preferential water flow has been shown to move along the rhizoplane surface in soil systems (Natsch et al. 1996), and bacteria can be carried in large numbers along root surfaces by this mechanism (Parke et al. 1986; Kluepfel 1993). Chao et al. (1986) demonstrated that the depth of rhizosphere colonization by bacteria was influenced by the volume of percolating water added to a root system. In field condition, bacteria inoculated on seed would also be distributed along roots by preferential flow from rainfall (Natsch et al. 1996), although this movement would also be influenced by other environmental factors.

Since the possibilities for directed movement of bacteria in soil are limited (Foster 1988), microbial cells upon introduction into soil are probably not immediately localized in protective soil sites. Specific action may be needed to localize as many inoculant bacteria as possible at protective sites. Postma et al. (1989) showed that manipulation of the introduction method affected the localization of inoculant bacteria in protective microhabitats in soil. In their experiments, recipient soil was dried to different moisture levels prior to addition of bacterial cells. The bacterial cells would reach the soil pore space with the water of the inoculum. In the case of moist soil at the moment of introduction, the water present in small, potentially protective pores is assumed to act as a physical barrier to the entry of cells, resulting in the localization of the inoculum in the relatively large pores. In initially dry soil, a greater proportion of smaller pores are not water filled and hence could be reached by the inoculant cells with the incoming water. This implies that bacterial cells introduced into initially wet soil would be accessible for predation to a larger extent than would cells introduced into initially dry soil (van Veen et al. 1997).

Water (soil moisture) is useful to the microorganisms in two ways, i.e., it serves as source of nutrients by supplying hydrogen/oxygen to the organisms and as solvent and carrier of other food nutrients to the microorganisms. Microbial activity and population proliferate best in the moisture range of 20–60 %, typically due to the increased diffusivity of substrates in soil solution and the activation of dormant microorganisms by available water. Under excess moisture conditions/water-logged conditions, due to lack of soil aeration (oxygen), anaerobic microflora become active and the aerobes get suppressed. While in the absence of adequate moisture in soil, some of microbes die out due to tissue dehydration, and some of them change their forms into resting stages/spores or cysts to tide over adverse conditions (Davidson 1993). One of the immediate responses of rhizobia to water stress (low water potential) concerns the morphological changes. Rhizobia from *Prosopis* (mesquite) (Shoushtari and Pepper 1985) and *Rhizobium meliloti* (Busse and Bottomley 1989) showed irregular morphology at low water potential. The modification of rhizobial cells by water stress will eventually lead to a reduction in infection and nodulation of legumes (Hunt et al. 1981). A reduction in the soil moisture from 5.5 to 3.5 % significantly decreased the number of infection threads formed inside root hairs and completely inhibited the nodulation of *Trifolium subterraneum* (Worrall and Roughley 1976). The decrease in water availability may cause a change in the physiology of microbes as they adjust to more desiccating conditions (Stark and Firestone 1995). For example, one microbial response to water stress is the accumulation of intracellular solutes for osmoregulation (Harris 1981). Microbial growth may be hindered biochemically by a high intracellular concentration of solutes, or growth efficiency may be lowered because of costs of osmoregulation (Harris 1981), because the efficiency with which microbes use carbon substrates to create biomass is directly related to the tendency to immobilize other mineral nutrients into biomass (Herron et al. 2009).

Oxygen limitation and reducing conditions were likely to be present in soil located immediately above the plow plan, where decomposing crop residues were apparent and the soil was water logged. Mascher et al. (2000) reported that abiotic stress resulting from a combination of oxygen limitation and reducing conditions had a role in the formation of viable but nonculturable (VBNC) cells of inoculated *Pseudomonas fluorescence* in field experiment. The exposure of *P. fluorescens* to a combination of these two stress factors or to a high intensity of a single type of stress in vitro resulted in mixed populations of culturable and nonculturable cells, and VBNC cells represented a significant proportion of the latter. Once in soil, the VBNC cells thus obtained did not display enhanced persistence compared with culturable cells, suggesting that their physiological state did not represent a successful adaptive response to adverse environmental conditions. Therefore, optimum soil moisture (range 20–60 %) must be there for better population and activity of microbes in soil.

Nutrient Components

In general, population sizes of PGPMs decline more or less rapidly following introduction into a natural soil, and growth of introduced populations in microbiologically undisturbed soils is a rare phenomenon. One of the main reasons is the scarcity of available nutrient sources to microbes in soil (van Veen et al. 1997). Soil concentrates C, N, P, Fe, essential metals, and various organic compounds above surrounding environmental levels. These essential nutrients may be acquired from soil mineral surfaces (Vaughan et al. 2002). Many important microbial processes can be influenced by minerals, including energy generation, nutrient acquisition, cell adhesion, and biofilm formation (Hochella 2002; Brown et al. 2008). The amount of organic matter that is stabilized in soils can be predicted from the total mineral surface area (Hassink 1997; Six et al. 2002). Soil organic matter (SOM) plays very significant roles in soil. According to a review by Bloem et al. (2006), SOM is usually considered to comprise two main pools: the "living" (e.g., roots and microbial populations) and the "nonliving" (nutrients of different ages and origins) pool. For the purpose of this review, microorganisms will be treated independently of SOM for simplification. SOM strengthens soil aggregates, increases water retention, buffers soil pH, chelates metals, interacts with xenobiotics, and retains cations and anions in soil systems. In doing so, SOM largely determines the physical, chemical, and biological qualities of a soil (Stevenson 1994; Etebu and Osborn 2012). The organic matter in soil being the chief source of energy and food for most of the soil organisms has great influence on the microbial population. Wididana (1990) conducted an experiment in which a 0.1 % aqueous solution of molasses was applied to soil. It resulted in a significant increase in the number of indigenous microorganisms in soil over that of the unamended control. Nakkeeran et al. (2005) mentioned that the performance of bioformulations can be increased by the incorporation of enrichment materials.

Microorganisms need nutrients from the environment for cell growth and division. Even though carbon is essential for microbial growth and survival, relatively few studies have examined the effects that carbon substrates have on the expression of microbial genes in a soil environment. Studies with *lac*Z as a reporter gene controlled by a carbon starvation-induced promoter in *P. fluorescens* in natural soil revealed that *lac*Z gene expression increased shortly after introduction of the cells into soil (van Overbeek et al. 2006). Apparently, the introduced cells indeed sensed carbon-limited conditions in soil, and this resulted in a physiological response which involved the induction of the carbon starvation-inducible reporter gene.

Nitrogen is an essential macronutrient required by microorganisms to support cell growth. Since nitrogen concentrations and form vary in soil environments, microbial formulations applied in the soil must be able to adapt in order to continue to grow and divide. In response to a decrease in cellular ammonia and consequently glutamine, transcription in the gene *gln*N encoding for the enzyme glutamine synthase increases. The increase in this enzyme allows for an enhanced ability to use small amounts of available ammonia. Concurrent with the transcriptional activation of *gln*N, other nitrogen regulatory genes are activated causing the formation of many enzymes and permeases responsible for utilizing other N sources (Magasanik 1996). Enzymatic activity studies coupled with gene expression determinations have also shown that nitrogen fixation is induced when other sources of nitrogen are not available (Bürgmann et al. 2003). Nitrogen fixation was induced through the provision of excess carbon source, causing nitrogen to be a limiting nutrient. This corresponded with the study of Jensen et al. (1998), where addition of excess carbon and phosphorus resulted in a 1,000-fold increase of bioluminescence, as measured by a nitrogen-starvation-inducible gene.

Phosphorus (P) in soil is subject to extensive physicochemical and biological reactions with only a small component of total soil P being in a form directly available for plant or microbial uptake. Nutritional constraints linked to scarcity of available P can shape soil microbial community composition in highly weathered soils (Cruz et al. 2009). It is known that every aspect of the process of formation of the N_2-fixing nodule is limited by the availability of P. Legumes like alfalfa, clover, common bean, cow pea, and pigeon pea show a high positive response to P supplementation (Gyaneshwar et al. 2002). In *Rhizobium*, P limitation can result in higher P transport rates and the induction of alkaline phosphatase (Al-Niemi et al. 1997). Moreover, while microorganisms may have capacity to directly solubilize P to meet their own requirements, subsequent benefits to plants may only occur following turnover of the microbial biomass, i.e., through mineralization as the bacteria die (Gyaneshwar et al. 2002). There is also a need to demonstrate that microorganisms in the rhizosphere have the ability to deplete inorganic forms of soil P over and above that by plants alone. There are also reports that inorganic phosphate represses antibiotic production by rhizobacteria posing a major concern about potential adverse effects of phosphate fertilizers commonly used in agriculture on not only introduced biocontrol agents but also on indigenous populations of antagonists (Slininger and Jackson 1992; Millner et al. 1996; Dutta and Podile 2010).

Iron is the fourth most abundant element on earth (Ma 2005); however, in aerobic soils, iron is mostly precipitated as hydroxides, oxyhydroxides, and oxides so that the amount of iron available for assimilation by living organisms is very low, ranging from 10^{-7} to 10^{-23} M at pH 3.5 and 8.5, respectively. Both microbes and plants have a quite high iron requirement (i.e., 10^{-5}–10^{-7} and 10^{-4}–10^{-9} M, respectively), and this condition is more accentuated in the rhizosphere where plant, bacteria, and fungi compete for iron (Guerinot and Ying 1994). Cellular iron deficiency in some bacteria induces the synthesis of low-molecular weight siderophores, molecules with an extraordinarily high affinity for Fe^{+3} (Ka ranging from 10^{23} to 10^{52}), as well as membrane receptors able to bind the Fe–siderophore complex, thereby allowing iron uptake by microorganisms (Neilands 1981). Plant iron nutrition can affect the structure of microbial communities in the rhizosphere. For example, transgenic tobacco that overexpresses ferritin and accumulates more iron than non-transformed tobacco has less bioavailable iron in the rhizosphere. As a consequence, the composition of the rhizosphere microbial community differed significantly when compared to non-transformed tobacco lines (Robin et al. 2006).

Duffy and Défago (1999) studied the environmental factors that modulate the biosynthesis of antibiotics and siderophores by the disease-suppressive strain *P. fluorescens* CHAO. The production of the antibiotic 2,4-diacetylphloroglucinol (DAPG) was stimulated by Zn^{2+}, NH_4Mo^{2+}, and glucose, and production of pyoluteorin was enhanced by Zn^{2+}, Co^{2+}, and glycerol and was repressed by glucose. The production of the siderophore pyochelin was increased by Co^{2+}, fructose, mannitol, and glucose. Comparison of strain CHAO with a genetically diverse collection of 41 *P. fluorescens* biocontrol strains indicated that the effect of some factors like the stimulation of DAPG by Zn^{2+} and glucose was strain dependent (Duffy and Défago 1999). Fe^{3+} and sucrose have been reported to increase the levels of DAPG in *P. fluorescens* F113, whereas in *P. fluorescens* Pf-5 and CHAO, DAPG was stimulated by glucose (Duffy and Défago 1999; Nowak-Thompsan et al. 1994). In *P. fluorescens* strain S272, the highest DAPG yield was obtained with ethanol as the sole source of carbon (Dwivedi and Johri 2003). Microelements, such as Zn^{2+}, Cu^{2+}, and Mo^{2+}, have been found to stimulate DAPG production in *P. fluorescens* CHAO (Notz et al. 2001). Fructose, mannitol, and a mixture of

Zn^{2+} and NH_4Mo^{2+} increased pyrrolnitrin production (Fernando et al. 2005). Apart from being an excellent nutrient supply to plants and microorganisms, organic matter makes the soil less compact, therefore providing better aeration and water movement, as well as protecting it from erosion. This is particularly important in hot and dry environments, whereby organic covering (mulch) on the soil prevents, rapid moisture loss, and excessive temperature increases are prevented (Miller and Donahue 1995).

Osmotic Stress

High concentrations of salt in soil can have a detrimental impact on its physical, chemical, and biological properties (Rengasamy 2006) and unbalance the ecosystem. Salts in soil cause osmotic stress to plants and soil biota by decreasing the osmotic potential of the soil solution. Fungi are more sensitive to salinity than bacteria (Pankhurst et al. 2001; Wichern et al. 2006). The study by Sardinha et al. (2003) found decreasing ergosterol concentration with increasing salinity which indicates fungal abundance decreased with salinity. Salinity can stress or even kill soil microorganisms which drive nutrient cycling (Wichern et al. 2006). Key ecological functions such as nitrogen transformation are reduced or delayed if a large proportion of a microbial population dies (Schimel et al. 2007). Salinity can also cause nutrient imbalance as sodium and chloride may compete with nutrient cations and anions in uptake and affect key metabolic functions. The reductions in soil microbial activity in saline soil result in poor plant growth, which reduces crop yields and organic matter input into the soil (Summer et al. 1998); these lead to environmental problems such as erosion and eutrophication (Rengasamy 2006). Mavi and Marschner (2013) reported that in the short term, addition of a readily available and easily decomposable source of energy improves the ability of microbes to tolerate salinity. The results also suggested that in saline soils, irrespective of the C substrate, N addition has no impact, or a negative impact, on microbial activity and growth.

Soil salinity and related problems generally occur in arid or semiarid climates where rainfall is insufficient to leach soluble salt from the soil or where surface or internal soil drainage is restricted. Another cause of increase in soil salinity is the application of fertilizers, for example, the use of potassium as fertilizer, which can form sylvite, a naturally occurring salt (Romheld and Kirkby 2010). The osmotic potential ($\Psi\pi$) which is the result of interactions of salts with soil water is important in saline soils (Papendick and Campbell 1981). Moreover, the reduction in soil water content by evapotranspiration that concentrates the salts in the remaining solution leads to the changes in osmotic potential of the soil solution. Water molecules are attracted to electrostatically charged cations such as hydrogen, sodium, potassium, and calcium. Therefore, increasing concentrations of salt in soil solution restrict the availability of the soil water to microorganisms and plant roots. In extreme cases, water moves from the microbial cells to the lower osmotic potential of the soil solution (Brady and Weil 1999). The effect is always more pronounced in the rhizosphere because of the increase in water absorption by the plants due to transpiration (Jiang et al. 2007).

To minimize the effect of low osmotic potential in saline soils, the tolerant microorganisms accumulate solutes in the cell to counteract the concentration gradient between cell cytoplasm and soil solution (Killham 1994). Polyhydroxyalkanoates (PHAs) are biopolymers synthesized by various bacteria as carbon, energy, and reducing power storage material in response to conditions of salinity stress (Arora et al. 2006; Khare et al. 2014). Synthesizing osmoregulatory compounds requires large amounts of energy: 30–110 ATP (adenosine triphosphate) compared to 30 ATP for cell wall synthesis (Oren 1999). Thus, high concentrations of salts can not only kill sensitive microorganisms but also reduce the activity of surviving organisms because of the metabolic burden imposed by the need for stress tolerance mechanisms (Schimel et al. 2007). Khare et al. (2011) reported reduction in production of plant growth-promoting and biocontrol metabolites with increase in osmotic stress. High salinity reduces the available phosphorus by sorption processes and results in low solubility of Ca-P complexes (Chookietwattana and Maneewan 2012). The effect of salinity on microbial biomass and community structure can be variable. The varying effects may be due to the salinity level, soil type, and microbial community composition as well as water content. Selection of salt-tolerant strains from arid or semiarid habitats can serve the purpose of inoculants in such habitats. Kumar et al. (1999) reported that biological nitrogen fixation and fertility of saline soil can be increased by planting salt-tolerant tree legumes along with inoculation with salt-tolerant rhizobia. Highly salt-tolerant strains of rhizobia have been reported by various workers from saline and arid soils (Singleton et al. 1982; Abd Alla and Abdel Wahab 1995; Kumar et al. 1999).

Mascher et al. (2000) reported that high osmotic stress in soil results in mixed populations of culturable, nonculturable, and VBNC cells of inoculated pseudomonads. High NaCl concentrations do not have the same effect on all Gram-negative bacteria. The culturability of cells of *P. fluorescens* AH9 was reduced after incubation in 1.7 M NaCl but not after incubation in 1 M NaCl (Jørgensen et al. 1994). In contrast, *Pseudomonas aeruginosa* PAO1 (Velasco et al. 1995) and *Escherichia coli* (Roth et al. 1988) were already affected at 0.7 and 0.8 M NaCl, respectively. This may, in part, reflect differences in osmotic potential between the habitats from which these bacteria originate. Interestingly, tolerance to high NaCl concentrations was suggested to be an important bacterial property for successful colonization of the root (Loper et al. 1985; Miller and Wood 1996). Once introduced in soil, VBNC cells did not display enhanced persistence compared with culturable cells, which suggests that this VBNC state did not represent a physiological strategy to improve survival under adverse conditions. Successful commercial application of microbial inoculants to improve crop growth and yield in saline soil implies that the inoculants are also salt tolerant, which highlights the potential of using microorganisms from saline habitats. It is important to conduct screening experiments with different soil types under controlled conditions and the consideration of water potential instead of electrical conductivity only.

Soil Acidity and Alkalinity

Soil pH has numerous effects on physiological, morphological, and metabolic responses of soil microorganisms, as well as the availability of mineral nutrients (Fierer and Jackson 2006). Agricultural residues and other plant materials can have a liming effect when added to soil in the absence of plants and leaching (Tang et al. 1999; Tang and Yu 1999; Sakala et al. 2004; Xu et al. 2006b). Increases in pH after the addition of residues are purported to occur due to the decarboxylation of organic anions (Yan et al. 1996; Tang and Yu 1999). Other studies have suggested that basic cations which are released during decomposition increase the pH (Pocknee and Sumner 1997; Noble and Randall 1999). Nitrogen (N) cycling is considered a main mechanism for pH change, with the conversion of organic N to NH_4^+ consuming H^+ and subsequent conversion to NO_3-releasing $2H^+$ (Xu and Coventry 2003; Xu et al. 2006a). Leaching of nitrate results in net acidification and would otherwise be balanced by nitrate uptake by the plant. Furthermore, association/dissociation reactions of H^+ with residue surfaces, organic compounds, and the soil matrix may also occur and contribute to the pH change. Soil alkalinity (pH>7) presents itself in soils where the material is calcareous or dolomitic. Sodic soils occur where there has been an accumulation of exchangeable sodium, naturally or under irrigation. Such soils have high concentrations of OH^- ions associated with high contents of bicarbonates and carbonates.

Soil nutrient availability is greatly influenced by soil pH. With the exception of phosphorus, which is most available within a pH range of 6–7, macronutrients (N, K, Ca, Mg, and S) are more available within a pH range of 6.5–8, while the majority of micronutrients (B, Cu, Fe, Mn, Ni, and Zn) are more available within a pH range of 5–7. Outside of these optimal ranges, nutrients are available to microbes and plants at lesser amounts (Haby 1993). With the exception of molybdenum (Mo), micronutrient availability decreases as soil pH values approach 8, due to cations being more strongly bound to the soil and not as readily exchangeable. Metals (Cu, Fe, Mn, Ni, and Zn) are very tightly bound to the soil at high pH and are therefore more available at low pH levels than high pH levels. Alkaline and especially sodic conditions cause deficiencies of copper and zinc and also of phosphorus (due to its low solubility). If the alkaline soil has high $CaCO_3$ content, potassium deficiency can occur because this can readily leach. Nitrogen may be deficient as well in alkaline soil due to the generally low organic matter content (Rowell 1994). Acidic pH reduces activity of many soil organisms, in extreme cases resulting in an accumulation of organic matter, reduced mineralization, and low availability of nitrogen, phosphorus, and sulfur (Rowell 1994). "Base" cations (Ca, K, Mg) are more weakly bound to the soil and are prone to leaching at low pH. Thus, in acidic soils, the availability of nutrients such as potassium (K), calcium (Ca), and magnesium (Mg) is reduced, while the availability of potentially toxic elements such as aluminum (Al), iron (Fe), and zinc (Zn) is increased (Haby 1993). Exposure of the organisms to Al and heavy metals may result into either stimulation or no change or inhibition of the enzyme activity, which depends upon the type of the metals used (Giller et al. 1998). Arora et al. (2009) reported that Al was deleterious for the enzymatic activities (nitrate reduction, nitrite reduction, nitrogenase, and uptake hydrogenase) of rhizobia. Cu showed inhibitory effect on growth and enzyme activities of *Bradyrhizobium* strain at all concentrations (Arora et al. 2009). However, in *Sinorhizobium meliloti*, all the tested enzymatic activities increased up to the concentration of 0.1 mM Cu (Arora et al. 2009). Dahlin et al. (1997) observed that there is no distinct threshold for metal toxicity, or for detrimental effects of other inputs, partly because the effects depend on site-specific characteristics such as climate and soil type.

Chin-A-Woeng et al. (1998) observed that at pH 5.7, the in vitro antifungal activity of phenazine-1-carboxamide was ten times higher than phenazine-1-carboxylic acid (PCA). PCA activity was completely abolished under near-neutral conditions. de Souza et al. (2003) proved that the activity of DAPG is more against *Pythium* species at acidic rather than at neutral to alkaline pH, in vitro conditions. Only some studies have examined gene expression, as affected by pH from pure cultures isolated from soil. Soto et al. (2004) described the effects of pH and calcium on *nod* gene expression in *S. meliloti* LPU63 and the acid-tolerant *Rhizobium* sp. LPU83. Bacterial cells were incubated with root exudates of alfalfa plants grown in either a pH of 7 or 5.6 with calcium concentrations of 0.7 or 6 mM. In *Rhizobium* sp. LPU83, the *nodC* gene expression was similar at both pH and calcium levels. However, the *nodC* expression in *S. meliloti* LPU63 was adversely affected at the lower pH and in the presence of low calcium concentrations. Soto et al. (2004) suggested that their findings may be due to low acidity causing the quantity or quality of alfalfa exudates to be suboptimal. An inherent problem in studying soil pH effects is its varied influence on multiple parameters. Experimentally manipulating the pH of soil may result in changes in several factors that are hard to separate (Rousk et al. 2009).

Temperature

Temperature is an important environmental factor regulating soil microbial biomass, activity, and community (Wu et al. 2010). Temperature directly affects the activity of the soil biota by determining the rate of physiological activity such as enzyme activity and indirectly by affecting physicochemical properties such as diffusion and solubility of nutrients, mineral weathering, evaporation rates, and so on. Organisms have a specific range of temperature at which their biological activity operates. So far, most effort has been made to understand how increased soil temperature influences the CO_2 efflux from soil. In comparison, physiology and composition of the microbial community were rarely studied. There is evidence that warming potentially alters decomposer physiology and accordingly the CO_2

efflux from soil (Zogg et al. 1997; Bradford et al. 2008; Balser and Wixon 2009; Allison et al. 2010). The mechanisms behind temperature adaptations of soil microbes could be physiological adaptations of single species (Malcolm et al. 2008) or species shifts within the microbial community. Lipson et al. (2009) showed that microbial winter and summer communities differed in growth kinetics, biomass-specific respiration rates, and respiration rates. Schindlbacher et al. (2011) reported that soil warming did not affect microbial biomass, nor did warming affect the abundances of most microbial groups. Warming significantly enhanced microbial metabolic activity in terms of soil respiration per amount of microbial biomass. To elucidate the effect of warming on the microbial community, Schindlbacher et al. (2011) analyzed soil from the soil warming experiment at Achenkirch, Austria. Soil of a mature spruce forest was warmed by 4 °C during snow-free seasons since 2004. Repeated soil sampling from control and warmed plots was done from 2008 until 2010. They observed that the 4 °C increase in soil temperature during the snow-free season had no influence on microbial community composition and biomass but strongly increased microbial metabolic activity and hence reduced carbon use efficiency.

Changing soil temperature especially warming has been shown to increase the mortality rate of soil microbes (Wu et al. 2010). High soil temperatures in tropical and subtropical areas are a major problem for biological nitrogen fixation of leguminous crops (Michiels et al. 1994). Temperature affects root hair infection, bacteroid differentiation, nodule structure, and the functioning of the legume root nodule (Roughley 1970). Nodule functioning in common beans (*Phaseolus* spp.) is optimal between 25 and 30 °C and is hampered by root temperatures above 30 °C (Piha and Munnus 1987). Piha and Munnus (1987) reported that bean nodules formed at 35 °C were small and had low specific nitrogenase activity. For most rhizobia, the optimum temperature range for growth in culture is 28–31 °C, and many are unable to grow at 37 °C (Graham 1992). However, 90 % of cowpea *Rhizobium* strains obtained from the hot, dry environment of the Sahel Savannah grew well at 40 °C (Eaglesham and Ayanaba 1984). Strain adaptation to high temperature has also been reported by Hartel and Alexander (1984) and Karanja and Wood (1988). The latter found that a high percentage of the strains that persisted at 45 °C lost their infectiveness at high temperature. They attributed these losses in infectiveness to plasmid curing. If nodules are formed at high temperatures, they were ineffective and plants did not accumulate nitrogen in shoots (Hungria and Franco 1993). Heat treatment of *Rhizobium phaseoli* at 35 and 37 °C resulted in mutant strains lacking a plasmid DNA implicated in the synthesis of melanin and is related to the loss of symbiotic properties of these bacteria (Beltra et al. 1988).

There are reports that low temperature near 15 °C caused inhibitory effect on *Bradyrhizobium japonicum*–soybean symbiosis (Antoun and Prevost 2005). It was shown that the presence of plant growth-promoting rhizospheric strains of *Serratia proteamaculans* 1–102 or *Serratia liquefaciens* 2–68 allowed a better plant growth and higher nitrogen fixation and nodule numbers by *Bradyrhizobium* at root zone temperatures of 15 and 25 °C (Zhang et al. 1996, 1997). Pietikäinen et al. (2005) measured the instantaneous total activity (respiration rate), bacterial activity (growth rate as thymidine incorporation rate), and fungal activity (growth rate as acetate-in-ergosterol incorporation rate) in soil at different temperatures (0–45 °C). Fungal and bacterial growth rates had optimum temperatures around 25–30 °C, while at higher temperatures, lower values were found. This decrease was more drastic for fungi than for bacteria, resulting in an increase in the ratio of bacterial to fungal growth rate at higher temperatures. A tendency towards the opposite effect was observed at low temperatures, indicating that fungi were more adapted to low-temperature conditions than bacteria. Fernando et al. (2005) reported that in *P. fluorescens* CHA0, the concentration of DAPG accumulated at 18 °C was twice the concentration at 30 °C. Little research has been conducted investigating temperature as a factor that affects microbial gene expression in soil. It is known that temperature plays an important role in processes which take place in the soil, including chemical reactions and biological interactions. The soil temperature varies in response to exchange processes that take place primarily at the soil surface. Therefore, it is thought that fluctuations in temperature will have a profound effect on microbial gene expressions within a soil environment. Little is known about the responses of soil microbes coming from different climatic regions to different incubation temperatures. Wu et al. (2010) showed that the majority of soil microorganisms coming from the hot regions are more adapted to high temperature (35 °C) compared to those from the cold area. The effect of incubation temperature on soil microbial community structure was inconclusive. However, genotype experiments are required to be done for the better understanding of fate of inoculant microbes from different climate regions with concerning temperature variation in soils.

Soil Amendment with Agrochemicals

Agriculture is directly dependent upon certain microbiological processes which build and maintain soil fertility. In relation to this, addition of agrochemicals to soil involves certain risks. Even small changes in the chemical structure of compounds can change their biological effects greatly and unpredictably (Niewiadomska and Klama 2005). For example, pesticides used in leguminous crops often show side effects, influencing the symbiotic efficiency and nitrogenase activity (Gaur 1980; Niewiadomska and Klama 2005). Application of lindane to soil at the rate 5 ppm and above inhibits nodulation and growth yield of chickpea, whereas lower concentrations are nontoxic. It was reported that dimethioate and monocrotophos reduced the germination, while carbofuran, formothion, and dimethioate reduced the plumule weight of green gram (Musarrat and Haseeb 2000). Also, treatment with malathion results in the reduced germination of *Vigna mungo*. It is speculated that certain pesticides exhibiting high affinity for plant lectins may interfere with the binding of *Rhizobium* spp. to lectins on the root surface of specific legumes. As a result, the interactions between the legume and *Rhizobium* from the analogous cross-inoculation groups will be inhibited (Musarrat and Haseeb 2000). Niewiadomska and Klama (2005) showed that the nitrogenase activity of *S. meliloti* in the presence of carbendazim and thiram dropped by 93 %, whereas in the presence of imazethapyr by 91 %, compared to the control. In testing procedures for the side effects of pesticides on soil microorganisms, it has long been recognized that effects are more likely to be seen on light-textured soils that are low in organic matter than on heavier soils, and it is therefore recommended to use at least two contrasting soil types for study (Somerville et al. 1987).

In modern agricultural production, herbicide application is a regular practice. Herbicides adsorb on cell surface, affecting ion transport. They affect the metabolism inside the cell by binding to amino and sulfide groups. In the course of these processes, changes take place in the oxidoreduction level of soil, and depending on the chemical composition and dose of herbicide, the microorganism concerned may be killed. Bioaccumulation mitigates the toxic effect of herbicides. Miloševiã et al. (2001) showed that, generally, herbicides tended to reduce the total number of soil microorganisms 7–30 days after application. Miloševiã et al. (1998, 2000) reported that soybean treatment with flumetsulam + trifluralin (Rival), imazethapyr (Pivot), clomazone (Command), and alachlor + linuron (Linuron S-50) reduced the total number of microorganisms in the period of 14 days after application by 15–27 %. After that period, the numbers of microorganisms in the treated variants reached the level of the control variant. The studies have shown that *Azotobacter* is most sensitive to herbicide application (Miloševiã et al. 1995; Miloševiã and Govedarica 2000). Herbicide application also inhibits the symbiotic bacterium *B. japonicum*. Herbicide application reduced nodulation rate, i.e., the number of nodules formed on soybean roots, by 5–21 % (Miloševiã et al. 2000). Studies have also shown that some herbicide applications stimulated the growth and development of actinomycetes. Dimethenamid and metolachlor applied to sugar beet (Govedarica et al. 2001) and prometryn applied to soybean and sunflower (Miloševiã and Govedarica 2001) increased the number of actinomycetes. Certain microbial species may decrease or increase the toxic action of herbicides. Most soil microorganisms are capable of decomposing herbicides, using them most frequently as sources of biogenous elements. The hormone herbicide 2,4-D is rapidly decomposed in the soil, as much as 3.4×10^6 mol in a single day. Various microorganisms decompose 2,4-D: *Mycoplana*, *Corynebacterium*, *Achromobacter*, *Rhizobium*, *Arthrobacter*, *Flavobacterium*, and some actinomycetes (Lynch 1983). According to Kosinkievicz (1984), the phenolic compounds produced by *Pseudomonas acidovorans* may increase or decrease the phytotoxicity of lenacil (Venzar), the mode of action depending on herbicide concentration. A detailed description of impacts of various pesticides on soil microbial communities is summarized in Table 19.1.

Farmers commonly report that the long-term use of synthetic fertilizers, especially anhydrous ammonia, leads to soil compaction and poor tilth (Francis et al. 1990). When bacteria increase and organic matter decreases, aggregation naturally declines because there is no more glue being produced to stick the soil particles together. Anhydrous ammonia used as fertilizer contains approximately 82 % nitrogen and is applied subsurface as a gas. Anhydrous ammonia speeds the decomposition of organic matter in the soil, leaving the soil to be more compact. The addition of anhydrous ammonia contributes acidity to the soil, requiring 148 lb of lime to neutralize 100 lb of anhydrous ammonia or 1.8 lb of lime for every pound of nitrogen contained in the anhydrous (Tisdale et al. 1985; Francis et al. 1990). Anhydrous ammonia initially kills many soil microorganisms in the application zone. Bacteria and actinomycetes recover within 1–2 weeks to levels higher than those prior to treatment (Francis et al. 1990). Soil fungi, however, may take 7 weeks to recover. During the recovery time, the bacteria are stimulated to grow and decompose more organic matter due to the high soil nitrogen content. This is why their numbers increase after anhydrous ammonia applications. Sodium nitrate, also known as Chilean nitrate or nitrate of soda, is a high salt fertilizer. The concern is that excessive sodium acts as a dispersant of soil particles, degrading aggregation.

Organophosphates are another group of fertilizers commonly used in agricultural lands, responsible for the decrease in N-mineralization process (Sato 1983). Mycorrhizal fungi have been consistently reported as being decreased by P fertilizer, but the extent to which this occurs may be dependent on species of fungus involved and level of plant available P in soil (Seymour 2002). The appliance of biopesticides in these chemical fertilizer-supported soils needs understanding of proper application timing and field trials. Despite the numerous efforts aimed at understanding the effects of agrochemicals on the soil ecosystem, it is difficult to comprehend the role of agrochemicals in perturbing soil environment due to divergent research findings reported in the literature. Generally, long-term application of agrochemicals can disturb biochemical equilibrium which can reduce soil fertility and productivity. Understanding the mechanisms underlying molecular responses in microbes in response to agrochemical application could be helpful in elucidating the risk assessment of agrochemical contaminations and its consequent adverse impacts on soil microbial diversity, enzymatic activities, and biochemical reactions.

Effect of Soil Biotic Factors

Soil Indigenous Microbes

In order to have a beneficial effect on a target plant, PGPMs are introduced in large number with the aim of having good root colonization, a prerequisite for the successful use of PGPM. Microbial inoculant colonization is affected by interactions with other soil microorganisms. Chanway and Holl (1992) provided anecdotal evidence to support the view that changes in soil microbial composition cause variability of plant growth responses. Little work has been done to investigate the relationship between PGPM efficacy and changes in the soil microbial

Table 19.1 Effect of agrochemicals on soil microorganisms

Pesticides	Effect on soil microbes	References
HCH, phorate, carbofuran	Increased soil microbial population including N_2-fixing bacteria and P-solubilizing microorganisms	Das and Mukherjee (2000)
Atrazine, isoproturon, metribuzin, sulfosulfuron	Adversely affected *Bradyrhizobium* sp.	Khan et al. (2006)
Mefenoxam, metalaxyl	Inhibited N-fixing bacteria	Monkiedje et al. (2002)
Carbofuran, ethion	Adversely affected soil microorganisms	Kalam and Mukherjee (2001)
Hexaconazole	Reduction in nitrogenase activity of *Anabaena doliolum*	Hammouda (1999)
Phorate	Decreased total number of bacteria and N_2-fixing bacteria	Gonzalez-Lopez et al. (1993)
Malathion	Increased number of denitrifying bacteria; nitrifying and fungal populations were not affected	Gonzalez- Lopez et al. (1993)
Captan	Decreased viable counts of rhizobia; fungi, nitrifying and N2-fixing bacteria significantly decreased, while denitrifying bacteria increased; nitrogenase activity decreased in soils under aerobic conditions	Kyei-Boahen et al. (2001) Martinez-Toledo et al. (1998)
DDT, 2,4-D	Inhibited the growth of *Rhizobium* sp.; inhibited *nod* expression by 32–90 % by disrupting plant–*Rhizobium* signaling	Fabra et al. (1997) Mclachlan (2001)
Isoproturon	Biodegradation of isoproturon-favored bacterial growth while suppressed actinomycetes and fungi	Nowak et al. (2004)
Butachlor	Stimulated anaerobic fermentative and sulfate-reducing bacteria while inhibited acetogenic bacteria in paddy soil	Min et al. (2001)
Ridomil gold plus copper	Increased N- and organic P-mineralization, ammonification	Demanou et al. (2004)
Carbendazim, imazetapir, thiram	Inhibited nodulation; reduced nitrogenase activity in rhizobia	Niewiadomska (2004)
Pentachlorophenol	Reduction in nitrification rates	Colores and Schmidt (2005)
Lindane	Nitrification and phosphate solubilization decreased	Ogunseitan and Odeyemi (1985)
Organophosphates	Denitrifying bacteria decreased thus N-mineralization decreased; decreased root colonization and/or spore number of mycorrhiza	Sato (1983) Seymour (2002)
N- fertilizers	Mycorrhizal fungi spore number and root colonization decreased; adversely affect soil bacteria	Jensen and Jakobsen (1980) Francis et al. (1990)

communities in which PGPMs function. One type is mutualism – interactions beneficial to both parties, such as those involving mycorrhizae, symbiotic nitrogen fixers, and many rhizosphere microbes. The second type of biotic interactions is competition and antagonism for limited food and habitat (Chapin et al. 2002).

Several reports from the Himalayan region have shown that the effects of seed inoculation on plant growth may be due to stimulation of already existing native microbial communities of the rhizosphere/rhizoplane. Pandey et al. (1998) have observed a two- to five fold increase in population of actinomycetes (a group known to fix nitrogen), following inoculation of maize with *Azotobacter chroococcum* and *Azospirillum brasilense* at the Mamlay watershed in Sikkim. Bacterial inoculation of tea rhizosphere resulted in the stimulation of the native bacteria, actinomycetes, and a group of bacteria capable of growing on N-free medium while suppressing the rhizospheric fungal population (Trivedi et al. 2005). Kumar et al. (2007) have also attributed the positive effect of bacterial inoculation on growth of maize through the stimulation of native microflora. However, Robleto et al. (1998) showed that a field release of a *Rhizobium etli* strain containing genes encoding trifolitoxin (an antibiotic peptide active against members of a specific group of α-proteobacteria that enhances the ability to compete trifolitoxin-sensitive strains) strongly reduces the diversity of trifolitoxin-sensitive members of α-proteobacteria in bean rhizosphere as shown by ribosomal intergenic spacer analysis (RISA), with little apparent effect on most microbes.

Soil is generally poor in nutrients and energy sources (estimated to be only about 5 % of its solid material), when compared with the conditions used for microbial growth under laboratory conditions (Ashman and Puri 2002; Nannipieri et al. 2003). Microorganisms are thus constantly competing for nutrient sources (Bridges 1998), mainly carbohydrates, nitrogen, phosphorus, and ions such as iron (Fe^{3+}) and calcium (Ca^{2+}) (Miller and Donahue 1995; Stengel and Gelin 2003). Various studies have shown that the ecologically competent bacteria can effectively establish in a plant environment as compared to nonnative strains. Further, the effect of a particular strain to promote plant growth is also dependent on its interaction with other beneficial microorganisms present in the vicinity of target plants (Trivedi et al. 2012). Latour et al. (2003) described a strategy to study traits involved in the rhizosphere competence. First, the diversity of indigenous populations associated with plant roots was compared with that of the uncultivated soils in order to identify traits discriminating between the two populations. Comparing a wild-type strain to mutants affected in the corresponding phenotypes allowed the determination of the involvement of the traits involved in rhizosphere competence. Finally, traits shared by populations adapted to the rhizosphere were identified by comparing the metabolism and the competitiveness in the rhizosphere of a collection of bacterial strains. The results obtained indicated that rhizosphere-competent pseudomonads are particularly efficient in using pyoverdine-mediated iron uptake system and in reducing nitrogen oxides (Latour et al. 2003).

The agonistic and antagonistic effects of soil microbes through various interactions of bacteria, fungi, and *Actinomycetes* on *Rhizobium* have profoundly influenced sustainable annual harvests (Parmar and Dufresne 2011). Nine *Streptomyces* spp. were isolated from the nodule surface of black gram which were found to inhibit fast- and slow-growing strains of cowpea and soybean rhizobia. The fast-growing strain of *Rhizobium* (both cowpea miscellany and soybean) was more sensitive to antibiosis as compared to slow-growing stains (Jayaraman et al. 1985; Pugashetti et al. 1992). Another study observed that 90 % of the *Actinomycetes* sp. isolated from soil obtained from field plots was antagonistic to *B. japonicum* (Pugashetti et al. 1992). Conn and Franco (2004) found a significant reduction in indigenous actinobacterial endophytes upon inoculation of soil with a commercial multi-organism product, compared with no change in diversity after inoculation with a single species. Most inoculants often rely on application of a single strain which might partially account for the recorded inconsistencies in the field. However, for utilization of multi-organism products, thorough investigation is necessary as the combination of inoculants will not necessarily produce an additive or synergic effect, but rather a competitive process, and, hence, growth enhancement could be reduced or eventually disappear (Trabelsi and Mhamdi 2013). Many researchers have given conclusion that in the rhizosphere, PGPMs (e.g., *Pseudomonas* and *Bacillus*) are involved in interactions with root-nodulating bacteria. Interactions of these rhizobacteria with *Rhizobium* may be antagonistic or synergistic (Guaiquil and Luigi 1992; Capper and Higgins 1993; Parmar and Dadarwal 1997).

It is now clear that the development of endo- or ectomycorrhizae cause rhizosphere microbial changes which can result in interactions among rhizosphere microbes (Bianciotto and Bonfante

Table 19.2 Signal molecules of the quorum-sensing systems identified from microorganisms

Autoinducers	Producing species	Phenotype(s) regulated	References
Acylated homoserine	Many Gram-negative bacteria	Bioluminescence, plasmid conjugal transfer, biofilm formation, motility, antibiotic biosynthesis, and production of virulence factors	Eberhard et al. (1981), Ryan and Dow (2008)
Quinolones	*Pseudomonas aeruginosa*	Elastase, rhamnolipid, pyocyanin, production, and biofilm formation	De'ziel et al. (2004), Diggle et al. (2006)
Autoinducer-2	Many Gram-negative, Gram-positive bacteria, and oomycetes	Bioluminescence, ABC transporters	Xavier and Bassler (2003), Vendeville et al. (2005), Kong et al. (2010)
Bradyoxetin	*Bradyrhizobium japonicum*	Nodulation	Loh et al. (2002)
Butyrolactone	*Streptomyces*	Morphological differentiation and secondary metabolite production	Chater and Horinouchi (2003)
Cyclic dipeptides (CDPs)	*Pseudomonas fluorescens*, *P. alcaligenes*	Influence N-AHL-dependent quorum sensing	Holden et al. (1999), Degrassi et al. (2002)
Fatty acid derivatives	*Ralstonia solanacearum*	Prokaryotic differentiation	Flavier et al. (1997)
Diffusible signal factor	*Xanthomonas campestris*	Endoglucanase production	Wang et al. (2004)
Indole and derivatives	Many Gram-negative and Gram-positive bacteria	Virulence, biofilm formation, drug resistance, sporulation	Kim et al. (2011)
Short peptides	Gram-positive bacteria	Genetic competence	Okada et al. (2005)

2002). Linderman and Paulitz (1990) reviewed the interactions between mycorrhizae and groups of bacteria such as nitrogen-fixing bacteria, PGPR including phosphate-solubilizing bacteria, and biocontrol agents. Bacteria associated to mycorrhizal fungi adhere to fungal spores and hyphal structures and thus spread to the rhizosphere (Bianciotto and Bonfante 2002). Recently, Bianciotto et al. (2004) observed strong evidence of a vertical transmission of endobacteria through the arbuscular mycorrhiza (AM) fungus vegetative generation. In a study using the nitrogen-fixer *A. chroococcum* and *P. fluorescens*, the chemotaxis of these two PGPRs towards the roots of mycorrhizal tomato plants (*Glomus fasciculatum*) was an important step of communication for root colonization (Sood 2003). It was found that *G. fasciculatum* alters the characteristics of root exudates which are chemoattractants specific for each PGPR, amino acids for *P. fluorescens*, and sugars for *A. chroococcum*. Silveira et al. (1995) showed in experiments with beans (*Phaseolus vulgaris*) that bacterial culture of fluorescent *Pseudomonas* co-inoculated with *Glomus etunicatum* increased root growth, nodulation, and N and P uptake. However, the interactions of PGPR with AM fungi are often contradictory and probably depend on the bacterium, the plant species, and the environmental factors.

Complex biotic factors such as plant species, plant age, host cultivar, and infection with the plant pathogen can significantly alter the expression of the gene *phlA*25 for DAPG production in fluorescent pseudomonads (Dwivedi and Johri 2003). Bacteria can perceive extracellular conditions using numerous sensor/kinase signal transduction pathways (Gross et al. 1989). Quorum sensing enables bacteria to regulate their gene expression in a population density-dependent way and adjust their physiology according to their environmental conditions and coordinate the behavior of the entire cell population. Many bacterial functions related to pathogenicity or symbiotic interactions with plants or animals are regulated by quorum-sensing systems. To date, several classes of extracellular signal molecule have been described in microorganisms (Table 19.2). Most QS signals are either small (1,000 Da) organic molecules or peptides with 5–20 amino acids (Chhabra et al. 2005; Williams et al. 2007). Gram-negative bacteria, for example, employ N-acylhomoserine lactones (AHLs), 2-alkyl-4(1H)-quinolones (AQs), long-chain fatty acids, and fatty acid methyl esters as well as autoinducer-2 (AI-2), a collective term for a group of interconvertible furanones derived from dihydroxypentanedione (DPD). AI-2 is also produced by some Gram-positive bacteria, although generally these organisms prefer linear, modified, or cyclic peptides such as the autoinducing peptides (AIPs) made by the staphylococci (Xavier and Bassler 2003). AI-2 production has also been reported in fungi (Kong et al. 2010). Signal molecules belonging to further structural classes such as indole and its derivatives have also been described (Kim et al. 2011). Since two or more bacterial species can use identical or the same class of signal molecules for interspecies communication, cross talk may occur between different signaling systems. Microcolonies on the root surface can indeed consist of more than one population or strain (Bloemberg et al. 2000; Dekkers et al. 2000). Khare and Arora (2011) reported that pyocyanin, an antibiotic from fluorescent pseudomonads at lower concentrations, acts as signaling molecules and modulate gene expression. The application of pyocyanin-producing pseudomonads together with rhizobia contributes to the enhancement of nodulation ability.

Phenotypic and genotypic approaches are now available to characterize the rhizobacterial community structure. Methods that rely on the ability to culture microorganisms include standard plating methods on selective media, community level physiological profiles (CLPP) using the BIOLOG system (Garland 1996), phospholipid fatty acid (PLFA; Tunlid and White 1992), and fatty acid methyl ester (FAME profiling; Germida et al. 1998). Culture-independent molecular techniques are based on direct extraction of DNA from soil and 16S rRNA gene sequence analysis, bacterial artificial chromosome, or expression cloning systems. These provide new insight into the diversity of rhizosphere microbial communities, the heterogeneity of the root environment, and the importance of environmental and biological factors in determining community structure (Smalla et al. 2001). Recently, microarray technique was adapted to elucidate the structure (phylogenetic oligonucleotide microarrays) and functions (functional gene arrays, FGA) of microbial community. Pyrosequencing has broken the barrier of sequence limitations in the study of bacterial diversity. The ability to generate megabases of sequences in a few hours allows intense exploration of species in any environmental sample (Edwards et al. 2006). These approaches can also be used to determine the impact of inoculation of plant growth-promoting rhizobacteria on the rhizosphere community.

Soil Fauna

Different types of soil microfauna (protozoa and free-living nematodes), mesofauna (collembola and mites), and macrofauna (earthworms and termites) play a significant role in soil biological processes (Coleman et al. 2004). Grazers and predators release plant-available nutrients as they consume microbes. Often, organisms specialize in one type of prey, such as either bacteria or fungi. Certain collembolans (springtails) even specialize on specific species of fungi. Other organisms are generalists and will feed on any microbial species they encounter (Hol et al. 2013). Several investigations from freshwater (Pernthaler et al. 1997; Jürgens et al. 1999; Posch et al. 1999) but also from soil systems show that protozoa do not indiscriminately ingest bacteria, but selectively feed on certain bacterial strains (Griffiths et al. 1999; Rønn et al. 2002; Rosenberg 2008). Ciliates and flagellates have distinct and constant cell shapes and feed on free-swimming soil bacteria, whereas amoebae have more plastic bodies that continually change shape and ingest organisms attached to soil particles (Hausmann and Hülsmann 1996; Clarholm et al. 2006). Bonkowski and Brandt (2002) demonstrated that grazing of soil protozoa on bacteria shifts the composition of the microbial community towards auxin-producing bacteria. Generally, feeding by flagellates and ciliates is size selective (Jürgens and Güde 1994; Jürgens et al. 1999; Hahn and Höfle 2001). Medium-size bacterial cells are most susceptible to predation by flagellates and ciliates, whereas smaller cells and large filamentous forms may be partly resistant to grazing (Jürgens and Güde 1994; Hahn and Höfle 2001). Hence, grazing may lead to a bidirectional shift in the relative distribution of bacterial size classes, with increases in the relative abundance of large filaments and small cells (Hahn and Höfle 2001).

A study using DNA fingerprinting (denaturing gradient gel electrophoresis [DGGE] analysis of PCR-amplified 16S rRNA gene sequences), CLPP, and phospholipid fatty acid analysis demonstrated that protozoan grazing changed the compositions of the microbial communities in soil microcosms towards Gram-positive bacteria (Griffiths et al. 1999). DGGE of PCR amplifications of 16S rRNA gene fragments then excising and sequencing of bands from the DGGE gels indicated that high G+C Gram-positive bacteria closely related to *Arthrobacter* spp. were favored by grazing, whereas the excised bands that decreased in intensity were related to Gram-negative bacteria (Rønn et al. 2002). As reported by Rønn et al. (2002), five different protozoan isolates (*Acanthamoeba* sp., two species of *Cercomonas*, *Thaumatomonas* sp., and *Spumella* sp.) had different effects on the bacterial communities. This suggests that the composition of protozoan communities is important for the effect of protozoan grazing on bacterial communities.

Among predatory prokaryotes, *Bdellovibrionaceae* species (small deltaproteobacteria) prey on other Gram-negative cells, using flagella to swim rapidly until they collide with a prey cell (Morgan et al. 2010). After collision, the predator cells then enter the periplasmic space of the prey cell, consume the host cell from within, elongate, and divide into new cells that are released upon host cell lysis (Lambert et al. 2006). Predation by myxobacteria, which are also δ proteobacteria but constitute a monophyletic clade divergent from the *Bdellovibrionaceae*, was also reported by several workers (Morgan et al. 2010). Myxobacteria are found in most terrestrial soils and in many aquatic environments as well (Dawid 2000; Velicer and Vos 2009) and exhibit predation on both bacteria and fungi (Morgan et al. 2010). Myxobacteria use gliding motility (Spormann 1999) to search the soil matrix for prey and produce a wide range of antibiotics and lytic compounds that kill and decompose prey cells and break down complex polymers, thereby releasing substrates for growth (Sudo and Dworkin 1972). There is evidence that *Myxococcus xanthus* employs chemotaxis-like genes in its attack on prey cells (Berleman et al. 2008) and that predation is stimulated by close contact with prey cells (McBride and Zusman 1996).

There are reports on root-knot nematodes as potential inhibitors of nodulation in legumes, particularly mung bean, chickpea, cowpea, and green gram. The presence of sugars such as *N*-acetylglucosamine, galactose, *N-acetylgalactosamine*, and mannose and/or glucose on the cuticle surface of plant-parasitic nematodes may play an important role in the interaction between nematodes and their hosts (Musarrat and Haseeb 2000). It has been demonstrated that the binding of rhizobia to nematode-free roots was inhibited only after pretreatment with certain sugars (Huang et al. 1984). Studies on the interference of nematodes with soybean lectin metabolism showed the reduced binding of rhizobia to *Heterodera glycines*-infected soybean roots, suppressing the nodule formation (Huang et al. 1984). Selective grazing can change bacterial competition (Pedersen et al. 2009). Bacterivores avoid *P. fluorescens* due to the secondary metabolite production by them. This can improve the competitive advantage of *P. fluorescens* over other bacteria (Jousset et al. 2008; Jousset 2012). *Pseudomonas* can produce hydrogen cyanide and thus repel bacterivorous nematodes. Also DAPG, an antibiotic compound produced by *P. fluorescens*, acts as nematicide (Neidig et al. 2011).

Decomposer soil invertebrates also affect the distribution of microorganisms among soil particle-size fractions, their activity, and, therefore, mobilization and stabilization processes of organic matter by mixing organic and mineral soil components during the gut passage (Scullion and Malik 2000; Marhan and Scheu 2006). A large number of soil invertebrates are involved in these processes, but due to their large size and dominance in numbers and biomass in many terrestrial ecosystems, earthworms are the most important drivers (Beare et al. 1995; Bardgett

2005). Fluorescence microscopy, genetic fingerprinting (T-RFLP analysis), and phospholipid fatty analysis suggests that the general structure of microbial communities remains little affected by the gut passage through earthworms (Egert et al. 2004; Marhan et al. 2007). However, by feeding on both organic matter-rich patches and mineral soil and intimately mixing them during the gut passage, the structure and activity of microbial communities are strongly affected (Devliegher and Vertraete 1995) and alter the distribution of microorganisms associated with soil macro- and micro-aggregates (Mummey et al. 2006). Further, despite the structure of microbial communities may be little affected, their activity may change markedly, e.g., it has been documented recently that the activity of denitrifiers is strongly increased in earthworm guts and fresh cast materials (Drake and Horn 2006; Horn et al. 2006). Macrofauna have been shown to reduce pathogen inocula (*Rhizoctonia solani*, *Fusarium* spp.), disperse beneficial organisms (e.g., *Rhizobium*, biocontrol bacteria), and release nutrients from crop residues (Stephens et al. 1993; Doube et al. 1994; Baker 2007).

Conclusion

On the application of bioformulation in soil, its immediate response varies considerably depending on plant species, soil type, soil biotic factors, inoculant density, and environmental conditions. This results in a progressive decline in the inoculant density which fails to elicit intended plant response. Soil texture is an important factor that influences distribution of minerals, organic matter retention, microbial biomass, and other soil properties (Scott and Robert 2006). Accordingly, pore space distribution plays a major impact on the abundance of bacteria and fungi and might be responsible for higher rates of carbon mineralization (Jensen et al. 1996; Raiesi 2006). The soils in good physical condition have better aeration and moisture content which is essential for optimum microbial activity. Fluctuations in nutrient availability, temperature, and water status greatly influence the growth, survival, and metabolic activity of soil microorganisms and plants and may quickly diminish the population of any microbe introduced into soil (Van Veen et al. 1997).

The soil being a heterogeneous, unpredictable environment, the inoculated microbe finds it often difficult to establish a niche for survival among the competitors and predators. Microorganisms interact with each other giving rise to antagonistic or symbiotic interactions. The association existing between one organism and another whether of symbiotic or antagonistic nature influences the population and activity of soil microbes to a great extent. On the other hand, the activities of some of the microorganisms are beneficial to each other. For instance, organic acids liberated by fungi, increase in oxygen by the activity of algae, and change in soil reaction favor the activity of bacteria and other organisms in soil. The predatory habit of protozoa and some myxobacteria which feed on other bacteria may also suppress or eliminate certain inoculants (Chapin et al. 2002). The interactions of microbial inoculants with soil abiotic and biotic (indigenous soil organisms) factors are likely to be complex, and a better mechanistic understanding is necessary to predict short- and long-term effects.

After being introduced into the soil, the inoculants have to compete with native soil microorganisms for the nutrient and habitable niche and have to survive against grazing protozoa. The inoculants can be amended with nutrients to improve the short-term survival of cells upon inoculation, which is essential to the success of any bioformulation (Trivedi et al. 2012). Such carrier materials that offer the available nutrient and/or habitable micropore to the inoculants will be desirable for the development of future bioformulations. In this sense, materials with microporous structure, such as soil aggregate and charcoal, will be good carriers for soil inoculants (Malus'a et al. 2012). Further, to make effective utilization of microbial inoculants, accurate and reliable methods for monitoring the fate of introduced PGPM in the rhizosphere/rhizoplane are required to enhance their efficacy under field conditions. Proteomics can be exploited as a powerful tool to understand the complex patterns of expression of genomes with respect to different environmental niches in which the bacteria adapt and survive. Metabolomics (the quantitative complement of all low-molecular-weight molecules present in the cells in a particular physiological or developmental state) techniques have a huge potential to be applied for strategies aimed at improving competitive abilities of biofertilization and biocontrol strains (Narasimhan et al. 2003; Trivedi et al. 2012). The chance of obtaining a successful inoculant in the end will be greatly enhanced when ecological principles are applied throughout the procedure of development of microbial inoculants. Only then, the major goal of inoculant formulation to provide more suitable microenvironment for survival of introduced PGPR in soil will be fulfilled.

Acknowledgments EK is grateful to Vice Chancellor, Chhatrapati Shahu Ji Maharaj University, Kanpur, India, for providing facilities and support while NKA to VC, BBA University, Lucknow.

References

Abd Alla MH, Abdel Wahab AM (1995) Survival of *Rhizobium leguminosarum* bv. viciae subjected to heat, drought and salinity in soil. Biol Plant 37:131–137

Adl SM, Acosta-Mercado D, Anderson TR, Lynn DH (2006) Protozoa. In: Carter MR, Gregorich E (eds) Soil sampling and methods of analysis, 2nd edn. CRC Press, Boca Raton

Allison SD, Wallenstein MD, Bradford MA (2010) Soil-carbon response to warming dependent on microbial physiology. Nat Geosci 3:336–340

Al-Niemi TS, Kahn ML, McDermott TR (1997) P metabolism in the *Rhizobium tropici*-bean symbiosis. Plant Physiol 113:1233–1242

Antoun H, Prevost D (2005) Ecology of plant growth promoting rhizobacteria. In: Siddiqui ZA (ed) PGPR: biocontrol and biofertilization. Springer, Dordrecht, pp 1–38

Arora NK, Singhal V, Maheshwari DK (2006) Salinity induced accumulation of poly-β-hydroxybutyrate in rhizobia indicating its role in cell protection. World J Microbiol Biotechnol 22:603–606

Arora NK, Khare E, Singh S, Maheshwari DK (2009) Effect of Al and heavy metals on enzymes of nitrogen metabolism of fast and slow growing rhizobia under explanta conditions. World J Microbiol Biotechnol 26:811–816

Ashman MR, Puri G (2002) Essential soil science: a clear and concise introduction to soil science. Blackwell Science Ltd., Oxford, p 208

Baker GH (2007) Differences in nitrogen release from surface and incorporated plant residues by two endogeic species of earthworms (Lumbricidae) in a red-brown earth soil in southern Australia. Eur J Soil Biol 43:S165–S170

Balser TC, Wixon DL (2009) Investigating biological control over soil carbon temperature sensitivity. Global Change Biol 15:2935–2949

Bamforth SS (1997) Protozoa: recycling and indicators of agroecosystem quality. In: Benckiser G (ed) Fauna in soil ecosystems. Dekker, New York, pp 63–84

Bardgett RD (2005) The biology of soil. Oxford University Press, Oxford

Bashan Y (1998) Inoculants of plant growth-promoting bacteria for use in agriculture. Biotechnol Adv 15:729–770

Beare MH, Coleman DC, Crossley DA Jr, Hendrix PF, Odum EP (1995) A hierarchical approach to evaluating the significance of soil biodiversity to biogeochemical cycling. Plant Soil 170:5–22

Beltra R, Del-Solar G, Sanchez-Serrano JJ, Alonso E (1988) Mutants of *Rhizobium phaseoli* HM Mel(2) obtained by means of elevated temperatures. Zentbl Mikrobiol 143:529–532

Berleman JE, Scott J, Chumley T, Kirby JR (2008) Predataxis behavior in *Myxococcus xanthus*. Proc Natl Acad Sci U S A 105:17127–17132

Bianciotto V, Bonfante P (2002) Arbuscular mycorrhizal fungi: a specialised niche for rhizospheric and endocellular bacteria. Antonie Leewvenhoek 81:365–371

Bianciotto V, Genre A, Jargeat P, Lumini E, Bécard G, Bonfante P (2004) Vertical transmission of endobacteria in the arbuscular mycorrhizal fungus *Gigaspora margarita* through generation of vegetative spores. Appl Environ Microbiol 70:3600–3608

Bloem D, Hopkins W, Benedetti A (2006) Microbiological methods for assessing soil quality. CABI Publishing, CAB International, Oxfordshire

Bloemberg GV, Wijfjes AH, Lamers GE, Stuurman N, Lugtenberg BJJ (2000) Simultaneous imaging of *Pseudomonas fluorescens* WCS365 populations expressing three different autofluorescent proteins in the rhizosphere: new perspectives for studying microbial communities. Mol Plant Microbe In 13:1170–1176

Bonkowski M, Brandt F (2002) Do soil protozoa enhance plant growth by hormonal effects? Soil Biol Biochem 34:1709–1715

Boyle M, Frankenberger WT Jr, Stolzy JH (1989) The influence of organic on soil aggregation and water infiltration. J Prod Agr 2:209–299

Bradford MA, Davies CA, Frey SD, Maddox TR, Melillo JM, Mohan JE, Reynolds JF, Treseder KK, Wallenstein MD (2008) Thermal adaptation of soil microbial respiration to elevated temperature. Ecol Lett 11:1316–1327

Brady NC, Weil RR (1999) The nature and properties of soil, 12th edn. Prentice-Hall Inc., Upper Saddle River

Brady NC, Weil R (2002) Elements of the nature and properties of soils. Prentice-Hall Inc., Upper Saddle River

Bridges EM (1998) World soils. Cambridge University Press, Cambridge

Brown GE, Trainor TP, Chaka AM (2008) Geochemistry of mineral surfaces and factors affecting their chemical reactivity. In: Nilsson A, Pettersson LGM, Norskov JK (eds) Chemical bonding at surfaces and interfaces. Elsevier, Amsterdam, pp 457–509

Bürgmann H, Widmer F, Sigler WV, Zeyer J (2003) mRNA extraction and reverse transcription-PCR protocol for detection of nifH gene expression by *Azotobacter vinelandii* in soil. Appl Environ Microbiol 69:1928–1935

Busse MD, Bottomley PJ (1989) Growth and nodulation responses of *Rhizobium meliloti* to water stress induced by permeating and nonpermeating solutes. Appl Environ Microbiol 55:2431–2436

Cabeda MS (1984) Degradação física e erosão, In: I Simpósio de manejo do solo e plantio direito no sul do Brasil, 1, Simpósio de conservação de solos do planalto, 3. 1984, Passo Fundo, Anais, Passo Fundo: PIVCS e UPF – Faculdade de Agronomia, pp 28–33

Capper AL, Higgin KP (1993) Application of *Pseudomonas fluorescens* isolates to wheat as potential biological control agents against take-all. Plant Pathol 42:560–567

Chanway CP, Holl FB (1992) Influence of soil biota on Douglas-fir (*Pseudotsuga menziesii*) seedling growth: the role of rhizosphere bacteria. Can J Bot 70:1025–1031

Chao WL, Nelson EB, Harman GE, Hoch HC (1986) Colonization of the rhizosphere by biological control agents applied to seeds. Phytopathology 76:60–65

Chapin FSIII, Matson PA, Mooney HA (2002) Principles of terrestrial ecosystem ecology. Springer, New York

Chater KF, Horinouchi S (2003) Signalling early developmental events in two highly diverged *Streptomyces* species. Mol Microbiol 48:9–15

Chhabra SR, Philipp B, Eberl L, Givskov M, Williams P, Ca' mara M (2005) Extracellular communication in bacteria. In: Schuz S (ed) Chemistry of pheromones and other semiochemicals 2. Springer, Berlin/Heidelberg, pp 279–315

Chin-A-Woeng TFC, Bloemberg GV, van der Bij AJ, van der Drift KMGM, Schripsema J, Kroon B, Scheffer RJ, Keel C, Bakker PAHM, De Bruijn FJ, Thomas-Oates JE, Lugtenberg BJJ (1998) Biocontrol by phenazine-1-carboxamide producing *Pseudomonas chlororaphis* PCL1391 of tomato root rot caused by *Fusarium oxysporum* f. sp. *radicis-lycopersici*. Mol Plant Microbe In 10:79–86

Chookietwattana K, Maneewan K (2012) Screening of efficient halotolerant phosphate solubilizing bacterium and its effect on promoting plant growth under saline conditions. World Appl Sci J 16(8):1110–1117

Clarholm M, Bonkowski M, Griffiths BS (2006) Protozoa and other protists in soil. In: Van Elsas JD, Jansson JK, Trevors JT (eds) Modern soil microbiology, 2nd edn. CRC Press, Boca Raton, pp 147–176

Coleman DC, Crossley DA Jr, Hendrix PF (2004) Fundamentals of soil ecology, 2nd edn. Elsevier, San Diego

Colores GM, Schmidt SK (2005) Recovery of microbially mediated processes in soil augmented with a pentachlorophenol-mineralizing bacterium. Environ Toxicol Chem 24:1912–1917

Compant S, Duffy B, Nowak J, Clement C, Barka EA (2005) Use of plant-growth promoting bacteria for biocontrol of plant diseases: principles, mechanisms of action, and future prospects. Appl Environ Microbiol 71(9):4951–4959

Conn VM, Franco CMM (2004) Effect of microbial inoculants on the indigenous actinobacterial endophyte population in the roots of wheat as determined by terminal restriction fragment length polymorphism. Appl Environ Microbiol 70:6407–6413

Cruz AF, Hamel C, Hanson K, Selles F, Zentner RP (2009) Thirty-seven years of soil nitrogen and phosphorus fertility management shapes the structure and function of the soil microbial community in a Brown Chernozem. Plant Soil 315:73–184

Dahlin S, Witter E, Martensson A, Turner A, Baath E (1997) Where's the limit? Changes in the microbiological properties of agricultural soils at low levels of metal contamination. Soil Biol Biochem 29:1405–1415

Das AC, Mukherjee D (2000) Soil application of insecticides influences microorganisms and plant nutrients. Appl Soil Ecol 14:55–62

Davidson EA (1993) Soil water content and the ratio of nitrous oxide to nitric oxide emitted from soil. In: Oremland RS (ed) Biogeochemistry of global change. Chapman & Hall, New York, pp 369–386

Dawid W (2000) Biology and global distribution of myxobacteria in soils. FEMS Microbiol Rev 24:403–427

de Souza J, Arnould C, Deulvot C, Lamanceau P, Pearson VG, Raaijmakers JM (2003) Effect of 2,4 diacetylphloroglucinol on *Pythium*: cellular responses and variation in sensitivity among propagules and species. Phytopathology 93:966–975

De'ziel E, Le'pine F, Milot S, He JX, Mindrinos MN, Tompkins RG, Rahme LG (2004) Analysis of *Pseudomonas aeruginosa* 4-hydroxy-2-alkylquinolines (HAQs) reveals a role for 4-hydroxy-2-heptylquinoline in cell-to-cell communication. Proc Natl Acad Sci U S A 101:1339–1344

Degrassi G, Aguilar C, Bosco M, Zahariev S, Pongor S, Venturi V (2002) Plant growth-promoting *Pseudomonas putida* WCS358 produces and secretes four cyclic dipeptides: cross-talk with quorum sensing bacterial sensors. Curr Microbiol 45:250–254

Dekkers LC, Mulders IH, Phoelich CC, Chin-A-Woeng TFC, Wijfjes AH, Lugtenberg BJ (2000) The *sss* colonization gene of the tomato-*Fusarium oxysporum* f. sp. *radicis-lycopersici* biocontrol strain *Pseudomonas fluorescens* WCS365 can improve root colonization of other wild-type *Pseudomonas* spp. Bacteria. Mol Plant Microbe In 13:1177–1183

Demanou J, Monkiedje A, Njine T, Foto SM, Nola M, Serges H, Togouet Z, Kemka N (2004) Changes in soil chemical properties and microbial activities in response to the fungicide Ridomil gold plus copper. Int J Environ Res Public Health 1:26–34

Devliegher W, Verstraete W (1995) Lumbricus terrestris in a soil core experiment—nutrient-enrichment processes (nep) and gut associated processes (gap) and their effect on. Soil Biol Biochem 27:1573–1580

Diggle SP, Cornelis P, Williams P, Camara M (2006) 4-Quinolone signalling in *Pseudomonas aeruginosa*: old molecules, new perspectives. Int J Med Microbiol 296:83–91

Doube BM, Stephens PM, Davoren CW, Ryder MH (1994) Earthworms and the introduction and management of beneficial soil microorganisms. In: Pankhurst C, Doube BM, Gupta VVSR, Grace PR (eds) Soil biota: management in sustainable farming systems. CSIRO, Melbourne, pp 32–41

Drake HL, Horn MA (2006) Earthworms as a transient heaven for terrestrial denitrifying microbes: a review. Eng Life Sci 6:261–265

Duffy BK, Défago G (1999) Environmental factors modulating antibiotic and siderophore biosynthesis by *Pseudomonas fluorescens* biocontrol strains. Appl Environ Microbiol 65:2429–2438

Dutta S, Podile AR (2010) Plant growth promoting rhizobacteria (PGPR): the bugs to debug the root zone. Crit Rev Microbiol 36(3):232–244

Dwivedi D, Johri BN (2003) Antifungals from fluorescent pseudomonads: biosynthesis and regulation. Curr Sci 85(12):1693–1703

Eaglesham ARJ, Ayanaba A (1984) Tropical stress ecology of rhizobia, root-nodulation and legume fixation. In: Subba Rao NS (ed) Current developments in biological nitrogen fixation. Edward Arnold Publishers, London, pp 1–35

Eberhard AA, Burlingamme L, Eberhard C, Kenyon GL, Nealson KH (1981) Structural identification of autoinducer of *Photobacterium fischeri* luciferase. Biochemistry 20:2444–2449

Edwards RA, Rodriguez-Brito B, Wegley L, Haynes M, Breitbart M, Peterson DM, Saar MO, Alexander S, Alexander EC Jr, Rohwer F (2006) Using pyrosequencing to shed light on deep mine microbial ecology. BMC Genomics 7:57

Egert M, Marhan S, Wagner B, Scheu S, Friedrich MW (2004) Molecular profiling of 16S rRNA genes reveals diet-related differences of microbial communities in soil, gut, and casts of Lumbricus terrestris L(Oligochaeta:Lumbricidae). FEMS Microbiol Ecol 48(2):187–197

Etebu E, Osborn AM (2012) A review of indicators of healthy agricultural soils with pea footrot disease suppression potentials. Sustain Agr Res 2:235–250

Evans J, Wallace C, Dobrowolski N (1993) Interaction of soil type and temperature on the survival of *Rhizobium leguminosarum* bv. *Viciae*. Soil Biol Biochem 25:1153–1160

Fabra A, Duffard R, de Duffard AE (1997) Toxicity of 2,4-Dichlorophenoxyacetic acid to *Rhizobium* sp. in pure culture. Bull Environ Contam Toxicol 59:645–652

Fang M, Kremer RJ, Motavalli PP, Davis G (2005) Bacterial diversity in rhizospheres of nontransgenic and transgenic corn. Appl Environ Microbiol 71:4132–4136

Fernando WGD, Nakkeeran S, Zhang Y (2005) Biosynthesis of antibiotics by PGPR and its relation in biocontrol of plant diseases. In: Siddiqui ZA (ed) PGPR: biocontrol and biofertilization. Springer, Dordrecht, pp 67–109

Fierer N, Jackson RB (2006) The diversity and biogeography of soil bacterial communities. Proc Natl Acad Sci 103:626–631

Flavier AB, Clough SJ, Schell MA, Denny TP (1997) Identification of 3-hydroxypalmitic acid methyl ester as a novel autoregulator controlling virulence in *Ralstonia solanacearum*. Mol Microbiol 26:251–259

Foster RC (1988) Microenvironments of soil microorganisms. Biol Fertil Soils 6:189–203

Francis CA, Flora CB, King LD (1990) Sustainable agriculture in temperate zones. Wiley, New York, p 487

Garbeva P, van Veen JA, van Elsas JD (2004) Microbial diversity in soil: selection of microbial populations by plant and soil type and implications for disease suppressiveness. Ann Rev Phytopathol 42:243–270

Garland JL (1996) Patterns of potential C source utilization by rhizosphere communities. Soil Biol Biochem 28:223–230

Gaur AC (1980) Effect of pesticides on symbiotic nitrogen fixation by legumes. Indian J Microbiol 20:362–370

Germida JJ, Siciliano SD, de Freitas JR, Seib AM (1998) Diversity of root-associated bacteria associated with field-grown canola (*Brassica napus* L.) and wheat (*Triticum aestivum*). FEMS Microbiol Ecol 26:43–50

Giller KE, Witter E, McGrath SP (1998) Toxicity of heavy metals to microorganisms and microbial processes in agricultural soils a review. Soil Biol Biochem 30(10–11):1389–1414

Gonzalez-lopez J, Martinez-Toledo MV, Salermon V, Rodelas B (1993) Studies on the effects of the insecticide phorate and malathion on soil microorganisms. Environ Toxicol Chem 12(7):1209–1214

Govedarica M, Miloševiã N, Konstantinoviã B (2001) Uticaj dimetenamida i metalahlora na mikrobiološka svojstva zemljišta pod šeãernom repom, V jugoslovensko savetovanje o zaštiti bilja, Zlatibor, pp 3–8, 12

Graham PH (1992) Stress tolerance in *Rhizobium* and *Bradyrhizobium*, and nodulation under adverse soil conditions. Can J Microbiol 38:475–484

Griffiths BS (1990) Comparison of microbial-feeding nematodes and protozoa in the rhizosphere of different plants. Biol Fert Soils 9:83–88

Griffiths BS, Bonkowski M, Dobson G, Caul S (1999) Changes in soil microbial community structure in the presence of microbial-feeding nematodes and protozoa. Pedobiologia 43:297–304

Gross R, Arico B, Rappuoli R (1989) Families of bacterial signal-transducing proteins. Mol Microbiol 3:1661–1667

Guaiquil VH, Luigi C (1992) Plant growth promoting rhizobacteria and their effect on rapeseed (*Brassica napus* L.) and potato seedlings. Microbiol Rev 23:264–273

Guerinot ML, Ying Y (1994) Iron: nutritious, noxious, and not readily available. Plant Physiol 104:815–820

Gyaneshwar P, Kumar GN, Parekh LJ, Poole PS (2002) Role of soil microorganisms in improving P nutrition of plants. Plant Soil 245:83–93

Haby VA (1993) Soil pH and plant nutrient availability. Ferti Gram, X(2)

Hahn MW, Höfle MG (2001) Grazing of protozoa and its effect on populations of aquatic bacteria. FEMS Microbiol Ecol 35:113–121

Hamarashid NH, Othman MA, Hussain MAH (2010) Effects of soil texture on chemical compositions, microbial populations and carbon mineralization in soil. Egypt J Exp Biol (Bot) 6(1):59–64

Hammouda O (1999) Response of the paddy field cyanobacterium *Anabaena doliolum* to carbofuran. Ecotoxicol Environ Saf 44:215–219

Harris RF (1981) Effect of water potential on microbial growth and activity. In: Parr JF, Gardner WR, Elliott LF (eds) Water potential relations in soil microbiology. Soil Science Society of America, Madison, pp 23–95

Hartel PG, Alexander M (1984) Temperature and desiccation tolerance of cowpea rhizobia. Can J Microbiol 30:820–823

Hassink J (1997) The capacity of soils to preserve organic C and N by their association with clay and silt particles. Plant Soil 191:77–87

Hattori T, Hattori R (1976) The physical environment in soil microbiology: an attempt to extend principles of microbiology to soil microorganisms. CRC Crit Rev Microbiol 4:423–461

Hausmann K, Hülsmann N (1996) Ecology of protozoa. In: Hausmann K, Hülsmann N (eds) Protozoology. Georg Thieme Verlag, Stuttgart, pp 272–300

Herron PM, Stark JM, Holt C, Hooker T, Cardon ZG (2009) Microbial growth efficiencies across a soil moisture gradient assessed using 13C-acetic acid vapor and 15N-ammonia gas. Soil Biol Biochem 41:1262–1269

Hochella MF (2002) Sustaining earth: thoughts on the present and future roles in mineralogy in environmental science. Mineral Mag 66:627–652

Hol WHG, Bezemer TM, Biere A (2013) Getting the ecology into interactions between plants and the plant growth-promoting bacterium *Pseudomonas fluorescens*. Front Plant Sci 4(81):1–9

Holden MT, Chhabra SR, de Nys R, Stead P, Bainton NJ, Hill PJ, Manefield M, Kumar N, Labatte M, England D, Rice S, Givskov M, Salmond GP, Stewart GS, Bycroft BW, Kjelleberg S, Williams P (1999) Quorum-sensing cross talk: isolation and chemical characterization of cyclic dipeptides from *Pseudomonas aeruginosa* and other Gram-negative bacteria. Mol Microbiol 33:1254–1266

Horn MA, Mertel R, Gehre M, Kaestner M, Drake HL (2006) In vivo emission of dinitrogen by earthworms via denitrifying bacteria in the gut. Appl Environ Microbiol 72:1013–1018

Huang JS, Barker KR, Van Dyke CG (1984) Suppression of binding between rhizobia and soybean roots by *Heterodera glycines*. Phytopathology 74:1381–1384

Hungria M, Franco AA (1993) Effects of high temperature on nodulation and nitrogen fixation by *Phaseolus vulgaris* L. Plant Soil 149:95–102

Hunt PG, Wollum AG, Matheny TA (1981) Effects of soil water on *Rhizobium japonicum* infection nitrogen accumulation and yield in bragg soybean. Agron J 73:501–505

Ingham ER, Trofymow JA, Ames RN, Hunt HW, Morley CR, Moore LC, Coleman DC (1986) Trophic interactions and nitrogen cycling in a semi-arid grassland soil II system responses to removal of different groups of soil microbes or fauna. J Appl Ecol 23:615–630

Jayaraman S, Rangarajan M, Jayachandran S (1985) Metabolism of antibiotic resistant mutants of *Rhizobium* sp. as influenced by different temperature regimes. Hindustan Antibiot Bull 27:38–41

Jensen A, Jakobsen I (1980) The occurrence of vesicular-arbuscular mycorrhiza in barley and wheat grown in some Danish soils with different fertilizer treatments. Plant Soil 55:403–414

Jensen LS, McQueen DJ, Ross DJ, Tate KR (1996) Effect of soil composition on N-mineralization and microbial-C and -N II: laboratory simulation. Soil Till Res 38:189–202

Jensen LE, Kragelund L, Nybroe O (1998) Expression of a nitrogen regulated lux gene fusion in Pseudomonas fluorescens DF57 studied in pure culture and in soil. FEMS Microbiol Ecol 25:23–32

Jiang H, Dong H, Yu B, Liu X, Li Y, Ji S, Zhang CL (2007) Microbial response to salinity change in Lake Chaka, a hypersaline lake on Tibetan plateau. Environ Microbiol 9(10):2603–2621

Jørgensen F, Nybroe O, Knøchel S (1994) Effects of starvation and osmotic stress on viability and heat resistance of *Pseudomonas fluorescens* AH9. J Appl Bacteriol 77:340–347

Jousset A (2012) Ecological and evolutive implications of bacterial defences against predators. Environ Microbiol 14:1830–1843

Jousset A, Scheu S, Bonkowski M (2008) Secondary metabolite production facilitates establishment of rhizobacteria by reducing both protozoan predation and the competitive effects of indigenous bacteria. Funct Ecol 22:714–719

Jürgens K, Güde H (1994) The potential importance of grazing-resistant bacteria in planktonic systems. Mar Ecol Prog Ser 112:169–188

Jürgens K, Pernthaler J, Schalla S, Amann R (1999) Morphological and compositional changes in a planktonic bacterial community in response to enhanced protozoan grazing. Appl Environ Microbiol 65:1241–1250

Kalam A, Mukherjee AK (2001) Influence of hexaconazole, carbofuran and ethion on soil microflora and dehydrogenase activities in soil and intact cell. Indian J Exp Biol 39:90–94

Karanja NK, Wood M (1988) Selecting *Rhizobium phaseoli* strains for use with beans (*Phaseolus vulgaris* L.)

in Kenya, Tolerance of high temperature and antibiotic resistance. Plant Soil 112:15–22

Khan MS, Zaid A, Rizvi PQ (2006) Biotoxic effects of herbicides on growth, nodulation, nitrogenase activity, and seed production in chickpeas. Commun Soil Sci Plant Anal 37:1783–1793

Khare E, Arora NK (2011) Dual activity of pyocyanin from *Pseudomonas aeruginosa*: antibiotic against phytopathogen and signal molecule for biofilm development by rhizobia. Can J Microbiol 57(9):708–713

Khare E, Singh S, Maheshwari DK, Arora NK (2011) Suppression of charcoal rot of chickpea by fluorescent pseudomonas under saline stress condition. Curr Microbiol 62:1548–1553

Khare E, Chopra J, Arora NK (2014) Screening for MCL-PHA producing fluorescent pseudomonads and comparison of MCL-PHA production under iso-osmotic conditions induced by PEG and NaCl. Curr Microbiol 68(4):457–462

Killham K (1994) Soil ecology. Cambridge University Press, Cambridge, p 242

Kim Y-G, Lee J-H, Cho MH, Lee J (2011) Indole and 3-indolylacetonitrile inhibit spore maturation in *Paenibacillus alvei*. BMC Microbiol 11:119

Kluepfel DA (1993) The behaviour and tracking of bacteria in the rhizosphere. Ann Rev Phytopathol 31:441–472

Kong P, Lee BWK, Zhou ZS, Hong C (2010) Zoosporic plant pathogens produce bacterial autoinducer-2 that affects *Vibrio harveyi* quorum sensing. FEMS Microbiol Lett 303(1):55–60

Kosinkievicz B (1984) Interaction between bacterial metabolites and some pesticides, I Interaction between the phenolic compounds produced by *Pseudomonas acidovorans* and the herbicide Venzar. Acta Microbiol Pol 33(12):103–110

Kumar H, Arora NK, Kumar V, Maheshwari DK (1999) Isolation, characterization and selection of salt tolerant rhizobia nodulating *Acacia catechu* and *A. nilotica*. Symbiosis 26:279–288

Kumar B, Trivedi P, Pandey A (2007) Pseudomonas corrugata: a suitable bioinoculant for maize grown under rainfed conditions of Himalayan region. Soil Biol Biochem 39:3093–3100

Kyei-Boahen S, Slinkard AE, Walley FL (2001) Rhizobial survival and nodulation of chickpea as influenced by fungicide seed treatment. Can J Microbiol 47:585–589

Lambert C, Morehouse KA, Chang CY, Sockett RE (2006) *Bdellovibrio*: growth and development during the predatory cycle. Curr Opin Microbiol 9:639–644

Latour X, Delorme S, Mirleau P, Lemanceau P (2003) Identification of traits implicated in the rhizosphere competence of fluorescent pseudomonads: description of a strategy based on population and model strain studies. Agronomie 23:397–405

Linderman RG, Paulitz TC (1990) Mycorrhizal-rhizobacterial interactions. In: Hornby D, Cook RJ, Henis Y, Ko WH, Rovira AD, Schippers B, Scott PR (eds) Biological control of soil-borne plant pathogens. CAB International, Wallingford, pp 261–283

Lipson DA, Monson RK, Schmidt SK, Weintraub MN (2009) The trade-off between growth rate and yield in microbial communities and the consequences for under-snow soil respiration in a high elevation coniferous forest. Biogeochemistry 95:23–35

Loh J, Carlson RW, York WS, Stacey G (2002) Bradyoxetin, a unique chemical signal involved in symbiotic gene regulation. Proc Natl Acad Sci U S A 99:14446–14451

Loper JE, Haack C, Schroth MN (1985) Population dynamics of soil pseudomonads in the rhizosphere of potato (*Solanum tuberosum* L.). Appl Environ Microbiol 49:416–422

Lugtenberg B, Kamilova F (2009) Plant-growth-promoting rhizobacteria. Annu Rev Microbiol 63:541–556

Lynch JM (1983) Microorganisms and enzymes in the soil. In: Marumoto T, Watanabe I, Satoh K, Kanazawa S (eds) Soil biotechnology, microbiological factors in crop productivity. Blackwell Science Publications, London, p. 185

Ma JF (2005) Plant root responses to three abundant soil minerals: silicon, aluminum and iron. Crit Rev Plant Sci 24:267–281

Madsen EL, Alexander M (1982) Transport of *Rhizobium* and *Pseudomonas* through soil. Soil Sci Soc Am J 46:557–560

Magasanik B (1996) Regulation of nitrogen utilization, Escherichia coli and Salmonella: cellular and molecular biology, 2nd edn. American Society of Microbiology, Washington, DC, pp 1345–1356

Malcolm GM, López-Gutiérrez JC, Koide RT, Eissenstat DM (2008) Acclimation to temperature and temperature sensitivity of metabolism by ectomycorrhizal fungi. Global Change Biol 14:1–12

Malus'a E, Sas-Paszt L, Ciesielska J (2012) Technologies for beneficial microorganisms inocula used as biofertilizers. Sci World J 2012:491206

Marhan M, Scheu S (2006) Mixing of different mineral soil layers by endogeic earthworms affects carbon and nitrogen mineralization. Biol Fertil Soils 32:308–314

Marhan S, Kandeler E, Scheu S (2007) Phospholipid fatty acid profiles and xylanase activity in particle size fractions of forest soil and casts of *Lumbricus terrestris* L. (Oligochaeta, Lumbricidae). Appl Soil Ecol 35:412–422

Marshall K (1975) Clay mineralogy in relation to survival of soil bacteria. Annu Rev Phytopathol 13:357–373

Martinez-Toledo MV, Salermon V, Rodelas B, Pozo C, Gonzalez-Lopez J (1998) Effects of the fungicide captan on some functional groups of soil microflora. Appl Soil Ecol 7:245–255

Mascher F, Hase C, Moënne-Loccoz Y, De'fago (2000) The viable-but-nonculturable state induced by abiotic stress in the biocontrol agent *Pseudomonas fluorescens* CHA0 does not promote strain persistence in soil. Appl Environ Microbiol 66(4):1662–1667

Mavi MS, Marschner P (2013) Salinity affects the response of soil microbial activity and biomass to addition of carbon and nitrogen. Soil Res 51(1):68–75

McBride MJ, Zusman DR (1996) Behavioral analysis of single cells of *Myxococcus xanthus* in response to prey cells of Escherichia coli. FEMS Microbiol Lett 137:227–231

Mclachlan JA (2001) Environmental signaling: what embryos and evolution teach us about endocrine disrupting chemicals. Endocr Rev 22:319–341

Michiels J, Verreth C, Vanderleyden J (1994) Effects of temperature stress on bean nodulating *Rhizobium* strains. Appl Environ Microbiol 60:1206–1212

Miller RW, Donahue RL (1995) Soils in our environment, 7th edn. Prentice-Hall, Inc., Upper Saddle River

Miller KJ, Wood JM (1996) Osmoadaptation by rhizosphere bacteria. Annu Rev Microbiol 50:101–136

Millner JL, Silo-Suh L, Lee JC, He HY, Clardy J, Handelsman J (1996) Production of kanosamine by *Bacillus cereus* UW85. Appl Environ Microbiol 62:3061–3065

Miloševiã N, Govedarica M (2000) Effect of some herbicides on microbial properties of soil. In: Albanis T (ed) Proceedings of the 1st European conferences on pesticides and related organic micropollutants in the environment, Ioannina, Greece, pp 61–62

Miloševiã N, Govedarica M (2001) Uticaj prometrina na mikrobiološka svojstva zemljišta pod sojom i suncokretom, V jugoslovensko savetovanje o zaštiti bilja, Zlatibor, pp 3–7

Miloševiã N, Govedarica M, Jarak M, Konstantinoviã B, Miletiã S (1995) Effect of herbicides on the number of microorganisms and dehydrogenase activity in soil under soybean, I Regional symposium: chemistry and environment, 25–29 Sept, Vr. Banja, Proceedings II, pp 551–554

Miloševiã N, Govedarica M, Jarak M, Đuriã S, Konstantinoviã B (1998) Uticaj herbicida na brojnost i dehidrogenaznu aktivnost mikroorganizama u zemljištu pod usevom soje. IV Jugoslovenski kongres o zaštiti bilja, Vrnjaåka Banja 9:21–26

Miloševiã N, Govedarica M, Konstantinoviã B (2000) Uticaj herbicida na nodulaciju soje i mikrobiološku aktivnost zemljišta, Šesti kongres o korovima. Zbornik radova, Banja Koviljaåa, pp 455–460

Miloševiã N, Govedarica M, Jarak M, DĐorðeviã S (2001) Pesticidi i mikroorganizmi. In: Konstatinoviã A (ed) Zaštita šeãerne repe od štetoåina, bolesti I korova. Stylos, Novi Sad, pp 109–149

Min H, Ye YF, Chen ZY, Wu WX, Yufeng D (2001) Effects of butachlor on microbial populations and enzymes activities in paddy soil. J Environ Sci Health B 36:581–595

Misaghi IJ, Olsen MW, Billotte JM, Sonoda RM (1992) The importance of rhizobacterial mobility in biocontrol of bacterial wilt of tomato. Soil Biol Biochem 24:287–293

Monkiedje A, Ilori MO, Spiteller M (2002) Soil quality changes resulting from the application of the fungicides mefenoxam and metalaxyl to a sandy loam soil. Soil Biol Biochem 34:1939–1948

Moore JC, Walter DE, Hunt HW (1988) Arthropod regulation of micro- and mesobiota in below-ground detrital food webs. Ann Rev Entomol 33:419–439

Morgan AD, MacLean RC, Hillesland KL, Velicer GJ (2010) Comparative analysis of *Myxococcus* predation on soil bacteria. Appl Environ Microbiol 76(20):6920–6927

Mummey DL, Rillig MC, Six J (2006) Endogeic earthworms differentially influence bacterial communities associated with different soil aggregate size fractions. Soil Biol Biochem 38:1608–1614

Musarrat J, Haseeb A (2000) Agrichemicals as antagonist of lectin-mediated *Rhizobium*–legume symbiosis: paradigms and prospects. Curr Sci 78(7):793–797

Nakkeeran S, Dilantha Fernando WG, Siddiqui WA (2005) Plant growth promoting rhizobacteria formulations and its scope in commercialization for the management of pests and diseases. In: Siddiqui ZA (ed) PGPR: biocontrol and biofertilization. Springer, Dordrecht, pp 257–296

Nannipieri P, Ascher J, Ceccherini MT, Landi L, Pietramellara G, Renella G (2003) Microbial diversity and soil functions. Eur J Soil Sci 54:655–670

Narasimhan K, Basheer C, Bajic VB, Swarup S (2003) Enhancement of plant-microbe interactions using a rhizosphere metabolomics driven approach and its application in the removal of polychlorinated biphenyls. Plant Physiol 132:146–153

Natsch A, Keel C, Troxler J, Zala M, Von Albertini N, De'fago G (1996) Importance of preferential flow and soil management in vertical transport of a biocontrol strain of *Pseudomonas fluorescens* in structured field soil. Appl Environ Microbiol 62:33–40

Neidig N, Paul RJ, Scheu S, Jousset A (2011) Secondary metabolites of *Pseudomonas fluorescens* CHA0 drive complex non-trophic interactions with bacterivorous nematodes. Microb Ecol 61:853–859

Neilands JB (1981) Iron adsorption and transport in microorganisms. Annu Rev Nutr 1:27–46

Niewiadomska A (2004) Effect of carbendazim, imazetapir and thiramon nitrogenase activity, the number of microorganisms in soil and yield of red clover (*Trifolium pretense* L). Pol J Environ Stud 13:403–410

Niewiadomska A, Klama J (2005) Pesticide side effect on the symbiotic efficiency and nitrogenase activity of rhizobiaceae bacteria family. Polish J Microbiol 54(1):43–48

Nimmo JR, Perkins KS (2002) Aggregate stability and size distribution. In: Dane JH, Topp GC (eds) Methods of soil analysis, Part 4: Physical methods: Soil Science Society of America Book Series No. 5. Soil Science Society of America, Madison, pp 317–328

Noble AD, Randall PJ (1999) Alkalinity effects of different tree litters incubated in an acid soil of NSW. Aust Agrofor Syst 46:147–160

Notz R, Maurhofer M, Schnider-Keel U, Duffy B, Haas D, Défago G (2001) Biotic factors affecting expression of the 2,4-diacetylpholoroglucinol biosynthesis gene *phlA* in *Pseudomonas fluorescens* biocontrol strain CHA0 in the rhizosphere. Phytopathology 91:873–881

Nowak A, Nowak J, Klodka D, Pryzbulewska K, Telesinski A, Szopa E (2004) Changes in the microflora and biological activity of the soil during the degradation of isoproturon. J Plant Dis Prot 19:1003–1016

Nowak-Thompsan B, Gould SJ, Kraus J, Loper JE (1994) Production of 2,4-diacetylphloroglucinol by the biocontrol agent *Pseudomonas fluorescens* Pf-5. Can J Microbiol 40:1064–1066

Ogunseitan OA, Odeyemi O (1985) Effect of lindane, captan and malathion on nitrification, sulphur oxidation, phosphate aerobic heterotrophic soil bacteria to the generic level by solubilization and respiration in tropical soil. Environ Pollut 37:343–354

Okada M, Sato I, Cho SJ, Iwata H, Nishio T, Dubnau D, Sakagami Y (2005) Structure of the *Bacillus subtilis* quorum-sensing peptide pheromone ComX. Nat Chem Biol 1:23–24

Orchard VA, Cook FG (1983) Relation between soil respiration and soil moisture. Soil Biol Biochem 15:447–453

Oren A (1999) Bioenergetic aspects of halophilism. Microbiol Mol Biol Rev 63:334–348

Pandey A, Sharma E, Palni LMS (1998) Influence of bacterial inoculation on maize in upland farming systems of the Sikkim Himalaya. Soil Biol Biochem 30:379–384

Pankhurst CE, Yu S, Hawke BG, Harch BD (2001) Capacity of fatty acid profiles and substrate utilization patterns to describe differences in soil microbial communities associated with increased salinity or alkalinity at three locations in South Australia. Biol Fert Soils 33:204–217

Papendick RI, Campbell GS (1981) Theory and measurement of water potential. In: Parr JF, Gardner WR, Elliott LF (eds) Water potential relations in soil microbiology, Soil Science Society of America Special Publication Number 9. Soil Science Society of America , Madison, pp 1–22

Parke JL, Moen R, Rovira AD, Bowen GD (1986) Soil water flow affects the rhizosphere distribution of a seed-borne biological control agent, *Pseudomonas fluorescens*. Soil Biol Biochem 18:583–588

Parmar N, Dadarwal KR (1997) Rhizobacteria from rhizosphere and rhizoplane of chick pea (*Cicer arietinum* L.). Indian J Microbiol 37:205–210

Parmar N, Dufresne J (2011) Beneficial interactions of plant growth promoting rhizosphere microorganisms. In: Singh A, Parmar N, Kuhad RC (eds) Bioaugmentation, biostimulation and biocontrol, vol 28, Soil Biology., pp 27–42

Pedersen AL, Nybroe O, Winding A, Ekelund F, Bjornlund L (2009) Bacterial feeders, the nematode *Caenorhabditis elegans* and the flagellate *Cercomonas longicauda*, have different effects on outcome of competition among the *Pseudomonas* biocontrol strains CHA0 and DSS73. Microb Ecol 57:501–509

Pernthaler J, Posch T, Simek K, Vrba J, Amann R (1997) Contrasting bacterial strategies to coexist with a flagellate predator in an experimental microbial assemblage. Appl Environ Microbiol 63:596–601

Pietikäinen J, Pettersson M, Bååth E (2005) Comparison of temperature effects on soil respiration and bacterial and fungal growth rates. FEMS Microbiol Ecol 52(1):49–58

Piha MI, Munnus DN (1987) Sensitivity of the common bean (*Phaseolus vulgaris* L.) symbiosis to high soil temperature. Plant Soil 98:183–194

Pocknee S, Sumner ME (1997) Cation and nitrogen contents of organic matter determine its soil liming potential. Soil Sci Soc Am J 61:86–92

Posch T, Simek K, Vrba J, Pernthaler J, Nedoma J, Sattler B, Sonntag B, Psenner R (1999) Predator-induced changes of bacterial size-structure and productivity studied on an experimental microbial community. Aquat Microb Ecol 18:235–246

Post E (1922) Etiquette in society, in business, in politics and at home. New York: Funk & Wagnalls. Bartleby. com, www.bartleby.com/95/37

Postma J, Walter S, van Veen JA (1989) Influence of different initial soil moisture contents on the distribution and population dynamics of introduced *Rhizobium leguminosarum* biovar *trifolii*. Soil Biol Biochem 21:437–444

Pugashetti BK, Angle JS, Wagner GH (1992) Soil microorganism antagonistic towards *Rhizobium japonicum*. Soil Biol Biochem 14:45–47

Raiesi F (2006) Carbon and N mineralization as affected by soil cultivation and crop residue in a calcareous wetland ecosystem in Central Iran. Agr Ecosyst Environ 112:13–20

Rengasamy P (2006) Soil salinity and sodicity. In: Stevens D, Kelly J, McLaughlin M, Unkovich M (eds) Growing crops with reclaimed wastewater. CSIRO Publishing, Collingwood, pp 125–138

Rieu M, Sposito G (1991) Fractal fragmentation, soil porosity, and soil water properties- I, theory. Soil Sci Soc Am J 55:1233–1238

Robin A, Mougel C, Siblot S, Vansuyt G, Mazurier S, Lemanceau P (2006) Effect of ferritin overexpression in tobacco on the structure of bacterial and pseudomonad communities associated with the roots. FEMS Microbiol Ecol 58:492–502

Robleto EA, Borneman J, Triplett EW (1998) Effects of bacterial antibiotic production on rhizosphere microbial communities from a culture-independent perspective. Appl Environ Microbiol 64(12):5020–5022

Romheld V, Kirkby E (2010) Research on potassium in agriculture: needs and prospects. Plant Soil 335:155–180

Rønn R, Mc Caig AE, Griffiths BS, Prosser JI (2002) Impact of protozoan grazing on bacterial community structure in soil microcosms. Appl Environ Microbiol 68:6094–6105

Rosenberg K (2008) Interactions in the rhizosphere of *Arabidopsis thaliana*: effects of protozoa on soil bacterial communities, Dissertation, Technical University of Darmstadt

Roth WG, Leckie MP, Dietzler DN (1988) Restoration of colony forming activity in osmotically stressed *Escherichia coli* by betaine. Appl Environ Microbiol 54:3142–3146

Roughley RJ (1970) The influence of root temperature, *Rhizobium* strain and host selection on the structure and nitrogen-fixing efficiency of the root nodules of *Trifolium subterraneum*. Ann Bot 34:631–646

Rousk J, Brookes PC, Bååth E (2009) Contrasting soil pH effects on fungal and bacterial growth suggest functional redundancy in carbon mineralization. Appl Environ Microbiol 75(6):1589–1596

Rowell DL (1994) Soil science: methods and applications. Longman, London, p 350

Ryan RP, Dow JM (2008) Diffusible signals and interspecies communication in bacteria. Microbiology 154:1845–1858

Sakala GM, Rowell DL, Pilbeam CJ (2004) Acid-base reactions between an acidic soil and plant residues. Geoderma 123:219–232

Sardinha M, Müller T, Schmeisky H, Joergensen RG (2003) Microbial performance in soils along a salinity gradient under acidic conditions. Appl Soil Ecol 23:237–244

Sato K (1983) Effect of pesticide, pentachlorophenol (PCP) on soil microflora. Plant Soil 75:417–426

Scheu S, Ruess L, Bonkowski M (2005) Interactions between microorganisms and soil micro- and mesofauna. In: Buscot F, Varma A (eds) Microorganisms in soils: roles in genesis and functions. Springer, Germany, pp 253–275

Schimel J, Balser TC, Wallenstein M (2007) Microbial stress-response physiology and its implications for ecosystem function. Ecology 88:1386–139

Schindlbacher A, Rodler A, Kuffner M, Kitzler B, Sessitsch A, Zechmeister-Boltenstern S (2011) Experimental warming effects on the microbial community of a temperate mountain forest soil. Soil Biol Biochem 43(7):1417–1425

Scott J, Robert J (2006) Soil texture and nitrogen mineralization potential across a riparian toposequence in a semi-arid savanna. Soil Boil Biochem 38(6):1325–1333

Scullion J, Malik A (2000) Earthworm activity affecting organic matter, aggregation and microbial activity in soils restored after open cast mining for coal. Soil Biol Biochem 32:119–126

Seymour NP (2002) Responses of linseed to vesicular-arbuscular mycorrhizae, phosphorus and zinc in a vertisol, Doctor of Philosophy, University of Queensland, p 263

Shoushtari NH, Pepper IL (1985) Mesquite rhizobia isolated from the Sonoran desert: competitiveness and survival in soil. Soil Biol Biochem 17:803–806

Silveira APD, Freitas SS, Silva LRC, Lombardi MLCO, Cardoso EJBN (1995) Interactions between arbuscular mycorrhizas and plant-growth promoting rhizobacteria in beans. Rev Bras Ciênc Solo 19:205–211

Silver WL, Neff J, McGroddy M, Veldkamp E, Keller M, Cosme R (2000) Effects of soil texture on belowground carbon and nutrient storage in a lowland Amazonian forest ecosystem. Ecosystems 3:193–209

Singleton PW, El Swaify SA, Bohlool BB (1982) Effect of salinity on *Rhizobium* growth and survival. Appl Environ Microbiol 44:884–890

Six J, Conant RT, Paul EA, Paustian K (2002) Stabilization mechanisms of soil organic matter: Implications for C-saturation of soils. Plant Soil 241:155–176

Slininger PJ, Jackson MA (1992) Nutritional factors regulating growth and accumulation of phenazine-1-carboxylic acid by *Pseudomonas fluorescens* 2–79. Appl Microbiol Biotech 37:388–392

Smalla K, Wieland G, Buchner A, Zock A, Parzy J, Kaiser S, Roskot N, Heuer H, Berg G (2001) Bulk and rhizosphere soil bacterial communities studied by denaturing gradient gel electrophoresis: plant-dependent enrichment and seasonal shifts revealed. Appl Environ Microbiol 67:4742–4751

Somerville L, Greaves MP, Domsch KH, Verstraete W, Poole NJ, van Dijk H, Anderson JPE (1987) Recommended laboratory tests for assessing the side effects of pesticides on soil microflora. In: Somerville L, Greaves MP (eds) Pesticide effects on soil microflora. Taylor and Francis, London, pp 205–219

Sood SG (2003) Chemotactic response of plant-growth-promoting bacteria towards roots of vesicular-arbuscular mycorrhizal tomato plants. FEMS Microbiol Ecol 45:219–227

Soto MJ, van Dillewijn P, Martı'nez-Abarca F, Jime'nez-Zurdo JI, Toro N (2004) Attachment to plant roots and nod gene expression are not affected by pH or calcium in the acid-tolerant alfalfa-nodulating bacteria *Rhizobium* sp. LPU83. FEMS Microbiol Ecol 48:71–77

Spormann AM (1999) Gliding motility in bacteria: insights from studies of *Myxococcus xanthus*. Microbiol Mol Biol Rev 63:621–641

Stark JM, Firestone MK (1995) Mechanisms for soil-moisture effects on activity of nitrifying bacteria. Appl Environ Microbiol 61:218–221

Stengel P, Gelin S (2003) Soil, fragile interface. Science Publishers, Inc., New Hampshire

Stephens PM, Davoren CW, Doube BM, Ryder MH, Benger AM, Neate SM (1993) Reduced severity of *Rhizoctonia solani* disease on wheat seedlings associated with the presence of the earthworm *Aporrectodea trapezoids* (Lumbricidae). Soil Biol Biochem 25:1477–1484

Stevenson FJ (1982) Humus chemistry. Wiley, New York

Stevenson FJ (1994) Humus chemistry: genesis, composition, reactions. Wiley, New York

Subba Rao NS (1993) Biofertilizers in agriculture and forestry. Oxford and IBH publishing Co. Pvt. Ltd., New Delhi, p 242

Sudo S, Dworkin M (1972) Bacteriolytic enzymes produced by *Myxococcus xanthus*. J Bacteriol 110:236–245

Sumathi T, Janardhan A, Srilakhmi A, Sai Gopal DVR, Narasimha G (2012) Impact of indigenous microorganisms on soil microbial and enzyme activities. Arch Appl Sci Res 4(2):1065–1073

Sumner ME, Rengasamy P, Naidu R (1998) Sodic soils: a reappraisal. In: Sumner ME, Naidu R (eds) Sodic soils: distribution, properties, management and environmental consequences. Oxford University Press, New York, pp 3–17

Suslow TV (1982) Role of root colonising bacteria in plant growth. In: Mount MS, Lary GH (eds) Phytopathogenic Prokaryotes, vol 2. Academic, London, pp 187–223

Sylvia DM, Fuhrmann JF, Hartel PG, Zuberer DA (2005) Principles and applications of soil microbiology. Pearson Education Inc., Upper Saddle River

Tang C, Yu Q (1999) Impact of chemical composition of legume residues and initial soil pH on pH change of a soil after residue incorporation. Plant Soil 215:29–38

Tang C, Sparling GP, McLay CDA, Raphael C (1999) Effect of short-term legume residue decomposition on soil acidity. Aust J Soil Res 37:561–573

Theng BKG (1979) Formation and properties of clay-polymer complexes. Elsevier, New York

Tisdale SL, Nelson WL, Beaton JD (1985) Soil fertility and fertilizers, 4th edn. MacMillan Publishing Company, New York, p 754

Trabelsi D, Mhamdi R (2013) Microbial inoculants and their impact on soil microbial communities: a review. BioMed Res Int. doi:10.1155/2013/863240

Treonis AM, Wall DH (2005) Soil nematodes and desiccation survival in the extreme arid environment of the antarctic dry valleys. Integr Comp Biol 45:741–750

Trivedi P, Pandey A, Palni LMS, Bag N, Tamang MB (2005) Colonization of rhizosphere of tea by growth promoting bacteria. Int J Tea Sci 4:19–25

Trivedi P, Pandey A, Palni LMS (2012) Bacterial inoculants for field applications under mountain ecosystem: present initiatives and future prospects. In: Maheshwari DK (ed) Bacteria in agrobiology: plant probiotics. Springer, Berlin/Heidelberg, pp 15–44

Tunlid A, White D (1992) Biochemical analysis of biomass, community structure, nutritional status, and metabolic activity of microbial communities in soil. In: Stotzky G, Bollag JM (eds) Soil biochemistry. Marcel Dekker, New York, pp 229–262

Turnbull GA, Morgan JAW, Whipps JM, Saunders JR (2001) The role of bacterial motility in the survival and spread of *Pseudomonas fluorescens* in soil and in the attachment and colonisation of wheat roots. FEMS Microbiol Ecol 36(1):21–31

van Elsas JD, Heijnen CE, van Veen JA (1991) The fate of introduced genetically engineered microorganisms in soil, in microcosms and the field: impact of soil textural aspects. In: MacKenzie DR, Henry SC (eds) Biological monitoring of genetically engineered plants and microbes. Agricultural Research Institute, Bethesda, pp 67–79

van Overbeek LS, van Elsas JD, van Veen JA (2006) *Pseudomonas fluorescens* Tn5-B20 mutant RA92 responds to carbon limitation in soil. FEMS Microbiol Ecol 24(1):57–71

van Veen JA, van Overbeek LS, van Elsas JD (1997) Fate and activity of microorganisms introduced into soil. Microbiol Mol Biol Rev 61(2):121–135

Vaughan DJ, Pattrick RAD, Wogelius RA (2002) Minerals, metals and molecules: ore and environmental mineralogy in the new millennium. Mineral Mag 66:653–676

Velasco R, Burgoa R, Flores E, Hernandez E, Villa A, Vaca S (1995) Osmoregulation in *Pseudomonas aeruginosa* under hyperosmotic shock. Rev Lat Am Microbiol 37:209–216

Velicer GJ, Vos M (2009) Sociobiology of the myxobacteria. Annu Rev Microbiol 63:599–623

Vendeville A, Winzer K, Heurlier K, Tang CM, Hardie KR (2005) Making sense of metabolism: autoinducer-2, LuxS and pathogenic bacteria. Nature Rev 3:383–396

Wang LH, He Y, Gao Y, Wu JE, Dong YH, He C, Wang SX, Weng LX, Xu JL, Tay L, Fang RX, Zhang LH (2004) A bacterial cell-cell communication signal with cross-kingdom structural analogues. Mol Microbiol 51:903–912

Wardle DA (2002) Linking the aboveground and belowground components. Princeton University Press, Princeton

Wichern J, Wichern F, Joergensen RG (2006) Impact of salinity on soil microbial communities and the decomposition of maize in acidic soils. Geoderma 137:100–108

Wididana GN (1990) Inducing disease suppressive soil through "effective microorganisms" (EM), M.S. Thesis. Department of Agriculture, University of the Ryukyus, Okinawa

Williams P, Winzer K, Chan W, Camara M (2007) Look who's talking: communication and quorum sensing in the bacterial world. Philos Trans R Soc Lond B Biol Sci 362:1119–1134

Worrall VS, Roughley RJ (1976) The effect of moisture stress on infection of *Trifolium subterraneum* L. by *Rhizobium trifolii* Dang. J Exp Bot 27:1233–1241

Wu Y, Yu X, Wang H, Ding N, Xu J (2010) Does history matter? Temperature effects on soil microbial biomass and community structure based on the phospholipid fatty acid (PLFA) analysis. J Soils Sediments 10(2):223–230

Xavier KB, Bassler BL (2003) LuxS quorum sensing: more than just a numbers game. Curr Opin Microbiol 6:191–197

Xu RK, Coventry DR (2003) Soil pH changes associated with lupin and wheat plant materials incorporated in a red-brown earth soil. Plant Soil 250:113–119

Xu JM, Tang C, Chen ZL (2006a) Chemical composition controls residue decomposition in soils differing in initial pH. Soil Biol Biochem 38:544–552

Xu JM, Tang C, Chen ZL (2006b) The role of plant residues in pH change of acid soils differing in initial pH. Soil Biol Biochem 38:709–719

Yan F, Schubert S, Mengel K (1996) Soil pH increase due to biological decarboxylation of organic anions. Soil Biol Biochem 28:617–624

Yeates GW, Bongers T, de Goede RGM, Freckman DW, Georgieva SS (1993) Feeding habits in soil nematode families and genera—an outline for soil ecologists. J Nematol 25:315–331

Zhang F, Dashti N, Hynes RK, Smith DL (1996) Plant growth promoting rhizobacteria and soybean [Glycine *max* (L.) Merr.] nodulation and nitrogen fixation at suboptimal root zone temperatures. Ann Bot 77:453–459

Zhang F, Dashti N, Hynes RK, Smith DL (1997) Plant growth promoting rhizobacteria and soybean [Glycine *max* (L.) Merr.] growth and physiology at suboptimal root zone temperatures. Ann Bot 79:243–249

Zogg GP, Zak DR, Ringelberg DB, MacDonald NW, Pregitzer KS, White DC (1997) Compositional and functional shifts in microbial communities due to soil warming. Soil Sci Soc Am J 61:475–481

Printed by Printforce, the Netherlands